Cryosols

Springer
Berlin
Heidelberg
New York
Hong Kong
London
Milan
Paris
Tokyo

John M. Kimble (Ed.)

Cryosols

Permafrost-Affected Soils

With 115 Figures

Springer

John M. Kimble
USDA-NRCS-NSSC
Fed. Bldg. RM 152,
MS 34, 100 Centennial Mall North
Lincoln, NE 685080-3866
USA

Email: John.kimble@usda.gov

ISBN 3-540-20751-1 Springer-Verlag Berlin Heidelberg New York

Library of Congress Control Number: 2004103664

Bibliographic information published by Die Deutsche Bibliothek
Die Deutsche Bibliothek lists this publication in the Deutsche Nationalbibliografie; detailed bibliographic data is available in the Internet at <http://dnb.ddb.de>.

Springer-Verlag Berlin Heidelberg New York
Springer-Verlag is a part of Springer Science+Business Media GmbH

springeronline.com

Printed in Germany

Camera ready by the editor

Cover design: E. Kirchner, Heidelberg

Printed on acid-free paper 30/2132/AO 5 4 3 2 1 0

Preface

Cryosols, as the *World Reference Base for Soil Resources* defines them, or *Gelisols*, as in *Soil Taxonomy*, are soils of the permafrost regions of the world. These regions are at high latitudes and also at high elevations. They cover approximately 25% of the world's surface and, in countries such as Canada, include up to 40% of the soils. Permafrost creates such unique conditions that the permafrost-affected soils have different physical, biological, and chemical properties than other soils.

In the last couple of decades, development in regions of permafrost has increased and so has interest in these regions' soils. Also, these areas are projected to undergo the greatest impact from global warming, which may have a major effect on soil-forming processes and on how the soils are used and managed. Thus there is an even greater need to better understand the distribution, soil-forming processes, ecological processes, classification, use, and management of permafrost-affected soils.

In the late 1980s, the International Permafrost Association and the International Union of Soil Sciences formed the Cryosol Working Group, which began efforts to improve the classification of soils with permafrost. Out of this effort came the *Gelisols* classification of *Soil Taxonomy* and the *Cryosols* classification of the *World Reference Base for Soil Resources.* Another major accomplishment included a Northern Circumpolar Soils Map and database. Under the leadership of Dr. Charles Tarnocai, the group decided to produce this book, *Cryosols – Permafrost-Affected Soils.*

The Working Group is made of people who have worked in the Arctic and Antarctic for many years. They are leaders in the study of permafrost-affected soils and long have recognized the ecological importance of the cold regions. They developed the outline for this book, identified leading experts in different regions to address specific topics, and met formally and informally in Russia, Denmark, the United States, Germany, Finland, Austria, France, and Thailand to review and revise the book's contents.

The idea behind the book was to compile as much knowledge as possible about permafrost-affected soils. Numerous papers on permafrost-affected soils and specific permafrost topics have appeared in many journals, proceedings, and conference publications, and much of that has been in Russian, as the largest extent of such soils is in Russia, but no single book has been dedicated to the soils of cold regions. Therefore much of the information in this book is being presented to a wide audience for the first time. We hope this book will serve as a resource for anyone interested in the soils affected by permafrost, no matter where those soils occur.

The book is divided into six sections. The first section is a history of research of polar soils, which began in the early 1900s. The second section covers the geographic distribution of Cryosols in all regions where the soils occur. The third section deals with the soils' properties and processes. The fourth section covers ecological processes of Cryosols. The fifth section deals with their classification. And the last section covers the management and use of such soils.

This book is the culmination of a great deal of effort by the Cryosol Working Group. The inspiring leadership of Dr. Charles Tarnocai kept the project moving along. Charles was the visionary who made the book become a reality. We are all deeply indebted to him for his ideas and leadership.

Charles and the late Dr. Kaye R. Everett were the first to introduce me to permafrost-affected soils. I know many members of the working group were influenced by Kaye's work in the Arctic and Antarctic, and we feel there are many parts of him in this book.

The lead authors for each of the sections are to be complemented for their efforts in making sure all of the chapters in their sections were completed and reviewed and revised as needed. I especially thank Gabriele Broll, for her help in all aspects of the book including working with the publisher, and Sergey Goryachkin, for his help with the many Russian papers. Sergey was also co-chair of the Working Group for many years. We also thank Maria M. Cadwallader, Ph.D. (formerly Maria M. Lemon) of The Editor Inc. for editing all of the papers for English usage and for completing the layout and preparing the camera-readies.

Thanks also are due to Springer for publishing this large amount of scientific information which will make it generally available to the scientific community.

John M. Kimble
for the members of the
Cryosol Working Group

CWG members, 1998-2003

Charles Tarnocai[1] (Canada)
Sergey Goryachkin[2] (Russia)
Arnold Arnoldussen (Norway)
Megan Balks (New Zealand)
Lothar Beyer (Germany)
Jim Bockheim (USA)
Gabriele Broll (Germany)
Jerry Brown (USA)
Iain Campbell (New Zealand)
David Gilichinsky (Russia)
Luo Guobao (China)
Bjarne Holm Jakobsen (Denmark)
John Kimble (USA)
Dimitri Konyushkov (Russia)
Zhao Lin (China)
Galina Mazhitova (Russia)
Eugeni Naumov (Russia)
Eva-Maria Pfeiffer (Germany)
Chien-Lu Ping (USA)
Marja Liisa Raisanen (Finland)
Scott Smith (Canada)
Ilya Sokolov (Russia)
David Swanson (USA)
Brigitte Van Vliet-Lanoe (France)
Markku Yli-Halla (Finland)

[1]Chair of the Cryosol Working Group of the International Union of Soil Science (1998–2002) and co-chair of the Cryosol Working Group of the International Permafrost Association.
[2]Chair of the Cryosol Working Group of the International Union of Soil Science (2002–) and co-chair of the Cryosol Working Group of the International Permafrost Association.

Contents

Contributing Authors

R.J. Ahrens, Director
National Soil Survey Center, USDA-NRCS
100 Centennial Mall North, Rm 152
Lincoln, NE 68508, USA
e-mail bob.ahrens@nssc.nrcs.usda.gov

T.V. Ananko, Senior Researcher
V.V. Dokuchaev Soil Science Institute
Pyzhevskii per. 7
Moscow, 109017 Russia
e-mail dkon@agro.geonet.ru

I. Archegova, Professor
Institute of Biology, Komi Science Center
Russian Academy of Sciences
28 Kommunisticheskaya St.
Syktyvkar, 167982 Russia
e-mail archegova@ib.komisc.ru

L. Beyer, Professor
Institute for Polar Ecology
Christian-Albrechts-University Kiel
Wischhofstr. 1-3, Bldg. 12
24148 Kiel, Germany
e-mail soil-science@web.de

H.-P. Blume, Professor
Institute for Plant Nutrition and Soil Science
Christian-Albrechts-University Kiel
Olshausenstr. 40
24098 Kiel, Germany
e-mail hblume@soils.uni-kiel.de

J.G. Bockheim, Professor
University of Wisconsin, Soils Department
1525 Observatory Dr.
Madison, WI 53706, USA
e-mail bockheim@facstaff.wisc.edu

M. Bölter, Professor
Institute for Polar Ecology
Christian-Albrechts-University Kiel
Wischhofstr. 1-3, Bldg. 12
24148 Kiel, Germany
e-mail mboelter@ipoe.uni-kiel.de

G. Broll, Professor
Division of Geology and Agroecolgy
ISPA - University of Vechta
P.O. Box 1553, 49364
Vechta, Germany
e-mail gbroll@ispa.uni-vechta.de

C.R. Burn, Professor
Department of Geography and Environmental Studies
Carleton University
1125 Colonel By Drive
Ottawa, Ontario, K1S 5B6, Canada
e-mail crburn@ccs.carleton.ca

I.B. Campbell, Research Scientist
Land and Soil Consultancy Services
23 View Mount
Nelson 7001, New Zealand
e-mail iaincampbell@xtra.co.nz

J. Chen, Professor
Institute of Soil Science
Chinese Academy of Sciences
East Beijing Road 71, P.O. Box 821
Nanjing 210008, People's Republic of China
e-mail jchen@issas.ac.com

G.C. Claridge, Research Scientist
33 Kotari Road
Days Bay
Eastbourne, New Zealand
e-mail graeme.claridge@xtra.co.nz.

M.H. Clark, Soil Scientist
USDA-NRCS
800 W. Evergreen Avenue, Suite 100
Palmer, Alaska 99645, USA
e-mail mclark@ak.usda.gov.

X.Y. Dai, Senior Research Associate
Dept. of Rangeland Ecology and Management
Texas A&M University
2126 TAMU
College Station, Texas 77843, USA
e-mail xdai@ag.tamu.edu

B. Elberling, Associate Professor
Institute of Geography
University of Copenhagen
Oster Voldgade 10
DK-1350
Copenhagen K., Denmark
e-mail be@geogr.ku.dk

D.G. Fedorov-Davydov, Senior Researcher
Institute of Physico-Chemical and Biological Problems of Soil Science
Russian Academy of Sciences
Institutskaya St.
Pushchino, Moscow region, 142290 Russia
e-mail gilichin@issp.serpukhov.ru

C.A. Fox, Research Scientist
Agriculture and Agri-Food Canada
Southern Crop Protection and Food Research Centre
1391 Sandford Street
London, Ontario, Canada, N5V 4T3
e-mail foxc@em.agr.ca

S.V. Goryachkin, Head
Laboratory of Soil Geography and Evolution
Institute of Geography, Russian Academy of Sciences
Staromonetny, 29
Moscow, 109017 Russia
e-mail pedology@igras.geonet.ru

R.G. Gracheva, Senior Researcher
Laboratory of Soil Geography and Evolution
Institute of Geography, Russian Academy of Sciences
Staromonetny, 29
Moscow, 109017 Russia
e-mail pedology@igras.geonet.ru

S.V. Gubin, Leading Researcher
Institute of Physico-Chemical and Biological Problems of Soil Science
Russian Academy of Sciences
Institutskaya St.
Pushchino, Moscow region, 142290 Russia
e-mail gubin@issp.serpukhov.ru

I.V. Ignatenko, Professor
Gertsen Russian State Pedagogical University
Moika, 48
Saint Petersburg, 191186 Russia

E. Kalk, Dr.
Institute for Plant Nutrition and Soil Science
Christian-Albrechts-University Kiel
Olshausenstr. 40
24098 Kiel, Germany
e-mail hblume@soils.uni-kiel.de

E. Kandeler, Professor
Institute of Soil Science
University of Hohenheim
70599 Stuttgart, Germany
e-mail kandeler@uni-hohenheim.de

N.A. Karavaeva, Leading Researcher
Institute of Geography, Russian Academy of Sciences
Staromonetny, 29
Moscow, 109017 Russia
e-mail pedology@igras.geonet.ru

J.M. Kimble, Research Soil Scientist
USDA-NRCS-NSSC
Fed. Bldg. RM 152 MS 34
100 Centennial Mall North
Lincoln, Nebraska 68508, USA
e-mail John.Kimble@usda.gov

D.Ye. Konyushkov, Head
Laboratory of Soil Evolution and Ecology
V.V. Dokuchaev Soil Science Institute
Pyzhevskii per. 7
Moscow, 109017 Russia
e-mail dkon@agro.geonet.ru

N. Kotelina, Senior Researcher
Institute of Biology, Komi Science Center
Russian Academy of Sciences
28 Kommunisticheskaya St.
Syktyvkar 167982, Russia
e-mail archegova@ib.komisc.ru

D. Kuhn, Dr.
Institute for Plant Nutrition and Soil Science
Christian-Albrechts-University Kiel
Olshausenstr. 40
24098 Kiel, Germany
e-mail hblume@soils.uni-kiel.de

O.V. Makeev, Professor (deceased)
Institute of Physico-Chemical and Biological Problems of Soil Science
Russian Academy of Sciences
Institutskaya St.
Pushchino, Moscow Region, 142290 Russia

G. Mazhitova, Senior Researcher
Institute of Biology, Komi Science Center
Ural Division, Russian Academy of Sciences
28 Kommunisticheskaya St.
Syktyvkar, 167982 Russia
e-mail mazhitova@ib.komisc.ru

S.V. Maximovich, Senior Researcher
Institute of Physico-Chemical and Biological Problems of Soil Science
Russian Academy of Sciences
Institutskaya St.
Puschino, Moscow region, 142290 Russia
e-mail svmax@ibbp.psn.ru

G.J. Michaelson, Research Associate
University of Alaska Fairbanks
Agricultural and Forestry Experiment Station
Palmer Research Center
533 E. Fireweed
Palmer, Alaska 99645, USA
e-mail pngjm@uaa.alaska.edu

Ye.M. Naumov, Leading Researcher
Dokuchaev Soil Science Institute
Pyzhevsky per. 7
Moscow, 109017 Russia
e-mail naumov@agro.geonet.ru

V. Ostroumov, Senior Researcher
Cryosol Laboratory
Institute of Physico-Chemical and Biological Problems of Soil Science
Russian Academy of Sciences
2 Institutskaya St.
Puchshino, Moscow region, 142290 Russia
e-mail vostr@issp.psn.ru.

C.-L. Ping, Professor
University of Alaska
Agricultural & Forest Experiment Station
533 E. Fireweed
Palmer, Alaska 99645, USA
e-mail pfclp@uaa.alaska.edu.

K. Pingpank, Dr.
Institute for Plant Nutrition and Soil Science
Christian-Albrechts-University Kiel
Olshausenstr. 40
24098 Kiel, Germany

E. Pospelova, Senior Research Assistant
Geographical faculty
Moscow State University
Vorobievy gory
Moscow, 119899 Russia
e-mail taimyr@orc.ru

G. Qiu, Dr. (deceased)
Cold and Arid Environmental and Engineering Research Institute
Chinese Academy of Sciences
Lanzhou, China

R.D. Seppelt, Professor Dr.
Institute of Arctic Biology
University of Alaska, Fairbanks, and
Australian Antarctic Division
Channel Highway, Kingston 7050
Tasmania, Australia
e-mail rod.seppelt@antdiv.gov.au

C.A.S. Smith, Research Scientist
Pacific Agri-Food Research Centre
Agriculture and Agri-Food Canada
Summerland, BC V2A 8Y2 Canada
e-mail smithcas@agri.gc.ca

I.A. Sokolov, Professor
V.V. Dokuchaev Soil Science Institute
Pyzhevskii per. 7
Moscow, 109017 Russia
e-mail dkon@agro.geonet.ru

D.K. Swanson, Ecologist
USDA Forest Service
P.O. Box 907
Baker City, Oregon 97814, USA
e-mail dkswanson@fs.fed.us

V. Targulian, Professor
Institute of Geography, Russian Academy of Sciences
Staromonetny, 29
Moscow, 109017 Russia
e-mail pedology@igras.geonet.ru

C. Tarnocai, Research Scientist
Agriculture and Agri-Food Canada
Research Branch (ECORC)
K.W. Neatby Building, Rm. 1135
960 Carling Avenue
Ottawa, Canada, K1A 0C6
e-mail tarnocaict@agr.gc.ca

J.C.F. Tedrow, Distinguished Professor
Dept. of Ecology, Evolution, and Natural Resources
Cook College, Rutgers University
14 College Farm Road
New Brunswick, New Jersey 08901-8551, USA

V. Telesnina, Junior Research Assistant
Faculty of Soil Science
Moscow State University
Vorobievy gory
Moscow, 11899 Russia
e-mail taimyr@orc.ru

B. Van Vliet-Lanoë, Director of Research
Processus et Bilan des Domaines Sédimentaires
UMR 8110 CNRS & University of Lille 1
59655 Villeneuve d'Ascq, France
e-mail Brigitte.Van-Vliet-Lanoe@univ-lille1.fr

V. D. Vassiljevskaja, Professor
Faculty of Soil Science, Moscow State University
Vorobievy gory
Moscow, 11899 Russia
e-mail vasilev@soil.msu.ru

H. Veldhuis, Pedologist
University of Manitoba
Land Resource Unit-Manitoba
Ellis Bldg., Room 362A
Winnipeg, Manitoba R3T 2N2, Canada
e-mail veldhuish@agr.gc.ca

D.M. White, Associate Professor
Department of Civil and Environmental Engineering
University of Alaska, Fairbanks
248 Duckering Bldg.
P.O. Box 755900
Fairbanks, Alaska 99775-5900 USA
e-mail ffdmw@uaf.edu

L. Zhao, Dr.
Cold and Arid Environmental and Engineering Research Institute
Chinese Academy of Sciences
Lanzhou, Gansu 730000
People's Republic of China
e-mail linzhao@ns.lzb.ac.cn

D.G. Zamolodchikov, Leading Researcher
Forest Ecology and Production Center of Russian Academy of Sciences
Profsouznaya ul., 84/32
Moscow, 117810 Russia
e-mail dzamolod@cepl.rssi.ru

Section 1. The History of Research of Polar Soil

The History of Research of Polar Soil: Introduction

S.V. Goryachkin[1]

[1]Head, Laboratory of Soil Geography and Evolution, Institute of Geography, Russian Academy of Sciences, Staromonetny, 29, Moscow, 109017 Russia; e-mail pedology@igras.geonet.ru.

Although geologists, botanists, and travelers provided earlier information on polar soils, soil research in high latitudes of the northern and southern hemispheres really began in the early 20th century, and the different regions of permafrost-affected soils each have their own histories of explorations. In eastern Siberia, Russia, soils with permafrost are used widely for agriculture, and they have been studied since 1906; but in most of the ice-free Antarctic, the existence of soils in a strict sense is still disputable and the Antarctic is not a common place for soil studies.

The three chapters in this section each deal with the explorations of a different region of permafrost-affected soils. Chapter 1 summarizes the history of soil research, including soil mapping, in Arctic Alaska, Greenland, and Antarctica, and its Table 1.1.1 shows how many of the original terms used for polar soils have changed over the years. Chapter 2 discusses the history of polar soil research in Eurasia, which has been extensive; its bibliography includes many works in Russian and German. Chapter 3 deals with the history of polar soil research in Canada and its impact on the development of soil science.

The authors of these chapters have studied soils in high latitudes since the 1950s and 1960s, and they witnessed field research blossom in the polar regions through the 1970s.

Chapter 1. Soil Research in Arctic Alaska, Greenland, and Antarctica

J.C.F. Tedrow[1]

[1]Distinguished Professor, Dept. of Ecology, Evolution, and Natural Resources, Cook College, Rutgers University, 14 College Farm Road, New Brunswick, New Jersey 08901-8551, USA.

1. Arctic Alaska

During the formative years of northern pedologic studies in Alaska, the term *Arctic* generally was reserved for the Brooks Range and land to the north. Some investigators, however, extended the definition to include some of the treeless sectors of western and central Alaska. Delineations generally were based on a combination of vegetation and temperature, plus the presence of permafrost. Permafrost is present in various patterns in virtually all of Alaska, except some land bordering the Gulf of Alaska (Brown et al., 1997). One rightfully could include as Arctic nearly all of Alaska, but, in this chapter, discussions are almost exclusively of the Brooks Range, Arctic Foothills, and Arctic Coastal Plain.

Kellogg and Nygard (1951) conducted the first soil investigations in northern Alaska. They delineated three soil zones: 1. Tundra soils of the Arctic Coastal Plain; 2. Tundra soils of the Arctic Plateau (Foothills); 3. Lithosol and Tundra soils of the northern mountains (Brooks Range). Continuing pedologic investigations found a spectrum of soils present within the areas designated as Tundra (Tedrow et al., 1958).

1.1. Spectrum of soils

On the elevated positions, Arctic brown, a well drained soil, is present in some locations (Tedrow and Hill, 1955; Drew and Tedrow, 1957; Hill and Tedrow, 1961). Tundra, a poorly drained mineral soil, is widespread (Douglas and Tedrow, 1960). Brown (1965, 1967) outlined the complex morphology of Tundra soil. Bog soils are also common throughout the Foothills and the Coastal Plain (Tedrow and Brown, 1968).

Tedrow and Brown (Figure 1.1.1) published a small-scale map of the major soil regions of northern Alaska. Reiger, Schoephorster, and Furbush (1979) published an informative map showing major soil regions in all of Alaska. In addition to the more common soils just named, small, isolated pockets of Rendzina soil occur in the Southern Foothills and in the Brooks Range (Ugolini and Tedrow, 1963; Holowaychuk et al., 1966). In the Porcupine Lake area of the Brooks Range, Shungite (black, organic-rich shales) soils occur (Ugolini, Tedrow and Grant, 1963). On the granite-rich till of the Jago and Okpilak Rivers area, Podzol-like soils occur (Brown and Tedrow, 1964; Brown, 1966). The bentonite deposits near Umiat contain Grumusol (MacNamara and Tedrow, 1966). In the Brooks Range, an altitudinal zonation of Arctic brown soil occurs, with the higher elevations having more shallow sola (Tedrow and Brown, 1962).

Aeolian-derived sands are widespread within the Arctic Coastal Plain, with wind erosion now occurring (Rickert and Tedrow, 1967; Everett, 1979, 1980). Patterned ground is ubiquitous throughout the Arctic Slope (Drew, 1957; Tedrow and Cantlon, 1958; Tedrow, 1962; Drew and Tedrow, 1962; Brown 1966; Brown et al., 1980; Everett and Brown, 1982; Ahrens et al., 1994; and many others).

The western sectors of Alaska, from Kotzebue Sound south to Kuskokwim Bay and beyond to the Aleutian Chain, have many volcanic ash deposits. Soils formed on these deposits are dark. Most of them have been described by the term *Arctic* (Rieger, 1964; Simonson and Rieger, 1967; Everett, 1971a; Ping et al., 1989).

With permafrost present in virtually all of Alaska, the soils could be placed within the order of Gelisols (Bockheim et al., 1994). This chapter does not address soils of those vast areas in central, southern, and eastern Alaska, but Rieger, Schoephorster, and Furbush's maps (1979) provide generalized information on the distribution of soils throughout Alaska.

1.2. Frost action

One of the most important factors associated with Arctic soil formation is frost action, which causes high-order displacement within the soil matrix, resulting in complex morphologies. Reiger (1983), Ahrens et al. (1994), Bockheim et al. (1994), and Bockheim and Tarnocai (1998) have reported on this in considerable detail.

1.3. Mapping

There has been very little detailed soil mapping in northern Alaska. MacNamara (1964), however, prepared a soils map of the Howard Pass area, showing the distribution of genetic soils.

The U.S. Department of Agriculture now uses a new method for classifying soils. The original volume (Soil Survey Staff, 1975) said very little about soils of the polar regions, but developments of more recent years, such as the work by Reiger (1983), Ahrens et al. (1994), Bockheim et al. (1994), and others, have made the system more complete, and it is being used in the United States and beyond. Table 1.1.1 shows how many of the original terms have been revised and redefined.

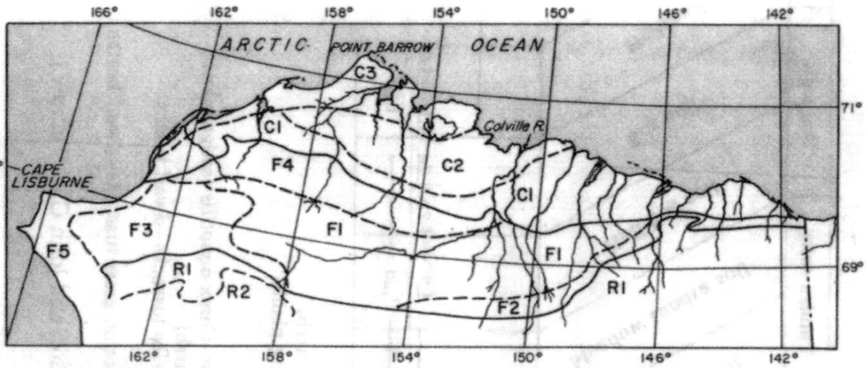

Figure 1.1.1. Soil distribution pattern of Northern Alaska. **C1**, Tundra soils of the Coastal Plain; **C2**, Tundra soils of the Coastal Plain with many sand dunes; **C3**, Tundra soils of the Coastal Plain with extensive Bog-like soils; **F1**, Tundra soils of the Foothills; **F2**, Tundra soils of the Foothills with many stony conditions; **F3**, Tundra soils of the Foothills with steep topography; **F4**, Tundra soils on fine sands; Foothills-Coastal Plain transition; **F5**, Tundra soils of the Foothills with dark-colored sola; **R1**, shallow soils of the mountains; **R2** Tundra soils of the Noatak Basin. (from Tedrow and Brown, 1967).

1.4. Special studies

Oil exploration and construction, including road building, has caused considerable terrain disturbance in northern Alaska and beyond. A number of investigators thus have studied terrain recovery (Gersper and Challinot, 1975; Everett et al., 1985; Walker et al., 1987; Everett, 1994).

The International Biological Program has given considerable attention to the problem of soil fertility, soil chemistry, and nutrient cycling in soils (Bilgin and Douglas, 1972).

2. Greenland

Hartz (1896) and Nordenskjold (1914), who focused their studies in the more southerly sectors of Greenland, described the presence of salt crusts on the soils, with the dominant salt being sodium sulphate. Meinardus (1930) provided valuable information on the soils of Greenland, particularly as related to mineral weathering and frost action.

Along the southern margin of Greenland, soils are rather poorly developed, but some show evidence of podzolization (Hansen, 1969). In the Taseriaq area of southwest Greenland, Holowaychuk and Everett (1965, 1972) recognized seven genetic soil varieties, including those of a hydric nature. In the Jakobshavn area (west coast ca. 69° N) Fredskild (1961) reported soils similar to those present in northern Alaska.

In the Mesters Vig area (east coast ca. 72° N), Ugolini (1966a) reported Arctic brown, Podzol-like, Tundra, Proto-ranker, and Tundra ranker soils. Further, Ugolini (1966b) recognized considerable cryogenic activity similar to that which Meinardus (1930) reported. Jakobsen (1989, 1992) reported on the general soil conditions in the mid-latitudes of east Greenland. Among his findings were gley soils and peats. Further, Jakobsen (1997) published a small map of Greenland showing soil distribution, including the desert-like conditions along the north coast.

Table 1.1. Comparison of the proposed great groups of Gelisols by Bockheim et al., 1994, to Tedrow's (1977) zonal system for polar soils (from Bockheim et al., 1994).

Proposed suborder and great group	Zonal system (Tedrow, 1977, et seq.)
Histels	
Folistels	Organic (bog) soils
Fibristels	Organic (bog) soils
Hemistels	Organic (bog) soils
Sapristels	Organic (bog) soils
Turbels	
Histiturbels	Meadow Tundra soils
Hydriturbels	Meadow Tundra soils
Ariditurbels	Ahumisols, Ahumic (subxerous) soils
Humiturbels	Upland Tundra soils, Soils of the Hummocky Ground, Grumusols
Ochriturbels	Some Polar Desert soils, Soils of the Soliflucton Slopes
Aquels	
Histaquels	Some Meadow Tundra soils
Andaquels	(none)
Humaquels	Some Upland Tundra soils
Psammaquels	Some Regosols
Ochraquels	Some Meadow Tundra soils
Aridels	
Gypsiaridels	Evaporite soils, Ahumic (ultraxerous, xerous) soils
Salaridels	Evaporite soils, Ahumic (ultraxerous, xerous) soils
Nitriaridels	Evaporite soils, Ahumic (ultraxerous, xerous) soils
Calciaridels	Ahumisols
Haplaridels	Ahumisols, Ahumic soils, Protorankers
Statels	
Spodistatels	Podzol and Podzol-like soils
Andistatels	Andosols
Humistatels	Arctic Brown soils, Rendzinas, Shungite soil, Subarctic Brown soils
Argistatels	Subarctic Grey Wooded soils
Psammistatigels	Regosols
Ochristatels	Arctic Brown (northern variant) soil, some Polar Desert soils

Passarge (1920) designated the highest northern latitudes of Greenland as *polar desert*. Most investigators (Nordenskjold and Mecking, 1928; Korotkevich, 1967; Walton, 1972; Charlier, 1978; Aleksandrova, 1988; Tedrow, 1991) used the terms *High Arctic* and *polar desert* somewhat synonymously.

Bocher, Holmen and Jakobsen (1959) described physical features of the north coast of Greenland. Davies (1961) designated ice-free land north of Thule (ca. 77°30' N) as polar desert. Goryachkin, Karavaeva, and Targulian (1998) outlined the problem of soil zonation in the Arctic. They also projected that the north coast of Greenland is the only place in the Arctic where cold desert soils may exist.

Fristrup (1952, 1952-53) described salt encrustations composed of magnesium and calcium chlorides and sulphates. Tedrow (1970) described the soils of Inglefield Land (ca. 78-79° N), a part of the polar desert. In addition to Polar desert soil, other soils such as gley soils and raw, earthy material of the solifluction slopes are present. The organic matter in the surface horizon of Polar desert soil yielded an age of 3,300 ± 110 yr. B. P. (Tedrow, 1970).

3. Antarctica

Antarctic soils form under a very dry, low-temperature environment which includes permafrost. Factually, conditions can be considered those of a desert or, more specifically, a cold desert. Except for special situations such as those of the penguin rookeries and occasional algal residues along lake edges and few other locations, no effective organic component is in Antarctic soils (Rudolph, 1966; Boyd, Staley and Boyd, 1966).

Opinions differ as to whether the term *soil* should apply on the continent, in the Dokuchaevian sense. Most global soil maps do not provide for conditions in the ice-free sectors of Antarctica, and textbooks on soil genesis and soil geography seldom mention conditions in Antarctica. In any case, mineral alteration, salt accumulation including carbonate formation, and horizon differentiation occur in the uppermost layers.

3.1. Initial studies

Jensen (1916) made the initial study of Antarctic soils. He described conditions on the alkali trachyte on Ross Island, where he found significant quantities of available calcium, potassium, and phosphorus in the soils. Next, Markov (1956) conducted studies in the ice-free land of the Mirnyy sector in East Antarctica (Figure 1.1.2).

Markov's work was important because it showed chemical migration within rocks and soils. The processes related to temperature amplitudes in a very dry environment. Markov generally receives credit for introducing the term *cold desert* for soils of Antarctica. Avisuk, Markov, and Shumskiy (1956) and Taubert (1956) provided further information on the chemical processes occuring in cold desert soils.

Glazovskaia (1958) recognized mineral alteration patterns in the soils of the Mirnyy sector. She showed that there was physical weathering and an enrichment of alteration products, particularly calcium, magnesium, iron, aluminum, and manganese, plus the more common forms of soluble salts.

Blakemore and Swindale (1958), Gibson (1962), and Voronov and Spiro (1964), among others, established that there are significant quantities of soluble salts and other altered mineral material present in Antarctic soils. Ugolini and Anderson (1972) showed that ions migrated within the unfrozen water films of otherwise frozen, dry soils of Antarctica. Voronov and Spiro (1964) concluded that there were also additions of soluble salts from atmospheric precipitation, particularly sea salts, in Antarctic soils. Extensive areas of salt-bearing deposits are present.

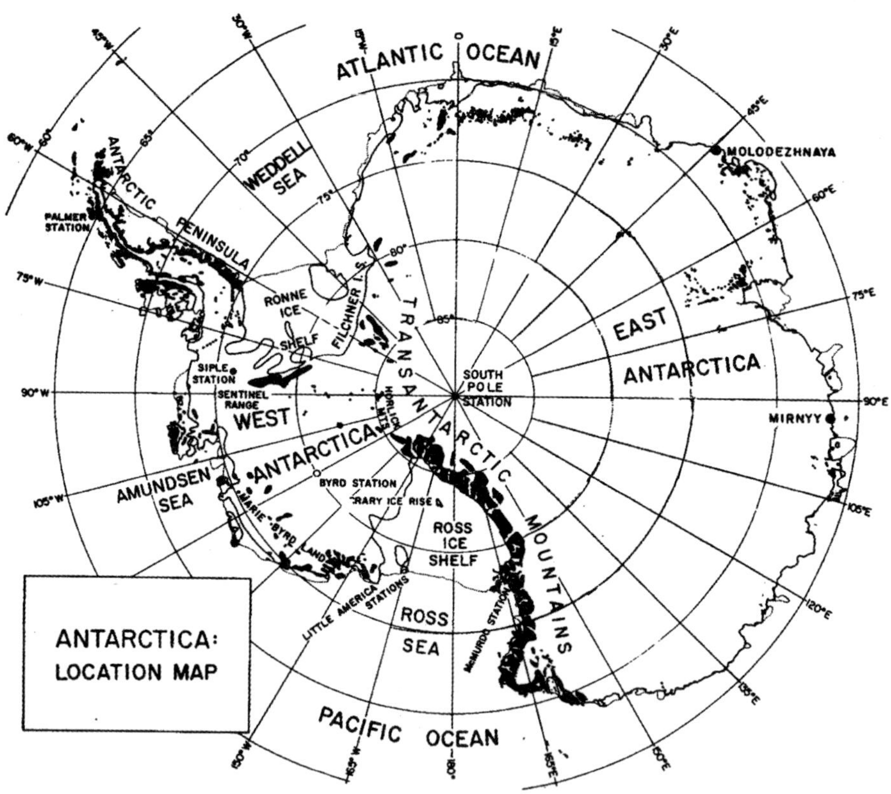

Figure 1.1.2. Map of Antarctica. The shaded areas are largely ice-free (courtesy of C.R. Bentley).

Soil clays in Antarctica consist mainly of illite-like structures, with smaller quantities of vermiculite, chlorite, kaolinite and montmorillonite present (Blakemore and Swindale, 1958; Claridge, 1965; Campbell and Claridge, 1987).

3.2. Soil genesis and classification

Considerable effort has focused on the problem of soil development and classification. McCraw (1960) was the first to develop a soils map of a sector of Antarctica. His map of Taylor Valley in Victoria Land consisted of separating lithologic-landform units, followed by soil descriptions within the delineations. McCraw also displayed his soils map at the 7th International Soil Science Congress.

Syroechovsky (1959) investigated soils of the penguin rookeries of Antarctica and showed that the organic matter accumulation from droppings precedes soil formation. He also introduced the term *ornithogenic* to describe soils of the penguin nesting grounds.

Tedrow and Ugolini (1966) developed a provisional scheme for classifying Antarctic soils. Campbell and Claridge (1969) followed with an improved and more complete method for soil classification. Korotkevich (1967) published an in-depth report on soil formation and classification possibilities in Antarctica.

Many details concerning the soils of Antarctica have been recorded during the past 30 or so years. Everett and Behling (1968) and Everett (1971b) characterized the rate of soil formation on variously aged glacial deposits of South Victoria Land. Ugolini (1970), Bockheim (1979, 1980), and Bockheim and Ugolini (1990) addressed the overall problem of soil development in Antarctica. Ugolini and Bull (1965), Campbell and Claridge (1967, 1968), Claridge (1968b), Everett, (1969) and Linkletter (1972), among others, outlined soil distribution, including patterned ground formation, in South Victoria Land. In the Molodezenya sector of East Antarctica, MacNamara (1969a,b) recorded prominent areas of reddish brown soils, heretofore unknown to exist on the continent.

Korotkevich (1967), Claridge and Campbell (1968a), Ugolini (1970, 1986), Everett (1976), Bockheim (1980), Tedrow (1991), and others have discussed linkage between the soils of Antarctica and those of the High Arctic.

Campbell and Claridge (1987) completed the most definitive work on Antarctic soils. This premier volume draws together virtually all existing information on Antarctic weathering and soil formation.

References

Ahrens, R.J., C.A.S. Smith, J.P. Moore, and L. Ratliff. 1994. Review of the classification of permafrost-affected soils in the United States and Canada. In: Proceedings of the meeting of the classification, correlation, and management of permafrost-affected soils - July, 1994. USDA, SCS, Lincoln, NE. pp. 3-15.

Aleksandrova, V.D. 1988. Vegetation of the Soviet Polar Deserts. trans. from Russian by D. Love. Cambridge Univ. Press. New York. 228 pp.

Avsiuk, G.A., K.K. Markov, and P.A. Shumsky. 1956. Cold deserts in Antarctica. Izv. Akad. Nauk SSSR. Ser. Geogr. No. 4:16-25. (In Russian.)

Beyer, L.J., G. Bockheim, I.B. Campbell, and G.G.C. Clardige. 1999. Review Genesis properties and sensitivity of Antarctic Gelisols. Antarc. Sci. 11:387-398.

Bilgin, A., and L.A. Douglas. 1972. Relationship of soils and hydrological nutrient cycling in a tundra ecosystem. Proc. 1972 tundra biome symposium. Univ. Wash. (Seattle). pp. 80-86.

Blakemore, L.C., and L.F. Swindale. 1958. Chemistry and clay mineralogy of a soil sample from Antarctica. Nature. 182:47-48.

Blume, H.P., and M. Bolter. 1993. Soils of Casey Station, Wilkes Land, Antarctica. Joint Russian-American seminar on Cryopedology and Global changes, Post-Sem.Proc. In: 1st Intern. Conf. on Cryopedology. Russian Academy of Sci. Pushchino, Russia. pp. 96-103.

Bocher, T.W., K. Holmen, and K. Jakobsen. 1959. A synoptic of the Greenland flora. Medd. om Gronland. 163(1):32 pp.

Bockheim, J.G. 1979. Relative age and origin of soils in eastern Wright Valley, Antarctica. Soil Sci. 128:142-152.

Bockheim, J.G. 1980. Properties and classification of some desert soils in coarse-textured glacial drift in the Arctic and Antarctic. Geoderma. 24:45-69.

Bockheim, J.G., and C. Tarnocai. 1998. Recognition of cryoturbation for classifying perma-frost-affected soils. Geoderma. 81:281-293.

Bockheim, J.G., and F.C. Ugolini. 1990. A review of pedogenic zonation in well drained soils of the southern circumpolar region. Quat. Res. 34:47-66.

Bockheim, J.G., C.L. Ping, J.P. Moore, and J.M. Kimble. 1994. Gelisols: A new proposed order for permafrost-affected soils. In: Proceedings of the meeting on the classification, correlation, and management of permafrost-affeted soils, July, 1994. USDA, SCS, Lincoln, NE. pp. 25-44.

Boyd, W.L., J.T. Staley, and J.W. Boyd. 1966. Ecology of soil microorganisms of Antarctica. In: Antarctic Soils and Soil-Forming Processes. Antarctic Res. Ser. AGU. Wash. D.C. 8:125-159.

Brown, J. 1965. Radiocarbon dating, Barrow, Alaska. Arctic. 18:37-48.

Brown, J. 1966. Soils of the Okpilak River region, Alaska. Cold Reg. Res. and Eng. Lab. (Hanover, N.H.) Res. Rept. 188. 49 pp.

Brown, J. 1967. Tundra soils formed over ice wedges, northern Alaska. Soil Sci. Soc. Amer. Proc. 31:686-691.

Brown, J., and J.C.F. Tedrow. 1964. Soils of the northern Brooks Range, Alaska: 4. Well-drained soils of the glaciated valleys. Soil Sci. 97:187-195.

Brown, J., O.J. Ferrians, Jr., J.A. Heginbottom, and E.S Melnikov. 1997. Circum-Arctic map of permafrost and ground-ice conditions. U.S. Geol. Surv. Wash. D.C. Map CP-45.

Brown, J., K.R. Everett, P.J. Webber, S.F. McLean, Jr., and D.F. Murray. 1980. The coastal tundra at Barrow. In: J. Brown, P.C. Miller, L.L. Tieszen, and F.L. Bunnell, eds. An Arctic Ecosystem: The Coastal Tundra at Barrow, Alaska. Dowden, Hutchinson and Ross. Stroudsburg, Pa. pp. 1-29.

Campbell, I.B., and G.G.C. Claridge. 1967. Site and soil differences in the Brown Hills region of the Darwin Glacier, Antarctica. N.Z. J. Sci. 10:563-577.

Campbell, I.B., and G.G.C. Claridge. 1968. Soils in the vicinity of Edisto Inlet, Victoria Land, Antarctica. N.Z. J. Sci. 11:498-520.

Campbell, I.B., and G.G.C. Claridge. 1969. A classification of frigic soils - The zonal soils of the Antarctic continent. Soil Sci. 107:75-85.

Campbell, I.B., and G.G.C. Claridge. 1987. Antarctica: Soils, Weathering Processes and Environment. Elsevier. New York. 368 pp.

Charlier, R.H. 1969. The geographic distribution of Polar desert soils in the Northern Hemisphere. Geol. Soc. Amer. Bull. 80:1985-1996.

Claridge, G.G.C. 1965. The clay mineralogy and chemistry of some soils form the Ross Dependency. N.Z. Jour. Geol. and Geophys. 8:186-220.

Claridge, G.G.C. 1968a. Some features of Antarctic soils and their relation to other desert soils. Tran. 9th Int. Congr. Soil Sci. 4:541-549.

Claridge, G.G.C. 1968b. Soils of the Shackleton Glacier region, Queen Maud Range, Antarctica. N.Z. J. Sci. 11:171-218.

Davies, W.E. 1961. Glacial geology of northern Greenland. Polarforschung (Kiel) 5:95-102.

Douglas, L.A., and J.C.F. Tedrow. 1960. Tundra soils of Arctic Alaska. Proc. 7th Int. Congr. Soil Sci. Vol.4, Comm V: 291-304.

Drew, J.V. 1957. A pedologic study of Arctic Coastal Plain soils near Point Barrow, Alaska. Ph.D. Thesis. Rutgers Univ. 117 pp.

Drew, J.V., and J.C.F. Tedrow. 1962. Arctic soil classification and patterned ground. Arctic. 15:109-116.

Drew, J.V., and J.C.F. Tedrow. 1957. Pedology of an Arctic brown profile near Point Barrow, Alaska. Soil Sci. Soc. Amer. Proc. 21:336-339.

Everett, K.R. 1969. Pedology of the Trinity Peninsula and offshore islands. Antarct. J. U.S. 4:138.

Everett, K.R. 1971a. Composition and genesis of the organic soils of Amchitka Island, Aleutian Islands, Alaska. Arctic and Alpine Res. 3:1-16.

Everett, K.R. 1971b. Soils of the Meserve glacier area, Wright Valley, South Victoria Land, Antarctica. Soil Sci. 112:425-438.

Everett, K.R. 1976. A survey of soils in the region of the South Shetland Islands and adjacent parts of the Antarctic Peninsula. Ohio State Univ. Inst. of Polar Studies. Rept. 58. 44 pp.

Everett, K.R. 1979. Evolution of the soil landscape in the sand region of the Arctic Coastal Plain as exemplified at Atkasook, Alaska. Arctic: 32:207-223.

Everett, K.R. 1980. Distribution and variability of soils near Atkasook, Alaska. Arctic and Alpine Res. 12:433-446.

Everett, K.R. 1994. Long-term active layer response to industrial disturbance at Prudhoe Bay, Alaska. Eos 75 (44): 76.

Everett, K.R., and R.E. Behling. 1968. Chemical and physical characteristics of Meserve Glacier morainal soils, Wright Valley, Antarctica: An index of relative age. Int. Symp. on Antarctic Geol. Exploration Proc. (Hanover, N.H.) pp. 459-460.

Everett, K.R., and J. Brown. 1982. Some recent trends in the physical and chemical characterization and mapping of tundra soils, Arctic Slope of Alaska. Soil Sci. 133:264-280.

Everett, K.R., B.M. Murray, D.F. Murray, A.W. Johnson, A.E. Linkins, and P.J. Webber. 1985. Reconnaissance observations of long-term natural vegetation recovery in the Cape Thompson region, Alaska, and additions to the checklist of flora. Cold Reg. Res. and Eng. Lab. (Hanover, N.H.) Report 85-11. 84 pp.

Fredskild, B. 1961. Floristic and ecologic studies near Jakobshavn, West Greenland. Medd. om Gronland 163(4). 82 pp.

Fristrup, B. 1952. High Arctic deserts. Int. Geol. Congr. Proc., Algiers. Sect. 7:91-99.

Fristrup, B. 1952-53. Wind erosion within the Arctic deserts. Geogr. Tidsskr. 52:51-65.

Gersper, P.L., and J.L. Challinor. 1975. Vehicle perturbation effects upon a tundra soil-plant system: I. Effects on morphological and environmental properties on soils. Soil Sci. Soc. Amer. Proc. 39:737-744.

Gibson, G.W. 1962. Geological investigations in southern Victoria Land, Antarctica. 8. Evaporite salts in the Victoria Valley region. N.Z. J. Geol. and Geophys. 5:361-374.

Glazovskaia, M.A. 1958. Weathering and primary soil formation in Antarctica. Nauk. Dok. Vyss. Sk., Geol. Geogr. Nauk (Moscow). 1:63-76. (In Russian.)

Goryachkin, S.V., N.A. Karavaeva, and V.O. Targulian. 1998. Geography of Arctic soils: Current problems. Eurasian Soil Sci. 31:467-476.

Hansen, K. 1969. Analyses of soil profiles in dwarf-shrub vegetation in south Greenland. Medd. om Gronland. 178(5). 33 pp.

Hartz, N.I. 1896. Ostgronlands vegetation, forhold. In: C.H. Ryder, ed. Medd. om Gronland: Den Ostgronland Expedition. 4:105-314.

Hill, D.E., and J.C.F. Tedrow. 1961. Weathering and soil formation in the Arctic environment. Am. J. Sci. 259:84-101.

Holowaychuk, N., and K.R. Everett. 1965. Soils in the Lake Tasersiaq area, Greenland. Microclimatological, pedological, and geomorphological studies in the western Tasersiaq area, Greenland, during summer, 1964. U.S. Army Natik Labs. PY. 4:28-56.

Holowaychuk, N., and K.R. Everett. 1972. Soils of the Tasersiaq area Greenland. Medd. om Gronland 186(6). 34 pp.

Holowaychak, N., J.H. Petro, H.R. Finney, R.S. Farnham, and P.L. Gersper. 1966. Soils of the Ogotoruk Creek watershed. In: Environment of the Cape Thompson Region, Alaska. U.S. Atomic Energy Comm. pp. 221-273.

Jakobsen, B.H. 1989. Evidence for translocations into the B horizon of a subArctic podzol in Greenland. Geoderma. 45:3-17.

Jakobsen, B.H. 1992. Preliminary studies of soils in northeast Greenland between 74° and 75° northern latitude. Geogr. Tidsskr. 92:111-115.

Jakobsen, B.H. 1997. A soil map of Greenland. Cryopedology '97: II Int. Congr. p. 43.

Jensen, H.L. 1916. Report on Antarctic soils. Brit. Antarct. Exped. 1907-1909. Heinemann. London. Sci. Invest. Rept. Geol. 6:89-92.

Kellogg, C.E., and I.J. Nygard. 1951. Exploratory study of the principal soil groups of Alaska. U.S. Dept. Agr. Monograph no. 7. 138 pp.

Korotkevich, Y.S. 1967. Polar deserts. Sov. Ant. Expd. Int. Bull. Eng. Tr. 6(65). 459-472.

Linkletter, G.O. 1972. Weathering and soil formation in the Dry Valleys of southern Victoria Land: A possible origin for the salts in soils. In: Antarc. Geol. and Geophys. Universitesforlaget (Oslo). pp.441-445.

MacNamara, E.E. 1964. Soils of the Howard Pass area, northern Alaska. Ph.D. thesis. Rutgers Univ. 196 pp.

MacNamara, E.E. 1969a. Active layer development and soil moisture dynamics in Enderbyland, East Antarctica. Soil Sci. 108:345-359.

MacNamara, E.E. 1969b. Soils and geomorphic surfaces in Antarctica. Biul. Peryglacjalny. 20:299-320.

MacNamara, E.E., and J.C.F. Tedrow. 1966. An arctic equivalent of the Grumusol. Arctic. 19:145-152.

McCraw, J.D. 1960. Soils of the Ross Dependancy Antarctica, a preliminary note. N.Z. Soc. Soil Sci. Proc. 4:30-34.

Markov, K.K. 1956. Some facts concerning periglacial phenomena in Antarctica. Preliminary report. Vest. Mosk. Univ. Ser. 11, Geogr. (1):139-148. (In Russian.)

Meinardus, W. 1930. Verteilung der Boden an der Erdoberflache und ihre Ausbildung. Handbuch der Bodenlehre (E. Blanck, ed.), J. Springer (Berlin) Vol. 3 (1) Boden der kalten Region, a) Arctische Boden. pp. 27-96.

Nordenskjold, O.G. 1914. Einige Zuge der physischen Geographie und der Entwickelungsgeschichte Sud-Gronlands. Geogr. Z. (Leipzig). 20:505-524.

Nordenskjold, O.G., and L. Mecking. 1928. The Geography of Polar Regions. Spec. Pub. no. 8. Amer. Geogr. Soc., N.Y. 359 pp.

Passarge, S. 1920. Die Grundlagen der Landschaftskunde. Hamburg. 588 pp.

Ping, C.L., S. Shojo, T. Ito, T. Takahashi, and J.P. Moore. 1989. Characteristics and classification of volcanic ash-derived soils in Alaska. Soil Sci. 148:8-28.

Rickert, D.A., and J.C.F. Tedrow. 1967. Pedologic investigations on some aeolian deposits of northern Alaska. Soil Sci. 104:250-262.

Rieger, S. 1964. Dark, well-drained soils of the tundra regions in western Alaska. J. Soil Sci. 17:264-273.

Rieger, S. 1983. The Genesis and Classification of Cold Soils. Academic Press. New York. 230 pp.

Rieger, S., D.B. Schoephorster, and C.E. Furbush. 1979. Exploratory soil survey of Alaska. U.S. Dept. Agr., Wash. D.C. 213 pp.

Rudolph, E.D. 1966. Terrestrial vegetation of Antarctica: past and present studies. In: Antarctic Soils and Soil-forming Processes. AGU, Wash., D.C. Antarctic Res. Ser. 8:109-124.

Simonson, R.W., and S. Rieger. 1967. Soils of the Andept suborder in Alaska. Soil Sci. Soc. Amer. Proc. 31:692-699.

Soil Survey Staff. 1975. Soil Taxonomy: A Basic System of Soil Classification for Making and Interpreting Soil Surveys. Agr. Handb. 436. U.S. Govt. Prtg. Off. Wash., D.C. 754 pp.

Syroechkovsky, E.E. 1959. The role of animals in primary soil formation under the conditions of the pre-polar region of the globe, as exemplified by the Antarctic. Zool. Zh. 38:1770-1775. (In Russian.)

Taubert, H. 1956. Sowjetische Untersuchungen in der Bunger – "Oase" (Antarktika). Petermanns Geogr. Mitt. (Gotha). 100:329-334.

Tedrow, J.C.F. 1962. Morphological evidence of frost action in Arctic soils. Biul. Peryglacjalny. 11:343-352.

Tedrow, J.C.F. 1970. Soil investigations in Inglefield Land, Greenland. Medd. om Gronland 188(3). 93 pp.

Tedrow, J.C.F. 1977. Soils of the Polar Landscapes. Rutgers Press. New Brunswick, N.J. 638 pp.

Tedrow, J.C.F. 1991. Pedologic linkage between the cold deserts of Antarctica and the polar deserts of the high Arctic. In: Contributions to Antarctic Research II. AGU. Antarctic Res. Ser. 53:1-17.

Tedrow, J.C.F., and J. Brown. 1968. Soils of Arctic Alaska. In: Arctic and Alpine Environments. Indiana Univ. Press. pp. 283-294.

Tedrow, J.C.F., and J. Brown. 1962. Soils of the northern Brooks Range, Alaska: Weakening of the soil-forming potential at High Arctic altitudes. Soil Sci. 93:254-261.

Tedrow, J.C.F., and J.E. Cantlon. 1958. Concepts of soil formation and classification in Arctic regions. Arctic. 11:166-179.

Tedrow, J.C.F., and D.E. Hill. 1955. Arctic brown soil. Soil Sci. 80:265-275.

Tedrow, J.C.F., and F.C. Ugolini. 1966. Antarctic Soils. In: Antarctic Soils and Soil-Forming Processes. Antarctic Res. Ser. AGU, Wash. D.C. 8:167-177.

Tedrow, J.C.F., J.V. Drew, D.E. Hill, and L.A. Douglas. 1958. Major genetic soils of the Arctic Slope of Alaska. J. Soil Sci. 9:33-45.

Ugolini, F.C. 1966a. Soils of the Mesters Vig district, northeast Greenland: 1. The Arctic brown and related soils. Medd. om Gronland. 176(1) 22 pp.

Ugolini, F.C. 1966b. Soils of the Mesters Vig district, northeast Greenland: 2. Exclusive of Arctic brown and Podzol-like soils. Medd. om Gronland. 176(2) 25 pp.

Ugolini, F.C. 1970. Antarctic soils and their ecology. In: Antarctic Ecology. Academic Press. pp.673-692.

Ugolini, F.C. 1986. Pedogenic zonation in the well-drained soils of the Arctic regions. Quat. Res. 26:100-120.

Ugolini, F.C., and D.M. Anderson. 1972. Ionic migration and weathering in frozen Antarctic soils. Soil Sci. 115:461-470.

Ugolini, F.C., and C. Bull. 1965. Soil development and glacial events in Antarctica. Quaternaria. 7:251-269.

Ugolini, F.C., and J.C.F. Tedrow. 1963. Soils of the Brooks Range, Alaska: 3. Rendzina of the Arctic. Soil Sci. 96:121-127.

Ugolini, F.C., J.G. Bockheim, and D.M. Anderson. 1973. Soil development and patterned ground evolution in Beacon Valley, Antarctica. Int. Conf. on Permafrost, Second Conf., North American Contribution, NAS, Wash. D.C. pp.246-254.

Ugolini, F.C., J.C.F. Tedrow, and C.L. Grant. 1963. Soils of the Brooks Range, Alaska: 2. Soils derived from black shale. Soil Sci. 95:115-123.

Voronov, P.S., and N.S. Spiro. 1964. Composition of soluble salts in Quaternary terrigenic sediments of the central and western sectors of East Antarctica. Probl. Arkt. Antarkt. 15:5-10. (In Russian.)

Walker, D.A., D. Cate, J. Brown, and C. Racine. 1987. Disturbance and recovery of Arctic Alaskan tundra terrain. Cold Reg. Res. and Eng. Lab. (Hanover, N.H.) Rept. 87-11. 70 pp.

Walton, G.F. 1972. The High Arctic environment and polar desert soils. Ph.D. thesis, Rutgers Univ. 479 pp.

Chapter 2. The History of Research of Eurasian Cryosols

S.V. Goryachkin[1], N.A. Karavaeva[2], and O.V. Makeev[3]

[1]Head, Laboratory of Soil Geography and Evolution, Institute of Geography, Russian Academy of Sciences, Staromonetny, 29, Moscow, 109017 Russia; e-mail pedology@igras.geonet.ru.
[2]Leading Researcher, Institute of Geography, Russian Academy of Sciences, Staromonetny, 29, Moscow, 109017 Russia; e-mail pedology@igras.geonet.ru.
[3]Professor (deceased), Institute of Physico-Chemical and Biological Problems of Soil Science, Russian Academy of Sciences, Institutskaya St., Pushchino, Moscow Region, 142290 Russia.

1. Introduction

The Eurasian study and development of concepts of permafrost-affected soils began over a century ago, in two centers, Russia and Germany. In Russia, permafrost-affected soils cover about 65% of the total area. Pioneers of Russian soil studies were faced with Cryosols. Unfortunately, for a long period, the results of these soil studies were published only in Russian and poorly known elsewhere. And since there are several thousand Russian publications on permafrost-affected soils, for this paper we have analyzed only monographs and some principal papers.

German scientists at the beginning of the 20th century developed mostly a geological approach to soil studies, and the major region for that was Spitsbergen. Tedrow (1977) wrote an early history of polar soil investigations, but we included pioneer soil studies in our analysis here. However, we did not include the deep experimental studies of freezing-thawing processes and their influence on soil material (see the summary by Ostroumov and Makeev, 1985), investigations of permafrost characteristics with no direct relationship to soil, or the permafrost-affected phenomena of Eurasian alpine regions, remote from the major areas of Cryosols (Caucasus, Alps, Himalayas, etc.).

Studies of Eurasian permafrost-affected soil can be broken into four periods: 1. First information and general ideas (before 1920); 2. First systematic studies and concepts (1920-1949); 3. Extensive field studies and new concepts (1950-1985); 4. Mapping, classification, and ecology of Cryosols (1986-2000). These periods are close to those Sokolov (1991) distinguished for soil study in East Siberia and the Far East.

2. First Information and General Ideas (before 1920)

During this period, separate data on the soils of the Eurasian permafrost regions came mostly from the studies of naturalists and geologists, for example, the famous traveler in Siberia, A. Middendorf (Middendorf, 1864). Later, such information was used for the first concepts in soil science (Dokoutchaew, 1900; Ramann, 1911).

V.V. Dokuchaev (1900) wrote that the soils of the Eurasian "Arctic zone or tundra" were cemented by permafrost and thawed only for the upper 0.3 to 0.5 m; they were strongly acid and bogged and rich in semi-decomposed organic matter, non-altered minerals, and Fe^{2+}. E. Ramann (1911) wrote that soils of high latitudes of Spitsbergen had arid features similar to those of the Central Asian highlands. In this period, large geographical studies also were conducted, including soil studies in the southern regions of Siberia and in the Far East, which looked for acceptable living conditions for a large number of colonists from Ukraine and the European part of Russia and made the first local descriptions of permafrost-affected soils (Shchusev, 1906). Even in these early publications, Russian pedologists tried to describe the phenomena of permafrost-affected soils, generalize the data obtained, and describe the role of permafrost and other soil-forming factors (Kvashnin-Samarin, 1911; Sukachev, 1911; Dranitsyn, 1914).

Almost at the same time, German pedologists obtained the first information on soil weathering, the forming of permafrost-induced patterned ground, and pedogenesis in Spitsbergen (Meinardus, 1912; Blanck, 1919) (Figure 1.2.1). Salt crusts and other pedogenic features were identified in this region of the High Arctic. The first indications of such specific permafrost-induced features as palsas came from northern Sweden (Fries and Bergström,1910).

3. First Systematic Studies and Concepts (1920-1939)

In this period, soil studies widened in European and Asian parts of the Arctic, and systematic investigations of almost all large permafrost-affected regions began. And at the same time, some previous studies were generalized.

Based on soil investigations in the vast territories of Siberia and the Far East carried out for the Migration Department (Glinka, 1921), researchers concluded that podzolized soils were widely distributed in the taiga zone; but Russian pedologists also realized that the Siberian soil cover did not fit concepts of the Russian plain (Gorshenin, 1926; Prasolov, 1927). The study of permafrost's influence on soil processes also was developing (Sumgin, 1931).

Figure 1.2.1. First German investigators of polar soils, W. Meinardus (left) and E. Blanck (right).

During this period of soil studies, the Eurasian high latitudes were surveyed widely. Systematic studies were made of the soils of the European sector of the Arctic and Subarctic (Grigoryev, 1925; Ivanov, 1933; Ivanova and Koposov 1937) (Figure 1.2.2a), the Ural Mountains (Neustruev and Gorodkov, 1923), and some northern regions of Siberia (Gorodkov, 1932). As a result, researchers distinguished the specific type of gley pedogenesis in tundra (Liverovskiy, 1934) (Figure 1.2.2b) and discovered widespread development of podzolization processes in the Arctic, even on clayey-loamy substrates (Gorodkov, 1939; Ratmanov, 1930).

Pedologists at this time first opposed the northern "glacial zone" (Ivanov, 1933) or "polar desert" (Gorodkov, 1939) to the southern (tundra) part of the Arctic. They divided the latter into four subzones: "saline polygonal and structural soils (Arctic tundra), gley mottled soils (typical moss-lichen tundra), peat gley mottled and hummocky soils (southern tundra), and peat podzolic gley soils" (Liverovskiy, 1934).

Pedologists from the United Kingdom (Elton, 1927) and Norway (Björlykke, 1928) joined their German colleagues in the 1920s and 30s to study the soils of Spitsbergen.

The systematic surveys of Nordic countries began then. The first mention of permafrost-induced features in mineral soils of Norway (Björlykke, 1928) and Sweden (Tamm, 1928) appeared.

Wide soil studies began in Mongolia; Cryosols occur in the northern half of this territory. These studies were generalized after World War II (Bespalov, 1951).

Figure 1.2.2. The advanced Russian investigators of permafrost-affected soils, E.N. Ivanova (left) and Yu.A. Liverovskiy (right).

4. Extensive Field Studies and New Concepts (1950-1985)

During this period, pedologists gathered the most important facts about permafrost-affected soils, and these facts provided the basis for contemporary conceptions.

Russian specialists from both old centers of soil science (Moscow and Leningrad) and the newly organized pedological departments in the Russian North, Siberia, and the Far East (Syktyvkar, Krasnoyarsk, Yakutsk, Ulan-Ude, Magadan, and many others) developed extensive studies. The government financed a large project, including soil surveys of remote areas, to compile the *Soil Map of the USSR (scale 1:1000000)*. But these studies resulted not only in maps but in much geographical, morphological, and chemical data on Cryosols, published in numerous books and papers.

In the northern part of European Russia, pedologists studied the permafrost-affected soil cover of the Komi Republic (Archegova, 1972; Zaboeva, 1975), Nenets district (Ignatenko, 1979), and northern part of the Ural Mountains (Firsova and Dedkov, 1983). On the Kola Peninsula, V.O. Targulian (1971) and the specialists from the Kola Scientific Centre studied the tundra and alpine Arctic desert soils and generalized their data later (Nikonov and Pereverzev, 1989).

Pedologists completed numerous studies over a wide spectrum of topics (geography, genesis, moisture and temperature regimes, biological features, chemical and mineralogical composition, etc.) in Siberia and then generalized their results in a series of books about soils of Siberia as a whole (Ivanova, 1962; Ilyin et al, 1977; Kovalev et al., 1978; Kovalev, 1979). But the soil data on the Asian part of Russia appeared systematized mostly in monographs on the different regions. Studies of the Cryosols of West Siberia covered both the boreal forest zone (Karavaeva, 1973) and tundra (Vasilievskaya et al.,1986).

The most extensive areas of permafrost-affected soils are in central and eastern Siberia. Many of them, especially in the southern and central parts of Yakutia, are of agricultural value. Soil studies here thus embraced the large areas of taiga (Makeev, 1959; Zolnikov et al., 1962; Petrova, 1971; Korsunov, 1978), tundra (Karavaeva, 1969; Vasilievskaya, 1980), and grasslands (Volkovintser, 1973). Some studies included both plains and mountainous soils (Ivanova, 1963; Elovskaya et al., 1979).

In the mountainous regions of Southern Siberia, soil investigations continued for a long period. B.F. Petrov (1952), M.P. Smirnov (1970), R.V. Kovalev and his colleagues (1973), and V.N. Gorbachev (1978) investigated the mountain systems of the western part of this region - Altai and Sayans. V.A. Nosin (1963) studied the mountains and valleys of Tuva, which have a more continental climate and soils under steppe vegetation, and others extensively studied the soils of mountainous systems in different parts of the Lake Baikal region (Nogina, 1964; Koposov, 1983; Kuzmin, 1988).

The Cryosols of northeastern Eurasia were the last for systematic studies, but here pedologists found the widest spectrum of permafrost-affected soils (Naumov and Gradusov, 1974; Sokolov, 1980; Ignatenko and Berman, 1980).

After World War II, soil studies of Spitsbergen continued. French investigations yielded approaches for both soil geomorphology (Corbel, 1954) and pedology (Fedoroff, 1966). N. Fedoroff is known not only for the first systematic studies of soils of Spitzbergen but also for the first use of the term *Cryosol*. Polish soil scientists conducted extensive and multidisciplinary soil studies of West Spitsbergen since the 1950s; S. Uziak (1992) summarized these results. Some U.S. pedologists studied soils of Spitsbergen in the context of general Arctic soil cover (Mann et al., 1986).

As for the Subarctic soil cover of Nordic countries, the first study was of permafrost-affected features in the soils of Iceland (Johannesson, 1960; Friedmann et al., 1971). Studies continued of palsas, the most pronounced permafrost-induced features in the soil cover of Northern Scandinavia, in Sweden (White et al., 1969) and Finland (Seppälä, 1972).

Russian and Mongolian soil scientists continued to study the Cryosols of Mongolia cooperatively, mainly using the Russian scientific school's approaches (Nogina, 1978; Ogorodnikov, 1981; Nogina and Dorzhgotov, 1982). Mongolia became the study area, besides Spitsbergen, for investigations of Cryosols and arid soils for Polish specialists, who cooperated with their Mongolian colleagues (Dorzhgotov and Kowalkowski, 1981). S.V. Maximovich (1993) later generalized the data on Mongolian Cryosols obtained in the 1960s through 1980s (see Section 2 of this book).

A.A. Sokolov (1977) investigated small areas of Cryosols in East Kazakhstan, basically in mountainous regions. I.A. Gorbunova (1986) studied the effect of cryogenesis, including the seasonal freezing-thawing processes on recent soils and paleocryogenic features, in soils of the plains of Central and East Kazakhstan. The studies of "High latitude" and "High altitude" permafrost-affected soils began in China (Qui Guo Qing and Cheng Guo Dong, 1993) in the 1950s.

During this period, pedologists combined the extensive field studies with generalizations and new concepts of soils and the soil cover of Eurasia and the circumpolar area. This first occurred while preparing the scheme of Soil-geographic regions for the whole territory of the USSR (Letunov, 1962). The authors gathered all the soil data, including from remote areas, and compiled a set of soil zones and subzones (distinguished by bioclimatic criteria), which they divided into provinces (distinguished by geology and some climatic features).

V.O. Targulian and N.A. Karavaeva (1964) suggested a new variant of zonation of the whole Arctic. They divided the Arctic into two zones: 1. that of Arctic and tundra soils, subdivided into the subzones of desert and soddy Arctic soils and tundra gley and tundra gley podzolized soils, respectively, and 2. that of acid unsaturated tundra gley soils.

I.S. Mikhailov (1970) divided the Arctic soil zone of Russia into three subzones: northern subzone of desert Arctic soils, middle subzone of typical Arctic soils, and southern subzone of tundra Arctic soils. They differ by the percentage of mature soils in soil cover. J. Tedrow (1977) distinguished three soils zones in the non-forested areas of the circumpolar area including Eurasia. They are polar deserts, subpolar deserts, and tundra.

Based on the data obtained in North America, N.A. Karavaeva and V.O. Targulian (1977) revised their first zonal scheme. They distinguished the following zones: 1. soddy Arctic saline and calcareous Arctic soils in the Arctic zone and northern Arctic tundra subzone; 2. homogeneous gley soils and podburs in the Arctic tundra subzone and most of the subzone of typical tundra; 3. differentiated gley, gley-podzolic, and podzol soils in the southern part of the typical and southern tundra and northern taiga.

I.A. Sokolov et al. (1982) analyzed the geography and factor ecology of well drained soils of Northern Eurasia. They categorized the territories of high latitudes (with the exception of Spitsbergen) as areas of primitive *cambizems* and *cryoxerozems*. In addition to humid areas of the tundra zone, they were first to distinguish *semihumid* and *semiarid* and *arid* areas with soddy soils and *tundra-steppe xerozems* in this zone.

V.O. Targulian (1971) proposed distinguishing the well drained and imperfectly drained soils in tundra and north taiga. He proved the common set of soil processes for a large spectrum of well drained permafrost-affected soils, which strongly differed in respect to their taxonomy earlier. He proposed naming these soils *podburs* (for soils without bleached horizons) and *Al-Fe-humic podzols* (with bleached horizons).

O.V. Makeev proposed recognizing the great role of cryogenic processes in soil-formation (Makeev and Kerzhentsev, 1974; Makeev, 1981). He believed that both permafrost and seasonal freezing play important roles in pedogenesis, and he included both in the term *cryopedosphere*.

All these generalizations and concepts of Cryosols' distribution and genesis were developed during the last period of studies of Eurasian permafrost-affected soils.

5. Mapping, Classification, and Ecology of Cryosols (1986-2000)

During this period, field work to study the distribution and genesis of Cryosols was reduced markedly, because many large projects were completed, Russia and Poland faced economic crises, and Nordic countries conducted very pragmatic soil science. However, the territories not investigated earlier because of military secrets (for example, Novaya Zemlya) were studied during this time (Goryachkin, 1998), as were the poorly investigated Cryosols of China (Qui Guo Qing and Cheng Guo Dong, 1993).

In the 1990s, international cooperation progressed. This led to an intensive exchange of ideas and data and organization of joint field research aimed to correlate the old classification systems of Northern soils and to create new ones (Mazhitova et al., 1993). Special orders for permafrost-affected soils appeared in the national taxonomies of the USA (Gelisols), France (Cryosols), and Russia (Cryozems). This book discusses these things in the Section on classification problems.

The most important events of this period were the completion of several soil maps for different areas of Eurasia and for the whole circumpolar region. The latter was the first soil map of the Arctic (Treshnikov, 1985). The *Soil Map of the Russian Federation, scale 1:2500000* (Fridland, 1988), and the *Soil Map of Northeastern Asia*, of the same scale but more detailed for cryologic information (Naumov, 1993), were published. The schematic *Soil Map of Scandinavia (scale 1:2mln.)* was developed (Rasmussen et al., 1991), as well as the more detailed map (1:1mln.) for the Svalbard area (Låg J. 1993). On the basis of earlier data, the *Map of Mongolia (scale 1:3mln.)* was compiled (Dorzhgotov and Nogina, 1990).

Paralleling the large work of compiling soil maps, important regional generalizations were made about soils of northeastern Eurasia (Naumov, 1993). The concepts of general principles of soil distribution in the Arctic (Goryachkin et al., 1998) and in the permafrost-affected circumpolar area (Sokolov and Konyushkov, 1998; Sokolov et al., 1994) also were published (see Section 2).

The last period of Cryosol study in Eurasia is the blossoming of the ecological investigation of these soils. It concerns mainly the problem of global change. A large number of papers were published on this subject.

6. Conclusion

The Cryosols of Eurasia have been studied for more than a century. The first general concepts of their genesis and geography appeared in 1930s. But the period of the most extensive studies and most advanced concepts was 1950-1985. A set of soil maps of the scales 1:1 mln. to 1:3 mln. exists for almost the whole permafrost-affected area of Eurasia. Recently, the wide spectrum of environmentally oriented studies of Eurasian Cryosols began.

References

Archegova, I.B. 1972. Specific features of humus formation in soils of Vorkuta tundra (Osobennosti gumusoobrazovania v pochvakh Vorkutinskoi tundry). Syktyvkar. 60 pp. (In Russian.)

Bespalov, N.D. 1951. Soils of Mongolian People's Republic (Pochvy Mongolskoi Narodnoi Respubliki). Moscow. Academy of Science. (In Russian.)

Björlykke, K.O. 1928. Soil types and their profiles in Norway. Proc. I Int. Congr. Soil Sci. 4:223-269.

Blanck, E. 1919. Ein Beitrag zur Kenntis arktischer Böden, insbesondere Spitzbergens. Chemie der Erde. 1:421-476.

Corbel, J. 1954. Les sols polygonaux: observations, experience, genese. Rev. Geomorphologie dynamique. 29:49-68.

Dokoutchaew, W.W. 1900. Zone naturelles des sols. St. Petersburg.

Dorzhgotov, D., and A. Kowalkowski. 1981. Fundamental regularities in the geographical zonality of soils in Mongolia. Quaestiones geographical. N7:21-34.

Dorzhgotov, D., and N.A. Nogina, eds. 1990. Soils, scale 1:3000000. In: Mongolian People's Republic. National Atlas. (In Mongolian and Russian.)

Dranitsyn D.A. 1914. On several zonal relief forms of the Far North. Pochvovedenie. N4:21-68. (In Russian.)

Elovskaya, L.G., E.I. Petrova, and L.V. Teterina. 1979. Soils of North Yakutia (Pochvy Severnoi Yakutii). Novosibirsk. Nauka. (In Russian.)

Elton, C.S. 1927. The nature and origin of soil polygons in Spitsbergen. Geol. Soc. London Quart. J. 83:163-194.

Fedoroff, N. 1966. Les sols du Spitsberg occidental. Cent. Nat. De la Rech. Sci. Lyon. pp. 111-228.

Firsova, V.P., and V.S. Dedkov. 1983. Soils of high latitudes of Mountainous Ural. Sverdlovsk. 95 pp. (In Russian.)

Fridland, V.M., ed. 1988. Soil map of Russian Soviet Federal Socialist Republic, scale 1:2500000. (In Russian.)

Friedmann, J.D., C.E. Johansson, N. Oskarsson, H. Svensson, S. Thorarinsson, and R.S. Williams. 1971. Observation on Icelandic polygon surfaces and palsa areas. Geograf. Annal. 53A:115-145.

Fries, T., and E. Bergström. 1910. Some observations on palsas and their occurrence in Northernmost Sweden. Geol. Fören. Förhandl. Stockholm. 32(1):195-205. (In Swedish.)

Glinka, K.D. 1921. Results of the Siberian expeditions of Migration Department of the Ministry of agriculture. Bull. III syezda pochvovedov. N1-2:10-46. (In Russian.)

Gorbachev, V.N. 1978. Soils of East Sayan (Pochvy Vostochnogo Sayana). Moscow. Nauka. 199 pp. (In Russian.)

Gorbunova, I.A. 1986. Cryogenesis of soils of steppe zone in Kazakhstan (Kriogenez pochv stepnoi zony Kazakhstana). Candidate dissertation. Moscow. Moscow State University. (In Russian.)

Gorodkov, B.N. 1932. Soils of Gydan tundra. Trudy Polarnoi komissii. 7:43-62. (In Russian.)

Gorodkov, B.N. 1939. Specific features of Arctic soil cover. Izvestia Gos. Geogr. Obshchestva. 71:1516-1532. (In Russian.)

Gorshenin, K.P. 1926. On the study of the soil cover of Siberia. Pochvovedenie. N2:76-83. (In Russian.)

Goryachkin, S.V. 1998. Soil cover of the plains and mountains of the western part of Southern island. In: P.V. Boyarsky, ed. Novaya Zemlya: Nature, History, Archeology, Culture. Moscow. pp. 149-175. (Novaya Zemlya: Priroda, Istoria, Arkheologiya, Kultura). (In Russian.)

Goryachkin, S.V., N.A. Karavaeva. and V.O.Targulian. 1998. Geography of Arctic soils: current problems. Eurasian Soil Science. 31(5):467-476.

Grigoryev, A.A. 1925. Soils of Subarctic tundras and forest-tundras of Eurasia in connection with observations in Bolshezemelskaya tundra in 1921. Pochvovedenie. N4:5-32. (In Russian.)

Ignatenko, I.V. 1979. Soils of East European tundra and forest-tundra (Pochvy Vostochnoevropeiskoi tundry i lesotundry). Moscow, Nauka. (In Russian.)

Ignatenko, I.V., and D.I. Berman, eds. 1980. Geography and genesis of soil of Magadan region (Geografia i genezis pochv Magadanskoi oblasti). Vladivostok. (In Russian.)

Ilyin, V.B., V.F. Altergot, I.M. Gadzhiev, V.P. Panfilov, and S.S. Trofimov, eds. 1977. Problems of Siberian soil science (Problemy Sibirskogo pochvovedenia). Novosibirsk, Nauka. 256 pp. (In Russian.)

Ivanov, I.M. 1933. Soil formation in the glacial zone. Trudy Arkticheskogo instituta. 12:183-202. (In Russian.)

Ivanova, E.N., ed. 1962. On soils of Ural, West and Central Siberia (O pochvakh Urala, Zapadnoi i Tsentralnoi Sibiri). Moscow, USSR Academy of Science Press. 212 pp. (In Russian.)

Ivanova, E.N., ed. 1963. On soils of East Siberia (O pochvakh Vostochnoi Sibiri). Moscow. USSR Academy of Science press. (In Russian.)

Ivanova, E.N., and N.A. Koposov. 1937. Soils of Khibiny tundras (Pochvy Khibinskikh tundra). Leningrad. USSR Academy of Sciences Press. (In Russian.)

Johannesson, B. 1960. The soils of Iceland. Rep. Ser. B. N 13. Dep. Agr. Reykjavik, Iceland.

Karavaeva, N.A. 1969. Tundra soils of Northern Yakutia (Tundrovye pochvy Severnoi Yakutii). Moscow. Nauka. (In Russian.)

Karavaeva, N.A. 1973. Soils of taiga of West Siberia (Pochvy taigi Zapadnoi Sibiri). Moscow. Nauka. (In Russian.)

Karavaeva, N.A., and V.O. Targulian. 1977. Autonomous soil formation in the North of Eurasia and America. In: V.A. Kovda, ed. Problems of soil science. Moscow. Nauka. pp. 301-309.

Koposov, G.F. 1983. Genesis of soils of the Pribaikalie mountains (Genezis pochv gor Pribaikalia). Novosibirsk. Nauka. 254 pp. (In Russian.)

Korsunov, V.M., ed. 1978. Forest soils of Angara-Yenisey economic region (Lesnye pochvy Angaro-Yeniseiskogo ekonomicheskogo regiona). Krasnoyarsk. (In Russian.)

Kovalev, R.V., ed. 1979. Specificity of soil formation in Siberia (Spetsifika pochvoobrazovania v Sibiri). Novosibirsk. Nauka. 294 pp. (In Russian.)

Kovalev, R.V., ed. 1973. Soils of Mountain-Altai autonomous region (Pochvy Gorno-Altaiskoi avtonomnoi oblasti). Novosibirsk, Nauka. 351pp. (In Russian.)

Kovalev, R.V., I.M. Gadzhiev, V.M. Kurachev, and V.A. Khmelev, eds. 1978. On soils of Siberia (O pochvakh Sibiri). Novosibirsk. Nauka. 231pp. (In Russian.)

Kuzmin, V.A. 1988. Soils of Pre-Baikal and Northern Trans-Baikal regions (Pochvy Predbaikalia i Severnogo Zabaikalia). Novosibirsk. Nauka. 174 pp. (In Russian.)

Kvashnin-Samarin, N.V. 1911. On the permafrost as a soil-forming factor in the western part of Amur region and in the Olekma watershed. Materials on the study of Russian soils. 20:61-88. (In Russian.)

Låg, J. 1993. Jordbunn Svalbard 1:1 mill. Nasjonalatlas for Norge. Statens kartverk.

Letunov, P.A., ed. 1962. Soil-geographical regioning of the USSR. (Pochvenno-geograficheskoe raionirovanie SSSR). Moscow. USSR Academy of Sciences press.

Liverovskiy, Yu.A. 1934. Tundra soils of the Northern region (Pochvy tundr Severnogo kraya). Leningrad. USSR Academy of Science press. (In Russian.)

Makeev, O.V. 1959. Sod taiga soils of the south of Central Siberia. (Dernovye pochvy yuga Srednei Sibiri). Ulan-Ude. (In Russian.)

Makeev, O.V. 1981. Facies of soil cryogenesis and specific features of organization of soil profiles in them (Fatsii pochvennogo kriogeneza i osobennosti organizatsii v nih pochvennykh profilei). Moscow, Nauka. (In Russian.)

Makeev, O.V,. and A.S. Kerzhentsev. 1974. Cryogenic processes in the soils of Northern Asia. Geoderma. 12:101-109.

Mann, D.H., R.S. Sletten, and F.C. Ugolini. 1986. Soil development at Kongsfjorden, Spitsbergen. Polar Research. 4(1):1-16.

Maximovich, S.V. 1993. Cryogenesis and paleocryogenesis in soils of Mongolia. In: D.A. Gilichinsky, ed. Joint Russian-American Seminar on Cryopedology and Global Change. Pushchino. pp. 75-83.

Mazhitova, G., C.L. Ping, J. Moore, S. Gubin, and S. Smith. 1993. Correlation of Russian, US, and Canadian classification systems for selected tundra and taiga soils (in Northeastern Russia). In: D.A. Gilichinsky, ed. Joint Russian-American Seminar on Cryopedology and Global Change. Pushchino. pp. 29-38.

Meinardus, W. 1912. Über einige characteristiche Bodenformen auf Spitzbergen. Sitzungsberichte Naturhistorischer Verein der Preussischen Rheinlande und Westfalens. Bonn. pp. 1-42.

Middendorf, A. 1864. Übersicht über die Natur Nord- und Ostsibiriens: 4. Die Gewächse Sibiriens. Sibirische Reisen. St. Petersburg. Vol. 4. N1.

Mikhailov, I.S. 1970. Soils. In: Soviet Arctic (Sovetskaya Arktika). Moscow. Nauka. pp. 236-249. (In Russian.)

Naumov, E.M. 1993. Soil map of the Northeast of Eurasia (Pochvennaya karta Severo-Vostoka Evrazii). Moscow.

Naumov, E.M. 1993. Soils and soil cover of Northeast Eurasia (Pochvy i pochvennyi pokrov Severo-Vostoka Evrazii). Doctoral Dissertation. Moscow. (In Russian.)

Naumov, E.M., and B.P. Gradusov. 1974. Specific features of taiga soil formation in the extreme North-East of Eurasia. Moscow. Kolos. 148 pp. (In Russian.)

Neustruev, S.S., and B.N. Gorodkov. 1923. Soil regions of the Ural area (Pochvennye raiony Uralskoi oblasti). Ekaterinburg. 203 pp. (In Russian.)

Nikonov, V.V., and V.N. Pereverzev. 1989. Soil formation in Kola Subarctic (Pochvoobrazovanie v Kolskoi Subarktike). Leningrad. Nauka. 168 pp. (In Russian.)

Nogina, N.A. 1964. Soils of Transbaikal region (Pochvy Zabaikalia). Moscow, Nauka. (In Russian.)

Nogina, N.A., ed. 1978. Soil cover of the main natural zones of Mongolia (Pochvennyi pokrov osnovnykh prirodnykh zon Mongolii). Moscow. Nauka. (In Russian.)

Nogina, N.A., and D. Dorzhgotov. 1982. Soil-geographic zoning of Mongolia. Soviet Soil Sci. 14(2):23-31.

Nosin, V.A. 1963. Soils of Tuva (Pochvy Tuvy). Moscow. USSR Academy of Science Press. 342 pp. (In Russian.)

Ogorodnikov, A.V. 1981 Soils of Mountain Forests in People's Republic of Mongolia (Pochvy gornykh lesov Mongolskoi Narodnoi Respubliki). Novosibirsk: Nauka. 144 pp. (in Russian.)

Ostroumov, V.E., O.V. Makeev. 1985. Temperature field of soils (Temperaturnoe pole pochv). Moscow. Nauka. (In Russian.)

Petrov, B.F. 1952. Soils of Altai-Sayan area (Pochvy Altaisko-Sayanskoi oblasti). Annals of Dokuchaev Soil Institute. 35:352 pp. (In Russian.)

Petrova, E.I. 1971. Soils of South Yakutia (Pochvy Yuzhnoi Yakutii). Yakutsk. (In Russian.)

Prasolov, L.I. 1927. South Transbaikal region (Yuzhnoe Zabaikalye). Leningrad. USSR Academy of Science press. (In Russian.)

Qui Guo Qing and Cheng Guo Dong. 1993. Development condition of Cryosols in alpine and plateau regions of Western China. In: D.A. Gilichinsky, ed. Joint Russian-American Seminar on Cryopedology and Global Change. Pushchino. pp. 59-75.

Ramann, E. 1911. Bodenkunde. Berlin. Springer.

Rasmussen, K., J. Sippola, L. Urvas, J. Låg, T. Troedsson, and M. Wiberg. 1991. Soil map of Denmark, Finland, Norway and Sweden, Scale 1:2000000. Oslo. Landbruksforlaget.

Ratmanov, G.E. 1930. Soils of the Novaya Zemlya Archipelago. Trudy Pochvennogo Instituta. 3-4:145-148. (In Russian.)

Seppälä, M. 1972. Pingo-like remnants in the Peltojärvi area of Finnish Lapland. Turku Univ. Geogr. Inst. Pub. Geogr. Ann. 54, Ser. A, 55:38-45.

Shchusev, S.V. 1906. Materials on soil study in Amur region. Pochvovedenie. 8:81-89. (In Russian.)

Smirnov, M.P. 1970. Soils of Western Sayan (Pochvy Zapadnogo Sayana). Moscow. Nauka. 236 pp. (In Russian.)

Sokolov, A.A. 1977. General peculiarities of soil-formation and soils of the Eastern Kazakhstan (Obshchie osobennosti pochvoobrazovaniya i pochv Vostochnogo Kazakhstana). Alma-Ata. Nauka. (In Russian.)

Sokolov, I.A. 1980. Hydromorphic non-gleyic soil formation. Pochvovedenie. N1:21-32. (In Russian.)

Sokolov, I.A. 1991. Some theoretical conceptions and problems in the investigations of East Siberian and Far East soils. Pochvovedenie. N5:131-145. (In Russian.)

Sokolov, I.A., and B.P. Gradusov. 1978. Soil-formation and weathering on basic rocks in condition of cold humid climate. Pochvovedenie. N2:5-17. (In Russian.)

Sokolov, I.A., and D.E. Konyushkov. 1998. Soils and the soil mantle of the northern circumpolar region. Eurasian Soil Science. 31:1179-1193.

Sokolov, I.A., Y.M. Naumov, and D.Y. Konyushkov. 1994. Soil cover of the North of Russia and its Cartographic Modeling. In: Proceedings of the meeting on the classification, correlation and management of Permafrost-affected soils - July, 1994. USDA, SCS. National Soil Survey Center, Lincoln, NE. pp. 118-133.

Sokolov, I.A., G.I. Bystryakov, A.O. Makeev, E.V. Kulinskaya, L.T. Kozitskaya, and G.I. Bystrov. 1982. Soils of the North: ecological, genetic, geographic and classification-nomenclature concepts (Geokhimiya landshaftov i geografiya pochv). In: Landscape geochemistry and soil geography. Moscow. Moscow University Press. pp. 145-172. (In Russian.)

Sukachev, V.B. 1911. On the permafrost influence on soil. Izvestiya Akademii nauk, seria VI, 5:83-97. (In Russian.)

Sumgin, M.I. 1931. Soil-forming conditions in the area of permafrost. Pochvovedenie. N3:5-17. (In Russian.)

Tamm, O. 1928. Die klimatischen Bodenregionen in Schweden. Proc. I Int. Congr. Soil. Sci. Washington. 1:269-285.

Targulian, V.O. 1971. Pedogenesis and weathering in cold humid regions (Pochvoobrazovanie i vyvetrivanie v kholodnykh gumidnykh oblastyakh). Nauka. Moscow. (In Russian.)

Targulian, V.O., and N.A. Karavaeva. 1964. Experience in soil-geochemical division of polar regions. In Problems of the Arctic: Nature (Problemy Arktiki: Priroda). Moscow. Nauka. pp. 213-224. (In Russian.)

Tedrow, J.C.F. 1977. Soils of the polar landscapes. Rutgers University. press.

Treshnikov, A.F., ed. 1985. Atlas of the Arctic (Atlas Arktiki). Moscow. (In Russian.)

Uziak, S. 1992. Polish pedological studies on Spitsbergen. A review. Geographia Polonica. 60:67-78.

Vasilievskaya, V.D. 1980. Soil formation in tundra of Central Siberia (Pochvoobrazovanie v tundrakh Srednei Sibiri). Moscow. Nauka. (In Russian.)

Vasilievskaya, V.D., V.V.Ivanov, and L.G. Bogatyrev. 1986. Soils of the North of West Siberia. (Pochvy Severa Zapadnoi Sibiri). Moscow, Moscow State University Press. 227 pp. (In Russian.)

Volkovintser, V.L. 1973. Steppe cryoarid soils (Stepnye krioaridnye pochvy). Novosibirsk. Nauka. (In Russian.)

White, S.E., G.M. Clark, and A. Rapp. 1969. Palsa localities in Padjelanta National Park, Swedish Lappland. Geogr. Ann. (Stockholm) 51:97-103.

Zaboeva, I.V. 1975. Soils and land resources of Komi ASSR (Pochvy i zemelnye resursy Komi ASSR). Syktyvkar. (In Russian.)

Zolnikov, V.G., L.G. Elovskaya, L.V. Teterina, and E.I. Chernyak. 1962. Soil of Vilui basin and their management (Pochvy Viluiskogo basseina i ikh ispolzovanie). Moscow. (In Russian.)

Chapter 3. Northern Soil Research in Canada

Charles Tarnocai[1]

[1]Research Scientist, Agriculture and Agri-Food Canada, Research Branch (ECORC), K.W. Neatby Building, Rm. 1135, 960 Carling Avenue, Ottawa, Canada, K1A 0C6; e-mail tarnocaict@agr.gc.ca.

1. Introduction

Northern Canada, the vast, sparsely populated region of Canada affected by permafrost, encompasses about half (5×10^6 km^2) of the country and includes the Arctic and Subarctic zones and portions of the Boreal zone. This region, which is composed of the Northwest Territories, Nunavut (formerly the eastern portion of the Northwest Territories), the Yukon, and the northern portions of most provinces, has a population density of only 2 people per hundred square kilometers.

Since most of this population is scattered in small settlements separated by great distances, roads are rare, except in the Yukon and the western portion of the Northwest Territories, and travel relies primarily on fixed wing aircraft (both commercial flights to limited areas and charter flights) and helicopters. Access to much of this region is still very difficult.

During the early part of the 20th century, most scientific activities in northern Canada were exploratory in nature. Geologists and naturalists who carried out these studies made some reference to soils, but no scientific studies of soils were carried out. Studies focusing on northern soils began in the 1940s and have continued through to the present day. This paper discusses the history of this research and the impact it has made on the development of soil science.

2. Pioneering Soil Studies: 1940-1969

Pedological research in northern Canada began in the 1940s, but these studies and those that followed throughout the 1950s and 1960s generally had a single purpose and were exploratory in nature. The purpose was an interest in the possible use of land for homesteading and agriculture. The scientists who carried out these studies viewed the northern soils simply as cold variants of the non-permafrost soils occurring in the more southerly areas of Canada.

Leahey (1943, 1947, 1953) undertook the first soil investigations in the permafrost regions of Canada, in the Subarctic and Boreal regions, north of the 60th parallel. These pioneering studies generally were carried out along existing roadways, such as the newly constructed Alaska and Mackenzie highways, and along the Mackenzie River.

Reconnaissance level soil surveys and pedological investigations in the southern Yukon and the western and southern portions of the Northwest Territories continued throughout the 1950s and 1960s (Day and Leahey, 1957; Day, 1962, 1966, 1968, 1972). Soil surveys of the Slave, Liard, and Takini River areas of the upper Mackenzie River valley in the western Northwest Territories generally were carried out by boat, with the soils being mapped, described, and sampled during daily foot traverses. One of the earliest soil studies in the High Arctic (Day, 1964) also was carried out during this period.

These soil surveys provided the first systematic soil inventory and study of spatial distribution of soils in these areas. In addition, Day and Rice (1964) provided information on the characteristics of permafrost soils in the Mackenzie Valley while Wright et al. (1959) provided information on the characteristics of some alluvial soils occurring elsewhere in the Northwest Territories.

Throughout this period, work also was carried out in the northern parts of the western provinces. The early soil surveys in northern British Columbia and Alberta during the late 1950s and early 1960s used packhorses for transportation (Figure 1.3.1). In the early 1960s, Daws Lindsay and his colleagues from the Alberta government used helicopters to carry out exploratory soil surveys in northern Alberta. This was probably the first time that helicopters were used to fly transects.

Figure 1.3.1. Field camp with packhorses. Soil survey in the Peace River area, Alberta (1950).

In the late 1950s, Dr. Walter Ehrlich and his colleagues from the Manitoba federal soil survey used rail cars and float planes to carry out similar soil surveys along the Hudson Bay Railway in northern Manitoba. In the early 1960s, Dr. Ehrlich also carried out a broad study of soils in northern Manitoba, using a float plane to crisscross the northern half of the province. The soils he described and sampled along the flight lines provided what was probably the first information collected on frozen soils in this province.

Everett (1968) and Tedrow et al. (1968) conducted the first soil mapping in the Arctic areas of Canada. Everett studied soil development on the Queen Elizabeth Islands in the High Arctic, while Tedrow and his associates (Tedrow and Douglas, 1964; Tedrow et al., 1968) concentrated on defining the soil geographic zones in the Arctic areas of the Northern Hemisphere and on developing a classification for these soils (Tedrow and Cantlon, 1958; Tedrow, 1968). Although this classification improved on the existing classifications of Arctic soils, it still did not focus on their cryogenic characteristics.

These soil surveys provided the first insight into the distribution and genesis of northern soils, especially those with permafrost. Although some of the distinctive features of these soils were recognized, they often were misinterpreted. Nevertheless, studies by these scientists, who worked for long periods in isolated and largely inaccessible areas, often under dangerous conditions, provided the basis for an understanding of these unique soils.

3. Development of New Concepts: 1970-1989

During the early 1970s, the scope and intensity of pedological studies changed. Although a great deal of exploratory work was still in progress (Rostad et al., 1977), increasing emphasis was placed on in-depth studies by multidisciplinary teams of scientists. Fieldwork in the north primarily relied on helicopters and fixed-wing aircraft, operating from remote field camps (Figures 1.3.2 and 1.3.3), and usually took place from mid-June to late August, with the scientists remaining in the field throughout this period.

Figure 1.3.2. Soil and paleoecology studies in the fossil forest area, Axel Heiberg Island, NWT (July, 1988). University of Saskatchewan and Canada Soil Survey scientists loading a Twin Otter aircraft with field gear.

3.1. Energy development studies

The primary impetus for the change to in-depth studies was the increasing pace of energy development in the north. With such development came the need to find suitable routes for oil and gas pipelines in both the western and eastern Arctic – routes that would cause minimal disruption to the land and its wildlife and peoples. The broad nature of this problem made it necessary for scientists from several disciplines (biologists, geologists, pedologists, hydrologists, geomorphologists, permafrost specialists) to work together (Figure 1.3.4).

Pedological studies were a vital component of these large-scale, multidisciplinary studies that provided baseline information for energy developments and generated data for the required environmental assessment studies. The studies began in 1970 and covered large areas in the Mackenzie Valley, the Beaufort Sea coast, the Keewatin, and the High Arctic islands in the Northwest Territories. Numerous publications summarize the soil information generated (Tarnocai, 1973, 1976, 1977, 1978a and 1978b; Zoltai and Pettapiece, 1973; Pettapiece, 1974; Zoltai and Tarnocai, 1974; Woo and Zoltai, 1977; Tarnocai and Zoltai, 1978).

Figure 1.3.3. Survey camp with fresh snow, north central District of Keewatin, NWT (August, 1976).

Figure 1.3.4. Scientists operating a power auger during a cooperative Geological Survey of Canada and Canada Soil Survey terrain and soil mapping study. North central District of Keewatin, NWT (August, 1976).

Figure 1.3.5. Retrieving soil temperature data from an eight-channel data logger by using a laptop computer. The cylindrical data logger is to the right of the computer. Inuvik area, NWT (July, 1991).

3.2. Monitoring studies

Long-term monitoring studies that concentrate on the collection of data on soil temperature, soil moisture, active layer depth, and surface subsidence began during the late 1980s and are ongoing. They provide information that helps characterize the soils and determine the effect of human activity and natural processes on Arctic soils and ecosystems.

In 1986, a soil climate monitoring program was established in the Mackenzie Valley, to characterize the soil climate and determine the effect of pipeline construction on permafrost (Cryosolic) soils (Tarnocai, 1995). As part of this program, soil and air temperatures (Figure 1.3.5), active layer thickness, and surface subsidence were monitored on 33 sites along a 1400-km transect in the Mackenzie River valley. The monitoring continued into the 1990s.

In addition, long-term ecosystem monitoring of the Ellesmere Island National Park Reserve (now Quttinirpaaq National Park) began in 1989, to determine the human impact on Arctic ecosystems and provide a method for evaluating the sensitivity of Cryosols in a High Arctic environment (Tarnocai et al., 1991b; Tarnocai and Gould, 1996). In 1995, a similar study began in the Pangnirtung Pass area of Auyuittuq National Park Reserve (now Auyuittuq National Park) on the southern part of Baffin Island (Tarnocai and Gould, 1998). In association with these studies, soil mapping was carried out (Tarnocai et al., 1991a; Tarnocai and Veldhuis, 1998) and monitoring sites were established in both parks.

3.3. Paleosol studies

During the 1980s, several paleosol studies were carried out in northern Canada. Some involved old glaciated and unglaciated surfaces in the Yukon (Smith et al., 1986; Tarnocai and Smith, 1989) and in the fossil forest area of Axel Heiberg Island in the High Arctic of the Northwest Territories (Tarnocai and Smith, 1991).

The purpose of these paleosol studies was to use pedological properties, especially those derived from cryogenic processes, to determine past climates. Other studies were carried out at archeological sites in the Northwest Territories (Tarnocai, 1997), not only to determine the past climates, but also to separate soil features derived from human activities from those resulting from pedological processes.

3.4. Characterization of permafrost soils

Soil investigations continued at a rapid pace throughout the 1980s. The federal Soil Survey both completed systematic mapping in the continental part of the Northwest Territories and carried out studies relating to soil genesis and characterization (Bradley et al., 1982; Tarnocai, 1980, 1984; Pettapiece, 1984).

Scientists from universities and other agencies focused on site-specific research relating to soil genesis and characterization. Soil characterization studies in the Northwest Territories were carried out by James (1970), who studied soils in the Rankin Inlet area of the Keewatin District; Cruickshank (1971), who studied the soils and terrain in the Resolute area of Cornwallis Island; and Walker and Peters (1977) and Lev and King (1999), who studied soils on the Truelove Lowland of Devon Island.

Other soil studies in the Northwest Territories included a correlation of soil development on various ages of glacial materials on the Cumberland Peninsula of Baffin Island (Bockheim, 1974) and a study of the mineralogy and chemistry of Arctic desert soils on Ellef Ringnes Island (Foscolos and Kodama, 1981). Burn and Smith (1988) studied the thermal properties of soils near Mayo in the Yukon, while Hendershot (1984) compared upland and valley soils in the Ungava area of Labrador.

Several scientists studied the micromorphological characteristics of permafrost soils in Canada's northern territories. Brewer and Pawluk (1975) and Pawluk and Brewer (1975a) studied soils from the Subarctic and southern Arctic, Pawluk and Brewer (1975b) studied those on Devon and King Christian islands, and Fox (1979, 1985) and Fox and Protz (1981) studied those in the Mackenzie Valley. In addition, Smith et al. (1991) used micromorphology to describe the formation of granular structure in fine-textured Turbic Cryosols on Herschel Island in the Yukon Territory.

3.5. Changing concepts

Interactions with permafrost scientists such as Dr. A.L. Washburn and Dr. J.R. Mackay, who studied cryogenic processes, furthered the understanding of the mechanisms involved in cryogenic processes and gave insight into the soil properties that might result from such processes.

3.6. Development of the Cryosolic Order

During the early stages of the energy development studies (1971-72), it became increasingly clear that none of the North American soil classifications available at that time were suitable for handling permafrost soils. None of the classifications fully reflected the importance of cryogenic processes and the unique morphologies these processes created.

These classifications, which were based on the soil profile, viewed northern soils, especially those with permafrost, as cryic versions of non-permafrost soils. Because of the complexity and variability of soil horizons in cryoturbated soils, the use of the soil profile as a basis for classification caused confusion, especially when soils were classified on the basis of very narrow soil profiles instead of pedons, which capture the inherent lateral and vertical variability of permafrost soils.

Recognition of these problems made it evident that a major revision of the Canadian soil classification system was necessary. In the new Cryosolic Order, both mineral and organic soils were included, cryogenic processes were recognized as soil-forming processes, and the pedon, on which the classification of these soils is based, was defined specifically for these soils (Canada Soil Survey Committee, 1978; Agriculture Canada Expert Committee on Soil Survey, 1987; Soil Classification Working Group, 1998).

These developments in the classification of permafrost soils were introduced to international soil scientists during soil tours held in northern Canada in connection with the 11th Congress of ISSS in 1978, the International Permafrost Conference in 1983, and the INQUA Conference in 1987 (Pettapiece et al., 1978; French and Heginbottom, eds., 1983; Morison and Smith, eds., 1987).

4. Recent Studies: 1990-2000

During the 1990s, soil research has continued in northern Canada and, as a result, the amount of soil information available has increased. This information has been used primarily for land use planning in national parks and for solving problems relating to transportation, the environment, urban planning, and agriculture

(Walmsley et al., 1987; Smith et al., 1989; Tarnocai et al., 1991a,b). Increasingly, however, the studies have focused on problems that are of interest not just nationally but also continentally and globally.

Long-term soil climate monitoring studies in the Mackenzie Valley, N.W.T., are continuing, and new studies have begun at benchmark sites in the Yukon (Tarnocai, 1995; Burgess and Tarnocai, 1997; Smith et al., 1998). Long-term ecosystem monitoring of the human impact on Ellesmere Island National Park Reserve also is continuing, and a similar study has begun for Auyuittuq National Park Reserve on Baffin Island (Tarnocai and Gould, 1996, 1998). Additional studies carried out in Auyuittuq National Park Reserve during this period include a study of soil genesis (Broll et al., 1998), a study of soil temperatures at a microscale on hummocky terrain (Mueller et al., 1998), a study of soil nutrient and vegetation relationships (Broll et al., 1999), and a study of soil biology (Mueller et al., 1999).

Mapping activities on the Arctic islands have continued (Tarnocai et al., 1991a; Tarnocai and Veldhuis, 1998) and the federal Soil Survey has completed the Soil Landscapes of Canada database (CLBRR, 1996). This uniform database covers the entire area of Canada and makes it possible to produce updated soil maps and soil-related derivative maps (Shields et al., 1991; Tarnocai and Lacelle, 1996).

4.1. International studies and the circumpolar view

4.1.1. Soil classifications

A field trip in northwestern Canada in 1991 provided an opportunity for US scientists to learn how to recognize cryogenic features in these frozen soils and to become familiar with the Canadian method of using the pedon approach to describe, sample, and classify these soils.

Scientists visited numerous sites in the Northwest Territories and Yukon (Tarnocai et al., 1993) during an international tour on the classification, correlation, and management of permafrost-affected soils that was held in July, 1993. During this tour, the International Committee on Permafrost-Affected Soils was formed to develop a proposal for the Gelisol Order for inclusion in US soil taxonomy (Bockheim et al., 1997). The Gelisol Order, which follows the Canadian concept in principle, was incorporated into *Keys to Soil Taxonomy* in 1998 (Soil Survey Staff, 1998). Also in 1998, the *World Reference Base for Soil Resources* included a revised Cryosol classification based, in part, on the Canadian and US classifications (ISSS Working Group RB, 1998).

4.1.2. Global change

Recent activities relating to global change also have focused on northern Canada. The 4°C warming predicted for this region (Woo et al., 1992; Mitchell et al., 1990) would cause marked changes in the thermal and moisture regimes of these northern areas. This would have a great impact on permafrost and Cryosols, and it is predicted that the area of Cryosols would be reduced from the current 35% to 17% of the soil area of Canada (Tarnocai, 1999).

As a result of this emphasis on global change, Canadian soil scientists have become increasingly involved in cooperative projects, such as the Northern Circumpolar Database and map and the Circumpolar Active Layer Monitoring (CALM) network.

5. Summary

During the past 60 years, soil research in northern Canada has progressed from exploratory surveys of limited areas, through multidisciplinary studies focusing on specific problems, to broad studies involving issues of international concern.

An understanding of the properties and nature of Cryosols has become more vital as people move northward and expand activities such as agriculture, resource development, and tourism, all of which place increasing demands on these often fragile soils and ecosystems.

References

Agriculture Canada Expert Committee on Soil Survey. 1987. The Canadian System of Soil Classification (2nd ed.). Agriculture Canada. Ottawa. Publication 1646. 164 pp.

Bockheim, J.G. 1974. Soils of Cumberland Peninsula, Baffin Island, N.W.T., Canada. Final Report of Parks Canada. Contract 74-112.

Bockheim, J., J.M. Kimble, C. Tarnocai, and C.L. Ping. 1997. Gelisols: a newly proposed order in Soil Taxonomy. Cryosols in Classification Hierarchy. Special issue, Nauka, Novosibirsk, Russia. Russian Academy of Sciences. ISBN 5-201-14350-4. pp. 11-14.

Bradley, S.W., J.S. Rowe, and C. Tarnocai. 1982. An ecological land survey of the Lockhart River map area, Northwest Territories. Environment Canada. Ottawa. Ecological Land Classification Series, No. 16. 152 pp.

Brewer, R., and S. Pawluk. 1975. Investigations of some soils developed in hummocks of the Canadian Subarctic and Southern Arctic regions. 1. Morphology and micromorphology. Can. J. Soil Sci. 55:301-319.

Broll, G., G. Mueller, and C. Tarnocai. 1998. Permafrost-affected soils in the Pangnirtung Pass area, Baffin Island, Canada. In: A.G. Lewkowicz and M. Allard, eds. Proceedings, Seventh International Conference on Permafrost. Centre d'études nordiques, Université Laval. Québec. pp. 89-94.

Broll, G., C. Tarnocai, and G. Müller. 1999. Interactions between vegetation, nutrients and moisture in soils in the Pangnirtung Pass area, Baffin Island, Canada. In: C. Tarnocai, R. King, and S. Smith, eds. Cryosols and Cryogenic Environments, special issue of Permafrost and Periglacial Processes (Vol. 10, No. 3). pp. 265-277.

Burgess, M.M., and C. Tarnocai. 1997. Peatlands in the Discontinuous Permafrost Zone along the Norman Wells pipeline, Canada. In: I.K. Iskandr, E.A. Wright, J.K. Radke, B.S. Sharratt, P.H. Groenevelt, and L.D. Hinzman, eds. Proceedings of the International Symposium on Physics, Chemistry, and Ecology of Seasonally Frozen Soils,

Fairbanks, Alaska, June 10-12, 1997. U.S. Army Cold Regions Research and Engineering Laboratory. Hanover, U.S.A. Special Report 97-10. pp. 417-424.

Burn, C.R., and C.A.S. Smith. 1988. Observations of the "Thermal Offset" in near-surface mean annual ground temperatures at several sites near Mayo, Y.T. Arctic. 41:99-104.

Canada Soil Survey Committee. 1978. The Canadian System of Soil Classification. Can. Dept. Ag. Ottawa. Publication 1646. 164 pp.

CLBRR (Centre for Land and Biological Resources Research). 1996. Soil Landscapes of Canada, v.2.2. Research Branch, Agriculture and Agri-Food Canada. Ottawa. (Digital database.)

Cruickshank, J.G. 1971. Soils and terrain units around Resolute, Cornwallis Island. Arctic. 24:195-209.

Day, J.H. 1962. Reconnaissance soil survey of the Takhini and Dezadeash Valleys in the Yukon Territory. Research Branch, Can. Dept. Ag. Ottawa. 78 pp. and soil map.

Day, J.H. 1964. Operation Hazen: Characteristics of soils of the Hazen Camp area, northern Ellesmere Island, N.W.T. Defence Research Board, Department of National Defense. Ottawa, Canada. D Phys R (G), Hazen 24. 28 pp.

Day, J.H. 1966. Reconnaissance soil survey of the Liard River Valley, Northwest Territories. Research Branch, Can. Dept. Ag. Ottawa. 71 pp. and soil map.

Day, J.H. 1968. Soils of the upper Mackenzie River area, Northwest Territories. Research Branch, Can. Dept. Ag. Ottawa. 77 pp. and soil map.

Day, J.H. 1972. Soils of the Slave River lowland in the Northwest Territories. Agriculture Canada. Ottawa. 60 pp. and soil map.

Day, J.H., and A. Leahey. 1957. Reconnaissance soil survey of the Slave River lowland in the Northwest Territories of Canada. Can. Dept. Ag. Ottawa. 44 pp. and soil map.

Day, J.H., and H.M. Rice. 1964. The characteristics of some permafrost soils in the Mackenzie Valley, N.W.T. Arctic. 17:223-235.

Everett, K.R. 1968. Soil development in the Mould Bay and Isachsen areas, Queen Elizabeth Islands, Northwest Territories, Canada. Ohio State Univ. Rsch Found. Columbus, Ohio. 75 pp. and maps.

Foscolos, A.E., and H. Kodama. 1981. Mineralogy and chemistry of Arctic desert soils on Ellef Ringnes Island, Arctic Canada. J. Soil Sci. Soc. Am. 45:987-993.

Fox, C.A. 1979. The Soil Micromorphology and Genesis of the Turbic Cryosols from the Mackenzie River Valley and the Yukon Coastal Plain. Ph.D. thesis. Univ. of Guelph. Guelph, Ont. 196 pp.

Fox, C.A. 1985. Micromorphology of an Orthic Turbic Cryosol — a permafrost soil. In: P. Bullock and C.P. Murphy, eds. Soil Micromorphology. Vol. 2. Soil Genesis. A B Academic Pubs. Berkhamsted, U.K. pp. 699-705.

Fox, C.A., and R. Protz. 1981. Definition of fabric distributions to characterize the re-arrangement of soil particles in the Turbic Cryosols. Can. J. Soil Sci. 61:29-34.

French, H.M., and J.A. Heginbottom, eds. 1983. Guidebook to permafrost and related features of the northern Yukon Territory and Mackenzie Delta, Canada. Fourth International Conference on Permafrost. Univ. Alaska. Fairbanks, Alaska. 186 pp.

Hendershot, W.H. 1984. A comparison of some upland and valley soils in the Ungava-Labrador Peninsula. Géographie physique et Quaternaire. 38:243-256.

ISSS Working Group RB. 1998. World Reference Base for Soil Resources. International Society of Soil Science (ISSS), International Soil Reference and Information Centre (ISRIC) and Food and Agriculture Organization of the United Nations (FAO). Report 84, Rome. 91 pp.

James, P.A. 1970. The soils of the Rankin Inlet area, Keewatin, N.W.T., Canada. Arct. Alp. Res. 2:293-302.

Leahey, A. 1943. Preliminary report on an exploratory soil survey along the Alaska Military Highway and the Yukon River system. Can. Dept. Ag. Ottawa. 16 pp.

Leahey, A. 1947. Characteristics of soils adjacent to the Mackenzie River in the Northwest Territories of Canada. J. Soil Sci. Soc. Am. 12:458-461.

Leahey, A. 1953. Preliminary soil survey of lands adjacent to the Mackenzie highway in the Northwest Territories. Can. Dept. Ag. Ottawa. 22 pp. and map.

Lev, A., and R.H. King. 1999. Spatial variation of soil development in a High Arctic soil-landscape: Truelove Lowland, Devon Island, Nunavut, Canada. In: C. Tarnocai, R. King, and S. Smith, eds. Cryosols and Cryogenic Environments, special issue of Permafrost and Periglacial Processes (Vol. 10, No. 3). pp. 289-307.

Mitchell, J.F.B., S. Manabe, V. Meleshko, and T. Tokioka. 1990. Equilibrium climate change – and its implications for the future. In: J.T. Houghton, G.J. Jenkins, and J.J. Ephraums, eds. Climate Change: The IPCC Scientific Assessment. Cambridge Univ. Press. Cambridge. Ch. 5, pp. 131-172.

Morison, S.R., and C.A.S. Smith (eds.). 1987. Guidebook to Quaternary Research in Yukon. XIIth INQUA Congress, Ottawa, Canada. Nat'l Rsch Council of Can. Ottawa. 110 pp.

Mueller, G., G. Broll, and C. Tarnocai. 1998. Soil temperature regimes and microtopographic contrasts, Baffin Island, N.W.T., Canada. In: A.G. Lewkowicz and M. Allard, eds. Proceedings, Seventh International Conference on Permafrost, Centre d'études nordiques. Université Laval. Québec. pp. 771-775.

Mueller, G., G. Broll, and C. Tarnocai. 1999. Biological activity as influenced by microtopography in a Cryosolic soil. Permafrost and Periglacial Processes. 10(3):279-288.

Pawluk, S., and R. Brewer. 1975a. Investigation of some soils developed in hummocks of the Canadian Sub-Arctic and Southern-Arctic regions. 2. Analytical characteristics, genesis and classification. Can. J. Soil Sci. 55:321-330.

Pawluk, S., and R. Brewer. 1975b. Micromorphological and analytical characteristics of some soils from Devon and King Christian Islands, N.W.T. Can. J. Soil Sci. 55:349-361.

Pettapiece, W.W. 1974. A hummocky permafrost soil from the subarctic of northwestern Canada and some influences of fire. Can. J. Soil Sci. 54:343-355.

Pettapiece, W.W. 1984. Some considerations of soil development in northwestern Canada and some ecological relationships. In: R. Olson, R. Hastings and F. Geddes, eds. Northern Ecology and Resource Management. Univ. of Alberta Press. pp. 3-17.

Pettapiece, W.W., C. Tarnocai, S.C. Zoltai, and E.T. Oswald. 1978. Soil, permafrost and vegetation relationships in the Yukon and Northwest Territories of northwestern Canada. Guidebook for Tour 18, 11th Congress of the International Society of Soil Science, Edmonton, Canada, June 1978. 165 pp.

Rostad, H.P.W., L.M. Kozak, and D.F. Acton. 1977. Soil Survey and Land Evaluation of the Yukon Territory. Saskatchewan Institute of Pedology Publications S-174. Saskatoon, Canada. 495 pp. and maps.

Shields, J.A., C. Tarnocai, K.W.G. Valentine, and K.B. MacDonald. 1991. Soil landscapes of Canada: procedures manual and user's handbook. Land Resource Research Centre, Agriculture Canada. Ottawa. Publication 1868/E. 74 pp.

Smith, C.A.S., C. Tarnocai, and O.L. Hughes. 1986. Pedological investigations of Pleistocene glacial drift surfaces in the central Yukon. Géographie physique et Quaternaire, 40:29-37.

Smith, C.A.S., C. Kennedy, A.E. Hargrave, and K. McKenna. 1989. Soil and vegetation survey of Herschel Island, Yukon Territory. Land Resource Research Centre, Research Branch, Agriculture Canada. Ottawa. 19 pp. and 2 maps.

Smith, C.A.S., C.A. Fox, and A.E. Hargrave. 1991. Development of soil structure in some Turbic Cryosols in the Canadian Low Arctic. Can. J. Soil Sci. 71:11-29.

Smith, C.A.S., C.R. Burn, C. Tarnocai, and B. Sproule. 1998. Air and soil temperature relations along an ecological transect through the permafrost zones of northwestern Canada. In: A.G. Lewkowicz and M. Allard, eds. Proceedings, Seventh International Conference on Permafrost, Centre d'études nordiques. Université Laval. Québec. pp. 1009-1016.

Soil Classification Working Group. 1998. The Canadian System of Soil Classification. (3rd ed.) Research Branch, Agriculture and Agri-Food Canada, Ottawa, Publication 1646. 187 pp.

Soil Survey Staff. 1998. Keys to Soil Taxonomy. USDA. Washington, D.C. 326 pp.

Tarnocai, C. 1973. Soils of the Mackenzie River area. Environmental-Social Committee, Northern Pipelines, Task Force on Northern Oil Development. Report No. 73-26. 136 pp.

Tarnocai, C. 1976. Soils of Bathurst, Cornwallis, and adjacent islands, District of Franklin. Geological Survey of Canada, Department of Energy, Mines and Resources. Ottawa. Paper 76-1B. pp. 137-141.

Tarnocai, C. 1977. Soils of the north-central Keewatin. Geological Survey of Canada, Department of Energy, Mines and Resources. Ottawa. Paper 77-1A. pp. 61-64.

Tarnocai, C. 1978a. Genesis of organic soils in Manitoba and the Northwest Territories. Proceedings of the Third York Quaternary Symposium, York Univ., Toronto, Ontario. pp. 453-470.

Tarnocai, C. 1978b. Distribution of soils in northern Canada and parameters affecting their utilization. 11th Congress of the International Society of Soil Science, Edmonton, Canada, June 1978. Transactions, Volume 3, Symposia Papers. pp. 332-347.

Tarnocai, C. 1980. Summer temperatures of Cryosolic soils in the north-central Keewatin, N.W.T. Can. J. Soil Sci. 60:311-327.

Tarnocai, C. 1984. Characteristics of soil temperature regimes in the Inuvik area. In: R. Olson, R. Hastings and F. Geddes, eds. Northern Ecology and Resource Management. Univ. of Alberta Press. pp. 19-37.

Tarnocai, C. 1995. Soil climates of the Mackenzie Valley. Centre for Land and Biological Resources Research, Agriculture and Agri-Food Canada, Ottawa. 169 pp.

Tarnocai, C. 1997. Paleosols of the northern part of North America: their features and significance as indicators of past climates. In: I.P. Martini, ed. Late Glacial and Postglacial Environmental Changes: Quaternary, Carboniferous-Permian, and Proterozoic. Oxford Univ. Press. New York. pp. 276-293.

Tarnocai, C. 1999. The effect of climate warming on the carbon balance of Cryosols in Canada. Permafrost and Periglacial Processes. 10(3):251-263.

Tarnocai, C., and J. Gould. 1996. Ecosystem and trafficability monitoring for Ellesmere Island National Park Reserve. (5-year evaluation) Centre for Land and Biological Resources Research. Ottawa. 139 pp.

Tarnocai, C., and J. Gould. 1998. Ecosystem and trafficability monitoring for Auyuittuq National Park Reserve. Prepared for the Auyuittuq National Park Reserve by the Eastern Cereal and Oilseed Research Centre (ECORC), Agriculture and Agri-Food Canada. Ottawa. 65 pp.

Tarnocai, C., and B. Lacelle. 1996. Soil Organic Carbon of Canada. Eastern Cereal and Oilseed Research Centre, Agriculture and Agri-Food Canada, Research Branch. Ottawa, Ontario, Canada. Map only.

Tarnocai, C., and C.A.S. Smith. 1989. Micromorphology and development of some central Yukon paleosols, Canada. Geoderma. 45:145-162. LRRC Contrib. No. 89-97.

Tarnocai, C., and C.A.S. Smith. 1991. Paleosols of the fossil forest area, Axel Heiberg Island. In: R.L. Christie, and N.J. McMillan, eds. Tertiary Fossil Forests of the Geodetic Hills, Axel Heiberg Island, Arctic Archipelago. Geological Survey of Canada Bulletin 403. pp. 171-187.

Tarnocai, C., and H. Veldhuis. 1998. Soils and trafficability of Pangnirtung Pass, Auyuittuq National Park Reserve. Prepared for the Auyuittuq National Park Reserve by the Research Branch, Agriculture and Agri-Food Canada. Ottawa. 78 pp.

Tarnocai, C., and S.C. Zoltai. 1978. Earth hummocks of the Canadian Arctic and Subarctic. Arct. Alp. Res. 10:581-594.

Tarnocai, C., D. Kroetsch, J. Gould, and H. Veldhuis. 1991a. Soils, vegetation and trafficability, Tanquary Fiord and Lake Hazen areas, Ellesmere Island National Park Reserve. Canadian Parks Service, Prairie and Northern Region, Winnipeg, Man. Report RM91-1/ELS. 269 pp.

Tarnocai, C., H. Veldhuis, D. Kroetsch, and J. Gould. 1991b. Trafficability monitoring program for Ellesmere Island National Park Reserve. Canadian Parks Service, Prairie and Northern Region. Winnipeg. Man. Report PG91-01/ELS. 79 pp.

Tarnocai, C., C.A.S. Smith, and C.A. Fox. 1993. International Tour of Permafrost-Affected Soils: The Yukon and Northwest Territories of Canada. Centre for Land and Biological Resources Research, Research Branch, Agriculture Canada. Ottawa. 197 pp.

Tedrow, J.C.F. 1968. Pedogenic gradients of the polar regions. J. Soil Sci. 19:197-204.

Tedrow, J.C.F., and J.E. Cantlon. 1958. Concepts of soil formation and classification in Arctic regions. Arctic. 11:166-179.

Tedrow, J.C.F., and L.A. Douglas. 1964. Soil investigations on Banks Island. Soil Sci. 98:53-65.

Tedrow, J.C.F., P.F. Bruggemann, and G.F. Walton. 1968. Soils of Prince Patrick Island. The Arctic Institute of North America. Washington, D.C. Research Paper No. 44. 81 pp.

Walker, B.D., and T.W. Peters. 1977. Soils of Truelove Lowland and Plateau. In: L.C. Bliss, ed. Truelove Lowland, Devon Island, Canada: A High Arctic Ecosystem. Univ. of Alberta Press. Edmonton. pp. 31-62.

Walmsley M.E., D.E. Maynard, and K.M. McKenna. 1987. Soil survey of the Klondike River Valley, Yukon Territory. Westland Resource Group. Victoria, B.C. 68 pp. and map folio (1:20 000 scale).

Woo, V., and S.C. Zoltai. 1977. Reconnaissance of soils and vegetation of Somerset and Prince of Wales Islands, N.W.T. Northern Forest Research Centre, Edmonton, Alberta. Information Report NOR-X-186. 127 pp.

Woo, M., A.G. Lewkowicz, and W.R. Rouse. 1992. Response of the Canadian permafrost environment to climatic change. Phys. Geog. 13:287-317.

Wright, J.R., A. Leahey, and H.M. Rice. 1959. Chemical, morphological and mineralogical characteristics of a chronosequence of soils on alluvial deposits in the Northwest Territories. Can. J. Soil Sci. 39:32-43.

Zoltai, S.C., and W.W. Pettapiece. 1973. Studies of vegetation, landforms and permafrost in the Mackenzie Valley. Environmental-Social Committee, Northern Pipelines, Task Force on Northern Oil Development. Report No. 73-4. 105 pp.

Zoltai, S.C., and C. Tarnocai. 1974. Soils and vegetation of hummocky terrain. Environmental-Social Committee, Northern Pipelines, Task Force on Northern Oil Development, Report No. 74-5. 86 pp.

Section 2. The Geography of Cryosols

The Geography of Cryosols: Introduction

C.A. Scott Smith[1] and S. Goryachkin[2]

[1]Research Scientist, Pacific Agri-Food Research Centre, Agriculture and Agri-Food Canada, Summerland, BC V2A 8Y2 Canada; e-mail smithcas@agri.gc.ca.
[2]Head, Laboratory of Soil Geography and Evolution, Institute of Geography, Russian Academy of Sciences, Staromonetny, 29, Moscow, 109017 Russia; e-mail pedology@igras.geonet.ru.

Cryosols (Gelisols) are some of the most widely distributed soils on earth, covering over 11 million km^2 or more than 8% of the earth's ice-free land surface (Soil Survey Staff, 1999). Cryosol is the predominant soil order in both Canada and Russia and occurs throughout northern Eurasia, northern North America, and western China (see the map, this section, Chapter 11).

These soils form under a wide range of temperature, moisture, and physiographic conditions, but they all have one factor in common: Cryosols develop in association with near-surface permafrost. They are the zonal soils of the polar and subpolar regions and higher elevations in more temperature regions of the world.

The Cryosols of sub-polar regions often contain vast amounts of organic matter incorporated into the soil profile by frost churning (cryoturbation), a process of mass displacement within the seasonally thawed portion of the soil profile. The Cryosols associated with polar deserts in both the northern and southern hemispheres largely are devoid of any significant organic matter. Some Cryosols are characterized by large volumes of frozen water, others are without water. Cryosols form under tundra, coniferous forest, and cold desertic vegetation.

While the classic soil-forming factors of climate, topography, vegetation, organisms, and time each play a role in the development of Cryosols, these soils are defined entirely by their thermal condition. The upper portion of the soil profile freezes and thaws periodically (the active layer); the lower portion remains perennially frozen (the top of permafrost). Disruption of this thermal balance, generally through surface disturbance, can alter the nature of the soil fundamentally, lower the permafrost table,and place the soil into a different taxonomic order.

This section's chapters describe the nature and distribution of Cryosols in various regions of Russia, Canada, Alaska, China, and Antarctica. They summarize, for the first time in English, much of the literature and data about the soils of northern Russia. The authors of chapters on Russian soils have used Russian taxons in their descriptions but, when possible, have included the equivalent World Reference Base (ISSS Working Group RB, 1999) classification. Other chapters use the Canadian or US taxonomies. For subgroup equivalencies for these systems, see the Tables in the section on soil classification, Section 5.

References

ISSS Working Group RB. 1998. World Reference Base for Soil Resources. International Society of Soil Science (OSSS), International Soil Reference and Information Centre (ISRIC) and Food and Agriculture Organization of the United Nations (FAO). World Soil Resources Report 84. FAO. Rome. 1 pp.

Soil Survey Staff. 1999. Soil Taxonomy: A basic system of soil classification for making and interpreting soil surveys. 2nd ed. USDA, NRCS. Ag. Hndbk. 436. U.S. Gov't Printing Ofc. Washington, D.C. 869 pp.

Chapter 1. Similarities and Differences in Arctic and Antarctic Soil Zones

S.V. Goryachkin[1], H.P. Blume[2], L. Beyer[3], I. Campbell[4], G. Claridge[5], J.G. Bockheim[6], N.A. Karavaeva[7], V. Targulian[8], and C. Tarnocai[9]

[1]Head, Laboratory of Soil Geography and Evolution, Institute of Geography, Russian Academy of Sciences, Staromonetny, 29, Moscow, 109017 Russia; e-mail pedology@igras.geonet.ru.
[2]Professor, Institute for Plant Nutrition and Soil Science, Christian-Albrechts-University Kiel, Olshausenstr. 40, 24098 Kiel, Germany; e-mail hblume@soils.uni-kiel.de.
[3]Professor, Institute for Polar Ecology, Christian-Albrechts-University Kiel, Wischhofstr. 1-3, Bldg. 12, 24148 Kiel, Germany; e-mail soil_science@web.de.
[4]Research Scientist, Land and Soil Consultancy Services, 23 View Mount, Nelson 7001, New Zealand; e-mail iaincampbell@xtra.co.nz.
[5]Research Scientist, 33 Kotari Road, Days Bay, Eastbourne, New Zealand; e-mail graeme.claridge@xtra.co.nz.
[6]Professor, Soils Department, University of Wisconsin, 1525 Observatory Drive, Madison, Wisconsin 53706, USA; e-mail bockheim@facstaff.wisc.edu.
[7]Leading Researcher, Institute of Geography, Russian Academy of Sciences, Staromonetny, 29, Moscow, 109017 Russia; e-mail pedology@igras.geonet.ru.
[8]Professor, Institute of Geography, Russian Academy of Sciences, Staromonetny, 29, 109017, Moscow, Russia; e-mail pedology@igras.geonet.ru.
[9]Research Scientist, Agriculture and Agri-Food Canada, Research Branch (ECORC), K.W. Neatby Building, Rm. 1135, 960 Carling Avenue, Ottawa, Canada, K1A 0C6; e-mail tarnocaict@agr.gc.ca.

1. Introduction

Soil scientists developed basic concepts of the geography of Arctic and Antarctic soils about 20 years ago (Tedrow, 1977; Karavaeva, Targulian, 1977; Sokolov et al., 1982). They acquired abundant information on soils and soil-forming factors, studied the soils of the Antarctic circumpolar region in detail, and suggested zonal patterns for the region (Campbell, Claridge, 1987; Bockheim, Ugolini, 1990; Blume et al., 1997). Their findings contain many discrepancies, however, and they demand reinterpretation of the spatial distribution of soils at high latitudes of the Northern Hemisphere and comparative analysis of soil covers of the Arctic and Antarctic.

This chapter has four objectives:

1. To analyze data on climatic and geogenic factors of soil formation and their role in spatial distribution of soil in the Arctic.
2. To incorporate the new soil data into the general concepts of the Arctic and Antarctic soil distribution.
3. To revise or update the existing schemes of soil zonation in the Arctic and Antarctic.
4. To determine similarities and differences in soil covers of high latitudes in the northern and southern hemispheres.

2. Soil Zones in the Arctic

Understanding the distribution of soils in the Arctic and the nature of soil zones requires understanding the spatial differences of soil-forming factors, both climatic and geogenic.

2.1. Temperature

The mean temperature of the warmest month decreases from south to north in treeless areas (for summer from ca. $+12^{\circ}$ to $<+2^{\circ}$), and the duration of growing seasons likewise decreases (from 120 to <60 days). Also, the relief-induced and coast-induced change in the temperature regime is cardinal, especially in the High Arctic. In Greenland, at the heads of fjords (50 to 100 km from the coastline), the temperature in summer could be 3 to $4^{\circ}C$ higher than on the coast (Jakobsen, 1991, 1992; Bay, 1997). The same trends exist in Arctic Canada, Alaska (Tedrow, 1977), and Siberia (Karavaeva, 1969).

2.2. Precipitation and evaporation

Precipitation in the Arctic decreases from south to north and from the coast inland. Different sources show closely similar values of precipitation data. Precipitation ranges from 600 to 900mm in maritime tundra to 100 to 200 mm in the High Arctic. The inland and continental parts of the Arctic receive two thirds to one half as much precipitation as the coastal zones of the same thermal level.

In spite of the data's consistency, different specialists have interpreted the humidity of the Arctic climate quite differently. Some consider the vast territories in the Arctic to be semiarid or arid (Tedrow, 1977; Sokolov et al., 1982; 1994; Ugolini, 1986). Using not only precipitation data but also evapotranspiration data, others consider the Arctic in general to be humid (Targulian and Karavaeva, 1964; Mikhailov, 1970). Evapotranspiration values range from 47 to 48 mm/year for Isachsen (Arctic Canada) and Rudolfa (Franz Joseph Land) (Isachenko and Shlyapnikov, 1989), and 80 mm/year for West Spitsbergen (Khromov and Petrosyants, 1994), to 300 mm/year in inland areas of South Greenland (Jakobsen, 1991). Evapotranspiration in every case is about a half or a third of that of the local precipitation.

However, in inland areas of the High Arctic islands and continental parts of eastern Siberia, precipitation is much lower, up to 25 mm/year in northern Greenland (Bay, 1997). In these conditions, higher temperature and orographic and catabatic winds increase evaporation (Brutsaert, 1982; Khromov and Petrosyants, 1994).

2.3. Geogenic factors

An analysis of geogenic factors indicates a drastic decrease in the percentage of plains and an increase of mountains and glaciers in the High Arctic (Ecoregions Working Group, 1989). Terrigenic calcareous stony and marine sediments dominate here in glacier-free territories (Treshnikov, 1985; Goryachkin et al., 1999).

In other words, in the High Arctic, soil cover occupies either small islands, which are open to Arctic winds, or narrow stripes of terraces and colluvial foothills among the areas covered with glaciers, mountains, and sea. Marine and stony terrigenic calcareous parent materials cover these stripes.

2.4. Soil formation

In these High Arctic conditions, the following processes markedly affect soil formation:

1. the input of salt aerosols from the sea
2. the eolian dust transport of the products of rock weathering (including calcareous rocks) (Dredge, 1992)
3. permanent catabatic winds coming from glaciers and concentrating in valleys and winds blowing from the pack ice (Khromov, Petrosyants, 1994) (they markedly increase the evaporation from adjacent landscapes)
4. the high moisture and thermal gradient observed from sea to periglacial inland areas (Jakobsen, 1988; 1990; 1992; Ba, 1997)
5. the minor duration of soil-forming processes and the high content of stony material in surface layers of the territories, recently freed from glaciation (Bockheim, 1979; Tedrow, 1986).

All this may lead to salinization and calcification of soils in local areas of the High Arctic.

An analysis of soil-forming factors in the Arctic as a whole leads to these main conclusions:

1. With respect to macroclimate, humidity and low temperatures are intrinsic to the whole Arctic. The soil-transforming ability of the climate decreases substantially in the northern direction, due to a shortened season of active pedogenesis, less liquid precipitation, and low temperatures. The kinetics of the majority of soil processes decrease, which decreases the transformation of soil substances. In more southern latitudes, where the transforming ability of the climate is higher, the soil processes level the pedodiversity which the relief and substrates (soil convergence) cause.
2. Geogenic factors determine the ratios between land and sea, mountains and plains, and a general dissection and complexity of relief. They are an important influence on the character of meso- and microclimate, which become more diverse in the High Arctic and in the most continental areas of northeastern Siberia (Sokolov and Konyushkov, 1998), and they thus increase soil divergence.
3. Geogenic factors lead to an increased spatial complexity of the relief-substrate matrix, which includes a variety of parent materials (shown by the content of salts, calcite, and crushed stones) and slopes (shown by aspect and gradient). All this also enhances the pedodiversity of soil cover.

During the last 25 years, new data were obtained which have not yet been incorporated into existing generalized concepts about the soil cover of the Arctic. These are data about soils of the earlier unexplored regions of Greenland (Jakobsen, 1988, 1991, 1992), Northern Canada (Tarnocai, 1976; Bockheim, 1979; Foscolos and Kodama, 1981; Wang et al., 1986; Tarnocai et al., 1991; Tarnocai and Gould, 1996), Alaska (Everett, 1981; Rieger, 1983; Ugolini et al., 1987; Ping, 1997), Spitsbergen (Mann et al., 1986; Uziak, 1992; Låg, 1993), the Taimyr peninsula (Vasilievskaya, 1980), the Chelyuskin peninsula at the north of Taimyr (Goryachkin et al., 1995), Novaya Zemlya (Goryachkin, 1998), and the vast region of northeastern Eurasia (Naumov, 1993).

An analysis of this new data makes it possible to describe, geographically and qualitatively, the diagnostic features of main pedogenic processes as they manifest themselves in the traditional units of zonal division of the Arctic (Goryachkin et al., 1998, 1999).

2.4.1. Cryoturbations

Cryoturbations of different kinds occur in every region of the Arctic. They are characteristic even for permafrost-free tundra regions of Europe (see the chapters on Cryosols of the Russian European North). Maximum cryoturbations appear in the tundra and decrease northwards.

2.4.2. Gley processes (redoximorphism)

The degree of gleyzation and the percentage of gley soils in the soil cover of the Arctic decrease markedly in the northern direction. This phenomenon more reflects lithological reasons (domination of stony, mostly calcareous substrates in high latitudes) than climatic ones (low temperatures and consequent decrease of microbiological Fe-reducing processes). If clayey derivatives of acid non-calcareous materials are present, the gley process is clearly pronounced in the soils of the high latitudes (Goryachkin et al., 1999) and is confirmed by redox potential (Eh) measurements (Uziak, 1992).

2.4.3. Peat accumulation

Peat accumulation can proceed over all the territory of the Arctic. Peaty soils with an organic horizon as thick as 1 m or more occur under conditions of over-moistening in many places, even in the High Arctic (Goryachkin et al., 1999). However, in contrast to the Low Arctic, where peatlands with a thick organic layer occur, soils with a thin (5-20 cm) peaty horizon usually prevail at higher latitudes.

In some cases, peaty soils occur at the feet of slopes, formed under conditions of additional lateral water and fine earth input. A substantial addition of mineral elements into peaty soils of the Arctic can occur with eolian fluxes. When soil develops under the influence of water that drains bird colonies, its trophic properties improve.

Peaty horizons as thick as 3-4 m, formed under such conditions, occur at Spitsbergen and Ellesmere (Låg, 1993). Therefore, in high latitudes, the lateral input of sources of mineral nutrition compensates for climatic conditions, which are unfavorable for peat growth.

2.4.4. Humus (mull-moder) accumulation

The accumulation of humus of the mull-moder (soddy arctic) type often occurs within the traditional zone of polar deserts and continental areas of the tundra zone. This process is frequent also in the humid sector of the Arctic and in the southern part of the tundra in the European and western Siberian sectors of the Arctic (Goryachkin et al., 1995). Soils with clear features of this process occupy drained areas, both the coldest ones with a thin snow cover (hill tops) and the warmest ones with a thick snow cover (insolated slopes of hollows and karst sink holes) (see the subchapter on Cryosols of the Russian European North).

2.4.5. Podzolization

Podzolization – the migration of organic sesquioxides compounds – is widespread in cold humid (Targulian, 1971) and even cold arid regions (Sokolov et al., 1982). This process forms the profiles of Podzols in Low Arctic tundra and podzol-like soils (soils with qualitatively pronounced features of podzolization and Al-Fe-humus migration, but which do not meet quantitative classification criteria) in the rest of the Arctic.

Data on such soils recently discovered at high latitudes is available (Goryachkin et al., 1999). As a rule, these podzolized soils differed from their southern analogues in that they have higher pH values and less pronounced differentiation in the distribution of sesquioxides, due to a lower production of acids by the plant communities of high latitudes. However, their acidity is sufficient to remove ferrous coatings from sandy quartz grains.

2.4.6. Calcification and salinization

Calcification and salinization in the soils of polar deserts appear in the form of calcareous pendings and white mottles and thin coatings of salts. Marine terrace sediments and eolian input from the sea are known salt sources. Vast outcrops of limestone crushed into dust by physical weathering in a cold climate (Dredge, 1992) can serve as a source of calcite.

Under polar desert conditions, precipitation is mostly solid, and the snow is removed from the top of the ground. This prevents the leaching of salts and carbonates, allowing salinization and calcification to occur.

These processes were observed in the "polar deserts" of North America and in the humid European sector of the Arctic (Goryachkin et al, 1999). However, salinization and calcification do not occur in all polar desert soils; the soils on non-marine substrates of Ellesmere (Tarnocai and Gould, 1996) have very low salt content.

The majority of research on the Arctic points to an active salinization dynamic (Tedrow, 1977). In general, the temporal and spatial distribution of salinization and calcification varies considerably in Arctic soils. The general tendency is toward an increased salt and calcite accumulation in the northern direction.

2.4.7. Formation of desert pavement

A desert pavement is a horizon of accumulation of stony material at the soil surface. It relates to the freezing out of stones and eolian output of fine earth. J. Tedrow (1977) denoted it as the process typical only for the zone of "polar deserts." Other numerous reports indicate that fragments of desert pavements occur in the tundra zone of Eurasia, in every area where substrates contain enough stones. At Novaya Zemlya, the formation of pavements proceeds due to the active eolian removal of fine earth from the surface by bora-orographic hurricanes (Goryachkin, 1998).

2.4.8. Textural (clay) differentiation

Clay differentiation occurs in forest tundra and in the south of the tundra zone. But it is less pronounced than the same process in boreal soils, as only "outposts" of textural differentiation appear in the Arctic.

2.5. Zoning of the Arctic

Figure 2.1.1 shows the general scheme of distribution of soil processes in the Arctic. All the soil processes, from podzolization to salinization, except textural differentiation, occur in every region of the Arctic. The figure uses both the names of the zonal divisions which we propose (see below) and widely used ones (Tedrow, 1977).

The same set of pedogenic processes and complex geogenic factors make zoning the Arctic extremely difficult. Thus several approaches are based on Arctic soil cover. Section 1 of this book reviews these schemes (see the chapter by Goryachkin, Karavaeva, and Makeev).

One approach considers "climatic moistening" to be the basic factor of soil geography in the Arctic and Subarctic (Sokolov and Konyushkov, 1998). In this division, soil regions change from west to east, and the north-south thermally induced trends are less significant.

All other approaches recognize the zonal character of the Arctic soil cover. Some of them put the thermal factor first (Mikhailov, 1970), but some zonal schemes consider a simultaneous decrease of temperature and humidity northward (Karavaeva and Targulian, 1977; Tedrow, 1977).

An analysis of soil-forming factors shows the macroclimate of the Arctic region to be principally humid, with semiarid "pedotopes" caused by the combination of geogenic factors and consequent transformation of the mesoclimate. Therefore the main climatic control of soil cover belongs to the thermal level. This principle fully corresponds with the zonal concepts of vegetation of soil scientists (Aleksandrova, 1988; Edlund, 1990; Bay, 1997) and plant biologists (Yurtsev, 1994) and physical geographers (Isachenko and Shlyapnikov, 1989), who relate the polar desert ("High Arctic tundra") to low summer temperatures and describe it on the humid and cold coasts but not in the more warm and dry inland areas of Greenland, for example (Bay, 1997).

We used soil data analysis and map and satellite images to set up zones of Arctic soil cover, based on the set of soils prevailing in the soil cover and the character of upper horizons. Table 2.1.1 and Figure 2.1.2 demonstrate the zonal pattern. The names of soils mix WRB names and informal ones, as it is very difficult to express briefly, in formal names of up-to-date taxonomic systems, the reality and relevant details of the soils in high latitudes.

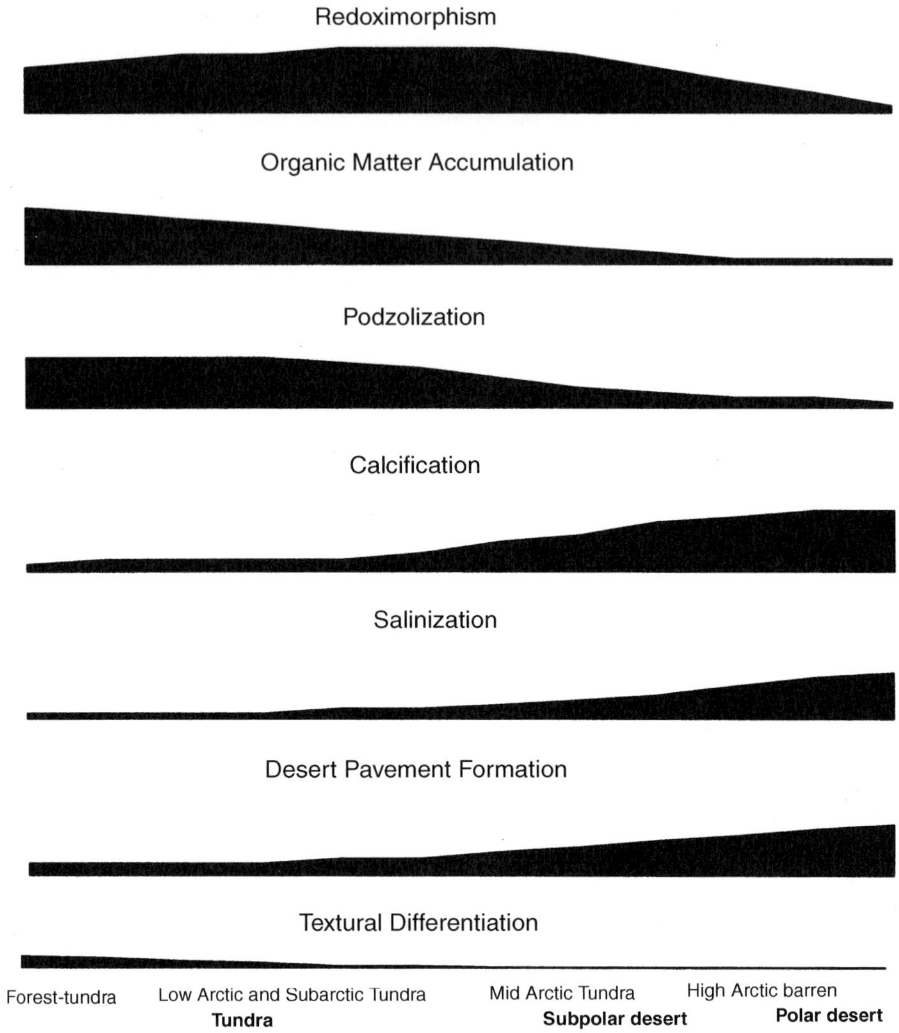

Figure 2.1.1. Pedogenic processes in the Arctic: qualitative estimation of their manifestation and frequency. Names of the zones are those the authors propose (see Table 2.1.1 and Tedrow, 1977).

As Table 2.1.1 and Figure 2.1.2 show, the boundaries differ from those of J. Tedrow (1977), especially in Eurasia. But the change of the soil zones' names is more principal. Proposed names reflect both Russian and North American traditional approaches (Treshnikov, 1985; Ecoregions Working Group, 1989) (see Table 2.1.1). We propose naming most of Tedrow's Subpolar desert "Mid-Arctic tundra," because in Eurasia this zone covers lowlands with a predominance of tundra gley (Gleyi-Turbic Cryosols) and bog soils (Histic Cryosols).

More exactly, we would name the southern part of Tedrow's Tundra zone "Low Arctic and Subarctic tundra," in accordance with both North American and Russian traditions (see Table 2.1.1). However, the northernmost zone is the most complicated case. We propose rejecting the term "desert" as not appropriate for this area; the term carries a landscape image of aridity which is a problem for anyone not involved in Arctic problems. In addition, landscapes with a combination of extremely low temperatures and non-disputable aridity, "cold deserts," do exist on earth (see below). Tedrow (1991) recently proved the great difference between "polar deserts" and "cold deserts." We would call the most northern unit of soil cover of the Arctic, "High Arctic barren."

3. Soil Zones in the Antarctic

In the Antarctic, the quantity of soils is low, because ice covers most of the continent and islands. From the Subantarctic Forests to the Antarctic Cold Desert, the mean annual temperature decreases from 6°C and more, to -18° and lower (Figure 2.1.3), whereas the mean temperature of the warmest month decreases from +9.5 (Falkland I.) to -3.4 (McMurdo). The precipitation decreases from 1300 mm to 36 mm (Table 2.1.2). The Cold Desert belongs to the xerous climate zone (< 250 m subxerous, > 1200 m ultraxerous; Bockheim 1997). The other areas are humid, including the Polar Desert, at least during the snow-free summer.

The soils of the Cold Desert are (except for some lichens in special positions) free of vegetation cover. The same is the case for soils of the Polar Desert which are not free of snow every summer, whereas lichens or mosses cover the other soils. Many soils of the Maritime Antarctic ("Subpolar Desert") are covered densely by grasses or mosses below 50 to 60 m a.s.l., whereas above 60 m is a thin cover of lichens only. Soils of the Subantarctic islands of the High Tundra (Figure 2.1.3) are covered by grasses, heath, or mosses that depend on water and nutrient conditions, whereas the Low Tundra of the Falkland Islands is characterized by tussock-grassland and peat mosses.

Acid and basic igneous and eruptive rocks, sandstones as well as moraine, solifluction, and fluvioglacial deposits of these rocks are the dominating substrates of soil formation. Limestone and till are rare (Campbell and Claridge, 1987).

3.1. Permafrost, cryoturbation, and cryoclastic weathering

Practically all soils of the Antarctic Cold, Polar, and Subpolar Deserts are soils with permafrost, as are some of the Subantarctic High Tundra (e.g., in South Georgia I.) (Campbell and Claridge, 1987; Bockheim, 1995). Most soils of ice-free parts of the Antarctic Polar Desert show a desert pavement as well as cryoturbation phenomena like stone circles and tonguing on soil surfaces, together with particle sorting inside. Higher silt or sand contents occur in the top soil, and higher clay contents occur farther down, just above permafrost. Vesicular structure appears in the first 1 to 10 cm of soil (Blume et al., 1997; Beyer et al., 2000). Dry permafrost dominates because of low temperatures in the Cold Desert (Beyer et al., 1999).

In the Subpolar Desert, cryoturbation phenomena are less pronounced and restricted to non-vegetated sites (although some slope creeping occurs). Permafrost phenomena combine with a strong cryoclastic weathering. This results in sandy to stony soils on leucogneiss and sandstone, but in loamy soils on basalt and schist (Blume et al., 1997).

Table 2.1.1. Arctic Soil Divisions.

Division	**Low arctic and subarctic tundra**	**Mid Arctic tundra**	**High Arctic barren**
Main soils	Histic, Gelic Gleysols, Podzols, elic Albeluvisols, Dystric, Gelic, Cryic Histosols; Turbic, Histic, Oxyaquic Cryosols, Gelic Cambisols in continental climate	Turbic, Histic, Spodic Cryosols, Podzol-like soils, Cryic Histosols	Turbic, Lithic, Gleyic, Oxyaquic Cryosols, dwarf Podzol-like and Rendzina-like soils, (Film-like) soils under algae
Features	Predominantly continuous soil cover, well developed litter, wide spread peats, peaty soils, palsas	Discontinuous soil cover, mull-moder A horizon, shallow, if any, litter	Contrast sporadic soil cover, soils with salic features adjacent to poorly drained peaty soils, eolian input of salt and calcite
Soil division by Tedrow, 1977	Tundra	Subpolar desert	Polar desert
Traditional Russian division (Treshniko, 1985)	Forest- tundra, Southern tundra	Typical tundra, Arctic tundra	Arctic tundra, Arctic (polar) desert
Division in Canada (Ecoregions WG, 1989)	Subarctic Low Arctic	Middle Arctic	High Arctic

3.2. Organic matter accumulation

Subantarctic Forest and Subantarctic Tundra are enriched with organic matter, as are soils with vegetative cover in the Antarctic Polar Desert. Imperfectly drained Histosols and Gleysols show 20 to more than 80 kg per m^2 in the Subantarctic Tundra and 5 to more than 20 kg in the Polar Desert (Beyer et al., 1998a; Beyer , 2000). Well drained Cambisols and Podzols of the Tundra as well as Gelic Podzols and Cryosols of the Polar Desert also can have higher amounts (Beyer et al., 1998b).

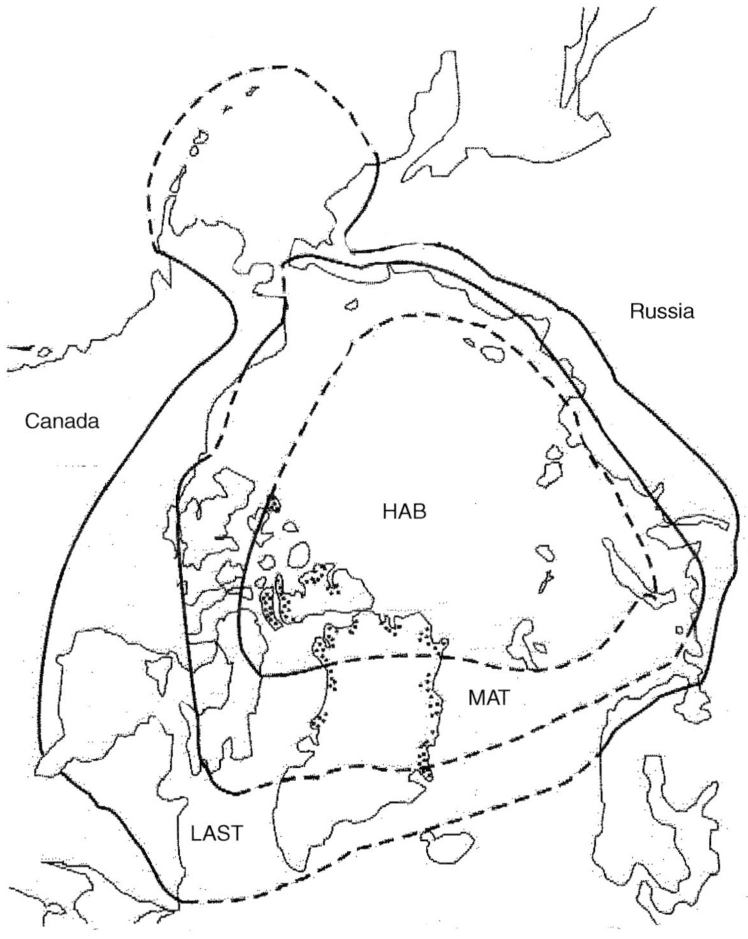

Figure 2.1.2. Soil zones of the Arctic. HAB - High Arctic barren, MAT - Mid Arctic tundra, LAST - Low Arctic and Subarctic tundra. Dotted areas are the territories, where local arid features ("pedotopes") mostly are developed.

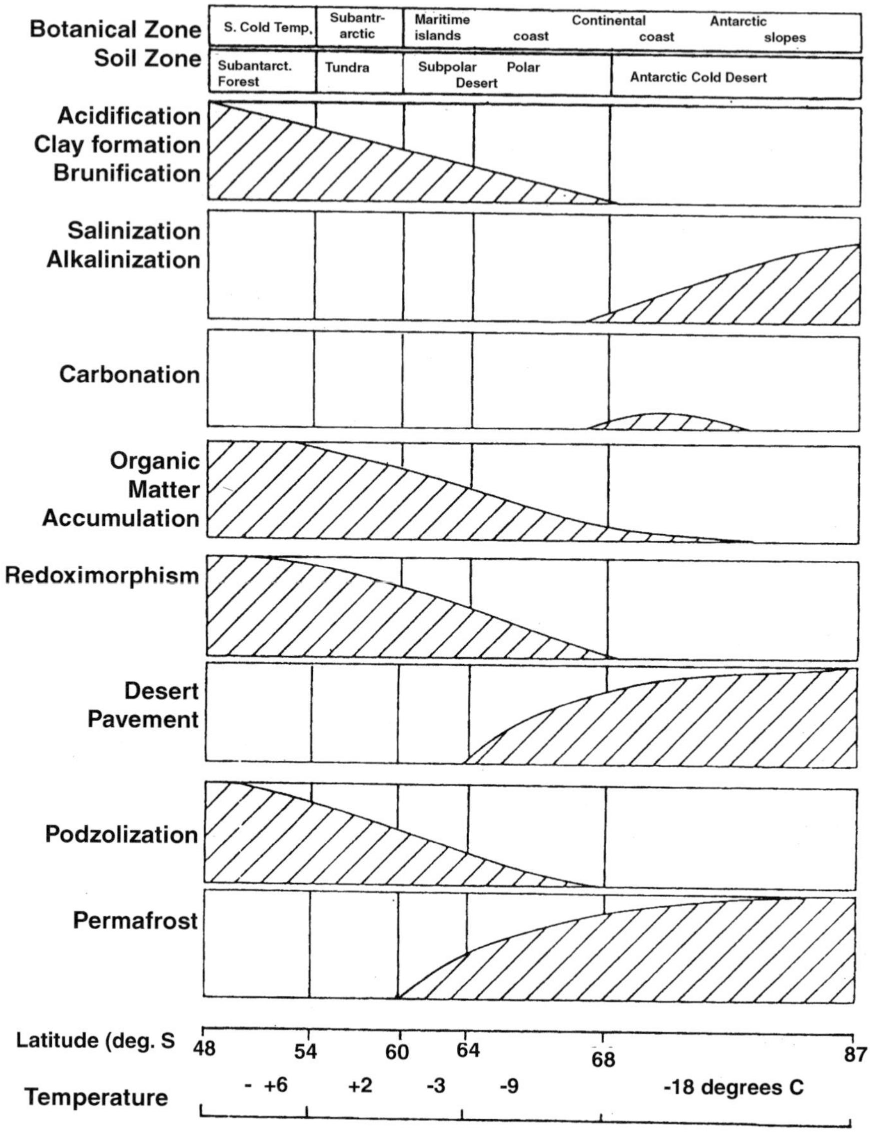

Figure 2.1.3. Changes in pedogenic processes in mineral soils along a latitudinal transect in the Southern Circumpolar Region (after Bockheim and Ugolini, 1990; changed and widened by Blume et al., 1997).

Table 2.1.2. Research sites and observed soils in the Southern Circumpolar Region (Blume et al., 1997; Bockheim, 1997; Kuhn, 1997; Beyer et al., 1999. Pedological zones after Bockheim and Ugolini, 1990. Soil classification after *WRB*, 1998).

Botanical zone	**Southern Cold Temperate**	**Subantarctic**		**Maritime Antarctic**	**Continental Coastal Antarctic**	**Antarctic**
Pedological zone	**Subantarctic Forest**	**Subantarctic**		**Subpolar Desert**	**Antarctic Desert, Polar Desert**	**Cold Desert**
		Low	**High**			
		Tundra				
Research places	South Chile, w. Cape Horn, Campbell I.	Falkland I., w. New I., Cartuss I., Falkland I.	South Georgia w., Gryviken Prion I., Salisbury Plain, Macquarie I.	S. Orkney I., S. Shetland, w. Elefant I., King George I., Decept. I., Antarc. Peninsul.	Wilkes Land	Mc Murdo Sound
Mean annual temperature	6.0	5.5	4.7 to 1.8	-1.8 to -5.0	-9.3	-18° C
Mean annual precipitation	>1000	670	1400 - 900	510 - 400	180	< 40 mm
Soils	Dystric Histosols, Dystric Cambisols, Dystric Leptosols	Dystric Histosols, Dystric Gleysols, Stagni-cambic Leptosols, Haplic Podzols (Ornithosols)*	Gelic Histosols, Dystric Histosols, Stagni-cambic Podzols, Dystric Regosols, Haplic Arenosols (Ornitosols)*	Gelic Histosols, Gelic Fluvisols, Gelic Umbrisols, Gelic Podzols; Lithic, Andic, Histic, Leptic, Mollic, Umbric, Turbic, Gleyic,Stagnic Cryosols (Ornithosols)*	Gelic Histosols, Gelic Podzols; Histic, Leptic, Lithic, Turbic Stagnic Cryosols, (Ornithosols)*	Aridi gelic, Solonchak, Lithic Cryosol, Aridi Salic Cryosol, Calci Salic Cryosol, Sali Turbic Cryosol, Eutri Turbic Cryosol

*recent penguin rookeries

Only soils without any vegetative cover and a short time without snow cover show low amounts of 1 to 2 kg in the Polar Desert, produced by snow and soil algae, and not more than 0.5 kg in the Cold Desert (Blume et al., 1997). This is the case because in cool climates the biomass production is absolutely low but is higher than the decomposing biological activity in the soils themselves; and, as organic matter is translocated from topsoil to subsoil by cryoturbation in loamy soils and podzolization in sandy soils, it is protected from decomposition directly above the permafrost. In addition, in many cases, penguins and other birds and animals translocated biomass from the sea to the soil (Blume et al., 1996; Beyer et al. , 2000).

3.3. Acidification, chemical weathering, and rubification

Many soils from carbonate free parent material of the Tundra and the Polar Desert not only strongly are weathered by frost but also are acidified (pH KCl values lower than 5 to 4) and slightly chemical weathered (Beyer et al. , 2000). Clay minerals and sesquioxides also formed. Soils from basalt, schist, and basic gneiss therefore have a loamy texture (Figure 3, Blume et al., 1997).

But the soils of the Cold Desert are neutral to alkaline, are chemically unweathered, and therefore have a sandy texture (Bockheim, 1997). Many subsoils are reddish brown, what Bockheim and Ugolini (1990) named "rubification." Relatively high amounts of oxalate and pyrophosphate extractable iron (in relation to dithionite extractable) show that this results from a formation of ferrihydrite and organic iron compounds in soils of the Tundra and the Polar Desert (Blume et al., 1997). In contrast, the red soils in the mountains of the Cold Desert seem to be relicts of a former warmer climate (Campbell and Claridge, 1987).

3.4. Podzolization

Some soils are podzolized, sandy to gravelly soils of the Polar Desert (Beyer and Bölter, 1999), as well as sandy and loamy soils of the Tundra (Blume et al., 1996). An albic E is very thin (1 to 2cm) or does not exist in many cases, but the B fits with the definition of a spodic horizon (Beyer et al., 1997; Beyer and Bölter, 1999), and thin sections show coatings of translocated metalorganic complexes (Blume et al., 1996).

Only podzolized soils of former penguin rookeries have a large albic E of 5 to 10 cm in the Polar Desert (Beyer and Bölter, 2000). In relation to Podzols of Central Europe, the C/N value of the spodic horizon is low and the content of humates with high aromaticity is high (Beyer et al., 1997).

3.5. Salinization and carbonatization

Most soils of the Cold Desert are enriched with airborne salts, due to xerous climatic conditions; chlorides dominate near the coast, and nitrates on old plateaus in the mountains (Campbell and Claridge, 1987; Bockheim, 1997). In the Polar Desert and the Tundra, recent penguin rookeries are enriched with some salts (Beyer and Bölter, 1999). Soils with secondary carbonates are rare in the Antarctic and restricted to the subxerous zone (Bockheim, 1997).

3.6. Redoximorphism

The subsoil of some loamy soils shows redoximorphic features in the form of mottles in the Tundra and also in the Polar Desert. Vivianite, which can exist only under reducing conditions, also occurs (Blume et al., 1997). In addition, peat accumulates in landscape positions of the Subantarctic Forest and Tundra and the Polar Desert, where well drained conditions can be expected in other parts of the world. A low evaporation of the precipitation, together with water stagnation, should be dominant reasons for these observations (Beyer et al., 1998c; Beyer and Bölter, 1999).

3.7. Penguins

Penguins form very special soils. Adelee penguins form their nests with translocated gravels and fill them each year with excrements of food from the sea as well as with sand and silt from the beach. Under these circumstances, 50- to150-cm-thick gravelly soils can form in the Polar Desert and in the Tundra. Recent rookeries have extremely eutrophic soils with high nutrient contents, especially N and P (Kuhn, 1997).

Formerly, penguin sites that were several thousands of years old became very acidic (perhaps influenced by a formation of nitric acid) and are enriched with Ca- and Fe-phosphates in the Polar Desert (Blume et al., 1997) and/or Al-phosphates in the Subpolar Desert and the Tundra (Myrcha and Tatur, 1991). Relatively large areas of the Tundra and the Polar Desert seem covered by current and abandoned penguin rookeries (also influenced by an uplift of more than 100 m during the last 10000 years; Myrcha and Tatur, 1991).

Soils with >2.0 of total P (or >0.20 of available P) are influenced by excrements of birds, in our opinion. We found that birds strongly influenced nearly all soils near Arctowski station, King George Islands, up to 75 m above sea level; only very young moraines didn't show this influence (Kuhn, 1997).

4. Similarities and Differences between Soil Zones in the Arctic and Antarctic

Tables 2.1.3 and 2.1.4 compare soil-forming factors, soil covers, soils, and some of their specific features. Table 2.1.3 uses our proposed names of Arctic zones. For the Antarctic (Table 2.1.4), the term "desert" remains only for the internal ice-free areas of the Antarctic – cold deserts. The subpolar and polar regions with humid climate are called "Low Antarctic and Mid Antarctic barrens," for the same reasons as for the Arctic

Table 2.1.4 shows that the cold deserts of Antarctica are very specific. They differ from other Arctic and Antarctic zones in their extra arid climatic conditions and the absence of liquid water, negative mean summer temperatures, much greater age (to 10^6 years), very poor and primitive plant cover, set of soils with saline features and high pH values, and very low organic matter content. Soil processes also distinguish cold deserts from other polar landscapes (Figures 2.1.1 and 2.1.3).

All the other zones of two high-latitude regions have both common and specific features. Antarctic zones differ from their northern analogues in their warmer winter and much colder summer and higher precipitation. The substrates are also different in the Arctic and Antarctic; they contain more fine earth and are more calcareous in the northern hemisphere. The land-ocean ratio differs, too, with very small islands and no large archipelagos in the Subantarctic and a very sharp transition to ice on the Antarctic continent.

These climatic and geogenic differences lead to clear vegetative and soil cover differences. They are much richer and continuous in the Arctic. The colder winter and looser parent material increase the role of permafrost and the proportion of Cryosols in the northern hemisphere. The paradox is that, in spite of such differences in soil-forming factors, the sets of soils in the Arctic and Antarctic, except in the cold deserts, are almost the same (see Tables 2.1.3 and 2.1.4).

That sameness could be because:

1. the coldness of the climate results mainly in discontinuity of soil cover but not in the set of soils; soil processes converge in very cold but humid regions
2. in critical conditions, life forms organize and concentrate in local shelters, and therefore here the same soil-forming processes occur as in warmer conditions
3. birds' input of sea organic matter can compensate to some extent for the climate-induced low productivity of terrestrial biota and thus results in the same set of soils.

Table 2.1.3. Some characteristics of soil-forming factors and soils of the Arctic.

	Arctic		
Soil Zone	**Low and Subarctic Tundra**	**Mid-Arctic Tundra**	**High Arctic barren**
Cold month temperature	-10-33	-14-32	-16-40
Warm month temperature	+11+13	+5+10	+1+5
Precipitation, mm	300-700	200-400	100-350
Relative humidity in summer, %%	70-90	70-90	70-90
Presence of liquid water	yes	yes	yes
Age of surface, yrs	$n*10^3$	$n*10^3$	$n*10^3$
Main substrates	loose non-marine	loose and stony	loose marine, stony terrigenic
Vegetation	shrubs, dwarf shrubs, herbs, mosses	dwarf- shrubs, herbs, mosses	lichens, mosses, herbs
Soil cover	continuous	discontinuous	Sporadic
Main soils	Histic, Gelic Gleysols, Podzols, Gelic Albeluvisols, Dystric, Gelic, Cryic Histosols; Turbic, Histic, Oxyaquic Cryosols, Gelic Cambisols	Turbic, Histic, Spodic Cryosols, Podzol-like soils, Cryic Histosols	Turbic, Lithic, Gleyic, Oxyaquic Cryosols, dwarf Podzol-like and Rendzina-like soils, (Film-like) soils under algae
Humus content, %	2-80	2-70	1-70
Salinity	low, rare	low, rare	Low-high, ephemeral
Calcite presence	rare	rare	locally
pH in upper horizons	4-6	4-7	4-9

Table 2.1.4. Some characteristics of soil-forming factors and soils of the Antarctic.

Soil Zone	**Subantarctic tundra**	**Antarctic Low Antarctic barren**	**Mid Antarctic barren**	**Antarctic cold desert**
Cold month temperature	+2 -7	-7 -11	-15 -19	-29 -30
Warm month temperature	+2+9	0 +0.4	-1 +0.3	-6. -0.8
Precipitation, mm	900-1400	400510	180-630	<40
Relative humidity in summer, %%	70-90	75-90	65-85	45-80
Presence of liquid water	yes	yes	yes	Almost no
Age of surface, yrs	$n*10^3$	$n*10^1$- $n*10^3$	$n*10^1$-$n*10^3$	$n*10^3$- $n*10^6$
Main substrates	moraines schists, sandstones	volc.+ metam. rock, moraine	gneiss, schist, moraine	stony, granite
Vegetation	herbs, mosses, dwarf shrubs	lichens, mosses, herbs (rare)	lichens, mosses, algae	Lichens (rare), algae
Soil cover	continuous to discont.	sporadic to discont.	Sporadic	sporadic
Main soils	Gelic Histosols Dystric Histosols Stagni-cambic Podzols Dystric Regosols Haplic Arenosols (Ornitosols)	Gelic Histosols Gelic Podzols Gelic Umbrisols Leptic, Turbic Lithic, Andic, Stagnic Histic, Leptic, Cryosols (Ornithosols)	Gelic Histosols Gelic Fluvisols Histic, Lithic Gelic Podzols, Mollic, Umbric, Turbic, Gleyic, Stagnic Cryosols (Ornithosols)	Aridi gelic Solonchak Lithic Cryosol Aridi Salic Cryosol Calcisalic Cryosol Salituric Cryosol Eutriturbic Cryosol
Humus content, %	2-70(4-64)[1)]	2-80(1-45)[1)]	0.4-60(.5-45)[1)]	.002-.24(.003-.7)[1)]
Salinity	low, rare [2)]	Low, rare [2)]	Low, rare [2)]	high, permanent
Calcite presence	rare	Rare	rare	everywhere
pH in upper horizons	3-7	3-7	4-7	7-10

[1)]carbon store in kg/m^2
[2)]mainly bird rookery

In any case, the sequence of soil zones, "tundra – subpolar desert – polar desert – cold desert," is not the common model for high latitude soil covers in both hemispheres. The cold desert of the Antarctic is a very dry landscape which radically differs from other polar landscapes. Subpolar and polar soil landscapes of the Arctic and Antarctic are not the same; they have essential differences, especially in discontinuity of soil covers. The Antarctic soil systems of humid sectors are much more connected to sea ecosystems than are their Arctic "analogs."

5. Conclusions

In high latitudes of both hemispheres, geogenic factors strongly control soil distribution. In the Arctic, they transform the macroclimate (principally humid), and local “arid” phenomena may occur in soil landscapes.

All the soil processes, from podzolization to salinization, can be distinguished qualitatively in every region of the Arctic and in the noncontinental Antarctic.

The continental cold desert of the Antarctic is a very dry landscape which radically differs from other polar soil landscapes in its set of soils and soil processes.

We propose three soil zones for the Arctic:

1. Low Arctic and Subarctic tundra
2. Mid Arctic tundra
3. High Arctic barren.

We propose four soil zones for the Antarctic:

1. Subantarctic tundra
2. Low Antarctic barren
3. Mid Antarctic polar barren
4. Antarctic cold desert.

Soil landscapes of the Arctic and Antarctic resemble each other in that the same set of soils and soil processes occur, except in the cold desert of the Antarctic. They differ in the more discontinuous and less permafrost-affected soil cover of the coastal Antarctic and the strong influence of sea birds on it.

References

Aleksandrova, V.D. 1988. Vegetation of the Soviet polar desert. Cambridge Univ. Press. Cambridge.

Bay, C. 1997. Floristical and ecological characterization of the polar desert zone of Greenland. J. of Veg. Sci. 8:685-696.

Beyer, L., H. Knicker, H.-P. Blume, M. Boelter, B. Vogt, and D. Schneider. 1997. Soil organic matter of suggested spodic horizons in relic ornithogenic soils of coastal contin. Antarctica in comparison with that of spodic soil horizons in Germany. Soil Sci. 162:518-527

Beyer, L., K. Pingpank, M. Bölter and R. Seppelt. 1998a. Small-distance variation of carbon and nitrogen storage in different Antarctic mineral Cryosols near Casey Station (Wilkes Land). Z. Pflanzenernähr. Bodenk. 161:211-220.

Beyer, L., K. Pingpank, M. Bölter, D. Schneider and H.-P. Blume. 1998b. Variation of carbon and nitrogen storage in soils of coastal continental Antarctica (Wilkes Land). Eurasian Soil Sci. 31:551-554.

Beyer, L., M. Bölter and H.-P. Blume. 1998c. Moisture conditions and organic matter accumulation at an ice-free oasis in East Antarctica (Wilkes Land). Abstracts 19th Annual SWS Meeting, Anchorage, 1998. p. 103.

Beyer, L., J. G. Bockheim, I. B. Campbell, and G. G. Claridge. 1999. Genesis, properties and sensitivity of Antarctic Soils. Antarc. Sci. 11:387-398.

Beyer, L., K. Pingpank, G. Wriedt and M. Bölter. 2000: Soil formation in coastal continental Antarctica. Geoderma. H, 95 (3-9):283-304.

Blume, H.-P., D. Schneider, and M. Boelter. 1996. Organic matter accumulation in and podzolization of Antarctic soils. Z. Pflanzenern. Bodenk. 159:411-412.

Blume, H.-P., L. Beyer, M. Boelter, H. Erlenkeuser, E. Kalk, U. Kneesch, U. Pfisterer, and D. Schneider. 1997. Pedogenic zonation in soils of the Southern circumpolar region. Advances in GeoEcology. 30:69-90.

Bockheim, J.G. 1979. Properties and relative age of soil of southwestern Cumberland peninsula, Baffin Island, N.W.T., Canada. Arctic and Alpine research. 11(3):289-306.

Bockheim, J. G. 1995. Permafrost distribution in the Southern circumpolar region and its relation to the environment. Permafrost a. Perigl. Processes. 6:27-45.

Bockheim, J. G. 1997. Properties and classification of cold desert soil from Antarctica. Soil Sci. Soc. America J. 61:224-231.

Bockheim, J.G., and F.C. Ugolini. 1990. A review of pedogenic zonation in well-drained soils of the Southern circumpolar region. Quaternary Research. 34:47-66.

Brutsaert, W.H. 1982. Evaporation into the atmosphere. D. Reidel Publishing Co. Dordrecht.

Campbell, I. B., and G.G.C. Claridge. 1987. Antarctica: Soils, Weathering Processes and Environment. Developments in Soil Sci. 16, Elsevier. Amsterdam.

Dredge, L.A. 1992. Breakup of limestone bedrock by frost shattering and chemical weathering, Eastern Canadian Arctic. Arc. Alp. Res. 24:314-323.

Ecoregions Working Group. 1989. Ecoclimatic Regions of Canada: First Approximation. Ecoregions Working Group of the Canada Committee on Ecological Land Classification. Ecological Land Classification Series, No. 23. Sustainable Development Branch, Canadian Wildlife Service, Conservation and Protection, Environment Canada. Ottawa, Canada. 118 pp. and map.

Edlund, S.A. 1990. Bioclimatic zones in the Canadian arctic archipelago. In: C.R. Harington, ed. Canada's missing dimension: Science and history in the Canadian Arctic Islands. Vol. 1. Canadian Museum of Nature. Ottawa. pp. 421-441.

Everett, K.R. 1981. The Soil Landscape Relations at Selected Sites Along Environmental Gradients in North Alaska. Ohio State Univ. Columbus, Oh.

Foscolos, A.E., and H. Kodama. 1981. Mineralogy and chemistry of Arctic desert soils on Ellef Ringnes Island, Arctic Canada. Soil Sci. Soc. Am. J. 45:987-993.

Goryachkin, S.V. 1998. Soil cover of the plains and mountains of the western part of Southern island. In: P.V. Boyarsky, ed. Novaya Zemlya: Nature, History, Archeology, Culture. (Novaya Zemlya: Priroda, Istoria, Arkheologiya, Kultura) Moscow. pp. 149-175. (In Russian.)

Goryachkin, S.V., R.I. Zlotin, and M.V. Glazov. 1995. Spatial and temporal organization of Arctic Ecosystems. In: Swedish-Russian Tundra Ecology Expedition - 94. A Cruise Report. Stockholm. pp. 334-339.

Goryachkin, S.V., N.A. Karavaeva, and V.O. Targulian. 1998. Geography of Arctic soils: current problems. Eurasian Soil Sci. 31(5):467-476.

Goryachkin, S.V., N.A. Karavaeva, V.O. Targulian, and M.V. Glazov. 1999. Arctic Soils: Spatial Distribution, Zonality and Transformation due to Global Change. Permafrost and Periglacial Processes. 10: 235-250.

Isachenko, A.G., and A.A. Shlyapnikov. 1989. Landscapes (Landshafty). Mysl. Moscow. (In Russian.)

Jakobsen, B.H. 1988. Soil formation on the peninsula Tugtuligssuaq, Melville Bay, North West Greenland. Geografisk Tidsskrift. 88:86-93.

Jakobsen, B.H. 1991. Aspects of soil geography in South Greenland. Folia Geographica Danica, XIX: 155-164.

Jakobsen, B.H. 1992. Preliminary Studies of soils in North-East Greenland between 74° and 75° northern latitude. Geografisk Tidsskrift. 92:111-115.

Karavaeva, N.A. 1969. Tundra soils of Northern Yakutia (Tundrovye pochvy Severnoi Yakutii). Nauka. Moscow. (In Russian.)

Karavaeva, N.A., and V.O. Targulian. 1977. Autonomous soil formation in the North of Eurasia and America. In V.A. Kovda, ed. Problems of soil science. Nauka. Moscow. pp. 301-309.

Khromov, S.P., and M.A.Petrosiants. 1994. Meteorology and climatology (Meteorologiya i klimatologiya). Moscow Univ. Press. Moscow. (In Russian.)

Kuhn, D. 1997. Genese, Ökologie und Soziologie einer Bodengesellschaft in einem Periglazialgebiet der King-George-Insel (West-Antarctis).Schriftenr. Institut Pflanzenernähr. Bodenk. Univ. Kiel 40:1-174 (with English summary.)

Låg, J. 1993. Humus accumulation and soil formation in Svalbard. In: D.A. Gilichinsky, ed. Joint Russian-American seminar on cryopedology and global change. Post-seminar proceedings. Pushchino. pp. 291-297.

Mann, D.H., R.S. Sletten, and Ugolini, F.C. 1986. Soil development at Kongsfjorden, Spitsbergen. Polar Research. 4(1):1-16.

Mikhailov, I.S. 1970. Soils. In: Soviet Arctic (Sovetskaya Arkika). Nauka. Moscow. pp. 236-249. (In Russian.)

Mircha, A., and A. Tatur. 1991. Ecological role of the current and abandoned penguin rookeries in the land environment of the Maritime Antarctic. Polish Polar Res. 12:3-24.

Müller, M.1987. Handbuch ausgewählter Klimastationen der Erde. Forschungsstelle Bodenerosion Univ. Trier, Heft 5.

Naumov, E.M. 1993. Soils and soil cover of the North-East Eurasia. (Pochvy i pochvennyi pokrov Severo-Vostoka Evrazii). Doctoral Dissertation. Moscow. (In Russian).

Ping, C.L. 1997. Characteristics of permafrost soils along a latitudinal transect in arctic Alaska. Agroborealis. 29(1):35-36.

Rieger, S. 1983. The genesis and classification of cold soils. Academic Press. New York.

Sokolov, I.A., G.I. Bystryakov, A.O. Makeev, E.V. Kulinskaya, L.T. Kozitskaya, and G.I. Bystrov. 1982. Soils of the North: ecological, genetic, geographic and classification-nomenclature concepts. In: Landscape geochemistry and soil geography (Geokhimiya landshaftov i geografiya pochv). Moscow Univ. Press. Moscow. pp. 145-172. (In Russian.)

Sokolov, I.A., and D.E.Konyushkov. 1998. Soils and the soil mantle of the northern circumpolar region. Eurasian Soil Science. 31(11):1179-1193.

Targulian, V.O. 1971. Pedogenesis and weathering in cold humid regions. (Pochvoobrazovanie i vyvetrivanie v kholodnykh gumidnykh oblastyakh). Nauka. Moscow. (In Russian.)

Targulian, V.O., and N.A. Karavaeva. 1964. Experience in soil-geochemical division of polar regions. In Problems of the Arctic: Nature. (Problemy Arktiki: Priroda). Nauka. Moscow. pp. 213-224. (In Russian.)

Tarnocai, C. 1976. Soils of Bathurst, Cornwallis, and adjacent islands, District of Franklin. Geological Survey of Canada Ottawa. Paper 76-1B.

Tarnocai, C., D. Kroetch, J. Gould, and H. Veldhuis. 1991. Soils, Vegetation and Trafficability, Tanquary Fiord and Lake Hazen areas, Ellesmere Island National Park Reserve. Canadian Parks Service, Prairie and Northern Region, Winnipeg. Man. Report RM91-1/ELS.

Tarnocai, C., and J.Gould. 1996. Ecosystem and trafficability monitoring for Ellesmere Island National Park Reserve. (5-year evaluation). Centre for Land and Biological Resources Research. Ottawa.

Tedrow, J.C.F. 1977. Soils of the polar landscapes. Rutgers Univ. press.

Tedrow, J.C.F. 1986. Factors affecting polar desert soil development in the High Arctic. Biuletyn Peryglacjalny. 31:275-282.

Tedrow, J.C.F. 1991. Pedogenic linkage between the cold deserts of Antarctica and the cold deserts of the High Arctic. Contributions to Antarctic research II. Antarctic research series, 53:1-17.

Tedrow, J.C.F., and C.C. Thompson. 1969. Chemical composition of polar soils. Biuletyn Peryglacjalny. 18:169-181.

Treshnikov, A.F., ed. 1985. Atlas of the Arctic (Atlas Arktiki). Moscow. (In Russian.)

Ugolini, F.C. 1986. Pedogenic zonation in the well-drained soils of the arctic regions. Quaternary Research. 26.100-120.

Ugolini, F.C., M.G. Stoner, and D.J. Marret. 1987. Arctic Pedogenesis: 1. Evidence for contemporary podzolization. Soil Science. 144:90-100.

Uziak, S. 1992. Polish pedological studies on Spitsbergen. A review. Geographia Polonica. 60:67-78.

Vasilievskaya, V.D. 1980. Soil formation in tundra of Central Siberia (Pochvoobrazovanie v tundrakh Srednei Sibiri). Nauka. Moscow. (In Russian.)

Wang, C., J.A. McKeague, and H. Kodoma. 1986. Pedogenic imogolite and soil environments: case study of spodosols in Quebec, Canada. Soil Science Society of America Journal. 50:711-718.

Yurtsev, B.A. 1994. Floristic division of the Arctic. Journal of Vegetation Science, 5:765-776.

Chapter 2. Cryosols in Alaska

Chien-Lu Ping[1], Mark H. Clark[2], and David K. Swanson[3]

[1]Professor, University of Alaska, Agric. & Forest Exp. Station, 533 E. Firewood, Palmer, Alaska 99645, USA; e-mail pfclp@uaa.alaska.edu.
[2]Soil Scientist, USDA - Natural Resources Conservation Service, 800 W. Evergreen Avenue, Suite 100, Palmer, Alaska 99645; e-mail mclark@ak.usda.gov.
[3]Ecologist, USDA Forest Service, P.O. Box 907, Baker City, Oregon 97814, USA; e-mail dkswanson@fs.fed.us.

1. Introduction

Permafrost, or perennially frozen ground, covers $9x10^5 km^2$ or 55% of the landmass of Alaska. The continuous zone of permafrost encompasses the physiographic subdivisions of the Arctic Coastal Plain, the Arctic Foothills, the northern Norton Sound Highlands, and the Brooks Range. The zone of discontinuous permafrost encompasses the physiographic subdivisions of the Interior Alaska Lowlands, the Interior Alaska Highlands, the Alaska Range, the Copper River Plateau, the Kuskokwin Highlands, the southern Norton Sound Highlands, and the Western Alaska Coastal Plains and Deltas (Wahrhaftg, 1965; Péwé, 1975) (see Figure 2.2.1).

The term *Cryosol* is used in this paper synonymously with *perennially frozen or permafrost-affected soils*. In the US soil classification system, *Soil Taxonomy* (Soil Survey Staff, 1999), Cryosols or permafrost-affected soils key out at the Gelisol Order. Gelisols are defined as soils having permafrost within 100 cm of the soil surface or having gelic materials within 200 cm of the soil surface. Gelic materials are defined as mineral or organic soil materials that show evidence of cryoturbation and/or ice segregation in the active layer and/or the upper part of the upper permafrost (Bockheim et al., 1997).

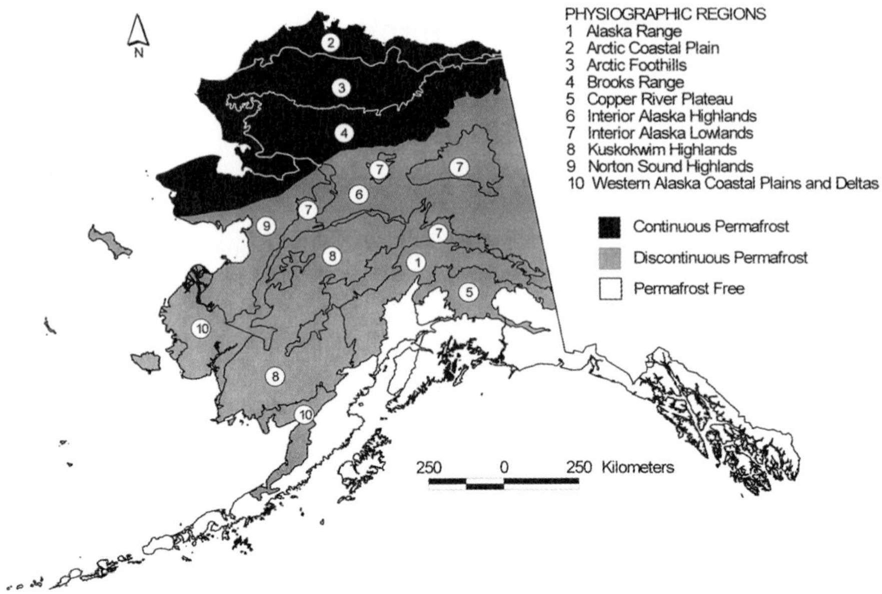

Figure 2.2.1. Distribution of Permafrost and Physiographic Regions with Permafrost in Alaska.

2. Cryosols in the Zone of Continuous Permafrost

2.1. Arctic Coastal Plain

2.1.1. Physical environment

This treeless area is characterized by multiple thaw lakes, many of them elongated in a north-northwesterly direction perpendicular to the prevailing winds. As much as 40 to 50% of the surface area is water. Low terraces, broad shallow depressions, and flood plains are typical of the landscape. Periglacial features appear, including frost polygons, hummocks, frost boils, and pingos. Thick permafrost (>600 m) underlies the entire area (Péwé, 1975). The drainage is poor or very poor. The dominant vegetation includes sedges (*Carex aquatilis*), cottongrass (*Eriophorum angustifolium*), and mosses (*Drepanocladus* spp. and *Scorpidium scorpoides*) (Walker, 1999).

2.1.2. Climate

The climate of the area varies with distance from the Arctic Ocean and elevation. In the Arctic Coastal Plain, the mean annual air temperature (MAAT) ranges from

-12.8° to -10.3°C, and mean annual precipitation (MAP) ranges from 12.5 to 14.2 cm, with 50% as snow (Haugen, 1982). The mean annual soil temperature (MAST) at 50 cm estimated from the temperatures of permafrost ranges from -7° to -9°C (Osterkamp and Romanovsky, 1996). The annual sum of growing degree days (GDD) (based on temperature of >0°C) measured at Franklin Bluff, at the southern Arctic Coastal Plain, is 121° days (L.D. Hinzman, unpublished data).

2.1.3. Soils

Two different soils are associated with low-center polygons on the coastal plain. The centers of the polygons are flat and poorly to very poorly drained, with water near or at the surface during most of the growing season. The dominant soils in the center of low-centered polygons and thaw lake basins are Typic Historthels (Hoefle et al., 1998; Ping et al., 1998). Table 2.2.1 provides a comparison among *US Soil Taxonomy* (Soil Survey Staff, 1999), the *World Reference Base* (FAO, 1998), and the *Canadian System of Soil Classification* (Soil Classification Working Group, 1998).

Typic Historthels consist of stratified peat and loamy sediments over medium-grained alluvium or lacustrine sediment that are gleyed with grayish colors (2.5Y 5/2, 2.5/2, and 5Y 2.5/1). The surface organic layer is generally 15 to 30 cm thick, with minimal cryoturbation. The active layer is about 35 cm thick, and the upper permafrost is ice-rich with thick ice lenses. The polygon rims are 20 to 40 cm above the polygon center. The soils on the polygon rim generally consist of 15 to 20 cm of organic material over 20 to 30 cm of an A horizon over a gleyed Bg horizon. The permafrost table is at 45 to 50 cm depth, and the upper permafrost (Cf) contains over 60% ice by volume.

Soil horizons in the polygon rim show a convex pattern, following the surface relief of the rim; thus the soils are cryoturbated and are classified as Aquic Molliturbels or Typic Histoturbels. Soils in the trough between polygons are very poorly drained and are characterized by 35 to 40 cm of organic material over an ice wedge. These soils are classified as Hemic Glacistels. Typic Aquorthels and Sulfuric Aquorthels occur along the river deltas and coastal estuaries (Ping, unpublished data).

Low-center polygons are common on the younger part of the Arctic Coastal Plain, whereas flat and high-centered polygons dominate the older part of this thaw-lake landscape. On flat-centered polygons, the soils are saturated throughout, during most of the growing season. Near the edge of the polygons, the organic layer is greater than 40 cm thick and overlays cryoturbated alluvium or lacustrine sediments with ice wedges. Thus, the soils are organic soils and are classified as Hemic Glacistels.

In the center of the polygons, the organic layer is sometimes thinner than 40 cm thick over mineral sediments, and the soils are classified as Typic Historthels; otherwise they are Histels (Soil Survey Staff, 1999) or Organic Cryosols (Soil Classification Working Group, 1998). On the better drained high-centered polygons are well developed "A" horizons. The subsoils show evidence of moderate cryoturbation, as manifested by broken and warped horizons. The permafrost table occurs at 45 to 50 cm, with organic matter frost-churned into the frozen "C" horizon; the "C" material is ice-rich, with about 60% ice by volume. These soils are either Aquic Molliturbels (Bockheim et al., 1998) or Ruptic-Histic Aquiturbels.

Soils formed in the Arctic Coastal Plain have reactions ranging from acidic to calcareous. Acidic soils in the Colville Delta formed in recently deposited alluvium on terraces (Jorgenson et al., 1996), and nonacidic soils are common on the broad alluvial fans (Ping et al., 1998). Soils on flood plains of major rivers throughout the coastal plains formed in carbonate rich parent materials, and marl

Table 2.2.1. A comparison of three approaches to classifying common soils in Alaska (subgroups).

US Soil Taxonomy **(Soil Survey Staff, 1999)**	***World Reference Base*** **(FAO, 1998)**	***Canadian System of Soil Classification*** **(Soil Classification (Working Group, 1998)**
Aquic Haplorthels	Haplic Gleyic Cryosols	Regosolic Static Cryosols
Aquic Haploturbels	Gleyic Turbic Cryosols	Gleysolic Turbic Cryosols Orthic Eutric Turbic Cryosols Orthic Dystric Turbic Cryosols Brunisolic Eutric Turbic Cryosols Brunisolic Dystic Turbic Cryosols
Aquic Molliturbels	Gleyic Turbic Cryosols	Gleysolic Turbic Cryosols
Glacic Fibristels	Gelic Glacic Histosols	Fibric Organic Cryosols
Glacic Hemistels	Gelic Glacic Histosols	Mesic Organic Cryosols
Hemic Glacistels	Gelic Glacic Histosols	Mesic Organic Cryosols
Humic Eutrocryepts	Mollic Cambisols	Orthic Sombric Brunisols
Ruptic-Histic Aquiturbels	Turbic Histic Cryosols	Gleysolic Turbic Cryosols
Sulfuric Aquorthels	Thionic Gleyic Cryosols	Gleysolic Static Cryosols
Typic Aquiturbels	Gleyic Turbic Cryosols	Brunisolic Turbic Cryosols Dystric Turbic Cryosols Eutric Turbic Cryosols Gleysolic Turbic Cryosols Orthic Turbic Cryosols
Typic Aquorthels	Haplic Gleyic Cryosols	Gleysolic Static Cryosols
Typic Dystrocryepts	Dystric Cambisols	Orthic Dystric Brunisols
Typic Eutrocryepts	Eutric Cambisols	Orthic Eutric Brunisols
Typic Fibristels	Fibric Gelic Histosols	Fibric Organic Cryosols
Typic Haplocryolls	Haplic Chernozems	Orthic Melanic Brunisols
Typic Haploturbels	Haplic Turbic Cryosols	Orthic Turbic Cryosols
Typic Historthels	Histic Cryosols	Gleysolic Static Cryosol
Typic Histoturbels	Turbic Histic Cryosols	Histic Eutric Turbic Cryosols Histic Dystric Turbic Cryosols Gleysolic Turbic Cryosols

commonly appears on the soil surface and as heavy coatings on vegetative parts below the water table and rock fragments (Ping et al., 1998).

2.2. Arctic Foothills and northern Norton Sound Highlands

2.2.1. Physical environment

The Arctic Foothills includes an area along the northern foot slopes of the Brooks Range and the northern Norton Sound Highlands that encompasses the northern part of the Seward Peninsula. These physiographic regions include broad sloping valleys separated by moderately steep to steep ridges, hills, and knolls. Elevations range from sea level to about 900 m, and these areas are within the zone of continuous permafrost.

There are two main types of vegetation in the region, moist nonacidic tundra and moist acidic tundra. Mixed dwarf shrub-sedge (*Dryas integrifolia, Salix reticulata, Carex biglowii*, and *Tomentypnum nitens)* dominates the former, and mixed cottongrass-shrub (*Eriophorum vaginatum*, *Betula nana, Salix pulchra*, and *Hylocomium splendens)* dominates the latter (Swanson et al., 1985; Walker, 1999).

The dominant parent material in valleys and on long foot slopes is loamy colluvium. On hills and ridges, parent materials include gravelly colluvium and residuum weathered from sedimentary rocks. Glaciated hills and plains of early to mid Pleistocene times (Anaktuvik and Sagavanirktok ages) are common elsewhere (Hamilton, 1987) and often are mantled with loess deposits a meter or more thick.

2.2.2. Climate

The mean annual air temperature (MAAT) of the Arctic Foothills is 2° to 4°C warmer than that of the Coastal Plain, due to the greater distance from the ocean (Zhang et al., 1996). The mean annual soil temperature (MAST) ranges from -4° to -7°C, and the mean annual precipitation (MAP) ranges from 14 to 27 cm, with 40% falling as snow (Haugen, 1982). The winter temperatures are higher in the southern portion of the Arctic Foothills because of atmospheric temperature inversions.

The growing degree-days (GDD) in this region range from 760° to 1125° days. Temperature-vegetation gradients on the Arctic Coastal Plain and the Foothills indicate that GDD accumulations linearly relate to the distance from the ocean. In general, MAP and diurnal temperature variations increase southward with distance from the coast.

2.2.3. Soils of the Arctic Foothills and northern Norton Sound Highlands

The Sagwon Hills encompass a major part of the Arctic Foothills. This landscape consists of hills dissected into Tertiary age gravelly outwash deposits of mixed mineralogy and blanketed with loess (Hamilton, 1987). Along the northern edge of this hilly upland, the vegetation is dominantly nonacidic tundra (Bockheim et al., 1998) characterized by dwarf shrub (*Dryas spp.* and *Lupinus spp.)*, in addition to other tundra types. Low, flat polygons about 40 to 100 cm across, with 15% bare soil in mudboils or frost scars, characterize the microrelief.

The soils associated with dwarf shrub vegetative types are strongly cryoturbated, as warped and broken soil horizons indicate throughout the entire profile. Under a discontinuous organic layer, portions of the profile have a high chroma color, indicating oxidation of iron minerals, while adjacent zones are gleyed.

The dominant soils are Aquic Molliturbels and Ruptic-Histic Aquiturbels (Ping et al., 1998). The former have an active layer more than 50 cm deep, and the later are interspersed with acidic tundra soils. Cryoturbated organic matter or humus scatters throughout the mineral horizons, and a second concentration of organic matter occurs directly above the permafrost table, at about 60 cm. This suggests the frost churning of humus and sequestration of carbon into the permafrost (Ping et al., 1997).

Soil structure is massive in the upper part of the active layer in mudboils and between tussocks, due to the constant wetness. Elsewhere, a strongly developed platy structure is common in the active layers of upland soils, as a result of lower relative wetness and higher exposure to annual freeze and thaw cycles. In the lower portion of the active layer above the permafrost table, the structure is reticular, as the result of vertical and horizontal vein-ice formation, due to frost cracks caused by freeze-back from the permafrost table in the fall (Zhestkova, 1982). The upper 30 to 40 cm of permafrost is very ice-rich, and soil aggregates appear suspended in ice; this has been identified as an “ataxitic” cryogenic fabric (Gasanova, 1963).

The ataxitic fabric forms (1) from freeze-back of the active layer from below, with the rise of the permafrost table due to vegetative succession and the formation of the O horizon (Shur, 1988) and (2) from gradual movement of water into the upper permafrost. The resulting structures of the thawed soil provide channels for soil water after the permafrost thaws; thus, biogeochemical weathering, mainly reduction-oxidation, can penetrate into the upper permafrost layer when it periodically thaws.

On similar landforms and loess parent materials farther to the south, the vegetation changes to acidic tundra. The region has patterned ground, with occasional frost boils and acidic tussock tundra. Soils associated with this land cover have a thick, but discontinuous, organic horizon commonly ranging from 15 to 30 cm. Cryoturbation is more pronounced than in the nonacidic soils. The permafrost table occurs at 45 to 50 cm, and cryoturbation has enriched the upper 20 cm with well humified organic matter. These soils are dominantly Ruptic-Histic Aquiturbels. In the acidic tundra there is better moss growth, hence a thicker organic horizon. In both soils, the upper permafrost layers have an ataxitic fabric, with ice content generally 60% or more by volume.

Soil texture suggests a uniform parent material for soils in the northern Arctic Foothills. The predominantly silt loam textures in both acidic and nonacidic tundra suggest a common source, loess. In general, the acidic tundra occurred to the south of the nonacidic tundra and the differentiation of the soil acidity was attributed to snow depth (Walker et al., 1998) and weathering (Ping et al., 1998).

The silt content in these soils ranges from 60 to 70% but is slightly higher in the nonacidic tundra because it is closer to the Sagavanirktok River (Walker and Everett, 1991). However, in the northern Norton Sound Highlands, the parent material is colluvium rather than loess (Reiger et al., 1979). Pebbles and gravel occur throughout the lower profile, but their position suggests that frost heave moved them upward in the soil profile. Sorted circles are common in areas of nonacidic tundra.

2.2.4. Soils of the southern Arctic Foothills

Glaciated rolling hills make up a large portion of the southern part of this region. The drift is Middle or Late Wisconsin in age (Hamilton, 1987) and contains coarse fragments from pebbles to boulders in size. Soil texture changes from silt loam or silty clay loam in the northern part of the area to sandy loam in the southern part, with an associated increase in coarse fragments. Soils of the moist acidic tundra in the southern Arctic Foothills have more strongly oxidized colors (higher chroma), i.e., are redder, indicating longer periods of oxidation than equivalent soils in the northern foothills.

Discontinuous surface organic horizons are common in the southern foothills, due to cryoturbation under tussock tundra and dwarf shrub. In addition, organic matter is frost-churned into the underlying mineral soil horizons and commonly concentrates in the upper permafrost. Interspersed within large areas of acidic tundra are small areas of nonacidic soils formed in calcareous drift and alluvium on till plains, fans, and alluvial basins (Ping, unpublished data).

Soils in the southern foothills have similar cryogenic structures as soils in the northern foothills, including ice lenses, reticular and blocky structures in the lower portions of the active layer, and ataxitic fabric in the upper permafrost. However, the layers with ataxitic fabric are generally thinner, ranging from 20 to 30 cm. This indicates a variety of causes: less fluctuation of permafrost tables occurs in the southern portion of the foothills, possibly due to thicker snow cover and better insulation; drier conditions result from the landscape position; or less time has gone by for soil development.

The dominant soils associated with tussock tundra and shrub tundra are Ruptic-Histic Aquiturbels. Histic Aquorthels are found along waterways. Glacic Fibristels and Glacic Hemistels are common in valley bottoms and depressions. Ridge tops and windswept valleys are exposed to a more severe climate and commonly contain mudboils. Typic Aquiturbels and Typic Haploturbels are common to these areas (Ping et al., 1998).

3. Cryosols in the Zone of Discontinuous Permafrost

3.1. Western Alaska

3.1.1. Physical environment

Western Alaska includes the Western Alaska Coastal Plains and Deltas, the southern part of the Norton Sound Highlands, and the southern part of the Kuskokwim Highlands physiographic regions (Figure 2.2.1). The Western Alaska Coastal Plains and Deltas are along the western Alaskan coast, north of the Alaska Peninsula, and south of the Seward Peninsula. They include deltas of the Yukon and Kuskokwim rivers and coastal plains and deltas along the north side of the Alaska Peninsula. The area is below 50 m elevation, of low relief with irregular microrelief, and has many lakes and ponds connected by a maze of waterways. It supports shrub and sedge tundra, alder and willow shrub, and, in a few places along streams, stunted spruce (*Picea glauca*) and paper birch (*Betula paperifera*) forest.

The southern Norton Sound Highlands is along the southern part of the Seward Peninsula and extends south along Norton Sound to the northern and eastern edge of the Yukon-Kuskokwim Delta. Elevation ranges from sea level to 900 m, with a few peaks as high as 1200 m. The area consists of steeply sloping hills and low mountains. Bedrock types include schist, dolomite, graywacke, mudstone, and sandstone (Beikman, 1980). Most of the area supports shrub and sedge tundra, with minor areas of spruce forests along lower slopes of inland valleys. Detailed soil information is generally unavailable for these physiographic regions. General soil descriptions here are based on information from Rieger et al. (1979).

3.1.2. Climate

The climate within these physiographic regions is influenced by the Bering Sea to the west. Cooler summer temperatures and the predominance of tundra vegetation are features that help distinguishes Western Alaska from Interior Alaska. Mean summer temperatures are significantly lower, 11.7 °C at Bethel, compared to 15.2°C for Fairbanks. Average annual temperatures range from -4°C in the north to more than 0°C in the south. Average annual precipitation ranges from 25 to 5 cm (Soil Conservation Service, 1981).

3.1.3. Soils of Western Alaska Coastal Plains and Deltas

The landscape is nearly level coastal plains and deltas, interspersed with small lakes and river channels. Soil occur in a complex pattern, depending on local drainage. Typic Fibristels are in meander scars and broad depressions and along the margins of shallow lakes with sedge (*Eriophorum* spp.) tussock or shrub tundra. Intermixed on micro-highs are Typic Historthels and Typic Histurbels, formed in nonacidic sandy and silty alluvium with shrub tundra. Ice-cored mounds, or *pingos,* occur in some areas (modified from Rieger et al., 1979).

3.1.4. Soils of the southern Norton Sound and southern Kuskokwim Highlands

Landforms include rolling to steep mountains and glaciated plains. Soil parent materials include glacial drift, as well as colluvium from argillite, shale, graywacke, sandstone, quartzite, conglomerate, and tuff (Beikman, 1980). In the Kuskokwim portion, parent materials are mantled with a silty mantle consisting of loess and volcanic ash. Steep high mountains with alpine shrub vegetation generally lack permafrost, and common soils are Typic Humicryepts and Typic Cryorthents.

More gently sloping mountains and plains have a complex array of soils with and without permafrost. Coarsely textured soils formed in silty volcanic ash over sandy and gravelly outwash, are generally free of shallow permafrost, are strongly acidic, and have bog birch (*Betula glandulosa*) shrub vegetation. These are classified as Typic and Andic Haplocryods. Adjoining broad valleys have soils formed in loamy and gravelly substratum and often are poorly drained with shallow permafrost. These soils have tussock or shrub tundra vegetation and are classified as Aquic Haploturbels and Typic Histurbels.

3.2. Interior Alaska

3.2.1. Physical environment

Interior Alaska lies between the Alaska Range on the south and the Brooks Ranges on the north, and between the Canadian Border on the east and the first occurrence of lowland tundra in the west (Figure 2.2.1). Interior Alaska consists of several broad, nearly level lowlands, with elevations mostly below 500 m, and rounded mountains with elevations up to about 2000 m asl (Wahrhaftig, 1965). Steep mountainous areas, though not extensive, occupy the southwestern part. The region conveniently divides into these three physiographic regions for discussion of its soils (Soil Conservation Service, 1981).

Soil morphological studies are few and far between in Interior Alaska, with most work concentrated near Fairbanks. Information here without citation comes from unpublished Natural Resources Conservation Service soil survey data and Rieger, et al. (1979).

3.2.2. Climate

This region has discontinuous permafrost (Brown et al., 1997). Mean annual soil temperatures at the top of the permafrost as low as -3.5°C have been recorded at valley-bottom stations, although most soils with permafrost have mean annual temperatures between -2°C and 0°C (Brown et al., 1997; Osterkamp and Romanovsky, 1999). Vegetation is dominantly boreal forest, with tundra at higher elevations; tree line occurs at about 900 m elevation over much of the region. Climate is continental, with mean temperatures at valley bottom stations ranging from 15 to 17°C in July, from -23 to -28°C in January, and -2.5 to -6°C annually (Owenby and Ezell, 1992).

Mean annual precipitation at valley bottom stations typically ranges from about 23 to 34 cm per year (Owenby and Ezel, 1992) but ranges as low as 17 cm per year in the most continental basin, Yukon Flats (National Climatic Data Center, 2000). While precipitation totals are low, the abundant surface water and predominance of acidic soils over most of the area indicate an excess of annual precipitation over evapotranspiration. Data from higher elevations are sparse, but they indicate warmer winters, cooler summers, and higher precipitation than at valley bottom stations.

3.2.3. Soils of the Interior Alaskan Lowlands

Soils with permafrost dominate the wet, nearly level lowlands of Interior Alaska. The parent material in these lowlands is mostly alluvium or loess over alluvium, with thick eolian sand or silt or glacial deposits in some areas (Péwé, 1975).

On active floodplains there is typically a sequence from well drained, permafrost-free soils on the most frequently flooded surfaces near the river, to progressively colder, wetter soils with permafrost on higher and less frequently flooded surfaces; surface organic horizons thicken along this same sequence (Péwé, 1948; Viereck, 1970; Viereck et al., 1983; Viereck et al., 1993). This perhaps counterintuitive toposequence (i.e., wettest soils at the highest positions) results from the greater age and different vegetation of the higher surfaces, both of which facilitate accumulation of a surface organic layer, cooling of the soil, formation of permafrost, and wetness due to perching of water on the permafrost.

Vegetation on the most frequently flooded, youngest surfaces without permafrost is typically deciduous shrubs (*Salix* spp.) or trees (*Populus balsamifera*). Older and less frequently flooded surfaces with permafrost often have forest of white or black spruce (*Picea glauca* or *Picea mariana*) on Aquic Haplorthels or Typic Aquorthels. The oldest and least frequently flooded surfaces typically have very sparse black spruce with low shrubs, sedges, and moss on Typic Histoturbels. Thermokarst is common on these older surfaces (Péwé, 1948), although patterned ground is not prominent, other than organic microtopography composed of tussocks and small peat mounds.

Organic soils (i.e., Histels [US], Organic Cryosols [Canadian]) form in nearly level surfaces with rare or no flooding and finely grained mineral substratum (silty alluvium or loess) in Interior Alaskan lowlands. The degree of thickness and decomposition of the peat vary greatly.

The most common peatland morphologic type is the peat plateau (Zoltai, 1972; Zoltai and Tarnocai, 1975). The plateaus consist of a nearly level surface, with active layers of about 0.5 m, interrupted by occasional thermokarst pits. The soils of the plateau surface are extremely acidic in the active layer (typically with pH in water of about 4.0), and the vegetation consists of acid-tolerant black spruce, ericaceous shrubs, mosses, and lichens.

Unlike most organic soils, these soils often have no water table, due to drainage of water into the thermokarst pits. Soils in the thermokarst pits are also organic but commonly have a water table near the surface and may lack permafrost entirely. Degrading peat plateaus in a few areas are less acidic and have been colonized by birch forest (Racine et al., 1998; Walters et al., 1998).

Some parts of the Interior Alaskan lowlands consist of plains with very thick loess deposits that accumulated primarily during the Pleistocene (Péwé, 1975). The topography is gently rolling, due to thermokarsting of sediments with tremendous quantities of ground ice. These soils typically have a thick surface organic horizon, mineral soil material that is dull gray due to the reducing condition, and permafrost at about 0.5 m depth. Profiles are acidic, with pH values (in water) as low as 4 in the organic horizons and 5 to 7 in the mineral soil material. These soils classify mostly as Typic Aquiturbels and Typic Histoturbels. Permafrost is typically ice-rich, and thermokarst is widespread. Distinct thermokarst depressions contain a wide variety of wet soils that mostly lack permafrost, while nearly level low areas have peat plateau soils as just described.

Some nonflooded terraces with coarsely grained substrata in the Interior Alaskan lowlands have soils that contain permafrost but are drier than the terraces' soils just described (Schoephorster, 1973). They support black spruce forest on Aquic Haploturbels and Typic Aquiturbels. Unpublished NRCS data suggests that soils with less than about 0.5 m of loamy mantle over sand and gravel are unlikely to have permafrost, while soils with more than 0.5 m of loamy mantle usually have permafrost if an intact organic horizon and late-successional vegetative community is present (see below for more on the effects of disturbance on these soils).

Portions of the Interior Alaskan Lowlands are covered with sand dunes or glacial moraines, the latter being extensive only near the Alaska and Brooks Ranges (Coulter et al., 1965). Soils with permafrost are present mostly in the depressions between dunes (Furbush et al., 1980) or between moraine ridges, and the taxa are mostly Typic Aquiturbels and Typic Histoturbels.

3.2.4. Soils of the Interior Alaskan Highlands and northern Kuskokwim Highlands

The Interior Alaskan Highlands and northern Kuskokwim Highlands consist of crystalline igneous and metamorphic rocks and non-calcareous sedimentary rocks (Beikman, 1980) and a loess cover that ranges from many meters thick near loess sources (primarily large braided glacial outwash rivers) to less than 0.5 m distant from loess sources (Péwé, 1975). Pleistocene glaciation was limited to a few high cirques (Coulter et al., 1965).

The most widespread soils with permafrost occur on finely grained loess and colluvium on lower slopes in valleys. These soils resemble those in the loess lowlands described above (Dyrness and Grigal, 1979; Swanson, 1996b). Permafrost is typically ice-rich, and thermokarst is widespread. Patterned ground generally is not strongly expressed, with earth hummocks the most common type. Large ice wedges often are hidden beneath the surface and become apparent only after disturbance of the surface (Péwé, 1954; 1982).

Vegetation consists of sparse, stunted black spruce (*Picea mariana*) trees with low shrubs, sedges, and moss (including *Sphagnum* spp. and others). Figure 2.2.2 shows the soil-landscape relationship of the Interior Alaskan Highlands.

Soils with permafrost also occur on sites that are warmer and drier than those just described, as a result of being higher on the slope or on a convex position, having south aspect, or having thinner finely grained surface mantle over bedrock. In such soils, surface organic layers are thinner, active layers are thicker, and soils are better drained, although still generally dull gray in color (Swanson, 1996b). Common classifications are Aquic Haploturbel and Typic Aquiturbel.

Soils with permafrost also occur on steep, north-facing slopes (Krause et al., 1959). These soils resemble those of the lowlands, in that they have a thick organic surface layer, dully colored mineral soil, and permafrost at about a 0.5-m depth, but they differ by having bedrock within the first meter below the surface. These soils are striking because their vegetation closely resembles that of lowland black spruce wetlands, but they can occur on slopes in excess of 30°.

Tundra soils occupy significant areas at higher elevations in the Interior Alaskan Highlands and southern Kuskokwim Highlands, but they have received little attention beyond two studies in the foothills of the Brooks Range (Drew and Shanks, 1965; Ugolini et al., 1981). Unpublished NRCS data suggest that soils are mostly stoney with a loamy matrix; surface organic layers are thinner and active layers thicker than those of lowland soils with permafrost. Patterned ground is common, consisting primarily of sorted circles, nets, stripes, and nonsorted circles. Reactions are mostly acidic, due to the mainly acidic nature of parent materials in this region.

Soils with permafrost in Interior Alaska can change markedly in response to forest fires or other disturbance and post-fire plant succession. Active-layer thickness often increases greatly after fire, from about 0.5 m to 2 m or more (Dyrness, 1982; Viereck, 1982), soils become drier, and oxidation of previously reduced soils may occur. The soils that show the greatest change after fires are those at the warmest and driest end of the continuum of permafrost soils before the fire.

For example, soils on terraces with 0.5 to 1.0 m of loamy material over gravel or soils on gentle sloping hill slopes at mid-slope, shoulder, or summit positions often thaw and dry markedly after fire, while soils with thick surface organic horizons in lowlands on thick silty deposits often thaw little after fire (Swanson, 1996a). In some cases, retreat of the permafrost table below a 2-m depth after disturbance can cause the soil classification to change greatly, from Gelisol to Inceptisol (US) or Cryosol to Brunisol (Canadian) (Ping, 1987).

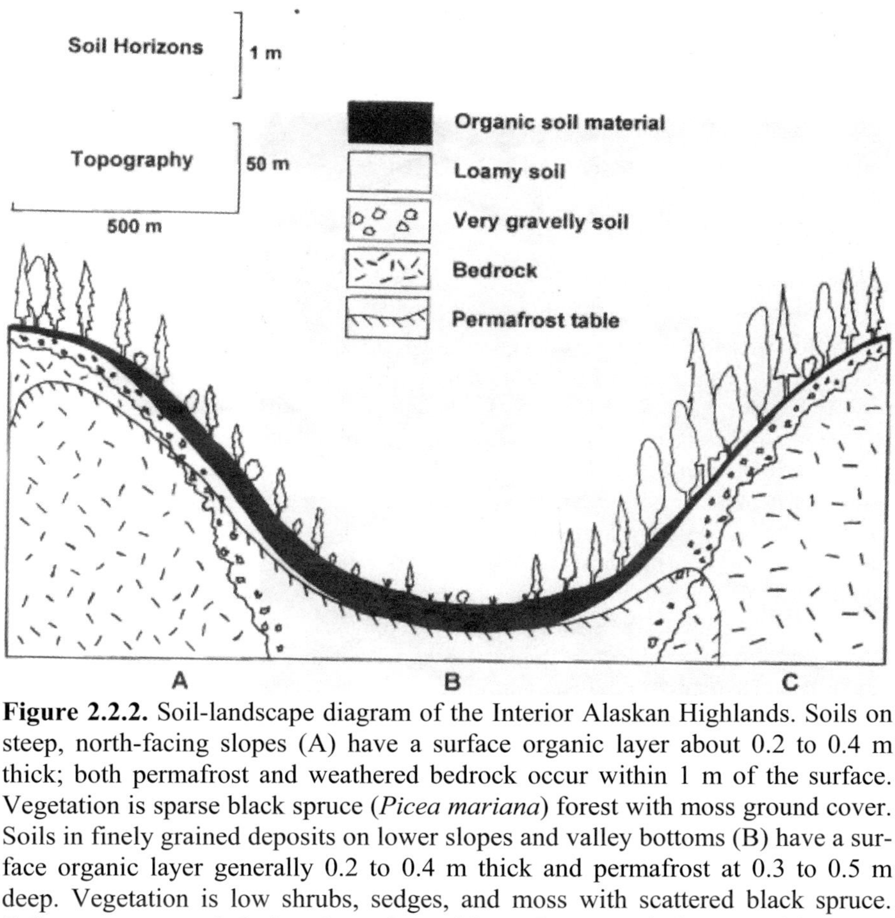

Figure 2.2.2. Soil-landscape diagram of the Interior Alaskan Highlands. Soils on steep, north-facing slopes (A) have a surface organic layer about 0.2 to 0.4 m thick; both permafrost and weathered bedrock occur within 1 m of the surface. Vegetation is sparse black spruce (*Picea mariana*) forest with moss ground cover. Soils in finely grained deposits on lower slopes and valley bottoms (B) have a surface organic layer generally 0.2 to 0.4 m thick and permafrost at 0.3 to 0.5 m deep. Vegetation is low shrubs, sedges, and moss with scattered black spruce. Soils on steep, south-facing slopes have thin surface organic layers and lack permafrost. Trees there are larger and may be white spruce (*Picea glauca*), birch (*Betula papyrifera*), or aspen (*Populus tremuloides*).

4. Alaska Range

4.1. Physical environment

The Alaska Range includes a north-trending mountainous arc that extends from the Alaska Peninsula on the west to the border of Canada on the east (Figure 2.2.1). Mountains consist of a variety of crystalline igneous, metamorphic, and sedimentary rocks (Beikman, 1980). The Alaska Range divides the Cook Inlet-Susitna Lowlands and Copper River Plateau regions to the south from the Interior Alaska and Western Alaska regions to the north and west.

North of the crest of the Alaska Range, the glaciers were essentially alpine in character; they filled the mountain valleys and in places spread as piedmont lobes in adjoining lowlands. South of the crest, glaciers originating in the mountains filled adjoining lowland basins, such as the broad Susitna Valley (Péwé, 1975).

A majority of the Alaska Range consists of alpine and sub-alpine shrub vegetation at elevations above about 550 meters, with less extensive areas of boreal forest below.

4.2. Climate

Along its western part, the Alaska Range divides the transitional maritime-continental climate to the south from the continental climate to the north. Along the eastern part, the climate is continental throughout.

This mix of climates, in conjunction with high mountainous topography, results in diverse microclimates with variable patterns of wind, temperature, and precipitation. Permafrost is generally absent where the climate is transitional maritime-continental and has a discontinuous distribution where continental conditions prevail. This discussion pertains to that portion of the Alaska Range that lies within the continental climate zone and thus the zone of discontinuous permafrost.

Long-term climatic data is available for the forested portion of the Alaska Range but is unavailable for the alpine zone. Mean annual air temperature at Denali Park Headquarters, located in the upper limit of the boreal forest zone, is -3.0°C with mean winter and summer air temperatures of -16.4°C and 11.3°C, respectively (National Climatic Data Center, 2000).

4.3. Soils of the Central Alaska Range

The core of the Alaska Range includes steep, rocky mountains, with elevation ranging from 900 to 6100 m asl. Slopes above about 1300 meters elevation are mostly barren, with scattered small islands of dwarf willow (*Salix reticulata* and *arctica*) and dryas (*Dryas* spp.) shrub. Below this is an assemblage of alpine and subalpine vegetation, including dwarf willow and dryas shrub, mixed ericaceous-bog birch (*Betula glandulosa*) shrub, and alder (*Alnus* spp.) shrub, interspersed with areas of barren rock and talus.

Lithology is diverse, with schist, shale, conglomerate, and diorite (Beikman, 1980) the major rock types and colluvium the dominant soil parent material. Based on initial soil survey information for Denali Park and Preserve and Delta River Areas (Clark and Duffy, in press) within the continental climate portions of this section, permafrost is of minor extent and limited primarily to the lower slopes.

A majority of soils lack shallow permafrost and are classified as Typic Dystrocryepts and Typic Eutrocryepts, Humic Eutrocryepts, and Typic Haplocryolls. These soils are devoid of surface organic mats and have gravelly surface textures and are well drained, and permafrost is absent within two meters of the surface. Though shallow permafrost is absent, mean annual soil temperatures (MAST) are below 0°C, and minimum winter soil temperatures are generally below -4°C. Seasonal soil temperature extremes measured in typical soils of the region ranged from maximums of about 9°C during summer to minimums of about -19°C during winter, with overall mean annual soil temperatures (MAST) from -1 to -3°C (Clark and Duffy, in press).

Consistent with the findings by French (1970), these gravelly soils with no appreciable organic mat at the surface are found to be warmer during summer than would normally be expected for the latitude. High thermal conductivity of this gravelly surface material (Jury et al., 1991) is thought to be an important contributor to the extreme seasonal temperatures and general lack of permafrost in these soils. Also important are winds that remove snow from the surfaces of these soils during winter, leaving the ground exposed to extreme winter air temperatures (MacKay and MacKay, 1974; Nichols and Moore, 1977).

Soils with shallow permafrost occupy less than about 15% of the landscape and are limited to benches, summits, saddles, and gently sloping foot slopes with convergent-plain to concave surface shapes. These landscape positions accumulate snow during winter and receive excess moisture as run-in from surrounding slopes during late spring and summer. Surface micro-relief is generally smooth but includes areas of turf hummocks with less than 30 cm relief.

Parent materials include colluvium from a host of rocks with a thin mantle of loess and eolian organic deposits. Eolian deposits accumulate each winter in snowdrifts and are deposited on the soil surfaces during the spring thaw.

The permafrost is at one to two meters depth, and Aquic Haplorthels and Histic Haplorthels are common soils. Vegetation is dwarf willow (*Salix reticulata*, *Salix Arctic*) and dryas shrub (*Dryas* spp.) and dwarf willow shrub-sedge (*Carex* spp.). Both Aquic Haplorthels and Histic Haplorthels have neutral to alkaline soil reactions in surface layers, from nutrient rich run-in water, compared to adjacent well drained Typic Distrocryepts and Eutrocryepts.

The upper mineral layers, or "A" horizons, in Aquic Haplorthels are black (10YR 2/2) silt loam with reactions that range from strongly through slightly acidic (pH 5.3 to 6.4). Histic Haplorthels have very dark brown (10YR 2/2) or black (10YR 2/1) surface organic mats and a thin underlying silt loam mineral layer with a reaction ranging from strongly acidic through neutral (pH 5.3 to 6.6). Substrata of both soils have variable color depending on lithology and have a strongly acidic through neutral reaction (pH 5.3 to 6.6).

4.4. Soils of the Outer Mountains, Glacial Plains, and Basins

Rounded glaciated mountains, glaciated plains and hills, and bedrock cored low mountains occupy much of the fringe of the Alaska Range at elevations from about 300 to 1300 meters. Common parent materials include gravelly till, loess, cryoturbate, and colluvium. Periglacial features, including gelifluction lobes, stripes, and steps, are common on this assemblage of landforms.

Equally represented on periglacial landforms are soils with and soils without permafrost. Common soils with shallow permafrost include Typic Histoturbels, and soils on periglacial features include Ruptic-Histic Aquiturbels. Aquic Haplorthels and Histic Haplorthels are common in basins. Alpine and subalpine vegetation comprises much of the area, with less extensive areas of boreal forest at elevations below about 550 m asl.

Soils lacking significant cryoturbation include Aquic Haplorthels and Histic Haplorthels. These soils are common on gently sloping mountain foot slopes, till plains, glaciated hills, and broad, mountain summits. Surface micro-relief is generally smooth but includes areas of turf hummocks with less than 30 cm relief. Vegetation is mixed bog birch (*Betula glandulosa*)-ericaceous shrub or alpine-mixed ericaceous shrub-sedge (*Carex* spp.) types.

These soils do not receive the nutrient rich run-in water described for similar soils in the "Steep Mountains" section and therefore are measurably more acidic in the upper soil horizons. Organic mats have very strongly or strongly acidic soil reaction (pH 4.3 to 5.5), and soil color is very dusky red to black (2.5YR 3/2 to 10YR2/1). Reaction in substrata is very strongly through moderately acidic (pH 4.7 to 6.0) and soil color is dark grayish brown to very dark gray (2.5Y 4/2 to 5Y 3/1) (Clark and Duffy, in press).

Highly cryoturbated soils, including Typic Histoturbels and Ruptic-Histic Aquiturbels, are common in area with periglacial features. They formed in frost-churned sediments from a variety of rocks and till. Cryoturbation resulted in multiple buried, mixed, and convoluted organic and mineral horizons. Hummocks and non-sorted steps are periglacial features most commonly associated with Typic Histoturbels. Vegetation on these soils includes alpine-ericaceous shrub or alpine-mixed ericaceous shrub-sedge types.

Nonsorted circles, another common periglacial feature, occur on mountain footslopes, toeslopes, and broad summits. These consist of a repeating pattern of convex circular barren soil areas about one to three meters in diameter encircled by concave troughs one to two meters across. Circle centers have well drained soils and permafrost at depths of 1.5 m or more and are often barren of vegetation. Soils in troughs have histic epipedons, shallow permafrost, saturated conditions above the ice, and bog birch (Betula glandulosa)-ericaceous shrub vegetation.

Soils with shallow permafrost within the boreal forest zone are common along mountain foot slopes and basins at elevations between 300 and 550 m asl. Soil materials include gravelly till, silty loess, and silty colluvium derived from loess. Distribution and kinds of soils found here are similar to those described in the "Interior Alaska Highlands" section.

5. Copper River Plateau

5.1. Physical environment

The Copper River Plateau refers to the Copper River Basin and adjacent uplands rimmed by the Alaska Range on the north, the Talkeetna Mountains on the west, the Chugach Range on the south, and the Wrangell Mountains on the east (Figure 2.2.1). This region includes the upper drainage basin of the glacially fed Copper River and tributaries at elevations from about 170 to 1100 m asl.

Vegetation is dominantly boreal forest, with alpine areas above about 700 m. Landforms of the basin floor include narrow river valleys entrenched tens of meters within finely textured lacustrine basin sediments. Broad, nearly level to undulating lacustrine terraces are the most extensive landform of the basin floor above the river canyons.

Extensive areas of soils formed in calcareous glaciolacustrine sediments are a unique feature of the Copper River Plateau. Lacustrine terrace sediments consist of clayey and loamy calcareous materials that were deposited within a large proglacial lake that occupied the basin during the Pleistocene from about 35,000 to 9,000 years ago (Ferrians et al., 1983). Scattered throughout the basin are sandy strandline or beach deposits associated with fluctuating lake levels during this glacial period. Skirting the basin to about 900 meters elevation are glacial landforms including drumlins, outwash plains, till plains, and hills.

Mountains bordering the Copper River Plateau consist of schist, greenstone, graywacke, shale, sandstone, and andesite rocks (Beikman, 1980). Many upland landforms are mantled with a thin layer of calcareous loess, ranging from a few meters thick on lacustrine terraces adjacent to the Copper River to a few millimeters on lacustrine, glacial, and mountain landforms more distant from the river (Clark and Kautz, 1990).

This region is included within the zone of discontinuous permafrost (Péwé, 1975), and permafrost underlies most of the plateau at varying depths, except on floodplains and under lakes. Permafrost is thought to have formed in the recent epoch or following the end of the last glaciation, since, during that period, the basin was under either the waters of the proglacial lake or glacial ice. Permafrost occurs as finely segregated ice crystals and thin discontinuous ice lenses.

In clayey lacustrine deposits, ice content can average as high as 30 to 60% of the dry weight of the soil (Nichols, 1956). Though massive ground ice has been observed in area soils, massive ground ice formations or thermokarsts are localized features and not extensive.

5.2. Climate

The climate of the Copper River basin is subarctic continental characterized by long cold winters and short warm summers, a climate similar to the Interior physiographic area. January temperature is -23°C at Glennallen and -17°C at Paxson.

Daily low temperatures of -46°C or less occur frequently at Glennallen during the winter and may last for two or more weeks (National Weather Service). Winter temperature inversions result in the lower average temperatures at Glennallen, in the basin bottom, than those measured for Paxson, which is at a higher elevation near the mountains.

Mean July temperature is 13°C at Glennallen and 12°C at Paxson; daily high temperatures on occasion exceed 30°C in Glennallen. Although the daily minimum temperature in summer averages less than 10°C, freezing temperatures have been recorded in every month.

Mean annual precipitation measurements for the plateau range from about 280 mm in the basin bottom at Glennallen to about 540 mm in upper elevations at Paxson. Though precipitation totals are low, abundant surface water on landscapes of the region suggests that evapotranspiration is less than the annual precipitation.

5.3. Soils

Permafrost is generally absent on low to mid-level flood plains and common on high flood plains and stream terraces. The sequence of soils and permafrost along high flood plains and stream terraces on the Copper River Plateau is similar to that previously discussed in the "Interior Alaska Lowlands" section.

Soils on lacustrine terraces, till plains, and hills consist of a dark grayish brown (2.5Y 4/2) through olive gray (5Y 5/2) lacustrine and till deposits mantled with dark brown (10YR 3/3) and black (10YR 2/2) loess of variable thickness. Permafrost at a shallow depth and poorly drained conditions are common but highly variable over short distances. Recurring wild fires have created a complex mosaic of permafrost states and soil drainage conditions, similar to those described previously for Interior Alaska.

In most places, across lacustrine terraces and till plains, soils potentially alter between shallow permafrost and a permafrost-free state (Clark and Kautz, 1990). Soils under late succession dwarf black (*Picea mariana*) and white spruce (*Picea glauca*)-forest and shallow permafrost have histic epipedons and shallow permafrost and are classified as Typic Historthels.

Reaction in soils is relatively high, compared to similar soils in Interior Alaska. In Typic Historthels on lacustrine terraces, reaction in surface organic mats ranges from strongly through slightly acidic (pH 5.1 to 6.2), slightly acidic to moderately alkaline (pH 6.1 to 8.4) in the loess mantle, and mildly or moderately alkaline (pH 7.4 to 8.4) in lacustrine materials. Free calcium carbonate occurs disseminated throughout soil materials and occasionally as local white concentrations. Soils lack appreciable cryoturbation and structure, with the exception of weak platy structure in loess mantles.

In areas of these soils recently affected by fire of moderate or severe intensity, vegetation is commonly altered to willow shrub and aspen or mixed white spruce-aspen forest types. Combustion and consolidation of the insulating organic mat results in warming of underlying soils, lowering of the permafrost table, and a transition to a well drained and permafrost-free state.

Rate of change from the permafrost rich, poorly drained condition to a well drained, ice-free state was observed at the Wilson Camp fire burn near Glennallen, which burned in July of 1981. Soils showed an increase in the depth of thaw the first year after the burn. Depth to permafrost in August, in similar unburned forest adjacent to the burn, averaged 40 cm below the mineral surface. In August of 1982, within the burned area, depth to permafrost averaged 80 cm; in August of 1983, depth to permafrost averaged 100 cm (Clark and Kautz, 1990).

This alteration results in a change in the classification of soils from Typic Historthels to Pachic Haplocryolls, Typic Haplocryolls, or Typic Eutrocryepts. Soil reaction in the surface organic ranges from moderately alkaline (pH 7.9 to 8.4) immediately following fire, decreasing to slightly acidic levels (pH 6.1 to 6.5) several years following fire. Reaction in subsurface layers is similar to that in preburn levels. Thawed soil materials exhibit minimal cryoturbation. Soil structure is platy in loess material, with strongly expressed subangular blocky structure in underlying lacustrine deposits and massive or weak platy structure in till materials.

Permafrost-affected soils of lesser extent include very poorly drained Typic Histoturbels in broad depressions and toeslopes on hills with cottongrass tussocks (*Eriophorum* spp.); very poorly drained Typic Hemistels with wet willow shrub (*Salix* spp.) vegetation; and, Ruptic-Histic Aquiturbels on mudboils or nonsorted circles with spruce (*Picea* spp.)-shrub bog birch (*Betula glandulosa*) woodland. The occurrence of fire and alteration of soil properties and vegetation is infrequent on these soils.

6. Summary

Five soil-forming factors, parent material, topography, climate, biota, and time, usually are used to define the state of the soil system. Though often described independently, they are related, and the relative importance of each factor varies with location.

In the high latitudes, climate plays an important and often overriding role in soil formation, since permafrost is a prevalent feature in landscapes. Permafrost, acting as a barrier for water movement, creates restricted drainage and often wet soil morphology. Most soils affected by permafrost have very poor to poor drainage and are classified within the "Aqui" Great group of the Gelisol Order.

In the zone of continuous permafrost, the cold ambient temperature, usually < -20°C during fall freeze-up, induces frost-churning or frost heave processes and cryoturbated soil profiles. These soils are keyed into the Turbic Suborder of the Gelisol Order.

In the Arctic Foothills, the soils are classified as Aquiturbels. Along the Arctic Coastal Plain, the cold environment in conjunction with level or concave surface relief contributes to the accumulation of organic matter into thick surface organic mats or Histic epipedons that qualify these soils to be classified as Histels or Typic Histoturbels. Cryosols lacking cryoturbation, or Orthels, are not extensive and only form on young landscapes, such as floodplains.

Soil temperatures in the zone of discontinuous permafrost are generally "warmer," compared with those of the zone of continuous permafrost. They range from a fraction of a degree below freezing to -2°C in most cases. Variation in soil temperature and the presence of permafrost often are associated with landscape position, parent material, and fire history.

On hills and mountains, incidental solar radiation and evapotranspiration differ significantly, based on exposure. Northerly exposures are often cooler, with relatively low rates of evapotranspiration, and have permafrost at a shallow depth and dwarf spruce forests. Histic Haplorthels and Typic Histoturbels are the common Subgroups represented.

Southerly exposures are generally warmer, drier, and permafrost free, with mixed hardwood and white spruce forests. These soils lack permafrost, and Typic Dystrocryepts and Typic Eutrocryepts are the common Subgroups.

The kind and thickness of parent material influences whether permafrost will form in soils. Generally, soils with < 0.5 meters of loamy loess or alluvium over coarse texture tills, alluvium, or bedrock have relatively warm summer soil temperatures and lack permafrost, due to high thermal conductivity properties of coarse texture materials. Well to excessively drained soils common to these parent material types include Typic Dystrocryepts, Typic Eutrocryepts, and Typic Haplocryods with spruce and dwarf birch vegetation.

Upland soils with 0.5 to 1.0 m of loamy material over gravel or soils on gentle sloping hill slopes at mid-slope, shoulder, or summit positions have permafrost that is sensitive to disturbance and have been observed to recede below 2 m and dry markedly after fire or land clearing. Recession of the permafrost table to below 2 m depth after major disturbance can cause the soil classification to change from the Gelisols before disturbance to Inceptisols in the years following. Soil subgroups and vegetation include Typic Historthels with dwarf spruce forest before fire and Typic Dystrocryepts and Typic Eutrocryepts with mixed forest several years following fire.

Soils with thick surface organic horizons over thick silty loess deposits have low thermal conductivity and often thaw little after fire (Swanson, 1996a). Thus some of the Typic Historthels may change to Typic Aquorthels, due to the loss of the organic horizon following fire.

Cryoturbation is less extensive in soils of the Subarctic zone, compared to those of the Arctic, especially in the boreal forest zone. This is attributed to the combined insulating effect of the mineral soil by a stable winter snow pack and thick organic horizons. Historthels are common in the boreal forest on flat or gently sloping areas. Cryoturbated soils are limited to steeper northerly exposures, where mixed horizons are generally attributed to downslope movement of saturated soil material over permafrost or gelifluction. Highly cryoturbated soils are also common on exposed alpine landscapes, where wind removes the snow cover (Ruptic-Histic Aquiturbels), and under sedge tussock and dwarf spruce-sedge tussock vegetative types (Typic Histoturbels) where standing water between tussocks freezes, causing differential lateral pressures and cryoturbation.

References

Beikman, H.M. 1980. Geologic map of Alaska. U.S. Dept. of Interior, Geological Survey. Scale 1:2,500,000.

Bockheim, J.G., C. Tarnocai, J.M. Kimble, and C.A.S. Smith. 1997. The concept of gelic materials in the new Gelisol order for permafrost-affected soils. Soil Sci. 162: 927-939.

Bockheim, J.G., D.A. Walker, L.R. Everett, F.E. Nelson, and N.I. Shiklomanov. 1998. Soils and cryoturbation in moist nonacidic and acidic tundra in the Kuparuk River Basin, Arctic Alaska, U.S.A. Arc. Alp. Res. 30: 166-174.

Brown, J., O.J. Ferrians, J.A. Heginbottom, and E.S. Melnikov. 1997. Circum-arctic map of permafrost and ground-ice conditions. U.S. Dept. of Interior, Geological Survey Map CP-45. Scale 1:10,000,000.

Clark, M.H., and M. Duffy. In press. Soil resources and ecosystems of Denali National Park, Alaska. National Park Service. Anchorage, AK.

Clark, M.H., and D.R. Kautz. 1990. Soil Survey of Copper River Area, Alaska. USDA, NRCS. US Gov't Prntg. Ofc. Washington, D.C.

Coulter, H.W., D.M. Hopkins, T.N.V. Karlstrom, T.L. Péwé, C. Wahrhaftig, and J.R. Williams. 1965. Extent of glaciation in Alaska. U.S. Dept. of Interior, Geological Survey Miscellaneous Geologic Investigations Map I-415, Scale 1:2,500,000.

Drew, J.V., and R.E. Shanks. 1965. Landscape relationships of soils and vegetation in the forest-tundra ecotone, upper Firth River valley, Alaska-Canada. Ecol. Mon. 35:285-306.

Dyrness, C.T. 1982. Control of depth to permafrost and soil temperature by the forest floor in Black Spruce/Feathermoss communities. U.S. Dept. of Ag., Forest Service, Research Note PNW-396. Pacific Northwest Forest and Range Experiment Station. Portland, OR

Dyrness, C.T., and D.F. Grigal. 1979. Vegetation-soil relationships along a spruce forest transect in interior Alaska. Can. J. of Bot. 57(23):2644-2656.

FAO. 1998. World Reference Base for Soil Resources. World Soil Resource Reports 84. Food and Agriculture Organization of the United Nations. Rome.

Ferrians, O.J., D.R. Nichols, and J.R. Williams. 1983. Copper River Basin, resume of Quaternary geology. In Guidebook to permafrost and Quaternary geology along the Richardson and Glenn Highway between Fairbanks and Anchorage, Alaska. Vol. 1. Dept of Natural Resources. Fairbanks, AK.

French, H.M. 1970. Soil temperatures in the active layer. Beaufort Plain. Arctic. 23:229-239.

Furbush, C.E., B.E. Koepke and D.B. Schoephorster. 1980. Soil survey of Totchaket area, Alaska. USDA, SCS. U.S. Gov't Prntg. Ofc. Washington, D.C. 68 pp.

Gasanova, S.S. 1963. Morfogeneticheskaya klassifikatsiya kriogennykh tekstur pykhlykh otlozherny. In: Trudy Severo-Vostochnogo Kompleksnogo Nauchno-Issledovatelskogo Instituta, VYP 3. Magadan, Russia. pp. 53-62.

Hamilton, T.D. 1987. Surfacial geologic map of the Philip Smith Mountains Quadrangle, Alaska. U. S. Geol. Surv. Misc. Field Stud. Map MF-879A. Scale 1:125,000. 1 pp.

Haugen, R.K. 1982. Climate of remote areas in north-central Alaska: 1975-1979 Summary. U.S. Army Cold Regions Research and Engineering Laboratory Rep. 82-35. Hanover, N.H. 114 pp.

Hoefle, C.M., C.L. Ping, and J.M. Kimble. 1998. Properties of permafrost soils on northern Seward Peninsula, northwest Alaska. Soil Sci. Soc. Am. J. 62(6):1629-1639.

Jorgenson, M.T., J.E. Roth, E.R. Pullman, R.M. Burgess, M.K. Raynolds, A.A. Stickney, M.D. Smith, and T.M. Zimmer. 1996. An ecological land survey for the Colville River Delta, Alaska. Final Report prepared for ARCO Alaska, Inc. Kuukpik Unit Owners. ABR, Inc. Fairbanks, AK.

Jury, W.A., W.R. Gardner, and W.H. Gardner. 1991. Soil Physics. John Wiley and Sons, Inc. New York.

Krause, H.H., S. Regier, S.A. Wilde. 1959. Soils and forest growth on different aspects in the Tanana watershed of interior Alaska. Ecol. 40(3):492-495.

MacKay, J.R., and D.K. MacKay. 1974. Snow cover and ground temperatures, Garry Island, N.W.T. Arctic. 27, 287-296.

National Climatic Data Center. 2000. Climatological Data Annual Summary, Alaska. NOAA. Asheville, NC.

Nichols, D.R. 1956. Permafrost and ground water conditions in the Glennallen area, Alaska. Preliminary report. U.S. Dept. of the Interior Geol. Survey. Fairbanks, AK.

Nichols H.M., and T.R. Moore. 1977. Pedogenesis in a subarctic iron-rich environment, Schefferville, Quebec. Can. J. Soil Sci. 57:35-45.

Osterkamp, T. E., and V.E. Romanovsky. 1996. Characteristics of changing permafrost temperatures in the Alaskan Arctic, U.S.A., Arc. Alp. Res. 28:267-273.

Osterkamp, T.E., and V.E. Romanovsky, V.E. 1999. Evidence for warming and thawing of discontinuous permafrost in Alaska. Permafrost and Periglacial Processes. 10:17-37.

Owenby, J.R., and D.S. Ezell. 1992. Monthly station normal of temperature, precipitation, and heating and cooling degree-days, 1961-1990: Alaska. Climatography of the United States No. 81. U.S. Dept. of Commerce, National Oceanic and Atmospheric Administration, National Climatic Data Center. Asheville, NC.

Péwé, T.L. 1948. Terrain and permafrost of the Galena Air Base, Galena, Alaska. U.S. Dept. of Interior, Geological Survey Permafrost Program Progress Report 7. 52 pp.

Péwé, T.L. 1954. Effect of permafrost on cultivated fields, Fairbanks, area, Alaska. U.S. Dept. of Interior. Geol. Survey Bull. 989-F. pp. 315-351.

Péwé, T.L. 1975. Quaternary geology of Alaska. U.S. Department of Interior, Geological Survey Professional Paper 835. U. S. Gov. Print. Ofc., Washington, D. C. 145 pp.

Péwé, T.L. 1982. Geologic hazards of the Fairbanks area, Alaska. Alaska Division of Geol. and Geophys. Surveys Special Report 15. College, Alaska. 109 pp.

Ping, C.L. 1987. Soil temperature profiles of two Alaskan soils. Soil Sci. Soc. Am. J. 51(4):1010-1018.

Ping, C.L., J.G. Bockheim, J.M. Kimble, G.J. Michaelson, and D.A. Walker. 1998. Characteristics of cryogenic soils along a latitudinal transect in arctic Alaska. J. Geophys. Res. 103(D22):917-928.

Ping, C.L., G.J. Michaelson, and J.M. Kimble. 1997. Carbon storage along a latitudinal transect in Alaska. Nutrient Cycling Agroecosyst. 49:235-242.

Ping, C.L., G.J. Michaelson, P.P. Overduin, and C.A. Stiles. 2003. Morphogenesis of frost-boils in the Galbraith Lake area, Arctic Alaska. Proceedings of the eighth international conference on permafrost, Zurich, Switzerland.

Racine, C.H., M.T. Jorgenson, and J.C. Walters. 1998. Thermokarst vegetation in lowland birch forests on the Tanana Flats, Interior Alaska, U.S.A. In: Proceedings of Permafrost: Seventh International Conference, June 23-27, 1998, Yellowknife, Canada. Centre d'etudes nordiques, Universite lavel. pp. 927-933.

Rieger, S., D.B. Schoephorster, and C.E. Furbush. 1979. Exploratory soil survey of Alaska. U.S. Dept. of Agriculture, Soil Conservation Service. U.S. Gov't Prntg Ofc. Washington, D.C. 213 pp.

Schoephorster, D.B. 1973. Soil survey of Salcha-Big Delta area, Alaska. U.S. Dept. of Agriculture, Soil Conservation Service. U.S. Gov't Prntg Ofc. Washington, D.C. 51 pp.

Shur, Y. L. 1988. The upper horizon of permafrost soils, in Proceedings of the Fifth International Conference on Permafrost, vol. 1. Tapir. Trondhiem, Norway. pp. 867-871.

Soil Classification Working Group. 1998. The Canadian system of soil classification. Agriculture and Agri-Food Canada Publication 1646 (Revised). 187 pp.

Soil Conservation Service. 1981. Land resource regions and major land resource areas of the United States. USDA Ag. Hndbk. 296. U.S. Gov't Prntg Ofc. Wash. D.C.

Soil Survey Staff. 1999. Soil Taxonomy: A basic system of soil classification for making and interpreting soil surveys. 2nd ed. USDA, NRCS. Ag. Hndbk. 436. U.S. Gov't Prntg Ofc. Washington, D.C. 869 pp.

Swanson, D.K. 1996a. Susceptibility of permafrost soils to deep thaw after forest fire in interior Alaska, U.S.A., and some ecologic implications. Arct. and Alp. Res. 28(2):217-227.

Swanson, D.K. 1996b. Soil geomorphology on bedrock and colluvial terrain with permafrost in central Alaska, USA. Geoderma. 71:157-172.

Swanson, J.D., M. Schuman, and P.C. Scorup. 1985. Range survey of the Seward Peninsula reindeer ranges, Alaska. USDA. SCS/ Anchorage, Alaska.

Ugolini, F.C., R. Reanier, G. Rau, and J. Hedges. 1981. Pedological, isotopic, and geochemical investigations of the soils at the boreal forest and alpine transition in northern Alaska. Soil Sci. 131:359-374.

Viereck, L.A. 1970. Forest succession and soil development adjacent to the Chena River in interior Alaska. Arct. and Alp. Res. 2(1): 1-26.

Viereck, L.A. 1982. Effects of fire and firelines on active layer thickness and soil temperatures in interior Alaska. Proceedings Fourth Canadian Permafrost Conference, Calgary, Alberta, March 2-6, 1981. pp. 123-135.

Viereck, L.A., C.T. Dyrness, and M.J. Foote. 1993. An overview of the vegetation and soils of the floodplain ecosystems of the Tanana River, interior Alaska. Can. J. Forest Res. 23:889-898.

Viereck, L.A., C.T. Dyrness, K. Van Cleve, and M.J. Foote. 1983. Vegetation, soils, and forest productivity in selected forest types in interior Alaska. Can. J. Forest Res. 13:703-720.

Wahrhaftig, C. 1965. Physiographic divisions of Alaska. U.S. Dept of Interior, Geological Survey Professional Paper 482. 52 pp.

Walker, D.A. 1999. A integrated mapping approach for northern Alaska. Int. J. Remote Sensing. 20:2895-2920.

Walker, D.A., N.A. Auerbach, J.G. Backheim, F.S. Chapin III, W. Eugester, J.Y. King, J.P. Mcfadden, G.J. Michaelson, F.E. Nelson, W.C. Oechel, C.L. Ping, W.S. Reeburg, S.

Regli, N.I. Shiklomanov, and G.L. Vourlitis. 1998. Energy and trace-gas fluxes across a soil pH boundary in the Arctic. Nature. 394:469-472.

Walker, D. A., and K. R. Everett. 1991. Loess ecosystems of northern Alaska: Regional Gradient and Toposequence at Prudhoe Bay, Ecol. Monogr. 61:437-464.

Walters, J.C., C.H. Racine, and M.Y. Jorgenson. 1998. Characteristics of permafrost in the Tanana Flats, Interior Alaska. In: Permafrost: Seventh International Conference, June 23-27, 1998. Proceedings. Yellowknife, Canada. Centre d'etudes nordiques, Universite lavel. pp. 1109-1114.

Zhang, T., T.E. Osterkamp, and K. Stamnes. 1996. Some characteristics of the climate in northern Alaska, USA., Arct. and Alp. Res. 28:509-518.

Zhestkova, T.N. 1982. Formirovaniye Krigennogo Stroyeniya Gruntov. Formation of cryogenic soil structure. Nauka. Moscow. (In Russian.)

Zoltai, S.C. 1972. Palsas and peat plateaus in central Manitoba and Saskatchewan. Can. J. Forest Res. 2(3):291-302.

Zoltai, S.C., and C. Tarnocai. 1975. Perennially frozen peatlands in the western arctic and subarctic of Canada. Can. J. Earth Sci. 12:28-43.

Chapter 3. Cryosols of Arctic Canada

Charles Tarnocai[1]

[1]Research Scientist, Agriculture and Agri-Food Canada, Research Branch (ECORC), K.W. Neatby Building, Rm. 1135, 960 Carling Avenue, Ottawa, Canada, K1A 0C6; e-mail tarnocaict@agr.gc.ca.

1. Introduction

The Arctic, which covers approximately 30% (2,375,000 km^2) of the land area of Canada, has a triangular shape. Its southern border is the Arctic tree line, the northern limit of trees. This border extends eastward from the Yukon-Alaska border (at approximately Lat. 69.5° N, Long. 142° W) across the middle of Hudson Bay (Lat. 57° N, Long. 80° W), to the Atlantic coast (at approximately Lat. 57.5° N, Long. 62° W). Its northernmost point is the northern tip of Ellesmere Island (Lat. 83° N, Long. 70° W).

Although early pedological investigations in the Arctic date back to the beginning of the 1940s (Leahey, 1947), most of the information available was collected during the last three decades. Many of these studies took place on the Arctic islands.

Soil information for the Tanquary Fiord and Lake Hazen areas of Ellesmere Island is in Tarnocai et al. (1991) and in Day (1964). Everett (1968) described the soils and soil development in the Mould Bay and Isachsen areas of Prince Patrick Island. Tedrow et al. (1968) also carried out soil investigations on Prince Patrick Island during this period, while Tedrow and Douglas (1964) earlier had carried out similar work on Banks Island.

Walker and Peters (1977) described the soils in the Truelove Lowland area of Devon Island, Cruickshank (1971) provided information on soils near Resolute Bay on Cornwallis Island, and Tarnocai (1976) provided information for Bathurst and Cornwallis islands. Woo and Zoltai (1977) mapped the soils of Somerset and Prince of Wales islands, Smith et al. (1989) mapped and described the soils on Herschel Island, and Tarnocai and Veldhuis (1998) mapped soils in the Pangnirtung Pass area of Baffin Island. Bockheim (1974) also described soils in selected locations on Baffin Island.

In the District of Mackenzie, the western portion of the continental Arctic, Zoltai and Tarnocai (1974) studied soils and vegetation along the Mackenzie Valley, including the Mackenzie Delta, while Bradley et al. (1982) mapped and described the soils in the Lockhart River area. To the east, in the District of Keewatin, Tarnocai (1977) described the soils in the north central portion of the district, and Tarnocai et al. (1976) described those in the northeastern portion and on Boothia Peninsula, while James (1970) described those in the Rankin Inlet area. Still farther east, Hendershot (1984) carried out soil investigations on the Ungava-Labrador Peninsula of northern Quebec.

In addition, broader scale studies dealt with soils throughout the various regions of the Arctic. These include Tarnocai (1978a and c), which focused on soils and land use, and Tarnocai (1978b) and Zoltai and Tarnocai (1975), which investigated peatlands. Information on soil development and characterization is in Lev and King (1999), Broll et al. (1998), Tarnocai et al. (1993), Smith et al. (1991), Tarnocai and Zoltai (1978), Brewer and Pawluk (1975), Pawluk and Brewer (1975a), and Day and Rice (1964).

The interactions between vegetation, nutrients, and moisture content of soils, as well as information on the biological activity, are in Broll et al. (1999) and Mueller et al. (1999). Information relating to mineralogy and micromorphology is in Foscolos and Kodama (1981), Fox (1985, 1979), Fox and Protz (1981), and Pawluk and Brewer (1975b), while information on soil thermal characteristics is in Mueller et al. (1998), Tarnocai et al. (1993), and Tarnocai (1980). Predictions about the effect of climate warming on Cryosols in Canada are in Tarnocai (1999).

This paper briefly describes the physical and ecological setting of the Canadian Arctic and gives information about the characteristics, genesis, and distribution of Cryosols, including detailed descriptions of selected pedons from the Arctic Ecoclimatic Province.

2. Description of the Area

The Canadian Arctic has been divided into various regions on the basis of either its ecoclimate (Ecoregions Working Group, 1989) or its physiography (Bostock, 1970). The ecoclimatic provinces and subprovinces are separated primarily by using vegetation as an indicator of climate, while distinctive terrain types, including bedrock and landforms, identify the physiographic regions. Ecoclimatically, the Arctic Ecoclimatic Province is divided into the High, Mid-, and Low Arctic ecoclimatic subprovinces (Figure 2.3.1), hereinafter referred to as the High Arctic, Mid-Arctic, and Low Arctic.

The physiography of the Canadian Arctic (Figure 2.3.2) includes a wide variety of terrain types. The *Canadian Shield Region*, the southernmost part of the Arctic, covers the eastern half of the continental Arctic and some of the adjacent islands, including a large part of Baffin Island. This region is composed primarily of undulating hills and valleys with many lakes, although the Baffin Island and Labrador portions are composed of rugged mountainous terrain with ice fields and glaciers.

The *Interior Plains Region* covers small areas along the coast of the Beaufort Sea and along the western margin of the continental portion of the Canadian

Shield. This region is composed of a series of plains, hills, and plateaus rising gradually from the Arctic coast.

The *Arctic Coastal Plains Region*, lying west of the Interior Plains Region, forms a very narrow band along the margins of the Arctic Ocean, the Beaufort Sea, and the inland channels on both the continental portion and the Arctic islands. It is generally a low area, often covered by wetlands.

The *Arctic Lowlands Region* covers the large central portion of the Arctic islands and consists of relatively gently rolling terrain with some hills and mountains.

Finally, the northernmost area, the *Innutian Region* of the High Arctic islands, is composed of rugged mountainous terrain with deep fiords, glaciers, and ice fields.

Permafrost is continuous in the Arctic and exists in all terrain types under all moisture conditions, except in newly deposited sediments where the climate has just begun to impose its influence on the soil thermal regime.

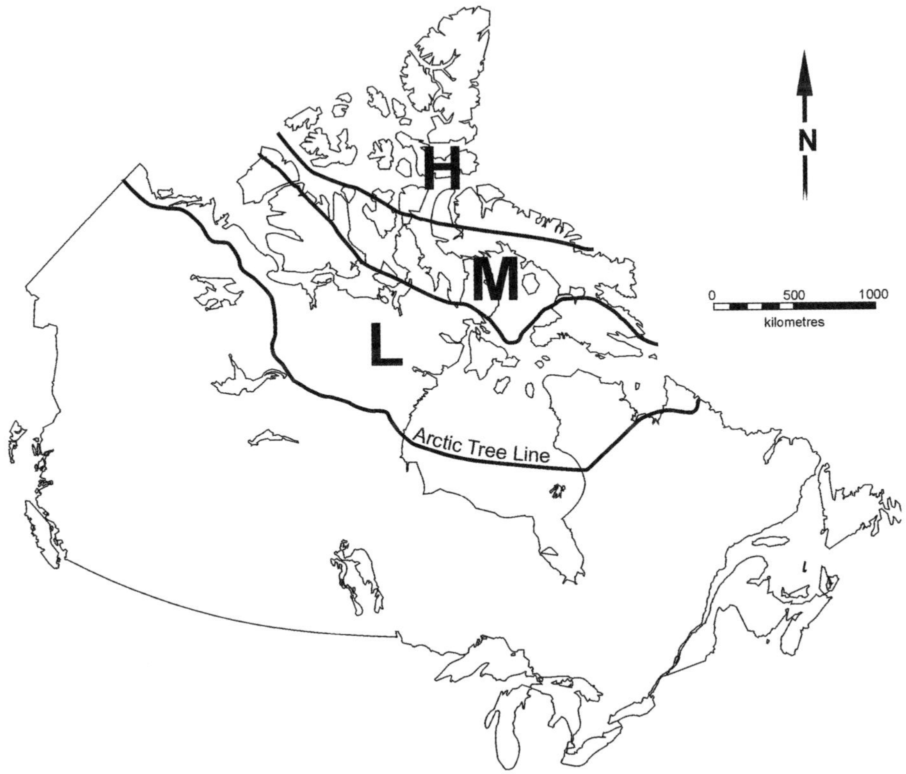

Figure 2.3.1. Ecoclimatic subprovinces of the Arctic – High (H), Mid- (M), and Low (L) Arctic (Ecoregions Working Group, 1989).

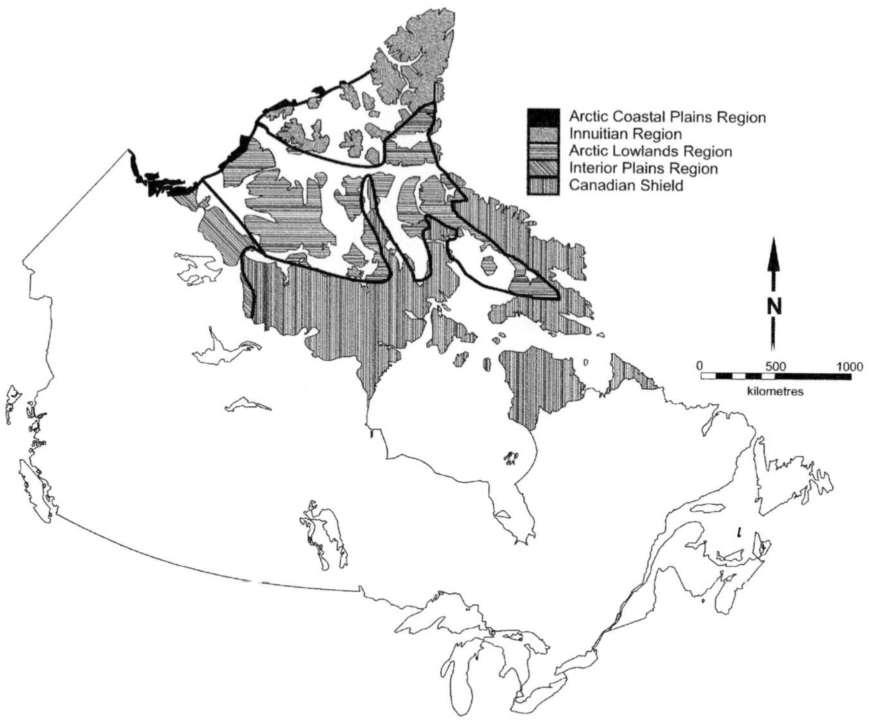

Figure 2.3.2. Physiographic regions of Arctic Canada.

3. Distribution of Cryosols

Approximately 63% (1,508,704 km^2) of the land area of the Arctic is covered by soils (Table 2.3.1 and Figure 2.3.3); rockland and glaciers cover the remaining 37% of the area. Cryosols cover 99% (1,493,483 km^2) of the soil area of the Arctic, with Turbic Cryosols, the most common Cryosols in the Arctic, covering approximately 79% of the soil area (Table 2.3.1). The remaining 20% of the area of Cryosols is covered by Static Cryosols (17%), found primarily in coarsely textured and gravelly materials, and Organic Cryosols (3%), which occur in peatlands. Brunisols and Organic soils occur in only small areas (1%) along the southern border of the Arctic. Regosols, found in recently deposited alluvial materials, also cover a very small area (<0.1%) (Table 2.3.1).

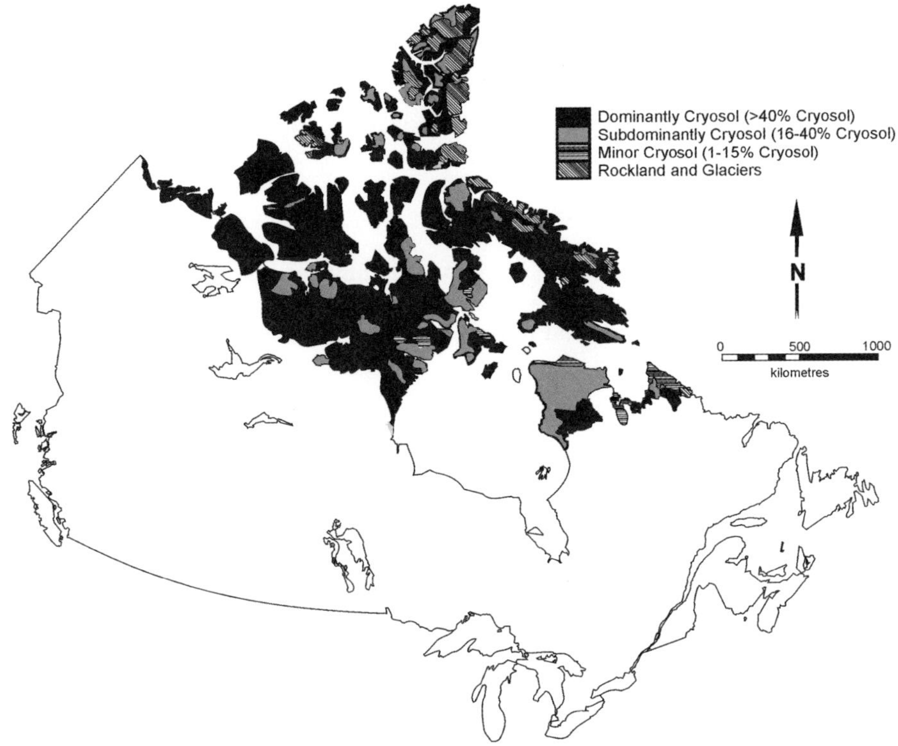

Figure 2.3.3. Distribution of Cryosols in Arctic Canada.

Table 2.3.1. Area and percent distribution of soils in Arctic Canada.

Soils Canada[1]	USA[2]	Area (km^2)	Area (%)
Turbic Cryosols	Turbels	1,192,659	79
Static Cryosols	Orthels	251,912	17
Organic Cryosols	Histels	48,912	3
Brunisols	Inceptisols	14,687	1
Regosols	Entisols	327	0.02
Organics	Histosols	207	0.01
TOTAL SOILS		1,508,704	100

[1]Soil Classification Working Group (1998).
[2]Soil Survey Staff (1998).

4. Regional Characteristics

Although Cryosols are the dominant soils in the Arctic, their characteristics differ greatly because of differences in parent materials, drainage, and ecoclimate.

4.1. Canadian Shield Region

Most of this region falls within the Low Arctic, although a sizeable portion lies in the Mid-Arctic. The only portion of this region in the High Arctic is on the north-eastern part of Baffin Island and in narrow bands on the eastern coasts of Devon and Ellesmere islands.

A large part of the soils developed on coarse-textured till and glaciofluvial materials occur in the Canadian Shield portion of the Arctic. These types of parent materials derived primarily from the granitic bedrock of the Canadian Shield. Most of the soils developed on till are Turbic Cryosols that have an active layer depth of less than one meter and are associated with nonsorted circle, stripe, step, and ice-wedge polygon types of patterned ground.

Tables 2.3.2-2.3.4 give examples of these Turbic Cryosols on till in the north-central District of Keewatin in the Low Arctic (pedon DK-1; see also cross section of soil profile in Figure 2.3.4) and on Baffin Island in the High Arctic (pedon 12-96-9) Arctic. Although these soils have a large sand fraction, they also contain considerable amounts of silt, which makes them extremely susceptible to cryoturbation. These soils are strongly to very strongly acidic (pH 4.5 to 5.2) and very low in nitrogen. The carbon content of the mineral horizons is usually low, except for the Ah horizon and some of the strongly cryoturbated horizons.

Vesicular structure, associated with a high water content, is very common in soils developed on till. Because of the high water content, this structure is very unstable, and the slightest disturbance (e.g., vibration) will cause the soil material to liquefy.

Cryoturbated soils similar to those described above occur on coarse-textured till in the southern portion of the Canadian Shield, but their active layers are deeper than two meters. As a result, these soils do not meet the criteria for Cryosols and are classified as Dystric Brunisols, cryoturbated phase. Tables 2.3.2-2.3.4 (pedon DP-3) give an example of this type of soil occurring in the Porter Lake area of the District of Mackenzie in the Low Arctic.

Soils developed on sandy glaciofluvial materials in the Canadian Shield region usually are noncryoturbated and have well developed B horizons. Tables 2.3.2-2.3.4 give characteristics of a well developed Brunisolic Static Cryosol (pedon 12-95-3) occurring on sandy glaciofluvial material on the High Arctic portion of Baffin Island.

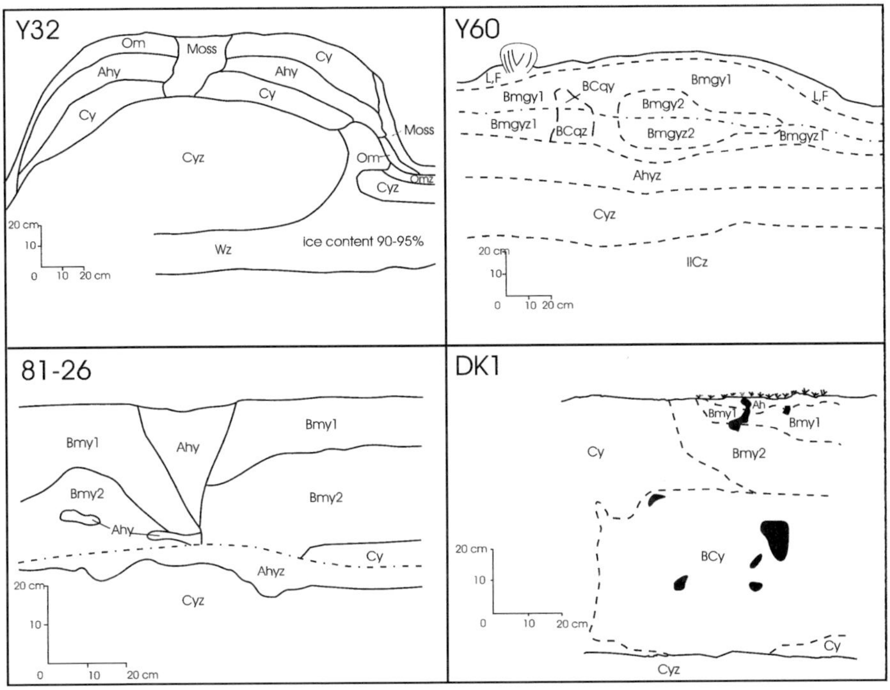

Figure 2.3.4. Cross-sections of cryoturbated soil profiles. Profiles Y32 and Y60 are associated with earth hummocks, while profiles 81-26 and DK1 are associated with nonsorted circles. Horizon designators follow the Canadian System of Soil Classification, where "y" indicates cryoturbated and "z" indicates perennially frozen (cryic) horizons. Locations are: Y32, Lat. 64° 36' N, Long. 126° 23' W; Y60, Lat. 69° 25' N, Long. 139° 59' W; 81-26, Lat. 76° 14' N, Long. 119° 04' W; DK1, Lat. 66° 06' N, Long. 94° 21' W (Tarnocai and Smith, 1993).

Table 2.3.2. Location of sites and site parameters of some Cryosols.

Parameter	Site										
	12–90–H1	12–90–T1	12–95–3	12–96–9	12–81–26	DC–3	DB–4	DK–1	Y60	DP–3	DB–3
Lat. (N)	81° 49' 48"	81° 23' 44"	66° 31' 05"	66° 41' 49"	76° 14'	93° 30'	75° 40'	66° 06'	69° 25'	62° 23'	97° 40'
Long. (W)	71° 17' 54"	76° 48' 00"	65° 29' 58"	64° 57' 16"	119° 20'	69° 57'	97° 40'	94° 21'	139° 59'	106° 05'	75° 40'
Landform1	CB	FT	EB	EVT	inclined	UM	rolling	UM	undulating	RM	EHCP
Slope (%)	5	7	20	5	5	3	0.5	1.5	3	–	0
Drainage	well	well	well	poor	mod. well	well	well	well	imperfect	well	imperfect
Vegetation[2]	D	Dryas tundra	LS	W	MS	ML	ML	LT	E	E	unvegetated
Parent material	colluvium	fluvial	eolian sand	eolian sand	colluvium	stony till	colluvium	stony till	colluvium	stony till	peat
Patterned ground[3]	IWP	EH	none	IWP	NC	EH	SP	NC	EH	NC	EHCP
Depth to permafrost (cm)	69	35	84	26	40	85	58	80	57	205 +	25
Soil Class. (Canada)[4]	RTC	RTC	BDSC	GSC	OETC	RTC	OETC	ODTC	ODTC	ODBcph	MOC
Soil Class. (US)[5]	TP	TP	TUO	TAO	TUT	TP	TUT	TUT	TUT	TOC	TH

[1]Landform: CB–colluvial blanket; EB–eolian blanket; EHCP–eroded high-centred lowland polygon; EVT–eolian veneer over till, inclined; FT–fluvial terrace; RM–ridged moraine; UM–undulating moraine.
[2]Vegetation: D–Dryas-sedge tundra; E–Ericaceous-sedge-moss-lichen tundra; LS–low shrub-lichen tundra; LT–lichen-shrub tundra; ML–moss-lichen-saxifrage tundra; MS–moss-sedge-lichen-willow tundra; W–arctic willow-sedge-Sphagnum moss tundra.
[3]Patterned ground: EH–earth hummocks; IWP–ice-wedge polygons; NC–nonsorted circles; SP–small (15–40 cm diam.) polygons.
[4]Soil Classification (Canada: Soil Classification Working Group, 1998): BDSC–Brunisolic Dystric Static Cryosol; GSC–Gleysolic Static Cryosol; MOC–Mesic Organic Cryosol; ODBcph–Orthic. Dystric Brunisol cryoturbated phase; ODTC–Orthic Dystric Turbic Cryosol; OETC–Orthic Eutric Turbic Cryosol; RTC–Regosolic Turbic Cryosol.
[5]Soil Classification (US: Soil Survey Staff, 1998): TAO–Typic Aquorthels; TDC–Typic Dystrocryepts; TH–Typic Hemistels; TP–Typic Psammoturbels; TUT–Typic Umbriturbels; TUO–Typic Umbrorthels.

Table 2.3.3. Morphological descriptions of some Cryosols. (Site information in Table 2.3.2.)

Horizon	Depth (cm)	Color (moist)	Texture[1]	Structure[2]	Ice content	Special features[3]
1. Pedon 12–90–H1 (High Arctic)						
Cky1	0–15	10YR 3/2m	L	gr	–	25% cf
Cky2	15–45	10YR 3/2m	L	gr	–	25% cf, 1.67 BD
Cky3	45–69	10YR 3/2m	SiL	pl	–	20% cf
Ckyz	69–85	10YR 3/2f	SiL	sl	25%	ice crystals
2. Pedon 12–90–T1 (High Arctic)						
Cky1	0–10	10YR 3/2m	L	gr	–	–
Cky2	10–35	10YR 3/1m	SiL	pl	–	–
Ckyz	35–65	10YR 3/1m	SiL	sl	71%	1.20 BD
Wz	65–187	–	–	–	100%	5% cf
3. Pedon 12–95–3 (High Arctic)						
C	0–5	10YR 4/2m	S	sg	–	–
2Ah	5–12	10YR 5/2m	S	gr	–	–
2Bm1	12–33	5YR 4/6m	S	gr	–	–
2Bm2	33–42	5YR 3/46m	S	gr	–	–
2BC	42–56	7.5YR 4/3m	S	sg	–	–
2C	56–84	10YR 4/3m	S	sg	–	–
2Cz	84–90	10YR 4/3f	S	sl	medium	ice crystals
4. Pedon 12–96–9 (High Arctic)						
Ah	0–16	5YR 2.5/2m	LS	gr	–	–
Cg	16–26	10YR 3/3m	LS	sg	–	–
Cgz	26 +	10YR 3/3f	–	sl	medium	ice crystals
5. Pedon 12–81–26 (High Arctic)						
Ahy	8–0	10YR 2/1m	–	gr	–	–
Bmy1	0–14	10YR 4/3m	SL	gr	–	–
Bmy2	14–55	10YR 4/2m	SL	gr	–	–
Cy	55–100	10YR 4/2m	SL	sg	–	org. intrusions
Ahyz	40–45	10YR 2/1f	–	sl	high	–
Cyz	45–80	10YR 4/2f	SL	sl	high	org. intrusions
6. Pedon DC–3 (Mid-Arctic)						
Om	0–15	10YR 2/1m	–	–	–	–
Cy	0–30	10YR 5/2m	L	sbk	–	–
Ahy	30–50	10YR 3/1m	SL	sbk	–	–
Ahyz	50–85	10YR 3/1m	SL	sl	high	vein ice
Cz	85–100	10YR 5/1f	SL	sl	high	ice lenses
7. Pedon DB–4 (High Arctic)						
Bmky	2–44	10YR 3.5/2m	SL	gr	–	vesicular
BCky	10–32	10YR 3/3m	SL	gr	–	vesicular
Ahkyz	1–15	10YR 2/1m	SL	sg	–	vein ice
Ckz	46–100	10YR 4/1f	SL	sl	–	ice crystals
8. Pedon DK–1 (Low Arctic)						
Ah	0–3	7.5YR 2/0m	S	gr	–	–
Bmy1	3–8	10YR 4/3m	LS	sg	–	–
Bmy2	8–30	10YR 3/3m	LS	sg	–	thix., vesicular
Bcy	30–80	10YR 5/2m	SL	sg	–	vesicular
Cy	0–80	2.5Y 6/3m	SL	sg	–	thix., vesicular
Cyz	80–100	2.5 6/2f	SL	sg	low	vein ice
9. Pedon Y60 (Low Arctic)						
L, F	3–0	10YR 3/1m	–	–	–	–
Bmgy1	0–13	10YR 5/4m	SiCL	gr	–	–
Bmgy2	13–25	7.5YR 4/2m	SiCL	gr	–	–
Bmgyz2	25–37	10YR 3/2f	SiCL	mss	–	–
Bmgyz1	37–43	10YR 3/1f	SiCL	mss	–	–
Bcgyz	11–25	2.5Y 4/0f	SiCL	mss	–	–
Ahyz	43–57	10YR 2/1f	SiL	sl	–	–
Cz	57–80	5Y 4/1f	SiL	sl	medium	vein ice
IICz	80–110	5Y 4/1f	SiL	sl	20%	vein ice

(continues)

Table 2.3.3. (cont.)

Horizon	Depth (cm)	Color (moist)	Texture[1]	Structure[2]	Ice content	Special features[3]
10. Pedon DP–3 (Low Arctic)						
Cy1	0–29	10YR 7/1m	SL	gr	–	thixotropic
Cy2	29–52	10YR 7/2 m	SL	gr	–	thixotropic
Ahb	52–73	10YR 2/2 m	SL	gr	–	–
Bmb	73–100	10YR 6/4 m	SL	gr	–	–
BC	100–205	10YR 7/3 m	SL	gr	–	–

[1]Texture: L–loam; SiL–silt loam; SL–sandy loam; S–sand; LS–loamy sand; SiCL–silty clay loam; SiC–silty clay.
[2]Structure: gr–granular; pl–platy; sl–structureless; sg–single grain; sbk–subangular blocky; mss–massive.
[3]Special features: cf–coarse fragments; BD–bulk density (g/cm^3); org.–organic; thix.–thixotropic.

4.2. Interior Plains Region

Soils in the western portion of this region developed on fine-textured till associated with earth hummocks, while those in the eastern part developed on loamy-textured calcareous till primarily associated with nonsorted circles. Most of the fine-textured till soils are strongly cryoturbated and contain a high amount of organic carbon in the mineral horizons (Tarnocai and Zoltai, 1978). Soils developed on calcareous till are also strongly cryoturbated.

4.3. Arctic Coastal Plains Region

This region stretches from the Mackenzie Delta area on the Beaufort Sea coast in the Low Arctic along the western coasts of the Arctic islands in the Mid- and High Arctic.

Most soils in the Arctic Coastal Plains Region, as well as in other areas that were affected by marine inundation, are clay and silty clay textures. These soils are highly cryoturbated and commonly are associated with earth hummocks (Figure 2.3.5, A and B). The B horizons of these soils usually have a granular structure resulting from cryodesiccation and cryoturbation, while the underlying soil horizons usually have a massive structure because of cryostatic pressure and cryodesiccation (Bockheim and Tarnocai, 1998; Van Vliet-Lanoë et al., in this book). Gray reduced colors and mottles are common redoximorphic features. These soils have high carbon and high ice content or pure ice layers in the subsoil. Tables 2.3.2-2.3.4 give an example of these soils occurring in the Malcolm River area on the North Slope of the Yukon in the Low Arctic (pedon Y60; see also cross section of soil profile in Figure 2.3.4).

Table 2.3.4. Physical and chemical characteristics of some Cryosols. (Site information in Table 2.3.2.)

Horizon	Depth (cm)	Total Sand (%)	Silt (%)	Clay (%)	pH	$CaCO_3$ equiv. (%)	Cond. (mmhos/cm)	C (%)	N (%)	C.E.C. (meq)	Exchangeable Cations meq/100 g Ca	Mg	K	Na
1. Pedon 12–90–H1 (High Arctic)														
Cky1	0-15	47.9	42.5	9.5	6.9	3.9	0.29	2.9	0.2	–	–	–	–	–
Cky2	15-45	51.5	28.0	20.5	6.5	3.5	0.22	1.0	Tr	–	–	–	–	–
Cky3	45-69	42.3	56.3	1.4	7.0	3.2	0.26	1.1	0.1	–	–	–	–	–
Ckyz	69-85	42.0	57.0	1.0	7.0	3.0	–	1.0	0.1	–	–	–	–	–
2. Pedon 12–90–T1 (High Arctic)														
Cky1	0-10	40.9	44.7	14.4	6.0	2.2	0.20	3.2	0.2	–	–	–	–	–
Cky2	10-35	44.9	54.1	1.0	7.0	5.6	0.29	4.9	0.5	–	–	–	–	–
Ckyz	35-65	–	–	–	–	–	–	–	–	–	–	–	–	–
Wz	65-187	–	–	–	–	–	–	–	–	–	–	–	–	–
3. Pedon 12–95–3 (High Arctic)														
C	0-5	94.5	4.0	1.5	5.0	–	0	0.2	tr	–	–	–	–	–
2Ah	5-12	92.9	4.8	2.3	4.8	–	0	1.8	0.2	–	–	–	–	–
2Bm1	12-33	91.7	2.7	5.6	4.9	–	0	0.5	0.1	–	–	–	–	–
2Bm2	33-42	95.8	1.9	2.3	4.9	–	0	0.5	0.2	–	–	–	–	–
2BC	42-56	97.3	1.6	1.1	5.0	–	0	0.3	tr	–	–	–	–	–
2C	56-84	98.3	1.0	0.7	5.0	–	0	0.3	0.2	–	–	–	–	–
2Cz	84-90	97.3	1.7	1.0	5.1	–	0	0.4	0.2	–	–	–	–	–
4. Pedon 12–96–9 (High Arctic)														
Ah	0-16	–	–	–	4.6	–	–	13.2	0.2	–	5.6	1.9	0.6	0.2
Cg	16–26	70.7	18.5	1.8	4.4	–	–	2.4	0.2	–	3.3	0.8	tr	0.
Cgz	26 +	–	–	–	–	–	–	–	–	–	–	–	–	–
5. Pedon 12–81–26 (High Arctic)														
Ahy	8–0	–	–	–	6.2	–	0.16	10.3	0.9	37.0	36.8	5.2	0.5	–
Bmy1	0–14	62.8	23.3	13.8	7.2	1.85	0.14	1.4	0.1	11.3	–	–	–	–
Bmy2	14–55	63.2	22.3	14.5	7.3	1.76	0.14	1.1	0.1	10.3	–	–	–	–
Cy	55–100	64.4	23.9	11.8	7.0	1.10	0.13	2.2	0.2	16.0	9.9	1.3	0.2	–
Ahyz	40–45	–	–	–	6.6	–	0.26	13.4	0.8	51.8	49.2	5.9	0.6	–
Cyz	45–80	59.0	29.3	11.7	6.9	–	0.19	2.4	0.2	19.8	11.1	1.8	0.2	–

(continues)

Table 2.3.4. (Cont.)

Horizon	Depth (cm)	Total Sand (%)	Silt (%)	Clay (%)	pH	$CaCO_3$ equiv. (%)	Cond. (mmhos/cm)	C (%)	N (%)	C.E.C. (meq)	Exchangeable Cations meq/100 g Ca	Mg	K	Na
6. Pedon DC–3 (Mid-Arctic)														
Om	0–15	–	–	–	7.0	–	–	33.6	1.8	–	20.9	9.0	1.4	–
Cy	0–30	37.1	47.7	15.2	7.6	30.2	0.55	0.7	0.1	7.2	88.6	25.5	0.7	–
Ahy	30–50	54.0	32.8	13.2	7.5	24.5	0.58	2.3	0.2	13.7	96.8	26.6	0.2	–
Ahyz	50–85	55.0	38.0	7.0	7.6	23.8	0.73	4.3	0.3	20.9	17.8	7.6	0.2	–
Cz	85–100	56.9	35.5	7.6	7.6	29.1	–	2.2	–	10.4	42.8	12.3	0.2	–
7. Pedon DB–4 (High Arctic)														
Bmky	2–44	72.3	16.0	11.7	7.4	7.5	0.3	1.7	0.1	–	–	–	–	–
BCky	10–32	75.3	14.0	10.7	7.2	4.6	0.4	0.2	0.1	–	–	–	–	–
Ahkyz	1–15	76.9	14.2	8.9	7.4	tr	0.6	5.5	0.3	–	–	–	–	–
Ckz	46–100	82.5	11.7	5.8	7.5	13.8	1.4	0.4	tr	–	–	–	–	–
8. Pedon DK–1 (Low Arctic)														
Ah	0–3	89.0	6.0	5.0	4.3	0.0	0.2	6.8	0.4	17.8	3.5	1.5	0.3	0.1
Bmy1	3–8	78.0	19.0	3.0	4.5	0.0	0.1	0.8	Tr	6.4	1.0	0.2	0.1	0.1
Bmy2	8–30	77.0	18.0	5.0	4.5	0.0	0.1	0.6	Tr	6.0	0.7	0.2	0.1	0.1
BCy	30–80	70.0	25.0	5.0	4.7	–	0.1	0.4	Tr	5.0	0.5	0.2	0.1	Tr
Cy	0–80	66.0	31.0	3.0	4.8	–	–	–	Tr	2.6	0.6	0.2	0.1	Tr
Cyz	80–100	74.0	21.0	5.0	5.1	–	0.1	–	0.1	3.3	0.8	0.4	0.1	Tr
9. Pedon Y60 (Low Arctic)														
L, F	3–0	–	–	–	–	–	–	50.9	0.9	63.2	14.1	17.2	0.7	Tr
Bmgy1	0–13	16.2	54.2	29.6	3.8	0.0	0.08	3.0	0.1	14.6	5.7	1.4	0.1	Tr
Bmgy2	13–25	16.8	47.3	35.8	3.7	0.0	0.12	6.1	0.3	25.8	11.9	3.2	0.1	Tr
Bmgyz2	25–37	17.8	51.8	30.4	3.7	0.0	0.11	5.7	0.3	21.0	10.9	3.1	0.2	Tr
Bmgyz1	37–43	16.2	51.9	31.8	4.1	0.0	0.12	6.6	0.3	24.8	12.5	5.3	0.3	Tr
BCgyz	11–25	17.1	57.3	25.6	4.4	0.0	0.10	1.2	0.2	15.3	6.2	2.1	0.1	Tr
Ahyz	43–57	14.5	59.1	26.3	5.9	0.0	0.34	14.8	0.7	43.2	31.1	5.8	0.3	0.1
Cyz	57–80	24.5	61.4	14.0	7.4	6.2	0.48	5.2	–	24.5	–	–	–	–
IICz	80–110	19.9	57.1	23.0	7.3	4.7	0.62	4.0	–	35.0	–	–	–	–
10. Pedon DP–3 (High Arctic)														
Cy1	0–29	54.0	43.0	3.0	5.2	–	–	0.2	tr	–	–	–	–	–
Cy2	29–52	55.0	43.0	2.0	4.8	–	–	0.2	tr	–	–	–	–	–
Ahb	52–73	55.0	38.0	7.0	4.6	–	–	1.6	tr	–	–	–	–	–
Bmb	73–100	71.0	25.0	4.0	4.7	–	–	0.5	tr	–	–	–	–	–
BCb	100–205	82.0	16.0	2.0	4.7	–	–	0.1	tr	–	–	–	–	–

Tr – trace

4.4. Arctic Lowlands and Innuitian regions

All of the islands composing the Innuitian Region lie in the High Arctic. The Arctic Lowlands Region, however, extends northward from the Arctic coast in the Low Arctic to Devon Island in the High Arctic. Loamy-textured soils dominate the large expanses of the Arctic islands, most of which lie within these two regions, and portions of the Arctic coast, including Boothia Peninsula in the District of Keewatin.

Soils developed on moderately calcareous materials usually have a thin B horizon that commonly has a dark brown color and is strongly cryoturbated. Examples of these soils occur on Bathurst Island (Tables 2.3.2-2.3.4, pedon DB-4) and at Mould Bay on Prince Patrick Island (Tables 2.3.2-2.3.4, pedon 12-81-26), both in the Innuitian Region. On very strongly calcareous materials, the soils are also strongly cryoturbated but usually do not have a B horizon, as a soil on Boothia Peninsula in the Mid-Arctic portion of the Arctic Lowlands shows (Tables 2.3.2-2.3.4, pedon DC-3).

On the High Arctic islands, soils are also cryoturbated but generally lack B horizon development. Two soils in the Innuitian Region on Ellesmere Island show this, one at Lake Hazen (Tables 2.3.2-2.3.4, pedon 12-91-H1), the other at Tanquary Fiord (Tables 2.3.2-2.3.4, pedon 12-90-T1). Most of these soils are calcareous, with a pH higher than 6.0, and usually have a salt crust on the soil surface. These soils also usually have higher conductivity, especially those developed on materials deposited or affected by marine inundation. Although these soils are sparsely vegetated, their organic carbon content is high.

4.5. Peatlands

In the Low Arctic, Organic Cryosols commonly occur in the form of high-centered lowland polygon bogs, low-centered lowland polygon fens (Figure 2.3.5, C and D) and peat mound bogs (National Wetlands Working Group, 1988). These soils range in thickness from less than one meter to several meters. Small patches of strongly eroded lowland polygon bogs associated with Organic Cryosols occur throughout the Mid- and High Arctic.

The basal peat in Arctic lowland polygon bogs yielded radiocarbon dates of 8120 $\pm$ 100 yrs. BP (Mackenzie Delta area of the Low Arctic), 9210 $\pm$ 170 and 8420 $\pm$ 80 yrs. BP (Bathurst Island in the High Arctic), and 6780 $\pm$ 390 yrs. BP (Tanquary Fiord, Ellesmere Island in the High Arctic). These dates indicate that these soils developed during the Hypsithermal interval. Later, when the climate cooled, peat deposition ceased, and they are now eroded remnants of the former lowland polygon bogs.

Most of these Organic Cryosols developed on mesic and humic peat in the Low Arctic, but some fibric peat also occurs either on the surface or as layers within the peat. Most of the eroded Mid- and High Arctic Organic Cryosols are associated with mesic peat. Tables 2.3.2 and 2.3.5 present an example of a strongly eroded Organic Cryosol on Bathurst Island in the High Arctic (pedon DB-3).

Figure 2.3.5 A, B. Common patterned ground types associated with Cryosols in the Canadian Arctic. Earth hummocks viewed from the air (A, top) and from the ground (B, bottom).

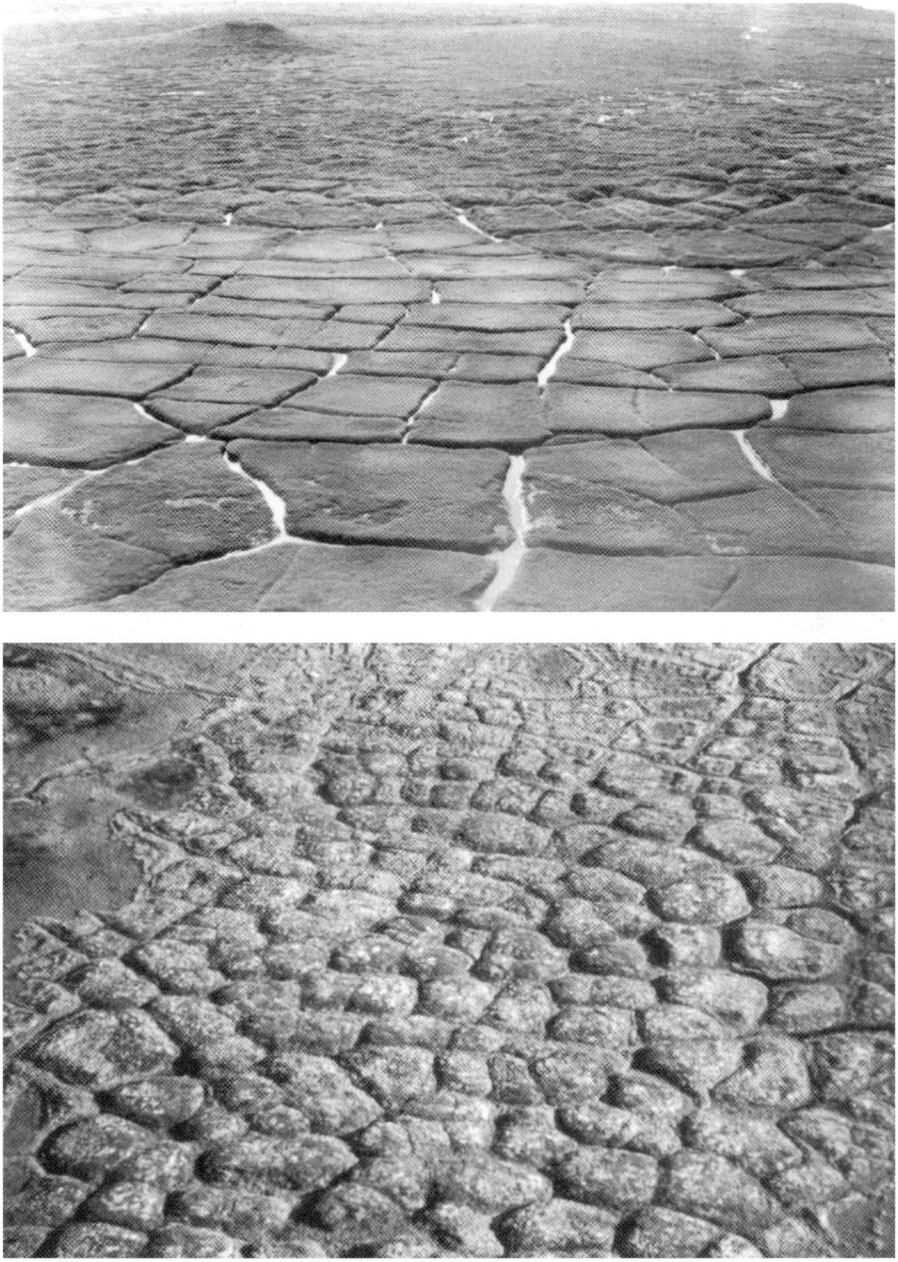

Figure 2.3.5 C, D. Common patterned ground types associated with Cryosols in the Canadian Arctic. Low-centered polygons (C, top) and high-centered polygons (D, bottom) viewed from the air.

5. Genesis of Cryosols

All soils are formed by the interaction of soil-forming factors, but because of the cold climate in the Arctic regions of Canada, cryogenic processes, which lead to the formation of Cryosols, dominate the soil formation (Figure 2.3.6A, B). The presence and mobility of unfrozen soil water, as it migrates toward the frozen front along the thermal gradient in the frozen system, drives these processes. The cryogenic processes that affect the genesis of Arctic soils are freeze-thaw, cryoturbation (frost churning), frost heave, cryogenic sorting, thermal cracking, and ice segregation. Other soil-forming processes that can leave an imprint on these soils include the gleyic process, brunification, eluviation, and salinization.

6. Discussion

Cryoturbation, a dominant process in the Arctic, results in irregular or broken soil horizons, involutions, organic intrusions, organic matter accumulation in the subsoil (often with concentration of this material along the top of the permafrost table), oriented rock fragments, silt-enriched layers, and silt caps (Bockheim and Tarnocai, 1998) (Figure 2.3.4). In the southern Arctic, it affects primarily clayey-textured materials or sandy loam materials. In the High Arctic, both of these soil materials are cryoturbated, as are some sandy materials, especially those with high moisture content.

Arctic soils generally have high moisture content, especially near the permafrost table, so gleying, associated grayish colors, and redoximorphic features occur even in better drained situations. In the western Arctic, soils associated with earth hummocks have characteristic reddish brown mottles and gray gleyed features. Soils of the High Arctic very often are saturated because of the shallow active layer, but gleying is not very common except on fine-textured materials. Gleying is likely lacking because most of the water entering the shallow active layer from rain and snow melt is oxygen-rich.

Thin eluvial or leached horizons commonly occur in Cryosols in the southern portion of the Low Arctic and have been observed on Cameron Island in the High Arctic. The brunification associated with coarse-textured Cryosols is usually associated with this eluvial horizon.

The active layer not only supports biological activity but also protects the underlying permafrost. The thickness of this layer is controlled by soil texture and moisture, thickness of the surface organic layer, vegetative cover, aspect, and latitude. In Arctic Cryosols, the active layer normally extends 40 to 80 cm below the soil surface.

In the Arctic, Cryosols situated on sloping terrain are usually subject to mass displacement of soil materials as a result of downslope movement and solifluction. Downslope movement is especially common in the Mid and High Arctic and can occur even on 1° slopes. Solifluction occurs throughout the Arctic.

Table 2.3.5. Morphological, physical and chemical characteristics of pedon DB–3, an Organic Cryosol. (Site information in Table 2.3.2.)

Horizon	Depth (cm)	Color (moist)	Fiber (%) Unrubbed	Fiber (%) Rubbed	Ice content	Special features	pH	Ash (%)	C (%)	N (%)	C.E.C. (meq)	Exchangeable Cations meq/100 g Ca	Mg	K	Na
Of	0-25	5YR 3/3 m	80	60			5.7	15	49.4	0.4	105	62	14	0.4	1.1
Omz1	25-78	5YR 2.5/2 f	72	60	high	ice crystals	5.9	12	52.5	0.3	105	63	19	0.3	0.9
Omz2	78-130	5YR 2.5/1 f	52	42	high	ice crystals	6.9	28	37.8	0.3	75	59	18	0.4	0.8
IICz	130-150	10YR 4/3 f			high	ice lenses	–	–	–	–	–	–	–	–	–

Figure 2.3.6 A. Photograph of a clay-loam-textured Brunisolic Turbic Cryosol associated with an earth hummock. Note the organic-rich subsurface horizon, the intrusions of organic materials and organic smears resulting from cryoturbation, and the high ice content layer at the top of the permafrost. Tuktoyaktuk Peninsula, Northwest Territories, Canada.

The particle-size distribution of Arctic Cryosols varies from silty clay to coarse gravelly sand (Table 2.3.4). The composition of the fine earth fraction commonly depends on the mode of deposition of the parent materials, because of their low rate of weathering combined with cryoturbation. The pH of Cryosols (Table 2.3.4) varies greatly and also depends on the chemistry of the parent materials. The pH is similar to that of the parent material partly because cryoturbation mixes soil materials not only between horizons but also with the parent material.

The nitrogen, potassium, and phosphorus contents of Arctic soils are generally low. Most of these nutrients are locked into compounds that form the surface organic matter (Broll et al., 1999). Salt crusts are common on soil surfaces on the High Arctic islands of Canada. These salt crusts develop by translocation of salts from the subsurface horizons and develop during dry periods in the summer because of increased evaporation from the soil surface. Calcareous crusts on the underside of rocks are a common phenomenon, especially in coarse-textured soils. Cryosols developed on fine-textured marine sediments usually have both high conductivity and high salt content.

Figure 2.3.6 B. Photograph of a sandy-textured Orthic Turbic Cryosol associated with non-sorted circles. Note the distorted and disrupted soil horizons resulting from cryoturbation. Bathurst Island, Nunavut, Canada.

Table 2.3.6. Carbon content and carbon masses of soils in Arctic Canada (Tarnocai, 2000).

Soils	Soil carbon content (kg/m^2)	Soil carbon mass (Gt)
Turbic Cryosol	27.77	33.12
Static Cryosol	21.53	5.43
Organic Cryosol	85.82	4.21
Brunisols	18.01	0.26
Regosols	5.17	0.0017
Organic	42.46	0.01
TOTAL SOILS		43.04

The total soil organic carbon mass in the Arctic is 43 Gt (Table 2.3.6) (Tarnocai, 2000). Turbic Cryosols contain 79% of this carbon, followed by Static Cryosols (17%), Organic Cryosols (3%), and other soils (Brunisols, Organic soils, and Regosols [1%]) (Tarnocai, 2000). The soil organic carbon contents of the Arctic soils (both total and surface contents) are greatest in the Low Arctic and decrease northward. Although the surface organic horizons become thinner and very patchy in the High Arctic, a high amount of carbon still is stored in both the surface and subsurface horizons.

In the Low and Mid-Arctic, both mineral and organic Cryosols are major soil organic carbon pools. In mineral Cryosols, this carbon commonly occurs as a surface organic layer or as organic-rich mineral layers. Considerable amounts of organic carbon also are stored in cryoturbated mineral horizons and cryoturbated subsurface organic horizons. In these regions, Organic Cryosols are associated primarily with low- and high-centered lowland polygons. The organic material is generally 2-2.5 m thick, especially in high-centered lowland polygons, but can be as thick as 10 m.

The unique thermal characteristic separating Cryosols from all other soils is the presence of a perennially frozen layer within two meters of the soil surface. Because of this frozen layer, Cryosols have a steep vertical temperature gradient. If these soils are associated with patterned ground, the lateral temperature gradient also can be large.

For example, in the case of Cryosols associated with earth hummocks in the Mackenzie Delta area, during summer months the soil temperature at the center of the hummock can decrease from 12°C at the surface to 0°C at the 50-cm depth. Soil temperatures at comparable depths under the interhummock depression, less than one meter away, can be 5° to 7°C lower (Tarnocai and Zoltai, 1978). Similar thermal characteristics also were found in Cryosols on Baffin Island (Mueller et al., 1998). Tarnocai (1980, 1994, 1993) demonstrated the dynamic thermal characteristics of Cryosols, showing that freeze-back from the permafrost table could occur several times during the spring and during cool, snowy periods in the summer.

7. Summary

Cryosols are major soils in the Canadian Arctic, with Turbic Cryosols being the dominant soils.

The characteristics of various Cryosols in the Arctic depend on their parent materials, drainage, and thermal conditions.

Cryogenic processes are the major soil-forming processes for Arctic Cryosols.

Parent materials control the pH and texture of these soils. The nutrient content (nitrogen, potassium, and phosphorus) of Arctic Cryosols is generally low. Carbon content, however, is generally high, and most Cryosols act as carbon sinks.

References

Bockheim, J.G. 1974. Soils of Cumberland Peninsula, Baffin Island, N.W.T., Canada. Final Report of Parks Canada. Contract 74-112.

Bockheim, J.G., and C. Tarnocai. 1998. Recognition of cryoturbation for classifying permafrost-affected soils. Geoderma. 81:281-293.

Bostock, H.S. 1970. Physiographic subdivisions of Canada. In: R.J.W. Douglas, ed. Geology and Economic Minerals of Canada, Dept. of Energy, Mines and Resources Canada, Geological Survey of Canada, Economic Geology Report No. 1. pp. 9-30.

Bradley, S.W., J.S. Rowe, and C. Tarnocai. 1982. An ecological land survey of the Lockhart River map area, Northwest Territories. Environment Canada, Ottawa. Ecological Land Classification Series, No. 16. 152 pp.

Brewer, R., and S. Pawluk. 1975. Investigations of some soils developed in hummocks of the Canadian Sub-arctic and Southern-Arctic regions. 1. Morphology and micromorphology. Can. J. Soil Sci. 55:301-319.

Broll, G., G. Müller, and C. Tarnocai. 1998. Permafrost-affected soils in the Pangnirtung Pass area, Baffin Island, Canada. In: A.G. Lewkowicz and M. Allard, eds. Proceedings, Seventh International Conference on Permafrost, Centre d'études nordiques, Université Laval, Québec. pp. 89-94.

Broll, G., C. Tarnocai, and G. Mueller. 1999. Interactions between vegetation, nutrients and moisture in soils in the Pangnirtung Pass area, Baffin Island, Canada. Permafrost and Periglacial Processes. 10(3):265-277.

Cruickshank, J.G. 1971. Soils and terrain units around Resolute, Cornwallis Island. Arctic. 24:195-209.

Day, J.H. 1964. Operation Hazen: Characteristics of soils of the Hazen Camp area, northern Ellesmere Island, N.W.T. Defence Res. Board, Dept. of Nat'l Defence. Ottawa, Canada. D. Phys. R. (G), Hazen 24. 28 pp.

Day, J.H., and H.M. Rice. 1964. The characteristics of some permafrost soils in the Mackenzie Valley, N.W.T. Arctic. 17:223-235.

Ecoregions Working Group. 1989. Ecoclimatic Regions of Canada: First Approximation. Ecoregions Working Group of the Canada Committee on Ecological Land Classification. Ecological Land Classification Series, No. 23, Sustainable Development Branch, Canadian Wildlife Service, Conservation and Protection, Environment Canada. Ottawa, Canada. 119 pp. and map.

Everett, K.R. 1968. Soil development in the Mould Bay and Isachsen areas, Queen Elizabeth Islands, Northwest Territories, Canada. Ohio State Univ., Res. Foundation. Columbus, Ohio. 75 pp. and maps.

Foscolos, A.E., and H. Kodama. 1981. Mineralogy and chemistry of Arctic desert soils on Ellef Ringnes Island, Arctic Canada. J. Soil Sci. Soc. Am. 45:987-993.

Fox, C.A. 1979. The soil micromorphology and genesis of the Turbic Cryosols from the Mackenzie River Valley and the Yukon Coastal Plain. Ph.D. Thesis. Univ. of Guelph. Guelph, Ont. 196 pp.

Fox, C.A. 1985. Micro morphology of an Orthic Turbic Cryosol – a permafrost soil. In: P. Bullock and C.P. Murphy, eds. Soil Micromorphology. Vol. 2. Soil Genesis. A B Academic Publishers. Berkhamsted, U.K. pp. 699-705.

Fox, C.A., and R. Protz. 1981. Definition of fabric distributions to characterize the re-arrangement of soil particles in the Turbic Cryosols. Can. J. Soil Sci. 61:29-34.

Hendershot, W.H. 1984. A comparison of some upland and valley soils in the Ungava-Labrador Peninsula. Géographie physique et Quaternaire, 38:243-256.

James, P.A. 1970. The soils of the Rankin Inlet area, Keewatin, N.W.T., Canada. Arct. Alp. Res. 2:293-302.

Leahey, A. 1947. Characteristics of soils adjacent to the Mackenzie River in the Northwest Territories of Canada. Soil Sci. Soc. Am. 12:458-461.

Lev, A., and R.H. King. 1999. Variation of soil development in a High Arctic Soil Landscape: Truelove Lowland, Devon Island, Nunavut, Canada. Permafrost and Periglacial Processes. 10(3):289-307.

Mueller, G., G. Broll, and C. Tarnocai. 1998. Soil temperature regimes and microtopographic contrasts, Baffin Island, N.W.T., Canada. In: A.G. Lewkowicz and M. Allard, eds. Proceedings, Seventh International Conference on Permafrost, Centre d'études nordiques. Université Laval. Québec. pp. 771-775.

Mueller, G., G. Broll, and C. Tarnocai. 1999. Biological activity as influenced by microtopography in a Cryosolic soil, Baffin Island, Canada. Permafrost and Periglacial Processes. 10(3):279-288.

Pawluk, S., and R. Brewer. 1975a. Investigation of some soils developed in hummocks of the Canadian Sub-Arctic and Southern-Arctic regions. 2. Analytical characteristics, genesis and classification. Can. J. Soil Sci. 55:321-330.

Pawluk, S., and R. Brewer. 1975b. Micromorphological and analytical characteristics of some soils from Devon and King Christian Islands, N.W.T. Can. J. Soil Sci. 55:349-361.

Smith, C.A.S., C.A. Fox, and A.E. Hargrave. 1991. Development of soil structure in some Turbic Cryosols in the Canadian Low Arctic. Can. J. Soil Sci. 71:11-29.

Smith, C.A.S., C. Kennedy, A.E. Hargrave, and K. McKenna. 1989. Soil and vegetation survey of Herschel Island, Yukon Territory. Land Resource Res. Centre, Res. Branch, Agriculture Canada. Ottawa. 19 pp. and 2 maps.

Soil Classification Working Group. 1998. The Canadian System of Soil Classification. (3rd ed.) Res. Branch, Agriculture and Agri-Food Canada. NRC Res. Press. Ottawa, Canada. Publication 1646. 187 pp.

Soil Survey Staff. 1998. Keys to Soil Taxonomy. USDA. Washington, D.C. 326 pp.

Tarnocai, C. 1976. Soils of Bathurst, Cornwallis, and adjacent islands, District of Franklin. Geological Survey of Canada, Dept. of Energy, Mines and Resources, Ottawa. Paper 76-1B, pp. 137-141.

Tarnocai, C. 1977. Soils of the north-central Keewatin. Geological Survey of Canada, Dept. of Energy, Mines and Resources. Ottawa. Paper 77-1A, pp. 61-64.

Tarnocai, C. 1978a. Soils and land use – Arctic soils. The Geosciences in Canada, 1977, Annual Report and Review of Soil Science, prepared by the Canadian Geoscience Council, Geological Survey Paper 78-6. pp. 28-32.

Tarnocai, C. 1978b. Genesis of organic soils in Manitoba and the Northwest Territories. Proceedings of the Third York Quaternary Symposium. pp. 453-470.

Tarnocai, C. 1978c. Distribution of soils in northern Canada and parameters affecting their utilization. 11th Congress of the International Society of Soil Science, Edmonton, Canada, June 1978. Transactions, Volume 3, Symposia Papers. pp. 332-347.

Tarnocai, C. 1980. Summer temperatures of Cryosolic soils in the north-central Keewatin, N.W.T. Can. J. Soil Sci. 60:311-327.

Tarnocai, C. 1994. Genesis of permafrost-affected soils. In: J.M. Kimble and R.J. Ahrens, eds. Proceedings of the Meeting on the Classification, Correlation, and Management of Permafrost-Affected Soils—July, 1994. USDA, SCS, NSSC. Lincoln, NE. pp. 143-154.

Tarnocai, C. 1999. The effect of climate warming on the carbon balance of Cryosols in Canada. Permafrost and Periglacial Processes. 10(3):251-263.

Tarnocai, C. 2000. Carbon Pools in Soils of the Arctic, Subarctic and Boreal Regions of Canada. In: R. Lal, J.M. Kimble and B.A. Stewart, eds. Global Climate Change and Cold Region Ecosystems. Advances in Soil Science. Lewis Publishers, Boca Raton, Fla. pp. 91-103.

Tarnocai, C., and C.A.S. Smith. 1993. The formation and properties of soils in the permafrost regions of Canada. In: Proceedings of the 1st International Conference on Cryopedology. November 10-14, 1992. Russian Academy of Sciences. Pushchino, Russia. pp. 21-42.

Tarnocai, C., and H. Veldhuis. 1998. Soils and trafficability of Pangnirtung Pass, Auyuittuq National Park Reserve. Prepared for the Auyuittuq Nat'l Park Reserve by the Res. Branch, Agriculture and Agri-Food Canada. Ottawa. 78 pp.

Tarnocai, C., and S.C. Zoltai. 1978. Earth hummocks of the Canadian arctic and subarctic. Arct. Alp. Res. 10:581-594.

Tarnocai, C., A.N. Boydell, J.A. Netterville and K.A. Drabinsky. 1976. Biophysical land classification of Boothia Peninsula and northeast Keewatin, N.W.T. Geological Survey of Canada Open File #390. 31 pp. (report and maps).

Tarnocai, C., D. Kroetsch, J. Gould, and H. Veldhuis. 1991. Soils, vegetation and trafficability, Tanquary Fiord and Lake Hazen areas, Ellesmere Island National Park Reserve. Canadian Parks Service, Prairie and Northern Region. Winnipeg, Man. Report RM91-1/ELS. 269 pp.

Tarnocai, C., C.A.S. Smith, and C.A. Fox. 1993. International Tour of Permafrost-Affected Soils: The Yukon and Northwest Territories of Canada. Centre for Land and Biological Resources Res., Res. Branch, Agriculture Canada. Ottawa. 197 pp.

Tedrow, J.C.F., and L.A. Douglas. 1964. Soil investigations on Banks Island. Soil Sci. 98:53-65.

Tedrow, J.C.F., P.F. Bruggemann, and G.F. Walton. 1968. Soils of Prince Patrick Island. The Arctic Institute of North America. Washington, D.C. Res. Paper No. 44. 81 pp.

Van Vliet-Lanoë, B., C.A. Fox, and S.V. Gubin. Micro morphology of Cryosols. In: J.M. Kimble, ed. Cryosols: Permafrost-Affected Soils. (This volume.)

Walker, B.D., and T.W. Peters. 1977. Soils of Truelove Lowland and Plateau. In: L.C. Bliss, ed. Truelove Lowland, Devon Island, Canada: A High Arctic Ecosystem. Univ. of Alberta Press. Edmonton. pp. 31-62.

Woo, V., and S.C. Zoltai. 1977. Reconnaissance of soils and vegetation of Somerset and Prince of Wales Islands, N.W.T. Northern Forest Res. Centre. Edmonton, Alberta. Information Report NOR-X-186. 127 pp.

Woo, M., A.G. Lewkowicz, and W.R. Rouse. 1992. Response of the Canadian permafrost environment to climatic change. Phys. Geog. 13:287-317.

Zoltai, S.C., and C. Tarnocai. 1974. Soils and vegetation of hummocky terrain. Environmental-Social Committee, Northern Pipelines, Task Force on Northern Oil Development. Report No. 74-5. 86 pp.

Zoltai, S.C., and C. Tarnocai. 1975. Perennially frozen peatlands in the western arctic and subarctic of Canada. Can. J. Earth Sci. 12:28-43.

Chapter 4. Cryosols of the Boreal, Subarctic, and Western Cordillera Regions of Canada

C.A. Scott Smith[1] and H. Veldhuis[2]

[1]Research Scientist, Pacific Agri-Food Research Centre, Agriculture and Agri-Food Canada, Summerland, BC V2A 8Y2 Canada; e-mail smithcas@agri.gc.ca.
[2]Pedologist, University of Manitoba, Land Resource Unit-Manitoba, Ellis Bldg., Room 362A, , Winnipeg, Manitoba R3T 2N2, Canada; e-mail veldhuish@agr.gc.ca.

1. Introduction

Cryosols are soils whose formation is affected by permafrost within the soil control section. In this chapter we define Cryosol as in the *Canadian System of Soil Classification* (Soil Classification Working Group, 1998) and we use the taxa from that system to describe the nature and distribution of Cryosols in three non-Arctic regions of Canada. The boreal, subarctic, and cordillera regions extend across the North American continent (Figure 2.4.1) to incorporate the zone of discontinuous permafrost and the southern portions of the zone of continuous permafrost.

The boreal region of Canada is a zone of closed-canopy coniferous or mixed forests that stretches from the Atlantic coast in the east to the foothills of the Rocky Mountains. Climatic conditions suitable for boreal forests define the limits of the region. Permafrost, and hence Cryosolic soils, are confined largely to the northern and western portions of this region, often referred to as the high boreal (Ecoregions Working Group, 1989). Cryosolic soils cover approximately 165,000 km^2 here (Table 2.4.1).

The subarctic region of Canada is a zone of open-canopy conifer woodlands with tundra patches. It stretches from the Atlantic to the Mackenzie and Richardson Mountains along the Yukon/Northwest Territories border. The region is an ecotone (i.e., transition) between the Arctic tundra to the north and the boreal forest to the south. It roughly coincides with much of the zone of discontinuous permafrost in the Canada (Heginbottom, 1995). Cryosolic soils form in both mineral and organic parent materials in this region.

Figure 2.4.1. Map of Canada showing the extent of the boreal, subarctic, and cordillera regions. Outlines of the cordillera and subarctic regions follow that defined for the respective ecoclimatic provinces (Ecoregions Working Group, 1989). The extent of the boreal region on the figure is confined to high boreal ecoclimatic regions where Cryosols are known to occur.

Physiography defines the extent of the cordillera region, here meaning, specifically, the western cordillera of Canada. It includes the mountain ranges of British Columbia and the Yukon Territory — the Coastal Ranges and the interior systems of the Rocky, Columbia, St. Elias, Mackenzie, Richardson, and British Mountains, as well as numerous smaller ranges and intervening plateaus. A wide range of climatic, vegetative, elevational, and permafrost conditions exists here. The distribution of Cryosols varies from isolated alpine patches in the south to over 90% of the landscape in the northernmost portions of the cordillera region (Table 2.4.1).

Table 2.4.1. Approximate areas of Cryosolic soils in the boreal, subarctic, and cordillera regions of Canada (values adapted from Tarnocai, 2000).

Region	Land Area (10^3 Km^2)	Area of Cryosols
Boreal [1]	10,851	165
Subarctic	1,720	700
Cordillera- boreal	710	40
Cordillera - subarctic	140	130

[1] includes only the high boreal portion of the region, the only part of the region where Cryosolic soils significantly occur.

2. Cryosols of the Boreal Region

The southern limit of permafrost runs through the boreal region over central and western Canada. Moderating air masses that move eastwardly onto the continent, from the Pacific Ocean, skew annual isotherms southeastward from the west toward Hudson Bay. Just east of the cordillera, the most southern distribution of Cryosols is at about 57° N, while near Hudson Bay in the center of the continent, this distribution extends south to 50°N latitude. From east of Hudson Bay, the southern boundary of permafrost follows the 51° N latitude to the Atlantic coast in southern Labrador (Heginbottom, 1995).

In much of the boreal region, Cryosols are associated mainly with peatlands with open cover of black spruce (*Picea mariana*) and ericaceous shrubs (*Ledum* and *Chamaedaphne* spp.) and lichens (*Cladina*, *Cladonia*, and *Cetraria* spp.) (Zoltai et al., 1988a). The presence of permafrost produces characteristic peatland types; Organic Cryosols are associated with peat plateau and palsa bogs (Figure 2.4.2). In the western portion of the region, peatlands with sporadic permafrost form in areas with mean annual air temperature <1°C and > 250 mm annual precipitation (Vitt et al., 1994). However an apparent disequilibrium exists between the southern limit of permafrost (occasionally found south of the 0°C isotherm) and the present climate in northwestern Canada (Halsey et al.1995).

Active layers that are 60 to 100 cm deep characterize the Organic Cryosols of the boreal region. The temperature of the permafrost is only just below freezing. Permafrost development is cyclical, over periods as short as 600 years, such that both aggrading and degrading permafrost can occur contemporaneously in the same peatland (Zoltai, 1993). Ice content of the permanently frozen materials underlying the peat plateaus is usually >50% by volume, and clear ice lenses, occasionally tens of centimeters thick, are common.

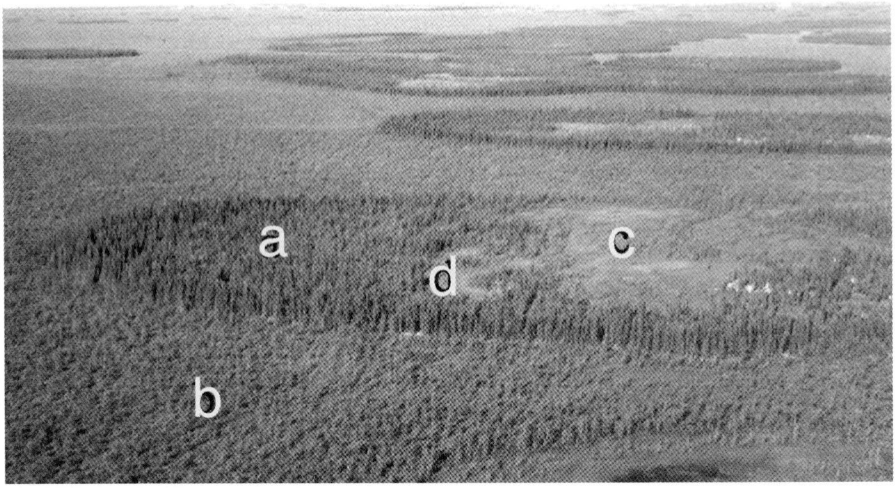

Figure 2.4.2. Peatlands in the subarctic region of Manitoba, Canada. The photo illustrates a peat plateau bog (a), a treed horizontal fen (b), a collapse fen (c), and a collapse bog (d). Permafrost is continuous or nearly continuous in peat plateau bogs but is not present in treed fens. The collapse fens and collapse bogs, which are thermokarst peatland forms, contained permafrost some time ago, but the underlying permafrost has receded to several meters in depth or has disappeared completely.

Most peat plateaus and palsa bogs initially developed as fens (Zoltai and Tarnocai, 1971) and, as a result, the deeper peat often derives from sedges and brown mosses with varying quantities of woody fragments. The deeper peat usually ranges from slightly acidic to weakly alkaline, but highly calcareous marl-rich layers may also be present. The peat can be layered in sequences of bryophytic, *Sphagnum*, and sylvic peat that reflect the cyclical nature of the ecological moisture conditions associated with permafrost aggradation and decay (Zoltai, 1993; Vitt et al., 1994). The upper part of the control section in these Fibric and Mesic Organic Cryosols often consists of weakly decomposed *Sphagnum* moss peat, very strongly to extremely acidic, developed during and after the transformation from fen to mature peat plateau or palsa bog.

With an increase in latitude, permafrost begins to appear in shallow peatlands (veneer bogs) (Mills et al., 1985) and in poorly to well drained clayey and fine loamy mineral deposits. Permafrost in the veneer bogs initially is confined to the larger *Sphagnum* moss hummocks and to parts with deeper peat within the veneer bog (Figure 2.4.3). The veneer bogs burn frequently, as wildfire is common in the boreal region. Moisture content of the surface peat and wind velocity control the impact of wildfire. When the surface peat has low moisture content, wildfire will consume a large part of the insulating moss and surface peat layer, frequently exposing the underlying mineral soil, thereby reducing the presence of permafrost.

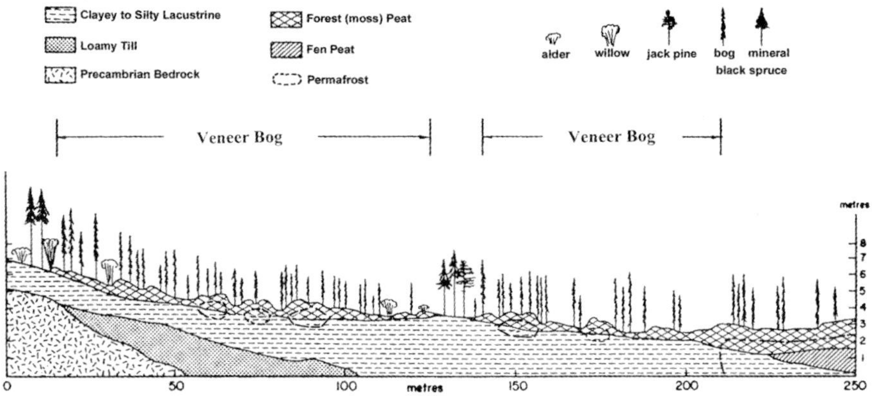

Figure 2.4.3. Cross-section of a veneer bog in the boreal region of western Canada. Cryosols occur in association with scattered shallow permafrost.

The extent of Cryosols increases with these peatlands' maturity and with increase in latitude. These usually are classified as Terric Fibric Organic Cryosols, with minor areas of Terric Humic Organic Cryosols. Thus, the main characteristic of these Cryosolic soils is their discontinuous occurrence within veneer bogs and their susceptibility to thermal alteration from surface disturbance. Taxonomically, these soils may change very quickly from the Organic Cryosol great group to taxa of other soil orders. Under similar conditions, Cryosols are dominant soils in the plains immediately east of Fort Nelson in northeastern British Columbia, where peat deposits >2 m thick support Fibric and Mesic Organic Cryosols with active layers around 100 cm thick in conjunction with Organic soils (where active layers are >100cm) over large areas of low-relief and inter-montane basins (Valentine, 1971).

The distribution of Cryosols in mineral soils in the northern portion of the boreal region depends on drainage, the presence or absence of shallow peat, and vegetative cover characteristics. Permafrost generally lies below the control section of mineral soils. Near-surface permafrost, in association with poor drainage and shallow peat development, produces Gleysolic Static Cryosols and, to a lesser extent, Gleysolic Turbic Cryosols.

Luvisolic soils have developed on moderately well drained clayey sites in the central portion of the boreal region. Permafrost will aggrade on these sites when mature, closed spruce (*Picea mariana*) stands, with thick forest floor layers, are present. The developing permafrost has no effect on the upper part of the soil profile, as ice content is low, and the active layer frequently exceeds 100 cm in depth. Evidence of cryoturbation in the control section is relic and relates to times when earth hummocks developed on these sites. When permafrost is present within 100 cm of the surface, these soils are classified as Luvisolic Static Cryosols. These soils revert back to Luvisolic soils after wildfire or other disturbance of vegetation.

In more northern areas of the boreal region under colder climate, permafrost distribution is not as strictly controlled by vegetative characteristics alone. The soils are usually classified as Brunisolic Eutric Turbic Cryosols, as cryoturbation is more evident (although it may be largely relic), active layers are shallower, upper permafrost ice content is higher, and Luvisolic features are weak (Veldhuis et al., 1990).

Cryosols are generally absent from well drained coarse loamy and sandy parent materials throughout the region, as both the depth of the moss and surface humus layers and stand density are insufficient to induce permafrost development. Cryosols generally are not found in boreal areas of eastern Canada, due to the insulating effects of deeper snow cover.

3. Cryosols of the Subarctic Region

The subarctic region forms the transition between the closed coniferous stands of the boreal region and the treeless tundra of the Arctic region. The subarctic region covers approximately 3.3 million km^2 and extends from Labrador on the Atlantic coast to the Mackenzie Delta on the coast of the Arctic Ocean (Figure 2.4.1). Latitudinally, the Subarctic region extends from 48° to 58° N in Labrador, from 53° to 55°30' N south of Hudson Bay, and from 62° to 69° N east of the cordillera. This vast region contains soil parent materials derived from various geological sources, with a resultant wide range of chemical and physical properties.

Table 2.4.2. Comparison of climatic variable for eastern (humid) and western (subhumid) portions of the subarctic region. (Adapted from Zoltai et al. 1988b).

Location	No. of stations	Mean annual temp (°C)	Mean daily July temp (°C)	Mean daily January temp (°C)	Mean annual snowfall (cm)
Western subarctic (subhumid)	5	-9.1	13.4	-30.1	130
Eastern subarctic (humid)	5	-5.4	11.5	-24.1	280

Climatic conditions of the subarctic region vary considerably east and west of Hudson Bay. Winters are colder, with less snow cover in the western subarctic (Table 2.4.2). These conditions are conducive to the formation and maintenance of near-surface permafrost, such that Cryosols form in both organic and mineral parent materials.

On the other hand, in the more humid eastern portion of the region, warmer winters with deep snow pack cool the soil less. Height of vegetation controls the distribution of Cryosols along the coast of eastern Hudson Bay. Near-surface permafrost is only maintained under low-growing vegetative communities of mosses and lichens that collect no more than 50 cm of snow during the winter (Allard et al., 1996; Menard et al., 1998). Cryosols are most common in peatlands locations (Payette et al., 1975). Mineral soils of the subarctic of eastern Canada tend to have Brunisolic or even Podzolic development.

While peatlands occur throughout the region, they are the dominant terrain cover in the lowlands southwest of the Hudson Bay and are widespread throughout the plains of the Mackenzie Valley in the extreme northwestern subarctic. Peatlands occupy less than 25% of the eastern subarctic of Quebec and Labrador, where Precambrian bedrock outcrops extensively (National Wetlands Working Group, 1986). In the southern part of the subarctic region, Fibric and Mesic Organic Cryosols are associated with peat plateau bogs, palsa bogs, and patterned fens (Zoltai and Tarnocai, 1975), while, in the northern portions, these soil types occur on polygonal peat plateau bogs and patterned fens (Zoltai et al., 1988b).

Permafrost is extensive in peatlands of the northern subarctic, where active layers are generally between 40 and 50 cm deep (Zoltai and Tarnocai, 1975). Soil temperatures of -0.7 °C (annual range from -0.1 to -7.5) at 50 cm and -1.6°C (annual range -0.3 to -6.0) at 100 cm have been reported for a polygonal peat plateau bog near Inuvik, adjacent to the Mackenzie River delta (Tarnocai, 1984; Tarnocai et al., 1993, site 1).

The mineral soil parent materials in the region derive from two distinct geological sources. Sandy and coarse loamy acidic parent materials derive from the igneous bedrock of the Precambrian Shield that underlies most of the region. The sandy and coarse loamy soils that lack significant silt generally do not support Cryosolic development, although permafrost is widespread. As cryoturbation is weak or lacking and active layers extend well below the control section depth of 100 cm, these soils are classified as Brunisols.

Coarse loamy soils having significant silt content show marked cryoturbation, are classified as Turbic Cryosols, and are associated with various forms of patterned ground. Also, Gleysolic Static and Turbic Cryosols are widespread in poorly drained areas with thin peaty surfaces. Calcareous, fine loamy, and clayey parent materials derive from the sedimentary geological strata that underlie the plains in the northwesternmost portions of the region.

Patterned ground (in the form of earth hummocks) is extensive on clayey and fine loamy materials, while circles and nets of various types are widespread on coarse and fine loamy materials. Ice wedge polygons are associated with coarse loamy and sandy deposits.

Earth hummocks, a feature of mineral soils underlain by near-surface permafrost, are common in the Mackenzie River valley in the western portion of the region (Tarnocai and Zoltai, 1978). Zoltai (1975) found that ground heave is periodic and speculated that it likely related to higher than normal fall temperatures and precipitation. The soils associated with these earth hummocks are Orthic and Brunisolic Eutric Turbic Cryosols; the latter largely are associated with low-activity earth hummocks. Also, Histic subgroups of Turbic Cryosols are present where activity has been low and an organic surface layer has developed.

On coarse to fine loamy textured till circles, stripes, steps, and nets occur. The associated soils are Regosolic, Eutric, and Dystric Turbic Cryosols, depending on the extent of differentiation of horizons and of parent material's characteristics. Orthic Turbic Cryosols occur along the edges of the circles, nets, and stripes where some B horizon is usually present although the centers of the circles, and stripes may lack any B horizon development and consist solely of C material.

In the Mackenzie Delta, deposition of fine loamy alluvium leads to the development of sequences of Cryosols according to relative elevation and age of deposition (Figure 2.4.4). Along active channels, permafrost is lacking and the soils are poorly drained and classified as Rego Gleysols. This strip is bordered by a strip of imperfectly drained Cumulic Regosols, with permafrost just below the control section.

Moving farther back from the active channel, the active-layer depth decreases to less than one meter and soils are classified as Regosolic Static Cryosols, with a typical mean annual soil temperature in the range of -3.5°C at 50 cm deep (Tarnocai et al., 1993, site 3). Soils on the oldest delta surfaces exist at about six meters above the late summer water level in the channel. These soils are typically well drained Regosolic Static Cryosols with a well developed litter layer. The permafrost has low ice content, the permafrost table occurs at a depth of around 35 cm, and the MAST at 50 cm is between -3.0 and -3.5°C.

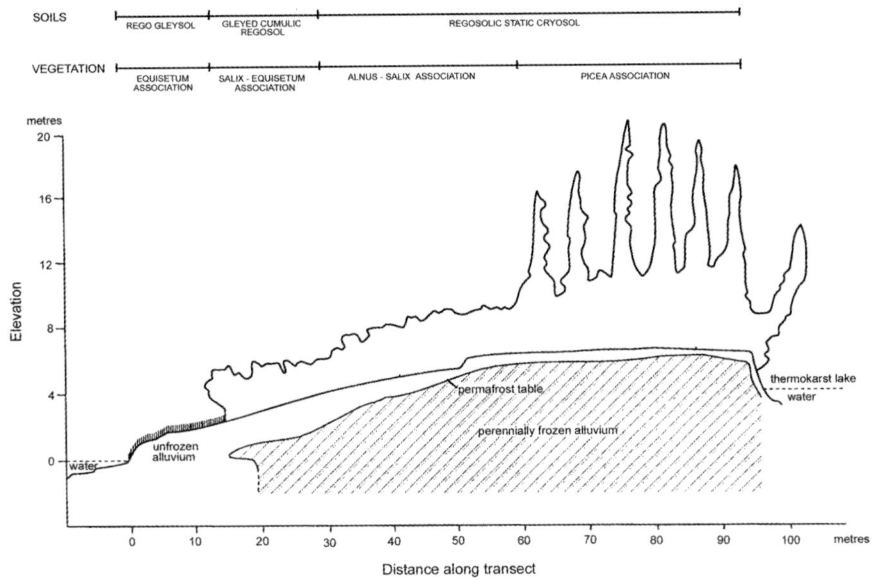

Figure 2.4.4. Cross-section showing vegetation, soil, and permafrost conditions for an ecosystem sequence in the Mackenzie River delta in the extreme northwestern portion of the subarctic region.

4. Cryosols of the Cordillera Region

In addition to macro-climate and site, elevation is a major determinant in the presence and distribution of permafrost and therefore Cryosolic soils within the mountainous region of western Canada. As a result of high elevations in western North America, permafrost occurs as far south as 35°N (International Permafrost Association, 1997), and pergelic soils (possibly equivalent to Cryosolic soils) occur at elevations >3500 m asl as far south as the state of Colorado at 39°N (Bockheim and Burns, 1991).

The western cordillera region of Canada extends from the Canada-U.S. border at 49°N latitude to the Arctic Ocean at approximately 70°N latitude. The total area of the region is approximately 1.2 million km^2; however, Cryosolic soils occur significantly only in the northern portion of the region, namely the northern boreal cordillera and the subarctic cordillera subregions (Ecoregions Working Group, 1989).

Permafrost underlies much of the northern half of the region from about 57°N latitude northwards. South of this latitude, isolated patches of permafrost exist in high elevation (>1600 m asl) alpine and sub-alpine environments (Heginbottom, 1995).

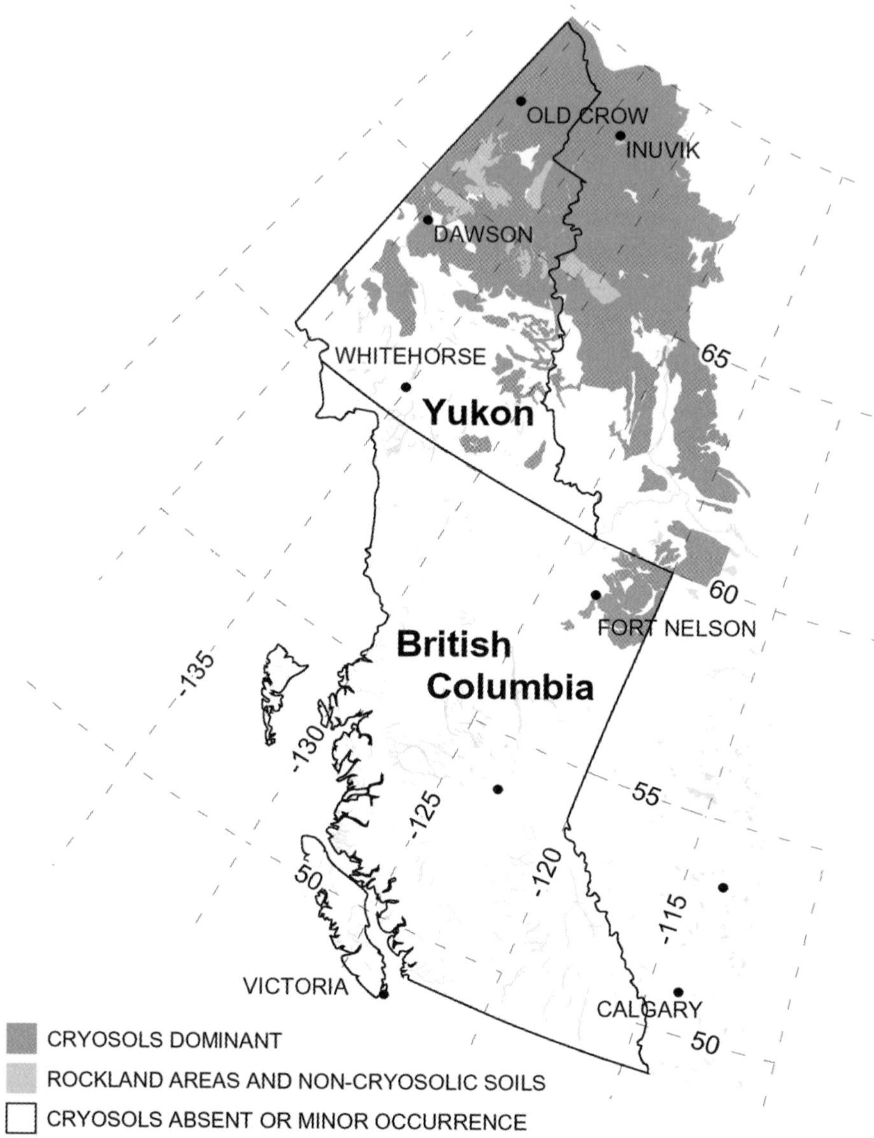

Figure 2.4.5. Map showing the soil landscapes in the cordillera region where Cryosols are the dominant form of soil development. Cryosols also occur as minor soils in the southern Yukon and northern British Columbia and occur sporadically at high elevations in the Rocky Mountains south of 55° N latitude.

The central portion of the cordillera region lies within the sporadic and extensive discontinuous permafrost zones, where Cryosolic soils occur to a limited extent at lower elevations under forested conditions as well as at higher elevations. Cryosols become the dominant soil on the landscape north of 64°N latitude under subarctic woodland and tundra vegetation (Figure 2.4.5). In the cordillera region, this latitude roughly corresponds to the southern limit of continuous permafrost.

4.1. Distribution of Cryosols in the southern portion of the region

Widely scattered Cryosols are associated with some of the isolated patches of alpine permafrost of the Interior Ranges of British Columbia and the Rocky Mountains of southern British Columbia and Alberta. These typically occur at elevations above 2000 m asl. Soils exhibiting evidence of active and relict solifluction (gelifluction) have been described in the Taseko Lakes area of the Coast Mountains (Valentine et al., 1987; Tarnocai and Valentine, 1989).

Farther east, in the Front Ranges of the Rocky Mountains in Banff and Jasper National Parks in southwestern Alberta, Achuff and Coen (1980) observed Cryosolic soils in localities within the upper subalpine forests (1880 to 2200 m asl) with annual mean temperatures of between -1 and -2°C. They attributed the scattered distribution of Regosolic and Brunisolic Static Cryosols in this area to low winter snow coverage, steep (40 to 60%) north-facing aspects, deep forest humus layers (up to 40 cm), and canopy coverage of subalpine fir (*Abies lasiocarpa*) and spruce (*Picea engelmannii*). Their description of strongly mounded surfaces and "a drunken forest appearance" would seem to indicate that Turbic Cryosols are also present in this environment.

Harris (1998) reports non-sorted circles from a xeric alpine site between 2250 and 2500 m asl on Plateau Mountain, also within the Front Ranges of the Rocky Mountains in extreme southwestern Alberta west of Calgary (Figure 2.4.5). While the soils associated with this patterned ground would have many of the properties of Turbic Cryosols, Harris reports active layers of 2.5 to 3.0 m in depth. Active layers within bedrock and surface unconsolidated sediments may exceed 10 m in thickness in many of the alpine areas in the southern portions of the region, which relegates these soils to non-Cryosolic status. Although mean annual air temperatures are quite cold (-2.3 to -2.5°C), these common xeric alpine sites tend not to support insulating humus layers which can preserve a permafrost table within the soil control section.

Strong outflow winds remove insulating snow cover, which leads to the development of Turbic Cryosols at latitudes between 53° and 54°N in some subalpine valleys at >1400 m asl in the Rocky Mountain foothills of Alberta (Brown, 1980). These Cryosols are associated with mineral soils of raised micro-relief, covered by both sedge (*Carex aquatilis, C. vesicaria)* and spruce (*Picea glauca x engelmannii*) vegetative communities in exposed flood plain locations.

4.2. Distribution of Cryosols in the northern boreal cordillera

Cryosolic soils have a common but subdominant presence on low elevation landscapes in the cordillera between about 58° and 63° N latitude. boreal forests composed of *Picea marianna*, *Picea glauca*, *Pinus contorta*, and *Populus tremuloides* cover most of the discontinuous permafrost zone in the region. Minor occurrences of mineral Cryosols occur throughout northern B.C. in the Rocky Mountains, Stikine Mountains, and Spatsizi Plateau areas (Luttmerdig et al., 1995).

Farther north, cooler temperatures and lower precipitation lead to more common occurrences of mineral Cryosols in the southern Yukon between 60° and 63° N latitude. Using geophysical techniques, Seguin et al. (1999) estimated that near-surface permafrost underlies 30% of subalpine and alpine areas (1200 to 1800 m asl) in a small mountainous watershed near Whitehorse (Figure 2.4.5). Active layers were thickest in the alpine zone, due to less insulating snow and surface cover. Much of the active layer is often within bedrock, and Cryosolic soils are intermittent in this zone. Turbic Cryosols are a minor component of most middle and lower elevation landscapes here (White et al., 1992). Most of the parent materials are calcareous, leading to high base status soils (Orthic Eutric Turbic Cryosols). Many of the Cryosols of southwestern Yukon are composed, at least in part, of unweathered Holocene tephra (Smith et al., 1999).

Mean annual soil temperatures at 50 cm deep in low elevation Cryosols of this zone vary from -0.6° to -1.5° C (Smith et al., 1998). Due to low annual precipitation (200 to 270 mm), and in sharp contrast to the boreal and subarctic regions of Canada, peatlands, and associated Organic Cryosols are largely absent from the northern boreal cordilleran region. Instead, marl-rich Gleysolic Turbic Cryosols dominate many wetlands. In these non-peaty wetlands, palsa-like formations occur in mineral soil and led Harris (1993) to propose the term “lithalsas” for these features (see also Seppälä, 1980). In the extreme southeastern Yukon, Fibric Organic Cryosols commonly are associated with bog landforms.

As elsewhere in the zone of discontinuous permafrost, wild fires are a major controlling factor where near-surface permafrost occurs on forested portions of the landscape (Swanson, 1996). Low precipitation in southern Yukon fosters extensive forest fires that indirectly control the thickness of active layers. In the Takhini Valley near Whitehorse, Cryosolic soils with an active layer of 1.5 m underlie unburned forest, while adjacent sites burned in 1958 had active layers 3.5 m deep (Burn, 1998). The presence of Turbic Cryosols on the plateau region of the central Yukon between 63° and 64° N latitude relates to surface organic horizons that are >10 cm thick and maintain summer moisture content >25% (gravimetric) (Williams and Burn, 1996) .

Aspect also can have a profound effect on distribution of Cryosols at this latitude (Figure 2.4.6). South- and west-facing slopes support non-Cryosolic soils (usually Brunisols), and north- and east-facing slopes support the development of Turbic Cryosols under thick moss carpets of *Sphagnum* and sedges (Tarnocai et al., 1993, site 28). Detailed soil mapping in the Klondike Valley near Dawson City revealed that all but the most active flood plains of major rivers supported Cryosolic soils with active layers as shallow as 30 cm under mature spruce (*Picea glauca* and *P. marianna*) forest (Tarnocai et al., 1993, site 25).

4.3. Distribution of Cryosols in the subarctic cordillera

Subarctic (taiga) woodlands of primarily open stands of *Picea mariana* characterize the subarctic vegetation of the western cordillera. The presence of this subarctic woodland corresponds closely to the southern limit of continuous permafrost in the region.

The subarctic vegetation extends from about 65° to 69° N latitude, just south of the coast of the Arctic Ocean (Figure 2.4.1). Mean annual air temperatures range between -5° and -10°C in the valley and lowland areas between major mountain ranges in the northern portion of this region. Cryosols are the dominant soils of the cordillera region north of 65°N latitude (Centre for Land and Biological Resources, 1996). Mean annual soil temperatures (at a 50-cm depth) of some monitored low elevation Cryosols in this region vary from -2.5 to -2.6°C (Smith et al., 1998).

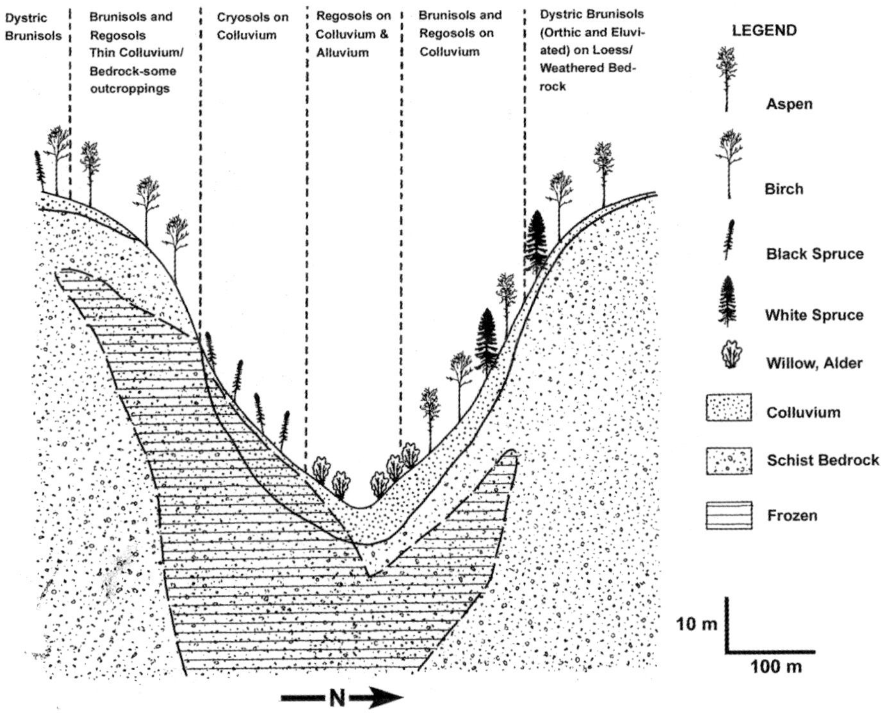

Figure 2.4.6. Landscape cross-section from the Klondike Valley near Dawson City, Yukon (64° 3' N, 139° 26' W), illustrating the influence of aspect on the distribution of Cryosols in the zone of discontinuous permafrost (adapted from Tarnocai et al., 1993).

Figure 2.4.7. A typical Histic Eutric Turbic Cryosol formed in finely textured parent material under a *Picea mariana* woodland, northern Yukon Territory. The profile consists of a 15-cm thick surface peat layer overlying highly cryoturbated Ahy and Bmgjy horizons. The markers illustrate the permafrost table, which undulates at about 40 cm deep. The tape measure is delineated in 10-cm intervals.

Tarnocai et al. (1993) and French and Heginbottom (1983) have summarized permafrost conditions and soil development in the mountains and plateaus of the northern Yukon and adjacent Northwest Territories. Orthic Eutric Turbic Cryosols predominate in the region (Figure 2.4.7). They formed on a variety of parent materials, many of which are calcareous. Active layer thicknesses are typically less than 100 cm and often less than 50 cm in mineral soils, depending on the thickness of surface organic layers.

Peatlands are common in the large inter-montane basins of northern Yukon (Zoltai et al., 1988b). Fibric Organic Cryosols are common, with active layers less than 50 cm. In the Ogilvie Mountains (Figure 2.4.8) at elevations >800 m asl between 64° and 65°N latitude in Yukon, palsa bogs, open system pingos, ice wedge polygons, seasonal frost mounds, thermokarst ponds, earth hummocks, sorted circles, and rock glaciers associated with Turbic and Regosolic subgroups of Cryosols occur (Kennedy and Smith, 1999).

Sorted patterned ground is widespread in alpine areas (Regosolic Static Cryosols) where active layers may be up to 1.5 m deep. Much of the bedrock is calcareous in this region, and most of the Regosolic Turbic Cryosols associated with non-sorted circles and related patterns are alkaline in reaction. Many of the alpine and subalpine soils of this region contain calcareous, organic matter-rich surface horizons that are Rendzina-like (Schrier and Lavkulich, 1985) and support unique assemblages of soil fauna (Smith et al., 1990).

In the northernmost sections of the region, mountain ranges and plateau remained unglaciated during the Pleistocene. Lauriol (1990) and Lauriol et al. (1997) described the formation of cryoplanation terraces in the ranges of the northern cordillera. Associated soils are usually classified as Turbic Cryosols with active layers often between 1 and 2 m deep. Well developed pediment surfaces (French and Harry, 1992) support finely textured Orthic Eutric Turbic Cryosols and Gleysolic Turbic Cryosols with shallow (< 50 cm) active layers (Welsh and Smith, 1989). Surface organic layers on pediment slopes tend to be extremely acidic (see Tarnocai et al., 1993, stop 12).

Figure 2.4.8. Vegetative communities illustrating the pattern of widespread discontinuous permafrost in the Ogilvie Mountains, Yukon. Active floodplains are permafrost free and support forests of *Picea glauca*. North-facing slopes (far side of the valley) and depressional valley bottom sites support tussock (*Eriophorum vaginatum*) tundra over shallow active layers. South- and west-facing slopes support mixed stands of *Picea glauca* and *Picea mariana* over active layers 75 to 200 cm deep. Sorted patterned ground predominates alpine areas.

5. Summary

As with all soils, development of Cryosols is a function of climate, relief, biological activity, and parent geological material. However, regions largely south of the tree line, such as those this chapter describes, show a particularly strong temporal component to their formation. Permafrost in these regions is warm, typically just below 0°C, and surface disturbances, most typically forest fires, influence the depths of active layers.

Soil properties such as texture, thickness of surficial organic material, drainage, and landscape (site) characteristics exert strong influences on the spatial and temporal distribution of Turbic and Static Cryosols in mineral soils, while cycles of aggradation and thaw in peatlands produce a similarly complex distribution of Organic Cryosols. In all but the most northerly reaches of the subarctic region, soils may oscillate in and out of the Cryosol order over time scales of several hundred to several thousands of years, due to natural surface disturbances (e.g., forest fires, erosion).

The Cryosols of the boreal, subarctic, and cordillera regions of Canada are highly susceptible to anthropogenic disturbances, including the effects of climate change, which has been projected to produce significant warming of these regions over the next century (Hengeveld, 2000).

6. Acknowledgments

We are grateful for the contributions Karen McKenna, Charlotte Mougeot, David Murray, and Nancy Steffan, who have mapped, described, and sampled soils over the last 20 years throughout the Yukon Territory, and to Chris Burn and his students for furthering our understanding of the nature and dynamics of permafrost in the region. We also are thankful for the valuable contributions of Gordon Mills and Robert Eilers in Manitoba. Finally we thank Charles Tarnocai for the guidance, training, and leadership he has provided to all of us working on the soils of northern Canada.

References

Achuff, P.L., and G. M. Coen 1980. Subalpine Cryosolic soils in Banff and Jasper National Parks. Can. J. Soil Sci. 60:579-581.

Allard, M., S. Caron, and Y. Bégin 1996. Climatic and ecological controls on ice segregation and thermokarst: The case history of a permafrost plateau in northern Quebec. Permafrost and Periglacial Processes. 7:207-227.

Bockheim, J.G., and S.F. Burns 1991. Pergelic soils of the western contiguous United States: Distribution and taxonomy. Arct. and Alp. Res. 23:206-212.

Brown, G. 1980. Palsas and other permafrost features in the lower Rock Creek valley, west-central Alberta. Arct. and Alp. Res. 12:31-40.

Burn, C.R. 1998. The response (1958-1997) of permafrost and near-surface ground temperatures to forest fire, Takhini River valley, southern Yukon Territory. Can. J. Earth Sci. 35:184-199.

Centre for Land and Biological Resources. 1996. Soil Landscapes of Canada, Version 2.2. Research Branch, Agriculture and Agri-Food Canada. Ottawa. (Digital map and database).

Ecoregions Working Group 1989. Ecoclimatic regions of Canada. Ecological Land Classification Series No. 23. Environment Canada. Ottawa. 118 pp.+ maps.

French, H.M., and D.G. Harry 1992. Pediments and cold-climate conditions, Barn Mountains, unglaciated northern Yukon, Canada. Geogr. Ann. 74A(2-3):145-157.

French, H.M., and J.A. Heginbottom (eds) 1983. Guidebook to permafrost and related features of the northern Yukon Territory and Mackenzie Delta, Canada.. 4th International Conference on Permafrost, Fairbanks, Alaska. 186 pp.

Halsey, L.A., D.H. Vitt, and S.C. Zoltai 1995. Disequilibrium response of permafrost in boreal continental western Canada to climate change. Climatic Change. 30:57-73.

Harris, S.S. 1993. Palsa-like mounds developed in a mineral substrate, Fox Lake, Yukon Territory in Proceedings of the Sixth International Conference on Permafrost, Beijing China. South Cina Univ. of Technology Press.

Harris, S.A. 1998. Nonsorted circles on plateau Mountain, SW. Alberta, Canada in Lewkowicz, A.G., and Alard M. (eds) Proceeding 7th International Conference on Permafrost, Yellowknife, Canada . Collection Nordicana, Université Laval. Ste Foy, Quebec. pp. 441-448.

Heginbottom, J. A. 1995. Permafrost. National Atlas of Canada 5th Edition. Map MCR 4177. Natural Resources Canada. Ottawa. 1:7,500,000 scale map.

Hengeveld, H.G. 2000. Projections for Canada's climate future. Meteorological Service of Canada Report CCD 00-01, Special Edition. Environment Canada, Ottawa. 27pp.

International Permafrost Association 1997. Circum-arctic map of permafrost and ground ice conditions. United States Geological Survey and Circum-Pacific Council for Energy and Mineral Resources. Map CP-45. 1:10,000,000 scale.

Kennedy, C.E., and C.A.S. Smith 1999. Vegetation, terrain and natural features in the Tombstone area, Yukon Territory. Department of Renewable Resources, Government of the Yukon and Agriculture and Agri-Food Canada. Whitehorse, Canada. 47pp.

Lauriol, Bernard. 1990. Cryoplanation terraces, northern Yukon. The Canadian Geographer. 34:347-351.

Lauriol, B. M., A.E. Lalonde, and V. Dewez 1997 Weathering of quartzite on a cryoplanation terrace in northern Yukon, Canada. Permafrost and Periglacial Processes. 8:147-153

Luttmerdig, H.A., E. Kenney, and A.J. Green 1995. Soil landscapes of Canada: British Columbia-north. Agriculture and Agri-Food Canada Publ 5280. Centre for Land and Biological Resources Research, Agriculture and Agri-Food Canada. Ottawa. 1:1,000,000 map and database.

Menard, É., M. Allard, and Y. Michaud. 1998. Monitoring of ground surface temperatures in various biophysical micro-environments near Umiujaq, eastern Hudson Bay, Canada. In: A.G. Lewkowicz and M. Alard, eds. Proceedings of the 7th International Conference on Permafrost, Yellowknife, Canada. Collection Nordicana, Université Laval. Ste Foy, Quebec. pp. 723-729.

Mills, G.F., H. Veldhuis, and R.G. Eilers. 1985. Preliminary results of soil climate and hydrology studies in the discontinuous permafrost zone near Thompson Manitoba. In Proc. 28th Annual Manitoba Society of Soil Science Meeting. Winnipeg, p.178-200.

National Wetlands Working Group 1986. Canada's Wetlands, Map Folio. Energy, Mines and Resources Canada and Environment Canada. Ottawa.

Payette S., H. Samson., and D. Lagarec. 1976. The evolution of permafrost in the taiga and in the forest-tundra, western Quebec-Labrador Peninsula. Can. J. Forest Res. 6:203-220.

Swanson, D.K. 1996. Susceptibility of permafrost soils to deep thaw after forest fires in interior Alaska, U.S.A. and some ecologic implications. Arct. and Alp. Res. 28:217-227.

Schrier, H., and L.M. Lavkulich 1985. Rendzina-type soils in the Ogilvie Mountains, Yukon Territory. Soil Sci. 139:2-12.

Sequin,M.K., J. Stein, O. Nilo, C. Jalbert, and Y. Ding. 1999. Hydrogeophysical investigation of the Wolf Creek watershed, Yukon Territory, Canada.. In: J.W. Pomeroy and R. J. Granger, eds. Wolf Creek Research Basin: Hydrology, Ecology and Environment. National Water Research Institute, Environment Canada. Saskatoon, SK. pp. 55-78.

Seppala, M. 1980. Stratigraphy of a silt-cored palsa, Atlin region, British Columbia, Canada. Arctic. 33:357-365.

Smith, C.A.S, C.L. Ping, C.A. Fox, and H. Kodama. 1999. Weathering characteristics of some soils formed in White River tephra, Yukon Territory, Canada. Can. J. Soil Sci. 79:603-613.

Smith, C.A.S., A.D. Tomlin., J.J. Millar, M.J. Tynen., and K.A. Coates. 1990. Large Enchytraeid (Annelida: Oligochaeta) worms and associated fauna from unglaciated soils of northern Yukon, Canada. Geoderma. 47:17-32.

Smith, C.A.S., C.R. Burn, C. Tarnocai, and B. Sproule. 1998. Air and soil temperature relations along an ecological transect through the permafrost zones of northwestern Canada. in Lewkowicz, A.G., and Alard M. (eds) Proceeding 7th International Conference on Permafrost, Yellowknife, Canada . Collection Nordicana, Université Laval. Ste Foy, Quebec. pp. 1009-1016.

Soil Classification Working Group 1998. The Canadian System of Soil Classification. 3rd edition. Agric and Agri-Food Canada Publ. 1646 (revised). Ottawa. 187 pp.

Tarnocai, C. 1984. Characteristics of soil temperature regimes in the Inuvik area. In: Olson et al., eds. Northern Ecology and Resource Management Univ. Alberta Press. Edmonton, Alberta. pp. 19-37.

Tarnocai, C. 2000. Carbon pools in soils of the Arctic, subarctic and boreal regions of Canada. In: R. Lal, J.M. Kimble and B.A. Stewart, eds. Global Climate Change and Cold Regions Ecosystems. Advances in Soil Science, Lewis Publishers. Boca Raton, Fla. pp. 91-103.

Tarnocai, C., and S.C. Zoltai. 1978. Earth hummocks of the Canadian Arctic and Subarctic. Arctic and Alpine Res. 10(3):581-594.

Tarnocai, C., and K.W.G. Valentine. 1989. Relict soil properties of the arctic and subarctic regions of Canada. Catena Supplement. 16:9-39.

Tarnocai, C., C.A.S. Smith, and C.A. Fox 1993. Guidebook to the international tour of permafrost affected soils: The Yukon and Northwest Territories of Canada. Centre for Land and Biological Resources Research. Agriculture Canada. Ottawa. 197 pp.

Valentine, K.W.G. 1971. Soils of the Fort Nelson Area of British Columbia. British Columbia Soil Survey Report No 12. Can. Dept. Ag. Vancouver, BC. 60 pp. + maps.

Valentine, K.W.G., W. Watt, and A.L. Bedwany 1987. Soils of the Taseko Lakes area, British Columbia. B.C. Soil Survey Report No 36. Ottawa. 131 pp. + maps.

Veldhuis, H., G.F. Mills, and R.G. Eilers. 1990. Thermal regimes of clay soils in Manitoba. Proc. 33rd Ann. Manitoba Soil Sci. Meet. Winnipeg. pp. 33-46.

Vitt, D.H., L.A. Halsey, and S.C. Zoltai 1994. The bog landforms of continental western Canada in relation to climate and permafrost patterns. Arct. and Alp. Res. 26:1-13.

Welsh, D.M., and C.A.S. Smith 1990. Soils of North Yukon (Ivvavik) National Park. In: Northern Yukon National Park Resource Description and Analysis. Natural Resource Conservation Section, Canadian Parks Service. Winnipeg. pp. 5-1 to 5-58.

White, M.P., C.A.S. Smith, D. Kroetsch, and K. McKenna 1992. Soil landscapes of Canada: Yukon. Agriculture Canada Publication 5281. Centre for Land and Biological Resources Research. Ottawa. 1:1,000,000 map and database.

Williams, D.J., and C.R. Burn 1996. Surficial characteristics associated with the occurrence of permafrost near Mayo, central Yukon Territory, Canada. Permafrost and Periglacial Processes. 7:193-206.

Zoltai, S.C. 1975. Tree ring record of soil movements on permafrost. Arct. and Alp. Res. . 7:331-340.

Zoltai, S.C. 1993. Cyclic development of permafrost in the peatlands of northwestern Alberta. Canada. Arct. and Alp. Res. 25:240-246.

Zoltai, S.C., and C. Tarnocai. 1971. Properties of a wooded palsa in northern Manitoba. Arct. and Alp. Res. 3:115-129

Zoltai, S.C., and C. Tarnocai. 1975. Perennially frozen peatlands in the western Arctic and Subarctic. Can. J. Earth Sci. 12:28-43.

Zoltai, S.C., and C. Tarnocai. 1981. Some nonsorted patterned ground types in northern Canada. Arct. and Alp. Res. 13:139-151.

Zoltai, S.C., S. Taylor, J.K. Jeglum, G.F. Mills, and J.D. Johnson 1988a. Wetlands of boreal Canada. In: Wetlands of Canada. Ecological Land Classification Series No 24. Environment Canada, Ottawa, and Polyscience Publications, Inc. Montreal. 452 pp.

Zoltai, S.C., C. Tarnocai, G.F. Mills, and H. Veldhuis 1988b. Wetlands of subarctic Canada. In: Wetlands of Canada. Ecological Land Classification Series No 24. Environment Canada, Ottawa, and Polyscience Publications, Inc. Montreal. 452 pp.

Chapter 5. Cryosols in the Russian Arctic Archipelagos

S.V. Goryachkin[1] and N.A. Karavaeva[1]

[1]Head, Laboratory of Soil Geography and Evolution, Institute of Geography, Russian Academy of Sciences, Staromonetny, 29, Moscow, 109017 Russia; e-mail pedology@igras.geonet.ru.
[2]Leading Researcher, Institute of Geography, Russian Academy of Sciences, Staromonetny, 29, Moscow, 109017 Russia; e-mail pedology@igras.geonet.ru.

1. Introduction

The Russian Arctic archipelagos, nearby isolated islands, and Spitsbergen are the main areas of High Arctic landscapes in the Eastern hemisphere. They include the large archipelagos of Novaya Zemlya, Severnaya Zemlya, and the New Siberian Islands, the northernmost but smaller Franz Josef Land archipelago, and several singular islands, including Wrangel, Vaigach, Belyi, Ayon, Kolguev Islands, and many smaller ones (Figure 2.5.1). Pedologists initiated their studies in the 1920s and visited almost all the archipelagos and large islands. These Russian northern islands later were mapped as part of a number of different small-scale soil mapping projects, compiled by extrapolating scarce field data, using remotely sensed images.

Tedrow reviewed a portion of the soil data from these regions during the 1970s (Tedrow, 1977). However, most data of that time were published only in Russian and were not known internationally. Soil investigations since then also were published mostly in Russian. However, the information on these soils is very important to understanding Arctic soil geography and for compiling circumpolar and global soil maps, databases, and classifications.

This chapter systematically reviews the soils and their distribution on the Russian Arctic archipelagos on the basis of published and unpublished field data. We discuss soil-forming factors, including parent material, landscape, and climate, and we describe the major components of the soil landscapes. We then describe the nature of the soil cover for each major island or archipelago in the region.

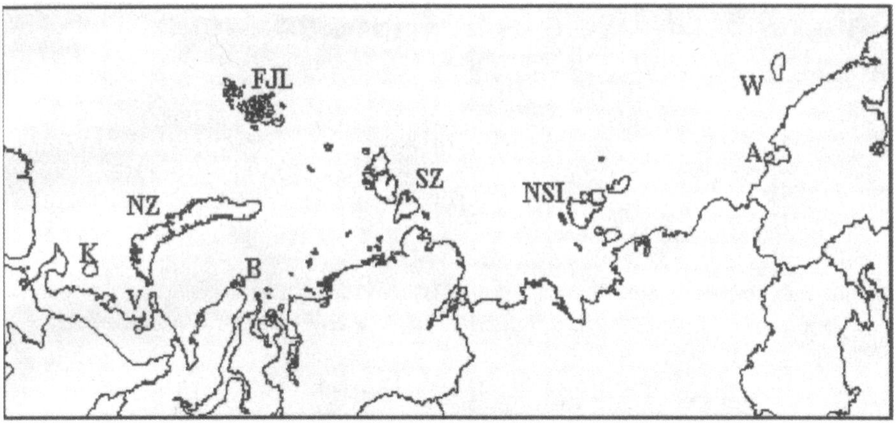

Figure 2.5.1. The geographic distribution of Russian Arctic archipelagos and islands. FJL – Franz Josef Land, NZ – Novaya Zemlya, K – Kolguev, V – Vaigach, B – Belyi, SZ – Severnaya Zemlya, NSI – New Siberian Islands, A – Ayon, W – Wrangel Island.

2. Soil-forming Factors in the Russian Arctic Archipelagos

The information here on soil-forming factors in the region is based on relevant monographs (Gerasimov, 1970; Goryachkin et al., 1994) and books (Ignatenko, 1979; Karavaeva, 1969) on soils of the archipelagos.

2.1. Landforms

With respect to geological and relief characteristics, all the islands of the Russian Arctic can be divided into 2 large groups:

1. Islands composed of mountains with large areas of ice caps and rock outcrops and relatively narrow strips of marine terraces along the coasts
2. Islands composed of plains with marine, glacial, eolian and lacustrine relief forms and sediments.

The islands of Novaya Zemlya, Severnaya Zemlya, and Wrangel and most of the Franz Josef Land islands belong to the first group. New Siberian Islands (with some exceptions), Kolguev, Belyi, Ayon, and some islands of the Franz Josef Land archipelago are in the second group. Vaigach and Kotelnyi Island (New Siberian archipelago) have the relief and substrate features of both groups; plains and rocky uplands occupy almost equal territory on each.

Periglacial landforms exist on all of the islands of the Russian Arctic. On mountainous islands, the sorted and non-sorted nets, stripes, and polygons are characteristic features. Palsas, low-center polygons, peat plateaus, solifluction terraces, hummocks, and patterned ground are widespread on islands dominated by plains. Very specific features of *baydzherakhs* (residual mounds, caused by the degradation of ice-wedge polygons) occur in New Siberian Islands (see below).

The parent materials of mountainous Arctic islands' soils are rock debris, stony and/or gravelly marine, and glacial sediments of different lithologies. Substrates derive mainly from schists, sandstones, limestones, and also basalts and dolerites. The unconsolidated surficial materials on the low-relief islands are mostly silty loams, but clayey substrates and sands also exist.

2.2. Climate

All the islands of the Russian Arctic have a cold, humid climate. But many climatic characteristics differ from south to north and from west to east. The attributes of summer climate vary mainly from south to north, while winter conditions vary more from west to east.

Mean annual temperatures on the Arctic Islands vary from -2.7°C (Kolguev Island) to -15°C (Kotelnyi Island, New Siberian islands). However, the difference in mean July temperatures between these two islands is only 4°C (6.6 and 2.6°C, respectively), while the difference between their January temperatures is much greater (-9 and -31°C, respectively). This is due to the strong influence of warm North Atlantic oceanic currents on the western part of the Russian Arctic and the effect of the very cold Siberian high-pressure systems on the New Siberian Islands during the winter.

The easternmost Arctic Islands are under the influence of air masses from the Pacific Ocean. For instance, Wrangel Island has a -23°C mean temperature in January and -10°C mean annual temperature. In general, summer (July) temperatures decrease from south to north, with July temperatures of 6.6°C for Kolguev Island and 1.0°C for the Severnaya Zemlya and Franz Jozef Land archipelagos.

The mean annual precipitation in the Arctic varies from 133 mm on the New Siberian Islands to 300 mm for Vaigach and Novaya Zemlya. These values are at least twice as large as the evapotranspiration rate for these regions. The evaporation is very low, due to low temperatures and high relative air humidity (70 to 90%). However, in high latitudes, most of the precipitation is snow, even in summer. The strong winds redistribute snow, resulting in snowless hills adjacent to deep accumulations of snow in valleys and depressions. Additionally, the Arctic islands are well known for their strong winds, especially Novaya Zemlya and Pevek Bay where Ayon Island is located, where winds can reach 70 to 80 m/sec.

Almost all the climatic data were gathered at meteorological stations situated along seacoasts. The interior regions of the larger islands are no doubt drier, with more pronounced continental effects (warmer in summer and colder in winter).

Russian Arctic islands lie within the zone of continuous permafrost. The depth of the active layer varies greatly from island to island and from site to site. Generally, active layers vary from about 30 to 80 cm deep, but at some ecotopes (sites) of Kolguev, Vaigach, and Novaya Zemlya, it can reach 1.5 to 2 m deep.

2.3. Vegetation

The vegetative cover of the Arctic Islands is one of tundra and High Arctic barrens. Dwarf-shrub-mossy tundra with widespread sedge-moss wetlands dominate Kolguev and Ayon Islands. Willow stands are typical for river valleys of Kolguev Island. Herbaceous and moss-herbaceous tundra are the vegetative covers of Vaigach, Novaya Zemlya (Southern Island), Belyi, Wrangel Island, and most of the New Siberian archipelago. Scarce and impoverished moss-lichen vegetation with sparse flowering plants is typical for the High Arctic barrens of Novaya Zemlya (Northern Island), Franz Josef Land, and Severnaya Zemlya.

2.4. Ecological zones

Soil-forming factors on the Arctic islands vary from place to place, due to the effects of aspect, wind and solar radiation, specific coastal influences, the fine-earth content in stony substrates, the possibility of eolian sea spray, and various other factors. The result is a great diversity of soils and complexity of soil distribution in the Arctic archipelagos and islands.

Based on climatic and vegetative parameters, all of the Russian Arctic islands can be related to three ecological zones:
1. High Arctic barrens
2. Mid Arctic tundra
3. Low Arctic tundra.

The islands of the Mid and Low Arctic also can be divided into three sectors: west, central, and east. Table 2.5.1 describes the distribution of island or archipelago by ecological zone and sector.

Table 2.5.1. Distribution of Russian Arctic Islands by ecological zones.

High Arctic Barrens	**Mid-Arctic Tundra** West	Central	East	**Low Arctic Tundra** West
Franz Joseph Land	Vaigach	New Siberian Islands	Wrangel	Kolguev
N. Island of Novaya Zemlya	S. Island of Novaya Zemlya	Ayon		
Severnaya Zemlya	Belyi			

3. Overview of Major Soil Units

The predominant components of most of the Arctic islands' soil landscapes are rock outcrops and rock debris; soil cover is discontinuous. However, various types of Cryosols have developed. Tables 2.5.2 and 2.5.3 present morphological and chemical data for several reference profiles. These profiles do not represent the complete variability of soils in the study area, and additional field studies and sampling are required to fully characterize the variables.

The soil taxa, typed in normal font within the text, are taken from the national Russian soil map, scale 1:2.5 M., Fridland (1988). Those in *italicized text* are taxa according to the World Reference Base (WRB) (ISSS Working Group RB., 1998.)

The Russian Arctic soils taxa include a large group of soils of high latitudes. They are well drained soils of variable textures with humic horizons 5 to 30cm deep. Mikhailov (1970) divided Arctic soils into Arctic desert, Arctic typical (low- and high-humic), and Arctic structure soils. They could correlate with the WRB subgroups *Humi-Turbic Cryosols*, *Umbri-Turbic Cryosols*, *Turbi-Leptic Cryosols*, and *Lithic Leptosols*.

The high latitude soils differ from soils of the more southerly regions of the Arctic, as they are well drained and show an absence of stagnic and gleyic features. Gleyic and stagnic features are also absent in Arctic hydromorphic non-gleyic soils (*Oxyaqui-Leptic Cryosols*) of northernmost archipelagos (Tables 2.5.2 and 2.5.3). The lack of reduction could result from low microbiological activity, oxygen-saturated melt water, high pH, and basaltic parent material. However, even in high latitudes, some ecotopes occur with the potential of water stagnation where the Arctic gleyzems (Gleyic Cryosols), with thin peat horizons over gleyic mineral masses, can form (Mikhailov, 1973; Karavaeva, 1969).

The soils that are widespread on all the Arctic islands, except the northernmost Franz Josef Land and Severnaya Zemlya, are Arctic tundra weakly gleyic humic soils (*Stagni-Turbic Cryosols*). On Vaigach Island are soils with a high content (1.6%) of water-soluble salts in the uppermost 5 cm of non-sorted circles (Ignatenko, 1967), while the profile under vegetation has a low pH and is salt-free and almost calcite-free (Table 2.5.3). These soils have thin litter, if any, and the dark humic horizon and grayish brown mineral horizons have weakly pronounced stagnic features.

Typically, the soil cover of the Arctic islands include weakly gleyic humic soils in complex with poorly drained Arctic gleyzems (*Gleyic Cryosols*) and Peaty bog gleyzems (*Turbi-Fibrihistic Cryosols* [*Gleyic*]) with histic horizons formed over gleyic gray-olive mineral subsoils. The transitional soils between them are Arctic tundra and tundra peat and mucky gleyzems (*Gleyi-Turbic Cryosols*) with thin peat or mucky horizons and underlying gleyic layers. Tundra peaty gleyzems – *Turbi-Fibrihistic Cryosols* (*Stagnic* or *Gleyic*) – are common soils for the tundra of the Eurasian continent.

Table 2.5.2. Morphological features of the major soils of the Russian Arctic islands and archipelagos.

Horizon	Depth (cm)	Color	Moisture Status	Texture	Structure	Pedofeatures	Roots
Arctic soil (Humi-Turbic Cryosol) – profile 63 (Mikhailov, 1967), Faddeevsky island, New Siberian Islands, flat surface, herbaceous community.							
A	0-9	BS-GR	Slightly M	SL	GR	No	M
AC	9-25	BR-GR	Slightly M	LS	GR	No	F
AC	25-36	GS-BR	M	LS	MA	No	VF
Ci	36-44	D-BR	Frozen	LS	MA	No	No
Arctic hydromorphic non-gleyic soil (Oxyaqui-Leptic Cryosol) – p.14-95 (Goryachkin), Southern Island, Novaya Zemlya, 335 m a s.l., mountainous impoverished ecosystems, hollow between sorted polygons, herbaceous-moss communit.							
H	0-4	10YR3/2 BS-BL	W	ST L	MA	No	M
C	4-10	10YR5/2 GS-YB	WS	ST L	MA	No	No
Arctic tundra weakly gleyic humic soil (Stagni-Turbic Cryosol [Humic]) – p. 70, 71 (Karavaeva, 1969) Bolshoy Lyakhovsky Island (New Siberian Islands), flat hill summit with patterned ground, herbaceous-polar willow-moss community.							
A	0-4	BS BL	M	CL	SB	No	M
AB	4-16	BS-BL	M	C	SB	No	C
Bg	16-32	BR	M	C	GR	RY mottles	F
B	32-44	Dull BR	M	C	MA	No	VF
Cri	44-55	GS-OL	Frozen	C	MA	No	No
Spot	0-4	BR	Dry	C	SB	No	F

(continues)

Table 2.5.2. (cont.)

Horizon	Depth (cm)	Color	Moisture Status	Texture	Structure	Pedofeatures	Roots
Arctic tundra weakly gleyic humic soil (Stagni-Turbic Cryosol) – p.39, 38 (Ignatenko, 1967), Vaigach Island, flat hill summit with patterned ground, herbaceous-sedge-polar willow-moss community, barren non-sorted circles –18-20%, troughs – 8% of the area.							
O	0-4	BR BL	W				M
Ag	4-10	BS GR	W	CL	MA	RY mottles	M
Bg	10-22	GS YB	W	CL	PL	BU mottles	F
BCg	22-36	GS YB	M	CL	MA	BU mottles	VF
Cgi	36-66	BR	Frozen 62 cm	CL	MA	BU mottles	VF
Circle	Brown loam, dry, salt efflorescences, limestone debris, permafrost at 75 cm						
Trough	Peat mat, depth 22 cm, table of permafrost in mineral mass at 25 cm						
Tundra podzolized Podbur (Spodic Cryosol) p. 20-95 (Goryachkin), Southern Island, Novaya Zemlya, hill slope, 10° steep, herbaceous-sedge-polar willow-dryas-moss community.							
O	0-2	7.7YR2/2 BS BL	M			No	M
AO	2-4	7.5YR3/1 BS BL	M	LS	MA	No	M
EA	4-6	10YR6/2 GR YB	M	LS	MA	No	M
Bhs	6-16	10YR3/3 Dk BR	M	LS	MA	No	M
BC	16-25	10YR4/2 GS YB	M	LS	MA	No	C
Dk	25-50	10YR5/1 BS GR	Frozen at Ca 95 cm	GR S	MA	No	No
Tundra mucky-calcareous soil (Calcari-Mollic Leptoosol) p.1-95 (Goryachkin), Southern Island, Novaya Zemlya, Barents Sea coast, wind shadow position, herbaceous community.							
Ak	0-10	10YR2/1 BL	M	L	GR	$CaCO_3$ pendents	M
Ak	10-20	10YR3/2 BR BL	M	GR L	SB	$CaCO_3$ pendents	C
Bk	20-28	10YR4/2 GS YB	M	GR L	SB	$CaCO_3$ pendents	F
Bk	28-55	10YR5/2 GR YB	M	GR L	SB	$CaCO_3$ pendents	VF

(continues)

Table 2.5.2. (cont.)

Horizon	Depth (cm)	Color	Moisture Status	Texture	Structure	Pedofeatures	Roots
Peaty bog Gleezem (Turbi-Histic Cryosol) p. 5 (Ignatenko, 1967), Vaigach Island, large flat depression, peat pool, sedge community.							
O	0-2	YB	WS			No	
H	2-12	YB	WS			No	M
H	12-28	Dull YB	W			No	VF
Cri	28-33	GS OL	Frozen	CL	MA	RY mottles	No
Ci	33-36	BR	Frozen	CL	MA	No	No
Peaty bog mesotrophic permafrost soil (Eutri-Cryic Histosol) p.8-95 (Goryachkin), Southern Island, Novaya Zemlya, slope flat-mound bog, summit of the peat mound, sedge-polar willow-moss (Sphagnum) community.							
O	0-3	10YR3/2 BS BL	W			No	M
H	3-7	10YR3/1 BS BL	W			No	M
H	7-10	10YR3/1	W			No	M
H	10-13	10YR3/1	WS			No	F
Hi	13-60	10YR3/1	Frozen			No	No

Abbreviations:

Color: BR – brown, BS – brownish, YB – yellowish brown, RY –reddish yellow, GR – gray, GS – grayish, BU – blue, BL – black, OL – olive. **Moisture status**: M – moist, W –wet, WS – wet, satiated. **Texture**: C – clay, L – loam, CL – clay loam, SL – sandy loam, LS –loamy sand, S – sand. **Texture modifiers:** ST – stony, GR –gravelly. **Structure:** MA – massive, GR – granular, SB – subangular blocky, PL – platy. **Roots (abundance):** M – many, C – common, F – few, VF – very few.

Table 2.5.3. Chemical properties of soils of Russian Arctic islands and archipelagos.

Horizon	Depth cm	Loss on ignition,%	Corg., %	pH H_2O	$CaCO_3$, %	Exchangeable cations, meq/100g soil			Base saturation,%	Oxalate extractable forms, %	
						Ca^{2+}	Mg^{2+}	H^+		Fe	Al
Arctic soil (Humi-Turbic Cryosol) – profile 63 (Mikhailov, 1967), Faddeevsky Island, New Siberian Islands.											
A	0-2	n.d.	3.04	6.3	n.d.	9.7	5.4	n.d.	n.d.	0.51	n.d.
A	4-9	n.d.	1.96	6.2	n.d.	8.2	4.5	n.d.	n.d.	0.70	n.d.
AC	15-20	n.d.	1.12	6.7	n.d.	6.1	5.0	n.d.	n.d.	0.72	n.d.
AC	30-35	n.d.	1.67	7.0	n.d.	8.2	5.0	n.d.	n.d.	0.74	n.d.
C	40-45	n.d.	0.99	7.2	n.d.	13.6	6.4	n.d.	n.d.	0.84	n.d.
Arctic hydromorphic non-gleyic soil (Oxyaqui-Leptic Cryosol) – p.14-95 (Goryachkin), Novaya Zemlya.											
H	0-4	36.45	12.21	7.1	n.d.	23.5	8.1	0	100	0.34	n.d.
C	4-10	n.d.	0.97	7.5	n.d.	2.7	1.6	0	100	0.11	n.d.
Arctic tundra weakly-gleyic humic soil (Stagni-Turbic Cryosol [Humic]) – p. 70, 71 (Karavaeva, 1969), Bolshoy Lyakhovsky Island, New Siberian Islands.											
A	1-4	15.27	4.35	5.5	0	20.0	9.5	0.9	97	0.56	0.25
AB	5-15	7.96	2.09	5.7	0	12.0	4.5	0.5	97	0.63	0.25
Bg	20-30	6.19	1.85	5.9	0	13.6	5.4	0.3	98	0.70	0.21
B	32-40	5.49	1.68	6.6	0	14.0	6.2	0.2	99	n.d.	n.d.
Cri	45-55	3.78	1.10	7.2	0	11.4	4.1	0.1	100	0.70	n.d.
Circle	0-4	15.69	1.74	5.5	0	13.9	6.2	0.4	98	0.77	n.d.
Circle	5-15	15.15	1.97	5.6	0	13.4	6.7	0.4	98	0.77	n.d.
Circle	20-30	17.42	1.85	6.1	0	15.3	7.7	0.2	99	0.70	n.d.
Circle	35-40	18.60	1.39	6.3	0	14.4	6.2	0.1	100	0.70	n.d.

(continues)

Table 2.5.3. (cont.)

Hori-zon	Depth cm	Loss on ignition,%	Corg., %	pH H_2O	$CaCO_3$, %	Exchangeable cations, meq/100g soil			Base satu-ration,%	Oxalate extract-able forms, %	
						Ca^{2+}	Mg^{2+}	H^+		Fe	Al
Arctic tundra weakly-gleyic humic soil (Stagni-Turbic Cryosol) – p.39, 38 (Ignatenko, 1967), Vaigach Island.											
O	0-4	74.03	n.d.	6.9	n.d.	74.2	13.6	n.d.	81	n.d.	n.d.
Ag	4-10	7.11	2.45	6.0	0.24	13.4	2.3	n.d.	66	0.11	0.04
Bg	12-20	3.28	0.93	5.7	0.36	10.1	2.3	n.d.	83	0.13	0.04
BCg	25-35	3.16	1.03	5.6	0.36	12.6	2.3	n.d.	94	0.11	0.04
Cg	55-65	4.09	1.53	7.2	0.65	15.4	2.3	n.d.	99	0.11	0.04
Circle	0-3	n.d.	1.92	7.9	4.15	n.d.	n.d.	n.d.	n.d.	n.d.	n.d.
Circle	3-13	n.d.	0.45	7.7	3.41	n.d.	n.d.	n.d.	n.d.	n.d.	n.d.
Circle	20-30	n.d.	0.35	7.6	2.91	n.d.	n.d.	n.d.	n.d.	n.d.	n.d.
Circle	45-55	n.d.	0.31	7.5	3.79	n.d.	n.d.	n.d.	100	n.d.	n.d.
Circle	75-85	n.d.	0.27	7.7	4.76	n.d.	n.d.	n.d.	100	n.d.	n.d.
Trough	0-4	84.56	n.d.	5.3	n.d.	n.d.	n.d.	n.d.	n.d.	n.d.	n.d.
Trough	4-10	56.44	n.d.	5.2	n.d.	45.9	20.4	n.d.	80	0.34	n.d.
Trough	10-22	49.41	n.d.	5.6	n.d.	26.4	16.6	n.d.	82	0.39	n.d.
Trough	22-34	n.d.	3.92	6.3	n.d.	13.6	8.8	n.d.	85	0.26	n.d.
Trough	34-40	n.d.	2.79	6.6	n.d.	14.6	8.4	n.d.	92	0.16	n.d.
Trough	50-60	n.d.	1.71	7.1	n.d.	13.5	8.2	n.d.	93	0.13	n.d.
Tundra podzolized Podbur (Spodic Cryosol) p. 20-95 (Goryachkin), Novaya Zemlya.											
O	0-2	75.79	n.d.	6.3	n.d.	37.4	13.9	4.9	92	n.d.	n.d.
AO	2-4	27.91	9.74	5.6	n.d.	16.9	6.42	1.4	95	0.15	0.05
EA	4-6	n.d.	3.78	5.8	n.d.	5.0	1.98	0.3	96	0.14	0.09
Bhs	6-16	n.d.	2.44	5.5	n.d.	2.1	1.09	1.3	73	0.20	0.18
BC	16-25	n.d.	1.64	5.4	n.d.	2.5	0.83	1.4	71	0.15	0.11
Dk	25-50	n.d.	n.d.	8.9	1.90	n.d.	n.d.	n.d.	n.d.	0.08	0.10

(continues)

Table 2.5.3. (cont.)

Horizon	Depth cm	Loss on ignition,%	Corg., %	pH H_2O	$CaCO_3$, %	Exchangeable cations, meq/100g soil			Base saturation,%	Oxalate extractable forms, %	
						Ca^{2+}	Mg^{2+}	H^+		Fe	Al
Tundra mucky-calcareous soil (Calcari-Mollic Leptosol) p.1-95 (Goryachkin), Novaya Zemlya.											
Ak	0-10	32.14	10.20	7.7	6.02	n.d.	n.d.	n.d.	n.d.	n.d.	n.d.
Ak	10-20	19.50	6.85	8.3	5.50	n.d.	n.d.	n.d.	n.d.	n.d.	n.d.
Bk	20-28	n.d.	1.31	8.6	2.14	n.d.	n.d.	n.d.	n.d.	n.d.	n.d.
Bk	28-55	n.d.	0.77	8.4	0.27	n.d.	n.d.	n.d.	n.d.	n.d.	n.d.
Stone	20-55	n.d.	n.d.	n.d.	62.0	n.d.	n.d.	n.d.	n.d.	n.d.	n.d.
Peaty bog Gleezem (Turbi-Histic Cryosol [Gleyic]) p. 5 (Ignatenko, 1967), Vaigach Island.											
H	2-12	93.61	n.d.	5.4	n.d.	36.9	9.8	n.d.	36	n.d.	n.d.
H	12-28	94.40	n.d.	5.1	n.d.	40.0	12.4	n.d.	40	n.d.	n.d.
Cri	28-33	n.d.	1.24	5.8	n.d.	14.3	5.6	n.d.	64	n.d.	n.d.
Ci	33-36	n.d.	1.65	5.8	n.d.	15.1	5.0	n.d.	56	n.d.	n.d.
Peaty bog mesotrophic permafrost soil (Eutri-Cryic Histosol) p.8-95 (Goryachkin), Novaya Zemlya.											
H	3-7	52.38	n.d.	6.2	n.d.	40.2	9.2	1.2	98	n.d.	n.d.
H	7-10	55.12	n.d.	6.4	n.d.	57.0	9.4	0.8	99	n.d.	n.d.
H	10-13	59.44	n.d.	5.8	n.d.	43.6	5.5	1.6	97	n.d.	n.d.
Hi	13-60	62.07	n.d.	5.7	n.d.	40.2	6.3	1.2	97	n.d.	n.d.

n.d. – not determined.

Well drained sandy soils occur in all three zones: the High Arctic barren zone, the Mid Arctic tundra zone, and the Low Arctic tundra zone. In the High Arctic barrens, they are classified as Arctic Soils, whereas in the Mid Arctic and Low Arctic Zones, they are classified as Podburs and Podzols (*Enti-Spodic Cryosols* or *Spodi-Turbic Cryosols*). These well drained sandy soils are distinctive for having a well developed spodic horizon, which could be under an albic horizon (Podzols) or directly under litter and an AO horizon (Podburs), the common situation when the transitional Podzolized podbur occurs (Table 2.5.2). These soils are well drained, but those with a shallow permafrost table, such as those found on Belyi Island, may have gleyic features because of water stagnation in the frozen ground.

As the Arctic islands consist mainly of sedimentary rocks, calcareous soils often are a component of the soil landscapes of Novaya Zemlya, Vaigach, Severnaya Zemlya, and especially Wrangel Island. According to formal criteria, these soils can belong to *Rendzic, Mollic*, and *Calcaric Leptosols*. They have a dark upper horizon which, depending on the local environmental conditions, can be humic (even mollic), mucky, or dry peat. Usually, these profiles include limestone debris, while the fine earth in mineral horizons can have a low calcite content (Table 2.5.3).

In spite of the fact that Arctic islands have many poorly drained soils, the Peaty-bog soils (*Histosols*) are not widespread here because of shallow peat horizons induced by low bioproductivity. The *Histosols* occur only on the warmest island, Kolguev Island, and in some landscape localities of the other archipelagos, where there is additional input of nutrients due to factors such as gradual slopes, valleys, and bird rookeries. All the *Histosols* of the Arctic belong to *Cryic* subgroups, including *Eutri-Cryic*, *Sapri-Cryic*, *Fibri-Cryic*, *Dystri-Cryic*, and *Foli-Cryic Histosols*.

Alluvial acid soils (*Dystric Fluvisols* and *Fluvic Cryosols)* and Alluvial bogged soils (*Histic Fluvisols*, *Gleyic Fluvisols*, and *Fluvi-Histic Cryosols*) are the common soils on flood plains of large rivers. Marshy saline soils (*Salic Fluvisols* and *Fluvi-Salic Cryosols*) occur widely in the tidal zone of the Arctic Ocean. In these coastal marine and riparian environments, non-permafrost soils (body of unfrozen ground) can substitute for Cryosols, as taliks are a common feature.

4. Soil Cover and Distribution on Russian Arctic Islands

4.1. Franz Josef Land

Franz Jozef Land is the northernmost archipelago of the Eastern hemisphere. It is mostly mountains of basalt, but some of its islands are denuded plains covered with Mesozoic quartzite sands. Glaciers occupy the major part of this archipelago. Much of the ice-free area is without soil development because of the recent age of marine terraces and glacial till. The majority of the ice-free area contains rocky outcrops, and/or stony material, and eolian sands. On sites with the most favorable conditions for vegetative growth, such as wind shadows, south oriented slopes, or rookeries, there is complete soil and vegetative cover.

Mean temperatures of the archipelago are -10°C for the year, +0.7 to 1.2°C for July, and -22 to -26°C for March. There are 200 to 300 mm of precipitation annually, mainly snow. Impoverished vegetation of the High Arctic barrens (polar desert) is characteristic.

Ivanov (Ivanov, 1933), Mikhailov (Mikhailov, 1970, 1973; Mikhailov and Govorukha, 1962), and Vostokova (Vostokova and Goryachkin, 1999) studied the soils of Franz Josef Land. All the pedologists classified these soils as Arctic soils. They have a 30- to 80-cm active layer, high organic carbon content (1.5 to 6%) for the whole depth of the active layer, high organic content of the total profile (15 to 16%), extractable forms of Fe due to basalt parent material, a lack of stagnic and gleyic features, and high base saturation of the soil exchange complex.

Depending on the character of the organic horizons, these soils can correlate to *Humi-Turbic Cryosols* and *Umbri-Turbic Cryosols*. The latter are the soils of the warmest site conditions mentioned above. If these soils are developed from rocky and stony substrates, they are *Leptosols*. Note that the Arctic soils of Franz Josef Land occur mainly under moss-lichen vegetation overlying cryogenic cracks. The soils are in complex with "film-like" soils under cryptogamic crusts and barren substrates.

Normally, mature soils occupy only 10 to 30% of the total surface of the plains and 1 to 5% of mountains and areas of eolian sands. The high organic matter content of Arctic soils occurs, in spite of the very low productivity of high latitudes, because of its accumulative character through the mass displacement of fine earth and organic remnants (residuals) from the spots (circles or nets) to cryogenic cracks (troughs) and their subsequent involvement in soil-forming processes (Vostokova and Goryachkin, 1999).

In addition to rock outcrops and Arctic Soils, the soil landscape of Franz Josef Land archipelago also contains Arctic hydromorphic non-gleyic soils (*Oxyaquic* and *Oxyaqui-Leptic Cryosols*), which are permanently over-saturated soils found in hollows containing oxygen-rich water from snow melt.

4.2. Novaya Zemlya

The mountainous archipelago, Novaya Zemlya, is in the European part of the Russian Arctic. Its two large islands (Northern and Southern) extend from north to south for approximately 1000 km. As a result, the northern and southern regions have different climates and vegetation. The mean July temperatures range from +2.5°C in the north to +6.6°C in the south. Mean January temperatures range from -20.4°C in the northeast to -14°C in the southwest. The precipitation is high both on the north (200 mm) and in the south (300 mm). Following the climatic gradient, the vegetation of Novaya Zemlya also differs: scarce plant communities of High Arctic barrens (polar deserts) in the north and Mid Arctic tundra vegetation on the plains of the south. Mountainous barrens, bordered with glaciers in the north and with mountainous tundra in the south, mainly cover the mountains.

For several decades, the archipelago was closed to civil soil surveys and, as a result, only data on early reconnaissance studies (Ratmanov, 1930) and recent data from case studies of soil and soil cover (Dobrovolsky and Filatova, 1993; Goryachkin, 1998) are available. During the period of extensive fieldwork and soil mapping of these islands (see Section 1), Novaya Zemlya was not mapped. According to available soil data, the soil landscapes of the northern and southern parts of the archipelago are different. The northern Novaya Zemlya soil landscape is similar to that found on Franz Josef Land, classified as Arctic soils and Arctic hydromorphic non-gleyic soils.

In the plains of the south, the important soil landscape feature is the solifluction stressed by rill erosion, giving a striped pattern on barren substrates and *Stagni-Turbic Cryosols*. Classic sorted and non-sorted nets occur only on very flat surfaces of less than one degree. The peat plateau bogs with Peaty bog mesotrophic permafrost soils (*Eutri-Cryic Histosols*) also occur at the base of long gradual slopes. The large flat wetlands are covered with Peaty bog gleyzems (*Turbi-Fibrihistic Cryosols* [*Gleyic*]).

The well drained soils of Novaya Zemlya southern island are *Humi-Turbic Cryosols* on loamy material or *Turbi-Leptic Cryosols* (*Humic*) on gravelly and cobbly substrates. *Calcari-Molli Leptosols* develop on limestones in the wind shadow, and the rock debris occupies the remaining limestone area. In the Tundra, podburs have bleached horizons (*Spodic Cryosols*), and the soils differ from their southern equivalents by having less Fe in the sandy material.

The specific soil landscape feature on the plains of the Novaya Zemlya is the occurrence of "polar desert" areas with low humus content and shallow leached soils in the tundra zone (*Calcari-Turbic Cryosols)* where katabatic winds (>40m/s) occur.

The soil formation in the mountains of Novaya Zemlya archipelago closely connects to small stream activity (recent and past) that produces localized fine earth, which gives the soil cover a pattern of linear forms. This pattern is composed of barren stones; *Oxyaqui-Leptic Cryosols* showing mucky and peaty horizons and lateral clay accumulations developed on stream beds, and *Leptic Cryosols* with moder humus developed in former stream beds (2.5.2).

Both the soils in the mountains and on the plains of Novaya Zemlya have a very low level of γ-radiation, except small areas totaling about 1 km^2 near the locations of past ground nuclear tests.

4.3. Vaigach Island

Vaigach Island is between the mainland and the Novaya Zemlya archipelago. Geologically, it is part of the Ural-Novaya Zemlya mountain system. However, the highest elevations here are only about 200 m asl. The Vaigach region includes the rocky calcareous highlands and plains of both marine and Quaternary glacial origins. The climate is similar to that of the southern part of Novaya Zemlya. The vegetation is Arctic tundra.

Ignatenko (1967, 1979) studied the soils of Vaigach. He identified the Tundra mucky-calcareous soils (*Molli-Leptic Cryosols* [*Calcaric*]) on limestones, which alternate with barren calcareous cryogenic spots. The loamy, well drained soils of Vaigach are *Stagni-Turbic Cryosols*, which form the polygonal pattern of cryogenic spots (circles/nets) with calcareous and saline (upper 5 cm) loamy substrates, with cryogenic cracks (troughs) characterized by peaty mat and shallow permafrost.

Tundra mucky gleyzems (*Gleyi-Turbic Cryosols*), which form a polygonal pattern of saline cryogenic spots (>2% of water-soluble salts in the upper 4 cm) and cryogenic cracks of Peaty gleyzems (*Turbi-Histic Cryosols* [*Gleyic*]), cover the areas with imperfect drainage.

The flat depressions are occupied by peat plateau bogs, where Peaty bog gleyzems (*Turbi-Fibrihistic Cryosols* [*Gleyic*]) are predominant. However, in some poorly drained ecotopes (sites), *Cryic Histosols* also occur.

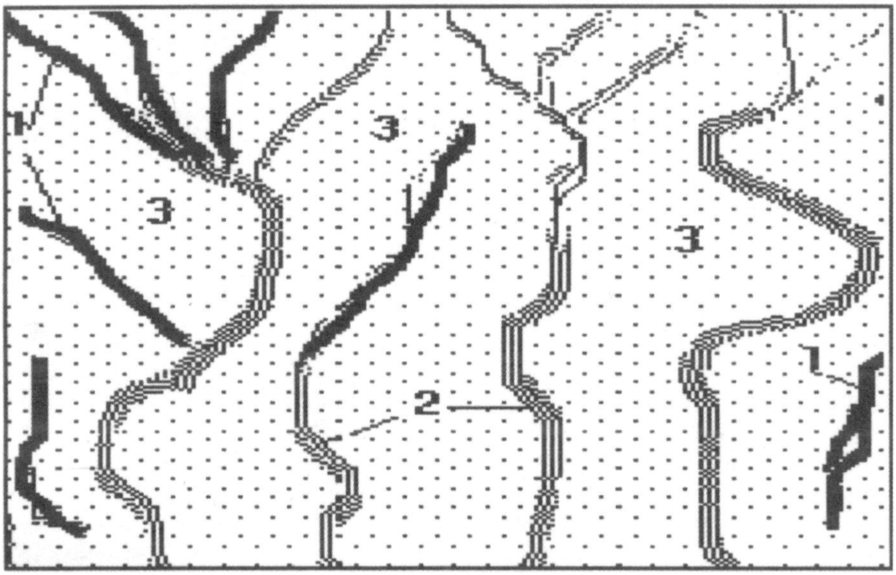

Figure 2.5.2. Linear pattern of soil cover in the mountains of Novaya Zemlya archipelago. 1. Arctic soils (*Leptic Cryosols*); 2. Arctic hydromorphic non-gleyic soils (*Oxyaqui-Leptic Cryosols*; 3. barren stones.

4.4. Kolguev Island

Kolguev is the southernmost island of the Russian Arctic. Quaternary glacial and marine deposits, mainly of loamy and partly of sandy texture, cover its surface. Kolguev is both hilly and gently undulating and Kolguev is the warmest of the

Arctic islands. The mean temperatures are -2.6°C annually, +6.6°C for July and -9°C for January. Annual precipitation is 300 mm. Kolguev's vegetation is dwarf shrub-herbaceous-mossy tundra with *Betula nana* and *Salix* spp in the warmest ecotopes.

No publications deal exclusively with the soils of Kolguev. However, Ignatenko (Fridland, 1988) mapped it, and Goryachkin has personal experience studying the soils of this island (Goryachkin et al., 1994, 1995).

The soils of Kolguev Island differ from those of other Arctic islands in the predominance of hummocky patterned ground and less area of polygonal ground. Arctic tundra weakly gleyed humic soils (*Stagni-Turbic Cryosols*) developed on loamy substrates in well drained sites. The most widespread soils occur on gentle slopes as Tundra peaty gleyzems (*Turbi-Fibrihistic Cryosols* [*Gleyic* or *Stagnic*]), showing pronounced hummocky patterned ground. Peat plateaus and palsa bogs also are common, with Peaty bog gleyzems (*Turbi-Histic Cryosols* [*Gleyic*]) and Peaty bog mesotrophic permafrost soils (*Eutri-Cryic Histosols*). Well drained sandy soils are Podburs and Podzols (*Spodi-Turbic Cryosols* and *Spodic Cryosols*).

Overgrazing by domestic reindeer has compacted and destroyed many surface horizons of Kolguev.

4.5. Belyi Island

Belyi Island is the former northernmost part of Yamal peninsula, West Siberia, which now is separated from the continent by Malygin Strait. The island has a low relief marine plain, with an altitude of 1 to 30 m asl. Some eolian forms such as dunes, barchans, and tafoni-like forms appear on the coast. Basically, the Quaternary sediments of Belyi are marine sandy and sandy-loam layered deposits. The mean January temperature is -23.4°C and the mean August temperature is +5.3°C. Annual precipitation is about 300 mm. The vegetation is classified as Arctic tundra subzone. Zonal plant communities are herbaceous-dwarf-shrub-mossy and lichen tundras.

The soil data for Belyi island are available only from the Russian national map (Fridland, 1988) and from a brief description of its soil landscape (Goryachkin et al., 1994). Peaty bog gleyzems (*Turbi-Histic Cryosols* [*Gleyic*]) with shallow (20 to 40 cm) permafrost, found in low-center ice-wedge polygons, occur widely on level marine and alluvial terraces. Active sand formations occur along the coastal areas. Sandy soils with a more stable surface are classified as Podburs and Podzols with gleyic features, which relate to saturation over the permafrost table (60 to 80 cm) (*Gleyi-Turbic Cryosols* [*Spodic*]). Hummocky tundra of sandy-loam layered substrata, supporting Peaty tundra gleyzems (*Turbi-Fibrihistic Cryosols* [*Gleyic* or *Stagnic*]), occupy higher surfaces.

4.6. Severnaya Zemlya

Severnaya Zemlya is the archipelago of the central part of the Russian Arctic and is mainly mountainous. About half its area is under glaciers. The mountains consist of calcareous and terrigenous rocks. Unconsolidated deposits on the plains are marine and glacial, cobbly, and gravelly loams. Periglacial forms, such as sorted nets and polygons, are widespread on the plains. The climate is as cold as Franz Josef Land in summer (0.8 to 1.6°C in July) and almost as cold as the New Siberian Islands (-28°C in March) in the winter. Precipitation is about 150 to 230 mm per year. Vegetation matches the criteria of High Arctic barrens (polar desert zone). The vegetative cover varies from sparse (1 to 2%) lichens on rocky substrates to nearly continuous mossy-herbaceous communities in warm sites with fine-earth accumulation.

Mikhailov (1962, 1970), Govorenkov (1981), and Aparin (1997) investigated and mapped the soils of Severnaya Zemlya. The predominant feature is mainly cobbly and rocky barren substrates. The most common soils for Severnaya Zemlya are Arctic soils (*Turbi-Leptic Cryosols* [*Humic*]) on cobbly parent material and *Humi-Turbic Cryosols* on loams. They form patterned ground with barren cryogenic spots in centers of polygons. On calcareous material, Arctic calcareous soils (*Rendric* and *Calcaric Leptosols*) occur. Arctic hydromorphic non-gleyic soils (*Oxyaqui-Leptic Cryosols*) occur widely in depressions with running melt water, similar to those on Franz Josef Land and Novaya Zemlya. In local sites under continuous vegetation cover and saturation, Cryosols with stagnic features (*Stagni-Leptic Cryosols*) occur.

4.7. New Siberian Islands

The New Siberian Archipelago is in Yakutia, that part of Asia with the most continental climate of the Northern Hemisphere. It consists of several large and small islands extending for more than 400 km from north to south. Most of these islands are plains of alluvial, lacustrine, and marine genesis, but some islands are mountainous with limestones, dolomites, sandstones, argillites, and basalts. In addition, the islands include a unique, very flat sandy area, Bunge Land, which has a wide distribution of different eolian forms. The micro-relief of Bunge Land relates to cryogenic (thermokarst, cryoturbation, solifluction) and eolian processes. Particular cryogenic forms on the plains of the New Siberian Islands are *baydzherakhs*, thermokarst mounds (2 to 3 m height) remaining after the ice wedges surrounding an ice-wedge polygon melt (Figure 2.5.3).

Figure 2.5.3. Baydzherakhs (thermokarst mounds) with excessively drained soils, Faddeevsky Island, New Siberian Islands. These are residual mounds formed after the melting of ice wedges and the subsidence of the surface of former ice wedge polygons.

The climate of the New Siberian Islands is very cold, especially in winter. It is humid but with the least precipitation (133 to 170 mm) of any of the Russian Arctic archipelagos. Vegetation of the islands belongs to the High Arctic tundra subzone and polar desert. The latter occurs in upper belts of mountains and on the northernmost small De Long Islands.

Mikhailov (1967), Karavaeva (1969), and the author (Goryachkin et al., 1994, 1995) studied the soil landscape of the New Siberian Islands. Targulian (1971) investigated soil development on rocky substrates. Soil landscapes of mountains, loamy plains, and sandy polar deserts are different.

In the mountains of the archipelago, rock outcrops predominate. Mountainous primitive soils (*Lithic Leptosols* and *Calcari-Lithic Leptosols*) and Arctic hydromorphic non-gleyic soils (*Oxyaqui-Leptic Cryosols*) are similar to those of Franz Josef Land, Novaya Zemlya, and Severnaya Zemlya. Podburs (*Spodi-Leptic Cryosols*) also occur on low rocky surfaces (Targulian, 1971).

Permafrost strongly influences pedogenesis on the plain, with the seasonally thawing active layer being 10 to 15 cm in peat, 20 to 40 cm in loams, and 50 to 60 cm in sands. Both Arctic soils (*Humi-Turbic Cryosols*) (Mikhailov, 1967) and Arctic tundra weakly-gleyic humic soils (*Stagni-Turbic Cryosols* [*Humic*]) (Karavaeva, 1969) were described as "zonal" well drained soils for loamy plains of New Siberian islands (Tables 2.5.2 and 2.5.3). Karavaeva (1969) proposed that the genesis of these two soil types relates to different stages of soil development, described earlier as connected with the development of baydzherakhs. Loamy lacustrine sediments were deposited under conditions of strong reduction.

Loamy soils have developed in association with level cryogenic polygons that were created after lake drainage. These soils have undergone a change in redox potential, having their gleyic properties substituted by stagnic ones, as expressed by the brown mottles on a grayish-olive matrix. Further thermokarst processes result in the formation of baydzherakhs (hillocks with excessive drainage). Soils on the upper positions develop under well aerated conditions, creating the brown loam that currently occurs on the tops of the mounds (Karavaeva, 1969, and personal observations). After denudation of the baydzherakhs, the soil begins to form on brown material, with saturation too weak and short for the stagnic features to reoccur. In such places, *Humi-Turbic Cryosols* develop. The water regime in the soils of the New Siberian Islands is adequate to preserve the stagnic features in strongly gleyed loams of the initial polygons. However, it is difficult to determine if the stagnic features are a result of the initial polygon or an eroded baydzherakh.

Bog gleyzems with very shallow peat layer (*Gleyic Cryosols*) are the typical soils of thermokarst depressions of New Siberian Land.

Active sand dunes and salinized boggy Marshy soils (*Sali-Fluvic Cryosols*) are common features for Bunge Land and the adjacent accumulative shore.

4.8. Ayon Island

The physiography of Ayon Island in Pevek Bay is an alluvial and lacustrine plain, with marine terraces and a complex of shore forms in the north. Layered loamy-sandy alluvial and lacustrine deposits are the dominant soil parent materials of Ayon Island, and Holocene marine sediments also occur as loose sands.

The mean January temperature of the island is -28.7°C, with the July mean temperature being +4.3°C. The climate is humid, with annual precipitation of 300 mm. The severe winds (up to 70 to 80 m/sec) cause strong deflation. Vegetation of Ayon Island is northern hypoarctic tundra and appears similar to that of the southern Arctic tundra.

The information on the soils of Ayon Island is available from papers (Yevseev, 1982; Goryachkin et al., 1994), maps (Fridland, 1988; Naumov, 1993), and the author's field experience. The soil cover of the island is a combination of three main types:

1. the sandy deposits of northern Ayon Island
2. the well drained soils of its central part
3. those with poor drainage in southern Ayon Island.

Active layers on Ayon vary from 20 to 40 cm in peat, to 40 to 60 cm in loams, and to 70 to 100 cm in sands.

Vast barren sand exposures occur on the northern coast of the island. Podzols and Tundra podzolized podburs (*Spodi-Turbic Cryosols*) occur in this region along with those with stagnic features developed immediately above the permafrost table (*Stagni-Turbic Cryosols* [*Spodic*]). These soils alternate with *Turbi-Histic Cryosols* in wetland areas. Locally, *Foli-Cryic Histosols* occur under herbaceous tussocks on peat ridges. These localities often are enriched by nitrogen and phosphorus, as Snowy Owls use them as nesting and eating places.

The well drained soil landscapes in the central part of the island are hummocky tundra with loamy substrata, and the two main soil types are Arctic tundra weakly gleyic humic soils (*Stagni-Turbic Cryosols* [*Humic*]) and Peaty tundra gleyzems (*Turbi-Histic Cryosols* [*Gleyic* or *Stagnic*]).

Peaty bog gleyzems (*Turbi-Histic Cryosols* [*Gleyic*]) developed in low-center polygons that are widely distributed on the low flat alluvial and marine terraces of southern Ayon Island.

Overgrazing of the tundra of Ayon Island is a striking ecological problem. In combination with severe winds, it results in deflation of well drained sandy soils and accumulation of eolian materials in the upper histic horizons of peaty soils.

4.9. Wrangel Island

Wrangel Island is in both the eastern and western hemispheres, as it spans the 180th meridian. Coastal plains are on the south and the north of its territory, while the central two-thirds is mountainous, with the highest elevations between 1000 to 1100 m asl. The mountains formed in the Paleozoic and were renewed in the Mesozoic epoch. Marine deposits overlap glacial, lacustrine, and alluvial deposits on the coastal plains. The mean January temperature on the island is -23°C and for July is +3.6°C, with an annual precipitation of 200 mm. The vegetation is Arctic tundra.

Svatkov (1958) and Oganesyan (1989) investigated the soils of Wrangel Island. The author also has experience studying the island's soil (Goryachkin et al., 1994, 1995).

Due to the substrate having a higher content of stony material, Wrangel Island differs from the Siberian islands described earlier by having less pronounced ice-wedge polygons. The main soils of the plains are Arctic tundra weakly gleyed humic soils (*Stagni-Turbic Cryosols* [*Humic*]) and Tundra gleyzems (*Gleyi-Turbic Cryosols* [*Humic*]). Peaty bog gleyzems are associated with small depressions on the northern and southern tundra plains. They differ by having a very shallow peat horizon, which is why its WRB equivalent is *Gleyic Cryosols*.

In the mountainous part of the island, the soil landscape is complex, due mostly to the influence of aspect, creating a situation where differently oriented slopes contain different soil types. On one slope, Mucky calcareous tundra soils (*Calcari-Leptic Cryosols*) with a 20-cm organic horizon can occur in complex with *Lithic Leptosols* and *Stagni-Turbic Cryosols*, depending on slope position and aspect. *Lithic Leptosols* can cover the opposite slope, which has the same gradient. In the wide valleys of the mountainous area, the *Stagni-Turbic Cryosols* (Humic) are the common soils. However, the predominant component of the surface in the mountains is rock outcrops, particularly at elevations above 350m.

5. Conclusion

The Russian Arctic archipelagos have a high diversity of climatic, geogenic, and resultant soil conditions. The islands belong to three soil zones – High Arctic barrens (Franz Jozef Land, northern part of Novaya Zemlya, and Severnaya Zemlya), Mid Arctic tundra (southern Novaya Zemlya, Vaigach, Belyi, Ayon, Wrangel, and New Siberian islands), and Low Arctic tundra (Kolguev). In the High Arctic and in the mountains of Mid Arctic tundra, the rock outcrops are the dominant element of the surface; they are combined with *Humi-Turbic* and *Oxyaqui-Leptic Cryosols*. Mainly *Humi-Turbic*, *Stagni-Turbic*, and *Turbi-Histic Cryosols* cover the Mid Arctic islands characterized by plains. Low Arctic Kolguev Island is covered by *Stagni-Turbic Cryosols* and *Cryic Histosols*. The podzol-like soils, *Spodi-Turbic Cryosols*, occur on sandy substrates on the islands.

6. Acknowledgments

The Russian Foundation for Basic Research partly provided project funding. Special thanks to the Swedish Polar Research Secretariat and P.V. Boyarski (leader, Integrated Marine Arctic Expedition) for arranging field work. We also thank Scott Smith for his the editorial help in preparing this chapter.

References

Aparin, V.B. 1997. Genetic features of soils on Bolshevik Island, the Severnaya Zemlya archipelago. In: Cryopedology 1997. II International conference. August 5-8, 1997. Syktyvkar. Russia. Abstracts. Syktyvkar. pp. 27-28.

Dobrovolsky, V.V., and E.V. Filatova. 1993. Geochemical features of soils of the Southern Island (Novaya Zemlya archipelago). Nauchnye doklady vysshei shkoly. Biologicheskie nauki. N2: 91-98. (In Russian.)

Fridland, V.M., ed. 1988. Soil map of Russian Soviet Federal Socialist Republic, scale of 1:2500000. PUBLISHER? PLACE? (In Russian.)

Gerasimov, I.P., ed. 1970. Soviet Arctic (Sovetskaya Arktika). Nauka. Moscow. (In Russian.)

Gorodkov, B.N. 1939. Specific features of Arctic soil cover. Izvestia Gos. Geogr. Obshchestva. 71:1516-1532. (In Russian.)

Goryachkin, S.V. 1998. Soil cover of the plains and mountains of the western part of Southern Island. In: P.V. Boyarsky, ed. Novaya Zemlya: Nature, History, Archeology, Culture (Novaya Zemlya: Priroda, Istoria, Arkheologiya, Kultura). PUBLISHER? Moscow. pp. 149-175. (In Russian.)

Goryachkin, S.V., R.I. Zlotin, and G.M. Tertitsky, eds. 1994. Diversity of natural ecosystems in Russian Arctic. Lund University. Sweden.

Goryachkin, S.V., R.I. Zlotin, and M.V. Glazov. 1995. Spatial and temporal organization of Arctic Ecosystems. In: Swedish-Russian Tundra Ecology Expedition -- 94. A Cruise Report. Stockholm. pp. 334-339.

Govorenkov, B.F. 1981. On pedogenesis in conditions of arctic deserts of October Revolution island (Severnaya Zemlya). Trudy NII Arktiki i Antarktiki. 367:132-141. (In Russian.)

Ignatenko, I.V. 1967. Soil complexes of Vaigach Island. Pochvovedenie. N9:86-99. (In Russian)

Ignatenko, I.V. 1979. Soils of East European tundra and forest-tundra (Pochvy Vostochno-evropeiskoi tundry i lesotundry). Nauka. Moscow. (In Russian.)

ISSS Working Group RB. 1998. World Reference Base for Soil Resources. International Society of Soil Science (OSSS), International Soil Reference and Information Centre (ISRIC) and Food and Agriculture Organization of the United Nations (FAO). World Soil Resources Report 84. FAO. Rome. 91 pp.

Ivanov, I.M. 1933. Soil formation in the glacial zone. Trudy Arkticheskogo instituta. 12:183-202. (In Russian.)

Karavaeva, N.A. 1969. Tundra soils of Northern Yakutia (Tundrovye pochvy Severnoi Yakutii). Nauka. Moscow. (In Russian.)

Mikhailov, I.S. 1960. Some features of sod Arctic soils on Bolshevik Island. Pochvovedenie. N6: 89-91. (In Russian.)

Mikhailov I.S. 1967. Soils. In: Y.Gakel, ed. New Siberian Islands (Novosibirskie ostrova). Gidrometeoizdat. Leningrad. pp. 146-150. (In Russian.)

Mikhailov, I.S. 1970. Soils. In: I.P. Gerasimov et al., ed. Soviet Arctic (Sovetskaya Arktika). Nauka. Moscow. pp. 236-249. (In Russian.)

Mikhailov, I.S. 1973. Structure of soil cover of arctic zone. In: V.M. Fridland. Structure of soil cover and methods of its study (Struktura pochvennogo pokrova i metody yeyo izucheniya). Annals of Dokuchaev Soil Institute. Moscow. pp. 119-125. (In Russian.)

Mikhailov, I.S. and L.S. Govorukha. 1962. Soils of Franz Josef Land. Vestnik Moskovskogo Universiteta, seria geograficheskaya. N6:42-48. (In Russian.)

Naumov, E.M. 1993. Soil map of the North-East of Eurasia (Pochvennaya karta Severo-Vostoka Evrazii). PUBLISHER? Moscow. (In Russian.)

Oganesyan, A.S. 1989. Soil and soil cover of Wrangel Island (Pochvy i pochvennyi pokrov ostrova Vrangelya). Candidate of Sciences (biology) Dissertation. Novosibirsk. (In Russian.)

Ratmanov, G.E. 1930. Soils of the Novaya Zemlya Archipelago. Trudy Pochvennogo Instituta. 3-4: 145-148. (In Russian.)

Svatkov, N.M. 1958. Soils of Wrangel Island. Pochvovedenie. N1:91-98. (In Russian.)

Targulian, V.O. 1971. Pedogenesis and weathering in cold humid regions (Pochvoobrazovanie i vyvetrivanie v kholodnykh gumidnykh oblastyakh). Nauka. Moscow. (In Russian.)

Tedrow, J.C.F. 1977. Soils of the polar landscapes. Rutgers Univ. Press. PLACE?

Vostokova, T.A., and S.V. Goryachkin. 1999. Soils of Alexandra Land Island, Franz Josef Land archipelago, and their place in zonal scheme of the Arctic. Abstracts. International Conference "Monitoring of Cryosphere." 20-23 April 1999. Pushchino. pp.137-138.

Yevseev, A.V. 1982. Arctic tundra soils of Ayon Island. Pochvovedenie. N1:31-33. (In Russian.)

Chapter 6. Soils and Soil Cover of Northeastern Eurasia

Ye. M. Naumov[1]

[1]Leading Researcher, Dokuchaev Soil Science Institute, Pyzhevsky per. 7, Moscow, 109017 Russia; e-mail naumov@agro.geonet.ru.

1. Introduction

Northeast Eurasia is a vast, remote territory with the most severe climate in the northern hemisphere. The lowest recorded temperature in the hemisphere (-71°C) was recorded in northeast Eurasia at Oimyakon. In this region, the role of cryogenesis in soil formation and soil distribution is most pronounced.

2. The History of Regional Soil Survey and Evolution of Soil Classification

The Dokuchaev Soil Science Institute has conducted special soil studies and soil survey expeditions since 1958, the Yakutsk Institute of Biology of the Russian Academy of Sciences, since 1968, and the Magadan Institute of Biological Problems of the North, Russian Academy of Sciences, since 1971. These studies and field expeditions were to determine the genesis and geography of soils of the region, to develop a suitable soil classification, to map and correlate soils, and to assess soil resources (Karavaeva and Targulian, 1960; Ivanova et al., 1961; Naumov, 1963, 1973; Naumov and Gradusov, 1974; Karavaeva et al., 1965; Ignatenko et al., 1982; Elovskaya et al., 1977; Naumov, 1996).

The study of the soil landscape in northeastern Eurasian began later than in other regions of the Russian north. Until the 1960s, soil studies in northeastern Eurasia were rather scarce. The All-Union soil maps compiled at that time showed hypothetical delineations of mapping units, based on analysis of physiographic conditions. For a long time, only non-systematized soils data existed for some parts of this vast territory. A.P. Vas'kovskii (1959) gave the first description of taiga soils in northeastern Eurasia and subdivided them into *svetlozems* (lightly colored soils) of continental regions and *podzols* of coastal regions.

The first comprehensive classification of the soils of northeastern Eurasia was developed for the southern part of this region on the basis of field investigations carried out from 1959 to 1963 and mapped on a scale of 1:1.5 M (Naumov, 1963). The map reflected specific features of the soil landscape of the coastal region: (a) the presence of Dry Peaty soils (*Folic Histosols*) under elfin forest (*Pinus pumila*) communities in coastal mountain ridges, (b) the development of tundra soils with barren unsorted circles on coastal plains as far south as 59°N, and (c) the presence of specific Ocherous Podzolic soils *(Andi-Spodic Cryosols*) in the areas with deposition of volcanic ash. The soils were grouped into six soil types distinguished by profile morphology; these in turn were placed in four major groups of soils (Targulian et al., 1970).

In 1967, large-scale (1:50,000) soil maps were compiled for the area of the state farms "Bilibino" and "Omsukchan" in the continental Kolyma region. These maps revealed the complexity of soil spatial patterns related to the effect of cryogenic processes and patterned ground formation. In the same year, a 1:1.5 M soil map was compiled for the taiga zone of northeastern Eurasia. It showed clearly the difference between the humid coastal region of the Sea of Okhotsk with a predominance of Al-Fe-humus soils (Podzols and Podburs) and the semi-arid and arid continental region (the Yano-Kolyma region) with its Pale soils, Cryozems, Steppe Cryoarid soils, and Podburs.

Later on, soil survey expeditions covered the whole territory of northeastern Eurasia. Their main goal was to compile the *State Soil Map (1:1 M)* for this vast region. Several sheets of this map were published in the 1970s, and the new data obtained during this period resulted in a more detailed subdivision of taiga soils of the region. Eight major types of taiga soils were distinguished (Naumov, 1973; Naumov and Gradusov, 1974). In 1974, the first draft of the *Soil Map of Northeastern Asia (1:2.5 M)* was presented to the participants of the 10th International Soil Science Congress.

In 1976, the *Soil Map of Magadan Region (1:5 M)* was published. The legend to this map contained 22 map units, in contrast to the 6 to 8 mapping units shown for this territory on the All-Union soil maps compiled in 1956 and 1960. The *Soil Map of the Russian Federation (1:2.5 M)*, published in 1988, using 22 map units, resulted in suggesting new soil classifications for the whole region (Ignatenko, 1977; Naumov, 1980) and for separate parts of northeastern Eurasia, in particular northeastern Yakutia (Elovskaya et al., 1977) and the Magadan Oblast (Ignatenko et al., 1982).

New field data from the 1980s and early 1990s led to further refinment of the maps, adding new soils to their legends. More attention was paid to the representation of the complexity of soil cover patterns in permafrost-affected regions.

The *Soil Map of Magadan Region (1:2.5 M)* published in 1990 displayed 53 map units. The most recent map encompassing the whole region was published in 1993 on a scale of 1:2.5 M (Naumov, 1993) and contains 22 different soil types in its legend.

A special feature of the legend to the latest map (Naumov, 1993) is the grouping of soils according to cryological conditions and types of soil microassociations (complexes); the names of map units include several soils composing soil cryocomplexes. Permafrost-affected soils are subdivided into groups with respect to the ice content in permafrost (ice-rich and dry permafrost), and depth of seasonal thawing (shallow <1 m and deep >1 m). The duration of preservation of frost in the soil profile is also indicated (both for the soils with permafrost and for seasonally freezing soils). Figure 2.6.1 shows a generalized (1:16 M) version of this soil profile.

3. Factors of Soil Formation

3.1. Relief

Northeastern Eurasia is a vast mountainous territory. The area of ridges, high uplands, and tablelands by far exceeds the area of lowland plains.

An extensive coastal plain occurs in the north of the region and consists of the Yana-Indigirka and Kolyma lowlands. Mountains that protect it from western and southwestern winds surround this plain on the west, south, and east. It is composed of ice-rich fine earth deposits (the "Yedoma" suite), subjected to active thermokarst during the Holocene and resulting in lake formation within the numerous thermokarst depressions. Polygonal, polygonal-ridge, and polygonal-trough cryogenic patterns are typical.

The several other lowland plains in the region are relatively small: the Anadyr, Chaun, Penzhina, Vankarem, Yama-Tauisk, and Cis-Verkhoyansk plains. The expansive Lena Delta is another special area of coastal plain.

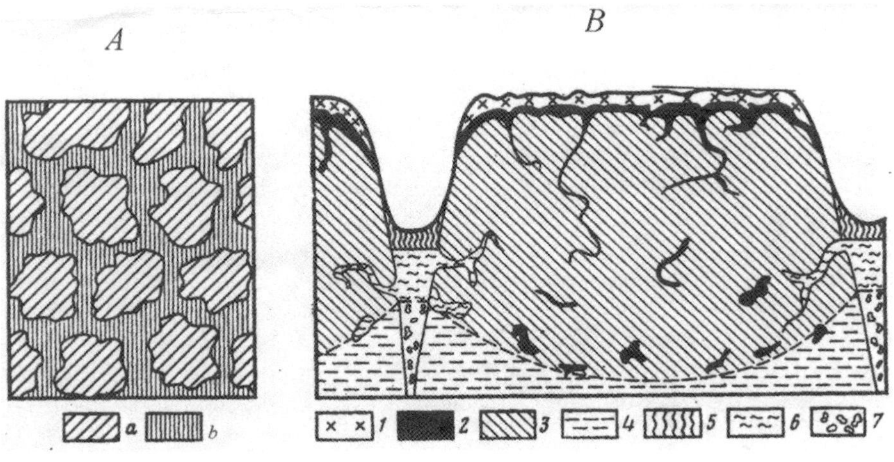

Figure 2.6.1. The scheme of polygonal-trough pattern of Cryozems. (A) Horizontal section: (a) polygons, (b) troughs. (B) Vertical section; soil horizons: (1) O, (2) AO, (3) Bjj (with fissures and mottles enriched in organic matter), (4) BCf, (5) Oi, (6) Oe, (7) ice wedge.

High tablelands occupy much of the region (the Yana, Elginsk, Yukagir, and Anadyr tablelands). River valleys dissect mountain systems, which include several intermontane depressions, e.g., the Oimyakon, Seimchan-Buyunda, and Upper Kupkinskaya depressions. The complexity of the relief governs the differentiation of climate, vegetation, soil parent materials, and soil development.

The presence of many high mountain ranges promotes the development of vertical soil zonation and aridity of the climate in interior areas, especially in intermontane depressions. These depressions have somewhat warmer summer temperatures and, as a result, have forest stands of higher quality, compared with those of surrounding territories. Differences in insolation and temperature conditions on southern and northern slopes are considerable. Thus, aspect is one of the most important factors of soil development in mountainous regions.

3.2. Climate

The climate of northeastern Eurasia is very severe. The vast latitudinal and longitudinal dimensions of the region and complex orography predetermine the differentiation of climatic conditions into three climatic zones: Arctic tundra, Subarctic tundra and forest-tundra, and Boreal coniferous forests (Klyukin, 1960). The tundra zone occupies the northern part of the territory and, owing to the cooling effect of cold seas, stretches along the eastern coast in the submeridianal direction. The southernmost patches of tundra are south of Magadan (approximately 57° N). The forest zone occupies the central part of northeastern Eurasia.

The eastern Siberian anticyclone (i.e., high pressure systems) governs the climate of continental areas, whereas Pacific cyclones (i.e., low pressure systems) affect the southeastern coastal regions. The difference in humidity between these two areas is considerable. The eastern and southeastern parts of the territory have a humid climate, with annual precipitation about 600 mm, and the precipitation-to-evaporation (P/E) ratio is above 1.3 to 1.5 in summer months. The rest of the territory has a semi-arid or even arid climate (except for high mountain regions).

The most arid conditions characterize intermontane depressions in the basins of the Kolyma, Yana, and Indigirka rivers. Summer precipitation in these intermontane depressions does not exceed 100 to 150 mm, and a droughty period lasts for 30 to 100 days. The P/E ratio in summer is about 0.3 to 0.5.

The differentiation of temperature conditions is also considerable, with the mean annual temperatures in the southeastern coast (Magadan) varying from -4.1 to -6.6°C and in interior regions (the Kolyma-Oimyakon Upland), from -12 to -16°C. In the northeastern part (the Bering Strait region), mean annual temperatures are about -8°C. Mean July temperatures in these regions vary from 11.6 to 13.4, 15 to16, and 4 to 5°C respectively; mean January temperatures vary from -20 to -25, -30 to -50, and approximately -22°C, respectively.

A comparison of different degree-day indices ($\Sigma t° > 0$, 5, 10°C; $\Sigma t° < 0$°C, etc.) in northern taiga, forest-tundra, and tundra zones of Eurasia shows the gradients and severity of the climate of northeastern Eurasia, resulting in very cold soil climates. Mean annual soil temperature in the profiles of taiga soils of northeastern

Eurasia is below zero; at a depth of 40 cm and more, soil temperature never rises above 10°C. The growing season is short, and all soils are subject to extended periods of freezing. Permafrost is continuous in the continental areas and discontinuous in the coastal zone of the Sea of Okhotsk.

Two kinds of permafrost, icy and relatively dry, exist. Their distribution depends on the composition of soil parent materials and drainage conditions. Ice-rich permafrost with huge ice wedges occurs in the valleys of large rivers and on coastal plains in the northern part of the territory. These specific loess-ice formations, the Yedoma suite, contain up to 80% by volume of ice.

In general, the differences in climatic conditions of relatively humid coastal areas and relatively arid interior regions dictate the differences in the direction of soil-forming processes.

3.3. Vegetation

The vegetation of northeastern Eurasia is mainly tundra, forest-tundra, and northern taiga communities. In agreement with the climatic pattern, zones of vegetation have both latitudinal (in the north) and longitudinal (in the east) directions. The tundra zone in the north extends from the Lena Delta to the Bering Strait; then it turns to the south and extends as a continuous zone to the mouth of Penzhina River and the Taigonos Peninsula. Discontinuous tundra alternating with larch stands and elfin (*Pinus pumila*) forests occupies the northern coast of the Sea of Okhotsk to the latitude of 59° N (Reutt, 1960; Treshnikov, 1985).

The forest-tundra zone between the tundra and the taiga zones is not always clearly seen. Some geobotanists do not distinguish it as a separate zone.

The taiga zone occupies the interior regions. Relatively sparse larch stands with *Pinus pumila* and dwarf birch in the undergrowth with mossy-lichen or mossy storey prevail. The quality of such stands is very low. Somewhat better stands of larch (*Larix dahurica*), poplar, and willow (*Chosenia arbutifolia*) occur on river terraces and high flood plains with deep permafrost or without permafrost at all.

3.4. Soil Parent Materials

Geological maps of northeastern Eurasia display a wide range of sedimentary rocks from the Proterozoic era to the Neogene. Permian, Jurassic, Triassic, and Cretaceous sandstone, siltstone, claystone, and slates prevail in the composition of the mountain systems of northeastern Eurasia. Calcareous rocks are less frequent. Igneous intrusions and basalt flows occupy smaller areas and occur as magmatic massifs among sedimentary rocks.

Large areas of igneous rocks are typical of mountain systems along the coast of the Sea of Okhotsk (granites and granodiorites of the Magadan batholith, basalts, and andesite-basalts). In mountainous areas, soils develop from the residuum (eluvium) of bedrock or colluvial deposits. Severe climate retards weathering processes; the soils develop from relatively coarse derivatives of bedrock with low clay content (Naumov and Gradusov, 1974). The genesis and composition of parent materials on plains are rather diverse.

The cryological status of the surface geological materials in the zone of continuous permafrost depends on soil texture, geomorphic position, and other factors. On the plains and gentle slopes, the ice content increases in the sediments with finer texture. Simultaneously, the depth of the active layer decreases. Shallow ice-rich permafrost favors the stagnation of water in the soil profile, the development of soil hydromorphism, and activation of cryogenic processes

Silty loams of alluvial, lacustrine-alluvial, and eolian genesis make up the coastal plains in the north. Loess-ice formations (the Yedoma suite) of the Late Pleistocene Age are widespread in these areas. The origin of loess-ice formations is widely discussed in the earth sciences literature and is controversial. Some researchers consider the loess-ice formations to be the result of alluvial sedimentation, whereas others believe them the result of eolian sedimentation in conditions of the cryoarid climate of the Late Pleistocene.

Alternating stages of eolian sedimentation (during cold periods with arid climate) and the development of thermokarst lakes (during warmer periods) have shaped a specific periglacial type of loess-ice deposits (Shilo and Tomirdiaro, 1970) typical of coastal plains in the north of northeastern Eurasia. During the Holocene evolution, thermokarst processes partly destroyed the Yedoma deposits; the development of thermokarst lakes led to the transformation of eolian deposits into lacustrine deposits.

Alluvial sediments are widespread in all parts of the region, and the composition of these sediments varies within wide limits. As a rule, alluvial sediments are of finer texture in low reaches of rivers and on high terraces. Along with mineralogical and textural properties, their cryological status is of primary importance for pedogenesis.

4. Classification, Diagnostics, and Properties of Soils

The manifestation of cryogenic phenomena in the soils of northeastern Eurasia is very diverse: from seasonal freezing, to very considerable cryogenic transformations of the soil mass, to the appearance of specific cryogenic features in permafrost-affected soils. Tables 2.6.1 to 2.6.6 give examples of soils with different degrees of manifestation of cryogenic features, from very vivid ones in Cryozems to relatively weak ones in Podzols with gleyic features above the permafrost table.

Soil taxa typed in normal font within the text come from the legend for the *Soil Map of Northeastern Eurasia*, published at a scale of 1:2.5 Mln., Naumov, (1993). Those in *italics* are taxa according to the *World Reference Base Classification System* (ISSS Working Group RB, 1998).

Table 2.6.1. The properties of major types of permafrost-affected soils in northeastern Eurasia.

Horizon	Depth cm	pH (H_2O)	Organic Matter %	Exchangeable, meq/100 g soil		Particle Size (mm), %		Total elements, %			
				Ca + Mg	H^+	<0.01	<0.001	SiO_2	Fe_2O_3	Al_2O_3	Fe_o**
Pit 1-1967. Typical Cryozem (Oxyaqui-Turbic Cryosols) (see Table 2.6.2)											
Oi	5-13	4.8	48.1	15.2	36.6	54	31	51.1	8.4	30.9	1.2
AOa	13-17	4.9	5.9	5.4	6.2	67	28	60.0	7.7	25.0	1.0
Bwjj	17-45	4.9	6.1	6.4	5.4	70	28	59.6	7.4	25.0	1.0
BCf	45-55	4.9	6.6	6.3	5.3	70	28	59.8	7.6	24.4	0.8
Pit 112-1967. Typical Cryogleyzem (Gleyi-Turbic Cryosols) (see Table 2.6.3)											
AOa	4-7	3.7*	4.9	11.3	0.9	–***	–	68.5	5.1	18.5	1.2
B1w	7-21	5.1*	1.6	16.2	0.2	53	24	68.3	5.0	19.4	1.2
Bwjj	21-53	4.9*	2.1	15.3	0.5	52	22	–	–	–	–
BCg	53-63	4.1*	5.6	13.6	0.6	51	21	69.0	5.1	17.7	1.1
Cf	63-73	4.4*	3.9	11.8	1.7	51	23	69.2	5.1	18.3	1.2
Pit 15-1958. Typical Pale Soil (Eutri-Turbic Cryosol) (see Table 2.6.4)											
A	2-10	4.3	7.1	6.0	5.0	33	10	71.1	5.3	13.1	0.3
B1w	10-30	6.0	1.3	8.0	0.1	31	7	73.3	5.2	15.6	0.2
B2w	30-75	6.1	0.9	7.7	0.0	28	5	74.6	4.7	14.2	0.2
BCf	75-85	6.3	–	7.9	0.0	30	8	73.6	4.7	15.1	0.2
Pit 24-1968. Pale Calcareous Soil (Calcari-Leptic Cambisols) (see Table 2.6.5)											
AOa	4-10	6.8	48.8	17.5	No	52	37	54.7	5.2	16.8	–
ABw	10-13	7.1	9.0	63.1	–	49	28	67.4	6.2	17.8	2.2
Bw	13-19	7.8	3.4	–	–	47	25	67.8	6.2	17.4	1.8
BCwk	1927	8.0	2.2	–	–	41	18	66.6	6.3	19.1	1.7
Ck	27-60	8.3	1.1	–	–	25	16	61.3	6.1	15.0	–
Rk	60-65	–	–	–	–	–	–	6.7	2.3	3.0	0.5
Pit 88-1960. Deep-Gleyic Podzol (Gleyi-Turbic Cryosols (Spodic)) (see Table 2.6.6)											
AOa	5-9	3.8	27.6	6.3	28.3	–	–	75.1	2.4	15.0	1.1
E	9-12	4.5	4.5	2.0	4.9	31	11	73.5	3.2	15.0	1.2
B1s	12-22	4.8	2.4	1.1	2.4	20	5	67.7	6.4	17.4	1.6
B2g	22-45	5.2	0.5	5.1	2.1	39	16	69.5	4.9	18.4	0.5
BCfg	45-70	5.3	0.3	10.8	1.1	41	24	68.3	5.9	17.2	0.5
Cf	70-85	5.1	0.5	12.8	0.3	41	21	67.5	6.2	17.6	0.5
Pit 72-1959. Typical Podbur (Entic Podzol)											
Oi	0-12	3.1	75.3	24.0	44.0	–	–	–	–	–	0.1
AOE	12-20	3.3	18.4	2.4	16.9	42	14	65.8	8.6	15.6	0.5
B1s	20-26	4.5	12.4	1.4	2.9	23	3	53.3	10.3	28.5	1.3
B2s	26-38	4.5	7.4	1.2	1.4	29	3	55.7	10.7	25.6	0.3
BC	38-50	4.8	2.3	1.9	0.2	14	0	55.1	10.3	27.5	0.5
C	50-85	–	–	–	–	–	–	55.3	10.5	24.2	–

* pH_{KCl}

** Oxalate-extractable Fe

*** not determined

Table 2.6.2. Genetic diagnostics of the main soil types of the North.

Soil: Genetic horizons	Processes and their morphological indices																			
	Organic matter accumulation		Eluvial			Illuvial (accumulation in B horizon)				Metamorphic						Cryogenic				
	peat, Oi	peaty litter, Oi/Oe	muck, Oa/AO	humus, A1	acidic hydrolysis, leaching	eluvial-gley bleaching	lessivage	amorphous Al-Fe-humus subst.	carbonates	silt, clay	amorphization	rubification	argillization	gleyzation	coagulative structuring	frost shattering and cracking	frost heave and cryoturbation	cryoseparation of coarse fractions	cementing by ice	platy structure
Podzols: O–E–Bhs–C	+	++	–/+	–	+++	–/+	+/–	+++	–	+/–	+++	–	++	–	+	+/–	–	+	–	–
Parapodzols (Podburs): O–Bhs–C	+	++	–/+	–	+++	–	+/–	+++	–	+/–	+++	–	++	–	+	+/–	–	+	–	–
Soddy Al-Fe-humus: O(A1)–Bhs–C	+	+++	+	+	++	–/+	+/–	+++	–	+/–	+++	–	++	+/–	–	–	–	+	–	–
Podbels: O–E–IIBt–C(f)	+	+	–/+	–	++	+++	–/+	++	–	+/–	+	–	+/–	+/–	+	–	–	–/+	–	++
Gley Podbels: O–Eg–IIBtg–C(g,f)	++	+	–/+	–	+	+++	–/+	+/–	–	+/–	++	–	+/–	+++	+	–	–	–	–	++
Granuzems: O(AO)–Bhsw–Bw–Cf	+	++	+	++	+	–	–	++	–	–	+++	–	++	–/+	+++	+/–	+/–	–	+	+
Pale unsaturated: O(AO)–Bw–C(f)	+/–	++	+	+/–	+/–	–	–/+	+/–	–	+/–	+/–	++	+/–	–/+	++	++	+/–	++	–/+	+/–
Pale typical: O(A1)–Bw–C(f)	–	+	+/–	+	–	–	–	–	+	–	–/+	+++	–/+	–	++	++	–	+/–	–/+	–/+
Pale-Cryozemic: AO–Bw–Bthix–Cf	+/–	++	+	+/–	–	–	–	–	+/–	–	+/–	++	–/+	–	++	++	+	+	++	+/–
Soddy: (O)–A1–AC–C	–	+/–	+	+++	+	–	–	+/–	–	–	++	–	+/–	+/–	++	+/–	–	+/–	–/+	–
Soddy-calcareous: A1–A1k–Bk–Ck	–	+/–	++	+++	–/+	–	–	–	+	–	–	–	–/+	+/–	+++	+/–	–	+/–	–/+	–
Rendzinas: AO–A1k–C(R)k	–	+/–	++	++	–/+	–	–/+	–	+	–/+	–	–	+/–	–	++	+/–	–	+/–	–	–

(continues)

Table 2.6.2. (Cont.)

Soil: Genetic horizons	Processes and their morphological indices																					
	Organic matter accumulation		Eluvial			Illuvial (accumulation in B horizon)				Metamorphic						Cryogenic				
	peat, Oi	peaty litter, Oi/Oe	muck, Oa/AO	humus, A1	acidic hydrolysis, leaching	eluvial-gley bleaching	lessivage	amorphous Al-Fe-humus subst.	carbonates	silt, clay	amorphization	rubification	argillization	gleyzation	coagulative structuring	frost shattering and cracking	frost heave and cryoturbation	cryoseparation of coarse fractions	cementing by ice	platy structure
Steppe cryoarid: O–A1–Bk–C(k)	–	+/–	+	++	–	–	–	–	++	–	–	+/–	–/+	–	+++	++	–	+/–	–	–
Soddy Gley: O–A1g–Bg–C(G)	+/–	+/–	+	+++	+/–	+/–	–	–	–	–	+	–	+/–	+++	++	–	–	–	+/–	–/+
Peaty Gley: Oi–BCg(G)–Cgf	+++	+/–	+	–	+/–	+/–	–	–	–	–	+/–	–	–	+++	+	–	–	–	+/–	–
Peaty Cryozems`: Oi–OB–Cf	+++	–	+	–	–	–	–	–	–	–	+/–	–	–	–/+	–	–	++	–	+++	–
Thixot. Cryozems: Oi(AO)–Bthix–Cf	+	++	++	+/–	–/+	–	–	–	–	–	–	+/–	+/–	–/+	+	+++	+++	+/–	+++	–
Surface-saline: Czm–C(z,f)	–	–	–	–	–	–	–	–	+	–	–	+/–	–	–	+/–	++	–	–/+	+	+/–

Note: The degree of manifestation of soil features – (+++) very distinct, (++) distinct, (+) indistinct, (–) not manifested.

Table 2.6.3. A typical cryogleyzem (Gleyi-Turbi Cryosol) on the right bank of the Malyi Anyui River, within a slightly undulating plain (30 m a.s.l), described on August 7, 1967, as Pit 112-1967. It was located within a taiga (larch) stand underlain by a moss and lichen ground cover, and with small shrubs (*Betula middendorffii*, *Vaccinium uliginosum*, *Ledum palustre*). The site is within a distinct hummocky microtopography.

Horizon*	Depth	Description
O	0-2 cm	Live lichen, needles.
Oi	2-4 cm	Slightly decomposed peat, dry, with insignificant (<10%) admixture of fine earth, clear transition.
AOa	4-7 cm	Dark brown dry loam mixed with highly decomposed plant residues, sharp transition.
B1wjjg	7-21 cm	Brown, with redox depletions, clayey, wet, cryoturbated, with loose granular foliated structure, slightly thixotropic, +5°C (20 cm).
B1B2 wjjg	21-35 cm	Darker and more compact, with gley features, clayey, with humus tongues, clear boundary.
B2wjjg	35-53 cm	Dark; compact; clayey; with foliated structure; humus streaks along the fissures; clear boundary.
BCjjg	53-63 cm	Dove-black; strongly gleyed; clayey; with inclusions of black peaty mass due to cryoturbation; with somc thixotropy.
Cf	63-73 cm	Frozen horizon with high ice content; ice lenses; -0.4°C (70 cm).

* Horizon designation follows Soil Survey Staff, 1999.

Table 2.6.4. The profile of a typical Pale soil (Eutri-Turbic Cryosol), as described from Pit 15-1958. The site is approximately 900 m asl on the southwestern slope (6°) of a ridge that serves as a divide between the Agayakan and Kuidusun Rivers (Verkhoyansk district). The relative height above the bottom of the Oimyakon Depression is about 200 m. The slope is covered by a larch stand with a shrub layer of dwarf birch, willow, and Pinus pumila. Red bilberry, moss (in microdepressions), and lichens (on small hummocks) compose the ground cover. The diameter of hummocks is about 0.5 to 1.0 m; their relative elevation is about 20 cm. The profile characterizes the soil of a hummock.

Horizon*	Depth	Description
Oi	0-2 cm	Dry dark brown litter of slightly decomposed lichens, needles, and leaves.
A	2-10 cm	Grayish pale, loamy, with a weak "curdled" structure; abundant roots, slightly dry; inclusions of fine channers (up to 20%); distinct transition.
B1w	10-30 cm	Grayish light pale, loamy, rather compact, slightly moist, few fine roots, up to 50% of channers, distinct transition.
B2w	30-75 cm	Brownish pale, loamy, wet to satiated wet, no roots, indistinct "curdled" structure, up to 70% of channers, underlain by permafrost.
BCf	>75 cm	Frozen loam filling the spaces between gravel and channers; fine earth contains ice crystals.

* Horizon designation follows Soil Survey Staff, 1999.

Table 2.6.5. Soil profile of a Pale Calcareous Soil as described from Pit 24-1968 on July 21, 1968, on the Yukagir Plateau. The site is on a steep (12°) southwestern slope at 200 m asl. The slope is covered by a sparse larch stand with dwarf willow and birch. Low shrubs are represented by bilberry. Lichens, green mosses, and grasses with some participation of horsetail compose the ground cover. The microrelief is not expressed. The soil develops from the pebbly residuum of Devonian limestone.

Horizon*	**Depth**	**Description**
Oi	0-4 cm	Litter layer.
AOa	4-10 cm	Mucky-humus horizon, black, well-decomposed organic remains constitute up to 50% of the volume, silty loamy fine earth, densely penetrated by roots, inclusions of fine limestone pebble, charcoal particles, gradual transition.
ABw	10-13 cm	Grayish dark brown, loamy, with granular structure, penetrated by roots, limestone pebble, no effervescence in the fine earth, sharp transition.
Bw	13-19 cm	Dark brown loam with granular-blocky structure, abundant roots, considerable amounts of limestone pebble, distinct transition.
BCwk	19-27 cm	Brown loam, slightly wet, up to 50% of limestone pebble, distinct effervescence of both pebble and fine earth, gradual transition.
Ck	27-60 cm	Brownish gray structureless loam admixed to limestone debris (70 to 80% of the mass of horizon).
Rk	>60 cm	Limestone debris. Permafrost is not found.

* Horizon designation follows Soil Survey Staff, 1999.

Table 2.6.6. A profile of a Deep-Gleyic Podzol, described in pit 88-1960, on a terrace of the Ola River on the northern coast of the Sea of Okhotsk under a larch stand with lichen-mossy ground cover. The soil surface has a hummocky topography.

Horizon*	**Depth**	**Description**
Oi	0-5 cm	Slightly decomposed peaty litter.
AOa	5-9 cm	Humus-mucky horizon, black-brown, wet, loamy, with indistinct platy structure and weak features of cryoturbation, distinct transition.
E	9-12 cm	Whitish loam with weak features of cryoturbation and gleyzation, loamy, with uneven boundary.
B1s	12-22 cm	Yellowish brown loam with admixture of gravel, slightly wet, with indistinct gleyic features and clear indices of cryoturbation, gradual transition.
B2g	22-45 cm	Transitional loamy horizon with increased content of gravel and distinct gley features in the form of ocherous and olive mottles.
BCfg	45-70 cm	Frozen brownish loam with considerable admixture of gravel material, rather compact, evident gleyic features, low-ice permafrost.
Cf	70–85 cm	Gravelly, with admixture of loamy fine earth and weak indices of gleyzation.

*Horizon designation follows Soil Survey Staff, 1999.

5. Major Soil Groups

5.1. Cryozems

Pedological literature introduced the term *Cryozem* in 1973 and suggested that Cryozems be defined as autonomous hydromorphic soils with icy permafrost and different degrees of gleyzation (Naumov, 1973). Later, this term was applied to autonomous hydromorphic soils without any features of gleyzation. Two soil types were distinguished in the group of Cryozems: Thixotropic Cryozems and Peaty (Homogeneous) Cryozems (Sokolov, 1980).

The subdivision of Cryozems at the type level is based on the proportion of organic and mineral parts of the soil profile. Thixotropic Cryozems *(Oxyaqui-Turbic Cryosols*) are the soils with relatively thin organic horizons underlain by cryoturbated mineral horizons with permanently high water content and thixotropic properties, but without gley features. The permafrost table is at a depth of less than 1.0 m from the soil surface. Peaty Cryozems *(Turbi-Histic Cryosols (Oxyaquic))* have a thick (20 to 35 cm) organic layer underlain by an organo-mineral horizon containing less than 30% (by weight) organic matter. Ice-cemented permafrost is typically at a depth of 30 to 50 cm.

Thixotropic Cryozems are widespread in taiga and tundra zones within leveled interfluves, gentle slopes, and terrace-like surfaces. They develop on clayey to loamy materials with low coarse fragment content and impeded internal drainage. The hummocky or polygonal-trough cryogenic microtopography is very typical of the areas with Thixotropic Cryozems. Patterned ground features are very distinct and dictate the complexity of the soil cover. The soils of hummocks occupy about 75 to 90% of the area and the soils of troughs, from 10 to 25% (Figure 2.6.1a). Cryozems are allocated to hummocks. In Russian taxonomic systems, the soils of the troughs between hummocks are distinguished separately. Thus, the combination of a soil on a hummock (Thixotropic Cryozem) and a soil in the surrounding troughs (usually, peaty soils with shallow permafrost) forms the cryogenic soil microcomplex (Figure 2.6.1b).

At pit 1-1967, the slope is covered by an open stand of larch with small shrubs and moss–lichen cover (Tables 2.6.1 and 2.6.2). Relatively elevated solifluctional stripes along the slope cover about 80% of the area and alternate with shallow hollows of meadow-wetland vegetation. Elevated land stripes (sorted or non-sorted striped patterned ground) are complicated by a hummocky microtopography that dictates the differentiation of soils and vegetation at the microlevel. Lichens predominate on hummocks, whereas mosses cover surrounding troughs. In the troughs around the hummock, the thickness of O horizons can reach >20 cm. Frozen, brown, clayey material with admixture of pebble and gravel underlies them.

The profile of a typical Thixotropic Cryozem consists of the following horizons: Oi-AO-(A)-Bwjj-Cf. The organic layer is thin (about 10 cm) and consists of peaty or mucky-peaty material. Sometimes, the soils contain a humus horizon. The mineral mass except for the A horizon is structureless, is water-saturated during the growing season, and displays thixotropic properties. Iron concentrations, as well as redox depletions, are not typical of Thixotropic Cryozems. However, some features of gleyzation occur in the intergrades between Thixotropic Cryozems and Gleyzems-Gleyic Thixotropic Cryozems *(Stagni-Turbic Cryosols*).

Ice-cemented permafrost occurs at a shallow depth (usually, about 50 to 80 cm from the soil surface). The deepest active layer location is typically at the central part of hummocks. Usually, Cryozems display evident features of cryoturbation. The fragments of organic horizons occur in the mineral mass of Cryozems.

There is no differentiation of mineral horizons by texture and bulk chemical composition. Soil reaction is acid in organic horizons and slightly acid to neutral in mineral horizons. Cryogenic involutions induce gradual mixing of the mineral mass with organic material; often, the fragments of plant tissues and humic mottles occur in deep horizons, just above the permafrost table, or even inside it. A diffuse distribution of finely dispersed charcoal particles is typical of mineral horizons of Thixotropic Cryozems (Naumov and Gradusov, 1974).

The degree of decomposition of organic remains in the organic layer and in the upper mineral horizon increases in Cryozems developing in extremely continental areas from relatively rich parent materials (loess-like loams). The humus content in the upper mineral horizon of such soils may reach 5 to 6% and more.

Peaty Cryozems consist of a thick Histic horizon (up to 30 to 35 cm) underlain by the mixture of organic and mineral materials with permafrost at a depth of about 50 cm. These soils are common in relatively humid cold continental regions, including the northeastern margins of Eurasia.

5.2. Gleyzems (Cryogleyzems)

A high degree of hydromorphism and the presence of gley features in the soil profile, in combination with cryoturbation, are the major characteristic features of Cryogleyzems. These soils develop both in humid coastal regions and in semi-arid interior regions of northeastern Eurasia. Usually, they occupy nearly level and concave parts of slopes, developing from loamy materials with high water retention capacity, icy underlying permafrost, and impeded drainage.

Typical Cryogleyzems *(Gleyi-Turbic Cryosols)* consist of a relatively thin organic surface horizon underlain by a cryoturbated mineral horizon with gley features. The distribution of these soils is much wider than that of Cryozems. Cryogleyzems occur in all parts of northeastern Asia with some predominance in humid coastal regions. The character and thickness of the organic surface horizon may vary, depending on local conditions; usually, this layer has a mucky or mucky-peaty character.

Cryoturbation is better pronounced in tundra and sparse forest areas with thin snow cover and deep soil freezing. The soils developed on river terraces under relatively dense forest stands are less susceptible to cryoturbation, which is connected with deeper snow cover and less active soil freezing. Typical Cryogleyzems are also associated with a hummocky cryocomplex. In the trough zone, the thickness of the surface organic layer is about 20 cm underlain by a saturated mineral horizon and a permafrost table at 25 cm deep.

The profile of Typical Cryogleyzems consists of the Oi-Bwjjg-Cf horizons (Table 2.6.3; see Pit 112 on Table 2.6.1). Often, a humus horizon is present under the surface peaty layer. The mineral mass of these soils is rather compact; after drying, it becomes indurated. Often, it displays thixotropic properties in lower horizons. Iron segregations and redox depletions occur throughout the solum.

The soil profile is underlain by ice-rich permafrost and is saturated with water, especially immediately above the permafrost table. No eluvial-illuvial differentiation features appear. Cryoturbation leads to mixing of the soil profile.

Physiochemical properties of Typical Cryogleyzems vary within a wide range of values. Organic horizons are strongly acid and have low base saturation. Soil pH in mineral horizons varies from strongly acid to neutral and even alkaline values, depending on the composition of parent rocks.

In the coastal zone of the Sea of Okhotsk, some Cryogleyzems display distinct differentiation of the profile; they are distinguished as Tundra eluvial-gley soils *(Gleyi-Turbic Cryosols (Albic)).* These soils are associated with gravelly morainal deposits developed in hummocky cryocomplexes with Typical Cryogleyzems and Peaty Soils of cryogenic troughs (Naumov et al., 1990).

Peaty Cryogleyzems *(Turbi-Histic Cryosols (Gleyic))* have a deep organic layer (<40 cm) underlain by a cryoturbated mineral horizon with gley features. These soils occur in complexes with Typical Cryogleyzems and Peaty Cryogenic soils.

5.3. Peat and Peaty Gley permafrost-affected soils and their cryogenic complexes

This group of soils encompasses a wide spectrum of hydromorphic soils developing in conditions of permanent saturation and the presence of ice-rich permafrost that serves as an aquiclude. These soils develop in all natural zones of northeastern Eurasia (northern taiga, forest-tundra, and tundra) under moss, sedge, and sedge-moss associations. They are common in the depressions but also occur on the interfluves, river terraces, flat-topped ridges, and footslopes, i.e., in all geomorphic positions with impeded drainage. Peat soils with shallow permafrost can develop even on steep northern slopes with ice-cemented stony rocks.

Peat soils are subdivided into oligotrophic, transitional, and eutrophic soils differing by the degree of decomposition of plant remains and chemical composition of feeding water. A detailed analysis makes it possible to reveal some differences in the properties of peat soils developed in different zones in mountainous and plain landscapes. The thickness of organic horizon varies from 40 to 80 cm and more. The depth of active layer ranges from 30 to 50 cm. Oligotrophic Peat Soils *(Fibri-Cryic Histosols)* – peaty soils of highmoor bogs – usually have acid reaction; Eutrophic Peat Soils *(Sapri-Cryic Histosols)* – peaty soils of lowmoor bogs – have a slightly acid or neutral reaction. Peat soils occur in complexes with other types of permafrost-affected seasonally freezing soils *(Fluvisols, Oxyaquic and Gleyic Cryosols*, etc.), which the soil maps reflect (Naumov, 1993).

5.4. Metamorphic Pale Soils

Pale (Palevye) soils develop in continental regions of northeastern Eurasia with semi-arid climate (the evapotranspiration exceeds precipitation). Insufficient precipitation retards the illuviation of Al-Fe-humus compounds; at the same time, soil leaching is sufficient to prevent the development of Ca-illuvial horizon. Thus, Pale Metamorphic soils *(Turbic Cryosols, Gelic Cambisols*) are considered the transitional group between Al-Fe-humus soils – Podzols *(Spodic Cryosols)* and Ca-illuvial soils – Cryoarid steppe soils *(Calci-Mollic Leptosols).*

Pale metamorphic soils develop from loamy and pebbly-loamy silicate-rich parent materials under northern taiga and tundra vegetation in relatively drained sites. There are two subgroups: Typical Pale Soils *(Eutri-Turbic Cryosols, Eutri-Gelic Cambisols)* with neutral reaction and high base saturation and Unsaturated Pale Soils *(Turbi-Leptic Cryosols [Dystric], Lepti-Gelic Cambisols [Dystric])* that were previously described under the names of Podburs and Pale Podburs (Naumov, 1980). Typical Pale soils have a thin organic horizon, a humus or mucky-humus horizon, and a metamorphic Bw horizon. The latter horizon bears the features of *in situ* soil weathering with the release of iron from crystal lattices of minerals; autochthonous films of iron oxides cover the soil particles. Typical Pale soils tend to predominate in more arid conditions, whereas Unsaturated Pale soils are widespread in semi-arid regions.

The absence of distinct horizon development and relatively monotonous brownish pale color of the soil mass characterize Typical Pale soils (Table 2.6.4). Active cryoturbation is not typical of them. Cryogenic phenomena show in cryogenic cracking, freezing out (sorting) of coarse particles and stony material, the development of platy (cryoplaty) soil structure, and the development of patterned ground.

Pale soils developed from calcareous parent materials may have carbonates in mineral horizons. These are Pale Calcareous Soils (*Calcari-Leptic Cambisols*) (Table 2.6.5).

Pale soils are the modal soils in interior regions of northeastern Eurasia.

5.5. Al-Fe-Humus Soils

Al-Fe-humus soils – Podzols and Podburs *(Podzols or Spodic Cryosols*) – are common in humid regions of northeastern Eurasia. Most of them are developed in the areas with deep permafrost or without permafrost at all. However, Deep Gleyic Podzols *(Gleyi-Turbic Cryosols (Spodic*)) are Podzols that have gley features in their lower horizon, which is underlain by an ice-rich frozen layer (either permafrost or persistant seasonal frost). Table 2.6.6 gives a profile description for this soil.

Similar soils, but without the E horizon, are Podburs *(Entic Podzols*) (pit 72-1959, Table 2.6.1). These soils develop under larch taiga with moss-lichen ground cover and form associations with Typical Podzols *(Haplic Podzols)* and Peat soils *(Cryic and Gelic Histosols).* They differ from Typical Podzols in their finer texture of bottom soil horizons, where water stagnation and the development of redoximorphic features occur. Cryoturbation phenomena are not very pronounced. In coastal areas, a thick peaty layer (Folic) horizon on the mineral surface of Podzols and Podburs develops. These soils are Dry-Peaty Podzols and Podburs *(Histi-Gelic Podzols, Histic Podzols).*

5.6. Fluvisols

Cryogenesis does not strongly affect Fluvisols, mineral soils developing on flood plains. Their properties (except for the temperature regime) do not differ much from the properties of Fluvisols developing beyond the permafrost zone. The permafrost table, if present, is at a depth of more than 1 m. However, the ice content in the permafrost may be considerable. Fluvisols developing from gravelly, loamy material display the features of cryogenic sorting of rock fragments. Cryoturbation is more pronounced on high flood plains not subjected to annual floods. In this case, the soil cover of flood plains is cryo-complexes of Soddy Gley Meadow Fluvisols *(Geli-Gleyic Fluvisols)* and Peaty Gley Fluvisols *(Gleyi-Histic Fluvisols [Gelic])* of wetland areas.

6. Soil Zones and Soil Distribution

Figure 2.6.2, which is a scheme of soil-geographic regions, shows the main geographic regularities of the soil landscape pattern of northeastern Eurasia. The taxonomic system of soil regionalization includes soil belts, regions, zones and subzones (on plains), provinces, and districts, distinguished on the basis of components of the soil landscape connected with particular agents of soil formation.

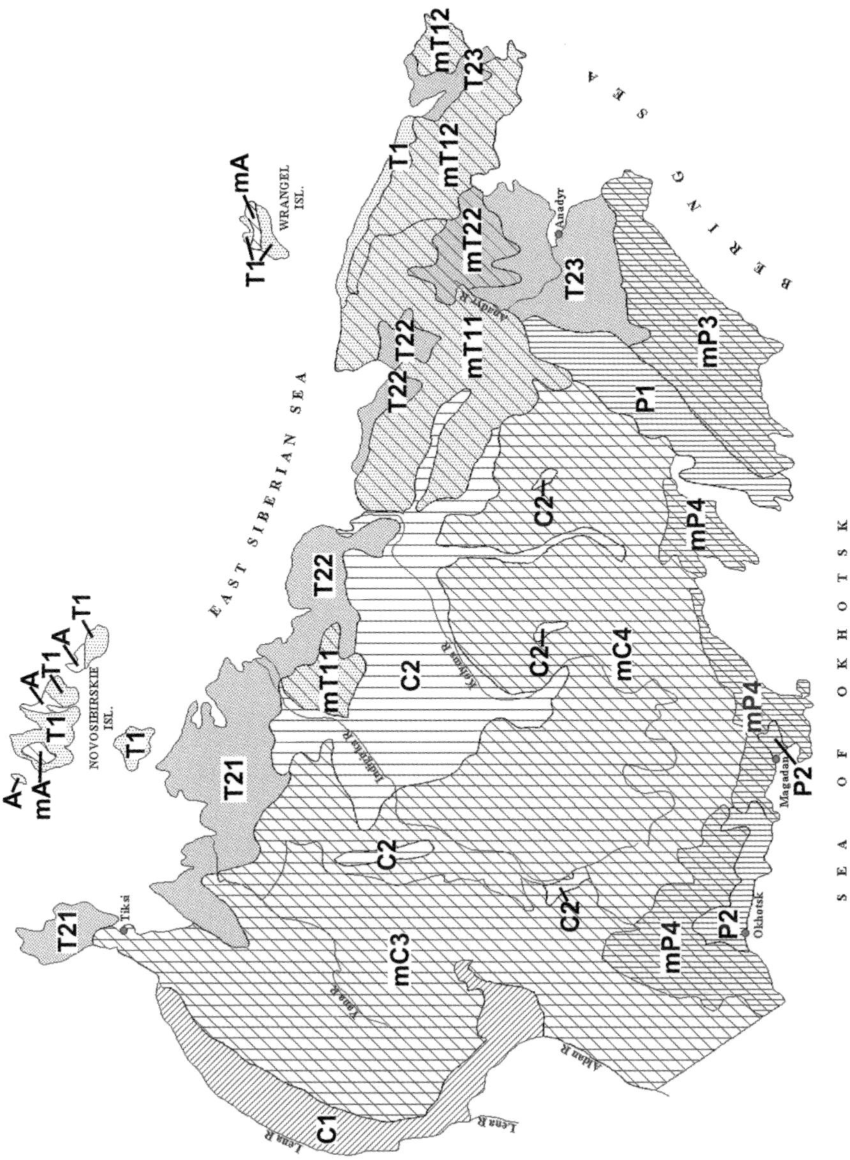

Figure 2.6.2. Map of northeastern Eurasia showing the major soil-geographic regions. The dark line separates the Polar belt (I) from the Boreal belt (II). The Polar belt is subdivided into the Arctic soil (A) and Tundra soil (T) zones. The Boreal belt is subdivided into the continental (C) and Pacific (P) zones. The prefix 'm' modifies Zone designations if the area is mountainous. Soil development, which this chapter describes, determines additional subdivisions.

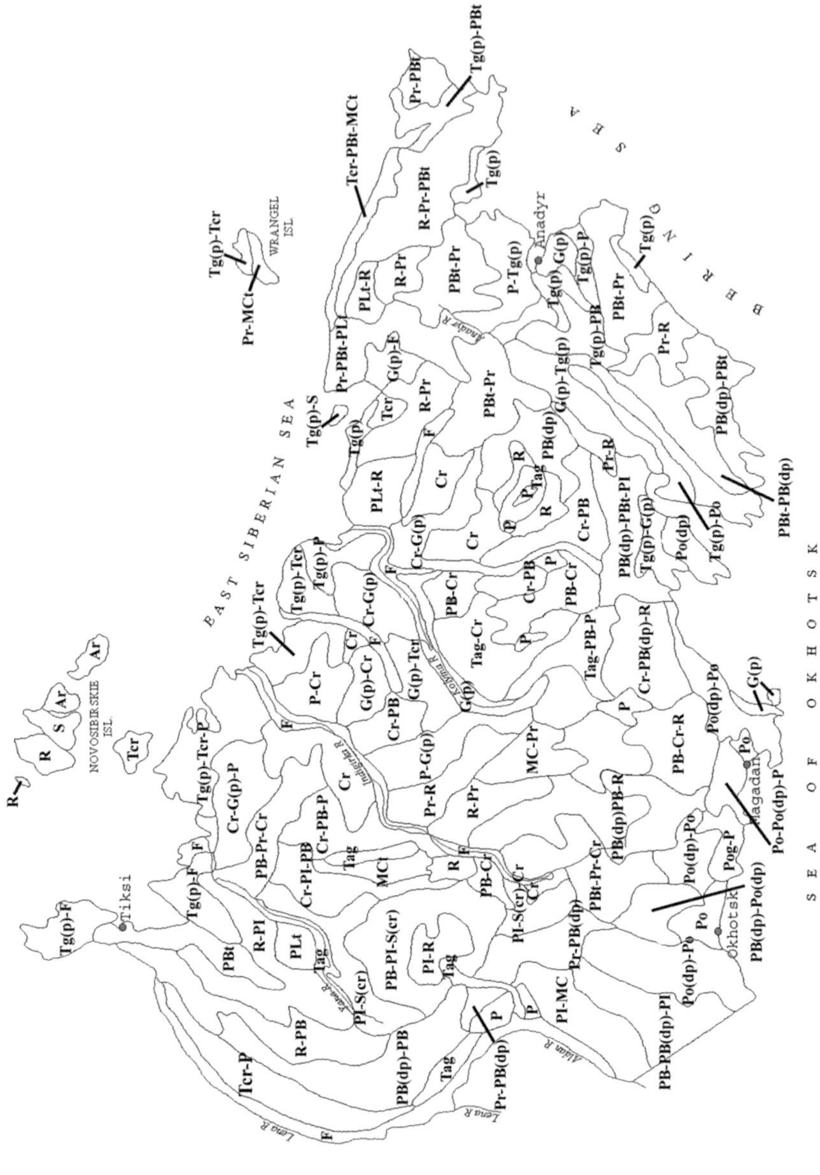

Figure 2.6.3. Soil map of Northeastern Eurasia. (See "Legend to Figure 2.6.3, next page.)

Legend to the map in Figure 2.6.3.

Index	Soils
Polygonal–trough and circle-shaped (spotty) cryocomplexes of arctic and tundra soils with shallow and ice-rich permafrost	
Ar	Arctic and Arctic-Gley Soils, Soils of Cryogenic Troughs, and Stony Polygons (Gleyi-Turbic Cryosols)
Tg^{p}	Peaty and Peat Tundra-Gley Soils (Turbi-Histic Cryosols(Gleyic))
Tcr	Tundra Cryozems (including gleyic varieties) (Oxyaqui-Turbic Cryosols)
Polygonal–trough and circle-shaped (spotty) cryocomplexes of tundra soils with deep and low-ice permafrost	
PBt	Tundra Podburs (Spodic Cryosols, Gelic Podzols)
PLt	Tundra Pale Soils (Gelic Cambisols, Turbic Cryosols)
MCt	Tundra Mucky-Calcareous Soils (Rendzinas, Soils of Circles, and Soils of Troughs) (Calcari-Turbic Cryosols, Lepti-Gelic Cambisols (Calcaric))
Taiga Soils (permafrost-affected, with dry permafrost, and without permafrost)	
PB	Taiga Podburs (Gelic Podzols, Entic Podzols)
PB^{dp}	Dry-Peaty Podburs (Lepti-Histic Podzols)
MC	Mucky-Calcareous Soils (Pale Rendzinas) (Calcari-Leptic Cambisols)
Po^{dp}	Dry-Peaty Humus-Illuvial Podzols (Histic Podzols)
Po	Podzols (including Ocherous, Humus-Illuvial, and Humus-Rich Varieties) (Podzols)
Pog	Peaty-Gley Podzols (predominantly, humus-illuvial podzols with gley above permafrost) (Gleyi-Gelic Podzols, Spodi-Gleyic Cryosols)
Pl	Pale Soils (including Mucky, Raw-Humus, Cryoturbated, and Calcareous varieties) (Gelic Cambisols, Calcaric Cambisols, Turbic and Haplic Cryosols)
Tag	Taiga Gley and Gleyic Humus/Peaty/Mucky Soils and their Cryocomplexes (Gelic Gleysols and Gleyi-Turbic or Histic Cryosols)
Cr	Cryozems (including Peaty-Mucky and Pale-Cryozems) and their Cryocomplexes (Oxyaquic Cryosols)
S^{cr}	Steppe Cryoarid Soils (Calci-Mollic Leptosols)
Bog and Flood-Plain Soils and Their Cryocomplexes	
P	Peat Soils (Cryic Histosols, Gelic Histosols)
G^{p}	Peat and Peaty Gley Soils of Bogs with Shallow Ice-Rich Permafrost (Cyric Histosols, Turbi-Histic Cryosols (Gleyic))
F	Flood-Plain Soils (including Soddy, Soddy-Gley and Swampy varieties) (Fluvisols, Gelic Fluvisols)
Miscellaneous	
Pr	Primitive Stony Soils (Lithic Leptosols)
R	Rock Outcrops
S	Sands

Climatic factors (temperature and the degree of moistening) dictate the separation of soil belts and regions. The vegetation cover (tundra, taiga, etc.) determines the separation of soil zones and subzones. Soil provinces on plains depend on the degree of climatic moistening and local lithological conditions; in mountains, on the vertical sequence of altitudinal soil zones. Soil districts (not shown on the map) reflect local geomorphic conditions.

The territory of northeastern Eurasia belongs to two soil-geographic belts (Figure 2.6.2): the Polar belt ($\sum T° > 10°$ is <600°C) and the Boreal belt ($\sum T° > 10°$ is 600 to 1400°C).

The Polar belt (I) (Figure 2.6.2) includes the zones of Arctic (A) soils and Tundra (T) soils. The Tundra zone includes the northern subzone (T1) with Cryozems, Cryoturbated Gleyic soils, and Podburs *(Oxyaqui- and Gleyi-Turbic Cryosols and Spodic Cryosols)* and the southern continental subzone (T2) with Cryozems, Cryoturbated Gleyic soils, and Pale soils *(Oxyaqui- or Gleyi-Turbic Cryosols, Turbic Cryosols).*

Areas of plains with typical Arctic soils under Arctic tundra and subpolar desert vegetation are mainly to the north of the continent, within the Novosibirskie Islands and the Wrangel Island (see the chapter on the soil landscapes of the Russian Arctic Archipelagos). Arctic Tundra Gleyic soils *(Gleyi-Turbic Cryosols)* in combination with Peat Tundra Gley soils *(Turbi-Histic Cryosols [Gleyic])*, Mucky-Calcareous Tundra soils *(Calcari-Leptic Cryosols)*, Peat soils (*Cryic Histosols*), and Tundra Podburs *(Spodi-Leptic Cryosols*), occur on a relatively small area of plains in the northern portion of the Chukotka Peninsula.

Mountainous areas with Arctic tundra vegetation occur in the Chukotka Peninsula (mT1) (Figure 2.6.2). In its western part, the soil landscape includes Primitive Stony soils *(Lithic Leptosols*) and Pale soils (*Leptic Cryosols*) alternating with rock outcrops ($mT1_1$). In the eastern part with more humid climate, Podburs and Dry-Peaty Podburs *(Histi-Gelic Podzols*) substitute for Pale soils ($mT2_2$). In places with calcareous bedrock, Mucky-Calcareous Tundra soils (Calcari-Gelic Leptosols) occur.

The main areas of tundra plains are on the vast coastal lowlands in the north. This is the area of cryogenic hummocky and polygonal-trough complexes of Peaty Tundra-Gley soils *(Turbi-Histic Cryosols (Gleyic))* (centers of hummocks) and Peat soils *(Turbi-Histic Cryosols*) (interhummock troughs). Complexes of Thixotropic Cryozems (*Oxyaqui-Turbi Cryosols)* and Peat soils occupy better drained sites.

In the western part between the Lena and Indigirka Rivers ($T2_1$) (Figure 2.6.2), the share of Cryozems in the soil landscape is less than in the eastern part of this zone ($T2_2$). Peat and Peat Tundra-Gley soils predominate in the Anadyr Lowland ($T2_3$). Isolated areas of Tundra-Gley soils *(Gelic Gleysols)* occur in the Penzhina River valley and along a narrow strip of coastal lowlands as far to the south as 59°N.

The Boreal belt (II) (Figure 2.6.2) includes two regions: (a) the East Siberian continental region (C) with permafrost-affected northern taiga soils – Pale soils *(Turbic Cryosols* and *Gelic Cambisols*), Podburs *(Enti-Gelic Podzols*), Steppe Cryoarid soils *(Calci-Mollic Leptosols*), Cryozems *(Oxyaqui-Turbic Cryosols)*, and Taiga Gley soils *(Gleyi-Turbic Cryosols)* and (b) the Pacific (P) coastal region of Podzols *(Podzols*), Podburs *(Entic Podzols*), and Eluvial-Gley soils *(Gleyi-Turbic Cryosols [Albic]*).

The continental region consists of a single zone of taiga soils (in the mountains is an altitudinal zonality with predominant areas under sparse thin forests and tundra vegetation). Plains characterize two provinces (Figure 2.6.2): C1, the Lena-

Aldan province (to the west of Verkhoyansk Ridge) of Taiga Gley soils *(Gleyi-Turbic Cryosols)*, Cryozems *(Qxyaqui-Turbic Cryosols*), Peat soils *(Cryic Histosols*), and Pale soils *(Turbic or Haplic Cryosols);* and C2, the Indigirka-Kolyma province of Cryozems, Taiga Gley soils, Podburs, and Peat soils. The latter province includes several districts within the Indigirka-Kolyma interfluve, separated by mountainous territories.

In turn, the mountainous territories include two vast provinces (Figure 2.6.3): mC3, the western province occupying Verkhoyansk Suntar-Khayata Ridges with Pale soils *(Turbic or Haplic Cryosols)*, Podburs *(Enti-Gelic Podzols)*, Cryoarid Steppe soils *(Calci-Mollic Leptosols*), Cryozems *(Oxyaqui-Turbic Cryosols)*, Primitive Stony soils *(Lithic Leptosols*), and rock outcrops; and mC4, the eastern province (Cherskiy, Yukarir, and Kolyma ridges) with the predominance of Podburs, Pale Soils, Cryozems, Primitive Stony soils and rock outcrops.

Within the Pacific region (Figure 2.6.3), Tundra eluvial-gley soils *(Gleyi-Turbic Cryosols [Albic]*) and Humus-Illuvial Podzols *(Haplic Podzols*) and Podburs *(Entic Podzols*) under tundra and northern taiga vegetation (P1) occupy the plains in the Penzhina River basin; Humus-Illuvial Podzols and Podburs, with minor participation of Gley soils (Gleysols,) cover the coastal plains near Magadan (P2), which have a coarser texture of rocks. In the mountainous territories of the Koryak Ridge (mP3) are the areas of Eluvial-Gley Taiga and Tundra soils *(Gleyi-Turbic Cryosols [Albic])*, Peaty Gley soils *(Turbi-Histic Cryosols [Gleyic*] or *Geli-Histic Gleysols*), and Primitive Stony soils *(Lithic Leptosols).* The southern (Magadan) mountainous province (mP4) is the area of Humus-Illuvial Podzols *(Haplic Podzols*) and Podburs *(Entic Podzols)* (including Dry Peaty variants), and Primitive Stony soils *(Lithic Leptosols*).

In general, the portion of permafrost-affected soils in the Pacific coastal region is low, except for its northern part. At the same time, the influence of cryogenesis on the pattern of soil development is evident everywhere, in the complexity of microtopography with distinct patterned ground features. Polygonal-trough and hummocky nonsorted patterns predominate in the areas with a thick mantle of loose loamy and clayey parent materials; sorted nets and polygons develop on stony substrates.

Aspect also complicates altitudinal zones in continental regions. In particular, the areas with steppe vegetation (with Steppe and Tundra-Steppe Cryoarid soils) are widespread on southern slopes, whereas taiga forests with Pale soils and Podburs cover the northern slopes at the same altitudes.

7. Conclusion

Northeastern Eurasia is a vast region with extremely diverse physiographic and permafrost conditions. The soil survey of this region started in the late 1950s. At present, the map of 1:2.5 M scale is available for the whole territory. There are some large-scale soil maps for key plots in the Magadan region, Kolyma Lowland, and some other areas.

The range of soil development of northeastern Eurasia is very diverse. Cryogenic processes affect nearly all soils. The extreme influence of cryogenesis on pedogenesis appears in hydromorphic heavily textured (clay loamy and clayey) soils with shallow (<1 m) permafrost and active cryoturbation in continental areas with semi-arid climate. These soils are distinguished as specific Non-Gley Cryohydromorphic soils, or Cryozems *(Oxyaqui-Turbic Cryosols)*. They have intergrades with Gley soils *(Gleyi-Turbic Cryosols)*. The minimum influence of cryogenesis on soil development appears in coarsely textured stony soils on southern slopes. These soils are Steppe Cryoarid soils *(Calci-Mollic Leptosols)* with well pronounced humus horizon and, often, with secondary carbonates in the B horizon. Usually, they do not contain permafrost.

The main differentiation factor in soil formation is climate, i.e., the degree of atmospheric moistening. The areas with humid climate occupy the western slope of Verkhoyansk Ridge and the eastern (Pacific) part of the territory, where Al-Fe-humus mesomorphic soils (Podzols and Podburs) and hydromorphic Gley and Peat soils predominate. The central part of northeastern Eurasia has an ultracontinental semi-arid and arid climate. Humus-accumulative Steppe Cryoarid soils and metamorphic Pale soils prevail in this region among xeromorphic and mesomorphic soils. Hydromorphic soils include Cryozems and Peat soils. The lithological factor, i.e., filtration properties and water retention capacity of parent materials, and regional soil drainage characteristics mainly control the development of mesomorphic or hydromorphic soils. Hydromorphic soils are typical of heavily textured, clay loamy materials with shallow, ice-rich permafrost.

8. Acknowledgments

The author thanks Dmitri Konyushkov of the Dokuchaev Soil Science Institute for translation assistance and Anne Hargrave and Scott Smith for their editorial and formating help.

References

Elovskaya, L.G., E.I. Petrova, and L.V. Teterina. 1977. Pochvy Severnoi Yakutii (Soils of Northern Yakutia), Novosibirsk. Nauka, 303 pp. (In Russian.)

Ignatenko, I.V., and A.A. Pugachev. 1980. Biological turnover in the biogeocenoses of the Okhotsk Mountainous Province. Pochvovedenie. 8:45-52. (In Russian.)

Ignatenko, I.V., Ye. M. Naumov, I.Ye. Bogdanov, G.G. Mazhitova, and B.A. Pavlov. 1982. Soil-geographic regionalization of the extreme northeast of the USSR, Pochvy ostrovov i priokeanicheskikh regionov Tikhogo okeana (Soils of Islands and Coastal Regions in the Pacific Basin). Vladivostok. pp. 44-96. (In Russian.)

ISSS Working Group RB. 1998. World Reference Base for Soil Resources. International Society of Soil Science (ISSS), International Soil Reference and Information Centre

(ISRIC) and Food and Agriculture Organization of the United Nations (FAO). World Soil Resources Report 84. FAO. Rome. 91 pp.

Ivanova E.N., N.N. Rozov, A.A. Erokhina, N.A. Nogina, V.A. Nosin, and K.A. Ufimtseva. 1961. New materials on the geography and classification of Polar and Boreal Soils in Siberia. Pochvovedenie. 11:7-23. (In Russian.)

Ivanova, E.N. 1965. Permafrost-affected taiga soils in northern Yakutia, Pochvovedenie. 7:1-14. (In Russian.)

Karavaeva, N.A., and V.O. Targulian. 1960. On the peculiarities of humus distribution in the tundra soils of northern Yakutia. Pochvovedenie. 12:36-45. (In Russian.)

Karavaeva, N.A., I.A. Sokolov, T.A. Sokolova, and V.O. Targulian. 1965. On the peculiarities of soil formation in tundra and taiga areas of Eastern Siberia and the Far East. Pochvovedenie. 7:26-37. (In Russian.)

Klyukin, N.K. 1960. Prikladnoi klimatologicheskii spravochnik Severo-Vostoka SSSR (Reference Book on the Climate of Northeastern USSR). Magadan. (In Russian.)

Naumov, Ye. M. 1963. Soils of the northern coast of the Sea of Okhotsk, Pochvy Vostochnoi Sibiri (Soils of Eastern Siberia). Nauka. Moscow. pp. 153-195. (In Russian.)

Naumov, Ye. M. 1973. The main types of genetic soil profiles and peculiarities of the soil cover in the taiga zone of the extreme northeast of Asia, Biologicheskie problemy Severa (Biological Problems of the North), part 1. Magadan. pp. 48-55. (In Russian.)

Naumov, Ye. M., and B.P. Gradusov. 1974. Osobennosti taezhnogo pochvoobrazovaniya na krainem severo-vostoke Evrazii (Peculiarities of Soil Formation in the Extreme Northeast of Eurasia). Kolos. Moscow. 148 pp. (In Russian.)

Naumov, Ye. M. 1980. Classification of the soils of northeastern Asia, All-Union Conf. "Forecast of Changes in Cryogenic Soils under the Impact of Economic Development of the Territory." Pushchino. pp. 24-27. (In Russian.)

Naumov, Ye. M., and I.V. Ignatenko. 1990. The Soil Map of the Magadan Region, 1:2.5 M. Moscow: GUGK. (In Russian.)

Naumov, Ye. M. 1993. The Soil Map of Northeastern Eurasia, 1:2.5 M. GUGK at the Council of Ministers of Ukraine. Kiev. (In Russian.)

Naumov, Ye. M. 1993. Soils and soil cover of the North-East Eurasia (Pochvy i pochvennyi pokrov Severo-Vostoka Evrazii). Doctoral Dissertation. Moscow. (In Russian.)

Reutt, A.T. 1960. Vegetation, Sever Dal'nego Vostoka (The Northern Part of Far East). Nauka. Moscow. pp. 335-337. (In Russian.)

Sokolov, I.A. 1980. On the diversity of hydromorphic non-gley soils. Pochvovedenie. 2:5-18. (In Russian.)

Soil Survey Staff. 1999. Soil Taxonomy. 2nd ed. USDA-NRCS, AH-436. Washington, D.C. 869 pp.

Targulian, V.O., N.A. Karavaeva, Ye.M. Naumov. I.A. Sokolov, and N.N. Rozov. 1970. Soils of the northeast of USSR, Sever Dal'nego Vostoka (The Northern Part of Far East). Nauka. Moscow. pp. 234-255. (In Russian.)

Treshnikov, A.F. (ed). 1985. Atlas of the Arctic (Atlas Arktiki). Moscow. (In Russian.)

Vas'kovskii, A.P. 1959. Genesis and geography of forest soils in the extreme northeast of the USSR. Kolyma journal. 1:15-23. (In Russian.)

Chapter 7. Cryosols of the Russian European North

S.V. Goryachkin[1] and I.V. Ignatenko[2]

[1]Head, Laboratory of Soil Geography and Evolution, Institute of Geography, Russian Academy of Sciences, Staromonetny, 29, Moscow, 109017 Russia; e-mail pedology@igras.geonet.ru.
[2]Professor, Gertsen Russian State Pedagogical University, Moika, 48, Saint Petersburg, 191186 Russia.

1. Introduction

This chapter summarizes the systematic information on the soils and soil cover of the Russian European North which before now has been known almost only by the Russian-speaking community of soil scientists.

The Russian European North is where researchers first obtained information on tundra soils and distinguished the specific features of tundra pedogenesis (Liverovskiy, 1934). However, the Cryosols of the Russian European North do not belong to the nucleus of the permafrost-affected soil concept.

Most of this tundra region includes non-permafrost areas and areas of sporadic and discontinuous permafrost. But the climate's alternate heavy freezing and thawing causes pronounced cryoturbation, heaving, and other processes specific to frozen ground and, because of the ground's seasonal freezing, these features are striking in northern Europe even in permafrost-free areas (Tamm, 1928; Polyntseva and Ivanova, 1936). Thus, information on soils of plains, where they gradually change from common Histosols, Podzols, and Luvisols into Gelic and Cryic suborder soils and Cryosols, is very important to our understanding of global and regional aspects of soil geography.

The soil cover of the Russian European North consists of naturally divided subregions (Figure 2.7.1): the Kola Peninsula, with the mountains of Khibiny; a small area of Koida tundra; the Kanin tundra of the Kanin Peninsula; the Malozemelskaya tundra to the west of the Pechora River; the Bolshezemelskaya tundra to the east of that river; the tundra of the Yugor Peninsula with the Pai-Hoi highland; and the Ural Mountains, where sporadic permafrost reaches 60° n.l. in the south.

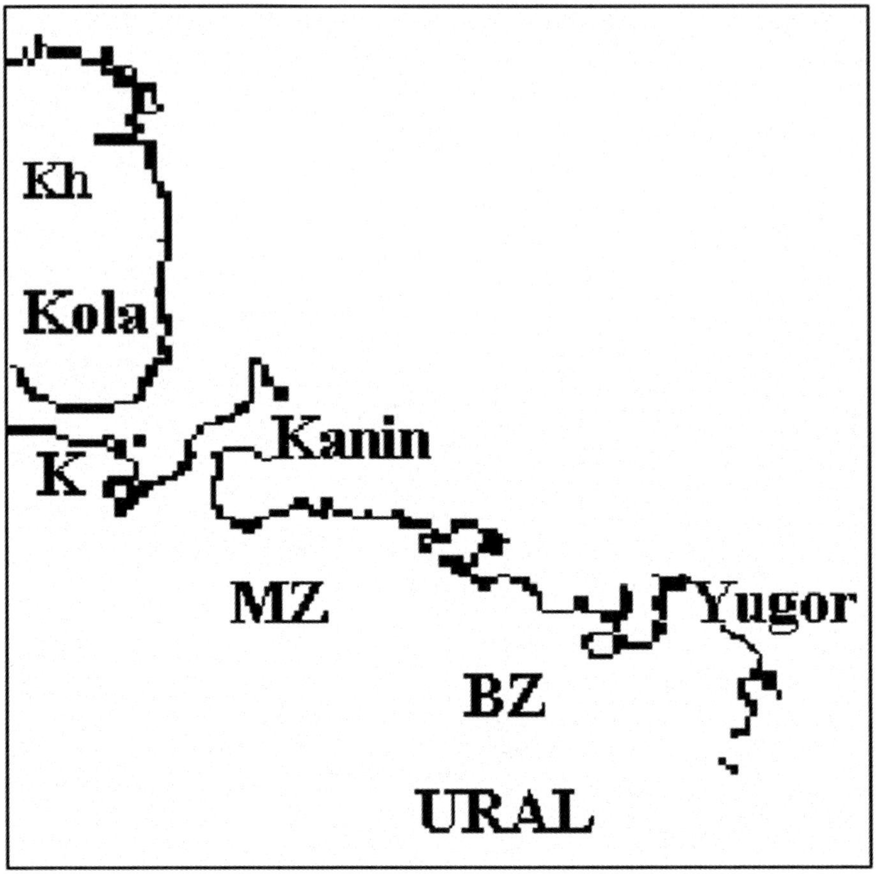

Figure 2.7.1. Regions of the Russian European North. Kola – Kola Peninsula; Kh – Khibiny mountains; K – Koida tundra; Kanin – Kanin tundra; MZ –Malozemelskaya tundra; BZ – Bolshezemelskaya tundra; Yugor – Yugor Peninsula; URAL – Ural Mountains.

The soils of the Russian European North have been studied for many decades, and information on the soil cover of the region was used to compile the national soil map of Russia (Fridland, 1988). In some regions of East European tundra, the very detailed studies of soil cover, characteristics of humus, physico-chemical and mineralogical composition, soil temperature, moisture regimes, and other dynamic parameters were organized. But soil information is very irregular about the whole territory of the Russian European North. The eastern part of the Kola Peninsula, the Koida tundra, and many parts of the Kanin and Malozemelskaya tundras have not been investigated well.

2. Soil-Forming Factors of the Russian European North

The information on soil-forming factors of the Russian European North comes from relevant monographs (Belov and Baranovskaya, 1969; Targulian 1971; Ignatenko, 1979; Nikonov and Pereverzev, 1989; Goryachkin et al., 1994).

2.1. Geological and relief characteristics

The Russian European North is mostly a plains area with a thick quaternary layer of loose sediments. Stony and rocky materials and deeply dissected topography occur only in Khibiny and other mountains and highlands of the Kola Peninsula, the Kanin and Timan highlands (which are very old mountains of Precambrian Baikal tectonic epoch and therefore very eroded), and the Ural mountains. All the territory was glaciated in the Quaternary period.

The Kola Peninsula is mainly a zone of glacial denudation, and its substrates are mostly very stony and gravelly sandy moraines upon granite and gneiss rocks. The other territory is a combination of glacial-marine, moraine, and limno- and fluvio-glacial deposits of loamy, clayey, and, to a lesser extent, sandy texture. The eastern part of the Bolshezemelskaya tundra is covered with loess-like silty loams. Marine and alluvial substrates occur on coasts and river terraces. The rocky material is nipheline syenit in Khibiny, granite and gneiss in Kola, and shale, sandstone, and limestone with small areas of basic rocks in Kanin, Timan, Pai-Hoi, and the Urals. Being derivatives of granites and gneiss of Kola, the glacial deposits of Koida tundra are brown and originated from shales. The loose sediments of the eastern regions often are brownish gray.

Permafrost-affected relief forms are spread out in the Russian European North. In mountainous areas, sorted and non-sorted nets, stripes, and polygons are characteristic features. Palsas, hummocks, sorted and nonsorted circles, and other different kinds of patterned ground are widespread in plains areas. All these permafrost-affected forms of relief result in the formation of a soil pattern, as it is called in the Russian literature, of soil cover complexity – regular alternation of different soils.

2.2. Climate

All the territory has a cold and humid climate. But many climatic characteristics differ, depending mostly on the distance from the seacoast; the nearer to the coast, the smaller the amplitude of winter and summer temperatures. The summer temperature increases from north to south and decreases from foothills to summits of the mountainous and highland areas. Thus the mean annual temperatures in tundra of the Russian European North vary from -0.8° C (Kola) to -7.6°C (Yugor Peninsula).

The mean July temperature is the highest in forest-tundra (+12°C) and the lowest in the coast of the Kara Sea of Yugor Peninsula (+5.8°C). In the mountains of Kola (900m a.s.l.), the mean July temperature is +8°C. The lowest mean January temperature is in the easternmost forest-tundra (-19.8°C), and the mildest winter is in the tundra of western Kola Peninsula (-5.5°C). This is because of the strong influence of the North Atlantic's warm oceanic currents on the western part of the Russian European North, and because of the effect of very cold Siberian and Arctic air in the eastern continental part in winter.

The mean annual precipitation in the area varies from 273 mm for the east of the Yugor Peninsula, to 450 to 550 mm for forest tundra on the plains, to 700 to 800 mm in Khibiny, to 1000mm in Ural Mountains. These values are at least twice larger than those of evapotranspiration in summer for the northern part of the territory and 1.5 times larger than those for the southern part of the territory. However, the larger portion of precipitation (50 to 75%) is snow. The strong winds redistribute snow, and its thickness depends on topography and vegetation. In bushy hollows in the southern tundra, 4m of snow can accumulate, while in higher latitudes the snow cover of hills is 10 to 30 cm.

2.3. Permafrost

The Russian European North is the area of different permafrost continuity and depth of the active layer. The Yugor Peninsula, the northern part of the Bolshezemelskaya tundra, and high altitudes (ca 500 to 600 m) of the mountains are in the zone of continuous permafrost. The depth of the active layer varies here from 20 cm in peat to 1 m in stony substrates.

The zone of discontinuous permafrost occupies the other part of tundra, except those of Kola. The active layer here increases from north to south (from 0.3 to 0.6 in palsas, 0.6 to 3.0 on loamy substrates, and 0.9 to 5.0 on sands).

The sporadic permafrost is in palsas of forests of the eastern part of the territory, in forest-tundra of the central part, and in the tundra of eastern Kola. The western Kola, both in tundra and forest-tundra, is permafrost-free.

2.4. Vegetation

The vegetation of the area is tundra and forest-tundra. Only on the high mountain tops of the Khibiny and Urals is the scarce and impoverished moss-lichen vegetation, with sparse flowering plants, characteristic – there are “goltsy” or alpine barrens here.

Arctic tundra (herbaceous, moss-herbaceous) is the vegetation of the northern part of the Yugor Peninsula. The rest of it and the northern parts of the Bolshezemelskaya and Malozemelskaya tundra belong to a dwarf-shrub-mossy tundra subzone, with widespread sedge-moss swamps. Willow standings are typical for river valleys here. And most of the region belongs to a dwarf-birch subzone of tundra with a large share of boreal species, Sphagnum bogs, and open woodlands in river valleys.

The transition to the boreal forests zone of forest-tundra differs in the western and eastern parts of the area. In the Kola Peninsula and the Koida tundra, the transition is mainly of low birch (Betula tortuosa) with an admixture of spruces. In the Kanin, Malozemelskaya, and Bolshezemelskaya tundra, the forest outposts are mainly spruce parklands with an admixture of birches and larches. The common element for alpine transition from forest to tundra is the gramen-herb meadow communities of the Ural Mountains. The meadow-like communities (*lugovinas*) are also typical for well drained slopes of hollows of plains tundras.

3. Major Soil Units

Because of the diversity of soil-forming factors, a wide spectrum of Cryosols, soils of Gelic suborders, and non-permafrost soils have developed in the Russian European North. Tables 2.7.1 and 2.7.2 present the morphological and chemical data of several reference profiles, but these profiles cannot represent all the varieties of soils in the studied area.

The coldest parts of the territory are covered with Arctic tundra weakly gleyic humic soils (Stagni-Turbic Cryosols) and Peaty bog gleyzems (Turbi-Histic Cryosols [Gleyic]), which are described in the chapter on the Russian Arctic Archipelagos.

Tundra mucky gleyzems (WRB equivalents: Gleyi-Turbic Cryosols, Stagni-Turbic Cryosols) are widespread soils on loamy and clayey substrate in the northern half of the territory. Their characteristic features are pronounced stagnic or gleyic properties and thixotropy. The upper horizons are the litter (if any) and the mucky horizon that is the mixture of well decomposed organic matter and mineral particles from lateral migration of fine earth from barren cryogenic spots. Sometimes the mucky horizon is like the saprihistic horizon of Histosols. These soils are strongly cryoturbated.

Tundra peaty gleyzems (WRB equivalents: Turbi-Histic Cryosols [Stagnic or Gleyic], Geli-Histic Gleysols, Stagni-Gelic Regosols [Histic], Histic Gleysols) are very widespread soils in the Russian European North. There are so many WRB equivalents because of the high amplitude of cryic conditions for their area. They are distinguished by a combination of gleyic or stagnic properties and a histic horizon 10 to 40 cm thick. This thickness fluctuates with hummocky microrelief. The features of different cryogenic disturbances are characteristic for these soils.
Tundra peaty-mucky gleyzems (WRB equivalents: Turbi-Histic Cryosols [Stagnic]), which have the features of the both soil types, also occur, especially in the northern part of the Malozemelskaya tundra.

Table 2.7.1. Morphological features of the major soils of the Russian European North.

Horizon	Depth	Color	Moisture Status	Texture	Structure	Pedofeatures	Roots
Tundra mucky gleyzem (Gleyi-Turbic Cryosol) – profile 28 (Ignatenko, 1963, 1979), Yugor Peninsula, hummocky sedge-mossy tundra, flat site							
O	0-4	BS-BL	Slightly M	No	No	No	M
A	4-9	BS-BL	W	SL	MA	No	M
Br	9-16	GS-OL	WS	SL	MA	No	C
Bg	16-25	GS-OL	W	SL	MA	RY mottles	C
BCg	25-59	BS-OL	M	SL	MA	RY mottles	VF
Cgi	60-70	BR	Frozen	L	MA	OL mottles	No
Tundra peaty gleyzem (Turbi-Histic Cryosol (Stagnic)) – profile 677 (Ignatenko, 1979), Yugor Peninsula, hummocky willow-sedge-mossy tundra, flat site							
O	0-5	YB	M	No	No	No	M
H	5-19	YB	W	No	MA	No	M
Br	19-50	GS-BU	WS	L	MA	RY mottles	C
BCgi	50-87	OL-BR	frozen	L	MA	BS mottles	VF
Cgi	87-125	OL-BR	frozen	L	MA	No	No
Tundra differentiated podzolized gleyzems (Albi-Stagnic Luvisols) – profile 710, (Ignatenko, 1979), south-eastern part of Bolshezemelskaya tundra, hummocky dwarf-birch-mossy tundra							
O	0-10	BR	M	No	No	No	M
Eg	10-41	L BS-GR	W	SL	PL-MA	BU mottles	F
E	41-60	L BS-GR	W	SL	SB	Silt coatings	VF
BCt	60-87	BR	WS	L	SN	Silt coatings	VF
BCtg	87-112	OL-BR	W	L	MA	Silt coatings, RY mottles	No
Cg	112-130, bore 230	OL-BR	W	L	MA	Nodules, BU mottles	No

(continues)

Table 2.7.1. (cont.)

Horizon	Depth	Color	Moisture Status	Texture	Structure	Pedofeatures	Roots
Tundra podbur with gleyic features (Spodi-Turbic Cryosol (Gleyic)) – profile 490 (Ignatenko, 1979), east part of Bolshezemelskaya tundra, irregular polygons, dwarf-shrub-mossy tundra							
O	0-2	BR BL	M	No	No	No	M
ABh	2-7	BS GR	M	LS	MA	Bleached grains	M
Bh	7-23	Dark BR	M	LS	MA	Fe-coatings	F
BC	23-87	L BR	M	S	MA	No	VF
BCgi	87-99	OL RB	Frozen	LS	MA	No	No
Cgi	99-127	L GR	Frozen	LS	MA	RY mottles	No
Podzol with gleyic features (Endogleyi-Gelic Podzol) – profile 158 (Ignatenko, 1979), eastern part of Bolshezemelskaya tundra, small depression in hummocky dwarf birch-lichen tundra							
O	0-4	BS BL	Dry	No	No	No	M
E	4-12	L GR	Slightly M	S	MA	No	F
Bh	12-31	Dark BR	M	S	MA	No	C
B	31-48	YB	M	S	MA	No	VF
Bs	48-96	RB	M	S	MA	Fe-coatings, weakly cemented	VF
Bsg	96-106	OL RB	W	S	MA	No	No
BCgi	106-123	OL YB	frozen	S	MA	No	No
Cgi	123-133	OL BR	frozen	S	MA	RY mottles	No

(continues)

Table 2.7.1. (cont.)

Horizon	Depth	Color	Moisture Status	Texture	Structure	Pedofeatures	Roots
Peaty bog degraded soils (Eutri-Cryic Histosol) – profile 8-93 (Goryachkin), southern part of Bolshezemelskaya tundra, hillocky bog, summit of palsa, dwarf-shrub-lichen community							
O	0-4	10YR5/6YB	M	No	No	No	M
O	4-14	10YR3/2BS BL	M	No	No	No	M
H	14-33	10YR2/3BS BL	M	No	No	No	C
H	33-42	10YR2/2BS BL	M	No	No	No	VF
Hi	42-50	10YR2/2BS BL	Frozen	No	No	No	No
Peaty bog mesotrophic soil (Eutric Histosol) – profile 7-93 (Goryachkin), southern part of Bolshezemelskaya tundra, near 8-93, hollow, willow-herbaceous-sedge community							
H	0-6	10YR3/2BS BL	W	No	No	No	M
H	6-36	10YR2/2BS BL	WS	No	No	No	M
H	36-41	10YR2/1 BL	W	No	No	No	M
Cr	41-56	7.5Y6/2GS OL	Frozen	CL	SB	RY mottles	VF

(continues)

Table 2.7.1. (cont.)

Horizon	Depth	Color	Moisture Status	Texture	Structure	Pedofeatures	Roots
Soil of tundra lugovina (meadow) (Ochric Regosol) – profile 11-93 (Goryachkin), southern part of Bolshezemelskaya tundra, north-oriented slope, 8-10%, gramen-herb meadow							
AO	0-3	10YR2/2 BS BL	M	L	GR	No	M
A	3-13	10YR2/2 BS BL	M	L	SB	Earthworm channels	M
A	13-20	2.5Y3/2 BS BL	M	L	SB	Earthworm channels	M
BC	20-38	2.5Y5/2 D GS Y	M	L	SN	Sandy layers	C
C	38-100	5Y5/2 GS OL	Slightly M	L	SN	Sandy layers	VF
Mountainous primitive soil (Lithic Leptosols) – profile 783 (Belyaev et al., 1965), Ural mountains, 64,5°n.l., 667 a s.l., small summit area, Betula nana, Cladonia rangiferina.							
O	0-2	BS BL	Dry	No	No	No	M
BC	2-4	GS BR	M	ST L	SB	No	M

Abbreviations:

Color: BR – brown, BS – brownish, Y – yellow, YB – yellowish brown, RY – reddish yellow, RB – reddish brown, GR – gray, GS – grayish, BU – blue, BL – black, OL – olive; L – light, D - dark

Moisture status: M – moist, W –wet, WS – wet, satiated;

Texture: C – clay, L – loam, CL – clay loam, SL – sandy loam, LS –loamy sand, S – sand,

Texture modifiers: ST – stony;

Structure: MA – massive, GR – granular, SB – subangular blocky, SN – nutty subangular blocky, PL – platy;

Roots (abundance): M – many, C – common, F – few, VF – very few

Table 2.7.2. Chemical properties of soils of Russian European North.

Horizon	Depth cm	Loss on ignition,%	Corg., %	PH H_2O	<0.0001 mm, %	Exchangeable cations, Meq/100g soil			Base satu-ration,%	Oxalate extractable forms, %	
						Ca^{2+}	Mg^{2+}	H^+		Fe	Al
Tundra mucky gleyzem (Gleyi-Turbic Cryosol) – profile 28 (Ignatenko, 1963, 1979), Yugor Peninsula											
O	0-4	84.22	n.d.	7.2	n.d.	71.3	12.3	26.3	75	n.d.	n.d.
A	4-9	36.54	14.10	6.9	16	67.4	6.6	10.4	88	0.39	0.15
Br	9-16	3.96	1.10	6.7	17	13.9	3.4	1.0	95	0.28	0.14
Bg	16-25	3.76	1.00	6.9	17	14.9	2.5	0.9	95	0.26	0.12
BCg	30-40	4.09	0.80	6.9	18	13.9	2.8	1.1	94	0.23	0.12
Cgi	60-70	7.55	0.83	6.6	18	14.2	8.0	1.1	94	0.25	0.12
Tundra peaty gleyzem (Turbi-Histic Cryosol (Stagnic)) – profile 677 (Ignatenko, 1979), Yugor Peninsula											
O	0-5	95.22	n.d.	4.0	n.d.	28.8	9.5	n.d.	n.d.	n.d.	n.d.
H	5-19	79.56	33.59	4.2	n.d.	11.2	8.2	89.0	18	n.d.	n.d.
Br	19-28	5.64	0.99	4.4	19	5.0	4.6	7.7	55	0.92	0.32
Br	28-35	4.19	1.09	4.3	17	5.9	3.3	7.9	54	1.83	0.46
Br	40-50	4.74	1.88	4.8	18	6.7	3.2	10.5	49	1.64	0.47
BCgi	50-60	5.18	1.79	5.0	20	7.1	3.0	12.6	45	1.52	0.50
BCgi	65-75	5.65	1.37	5.1	19	7.4	3.8	10.2	51	1.25	0.42
BCgi	75-85	5.03	1.14	5.3	19	7.4	3.6	1.4	89	0.71	0.48
Cgi	90-99	4.88	0.34	5.8	20	7.4	3.6	0.8	93	0.48	0.20

(continues)

Table 2.7.2. (Cont.)

Horizon	Depth cm	Loss on ignition,%	Corg., %	PH H_2O	<0.0001 mm, %	Exchangeable cations, Meq/100g soil			Base saturation,%	Oxalate extractable forms, %	
						Ca^{2+}	Mg^{2+}	H^+		Fe	Al
Tundra differentiated podzolized gleyzems (Albi-Stagnic Luvisols) – profile 710, (Ignatenko, 1979), southeastern part of Bolshezemelskaya tundra											
O	0-5	95.70	n.d	4.3	n.d.	18.5	8.3	n.d.	n.d.	n.d.	n.d.
O	5-10	76.40	n.d	3.9	n.d.	11.7	6.6	135	12	n.d.	n.d.
Eg	10-21	6.43	2.44	3.7	13	2.4	1.2	26.1	12	0.90	0.22
Eg	25-35	3.09	0.95	4.3	13	2.8	2.5	11.8	31	0.63	0.18
E	45-55	2.86	0.18	4.9	16	4.0	2.8	6.5	51	0.48	0.15
BCt	65-75	2.94	0.21	5.1	21	4.3	3.0	5.4	57	0.55	0.13
BCtg	90-100	3.11	n.d.	5.3	29	4.9	3.4	4.7	64	0.63	0.16
Cg	120-130	3.26	0.22	5.4	29	5.3	3.7	2.7	77	1.02	0.17
Tundra podbur with gleyic features (Spodi-Turbic Cryosols (Gleyic)) – profile 490 (Ignatenko, 1979), east part of Bolshezemelskaya tundra											
O	0-2	83.54	n.d.	4.6	n.d.	13.8	5.3	n.d.	n.d.		
ABh	2-7	7.61	2.81	4.0	9	3.6	0.9	10.2	31	0.34	0.20
Bh	7-15	2.28	1.00	4.4	9	3.1	1.6	9.4	33	0.36	0.20
Bh	15-23	2.34	0.66	4.8	7	2.8	0.9	7.5	33	0.23	0.21
BC	25-35	n.d.	0.35	5.0	5	2.5	0.7	7.3	30	n.d.	n.d.
BC	35-45	1.93	0.23	5.1	4	2.6	0.7	7.2	31	0.21	0.09
BC	50-60	1.78	0.13	5.2	4	3.0	1.0	3.5	53	0.19	0.09
BC	75-85	1.69	0.09	5.4	4	3.7	1.4	2.4	68	0.29	0.09
BCgi	87-99	1.84	0.14	5.5	7	4.3	2.7	2.5	74	0.61	0.08
Cri	99-115	1.71	0.09	5.7	2	2.6	1.2	0.7	83	0.20	0.08

(continues)

Table 2.7.2. (Cont.)

Horizon	Depth cm	Loss on ignition,%	Corg., %	PH H_2O	<0.0001 mm, %	Exchangeable cations, Meq/100g soil			Base satu-ration,%	Oxalate extractable forms, %	
						Ca^{2+}	Mg^{2+}	H^+		Fe	Al
Podzol with gleyic features (Endogleyi-Gelic Podzol) – profile 158 (Ignatenko, 1979), eastern part of Bolshezemelskaya tundra											
O	0-4	98.07	n.d.	4.0	n.d.	3.9	0.8	31.2	12	n.d.	n.d.
E	4-12	6.06	1.92	4.0	3	0.7	0.2	10.4	8	0.03	0.08
Bh	15-25	2.69	0.89	4.4	2	0.7	0.2	9.7	8	0.25	0.58
B	35-45	0.93	0.14	4.8	3	1.6	0.2	3.2	36	0.78	0.71
Bs	55-65	1.04	0.14	5.3	3	1.6	0.2	2.9	38	1.50	0.68
Bs	80-90	1.20	0.15	5.4	3	1.8	0.2	2.8	42	2.15	0.66
Bg	96-106	1.20	0.05	5.3	8	2.1	0.5	2.5	51	1.58	0.57
BCgi	110-120	1.12	0.12	5.3	n.d.	2.5	0.6	2.6	54	1.30	0.25
Cgi	123-133	1.22	0.13	5.5	10	4.8	1.0	2.1	73	0.08	0.30
Peaty bog degraded soils (Eutri-Cryic Histosol) – profile 8-93 (Goryachkin), southern part of Bolshezemelskaya tundra											
O	0-4	97.81	n.d.	4.3	n.d.	11.0	5.9	17.7	49	n.d.	n.d.
O	4-14	97.85	n.d.	4.0	n.d.	21.5	4.2	16.5	61	n.d.	n.d.
H	14-33	92.86	n.d.	4.4	n.d.	36.6	0.8	18.2	67	n.d.	n.d.
H	33-42	95.13	n.d.	4.4	n.d.	59.1	1.1	7.9	88	n.d.	n.d.
Hi	42-50	96.10	n.d.	4.7	n.d.	53.6	4.2	2.4	96	n.d.	n.d.
Peaty bog mesotrophic soil (Eutric Histosol) - profile 7-93 (Goryachkin), southern part of Bolshezemelskaya tundra											
H	0-6	85.49	n.d.	5.6	n.d.	48.9	11.2	2.0	97	n.d.	n.d.
H	6-36	86.97	n.d.	5.6	n.d.	50.6	9.2	1.7	97	n.d.	n.d.
H	36-41	44.49	n.d.	5.3	n.d.	38.8	5.3	5.8	88	n.d.	n.d.
Cr	41-56	n.d.	1.07	4.5	n.d.	9.2	2.6	5.1	69	n.d.	n.d.

(continues)

Table 2.7.2. (Cont.)

Horizon	Depth cm	Loss on ignition,%	Corg., %	PH H_2O	<0.0001 mm, %	Exchangeable cations, Meq/100g soil			Base satu-ration,%	Oxalate extract-able forms, %	
						Ca^{2+}	Mg^{2+}	H^+		Fe	Al
Soil of tundra lugovina (meadow) (Ochric Regosol) – profile 11-93 (Goryachkin), southern part of Bolshezemelskaya tundra											
AO	0-3	32.40	n.d.	5.6	n.d.	20.5	5.5	1.9	93	0.33	0.09
A	3-13	n.d.	3.92	4.8	n.d.	9.4	3.2	1.7	88	0.42	0.10
A	13-20	n.d.	1.59	5.1	n.d.	13.7	4.3	0.5	97	0.36	0.09
BC	20-38	n.d.	0.94	6.6	n.d.	15.1	4.4	0	100	0.36	0.06
C	38-100	n.d.	0.56	7.8	n.d.	13.5	3.5	0	100	0.55	0.05

n.d. – no data

Tundra Podburs (WRB equivalents: Spodi-Turbic Cryosols, Enti-Gelic Podzols, Entic Podzols) are predominant soils of the Russian European tundra on sandy and well drained stony material. They are distinguished for their lack of a (or very thin) bleached horizon over a spodic horizon. The WRB equivalent depends on the cryogenic status of the concrete profile. If the permafrost is shallow, the gleyic features related to cryogenic aquiclude are common in Podburs of the northern part of the area. These soils also are cryoturbated in both permafrost and non-permafrost conditions.

Podzols (WRB equivalents: Gelic Podzols, Haplic Podzols, Spodi-Turbic Cryosols [Albic]) are soils very like Podburs. They differ by the presence of an Albic horizon, but they usually occupy warmer positions in landscapes or more southern areas than Podburs and have a deeper active layer. Thus they correlate mostly with WRB Podzols and less frequently with Cryosols.

The most pronounced permafrost-affected features of the Russian European North are palsas-peat mounds, which are distributed from eastern Kola to the boreal forest zone of Ural. Usually, palsas are circled by wet swampy pools. Peaty bog degraded soils (WRB equivalents: Cryic Histosols) occur on the palsa and Peaty bog mesotrophic soils (WRB equivalents: Eutric Histosols) occur without (or very deep) permafrost in the pools. More detailed characteristics of bog permafrost-affected soils are given in the chapter on West Siberian Cryosols.

Loamy Tundra differentiated podzolized gleyzems (WRB equivalents: Albi-Stagnic Luvisols, Albi-Stagnic Luvisols [Gelic]) occur in the southern part of the Russian European North. They have some features of Podzolic texturally differentiated soils (WRB equivalents: Albeluvisols, Luvisols). In Russian papers they often are called "eluvial-gleyic soils." They have analytically pronounced textural differentiation but no albeluvic tonguing and small evidence (if any) of lessivage because of high moisture content, thixotropy, and stagnic features. They could occur with permafrost (1 to 2 m) and without or with very deep permafrost. The other soils near the tree line are Podzolic soils of the southern tundra, with permafrost within 2 m (Gelic Albeluvisols) and with a pronounced Argic horizon and albeluvic tonguing (See the section in this paper on Koida tundra).

Soils of tundra lugovinas (meadows) (WRB equivalents: Ochric Regosols, Haplic Phaeozems) are soils with a well developed sod and humic horizons under grassy vegetation in snowy warm relief positions – well drained slopes of tundra hollows. The soil of the profile 11-93 (Tables 2.7.1 and 2.7.2) requires only 3 cm of humic horizon thickness to be classified as Phaeozem, so some of these soils could match the criteria of Phaeozems but were not specially studied.

Mountainous soils of the Urals and Kola are close in genesis to those of the plains and have the same names, in accordance with the Russian approach (WRB equivalents: Podburs, Podzols, Gleyzems), but WRB names usually differ because of the Leptic suborder (e.g., mountainous Podburs equal Turbi-Leptic Cryosols [Spodic]). But there are also specific soils on rocky substrates, which are widely distributed. They are Mountainous primitive soils (WRB equivalents: Lithic Leptosols). Their profiles are characterized by litter and a shallow fine earth horizon (Table 2.7.1).

Rarer soils occur in the upper belts of the Ural Mountains near the southernmost outposts of permafrost. They are Mountainous forest-meadow soils (Ochri-Gelic Leptosols, Geli-Umbric Leptosols) with pronounced sod and usually are unsaturated by cations with humic horizons, overlaying lithic and paralithic contact. They developed under grassy vegetation within the belt of birch parklands.

Alluvial acid soils (Dystric Fluvisols) and Alluvial bogged soils (Histic Fluvisols, Gleyic Fluvisols and Fluvi-Histic Cryosols) are the common soils for flood plains of large rivers and Marshy saline soils (Salic Fluvisols) occur widely in the tidal zone of the White and Barents Seas. In these riparian positions, non-permafrost soils are common, as taliks (bodies of unfrozen ground) occur here.

4. Soil Covers

4.1. The Kola Peninsula

The Kola Peninsula is the westernmost and therefore the warmest part of the Russian European tundra. It is one of the rare places where tundra soils are permafrost-free.

The climate of the tundra part of Kola is characterized by a mean annual temperature of -0.8 to -1.5°C, a mean January one of -6 to -10°C, and a mean July one of +7.5 to +11°C. At higher altitudes (ca 900 to 1000 m), the mean annual temperature is lower, -3.6°C, as the winter temperatures are lower (mean January, -12.6 °C), while the summer ones are the same (+8.8°C).

Scientists have studied the soil cover of the Kola Peninsula since the 1930s (Polyntseva and Ivanova, 1936; Ivanova and Koposov 1937; Belov and Baranovskaya, 1969; Targulian 1971; Nikonov and Pereverzev, 1989; Goryachkin et al., 1994). The most noticeable characteristic of all the Kola tundra soil cover (both latitudinal and alpine) is the well pronounced evidence of intensive cryoturbations in non-permafrost areas. This is related to the shallow bedrock, which plays a role of cryogenic aquiclude, locking the soil profile from the bottom.

The main tundra soils of the Kola Peninsula are Podburs (Entic Podzols on sandy materials and Lepti-Entic Podzols on stony material with shallow lithic contact) and Podzols (consequently, Haplic Podzols on sands and Hapli-Leptic Podzols with shallow bedrock). The Podburs lack the Albic horizon because of both Fe-rich parent materials (gabbro and basalts of the some alpine tundras) and cryogenic disturbances, which prevent the E horizon's formation (Targulian, 1971).

The imperfectly drained soils are Gleyed podzols (Gleyic or Stagnic Podsols). In some parts of eastern Kola tundra are loamy Tundra gleyzems (Dystric

Gleysols), partially with deep permafrost (Dystri-Gelic Gleysols) (Belov and Baranovskaya, 1969; Goryachkin et al., 1994).

The slopes and small summits adjacent to the sea often have no soils at all, as only bare rock outcrops and Mountainous primitive soils (Lithic Leptosols) occur there. As a rule, they are very small in area, compared to rock outcrops.

Genuine permafrost-affected soils appear only in the eastern half of the Kola tundra zone and only in peat lands — Peaty bog degraded soils (Cryic Histosols) on the palsa, and, in high altitudes, Mountainous primitive soils (Lithic Leptosols) or Embriozems (Nikonov and Pereverzev, 1989). The soils of the peat land of the permafrost-free western Kola tundra are Eutric and Dystric Histosols.

4.2. Koida Tundra

The Koida tundra is a small tundra area on the coast of the White Sea, stretched out for 30 to 40 km from the seashore into the continent. It is very important to the soil geography in the Russian European North because part of it has the same geogenic parameters (relief and substrates) as large boreal plains to the south: glacial plains with loose moraine and limno-glacial deposits – sands and loams with an admixture of gravels, stones, and boulders.

The parent materials are Scandinavian glacial sediments, originally brown, unlike the parent materials of the eastern parts of the Russian European tundra, where brownish gray predominates (see above). Koida tundra also has karst topography (not the same as *thermokarst*) because of relatively shallow, extremely karstified hard gypsum bedrock.

The climate is the same as in eastern Kola. The vegetation is dwarf birch southern tundra, with outposts of low birch (*Betula tortuosa*) parklands and extensive areas of palsa mossy bogs.

No reliable soil information on Koida tundra soils exists, and this territory was mapped from extrapolation from the nearest tundra sites (Fridland, 1988). When we studied this region, we found Koida tundra covered mainly by extremely cryoturbated Podzols (Haplic Podzols) and Gleyed podzols (Histi-Gleyic Podzols) (Figure 2.7.2). But there was no evidence of permafrost in sandy material, at least to the depth of 3 m. Permafrost is present in loamy and organic material.

We identified Podzolic gleyed upon cryogenic aquiclude soils (Stagni-Gelic Albeluvisols [Ruptic]) on sandy-loamy (with discontinuity in the parent material) moraine substrates on well drained hill summits and slopes.

Peaty bog degraded soils (Cryic Histosols) occurred on the palsas and peat plateaus, alternating with Peaty bog mesotrophic and oligotrophic soils (Eutric and Dystric Histosols), common for vast flat limno-glacial areas of Koida tundra. String bogs with Dystric Histosols substitute for them to the south. Karst landscapes are well drained and the cryoturbated Haplic Podzols are widespread. Soil cover differs because of the orientation of sinkholes' slopes; Haplic Podzols with Ochric horizons occur on south-oriented slopes, and Haplic Regosols with pronounced solifluction terraces occur on north-oriented ones.

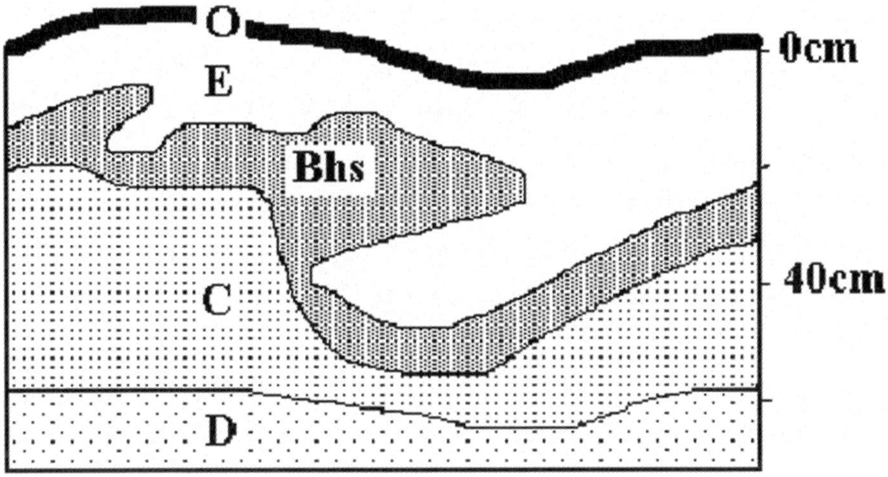

Figure 2.7.2. Cryoturbation features in a non-permafrost Haplic Podzol of Koida tundra under dwarf-shrub-mossy vegetation (Profile 20-86; Goryachkin). Microrelief is pronounced. Permafrost was not found at a depth of 2.5 meters.

4.3. Kanin, Malozemelskaya, and Bolshezemelskaya Tundras

The Kanin, Malozemelskaya, and Bolshezemelskaya tundra regions form the main body of the Russian European North, stretching for about 800 km from west to east and about 250 km from north to south. But the soil cover of these 3 regions has much in common, due to similar geogenic and climatic conditions.

The main territory of this macroregion alternates low plains and post-glacial uplands. Glacial-marine, moraine, and limno- and fluvio-glacial deposits of loamy, clayey, and sandy texture are widespread. The eastern part of the Bolshezemelskaya tundra is covered with loess-like silty loams. Rocky material occurs only in very small areas of the Kanin and Timan highlands.

Climatic conditions vary because of the large extent of the territory. The mean annual temperature varies from -1.8^{o} in the White Sea coast area of the Kanin tundra to -5.4^{o} in the continental and easternmost Vorkuta. Winter temperatures (in January, respectively, -11° and -19°) most affect the annual mean, as in summer the mean temperature rises from 8° in the north to 12^{0} in the south. Precipitation is about 300 to 400 mm on the whole territory, so the difference in the temperature regime causes the gradient of cryogenic conditions and vegetation.

Continuous permafrost (the depth of the active layer is 20 to 40 cm in peat, 40 to 60 cm in loam, and 60 to 80 cm in sand) occurs only in the north of easternmost Bolshezemelskaya tundra. To the north of ca 67° n.l., permafrost covers 70 to 90% of the area (the active layer is 40 to 60 cm in peat, 60 to 100 cm in loam, and 90 to

150 cm in sand). The rest of the territory of the region has discontinuous and sporadic permafrost: 30 to 40% of the area, up to 70% in the northern part (the active layer in peat mounds is 40 to 60 cm; in loams, 2 to 3 m; in sands 3 to 5 m; swampy pools are permafrost free).

The northern parts of the Bolshezemelskaya and Malozemelskaya tundra belong to the dwarf shrub-mossy tundra subzone, with widespread sedge moss swamps, and the rest of the region belongs both to the dwarf birch subzone of tundra with Sphagnum bogs and with open spruce-birch woodlands in river valleys, and to the subzone of forest tundra.

Scientists have studied this vast region much for soil distribution and genesis (Ivanova and Polyntseva, 1952; Kreida, 1958; Archegova, 1967; Parshevnikov, 1967; Gribova and Ignatenko, 1970; Ignatenko 1967, 1971, 1972, 1979; Rusanova, 1994; Goryachkin et al., 1994, 1995). But most of the studies were done for the eastern part – Bolshezemelskaya tundra – and other areas remain poorly examined.

The most widespread types of patterned ground of mineral soils in this region are hummocks, developed due to cryogenic heaving (Figure 2.7.3a). Strong winds erode the surface of a hummock with ice crystals, and this type of patterned ground turns into "spotty tundra" (Figure 2.7.3b), tundra with non-sorted circles (Ivanova and Polyntseva, 1952). Hummocks and spotty tundra can occur in areas of different soils.

In the northern part of the region on loamy materials, Tundra mucky gleyzems (Gleyi-Turbic Cryosols, Stagni-Turbic Cryosols) occur, while Tundra peaty gleyzems (Turbi-Histic Cryosol [Stagnic or Gleyic], Geli-Histic Gleysols) and Tundra peaty-mucky gleyzems (Turbi-Histic Cryosols [Stagnic]) occur in the central part. In the southernmost parts of the tundra and in forest tundra, hummocky and "spotty" tundras occur within areas of Tundra differentiated podzolized gleyzems (Albi-Stagnic Luvisols, Albi-Stagnic Luvisols [Gelic]). These types of patterned ground are also widespread on sandy substrates: within areas of Tundra Podburs (Spodi-Turbic Cryosols, Enti-Gelic Podzols) and Podzols (Gelic Podzols, Haplic Podzols, Spodi-Turbic Cryosols [Albic]).

The flat areas and depressions in the interior of undulated and hilly topography have palsa and peat plateau bogs which consist of two organic soils, Peaty bog degraded soils (Cryic Histosols) on a palsa or a peat plateau and Peaty bog mesotrophic soils (Eutric Histosols) without (or very deep) permafrost in the pool. Only in the northernmost parts of the territory with the coldest climates are the swampy flat areas occupied mainly by Peaty bog gleezems (Turbi-Histic Cryosols [Gleyic]).

Figure 2.7.4 shows the example of soil cover of the southern part of Bolshezemelskaya tundra with alternating of "spotty," hummocky tundra and palsa bog.

Soils of tundra lugovinas (meadows) (Ochric Regosols) also occur as a component of soil cover of Kanin, Malozemelskaya and Bolshezemelskaya tundra. They occupy the well drained slopes of hollows and river valleys. Usually, they have deep permafrost, if any, because of warmer conditions – south-oriented slopes and accumulations of snow.

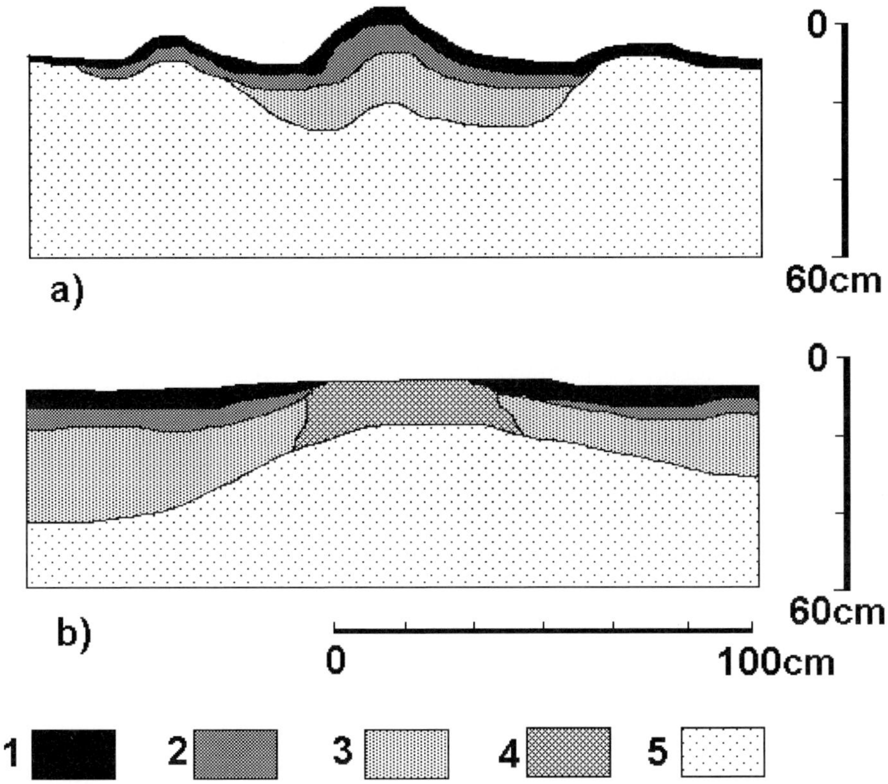

Figure 2.7.3. The most widespread types of soil-microrelief complexes (patterned ground) of Kanin, Malozemelskaya, and Bolshezemelskaya tundras: a) hummocky tundra, b) "spotty tundra" – tundra with non-sorted circles. Soil horizons (FAO): 1- O; 2 – AO; 3 – Br; 4 – Bg; 5 – BC.

4.4. Yugor Peninsula

The Yugor Peninsula is the most northeastern part of the Russian European North, and that is why it is the coldest, especially on the shore of the Kara Sea. The mean annual temperature of Amderma is -7.6°C, the mean in July is 5.8°C, and the mean in January is -19.5°C.

The Pai-Hoi range, with its schists and limestone, occupies the center of the peninsula, while marine deposits containing water-soluble salts in deeper layers make up the peninsula margins. Loamy and clayey substrates predominate here, but sandy and stony materials also play the role of parent substrata for soils.

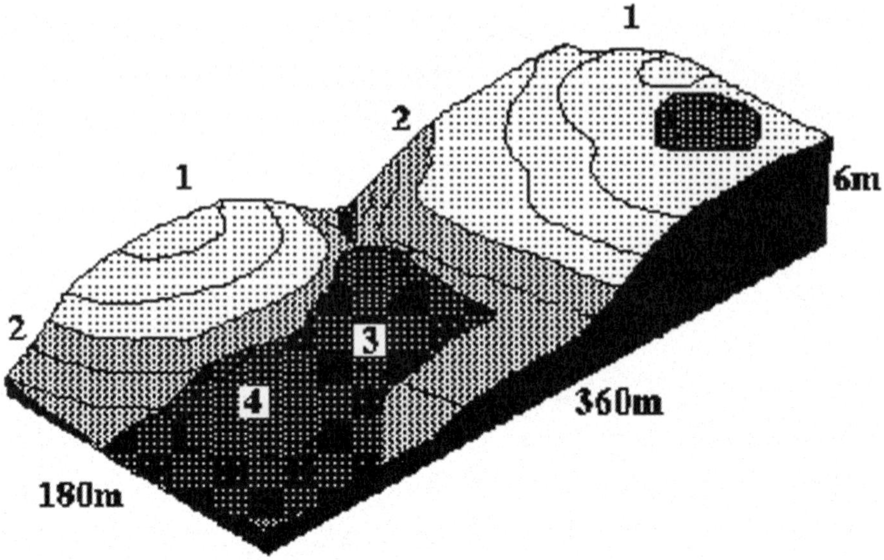

Figure 2.7.4. Soil landscape of undulated plain in the southern part of the Bolshezemelskaya tundra. Soils: 1. Albi-Stagnic Luvisols (spotty tundra – tundra with non-sorted circles), 2. Histic Gleysols (hummocky tundra), 3. Eutric Histosols, 4. Cryic Histosols (palsas).

Botanists consider the territory of the Yugor Peninsula to be the southern variant of Arctic tundra. Sedge-herbaceous-polar willow communities occupy the well drained positions and sedge-cotton-grass-mossy ones are typical for poorly drained flat areas.

Polygon tundras with Arctic tundra weakly gleyic humic soils (Stagni-Turbic Cryosols) with cryogenic circles having salic features (2.2% of salts in upper 2 cm) and Tundra peaty gleyzems (Turbi-Histic Cryosol [Stagnic or Gleyic]) in depressions are the typical soil complex for the Yugor Peninsula. Liverovsky (1934) conducted the first studies of Yugor soils, but Ignatenko (1963) systematically investigated them.

Tundra mucky gleyzems (Gleyi-Turbic Cryosols) are the soils of poorer drainage where hummocky tundras prevail. Tundra Podburs (Spodi-Turbic Cryosos) occupy sandy, well drained positions, and Podburs on stony substrates (Turbi-Leptic Cryosols [Spodic]) are common for the Pai-Hoi highland. Peaty bog gleezems (Turbi-Histic Cryosols [Gleyic]) occupy mainly the swampy flat areas, but in more favorable conditions Peaty bog degraded soils (Cryic Histosols) occur on the small palsas and Peaty bog mesotrophic soils (Eutri-Cryic Histosols) occur elsewhere.

Tundra mucky calcareous soils (Rendzic and Calcaric Leptosols) are well drained soils on limestone under dryas communities. Mountainous primitive soils (Lithic Leptosols) and rock outcrops occur in some places of Pai-Hoi tops.

4.5. Ural Mountains

The Urals are a mountainous chain stretched about 1000 km. along the meridian from the Yugor Peninsula to the steppe zone far in the south. The altitude here is about 800 to 1000 m, up to 1895 m. Such a long chain crosses several landscape zones (tundra and taiga) and subzones. Some alpine belts also occur in the Urals: tundra, alpine tundra, alpine barren in the north; taiga, meadow-birch parklands, alpine tundra, alpine barrens in the south. The parent materials of the Ural soils are shallow rock debris, mostly of basic rocks in the north and shales in the south.

Several specialists studied the soils of the Urals (Belyaev, et al., 1965; Zaboeva, 1975; Firsova and Dedkov, 1983). Permafrost-affected soils predominate in the north (to the north of Arctic Circle), but in the south they mostly occupy only the highest altitudes. Certainly, the predominant element of soil cover of high altitudes is rock and rock debris outcrops.

Mountainous primitive soils (Lithic Leptosols) are also common soils in every alpine belt. In tundra and alpine tundra belts, well drained Podburs and Podzols (Turbi-Leptic Cryosols [Spodic]) are common for mature soils. The soils of meadow birch parklands are Mountainous forest meadow soils (Ochric Leptosols, Umbric Leptosols). On gradual slopes in places of fine earth accumulation, Peaty gleyzems (Turbi-Histic Cryosols [Gleyic]) occur both in tundra and parkland belts, as do Cryic Histosols. The bog soils without permafrost (Eutric and Dystic Histosols) also occur in depressions and valleys of this zone (Belyaev et al., 1965).

5. Conclusions

The Russian European North is the westernmost place on the continent of Eurasia where permafrost-affected soils occur in the plains. That is why the soil cover of this area consists both of Cryosols and permafrost-free soils. The percentage of Cryosols in the soil cover increases from west to east and from south to north. However, only in the northeastern part is permafrost-affected soil cover continuous.

The specific feature of the territory is extensive and intensive cryoturbations in both Cryosols and seasonally frozen tundra soils. The dominant soils here are Gelic Gleysols and Turbi-Histic Cryosols (Gleyic or Stagnic) on loams, Haplic, Entic, and Gelic Podzols and Spodi-Turbic Cryosols on sands, and Eutric, Dystric, and Cryic Histosols on poorly drained flat sites. The highlight of the territory is the wide distribution of texturally differentiated soils (Albi-Stagnic Luvisols, Albi-Stagnic Luvisols [Gelic] and Gelic Albeluvisols) in the southern part of the tundra zone.

6. Acknowledgments

The Russian Foundation for Basic Research provided partial project funding. Special thanks to the Swedish Polar Research Secretariat and N.P. Soltseva for arranging field investigations. We also appreciate very much Scott Smith's great help and consultations in preparing this paper.

References

Archegova, I.B. 1967. Soil cover and elements of microrelief of Vorkuta tundra. Izvestia Komi filiala Vsesoyuznogo geograficheskogo obshchestva. 2(I):55-63. (In Russian.)

Belov, N.P., and A.V Baranovskaya. 1969. Soils of Murmansk region (Pochvy Murmanskoi oblasti). Nauka. Leningrad. (In Russian.)

Belyaev, S.V., I.V. Zaboeva, V.A. Popov, and D.M. Rubtsov. 1965. Soils of Pechora industrial region (Pochvy Pechorskogo promyshlennogo raiona). Nauka. Moscow-Leningrad. (In Russian.)

Firsova, V.P., and V.S. Dedkov. 1983. Soils of high latitudes of Mountainous Ural. Sverdlovsk. 95 pp. (In Russian.)

Fridland, V.M., ed. 1988. Soil map of Russian Soviet Federal Socialist Republic, scale 1:2500000. (In Russian.)

Goryachkin, S.V., R.I. Zlotin, and G.M. Tertitsky, eds. 1994. Diversity of natural ecosystems in Russian Arctic. Lund University. Sweden.

Goryachkin, S.V., R.I. Zlotin, and M.V. Glazov. 1995. Spatial and temporal organization of Arctic Ecosystems. In: Swedish-Russian Tundra Ecology Expedition - 94. A Cruise Report. Stockholm. pp. 334-339.

Gribova, S.A., and I.V. Ignatenko. 1970. The features of vegetation and soil cover of Korotaikha basin. In: B.A. Tihomirov, et al., eds. Biological basis for use of the nature of the North (Biologicheskie osnovy ispolzovaniya prirody Severa). Syktyvkar. pp.119-129. (In Russian.)

Ignatenko, I.V. 1963. Soils of Arctic tundra of Yugor Peninsula. Pochvovedenie. N5:26-40. (In Russian.)

Ignatenko, I.V. 1967. On swampy and bog soils of East European tundra. Pochvovedenie. N2:33-49. (In Russian.)

Ignatenko, I.V. 1971. Soils of Kara basin and their zonal position. Pochvovedenie. N2:3-16. (In Russian.)

Ignatenko, I.V. 1972. The pattern of soil cover of the East European forest-tundra. In: Soils and vegetation of East European forest-tundra. Nauka. Leningrad. pp. 64-101. (In Russian.)

Ignatenko, I.V. 1979. Soils of East European tundra and forest-tundra (Pochvy Vostochnoevropeiskoi tundry i lesotundry). Nauka. Moscow. (In Russian.)

Ivanova, E.N., and N.A. Koposov. 1937. Soils of Khibiny tundras (Pochvy Khibinskikh tundr). USSR Academy of Sciences Press. Leningrad. (In Russian.)

Ivanova, E.N., and O.A. Polyntseva. 1952. Soils of European tundras. In. E.N. Ivanova, ed. Annals of Komi affiliation of USSR Academy of Sciences (Trudy Komi filiala Akademii nauk SSSR. Seriya geograficheskaya. Vypusk 1). pp. 72-122. (In Russian.)

Kreida, N.A. 1958. On soils of East European tundras. Pochvovedenie. N1:62-67. (In Russian.)

Liverovskiy, Yu. A. 1934. Tundra soils of the Northern region (Pochvy tundr Severnogo kraya). USSR Academy of Science press. Leningrad. (In Russian.)

Nikonov, V.V., and V.N. Pereverzev. 1989. Soil formation in Kola Subarctic (Pochvoobrazovanie v Kolskoi Subarktike). Nauka. Leningrad. 168 pp. (In Russian.)

Parshevnikov, A.L. 1967. Soils of near-tundra forests of Arkhangelsk region. In: Vegetation of forest-tundra and ways of its colonization (Rastitelnost lesotundry i puti ee osvoeniya). Nauka. Leningrad. pp. 133-139. (In Russian.)

Polyntseva, O.A., and E.N. Ivanova. 1936. Complexes of spotty tundra of Khibin massive and their evolution related to the evolution of soil and vegetation cover. Annals of Dokuchaev Soil Institute (Trudy Pochvennogo instituta im. V.V.Dokuchaeva). XIII: 213-265. (In Russian.)

Rusanova, G.V. 1994. Natural and disturbed soils. In: M.V. Getsen and S.K. Nazarova, eds. Structure-functional organization of phytocenoses of Far North (Strukturno-Funktsionalnaya organizatsiya fitozenozov na Krainem Severe). Syktyvkar. pp.24-43. (In Russian.)

Tamm, O. 1928. Die klimatischen bodenregionen in Schweden. Proc. Int. Congr. Soil. Sci. Washington, D.C. 1:269-285.

Targulian, V.O. 1971. Pedogenesis and weathering in cold humid regions (Pochvoobrazovanie i vyvetrivanie v kholodnykh gumidnykh oblastyakh). Nauka. Moscow. (In Russian.)

Zaboeva, I.V. 1975. Soils and land resources of Komi ASSR (Pochvy i zemelnye resursy Komi ASSR). Syktyvkar. (In Russian.)

Chapter 8. Cryosols of Western Siberia

N.A. Karavaeva[1]

[1]Leading Researcher, Institute of Geography, Russian Academy of Sciences, Staromonetny, 29, Moscow, 109017 Russia; e-mail pedology@igras.geonet.ru.

1. Introduction

Cryosols are widespread in the northern and central regions of western Siberia, from $73^0 30'$ to 62^0 N latitude, where they formed under a vast spectrum of conditions, from tundra to taiga, including swampy areas (Figure 2.8.1). Between 62^0 and 60^0 N latitude, these soils occur only within the coldest ecotopes of forest areas. The features, diversity, and geography of Cryosols depend on both recent and paleoclimatic factors that control the presence and properties of permafrost (temperature, content of ice, depth, and water and heat regimes of the active layer).

2. Factors of Soil Formation

2.1. Climate

The western Siberian plain holds a transitional place between two climatic regions, the Atlantic climate of Europe and the extremely continental climate of Siberia. Table 2.8.1 gives climatic data for different natural zones and sub-zones. Climate continentality increases from northwest to southeast. Annual precipitation increases from north to south, and its largest portions fall in summer (55% in the north and 70% in the south) and in autumn. Precipitation as snow (80 to 85%) prevails in the north, and rains predominate (70%) in the south tundra and in forest-tundra. The snow cover lasts from 220 to 240 days in tundra and forest-tundra and from 170 to 220 days in taiga. Its thickness varies from several meters in depressions to only 20 to 30 cm on uplands, because of the strong winds typical for this territory.

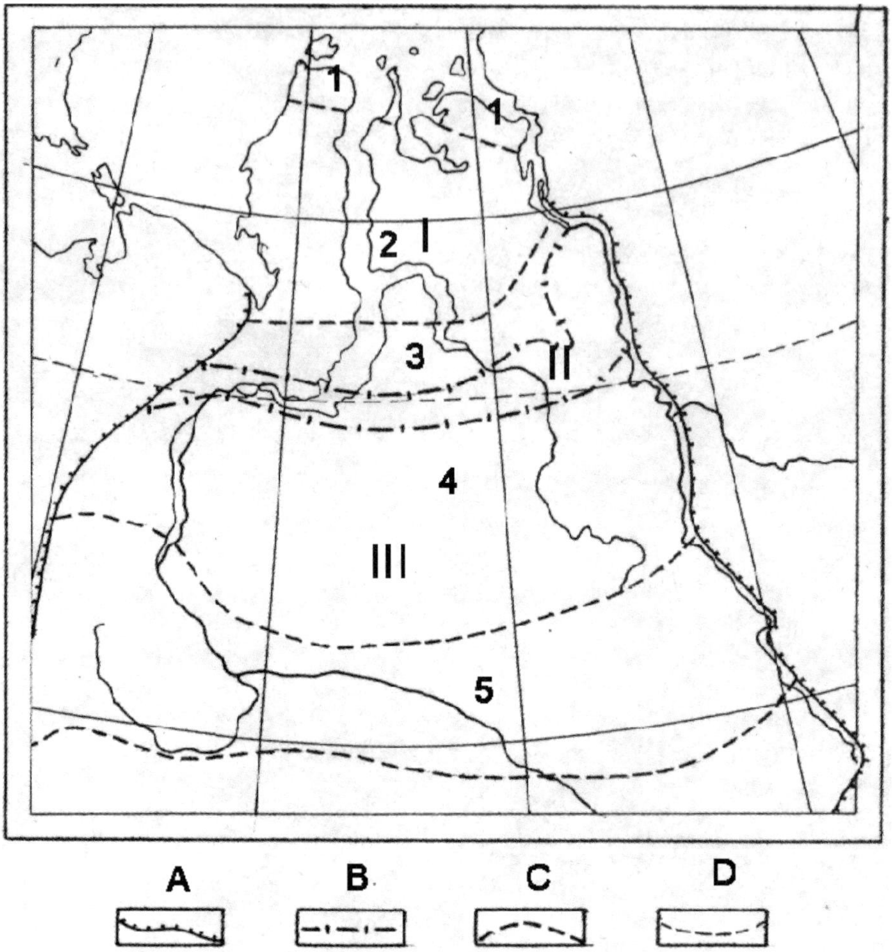

Figure 2.8.1. Natural Zones and Sub-zones in western Siberia. Boundaries. A. Western and eastern boundaries of West Siberia; B. Zonal boundaries; C. Subzonal boundaries; D. Arctic Circle boundaries. Zones and Sub-zones: I. Tundra, 1.Arctic (Sedge-Lichen-Moss), 2. Dwarf Shrub-Lichen-Moss, 3. Shrub; II. Forest-Tundra; III. Taiga, 4. Northern (Spruce-Larch), 5. Central (Siberian cedar-Spruce-Fir).

2.2. Parent materials and topography

Marine, glacial-marine, and fluvioglacial deposits of the middle and upper Quaternary ages cover the western Siberian plain. These deposits show significant spatial variation in texture, from sands to loams and clays, with inclusions of a buried peat. The substrates usually are stratified, even within soil profiles.

Table 2.8.1. Climate data for natural zones and sub-zones in western Siberia.

Zone/sub -zone.	Mean temperature, ^{0}C			Frost-free Days	Precipitation, mm			Snow Cover, Cm
	Yr.	Jly.	Jan.		Yr.	Sum.	Win.	
Tundra								
Arctic	-9.6	5.7	-23.5	40-52	285	115	170	50-60
Dwarf Shrub-moss	-11.2	8.1	-24.7	63	325	195	130	60-70
Shrub	-9.2	10.0	-25.2	77	400	240	160	40-50
Forest tundra	-6.0	12-14	-23 to -27	80-90	400	240	160	60-70
Taiga								
North	-4.9	15-16	-22 to -26	80-100	450-525	350-400	100-125	70-80
Central	-3.4	17	-19 to -23	95	450-525	350-400	100-125	80-100

In the middle part of the plain (between the Ob and Enisei Rivers), sands and loamy sands prevail. In its western part (between the Urals and the Ob River), as well as within the Yamal Peninsula, loamy sandy and loamy clayey deposits have equal extent. The texture of the deposits on the Gydan Peninsula becomes finer from south to north, and, on the east Kara Sea coast, montmorillonite clays are widespread. Middle and Late Holocene peat deposits are widespread parent materials. They cover 30 to 50% of the given territory.

The absolute height of the western Siberian plain is from 0 to 5 m asl (within the Kara Sea coast) to 150 to 200 m asl; altitudes of <100 m prevail. Upland territories constitute a minor portion of the area, 7% of the tundra zone and 20% of the taiga zone. The depth of erosional dissection within lowlands does not exceed 5 m; slopes of <1.5 degrees predominate. Thermokarst, which is an additional factor of mezo- and microrelief differentiation, occurs extensively.

Thus, the characteristic lithologic-geomorphologic features of the territory are level and low topography, poor drainage, and peaty and stratified sedimentary parent materials.

2.3. Permafrost and the active layer depth

Judging from the character of the heat exchange between the ground and the atmosphere, three permafrost zones (continuous, discontinuous, and sporadic) occur from north to south (Shpolyanskaya, 1973).

The continuous permafrost zone includes the whole tundra and the northern part of forest tundra up to 67^0 N latitude. The active-layer depth depends almost totally on the snow cover's thickness; the role of vegetation depends on its ability to favor snow's accumulation. The depth of thawing directly connects with the presence and thickness of the histic horizon and indirectly with the texture of the soil (Table 2.8.2).

Table 2.8.2. The depth (cm) of the active layer in soils of the tundra sub-zones in the continuous permafrost zone on different parent material deposits.

Sub-zones of tundra	Types of deposits		
	Loam and clay	**Sand**	**Peat**
Arctic	30-50	60-75	<30
Dwarf shrub..moss	30-70	50-110	20-40
Shrub	65-100	80-100	30-50

The discontinuous permafrost zone includes a greater part of forest tundra and the north of the taiga, up to about 65^0 N latitude. In general, the thickness of the active layer varies significantly in forest tundra and taiga zones, depending on landscape conditions (lithology, moisture or ice content in soil, aspect and gradient of slope, and type of vegetation). Under peat bogs, permafrost begins from a depth of less than 1 m, usually 40 to 50 cm. Under forests (often swamped), the same depth of thawing occurs in soils formed on loamy clayey deposits and on layered sandy loams and loams in conditions of poor drainage (Targulian et al., 1972; Ovchinnikov et al., 1973; Koposov, 1976). In well drained soils on these parent materials, the active layer of soils may be 1.5 to 2 m thick or more. Deeper active layers occur in soils on loamy sands and sands (Tyrtikov, 1973).

The sporadic permafrost zone spreads up to ~62^0 N latitude in the center of the West Siberian Plain, up to ~60^0 N latitude in its eastern part, and up to the polar circle in the vicinity of the Urals (Figure 2.8.1). Cryosols have small local areas, confined to elevated topographic elements within mossy dark coniferous (usually associated with poorly drained soils) forests on clays and within oligotrophic bogs.

In the discontinuous and sporadic cryolithozones, the modern spatial dynamics of permafrost occur. At present, frozen ground appears under forests as they begin to waterlog. Under swamped forests and young peat bogs, a seasonal frost turns into permafrost. Further accumulation of peat and water in bogs degrades the permafrost (Tyrtikov, 1973; Karavaeva, 1982).

2.4. Vegetation

The territory of western Siberia includes three geobotanic zones (tundra, forest tundra, and taiga), two of which (tundra and taiga) include sub-zones that reflect differences in the composition of plant communities.

The tundra zone extends over 750 km from the north to the south. Its northern sub-zone, the Arctic tundra, is mostly on the Yamal and also on the Gydan Peninsula, up to ~72^0 north latitude. In this sub-zone, hummocky and tussocky phytocenoses composed of dwarf shrubs, sedges, and mosses occupy loamy watersheds; many mud circles and cryogenic cracks appear. On watersheds of sands and loamy sands, lichens become a significant component of plant communities; bare circles are 1.0 to 1.5 m in diameter and cover nearly 50% of the surface. Within depressions and coastal areas, low-center polygon bogs usually occur. Mosses, herbaceous plants, and dwarf shrubs mostly occupy shoulders of polygons, while sedges, arctophila, dupontia, and mosses are confined to low centers and cryogenic cracks.

The next sub-zone, the dwarf shrub-lichen-moss (sub-Arctic) tundra, includes greater parts of the Yamal and Gydan Peninsulas and has a southern boundary at about 68^0 N latitude. Its characteristics are an abundance of dwarf shrubs from 20 to 25 cm tall; a high canopy density; willows in river valleys; peat plateau bogs and palsa bogs (along with low-center polygon ones); and large areas of swamped tundra with a tussocky microrelief and a prevalence of sedges, moss, and lichen species.

In the southern sub-zone, birch, willow, and alder and considerable numbers of tussocky phytocenoses and bare circles dominate the shrub tundra. Near its southern boundary, 67^0 N latitude, rare larch trees grow; areas of low-center polygon bogs reduce, and those of peat plateau and palsa bogs extend.

The forest tundra zone has a width of 100 to 250 km (Vasilievskaya, et al., 1986). Open woodlands (spruce and larch) occur only on flat watersheds and gentle slopes with thick snow cover. Tundra vegetation grows in windy places with a thinner snow cover. Only small areas of low-center polygon bogs occur, whereas peat plateau and palsa bogs occupy more than 50% of the forest tundra and northern taiga. Micro- and mezorelief of these bogs is very distinct, due to an intensive thermokarst. The vegetative cover of many frost mounds, especially hillocks, often has been destroyed mechanically. Depressions are wide and deep; water accumulation and peat formation occur in most of them.

The taiga zone has extensive areas of poorly drained soils (50 to 75% of the territory), especially in the north. Coniferous spruce, Siberian cedar, and fir forests with mossy forest floors spread up to 60^0 N latitude within relatively drained areas and on loamy clayey materials, where permafrost occurs within a 1-m depth; these forests correspond to the southernmost positions of Cryosols. Pine and pine-larch forests occur on sandy loams and sands that have a sporadic permafrost at a depth of over 2 to 3 m. Large areas of oligotrophic string fen bogs occur. Permafrost can form within 1 m at the highest points of these bogs.

3. Paleogeography of the Holocene

Several climatic changes took place during the Holocene in the northern and central regions of western Siberia that caused shifts of natural zones and transformations of permafrost (Kats, 1957; Hotinsky, 1977). During the Boreal period (8.0 to 9.2 thousand years ago), forests invaded the tundra beyond the polar circle. The Atlantic period (4.6 to 8.0 thousand years ago) was warmest and most humid, and the northward shift of the forest belt ranged from 200 to 400 km; peat lands actively formed and extended to surrounding landscapes.

During the Subboreal period (2.3 to 4.6 thousand years ago), the mean annual temperature decreased by 3 to 7^0C, the climate's continentality increased, and spruce forests retreated to 3 to 4^0 southward. The Subatlantic period had the coldest and most continental climate, which caused a further retreat of coniferous forests and an increased growth of peat bogs.

4. Main Soil Groups and their Geography

On the basis of literature and cartographic data, the permafrost-affected soils in western Siberia can be grouped as described below. Conventional Russian taxonomy for the region appears first, followed by the *World Reference Base* (WRB) (ISSS Working Group RB, 1998) equivalent taxa, in brackets.

4.1. Tundra Gleyzem (Turbi-Histic Cryosols [Stagnic])

Soils of this group have the surface histic or O horizon of 5 to 15 cm, mineral horizons with cryogenic and gleyic features, and permafrost from a depth of 50 to 100 cm. They occur in tundra and forest-tundra on elevated topographic positions. Judging from the drainage and temperature conditions and the amount of plant residues, this group includes Tundra Mucky Gleyzems (Turbi-Saprihistic Cryosols [Stagnic]), Tundra Peaty Gleyzems (Turbi-Fibrihistic Cryosols [Stagnic]), and Tundra Differentiated and Podzolized Gleyzems (Turbi-Fibrihistic Cryosols [Endogleyic and Albic]).

The first spread mainly in the Arctic tundra and lichen-moss tundra. The second occur throughout tundra and forest tundra. The third occur in the shrub tundra, forest tundra, and northern taiga, because temperature conditions, including the duration of a warm period and the depth of thawing, control the development of textural differentiation.

4.2. Tundra Mucky Gleyzems (Turbi-Saprihistic Cryosols [Stagnic])

These soils' properties illustrate changes in soil-forming conditions, in comparison with those in Europe, which the more continental climate in western Siberia causes (Ivanova, 1962). These changes result in better decomposition of plant residues, a smaller thickness of the histic or O horizon, a partial aeration of the upper horizons, the maximum development of gleyic features over the permafrost, and the small depth of an active layer.

In western Siberia, such soils often occur on hummocks in the mud circles tundra, which has a hummocky tussocky pattern ground. Hummocks are from 40 to 60 cm in diameter and 5 to 15 cm tall; microdepressions between them are 10 to 15 cm wide; circles are 20 to 30 cm in diameter and usually occur on elevated positions. All these dimensions increase from the north to the south of the tundra. (See Example 2.8.1.)

An acid reaction and an unsaturated exchange complex are typical for all soil horizons (Table 2.8.3). The humus profile shows, by their high concentration, a deep penetration of labile humus substances into the mineral part of the soil. Usually, there are two peak concentrations of the illuvial humus, one in the A1Bh horizon and the other near the contact with permafrost. Humin represents almost half the organic matter. The C/N ratio is very wide, especially in the O3 horizon. Fulvic acids prevail over humic ones; the Cha/Cfa in the O horizon is 0.27.

No differentiation occurs in distribution of soil particles <0.001 and <0.01 mm or of the total content of Al_2O_3, K_2O, and Na_2O (Tables 2.8.3 and 2.8.4). However, a weak differentiation in bulk Fe occurs; the minimum is in the Br1, and maxima are in the Bg and Br3 horizons. Simultaneous downward and upward flows of iron-containing solutions during the soil's freezing cause this distribution. An upward migration results in the accumulation of iron in the form of oxides in the Bg horizon. A downward migration leads to an increase in iron concentration in the permafrost horizon; however, anaerobic conditions prevent the oxidation.

The soils of a mud circle, Ochri-Gleyic Turbic Cryosol, within such a pattern ground, have a dry thin surface crust <1 cm thick, which is often glossy due to the presence of iron-humus compounds. The ABhg horizon underlies this crust. It resembles the upper mineral horizon formed under vegetation but is gleyed. Beneath it, the horizonation is the same as in the Turbi-Saprihistic Cryosol.

The specific features of Ochri-Gleyic Turbic Cryosol are the absence of the surface O horizon, an upward convexity of boundaries between horizons, their cryogenic break, and the burying of fragments of the upper layers into the lower ones.

4.3. Tundra Peaty Gleyzems (Turbi-Fibrihistic Cryosols [Stagnic])

Turbi-Fibrihistic Cryosols (Stagnic) occupy microdepressions between hummocks. These soils occur in different types of patterned ground and often (in cases of a poor drainage) are dominant components. Example 2.8.2 describes such soil.

An unsaturated exchange complex within a 50-cm deep layer and an acid pH increasing with depth are quite typical (Table 2.8.3). The profile distribution of labile humus substances, which penetrate down to the permafrost layer, is uniform (2 to 3%). The Cha/Cfa ratio is about 1.0 in upper horizons and from 0.4 to 0.5 in deeper ones.

Considerable seasonal variations occur in the humus composition. In the upper horizons, the content of humic and fulvic acids decreases along with an increase in the proportion of humin during freezing. At the same time, an increase in the content of humic acids and a decrease in the content of fulvic acids take place in lower horizons. Upon thawing, the concentration of fulvic acids becomes higher, and that of humin gets lower. Such a cyclic cryogenic transformation of the humus composition probably reduces the chemical action of humus (Dergacheva and Dedkov, 1977).

The contents of clay (<0.001 mm) and of physical clay (<0.01 mm) are uniform to a depth of 94 cm (Table 2.8.3). Bulk concentrations of elements in the total mass and in the clay fraction (<0.001 mm) are the same (Table 2.8.4). Reduction conditions and a stagnant water regime that hinders the removal of iron compounds from the profile explain the relatively high content of non-silicate iron. The clay fraction is mostly hydromica, mixed-layer silicates, kaolinite, and chlorite (Vasilievskaya, et al., 1986).

In complexes with Turbi-Fibrihistic Cryosols (Stagnic), soils of mud circles have a homogenous gleyed profile with an indistinct horizonation. Soils of microdepressions between hummocks are Boggy Peaty Gleyzems (Turbi-Fibrihistic Cryosols [Gleyic]) with an H horizon up to 50 cm thick, underlain by permafrost.

Example 2.8.1. Pit No. 116 (Tonkonogov, 1979). Arctic hummocky tundra with a sedge-dwarf shrub-moss vegetation, on the Yamal Peninsula.

O, 0 to 3 cm. Dark brown-gray Saprihistic horizon; the upper 1 cm is peaty; an average thickness is up to 10 cm.

ABh, 3 to 5 cm. Brown loam: impregnated with an illuvial humus; contains humus formed in situ as a result of root decomposition; subangular blocky; loose; non-gleyed.

Bg, 5 to 7 cm. Ochric loam; a bright stripe at the contact with the lower gleyed horizon; weakly compacted; weakly structured in small angular blocks.

Br, 7 to 35 cm. Bluish coarsely textured loam; thixotropic; contains soft Fe-Mn concentrations.

Cri, > 35cm. Frozen dark bluish coarsely textured loam; a considerable content of ice; the maximum depth of thawing is 50 to 60 cm.

Example 2.8.2. Tundra peaty Gleyzem from pit No. 101 (Vasilievskaya at al., 1986). Forest-tundra zone in an area of circle-hummocky dwarf shrub-lichen tundra, on the Yamal Peninsula.

O, 0 to 14 cm. Brown fibrihistic horizon; its thickness varies from 8 to 20 cm.

Bg1, 14 to 19 cm. Heterogeneous in color; bluish and ochric mottles on the brown background; loamy; loose; weakly structured in small subangular blocks.

Bg2, 19 to 33 cm. Heterogeneous in color; ochric and brown mottles and lenses and an illuvial humus on the bluish background; loamy; thixotropic; a bright ochric stripe along the lower boundary.

Br, 33 to 56 cm. The same color; loamy; thixotropic; wet; contains soft Fe-Mn concentrations.

BCr1, 56 to 71 cm. The same color; loamy; compact; not thixotropic.

BCr2, 71 to 94 cm. Homogenous brownish olive color; loamy; less compact.

2Cg, 94 to 100 cm. A sharp change to a sandy texture; bluish brown with rusty lenses; soft Fe-Mn concentrations; frozen from 100 cm.

Table 2.8.3. Some basic chemical properties of Turbi-Saprihistic (Stagnic) and Turbi-Fibrihistic (Stagnic) Cryosols.

Horizon depth, cm	pH H^2O	Exchangeable cations, meq/100g soil			Base saturation, %	Humus %	Particle size, mm; fractions, %	
		Ca^{2+}	Mg^{2+}	H^+			<0.01	<0.001
Turbi-Saprihistic Cryosol (Stagnic), pit No.116 (Tonkonogov, 1979).								
O, 0-3	4.5	7.2	6.2	15.6	46	n.d.	n.d.	n.d.
ABh, 3-5	4.6	1.8	3.00	6.1	44	7.0	34	19
Bg, 5-7	4.5	1.7	2.7	3.9	53	5.7	30	17
Br1, 7-10	5.0	3.2	4.5	2.2	78	5.6	30	16
Br2, 10-25	4.6	1.0	2.8	4.6	51	5.6	27	15
Br3, 25-35	4.2	1.5	2.3	4.1	48	6.5	28	16
Cri, 35-45	4.4	2.0	3.0	4.2	54	7.2	28	19
Turbi-Fibrihistic Cryosol (Stagnic), pit No.101 (Vasilievskaya et al., 1986).								
O, 0-14	4.9	3.2	0.9	76.0	5	28.7	n.d.	n.d.
Bg1, 14-19	5.0	2.0	0.9	17.3	14	3.1	29	17
Bg2, 19-33	4.9	2.1	0.4	17.3	13	2.5	28	17
Br, 33-56	5.0	1.5	0.5	5.9	26	2.1	27	17
BCr1, 56-71	5.6	1.3	0.8	2.0	50	2.0	29	18
BCr2, 71-94	6.2	2.0	1.4	0.8	81	2.5	27	15
2Cg, 94-100	6.4	0.5	0.3	0.3	77	0.1	4	2

Table 2.8.4. Total elemental analysis of Turbi-Saprihistic (Stagnic) and Turbi-Fibrihistic (Stagnic) Cryosols.

Horizon, depth, cm	Wt loss on ignition (%)	SiO_2	Al_2O_3	Fe_2O_3 Total	non silicate	CaO	MgO	K_2O	Na_2O
Turbi- Saprihistic Cryosol (Stagnic), pit No. 116 (Tonkonogov, 1979).									
ABh, 3-5	11.50	76.06	14.29	3.56	n.d.	2.26	1.00	2.13	1.65
Bg, 5-7	8.31	77.38	13.30	4.20	n.d.	1.34	0.84	2.15	1.86
Br1, 7-10	6.97	78.78	12.85	2.63	n.d.	1.32	0.95	2.18	1.91
Br2, 25-35	7.86	78.82	12.45	3.23	n.d.	1.33	0.89	2.18	1.97
Cri, 35-45	8.81	78.33	12.60	2.88	n.d.	1.52	1.09	2.16	1.95
Turbi-Fibrihistic Cryosol (Stagnic), pit No.101 (Vasil'evskaya et al., 1986).									
0, 0-14	28.73	77.05	15.31	3.51	1.85	1.15	0.99	2.34	n.d.
Bg1, 14-19	6.42	76.51	15.84	3.44	1.77	0.62	0.74	1.51	n.d.
Bg2, 19-33	2.55	76.34	16.35	3.05	2.05	0.82	0.53	1.48	n.d.
Br, 33-56	1.95	75.81	16.39	3.76	1.93	0.68	0.85	1.23	n.d.
BCr1, 56-71	1.35	75.94	16.48	3.49	1.61	1.47	0.71	1.47	n.d.
BCr2, 71-94	0.84	77.93	15.22	3.51	1.07	0.92	0.45	1.34	n.d.
2Cg, 94-100	0.35	87.31	5.83	3.18	n.d.	1.52	0.51	1.59	n.d.

4.4. Taiga Gleyzems (Endogleyi-Histic Cryosols)

This group resembles the Tundra Peaty Gleyzems. It differs in its thinner histic horizon (<10 to 5 cm), more decomposed organic matter (saprihistic or mollic), an appearance of distinct stagnic features from a depth of over 50 cm, an alternating oxidation-reduction regime within the upper 50 cm, occasional turbic features, and the presence of permafrost from about 100 cm. These soils occur only in taiga over elevated drained topographic elements consisting of loamy and clayey deposits.

The subdivision within this group reflects the presence or absence of the albic horizon: Taiga Peaty-Mucky Gleyzems (Endogleyi-Saprihistic Cryosols) and Taiga Differentiated and Podzolised Gleyzems (Endogleyi-Histic Cryosols [Albic]). Soils with similar diagnostic features and horizons, but with permafrost at 1 to 2 m and deeper or just with a seasonal frost, occur in forest-tundra and especially in taiga. In accordance with the WRB taxonomic system (ISSS Working Group RB, 1998), such soils should be classified as Histic Gleysols (including Albic and Gelic subgroups).

4.5. Tundra and Taiga Differentiated and Podzolized Gleyzems (Endogleyi-Histic Cryosols [Albic])

These soils occur only in conditions of a long (more than three-month) warm season and at a considerable depth of thawing. Hence, they appear from the shrubby tundra to taiga up to 60^0 N latitude. (See Example 2.8.3.)

A weakly acid pH of the whole profile, few exchangeable cations, and an insignificant saturation of the exchange complex (Table 2.8.5) characterize this soil. The organic profile includes the fibrihistic horizon (the loss on ignition exceeds 70%) and labile humus, which is impregnated into the Eg and Bgs horizons (to a depth of 35 cm) in concentrations from 1 to 2%. All humus is fulvate; the Cha/Cfa values in the O horizon, at 50 to 60 cm, and over the permafrost, are 0.5, 0.3, and 0.1, respectively.

The clay content is nearly the same (3%) in the eluvial and illuvial horizons, and its increase above the permafrost layer is insignificant (Table 2.8.5). Bulk concentrations of sesquioxides in the Eg and Bgs horizons are similar (Table 2.8.6). At the bottom of the active layer, the eluvial gley removal of Fe_2O_3 and sometimes of Al_2O_3 occurs. The bulk composition of clay and the oxalate-soluble forms better show the R_2O_3 redistribution within the upper part of the profile. Bleaching of the albic horizon implies the removal of iron films from coarse particles, of iron hydroxides from thin fractions, and of Al-humus compounds. The eluviation of SiO_2, R_2O_3, and humus from the whole profile accompanies these eluvial gley processes.

Example 2.8.3. Profile No. 9, (Karavaeva, 1982). Under taiga forest of fir and siberian cedar with a ground cover of mountain cranberry, grass, and mosses, on the right side of the Ob River near the Vakh River mouth, close to the southern limit of distribution of these soils.

O, 0 to 11 cm. Dark brown; peaty (fibrihistic).
Eg, 11 to 16 cm. Whitish light brown with small rusty spots; loam; subangular blocky; loose.
Bgs, 16 to 35 cm. Bright rusty brown with bluish spots; loam; subangular blocky.
B, 35 to 58 cm. Brown; loam with a cryogenic stratification.
Bg, 58 to 81 cm. Bluish brown; loam; a bluish color between cryogenic microlayers.
BCg, 81 to 92 cm. Loam with a bright rusty color of microlayers; seasonally frozen.
Cgi, 92 to 240 cm. Rusty and bluish spots; loam; horizontal layers with an angular blocky structure; thick iron oxide films with diffuse boundaries and whitish skeletans between the layers; permafrost is represented by very fine crystals of ice.

Table 2.8.5. Some basic chemical properties of Endogleyi-Histic (Albic) and Ochri-Turbic (Stagnic) Cryosols.

Horizon, depth cm	pH (H_2O)	Exchangeable cations (meq./100g)			Base saturation	Humus,	Particle size, mm. Fractions		Al_2O_3 Fe_2O_3 Oxalate-extractable,	
		Ca^{2+}	Mg^{2+}	H^+	%	%	<0.01.	<0.001	%	%
Endogleyi-Histic Cryosol (Albic), pit No.9 (Karavaeva, 1982).										
O, 0-11	4.5	16.5	3.0	18.4	52	73.2*	n.d.			
Eg, 11-16	4.8	4.2	2.1	12.3	34	2.5	37	19	0.50	0.72
Bgs, 16-35	5.0	5.1	2.4	9.0	46	1.0	39	22	0.69	0.91
B, 35-58	5.0	4.0	1.8	8.0	42	0.6	35	21	0.37	0.37
Bg, 58-81	4.8	3.3	1.2	7.4	38	0.5	32	20	0.53	0.64
BCg, 81-92	4.9	4.8	1.5	6.7	49	0.4	36	23	0.53	0.92
Cgi, 92-102	5.2	7.6	1.7	4.8	66	0.2	30	19	0.37	0.37
Ochri-Turbic Cryosol (Stagnic), pit No.17 on mud circle (Vasilievskaya at al., 1986).										
Au, 0-0.5	4.3	2.5	1.0	84	4	18.2	*	n.d.	n.d.	
B1, 0.5-10	4.5	0.8	0.4	13.8	8	1.0	33	18	n.d	
B1g, 10-30	4.5	0.6	0.4	19.0	6	0.3	32	18	n.d.	
B2g, 30-55	4.9	1.1	0.2	11.7	10	0.2	33	21	n.d.	
B3g, 55-85	5.0	1.5	1.7	8.2	28	0.05	36	21	n.d.	
BCgi, 85-100	5.9	3.2	1.6	1.6	75	0.0	34	20	n.d.	

*Loss on ignition, %.

Table 2.8.6. Total elemental analysis of Endogleyi-Histic (Albic) and Ochri-Turbic (Stagnic) Cryosols.

Horizon, depth, cm	Wt Loss on Ignition (%)	SiO_2	Al_2O_3	Fe_2O_3	CaO	MgO
Endogleyi-Histic Cryosol (Albic), pit No.9 (Karavaeva, 1982).						
Eg, 11-16	8.54	76.55	13.30	4.26	1.93	0.71
Bgs, 16-35	6.83	75.82	13.56	4.69	2.00	0.80
B, 35-58	6.05	75.39	13.75	4.60	2.09	0.73
Bg, 58-81	5.43	75.64	13.00	4.43	1.96	0.87
BCg, 81-92	6.18	74.11	13.57	5.74	1.82	1.22
Cgi, 92-102	4.96	75.94	12.58	3.73	1.96	0.94
Ochri-Turbic Cryosol (Stagnic), pit No.17 on mud circle (Vasilievskaya et al., 1986).						
Au, 0-0.5	18.18	78.34	14.63	4.61	1.49	0.29
B10.5-10	2.14	79.41	14.01	3.34	1.24	0.19
B1g, 10-30	2.31	79.12	14.80	3.51	0.91	0.12
B2g, 30-55	3.08	78.39	14.52	3.14	0.82	0.71
B3g, 55-85	3.11	78.04	15.31	3.18	0.99	0.81
BCgi, 85-100	1.94	78.09	15.84	4.21	0.56	0.88

Endogleyi-Histic Cryosols Albic can have a surface pattern of irregular circles (Vasilievskaya et al., 1986). Thawing within circles is from 15 to 20 cm deeper, so soils there can be classified as Ochri-Turbic Cryosols (Stagnic) as well as Ochri-Turbic Gleysols (Cryic). They have a surface crust (Au horizon) of up to 0.5 cm thick but no O or Eg horizon (Tables 2.8.5 and 2.8.6). The Bg horizon, found under the crust, is oxidized more strongly and contains more Fe-Mn concentrations than the same horizon under vegetation. This can result from an irregular oxidation and sharp changes in redox conditions.

Compared to soils developed under vegetation, soils of circles have a more acidic pH within the upper 30 cm, a lower base saturation of the exchange complex, and lower contents of the illuvial humus and of bulk R_2O_3 (especially non-silicate Fe_2O_3) and RO in a whole profile. Their surface crust and the upper 10 cm of the Bg horizon replace the eluvial part of the profile (i.e., Eg horizon), which was destroyed during the circle's formation.

4.6. Boggy Peaty Gleyzems (Turbi-Fibrihistic Cryosols [Gleyic])

An histic horizon of up to 50 cm, gleyed mineral horizons, and permafrost from a depth of 50 to 100 cm characterize these soils. They occur all over western Siberia, within depressions and poorly drained, level areas. They are subdivided by the thickness of the histic horizon (<30 cm and 30 to 50 cm) and by the peat quality into Eutrophic, Mesotrophic, and Oligotrophic. Similar properties (except that permafrost is absent or formed at a greater depth) and lithological-geomorphological conditions of soil formation are typical for Histic Gleysols (ISSS Working Group RB, 1998).

4.7. Tundra Podburs Gleyed (Endogleyi-Turbic Cryosols [Spodic])

The main characteristics of this group are a 5-cm thick O horizon, a shallow (from 2 to 7 cm) mottled albi-spodic horizon, the spodic horizon of about 5 cm (rarely up to 30 cm) thick with turbic features and often with buried peat fragments, a mottled gleyed horizon usually at a depth of more than 50 cm, and permafrost at 70 to 100 cm.

These soils formed under tundra vegetation on well drained sands, loamy sands, and stony loamy parent material deposits on plains and within small depressions. Strongly and weakly developed humus-illuvial processes distinguish Dark from Light Podburs. In forest-tundra and taiga, Haplic and Gleyic Podzols, which develop on the same parent deposits, replace them. Their surface pattern is usually hummocky-tussocky; mud circles appear only in the northern tundra. Endogleyi-Turbic Cryosols (Spodic) are confined to hummocks; their neighbors on lower elements of microrelief are Turbi-Fibrihistic Cryosols (Stagnic), described previously. Example 2.8.4 gives a profile description of an Endogleyi-Turbic Cryosol (Spodic).

Usually, turbic features such as distorted horizon boundaries and buried fragments of organic matter occur at a depth from 20 to 40 cm. These soils have an acid pH all over the profile (Table 2.8.7) and an eluvial-illuvial type of humus distribution (i.e., the minimum humus content [1.2%] in the EBh horizon and the maximum [5.3%] in the Bhs1 horizon); labile humus penetrates to a depth of 50 cm. They also show a distinct eluvial-illuvial distribution of bulk R_2O_3, especially Fe_2O_3 (Table 2.8.8).

Features of humus illuviation are better developed in the Arctic tundra than in areas to the south of it (Tonkonogov, 1979). The action of ground water above a shallow permafrost and the presence of buried organic fragments in the Bhs1 horizon, which are hardly distinguishable because of an intensive background color, explain this. In the forest-tundra and the northern part of the taiga zone, Podzols, often gleyic replace these soils.

Example 2.8.4. From pit No. 117 of (Tonkonogov, 1979), under hummocky, dwarf shrub-moss-lichen tundra vegetation, on the Yamal Peninsula.

O, 0 to 3 cm. Saprihistic, almost black, with an admixture of bleached sandy grains.
EBh, 3 to 10 cm. Brown-light gray sand forming lenses of 2 to 3 cm thick, without gleyic mottles.
Bhs1, 10 to 40 cm. Brown sand colored by illuvial humus, without gleyic mottles.
Bhs2, 40 to 50 cm. Ochric-light brown sand; the color intensity decreases with depth.
BCg, 50 to 75 cm. Yellow-brown sand with rusty and blue gleyic mottles.
Cgi, >75 cm. Bluish sand with rusty mottles; frozen, with a low ice content.

Table 2.8.7. Some basic chemical properties of Endogleyi-Turbic Cryosols (Spodic) and Histi-Gleyic Podzols (Placic).

Horizon depth, cm	pH H_2O	Exchangeable cations, meq./ 100 g soil			Base satura-tion,%	Humus, %	Particle size, mm; fractions, %.	
		Ca^{2+}	Mg^{2+}	H^+			<0.01	<0.001
Endogleyi-Turbic Cryosol (Spodic), pit No. 117 (Tonkonogov, 1979).								
O, 0-3	4.6	14.6	1.3	1.0	86	n.d.	n.d.	n.d.
EBh, 3-10	5.2	0.9	<0.1	<0.1	n.d.	1.2	4	3
Bhs1, 10-40	4.6	6.5	<0.1	<0.1	n.d.	5.3	10	8
Bhs2, 40-50	4.8	2.7	<0.1	<0.1	n.d.	1.4	6	4
BCg, 50-75	5.3	0.6	<0.1	<0.1	n.d.	0.5	1	1
Histi-Gleyic Podzol (Placic and Entic), pit No.T-1 (Vasilievskaya et al., 1986).								
O1, 0-4	3.6	14.0	2.6	2.3	12	91.6*	n.d.	n.d.
O2, 4-5	3.9	5.2	0.8	1.0	15	26.0*	n.d.	n.d.
E , 5-7	3.6	0.4	2.2	0.7	21	4.4	17	4
B1, 7-15	4.1	0.2	1.1	0.4	22	1.3	23	10
B2, 15-23	4.2	0.1	0.5	<0.1	11	1.2	24	11
Bg, 23-35	4.4	0.1	0.5	<0.1	8	0.9	21	10
2BCg, 35-56	5.3	0.1	0.4	<0.1	7	0.4	4	3
2Cms, 56-92	5.5	0.1	<0.1	<0.1	10	0.3	2	2

*Loss on ignition, %.

Table 2.8.8. Total elemental analysis of Endogleyi-Turbic Cryosols (Spodic) and Histi-Gleyic Podzols (Placic).

Horizon, depth, cm	Loss on ignition, %	SiO_2	Al_2O_3	Fe_2O_3	CaO	MgO
Endogleyi-Turbic Cryosol (Spodic), pit No.117 (Tonkonogov, 1979).						
Ebh , 3-10	1.60	92.69	3.80	0.62	1.25	0.30
Bhs1 , 10-40	8.01	89.18	5.21	2.40	0.98	0.35
BCg, 50-75	1.18	90.53	5.57	0.96	0.90	0.22
Histi-Gleyic Podzol (Placic and Entic), pit No.T-1 (Vasilievskaya et al., 1986).						
E, 5-7	7.41	87.33	6.29	1.41	0.82	0.63
B1, 7-15	6.10	80.73	10.04	3.34	1.69	0.62
B2, 15-23	5.58	81.40	9.89	3.00	1.01	1.40
Bgm, 23-35	5.03	88.64	4.10	3.10	1.05	0.58
2BCg, 35-56	3.08	91.34	3.96	0.90	0.92	0.83
2Cms, 56-92	3.00	93.78	3.19	0.45	0.57	0.81

4.8. Podzols

These soils are widespread in forest tundra and taiga on well drained sands and loamy sands. Permafrost occurs at a depth of 150 to 250 cm in forest tundra and is absent in taiga. According to the WRB (1988), these are Haplic Podzols (sometimes Gelic or Entic). Following the Russian soil classification, we distinguish iron-illuvial and humus-illuvial (poor or rich by illuviated organic carbon) Podzols.

4.9. Peaty Gleyed Podzols (Histi-Gleyic Podzols [Gelic and Densic])

These soils occur within the same area as those of the preceding group but in sites with the groundwater table at a depth of 1 to 2.5 m on flat watersheds, middle and lower parts of slopes, and shallow depressions. There is no permafrost, so these are seasonally frozen or non-permafrost soils.

Significant variations in the thickness and character of diagnostic (albic and spodic) horizons depend on the ground water regime and its chemical composition. The albic (E) horizon has a thickness from several to tens of centimeters (up to 50 cm) and usually is gleyed. The spodic (B1 and B2) horizon is from black to red in color and has a varying degree of cementation (up to Placic and Densic). The fibrihistic horizon is from 10 to more than 20 cm thick (pit No.T-1, Tables 2.8.7 and 2.8.8).

4.10. Organogenic (Boggy) soils (Cryic Histosols)

Bog soils cover up to 50% of the total area of western Siberia, especially in the north. Their profile consists of one or two diagnostic horizons. The first, and often the only, one is organogenic (formed of layers differentiated by the peat quality), and the second is a gleyic mineral one. The depth of permafrost depends on the geographic location and on specific site conditions.

The classification of bog soils reflects (1) the degree of organic matter decomposition, (2) the thickness of the peat layer, (3) the peat quality that is different in upland, transitory, and lowland bogs, (4) the type of microrelief, which determines parameters of the soil cover pattern. These criteria depend on the genesis and age of bogs and modern conditions that may favor the accumulation of peat or the adverse process of peat mineralization and degradation.

In the northern part of western Siberia, four types of bogs with specific soils and ground patterns occur: (1) low-center polygon bogs, (2) peat plateau bogs, (3) palsa bogs, and (4) leveled bogs. The first three types are most common.

4.10.1.Soils of low-center polygon bogs

These bogs spread mostly in the northern half of the tundra zone. To the south, their areas become smaller, but they do occur up to the northern forest-tundra.

The polygonal pattern forms due to the cryogenic cracking of moist ground, followed by bulging along the cracks filled by wedge ice (Figure 2.8.2). The peat formation on these bogs is slow because of a heat deficiency.

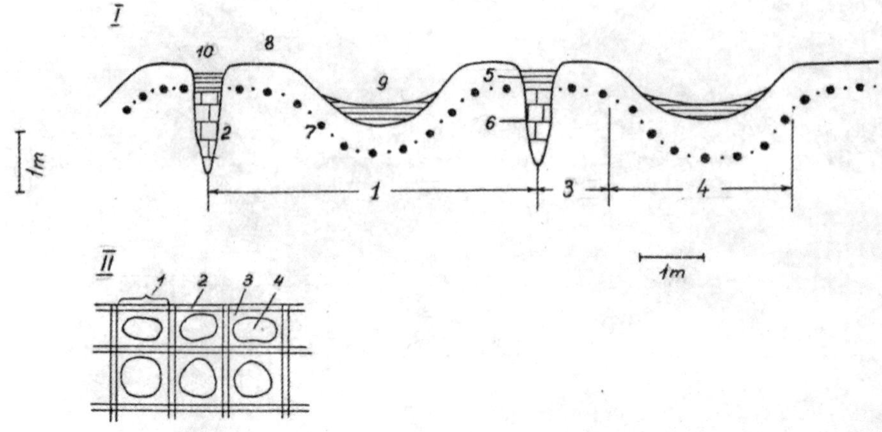

Figure 2.8.2. Low-center polygon bog and its soil association in dwarf shrub-lichen-moss tundra. I. cross-section, II. view from above. 1. low-center polygon, 2. trough, 3. shoulder, 4. low center, 5. peat in crack, 6. ice-wedge, 7. permafrost table. Soils: 8. Turbi-Saprihistic Cryosol (Stagnic), 9. Fibrihisti-Gleyic Cryosol, 10. Glaci-Histic Cryosol.

In western Siberia, polygons are typically 10 to 20 m in perimeter, and their periphery of 0.4 to 0.5 m in width heaves up to 0.3 to 0.7 m above the center of the polygon. The height of the shoulder corresponds to the depth of the trough (crack); the width of trough depends on the intensity of the thermokarst and on erosion.

Soil complexes that developed on such bogs include three components. Cryosols (Saprihisti-Gleyic) are confined to shoulder segments of the patterned ground and are underlain by permafrost at about 50 cm. Fibrihisti-Gleyic Endostagnic Oxiaquic Cryosols form within low centers of polygons with permafrost at 50 to 70 cm and a 30-cm layer of Dystric-Euthric peat and a constantly gleyed horizon. Fibrihisti-Glacic Cryosols in cryogenic cracks have a peat layer <20 cm thick, underlain by ground (ice wedge) ice.

4.10.2. Soils of peat plateau bogs

The area of peat plateau bogs lies between the shrub-lichen-moss tundra and the north of the taiga. The genesis of the peat plateau begins with the cryogenic polygonal cracking of a peat bog with an initially flat surface. Then, cracks grow in depth and width, due to thermokarst and erosion. The depth of thawing within cracks increases. In the taiga and forest-tundra, the ice wedges (permafrost in cracks) disappear completely. The heat regime within polygonal-hillocks, which are from 0.5 to several meters in height, is worse. Soil freezing begins earlier and goes deeper and is accompanied by moisture's accumulating under the polygon which causes heaving; permafrost rises to a depth of 50 to 70 cm below the soil surface.

The development of such patterns on bogs apparently began with cooling after the climatic optimum in the Holocene (peat contains the remains of plants that now have moved to the south). Bogs affected by cracking differ by age and peat quality and thickness. We thus now can observe thick (1.5 to 2.5 m) as well as thin layers of peat (corresponding soils are classified as Histic and Histi-Gleyic Cryosols) formed in upland (Dystric) as well as lowland bogs (Euthric).

The modern accumulation of peat occurs within depressions and on low peat plateaus. Vegetation that does not form peat occupies high peat plateaus, and they have disturbance features on the surface and a layer of mineralized peat from 5 to 15 cm thick. In general, the surface pattern of a peat plateau bog causes the development of a two-component soil complex: Fibrihisti-Gleyic Dystric Cryosol or Foli-Fibrihisti-Gleyic Cryosol on the plateaus and Histic Dystri-Eutric Cryosol (with a more thick histic horizon and a deeper or no longer extant permafrost) between them.

4.10.3. Soils of palsa bogs

These bogs spread mostly in forest tundra and in the northern part of the taiga, but their small areas can occur up to the Arctic tundra. Given the genesis of palsas, the age of these geomorphological elements differs from the age of the peat that constitutes them; palsa formation hinders and then stops the accumulation of peat (Figure 2.8.3). While peat formed throughout the Holocene, palsas developed during the Post-Atlantic cooling period leading up to the present (although some appeared during the Boreal Period, i.e., 8000 years ago).

A palsa thus can be 1) relic by composition and morphology, 2) relic by composition but morphologically modern, or 3) modern by both characteristics (the terms *relic* and *modern* are assigned to features formed before and after the Holocene optimum, respectively). Taking into account the active physical destruction of palsas, we can suppose that geomorphologically modern palsas also prevail.

As the active-layer depth decreases, the dimensions of palsas decrease. From forest-tundra to the north, their height decreases from 10 to 15 to 5 to 6 m; their diameter decreases from 15 to 20 to 0.7 to 1.0 m (Ivanova, 1962).

The soil complex of palsa bogs is formed by Foli-Histic Cryosols (on palsa) and Fibrihistic Dystric Cryosols (in depressions), with a spatial dominance of the latter. The depth to permafrost within a palsa is shallower than within depressions and lies at about 30 cm in the Arctic tundra and deeper than 2 m (or is absent) in forest tundra and taiga.

Some hydrolaccoliths (pingo) consist of mineral material. Their soils resemble those in areas of hummocky and mud circle tundra. However, pingo soils are better drained and, therefore, contain more acidic pH and show a stronger development of such processes as the oxidation of mineral horizons, organic matter decomposition, leaching, and the vertical differentiation of the profile by the content of clay (<0.001 mm) and by bulk concentrations of some components (Vasilievskaya, 1986). Podzols probably can form on sandy pingos in forest tundra and in taiga.

4.10.4. Alluvial Permafrost Soils (Fluvi-Gleyic Cryosols)

Diagnostic features of these soils include stratified parent deposits, an histic or mollic horizon of 10 to 20 cm, gleyic features in all mineral horizons, and permafrost not deeper than 1 m. These soils formed within floodplains and on the seacoast in the north of the tundra. Farther to the south, permafrost in such soils can occur at more than 1 m (Gelic Fluvisols) or just disappears.

Two specific groups of soils occur only in complexes of cryogenic pattern ground and do not form individual areas — Ochri-Gleyic Cryosols (Turbic and Oxiaquic) on circles and Glaci-Histic Cryosols within cryogenic cracks. The formation of soil complexes in connection with cryogenic processes is typical for tundra, forest tundra, and taiga areas with a continental climate. In western Siberia, soil complexes develop almost everywhere in tundra and forest tundra, over elevated topographic elements and bogs in the north of the taiga, and only within bogs (where they have a biogenic but not a cryogenic genesis) in the southern part of the taiga.

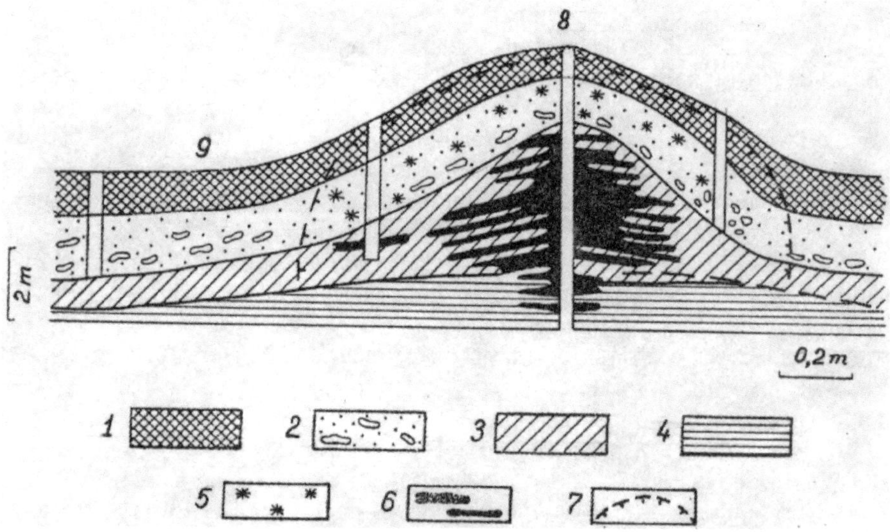

Figure 2.8.3. Palsa bog and its soil association in dwarf shrub-lichen-moss tundra. Lithology: 1. turf, 2. sand with pebbles, 3. layered loam, 4. clay, 5. segregations of ice, 6. ice layers and lenses, 7. permafrost table. Soils: 8. Foli-Cryic Histosol (Turbic), 9. Fibri-Dystric Histosol.

5. Summary

In the northern part of western Siberia, features of soils and of the soil cover are very special. Some of them are typical for western maritime and moderately continental regions, and others are typical for continental Siberian regions. This is mainly because of the transitional character of the climate and also the specific lithological and geomorphological conditions (level topography, stratified and saturated unconsolidated deposits, and generally poor soil drainage).

The distinguishing features of the Cryosolic soils of western Siberia are 1) a relatively high degree of organic matter decomposition and a prevalent formation of saprihistic horizons; 2) partial aeration and oxidation of upper mineral horizons, and constant gleyzation and thixotropy within middle and lower portions of the profile; 3) translocation of accumulated substances on the permafrost table by a downward filtration of solutions and by a cryogenic migration toward the lower front of freezing; 4) the gleyic differentiation of soil profiles on loamy-clayey parent materials within the region from the southern tundra to the middle taiga; 5) the development of Al-Fe-humus podzolization in soils on sands and loamy sands even under Arctic tundra; and finally, 6) the considerable illuviation of humus within these soils. The first four features also occur in Cryosols of the eastern continental regions of the Asian North; the last two features represent similarities to-Cryosols of the European North.

In general, specific characteristics of western Siberian Cryosols are better expressed in loamy and clayey textured materials rather than in sandy textured materials since the latter are less sensitive to changes in the soil-forming conditions.

The nature and distribution of soils (soil cover pattern) in western Siberia closely connects to the presence of periglacial patterned ground, which determines genetic differences between components of soil complexes. In tundra environments, the relative height of elements of microrelief is insignificant, genetic differences are small between soils on hillocks and on spots (mud circles) because of like conditions of soil freezing, and circles form and disappear relatively easily. These particularities may relate to the generally low relief of the territory, high soil moisture content, and high humidity of air masses (especially on the Yamal and Gydan Peninsulas surrounded by the Kara Sea). Yet in the southern (shrub) tundra, the hummocky circle pattern is quite distinct, and, therefore, genetic differences between soils of the microcomplex are better expressed.

Soil cover patterns within extensive bog areas of western Siberia show a particular diversity, which the presence of many types of bogs each with their own specific ground patterns at different stages of formation (from young to mature and degrading) express. Low-center polygon bogs are extensively developed in the tundra environment, and small areas of development also occur in the forest-tundra zone. South of the tundra, peat plateau bogs gradually replace low center polygon bogs; these peat bogs are widespread throughout forest-tundra up to the northern taiga and have the greatest variety of geomorphological and soil parameters. Palsa patterns occur mostly in forest-tundra and penetrate quite far to the south, into taiga, but singular palsas occur even in the Arctic tundra.

The relic features formed at earlier stages of the Holocene (especially during the climatic optimum) do not occur in mineral soils. Climatic fluctuations probably were buffered due to poor drainage and an over-saturation of soils. Some relic features occur in bog areas. For instance, palsa patterns in the Arctic tundra and dwarf shrub-lichen-moss tundra appear to have formed under warmer conditions when soil thawing was deeper. Peat plateau bog patterns have survived most cold stages of the Holocene. Now they remain in the north of their area, where the permafrost table is shallow, but peat plateau bogs are subject to active destruction in the southern portion of the region, due to erosion and the development of thermokarst. The patterns of low-center polygon bogs in forest-tundra also may be relict.

From the north to the south, changes in properties of soils and of the soil cover are very gradual, even at the zonal boundaries. Lithological and geomorphological conditions and the generally low relief and poor soil drainage of western Siberia cause this.

References

Katz, N.Ya. 1971. Swamps of the Earth (Bolota zemnogo shara). Nauka. Moscow. 296 pp. (In Russian.)

Dergacheva, M.I., and V.S Dedkov. 1977. The effect of freezing-thawing on the organic matter of soils in Obs tundra. Ecol. 2:17-30. (In Russian.)

Hotinsky, N.A. 1977. Holocene of the Northern Eurasia (Golocen severnoi Evrasii). Nauka. Moscow. 200 pp. (In Russian.)

Ivanova, E.N. 1962. Some regularities of the soil cover organization in the tundra and in the forest-tundra on the shore of the Obs gulf. On soils of Ural, West and Central Siberia. USSR Academy of Science Press. Moscow. pp. 49-116. (In Russian.)

ISSS Working Group RB. 1998. World Reference Base for Soil Resources. International Society of Soil Science (ISSS) and International Soil Reference and Information Centre and the Food and Agricuture Organization (FAO) of the United Nations. World Soil Resources Report 84. FAO, Rome. 91 pp.

Karavaeva, N.A. 1982. Bogging and the evolution of soils (Zabolachivanie i evolutsia pochv). Nauka. Moscow. 296 pp. (In Russian.)

Koposov, G.F. 1976. Soils of the river Taz upper reaches. Pochvovedenie. 4:28-39. (In Russian.)

Ovchinnikov, S.M., T.A. Sokolova, and V.O. Targulian. 1973. Clay minerals in the loamy soils of the taiga and the forest-tundra of the West Siberia. Pochvovedenie. 12:28-39. (In Russian.)

Shpolanskaya, N.A. 1973. Permafrost of the West Siberia and its connection with the actual heat-circulation between a ground and an atmosphere. The natural environment of the West Siberia. 3:127-149. (In Russian.)

Targulian, V.O., A.N. Gudina, G.M. Tereshkov, and V.T. Trofimov. 1972. The natural environment and economic development in the Taz oil-gaz region. Nauka. Moscow. pp. 134-142. (In Russian.)

Tonkonogov, V.D. 1979. Peculiarities of Gleyzems in Yamal-Gydan tundra. Peculiarities of the soil formation in Siberia. Nauka. Novosibirsk. pp. 45-52. (In Russian.)

Tyrtikov, A.N. 1973. Thawing of grounds in the West Siberia tundra. The natural environment of the West Siberia. 3:160-169. (In Russian.)

Vasilievskaya,V.D., V.V. Ivanov, and L.G Bogatyrev. 1986. Soils of The North of West Siberia (Pochvi Severa Zapadnoi Sibiri). Moscow State Unive. Press. Moscow. 227 pp. (In Russian.)

Chapter 9. Cryosols of the Mountains of Southern Siberia and Far Eastern Russia

Raissa G. Gracheva[1]

[1]Senior Researcher, Laboratory of Soil Geography and Evolution, Institute of Geography, Russian Academy of Sciences, Staromonetny, 29, Moscow, 109017 Russia; e-mail pedology@igras.geonet.ru.

1. Introduction

Southern Siberia and the far eastern parts of Russia run west to east, from 49 to 55 (60)° N latitudes and 84 to 140° E longitude and to the extent of approximately 5000 km, from western Siberia to the Pacific coast, along the Russian-Mongolian and Russian-Chinese borders. This region includes the vast southern Siberian mountain systems of the Altay, the West and East Sayans, Tuva, the Baikal regions, and the part of the Russian far east called "Priamurye" (Figure 2.9.1).

1.1. Climate

In winter, the Asian anticyclone, which causes intensive cooling of the land surface, influences the climate of southern Siberia; in summer, the Atlantic and Arctic air masses do. The climate is mostly strongly continental, with very low mean winter temperatures (-21° to -32°), a relatively warm but short summer, and low humidity, as annual evaporation exceeds precipitation. The most severe extra-continental climatic conditions are characteristic of the eastern and southern parts of the Altay-Sayans mountain system. Priamurye has a monsoon climate with continental features and with a cold, dry winter and humid, cool summer (Gerasimova, 1973).

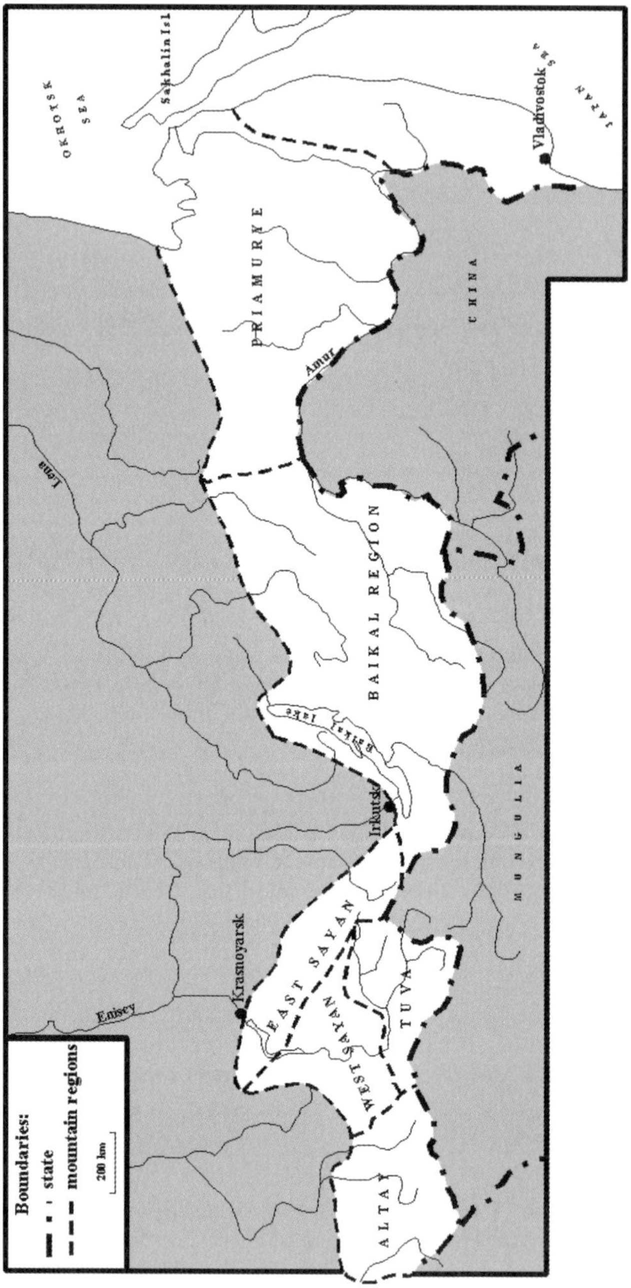

Figure 2.9.1. Mountain regions of southern Siberia and Priamurye.

1.2. Topography and vegetation

Regional and local climatic variability depends on the topography, which predominantly formed under Quaternary tectonic uplifts and glaciation. The main feature is the combination of sharp rugged ridges, ancient surfaces of denudation, and intermontane kettle depressions. The main vegetative altitude belts are mountain steppes, taiga, alpine meadows, and tundra, with bogs widespread on flat surfaces (Mikhailov, 1961).

1.3. Permafrost

Shallow snow cover and a short warm period promote the conservation of permafrost within soil profiles in southern Siberia and Priamurye. The permafrost has a discontinuous or sporadic expansion in these areas, except for the northern part of Priamurye, where it is continuous (Kudryavtsev, 1978; Romanovskyi, 1993). The thickness of frozen grounds varies from dozens of meters in the western parts of southern Siberia to 100 to 300 meters in the extra-continental eastern and southern parts of the Altay-Sayans and in northern Priamurye. The depth of the active layer ranges from 30 to 50 cm to several meters, depending on climatic conditions. In all of these regions, the permafrost-affected soils occur predominantly on high elevations, on the northern slopes and in the bogged depressions.

1.4. Soil studies

Cryosols of some regions of southern Siberia and far eastern Russia have not been studied in detail before now. The widest soil investigations were carried out in 1950 through 1970, when general regularities of soil cover were established and main properties of soils were examined.

2. The Altay

The Altay is the western and highest mountain system of southern Siberia. The absolute altitudes of alpine ridges, mainly situated in the central and eastern Altay, are more than 3000 to 4000 m, with the highest top, Beluha, at 4506 m. The main ridges consist of Paleozoic sedimentary and volcanic rocks (schist, sandstone, aleurolite, crystalline limestone, quartz rocks, effusive rocks), partly intruded into by granites. Deluvial, proluvial, alluvial, and glacial deposits fill the depressions (Devyatkin, 1965).

The climate of the permafrost-affected areas of the central and southeastern Altay is strongly continental. In the "anticyclone" provinces of the Altay, the mean annual temperatures (°C) are -1.1°, -4.2°, and -8.5° at 800, 1200, and 2200 m above sea level, respectively. Mean winter temperatures vary with the altitude gradient from -19° to -32°, to -60° in the southeastern Altay. Summer means are from 16° to 8°, rising to 31° in depressions. Usually the winter climatic conditions in intermontane depressions are essentially more severe at higher elevations, due to temperature inversions.

The annual precipitation is 345 to 509 mm in the central Altay and 110 to 300 mm in the southeastern areas, with the lowest precipitation in winter, 35 to 100 mm and 20 to 60 mm, respectively. Snow cover thickness is low and varies from 20 to 30 cm in the central Altay to 10 to 12 cm in the southeastern Altay.

The three main vertical vegetation belts follow the increasing altitudes: steppes (to 300 to 600 m); coniferous forests, consisting of fir (*Abies sibirica*), cedar pine (*Pinus sibirica*), and larch (*Larix sibirica*) (300 to 500, to 1700 to 2000 m); and alpine meadows, tundra, and bogs (1700 to 2000, to 2600 to 3000m). In addition, the specific steppe vegetation of high kettle depressions is widespread. Within southeastern Altay, which is the highest, coldest, and driest part of the Altay, the forest belt is fragmentary or absent, and tundra borders on the dry or meadow steppes (Kuminova, 1960). Figure 2.9.2 shows the distribution of soil according to macro topography within different parts of the Altay.

Permafrost has mostly discontinuous distribution, occupying 75 to 90% of the total area of the central and southeastern Altay. The annual temperatures of frozen ground are from +0.5° to -2°C (Romanovskyi, 1993).

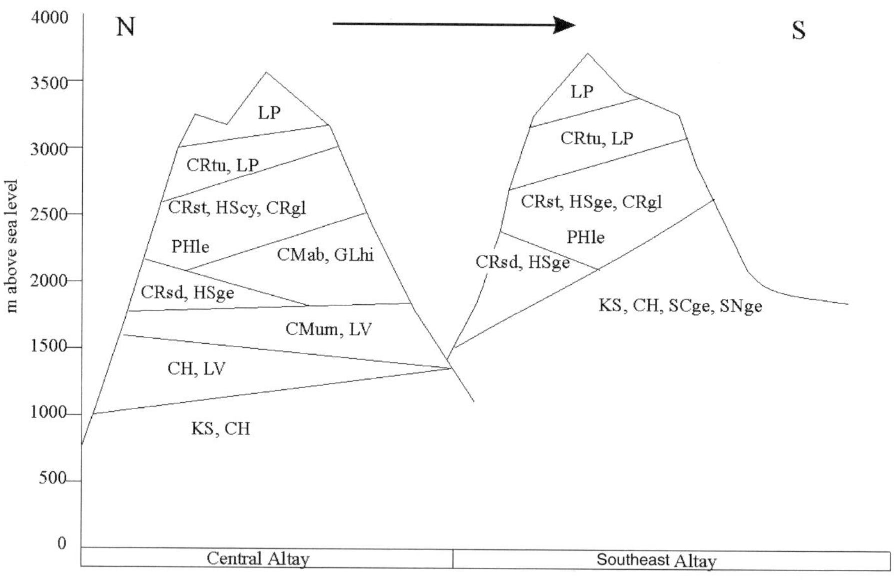

Figure 2.9.2. Distribution of soils depending on altitude and aspect in the Altay mountains. Soils: CR – Cryosol (gl – Gleyic, st – Stagnic, sd – Spodic, tu – Turbic); HS – Histosol (cy – Cryic, ge – Gelic); CM – Cambisol (ab – Albic, um – Umbric); GLhi – Histic Gleysol; PHle - Leptic Phaeozem; SCge – Gelic Solonchak; SNge - Gelic Solonetz; LP – Leptosol; LV – Luvisol; CH – Chernozem; KS – Kastanozem.

Permafrost-affected soils are distributed widely on high levels and, to a lesser extent, in extra-continental intermontane depressions. The list of these soils includes **Soils of cryogenic circles** (Turbi-Lithic and Turbi-Leptic Cryosols), **Tundra peaty gleyzems** (Stagni-Fibrihistic Cryosols), **Peaty bog gleyzems** (Gleyi-Histic Cryosols and Cryic Histosols), **Taiga peaty podzol gleyic soils** (Stagni-Fibrihistic Cryosols [Spodic]), **Solonchacks** (Geli-Sodic Solonchaks [Sulphatic]), and **Meadow Solonetzes** (Geli-Salic Solonetzes) (Petrov, 1952; Kovalev, 1973).

In the most arid and cold high mountains of the southeastern Altay, the soil cover of alpine tundra is sporadic, because of frequent rock outcrops and rock debris. In this area under the thin lichen or grass cover, the permafrost lies on the surface or at the depth of 20 to 30 cm. Cryogenic processes are intensive and have formed stone rings, stone polygons on the flat surfaces, and stone stripes and solifluction steps on the slopes. The soil-cover complex consists of **Soils of cryogenic circles** and **Mountainous primitive soils** (Turbi-Leptic Cryosols). They have much in common with such soils of the eastern Sayans (see below).

2.1. Tundra peaty gleyzems

The most widespread cryogenic soils are **Tundra peaty gleyzems**. They occur in tundra belt at levels of 1700 to 2500 (3000) m, up to the nivation belt, and are dominant soils on the flat surfaces neighboring the permafrost-free **Mountainous stony organogenic soils** (Histic Leptosols) and **Tundra soddy soils** (Leptic Phaeozems). The largest areas of these soils occur on the main Altay mountain ridges and high plateaus. The parent materials are coarse sandy loamy unconsolidated deposits such as eluvium and eluvium-deluvium of schists, mudstones, and sandstones.

Tundra peaty gleyzems occur under vegetation where peatmoss (*Sphagnum compactum, S. Fuscum*), different sedges (*Carex*), cotton grass (*Erophorum angustifolium*), and ground birch (*Betula rotundifolia*) are dominant. In the soil profile under the living mosses lies the fibrihistic layer, with a thickness of 25 to 30 cm and more. A mineral gray rust colored, mottled stagnic horizon underlies it. Ice-rich permafrost lies at a depth of 70 to 90 cm, and suprapermafrost water forms in the warm season, resulting in the development of reduction processes in the soil profile. Table 2.9.1 shows the soil properties. The specific feature of **Tundra peaty gleyzems** is high base saturation, caused by water and biophile elements' being retained in the permafrost barrier and by calcium-saturated soil water's neutralizing the acid organic products. Usually, the labile organic matter, in the form of chelates, migrates and accumulates directly above the frozen layer.

Groundwater-induced, permafrost-affected **Peaty bog gleyzems** (Gleyi-Fibrihistic or Gleyi-Saprihistic Cryosols) occur in the tundra and taiga alpine belts, forming narrow and small patterns on the river terraces or depressions within high intermontane lowlands. In these areas, **Peaty bog gleyzems** often are deformed by frost jacking and include ice-cored peat hillocks. The peat layer is 30 to 40 cm thick and covers the black-blue-gray humous-gleyic horizon, usually loamy-sandy layered. The permafrost table lies at 50 to 100 cm deep and in some places underlies the peat layer. Soils are acid and unsaturated, and the migration of organic matter and iron is not pronounced (Kovalev, 1973; Nosin, 1963).

2.2. Taiga peaty podzol gleyic soils

Within the upper part of the taiga alpine belt, at altitudes of 2000 to 2400 m, the **Taiga peaty podzol gleyic soils** (Stagni-Fibrihistic Cryosols (Spodic) occur. They form long but narrow areas, mostly on the cold, gentle, north-oriented slopes, where the permafrost is 30 to 40 cm deep. These soils occur under cedar pine-larch and larch open woodlands with thick sphagnum-moss cover, on loamy eluvia and deluvia of hard crystalline rocks.

The profile consists of these horizons: H (0 to 30 cm) – brown peat; E (30 to 35 cm) – bleached light gray loamy-sand; Bhi (35 to 50 cm) – dark brown stony loam, with ice lenses; Bgi (50 to 65 cm) – brown, mottled, stony, with ice lenses. Ice-rich permafrost underlies the lowest horizon at 50 to 60 cm deep.

In comparison with tundra Cryosols, the forest Cryosols are not affected by long-term over-moistening. This results in the formation of an eluvial horizon, the migration and illuviation of organic-mineral compounds, the acid pH condition, and base unsaturation. Table 2.9.1 shows the soil properties.

Table 2.9.1. Properties of Cryosols of Altay (Kovalev, 1973).

Horizon	Depth, cm	Corg., %	pH H_2O	Base saturation, %	Oxalate Fe_2O_3, %	Particles<0.001 mm, %
Central Altay (Terekhtinskiy ridge). Stagni-Fibrihistic Cryosol. P.767.						
H	0-8	29.2	6.9	100	0.41	-
O	8-34	5.0	6.7	100	1.15	9.9
AB	34-45	4.2	6.5	99	0.80	12.3
Bg	45-65	4.6	6.4	99	0.54	12.2
Bgi	65-80	5.5	6.5	99	0.48	8.3
East Altay (Chulyshmanskiy ridge). Stagni-Fibrihistic Cryosols (Spodic). P.896.						
H	0-30	28.8	4.1	-	-	-
E	30-35	2.4	4.5	31	0.5	8.8
Bhi	35-50	1.9	4.9	20	0.7	9.2
Bgi	50-65	0.4	5.1	62	0.2	9.0
East Altay (Chuyskaya depression). Geli-Sodic Solonchak (Sulphatic). P.5-70.						
Layer I	0-1	-	9.8	-	-	-
Layer II	1-8	2.8	8.7	-	-	10.8
Layer IIIa	8-28	2.2	9.3	-	-	14.6
Layer IIIb	28-45	1.4	9.4	-	-	9.2
Layer IY	45-90	1.3	9.2	-	-	4.2
East Altay (Chuyskaya depression). Geli-Salic Solonetz. P.6-70.						
Az	0-2	1.0	8.4	-	-	7.8
Bk	2-10	1.1	8.9	-	-	7.1
Bk	10-50	0.5	9.1	-	-	7.6
BCk	50-100	0.1	9.0	-	-	6.2

2.3. Forest soils

When the permafrost table lies at 100 to 200 cm deep, it does not affect the properties of well drained soils. Thus, **Brown forest soils** (Gelic Cambisols) and **Brown forest podzolized soils** (Albi-Gelic Cambisols) within the areas of sporadic permafrost do not differ from permafrost-free **Brown forest soils** (Dystric Cambisols). In contrast to its action on well drained soils, deep permafrost has an important soil-forming role in poorly drained conditions, where it acts as an aquiclude.

2.4. Gelic soils

Gelic Solonchaks and Gelic Solonetses occur in intermontane depressions under an extra-continental arid climate. Permafrost patches occur in these (Chuiskaya) depressions due to strong frost; the winter air temperatures decrease to -60°, and snow depth is only 10 to 12 cm.

Meadow and Typic Solonchaks (Gelic Solonchaks) and **Meadow Solonetzes** (Gelic Solonetzes) are spread at the altitude of 1700 m throughout Kastanozems areas, under steppes and desert-steppe vegetations. They occur in relief depressions on the mixed lacustrine-glacial and alluvial deposits under the influence of saline water, flowing above the cryogenic aquiclude. The salt sources are predominantly the bedrocks of the slopes from which the salts are released under weathering and transported with lateral flows in accumulative positions of relief.

The soil surface is covered sparsely by different kinds of alkali grass, wormwoods (*Artemisia frigida, A. argyrophylla*), lyme grass (*Aneurolepidium sp.*), *Statice gmelin*, and *Lasiagrostis splendens*. The latter forms the round giant (5 to 6 m in diameter) tussocks rising above the surface. The Solonchaks occur under the tussocks, while on the barren periphery of the tussocks are the **Solonchaks** with a saline crust. **Solonetzes** occur on the micro-highs in the **Solonchaks** areas and on the micro-lows throughout the permafrost-free Salic and Natric Kastanozems (Kovalev, 1973).

The Geli-Sodic Solonchak (Sulphatic) (P. 5-70) and Geli-Salic Solonetz (P. 6-70), which Examples 2.9.1 and 2.9.2 describe, occur under a pasture on the first flood-plain terrace of the Chuya river, 1700 m a.s.l. Parent rocks are sandy-loamy-gravelly alluvial deposits.

Example 2.9.1. P. 5-70. Micro-high, covered by alkali grass, Statice gmelin.

Layer I. 0 to 1(2) cm. Soft salt crust.

Layer II. 1(2)-8 cm. Wet, brownish blue-gray, structureless, loamy, penetration resistance is high. Includes the roots and rock debris. Strongly effervescent. Distinctness is gradual.

Layer III. 8 to 28 cm. Wet, pale blue-gray, structureless, clay loamy, penetration resistance is high, sticky. Includes very few roots, many pebbles and rock debris. Strongly effervescent. Distinctness is clear.

Layer IY. 28 to 90 cm. Loamy gravel-sand layers, blue-gray with common rust-colored mottles. Includes few roots. Strongly effervescent. Below – suprapermafrost water.

Example 2.9.2. P. 6-70. Microhigh close to P. 5-70.

Az. 0 to 2 cm. Dry, gray whitish, soft, loamy, with salt efflorescence.
Bk1. 2 to 10 cm. Dry-wet, brown, columnar structure, loamy. Includes few roots and gravel rock fragments. Strongly effervescent. Distinctness is clear.
Bk2. 10 to 50 cm. Wet, yellow-brown, structureless, clay loamy gravelly, includes roots and salt efflorescence – few and very few. Strongly effervescent. Distinctness is clear.
??k. 50 to 100 cm. Wet, rust-colored, loamy gravelly, includes very few roots. Strongly effervescent. Suprapermafrost water appears below 100 cm.

Tables 2.9.1 and 2.9.2 show soil properties. Solonchak and Solonetz are alkaline soils, and sodium is dominant among exchangeable cations. In the salt crust of the Geli-Sodic Solonchak (Sulphatic), sodium chloride and sodium sulfate prevail. In the Geli-Salic Solonetz, the exchangeable sodium-rich horizon (solonetz horizon) lies at 2 to 10 cm deep, and highest salt concentrations are marked above and under it.

3. Sayans

Sayans includes two mountains systems, the West and East Sayans. The nature of this region is essentially different from other parts of southern Siberia. Figure 2.9.3 shows the structure of the altitude soil belts of the West and East Sayans.

The West Sayan system is oriented partly west-east, and partly from southwest to northeast, to the extent of 650 km between 58°40' to 51°20' N latitudes and 87°40' to 97°E longitude. The absolute elevations of the highest ridges reach 2400 to 3000 m. Topography is of tectonic, erosive, and glacial origin. Recent landscapes preserve the features induced by Quaternary glaciation and periglacial environment, glacial deposits, and permafrost. Intermontane depressions occur on lower positions (300 to 1200 m). The ridges are underlain mainly by metamorphic (chlorite and quartz schist) and sedimentary (sandstone, conglomerate, limestone) rocks, from which weathering produces loamy and clay loamy gravelly slope deposits. Moraine occurs in highlands. The climate reflects the great distance from oceans and complicated topography, and it changes, with the altitude, from cool to very cold.

The southern province is under the great influence of the Siberian anticyclone and is significantly drier than the northern one (210 to 450 mm per year, versus 1000 to 1200 mm in the north) and has a colder winter. Thus, permafrost occurs predominantly in the southern part of the West Sayan. In the southern climatic province, the winter (January) temperatures (°C) are very low, from -22.7° to -33.9°, and the summer (July) ones are relatively high, 17.2 to 19.9°, rising sometimes to 32° in steppe depressions. The mean annual temperatures, depending on elevation, vary from -0.6° (440 m) to -5.4° (900 m) and -8.4° (2000 m). Winter precipitation is 10 to 20% of the annual amount, and snow thickness is not more than 19 to 45 cm.

Table 2.9.2. Chemical composition of water extracts from a Gelic Solonchak and Gelic Slonetz from Altay (from Kovalev, 1973).

Depth, cm	Insoluble residue, %	CO''_3	HCO_3^{-1}	Cl^{-1}	SO_4^{-2}	Ca^{2+}, %	Mg^{2+}, %	Na^{+}, %	Σ, %
Geli-Sodic Solonchak (Sulphatic). Profile 5-70.									
0-1	49.17	0.540	0.915	7.064	23.59	0.011	0.067	16.49	48.69
1-8	2.89	0	0.071	0.844	0.852	0.045	0.028	0.898	2.738
8-28	0.83	0	0.142	0.174	0.201	0.004	0.001	0.257	0.779
28-45	0.36	0	0.107	0.043	0.068	0.004	0.001	0.094	0.317
45-90	0.20	0	0.071	0.017	0.035	0.004	0.002	0.048	0.177
Geli-Salic Solonetz. Profile 6-70.									
0-2	0.941	0	0.004	0.175	0.418	0.119	0.032	0.153	0.901
2-10	0.450	0	0.059	0.023	0.224	0.006	0.003	0.134	0.449
10-20	1.376	0	0.046	0.168	0.692	0.018	0.015	0.421	1.360
30-40	0.895	0	0.032	0.295	0.247	0.009	0.015	0.282	0.890
50-80	0.504	0	0.029	0.200	0.095	0.007	0.016	0.159	0.506

Altitudinal vegetation belts are: <300 to 400 m in the north and <700 to 800 m in the south, steppe vegetation; <1600 to 1800 m, fir and cedar pine forests in the north and, in the south, larch forests with different ground vegetation — lingonberry (*Vaccinium vitis-idaea L.*) and lingonberry-herbaceous communities on the low mountain levels and ledum (*Ledum palustre L.*)-mossy and lingonberry-mossy ones on highlands; >1800 m in the north and > 2000 to 3000 m, subalpine and alpine meadows, followed by dryad (Dryas punctata Juz.), moss-lichen, and dwarf-shrub tundra (Mikhailov, 1961).

Permafrost has sporadic distribution (10 to 30% of total area), and its temperature is -1° or -2°C (Romanovskyi, 1993). It occurs widely on high elevations of all large ridges. Within the intermediate mountain belts, it occurs in the southern province, on northern slopes, usually under moss-peat cover. Permafrost is also common in the bottoms of boggy lowlands. Figure 2.9.3 shows the main features of soil distribution according to elevation and aspect of macroslopes.

The main permafrost-affected soils are **Soils of cryogenic circles** (Turbic Cryosols), **Peaty bog gleyzems** (Gleyi-Histic Cryosols), **Taiga peaty gleyzems** and **Taiga mucky gleyzems** (Stagni- Fibrihistic and Stagni-Saprihistic Cryosols), and **Humus-illuvial podzol** (Saprihisti-Gelic Podzol).

3.1. Turbic Cryosols

Soils of cryogenic circles under suppressed lichen vegetation are widespread in the tundra belt, combining with rock outcrops and stony fields. Soils are strongly deformed by cryogenic processes, forming such specific microrelief as stony rings and stony sorted polygons. The properties of the Turbic Cryosols of the West Sayan are similar to these of the East Sayan (see below) (Smirnov, 1970).

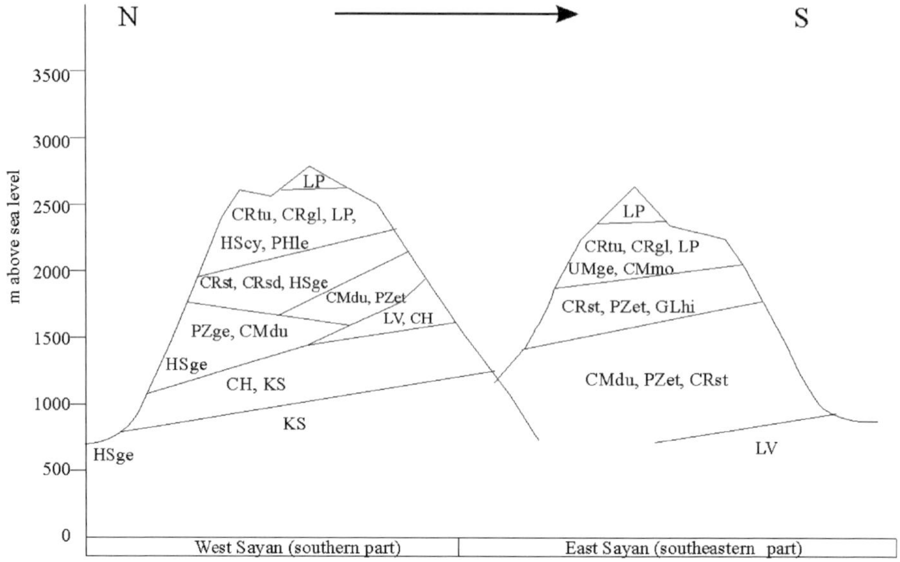

Figure 2.9.3. Distribution of soils depending on altitude and aspect in the Sayan mountains. Soils: PZ – Podzol (et – entic, ge – gelic); CM – Cambisol (dy – dystric, mo - mollic); UMge – Gelic Umbrisol; CR (gl, st, sd, tu), HS (cy, ge), GLhi; PHle, LP, LV, CH, KS (See Figure 2.9.2).

3.2. Peaty bog gleyzems

Permafrost-affected **Peaty bog gleyzems** (Gleyi-Histic Cryosols) occupy the large areas within the tundra, taiga, and steppe alpine belts (Table 2.9.3). **Peaty bog gleyzems** of the taiga occur at different elevations within the forest belt, in the depressions under the bog vegetation. The thickness of the peat layer varies from 30 to 200 cm and more. Usually the peat lies directly on frozen deluvial or alluvial deposits, which play a role of cryogenic aquiclude. Example 2.9.3 gives their soil description.

In the steppe belt, the cryogenic Peaty bog soils occur mainly in kettle depressions, along the border with the Tuva province, where permafrost occurs in isolated patches along rivers valleys. The profile includes: moss-sedge litter (0 to 5 cm), wet brown-black peat up to 15 cm; grayish blue-black almost liquid peat-clay mass (15 to 30 cm); from 30 to 80 cm, water saturated fine gravel and sand; below 80 cm, ice permafrost. These soils, unlike tundra and taiga soils, have high pH values, which weakly decrease from peat layers (7.5) to mineral horizons (6.9). The weak carbonate and sodium alcalinization occur due to lateral and surface flows of waters with high salt content (Nosin, 1963).

Table 2.9.3. Some properties of Cryosols and related soil of West Sayan (from Smirnov, 1970).

Horizon	Depth, cm	Corg., %	pH_{H2O}	Base saturation, %	Oxalate Fe_2O_3, %	Particles <0.001 mm, %
Northern part (Borus ridge). Tundra Gleyi-Histic Cryosol. P.256.						
H	0-8	84.0*	4.4	3	0.2	-
O	8-30	20.4	4.6	25	1.1	36.5
Bh	30-35	8.4	4.8	44	1.3	28.0
Bgi	35-60	5.7	5.2	32	1.8	25.6
Eastern part (Ergak-Torgak-Taiga ridge). Taiga Gleyi-Histic Cryosol. P.435.						
H	0-10	73.8*	5.3	98	0.2	18.0
O	10-30	57.5*	6.3	95	0.9	29.6
Cig	30-35	25.8	6.2	90	0.8	19.4
Central part (Sayanskiy ridge). Stagni-Fibrihistic Cryosol. P. 328.						
H	0-24	85.9*	5.2	28	0.2	-
Big	24-35	11.2	5.7	67	0.9	27.9
Big	35-48	10.5	5.6	70	0.8	29.4
Cg	48-58	9.9	5.5	70	0.7	27.2
Eastern part (Ergak-Torgak-Taiga ridge). Stagni-Saprihistic Cryosol. P.445.						
H	0-22	69.7*	4.6	28	0.3	-
O	22-33	22.4	5.0	75	0.8	49.9
Bg	33-50	3.7	5.0	75	0.9	27.2
Big	50-80	2.2	5.4	74	0.9	36.3
Ci	80-100	3.8	5.2	80	0.9	30.2
Southern part (Ridge Mirskiy). Saprihisti-Gelic Podzol. P. 176.						
H	0-9	83.5*	5.2	30	0.2	-
O	9-25	82.9*	4.2	13	0.7	-
Eh	25-43	12.8	4.6	29	1.9	24.4
Bih	43-62	5.1	5.9	54	3.3	34.4
BCi	62-95	0.3	6.7	84	1.6	13.0

* Loss of ignition.

Example 2.9.3. P.435. Peaty bog gleyzem (Gleyi-Histic Cryosol).

Watershed of Kuchar river, the northern flat slope, 900 m a.s.l. Spruce-larch-cedar pine forest with ledum-sphagnum cover. Well expressed cryogenic features of microtopography, such as micro-lows, micro-highs, moss hummocks. Parent materials are clay loamy deluvial deposits.

10 to 0 cm. Living mosses.

H. 0 to 24 cm. Wet, dark brown-black peat.

OA. 24 to 30 cm. Wet, dark brown, loamy clay, many roots.

Cig. 30 to 35 cm. Wet, brownish blue-gray, clayey loam, with lenses of ice; intensive water flowing from pit walls; very few roots.

Below 35 cm. Frozen ground.

As distinct from **Peaty bog gleyzems** in tundra, in taiga these soils are weakly acid and base-saturated. Mineralized ground water neutralizes the organic acids produced in the peat layer. The content of oxalate-extractable iron is relatively low and weakly rises in the mineral horizon (Table 2.9.3).

3.3. Taiga peaty gleyzems and Taiga peaty-mucky gleyzems

Taiga peaty gleyzems and **Taiga peaty-mucky gleyzems** (Stagni-Fibrihistic and Stagni-Saprihistic Cryosols) occur in the forest belt in the southern part of the West Sayan. They occur under severe climate and a shallow permafrost table, which results in an impermeable moisture regime. Atmospheric precipitation and ice melting, due to a short and cool vegetative period, moistens the active layer extensively, and weakly decomposed organic matter accumulates. This moss-peat layer prevents the permafrost within the soil profile from thawing.

Taiga peaty gleyzems dominate at high elevations on the clayey textured substrata, under ledum-moss larch and cedar pine depressed forests. **Taiga peaty-mucky gleyzems** are widespread in the areas of permafrost-free **Taiga soddy soils** (Dystric Cambisols and Dystri-Albic Cambisols) on the shadow slopes and the flat piedmonts at moderate elevations (900 to 1000 m), under larch, larch-cedar pine, and cedar pine-spruce forests with lingonberry-ledum-sphagnum and sphagnum-moss cover. These soils occur in warmer conditions, and organic matter is more decomposed.

The important feature of these soils is the formation of specific light humus which is morphologically invisible but chemically detectable (Table 2.9.4). Examples 2.9.4 and 2.9.5 give their typical profiles (Smirnov, 1970).

3.4. Podzols

Permafrost-affected Podzols occur under the larch and cedar pine moss and shrub-moss forests, too, but in a warmer climate than gleyzems. The thickness of the active layer is 100 and more cm, and cryogenic forms are not expressed on the soil surface. Nevertheless, the seasonally frozen ground may be preserved in the soils during summer. These soils occur on the northern slopes of ridges in the southern and central parts of the West Sayan. Humus-illuvial Podzols are strongly acid unsaturated soils. They are enriched by organic matter and the migration of organic-iron matter within the profile (Example 2.9.6) is clearly expressed (Table 2.9.3).

Table 2.9.4. Some properties of two Cryosols from East Sayan (from Gorbachev, 1978).

Horizon	Depth, cm	Corg., %	pH_{H2O}	Base saturation, %	Oxalate Fe_2O_3, %	Particles <0.001 mm, %
Central part (Manskoye belogorye). Turbic Cryosol. Profile 465.						
Layer I	0-20	12.0	4.5	24	2.2	10.4
Layer II	40-60	3.5	4.6	22	2.2	12.0
Layer III	85-95	1.7	4.7	50	1.9	8.2
Southern part (Vala ridge). Stagni-Saprihistic Cryosols. Profile 516.						
O	5-16	62.6*	3.9	18	0.6	-
Oi	16-25	28.8*	4.0	14	0.9	-
OCi	25-40	11.0*	3.7	16	1.6	16.6

* Loss of ignition.

Table 2.9.5. Some properties of Cryosols and related soils of the Priamurye (from Shapiro, 1984).

Horizon	Depth, cm	Corg., %	pH_{H2O}	Base saturation, %	Oxalate Fe_2O_3, %	Particles <0.001 mm, %
Lepti-Histic Cryosol. Profile 202-81.						
O1	3-15	80.9*	3.7	18	-	-
O2	15-26	38.2*	4.1	11	0.38	-
BCi	26-33	2.9	4.4	9	0.33	4.4
Lepti-Histic Cryosol (Spodic). Profile 89-78.						
O	2-15	45.3*	3.9	17	0.9	11.2
E	15-25	1.6	3.8	13	0.6	1.3
Bs	25-40	1.4	4.5	16	0.9	4.6
BCi	40-50	1.0	4.5	11	0.6	4.1
Taiga Histi-Gelic Podzol. Profile 37-79.						
O	0-5	24.2*	4.5	20	0.4	-
E	5-10	0.6	4.2	9	0.1	3.1
Bs	10-34	2.8	4.8	26	1.2	9.8.
BC	34-54	1.1	4.9	30	0.2	3.2
C	54-100	0.8	5.2	48	0.1	1.1
Tundra Histi-Gelic Podzol (Entic). Profile 127-81.						
H	0-4	53,4*	3.9	17	-	-
O	4-15	18.3*	4.3	10	0.6	8.6
Bsh	15-25	3.6	4.4	36	1.1	8.4
BC	25-40	1.6	4.8	48	0.2	6.2
C	60-100	0.9	4.8	52	0.1	2.0

* Loss of ignition.

Example 2.9.4. P. 328. Taiga peaty gleyzem (Stagni-Fibrihistic Cryosol).

Ridge Sayanskiy, northeast slope, 1750 m above sea level, hummocky microrelief. In micro-lows, the well sorted, fine-earth matter accumulates due to transportation of surface suprapermafrost waters. In the ledum-sphagnum-lichen larch open woodland, the larch trees form additional roots because of the close permafrost table. Parent rock is sandstone deluvium.

9 to 0 cm. Living lichens and mosses.

Hi. 0 to 24 cm. Wet dark brown peat, with many roots and ice pieces.

Big1. 24 to 35 cm. Wet, dirty blue-gray, with rust-colored spots, silty clay loam, gravelly, structureless, frozen, iced. Roots are very few.

Big2. 35 to 48 cm. Moist, blue grayish brown, silty clay loam, gravelly, structureless, frozen.

Cig. 48 to 58 cm. Blue grayish brown frozen iced clayey stony. Water table appears in one hour.

Example 2.9.5. P. 445. Taiga peaty-mucky gleyzem (Stagni-Saprihistic Cryosol).

Ridge Ergac-Torgac-Taiga, northern flat slope, 950 m above sea level. The surface topography is hillocky hollow. Vegetation: cedar pine-spruce forest with sedge, ledum, moss, sphagnum and lingonberry. Parent rock is red sandstone.
7 to 0 cm. Living mosses.
H. 0 to 22 cm. Wet, brown peat, with fair amount of roots. Includes charcoals.
OA. 22 to 33 cm. Wet, dark brown, clayey, weakly aggregate structure, soft; many roots.
B1. 33 to 50 cm. Wet, brown, clay loamy, thixotropic. Penetration resistance is high. Many stony debris.
Big2. 50 to 80 cm. Wet, brown-blue-gray with rust-colored spots, thixotropic; ice lenses.
Ci. 80 to 106 cm. Yellow-brown, clay loamy, frozen.

Example 2.9.6. P. 176. Humus-illuvial Podzol (Saprihisti-Gelic Podzol).

Ridge Mirskyi, northeastern steep slope, 1500 m a.s.l. Cedar pine moss and herbaceous-moss forest with spruce and fir. Parent rock is deluvium of chlorite schist.
H. 0 to 9 cm. Peat layer.
OA. 9 to 25 cm. Wet, dark brown, many roots, decomposed wood remains, distinctness is abrupt.
E. 25 to 43 cm. Dry, light gray, gravelly loamy clay, structureless, few roots, distinctness is clear.
Bih. 43 to 62 cm. Wet, brown-ocherous, gravelly clayey loam, structureless, seasonally frozen, distinctness is gradual.
BC. 62 to 95 cm. Brown, very gravelly loam, structureless, dense, very cold.

4. East Sayan

The East Sayan is northeast of the West Sayan, runs to 1350 km from north-west to southeast, and is 200 to 350 km wide. The elevations of ridges vary from 2000 to 2500 to 3000 m, and the main elements of topography are of glacial origin. Ancient interior plateaus are well preserved. They are situated at the elevations from 1200 to 1300, to 2700 to 2800 m and are isolated by high mountain ridges. Igneous and metamorphic rocks occur more widely in the East Sayan than in the West Sayan; thus stony, loamy sandy, and loamy eluvium and deluvium of granite, diorite, quartzite, marble, quartz sandstone, and schist play an important role as soil parent materials.

The winter Asian anticyclone air mass strongly affects the climate of East Sayan, which on the whole is more continental than West Sayan's. The most severe and continental climate conditions are characteristic for the southeastern part of the East Sayan. The winter in this region is long and cold (mean January temperature varies from -28 to -38°C), and amplitude between maximum and minimum temperatures reaches 80 to 90°C. Annual precipitation is not more than 200 to 250 mm in intermontane depressions and 400 to 450 mm on the mountain slopes. The winter part of annual precipitation is only 2 to 5%, and snow thickness is not more than 6 to 20 cm.

Climate-induced isolated patches of permafrost are characteristic for western macroslopes, and sporadic and extensive discontinuous permafrost occurs within the eastern macroslope and southeastern part of East Sayan. The annual temperatures of frozen grounds vary from -0.5° to -3° C in the first region, and to -5°C in the second (Romanovskiy, 1993).

The general features and changes of belts of vegetation, especially on the western slope, are not like those of the West Sayan. In the East Sayan, the steppe belt is not clearly expressed, and usually pine-birch wooded steppe occupies the piedmonts up to 400 to 500 m, and larch dominates in all mountainous forests.

In spite of the wide distribution of permafrost, genuine Cryosols occur only in the tundra belt and upper parts of shadow forested slopes. In other permafrost areas, the thickness of thawed ground is more than 200 cm.

The main permafrost-affected soils are: **Soils of cryogenic circles** (Turbic Cryosols), **Tundra peaty** and **Tundra mucky bog gleyzems** (Gleyi-Fibrihistic and Gleyi-Saprihistic Cryosols), **Mountainous meadow soddy soils** (Gleyi-Gelic Umbrisols), and **Taiga peaty-mucky gleyzems** (Stagni-Saprihistic Cryosols) (Targulian, 1971; Gorbachev, 1978).

4.1. Turbic Cryosols

Turbic Cryosols occur within stony lichen tundra on the flat surfaces of planation, high interior plateaus and the upper parts of smooth northern slopes at elevations from 1400 to 2200 to 2400 m. The surface cryogenic processes in these areas are very intensive and form such specific elements of microrelief as sorted circles and low-center polygons of 0.6 to 1.5 m in diameter. Usually soils of the central parts of circles and polygons have a monotonous profile, with light brown, friable, loamy fine gravelly mineral matter. The cryogenic layered structure may be expressed weakly. The permafrost lies at 80 to 90 cm deep (Gorbachev, 1978). Table 2.9.4 lists some properties of Turbic Cryosol.

4.2. Tundra peaty and Tundra mucky bog gleyzems

Tundra peaty and **Tundra mucky bog gleyzems** occur widely on flat surfaces and depressions under excessive moistening and imperfect drainage. They have much in common with the analogous soils of the West Sayan.

4.3. Mountain meadow soddy soils

As distinct from the Altay and the West Sayan, in the East Sayan, permafrost-affected soils occur in subalpine and alpine meadows belts.

The cryogenic **Mountain meadow soddy soils** (Gleyi-Gelic Umbrizems) occur on elevations of 1400 to 1900 m, both on flat and steep slopes, under dense grass cover, including violet (*Viola altaica*), geranium (*Geranium albiflorum*), gentian (*Gentiana altaica*), and monkshood (*Aconitum sajanense*). The parent rocks are usually loamy or clay loamy mixed deluvium of granite and gneiss, and ancient alluvial deposits. However, on stony eluvium, especially on limestone and marble rocks, the shallow (40 to 70 cm) and well drained non-cryogenic **Mountain meadow soddy soils** (Mollic Cambisols) are formed.

Specific features of Gleyi-Gelic Umbrisols are a good drainage of topsoil and accumulation of well decomposed organic matter (acid mull humus), on the one hand, and the action of lateral flows of suprapermafrost water and reduction-oxidation processes in subsoil, on the other hand. The profile of Gleyi-Gelic Umbrisols is relatively deep and consists of these horizons:

Dense sod cover (0 to 3 cm).

A (3 to 15 cm). Gray, blocky granular, silt loam, with gradual distinctness.

ABg (15 to 40 cm). Blue grayish gray, subangular blocky, friable, silt loam, with clear distinctness.

Bg (40 to 110 cm). Grayish blue-gray with red and rust-colored mottles, clay loamy, structureless.

BCg (110 to 180 cm). Water-saturated, grayish blue-gray and yellow-brown, mottled, loam with few fine gravel. Cryogenic aquiclude lies at the depth of 180 cm, and suprapermafrost water appears at the depth of 140 to 160 cm.

Soils are strongly or extremely acid and are base-unsaturated. Humus content is highest in the A (13%) and AB (10.6%) horizons and slowly declines with further depth, up to 1.5% at the bottom of the soil. The fulvic acids dominate in humus composition, and humus' enriching of mineral horizons occurs both due to organic accumulation *in situ* and migration of soluble organic matter. Oxalate-extractable iron content varies from 1.0 to 2.8% and exhibits two peaks, in topsoil and in the Bg horizon, and shows clear evidence of removal from the ABg horizon. The biogenic accumulation of calcium and magnesium is characteristic in humus horizons when active leaching of the profile occurs (Gorbachev, 1978).

4.4. Taiga peaty-mucky gleyzems

Taiga peaty-mucky gleyzems (Stagni-Saprihistic Cryosols) (see "West Sayan") occur in the coolest positions of the taiga belt, usually on the upper parts of northern slopes, under wet spruce-larch moss depressed forests, bordering with the Podzols and Taiga Soddy Soils (Dystric Cambisols), which occupy warmer sites. In the southeastern region, cryogenic Gleyzems occur more widely, forming relatively small areas in shadowed slopes, on mountain saddles, and in narrow valleys and canyons, where the active layer does not exceed 40 to 50 cm. Table 2.9.4 lists their properties.

5. Tuva

Tuva is a complex of mountain ranges and intermontane depressions located between the arc of the Sayan mountain system on the north and Mongolia on the south. The main ridges rise to 2300 to 3200 m; the intermontane depressions lie at 500 to 1100 m (Tuvinskaya) and higher, at 800 to 1200 m (Todjinskaya). The climate is continental and extra-continental, with mean winter temperatures of -25 to -34°C and summer means of 19 to 20°C, rising sometimes to 35 to 40°C, especially in the eastern part of Tuva, which has the most severe climate.

Tuva lies in the "rain shadow" of the Altay and the West Sayan, and annual precipitation is very low – from 100 to 180 to 300 mm. Snow cover thickness is 10 to 30 cm almost everywhere, and permafrost occurs in highlands, on northern slopes, and in intermontane depressions, where it occurs in small sites on lowlands and in river valleys. The permafrost table lies at 0.2 m to a few meters, and its annual temperatures are -2 to -5°C (Romanovskiy, 1993).

Vegetation belts follow humidity and elevation: desert steppes, steppe, forests (from 900 to 1300 m to 2000 m and, in semiarid southern regions, from 1700 to 2400 m), alpine meadows, and tundra. Soil cover is more complicated than in the Sayans and includes a wide spectrum of desert-steppe and steppe soils (Brown Desert-Steppe Soils [Calcisols]), Kastanozems, Chernozems) and forest and tundra soils. The slope aspect plays an important role in soils' distribution. On the southern slopes of the Tannu-Ola ridge, Kastanozems occur under dry steppes, and on the northern slopes, under cedar pine and larch forests, Entic Podzols and Stagni-Histic Cryosols prevail.

The main areas of cryogenic soils occur in tundra and taiga belts in northeastern and eastern Tuva, close to the East Sayan, and partly in extracontinental intermontane depressions. They include the same soils as in the permafrost areas of the East Sayan: **Soils of cryogenic circles** (Turbic Cryosols), **Peaty and mucky gleyzems** (Gleyi-Fibrihistic and Gleyi-Saprihistic Cryosols), **Mountain meadow soddy soils** (Gleyi-Gelic Umbrisols), and **Taiga peaty-mucky gleyzems** (Saprihistic Cryosols (Stagnic). The properties of the Cryosols of Tuva are similar to those of the East Sayan (Table 2.9.4). In the intermontane depressions under steppe vegetation occur **Peaty bog gleyzems**. Their properties are practically identical to those of **Peaty bog gleyzems** of the Turano-Uyukskaya kettle depression in the West Sayan (Table 2.9.3).

6. Priamurye

Priamurye is a part of the Russian far east, situated between the Stanovoy range and the Amur river and extended to the east to Okhotsk and the Japan sea, covering the vast left bank of the Amur basin and the river valley on the right bank. Priamurye includes several ridges (the southern edge of Stanovoy, Yankan, Tukuringra, Djagdy, and Bureinskiy) with mean elevations from 1000 to 1500 m and lowland provinces. Mountain ranges have wide massive forms of ridges, with separate alpine peaks, and mainly Proterozoic and Paleozoic metamorphic rocks and granite intrusions underlie them. In the lowlands, the alluvial deposits occur widely.

The Priamurye is the most humid part of the southern sector of the cryopedosphere in Russia. The climate is monsoon, with a cold dry winter and a humid summer. In the northwestern part of the area, the continental features of climate are strong. The mean summer and winter temperatures on the piedmont of the Tukuringra ridge (Zeya river basin, northern part of Priamurye) are 14 to 18°C and -30 to -32°C, respectively. Temperature inversions occur in valley bottoms, due to the flow of cold air from the slopes, so the landscapes of valleys are more severe than those of slopes. The annual precipitation varies from 550 to 650, to 800 to 1000 mm, increasing with elevation and toward the sea coast. Monsoon air circulation keeps the winter portion of precipitation at only 5 to 20% of annual quantity. Snow cover is 20 to 30 cm thick, increasing in high mountains to 50 cm and more.

The vegetation changes with altitude and from west to east. On the piedmonts and slopes of western mountain ridges, up to 900 to 1100 m, larch (*Larix dahurica*) forests with rhododendron (*Rhododendron aureum*), lingonberry, and ledum are dominant. To 1200 to 1300 m, forests of spruce (*Picea jezoensis*) and the narrow ecotone belt of mountain pine (*Pinus pumila*) form between forest and dwarf-shrub and lichen tundra belts. Tundra occupies the wide mountain ridges and tops at elevations of 1300 to 1400 m, with barrens on the highest ridges and peaks. To the east, spruce-Khingam fir (*Abies Nephrolepis*) forests occupy the large areas, and the larch lichen and ledum forests occur in cold and wet positions. The boundaries of vegetation belts are lower to the east, and, on the hills of the sea coast, tundra lies at 200 to 400 m, bordering with thin larch forests and bogs. The bogs occur very widely, and specific moss peat bogs with suppressed larch, ledum, and *Betula middendorffii*, so-called "maris," are characteristic for alluvial lowlands and wide mountain valleys (Nikolskaya, 1981).

The most widely distributed soils in tundra and taiga are Podzols and Podburs (Entic Podzols), which Brown Forest Soils (Eutri-and Dystri-Umbric Cambisols) replace in the warmest positions. The bog soils occur widely both in alluvial plains and mountain valleys.

The permafrost occurs continuously in the northwestern and northern parts of Priamurye, with a thickness of 100 to 300 m and annual temperatures from -2 to -5°C. The thawing layer varies from 20 to 40 cm to 150 to 200 cm and more. In the lowlands and on the seacoast, permafrost distribution is intermediate (not more than 60 to 65%) and sporadic (5 to 30%), and its table lies at 5 to 10 and more meters deep (Nekrasov, Klimovskiy, 1978; Romanovskiy, 1993).

Permafrost-affected soils occur mainly in areas with continuous permafrost, but soils with expressed cryogenic features occur widely in mountain valleys and lowlands, due to deep (2 to 2.5 m) and long (6 to 7 months) seasonal freezing. The evidences of cryogenic processes — frost mounds, cryoturbations, and frost wedging — are common in the sites with additive moistening and fine soil texture, especially in bog (maris) landscapes.

The main cryogenic soils are **Soils of cryogenic circles** (Turbic Cryosols), **Mountainous stony organic soils** (Lepti-Histic Cryosols), **Tundra podburs** and **Podzols** (Lepti-Histic Cryosol [Spodic]), **Taiga podburs** and **Podzols** (Histi-Gelic Podzols), and **Peaty bog gleyzems** (Histic Cryosols) (Shapiro, 1984).

The specific features of cryogenic soils of Priamurye, excluding bog soils, are coarse texture and high permeability.

6.1. Turbic Cryosols

Turbic Cryosols are similar to those described earlier.

6.2. Lepti-Histic Cryosols

Lepti-Histic Cryosols are the most widely distributed cryogenic soils in Priamurye. They occur on elevations higher than 1000 m, under coniferous forests, mountain pines, and tundra vegetation with dense moss cover. They occur in depressions of mountain relief, on massive ridges and flat slopes, among permafrost-free **Mountainous stony organic soils** and **Tundra podburs**, which have a thicker (to 60 to 80 cm) profile. In cryogenic **Mountainous stony organic soils**, the peat cover is usually 25 to 30 cm and more, and the permafrost table lies at 40 to 100 cm as aquiclude. (See Example 2.9.7.)

Example 2.9.7. P. 202-81. Lepti-Histic Cryosols.

Tukuringra ridge, flat saddle, 1180 m a.s.l. Mountain tundra with predominantly bog ledum-dwarf-shrub-sphagnum vegetation. Parent rock is gneiss.

H. 0 to 3 cm. Wet moss and lichen litter.

O1. 3 to 15 cm. Wet dark brown peat.

O2. 15 to 26 (30) cm. Very wet, black-dark brown peat with few gneiss debris and clear distinctness. Water appears from the walls of the pit.

BCi. 26 (30) to 33 cm. Wet, brownish gray, sandy, gravelly, structureless, with ice lenses. Water comes intensively from the pit wall.

Soils do not manifest gleyic features in spite of over-moistening, due to their coarse texture and the occurrence of flow channels. At the same time, the cryogenic aquiclude slows the migration of organic matter and iron elements through the soil profile. Table 2.9.5 shows properties of Lepti-Histic Cryosols of tundra and taiga.

6.3. Tundra podburs and Podzols

Cryogenic **Tundra podburs** and **Podzols** (Lepti-Histic Cryosol (Spodic) occur on well drained sites of watersheds and north-oriented slopes, under moss-lichen tundra vegetation. In the Podbur profile, there is a dark brown and brownish black peaty layer (0 to 15 cm) above the illuvial horizon Bh (15 to 40 cm), which is dark brown, structureless, sandy, gravelly, and stony, with black coatings on debris surfaces. The ice rich frozen horizon BCi (40 to 50 cm) is yellowish brown, structureless, sandy, and stony.

In **Podzols**, the light gray eluvial E horizon with a thickness of 3 to 5 cm underlies the peat layer. The cryic horizon acts as an aquiclude during the relatively short spring. The subsoil is wet but not gleyic. The very fine ice crystals do not deform the soil matter, and cryoturbations are not characteristic. Soils are strongly acid, and migration of organic matter and iron and aluminum is well pronounced (Table 2.9.5).

Cryogenic Tundra and **Taiga podzols** (Histi-Gelic Podzols) and Podburs (Histi-Gelic Podzols (Entic) do not occur widely. They occupy well drained sites in tundra and the upper part of the coniferous belt, on small, smooth mountain tops, mounds, and convex slopes. The properties of cryogenic **Podzols** and **Podburs** of tundra and taiga belts are very similar (Table 2.9.5). (See example 2.9.8.)

Example 2.9.8. P. 37-79. Taiga podzol (Histi-Gelic Podzols).

Tukuringra ridge, 950 m, flat terrace on the northern slope, larch taiga with ledum-lingonberry-moss cover. Parent rock is gneiss.

O. 0 to 5 cm. Dark brown-black peat.

E. 5 to 10 cm. Light gray, structureless, sandy loamy, with gneiss debris. Many roots.

Bhs. 10 to 34 cm. Yellow-brown, subangular, fine sandy loamy, very gravelly; debris coated by sporadic dark brown coatings; common roots, clear distinctness.

BC. 34 to 54 cm. Wet, yellow-brown, structureless, sandy loamy, extremely gravelly, with sporadic light brown coatings, few roots, gradual distinctness.

Ci. 54 to 100 cm. Wet, grayish olive brown, sandy, fine gravelly, stony, with ice lenses in lowest part. Below, the frozen cemented sandy gravelly ground.

6.4. Peaty bog soils

Peaty bog soils (Cryic Histosols) occur in a wide elevation range, occupying poorly drained positions on mountain saddles, wide mountain depressions, and flat terraces in all alpine belts, and cold sites on steep northern slopes, especially under larch ledum-sphagnum bog vegetation (maris) and tundra bog vegetation. The lowest part of the peat layer is frozen at 20 to 30 or 40 to 50 cm deep.

In the Fibri-Cryic Histosols, the layer of living moss (thickness is 10 to 12 cm) is underlain by dark brownish black peat with fine ice lenses. The peat is fully frozen at >20 cm deep.

In Sapri-Cryic Histosols, under the brown peaty layer (35 cm) lies well decomposed black peat, which at 50 to 55 cm deep is frozen and impermeable for percolating water. Cryoturbation processes are well expressed on surfaces and specific palsa bogs often form.

7. Conclusion

The cryogenic soils of the mountains of southern Siberia and far eastern Russia occur widely in the central and southeastern Altay, the southern part of the West Sayan and southeastern parts of the East Sayan, northeastern and eastern Tuva, and the northern part of the Priamurye.

Cryosols occur extensively under the severe climate of high tundra alpine belts and the upper parts of northern and shadowed forested slopes, and they occupy significant boggy areas through all elevations. Sporadic and isolated areas of cryogenic soils are specific for intermontane depressions and the alluvial plains of Priamurye.

The main permafrost-affected soils are the Turbic, Histic, Stagnic, and Gleyic Cryosols, as well as the Gelic Podzols in Priamurye. Weakly decomposed accumulation of organic matter, high acidity, and base unsaturation are characteristic for Cryosols in most cases, and, in the southern intermontane depressions under extra-continental arid climate, Gelic Solonchaks and Gelic Solonetzes form in poorly drained sites.

References

Devyatkin, E.V. 1965. Quartenary sediments and neotectonic geology of southeastern Altay (Chetvertichnye otlozheniya i neotektonika Yugo-Vostochnogo Altaya). Nauka. Moscow. (In Russian.)

Gerasimova, A.C. (Ed.). 1973. Reference Book for climate of USSR. (Spravochnik po klimatu SSSR). Krasnoyarsk. (In Russian.)

Gorbachev, V.N. 1978. Soils of Eastern Sayan. (Pochvy Vostochnogo Sayana). Nauka. Moscow. (In Russian.)

Kovalev, R.V. (Ed.). 1973. Soils of Mountain Altay Autonomous District. (Pochvy Gorno-Altayskoy avtonomnoy oblasti). Nauka. Novosibirsk. (In Russian.)

Kudryavtsev, V.A. (Ed.). 1978. General cryology. (Obshchee merzlotovedenie) Moscow Univ. (In Russian.)

Kuminova, A.V. 1960. Vegetation cover of Altay. (Rastitelnyi pokrov Altaya). Novosibirsk. (In Russian.)

Mikhailov, N.I. 1961. Mountains of southern Siberia. (Gory Yuzhnoy Sibiri). Moscow. (In Russian.)

Nekrasov, I.A., and I.V. Klimovskiy, 1978. Permafrost of BAM zone. (Vechnaya merzlota zony BAM). Nauka. Novosibirsk. (In Russian.)

Nikolskaya, V.V. 1981. Physical geography of Far East. (Fizicheskaya geografiya Dalnego Vostoka). Vysshaya Shkola. Moscow. (In Russian.)

Nosin, V.A. 1963. Soils of Tuva. (Pochvy Tuvy). Nauka Publishers, Moscow (In Russian.)

Petrov, B.F. 1952. Soils of Altay-Sayan region (Pochvy Altae-Sayanskoi oblasti). Proc. Dokuchaev Soil Science Institute, 35. Moscow. (In Russian.)

Romanovskiy, N.N. 1993. Principles of cryogenesis of lithosphere (Osnovy kriogeneza litosfery). Moscow Univ. (In Russian.)

Shapiro, M.B. 1984. Structure of soil cover of mountain ranges of central part of BAM zone. PhD thesis. Dokuchaev Soil Institute. Moscow. (In Russian.)

Smirnov, M.P. 1970. Soils of Western Sayan.(Pochvy Zapadnogo Sayana). Nauka. Moscow. (In Russian.)

Targulian, V.O. 1971. Soil formation and weathering in cold humid regions (Pochvoobrazovanie i vyvetrivanie v kholodnykh gumidnykh oblastyakh). Nauka. Moscow. (In Russian.)

Chapter 10. Geography and Ecology of Cryogenic Soils of Mongolia

Stanislav V. Maximovich[1]

[1]Senior Researcher, Institute of Physico-Chemical and Biological Problems of Soil Science, Russian Academy of Sciences, Institutskaya St, Pushchino, Moscow Region, 142290, Russia; e-mail svmax@ibbp.psn.ru.

1. Introduction

1.1. Area

Mongolia stretches for nearly 1200 km north to south and covers over 10 latitudinal degrees (from 52°04′N to 41°48′N), with forested mountains in the north and desert depressions in the south. Naturally, the latitudinal zonality influences the distribution of permafrost-affected soils; they decrease from north to south. Relevant literature (Melnikov, 1974; Lonzhid, 1966; Tumurbator, 1975, and others) describes geocryological conditions of Mongolia and provides information on historical records of studies, actual and paleocryological conditions, and the division of the area into geocryological regions.

1.2. Climate

Investigations of the heat balance of the seasonal freeze and thaw cycles of Mongolian soils show that general climatic conditions favor the complete summer thawing of ground frozen in winter. However, part of the ground may stay frozen, usually at depths greater than 100 cm, because of climatic conditions that megafactors (absolute altitude, slope aspect, etc.) and mesofactors (topography, degree of forestation, etc.) determine.

Much of the permafrost in Mongolia is a Pleistocene relict which deep winter freezing maintains, due to the strong continental climate of the country. Calculations based on data from the literature (Melnikov, 1974) show that the perennially frozen ground covers 10-15% of Mongolia.

1.3. Soil science history

Historical records of Mongolian soils (Gerasimov and Nogina, 1984) show soil scientists focused almost exclusively on steppe and desert territories, and they did not specifically deal with cryogenic soils. Thorough studies of cryogenic soils began in the 1960s and early 1970s and made it possible to compile a soil map of Mongolia at a scale of 1:2 500 000 (Nogina, 1980), to publish monographs and papers which characterized various cryogenic soil types, to systematize results of regular (including winter) field studies, and to define the area's regions of permafrost soils (Bannikova et al., 1980; Lavrenko and Bannikova, 1983; Maximovich and Mirkin, 1980; Maximovich, 1993; Ogorodnikov, 1981; Gerasimov and Nogina, 1984; Nogina, 1978; Grubov, 1980).

This chapter is based on both a review of the published literature and on still unpublished data I collected. It gives an overview of soil-forming factors, including cryogenic factors, and provides the basis for a systematic description of soils, followed by the delineation of the regions of permafrost-affected (but not generally Crysolic) soils in Mongolia.

2. Soil-forming Factors

Mongolia is near the center of the Eurasian continent, quite distant from maritime influence. High mountain systems, which prevent the penetration of humid air masses, surround it. The altitude ranges from 565 to 4362 m, with an average of 1500 m above sea level. These factors produce an extremely continental climate. Table 2.10.1 gives climatic data for the main zones of Mongolian permafrost-affected soils.

Table 2.10.1. The climatic data (1975-1980) for some of the belts (zones) containing permafrost-affected soils in Mongolia: high mountainous meadows and mountainous taiga. Data from the Tsetserleg meteorological station in the steppe zone is included for comparison.

Parameters	Steppe	Mountainous Taiga	High mountainous meadows
Altitude above sea level, m	1695	2050	2507
Mean annual temperature, °C	0.4	-1.1	-4.2
Mean July temperature, °C	14.4	11.8	8.9
Mean January temperature, °C	-15.0	-14.7	-19.3
Absolute air temperature maximum, year, °C	30.1	28.9	26.4
Absolute air temperature minimum, °C	-35.1	-34.8	-30.5
Annual air temperature amplitudes, °C	65.2	58.0	49.9
Daily air temperature amplitudes, °C	25.4	17.9	16.9
Relative air humidity, %	55	69	65
Atmospheric precipitation, mm	306	342	437

Autumns in Mongolia are dry and sunny. Normally, freezing of soils begins in late October, when there is little or no snow, and occurs at nearly the same rate within all altitudinal belts with grass vegetation, being delayed only in the taiga belts. The freezing goes deep, up to three meters and more. According to our data, in soils with active layers, ground up to 200 cm deep will refreeze and fuse with permafrost before the end of November.

Winters in Mongolia are cold, sunny, and with little snow. Continuous snow cover (25-30 cm, sometimes up to 50-60 cm) forms under forest communities of north-oriented mountainous slopes. In the high-mountain zone, winds often carry the snow away, and the snow cover is highly irregular. On windward sites, it is very thin (3-5 cm), and the height of dry grasses determines its thickness. During winters with abundant snowfalls, a continuous snow cover also can form on steppe plains.

Thawing of frozen soils is extremely slow in the spring, particularly in high altitude belts; it accelerates with the beginning of summer rains and proceeds rapidly with heavy rains. Sometimes the rainwater temporarily is frozen in the soil profile, forming a frozen icy screen. Thawing of taiga soils with insulating moss cover is delayed during the dry summers, as it is in high-mountain soils, where rain nearly exclusively determines the thaw rate.

The decrease of air temperatures and increase of humidity at higher altitudes leads to forest and tundra vegetation with underlying permafrost-affected soils. At these high altitudes, taiga (subdivided into lower-level, middle-level, and upper-level taiga) substitutes for open grassy woodlands. The mossy pre-tundra (i.e., upper subalpine) forests and alpine tundra are at elevations >2500m asl, where the amount of heat is insufficient to develop full forest vegetation. However, this sequence occurs only on north-facing slopes of mountains.

On the warmer and drier south-facing slopes, the forest belts disappear almost everywhere, giving way to the steppe zones which extend upwards until their contact with alpine vegetation. The soils of these slopes are permafrost-free. Alpine tundras spread over flat sites of high mountains. In the near-apical part of the southern slopes, the tundra does not border directly on the alpine steppes. The belt of low-grass alpine meadows dominated by *Cobresia* and *Carex* species, occurring over cryogenic soils, separates the alpine tundra and alpine steppe belts.

The forests covering the north-oriented slopes of mountains do not spread down to the foothills and intermontane valleys, where there is not enough heat for forest stands because of temperature inversions and cold air stagnation. These are niches for tall-grass meadows on permafrost-affected soils. Wetlands with cryogenic soils occupy the lower levels.

In the steppe belts, the soils with permafrost are confined to habitats with additional moistening near springs and saline depressions. Deep permafrost with active layers generally >200 cm also occur under the bottoms of some dried saline lakebeds.

3. Regions of Permafrost-affected Soils

Figure 2.10.1 shows Mongolia's division into five regions, on the basis of the presence of permafrost-affected soils and paleocryogenic features.

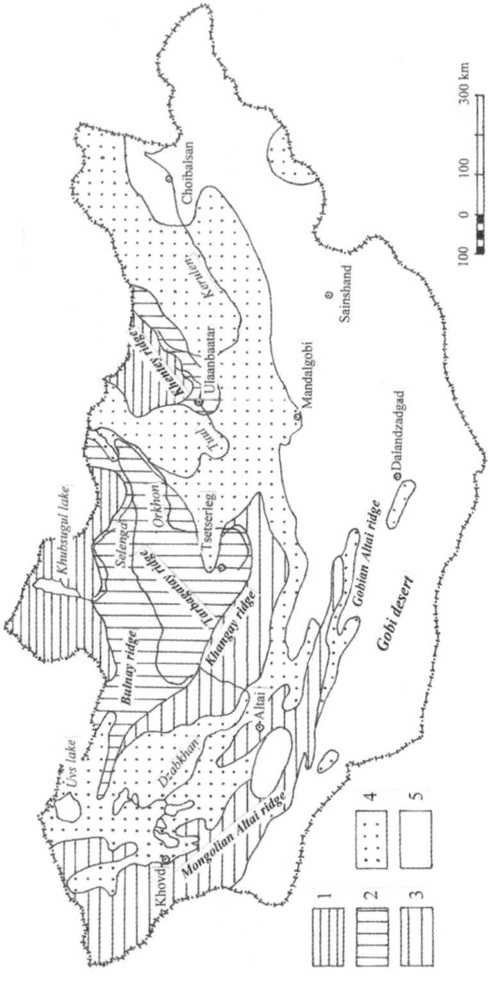

Figure 2.10.1. Permafrost-affected soil regions of Mongolia. 1. discontinuous areas of permafrost-affected soils; 2. sporadic areas of permafrost-affected soils; 3. isolated patches of permafrost-affected soils; 4. rare patches of permafrost-affected soils and pronounced paleocryogenic features; 5. non-permafrost soils and areas of soils with paleocryogenic features.

3.1. Khubsugul Mountain region and the northern slope of Khentey

The Khubsugul Mountain region and the northern slope of Khentey are the most humid regions in Mongolia, and the sequence of altitude belts is of the humid type. Forests cover all mountain slopes, except for the steepest south-facing ones. High mountains are humid.

The discontinuous occurrence of these permafrost-affected (but non-Cryosolic) soils characterize this region: Mountain frozen taiga soils (*Gelic Leptosols*), Mountain soddy-taiga deeply frozen soils *(Humi-Gelic Regosols),* Frozen-swampy soils *(Gelic Gleysols),* and Alpine-tundra soils – Soddy *(Lepti-Gelic Regosols, Gelic Leptosols*), Mucky-peaty-gley *(Geli-Saprihistic Gleysols [Leptic]).*

Permafrost often occurs in the form of palsas in wetlands of intermontane valleys at depths of 1 to 1.5 m. The taiga and high mountain soils thaw under the influence of rains in the beginning of summer.

3.2. Northern macroslope of Khangay and southern macroslope of Khentey

Khangay is in the wind shadow of mountains north of it, while the southern slope of Khentey is the wind shadow of the northern slope. Here, there is less precipitation than in the Khubsugul Mountain region and northern slope of Khentey. The sequence of altitude belts is of the subhumid type. Forests cover only the north-facing mountain slopes, while steppe vegetation dominates the south-facing slopes.

The high-mountain meadows contain steppe elements and even are sometimes transformed into high-mountain meadow steppes. The area of lichen-moss and dwarf shrub-mossy tundra decreases.

In summer, frozen horizons in the soil profile occur in wetland meadow soils of intermontane valleys and under forests of northern slopes. The high-mountain soils, which receive more precipitation, thaw in the first half of summer.

The sporadic cover of cryogenic soils are characteristic for Alpine-tundra soddy soils *(Lepti-Gelic Regosols, Gelic Leptosols*), High-mountain meadow soils *(Lepti-Gelic Phaeozems),* Mountain permafrost taiga soils *(Gelic Leptosols*), Mountain soddy-taiga deep permafrost soils *(Humi-Gelic Regosols*), Meadow deep permafrost soils *(Gelic Phaeozems* or *Gelic Chernozems),* and Permafrost-swampy soils *(Gelic Gleysols*).

3.3. Southern macroslope of Khangay and Mongolian and Gobian Altai

In the Southern macroslope of Khangay and Mongolian and Gobian Altai, the general dryness of climate is enhanced. The elevational stratification is subarid (Khangay) and arid. Forests disappear from northern slopes, while steppes and deserts extend upward to very high altitudes. Aridization of the high-mountain belt is enhanced. High-mountain meadows transform into high-mountain meadow steppes. Humid mossy tundra disappears, giving way to rocky tundra, the herbage of which contains plants capable of forming cushions.

These are regions of isolated patches of Alpine-tundra soddy *(Lepti-Gelic Regosols, Gelic Leptosols)* and High-mountain meadow *(Lepti-Gelic Phaeozems)* cryogenic soils in which permafrost occurs. Cryogenic soil-forming features are also evident in wetlands of intermontane valleys covered with frost mounds containing even hydrolaccolites and Permafrost-swampy soils *(Gelic Gleysols)*.

3.4. Low mountains, plains, and depressions in the northern half of Mongolia

Low mountains, plains, and depressions in the northern half of Mongolia are steppe, steppe-desert, and desert territories. These regions have rare isolated patches of permafrost-affected soils and clearly manifest paleocryogenesis. Permafrost sometimes occurs under saline meadows and the bottoms of saline lakes (Permafrost solonchaks-solonetzes *[Gelic Solonchaks, Gelic Solonetzes]*).

3.5. Dry steppe, desert-steppe, and desert areas in the southern half of Mongolia

The dry steppe, desert steppe, and desert areas in the southern half of Mongolia are regions with seasonally freezing soils and frequent relicts of paleocryogenesis.

4. Systematic Description of Cryogenic Soils

Permafrost-affected soils of Mongolia generally are confined to alpine tundras and high-mountain meadows, taiga and open woodlands of north-oriented slopes, tall-grass meadows and wetlands in high-altitude intermontane valleys, and saline depressions in the steppe belts.

Taxonomic names that here differentiate the cryogenic soils in these belts come from the soil map (Nogina, 1980) and the relevant monograph (Gerasimov and Nogina, 1984) published for the country (Table 2.10.2). Correlated taxonomic equivalents within the *World Reference Base* (WRB) (ISSS Working Group RB, 1998) system appear here in italics. Detailed descriptions follow for each group of cryogenic soil, the general locations, and associations with cryogenic features and characteristics.

Table 2.10.2. Cryogenic Soils of Mongolia.

1. Alpine tundra:
   - Alpine-tundra soddy soils *(Lepti-Gelic Regosols, Gelic Leptosols)*
   - Alpine-tundra mucky-peaty-gley *(Geli-Saprihistic Gleysols [Leptic])*
   - Alpine-tundra destructive soils *(Stagni-Gelic Regosols)*
2. High-mountain meadows:
   - High-mountain meadow soils *(Lepti-Gelic Phaeozems)*
3. Upper-level taiga:
   - Mountain permafrost taiga peaty-mucky-gley soils *(Turbi-Histic Cryosols [Gleyic])*
4. Middle-level taiga:
   - Mountain permafrost taiga soils *(Gelic Leptosols)*
5. Lower-level taiga:
   - Mountain soddy-taiga deep permafrost soils *(Humi-Gelic Regosol)*
6. Subtaiga open woodlands:
   - Mountain meadow-forest deep permafrost soils *(Gelic Phaeozems)*
7. Meadows and wetlands in intermontane valleys:
   - Meadow deep permafrost *(Gelic Phaeozems or Gelic Chernozems)*
   - Permafrost-swampy soils *(Gelic Gleysols)*
8. Near-spring and saline depressions around drying saline lakes in the steppe belt:
   - Meadow deep permafrost soils Permafrost solonchaks–solonetzes (*Gelic Solonchaks, Gelic Solonetzes)*
9. River flood-plains
   - Flood-plain boggy and meadow soils *(Geli-Saprihistic and Geli-Gleyic Fluvisols)*

4.1. Alpine Tundra Soils

4.1.1. Alpine-tundra soddy soils

Alpine-tundra soddy soils *(Lepti-Gelic Regosols, Gelic Leptosols)* form on well drained sites covered with dryad (*Dryas sp.*) and cobresial-dryad (*Cobresia, Dryas*) and, less frequently, lichen-sedge communities. They can occupy relatively large areas on high flat surfaces where snow blows away. At the same time, soddy soils also form the microcomplex pattern of polygonal and hummocky tundras.

Despite their well drained nature, solifluction phenomena and frost-sorting characterize them. The soil profile does not undergo gleyzation or has only poorly pronounced stagnic features. No frost is evident in these soils in July or August, but the depth of soil profiles did not exceed 90 cm because of shallow lithic contact.

The thickness of the solum of these soils ranges from approximately 10 to 40 cm. Horizons are often discontinuous, due to cryoturbation. However, the humus (A) horizon of these soils stands out distinctly, with its thickness ranging from 5 to 17 cm and its upper part enriched in plant roots (root mat, 1 to 2 cm) and humus.

Tables 2.10.3 through 2.10.5 present the analytical characteristics of these soils. This site information applies to the Alpine tundra soddy soils profiles: Profile 646, Tarbagatay, near the Solongot Mountain pass, 2600 m asl; Profile 731*a* and Profile 731*b* (*a*, cobresial patch; *b*, lichen patch), Gobian Altai, Ikh-Bogd Mountain,

3629 m asl; Profile 784, central Khangay, Tsagan-Bitu Mountain, 3050 m asl; Profile1217, western Khangay, Otgon-Tenger Mountain, 3200 m asl.

All of the A horizons are rich in organic matter, and humic acids prevail in the uppermost part of the horizon, but the share of fulvic acid increases with the depth. Humus is mobile and migrates down the soil column. Iron and silicon are also mobile. Despite their loamy texture and the abundance of humus, these soils are not well aggregated or microaggregated. The organic matter of these tundra soddy soils has undergone active humification only in the uppermost part of the A horizon, as low values of C:N ratio indicate, but at a depth of 5 to 10 cm this ratio increases more than two-fold. The soils are acidic over the entire profile, though they can have neutral pH near the southern limit of their distribution in the Gobian Altai.

Table 2.10.3. Particle-size and microaggregate analysis of soil complex associated patterned ground on alpine tundra.

Profile	Horizon	Depth (cm)	Coarse Fragments (>1 mm)	Fractions (mm), fine earth (<1 mm) (%)				Aggregation degree (%) according to Beiver and Roades
				< 0.001	< 0.01	>0.05	Aggregates >0.05	
731a	A	0-10	58	13	35	47	74	36.5
731b	A	0-10	54	9	25	54	67	19.4
731c	A	0-10	65	21	45	25	42	40.5

Table 2.10.4. Base saturation, pH, and humus properties of soils of alpine tundra.

Profile	Horizon	Depth (cm)	pH		Base Saturation (%)	Organic matter content (%)		C:N	C_{ha}:C_{fa}¶
			H_2O	KCl		Total	Water-soluble		
646	A	1-11	5.90	4.80	63.2	8.90	-	-	-
	B	20-30	6.50	4.60	-	1.95	-	-	-
	BC	55-65	6.60	4.70	26.3	1.25	-	-	-
647	O	0-5	5.40	4.40	-	77.3*	-	-	-
	AO	5-15	5.60	4.70	52.2	28.5*	0.090** (0.3)	-	-
731a	A	0-10	7.10	6.00	90.5	7.65	0.045 (0.6)	-	-
731b	A	0-10	6.83	5.87	89.6	8.30	0.039 (0.5)	-	-
731c		0-10	7.61	5.82	91.6	1.40	0.019 (1.4)	-	-
784	A	0-5	7.10	5.62	-	18.30	0.065 (0.4)	-	-
819	O	0-9	5.15	4.20	-	83.8*	-	-	-
	AO	10-20	5.90	4.49	-	29.1*	0.058 (0.2)	-	-
	Ag	30-40	5.78	4.22	-	9.45	0.045 (0.5)	-	-
	Bg	60-70	6.12	4.78	-	4.65	0.017 (0.4)	-	-
1217	AO	2-5	5.60	4.80	69.6	10.61	-	11	1.51
	B	5-9	5.70	4.50	71.6	3.94	0.010 (0.2)	29	0.82
	BC	50-60	5.95	4.50	77.0	1.05	-	-	-

* Value based on loss on ignition method. ** Figures in parentheses indicate % of total humus.
¶ Ratio of humic to fulvic acid carbon. - = was not determined.

Table 2.10.5. Labile compounds in soils of alpine tundra.

Profile	Horizon	Depth (cm)	Exchangeable Cations (1M KCl) (meq/100g)		Oxalate extactable (%)		Water extractable (mg/100g)		
			H^+	Al^{+3}	Fe_2O_3	Al_2O_3	Al_2O_3	SiO_2	Fe
646	A	1-11	0.1	0.3	0.40	0.20	-	-	-
	BC	55-65	0.1	1.7	-	-	7.56	-	-
819	O	0-9	0.60	0.30	-	-	-	-	-
	AO	10-20	0.40	0.35	-	-	-	-	-
	Ag	30-40	0.60	0.55	-	-	-	-	-
	Bg	60-70	0.30	0.20	-	-	-	-	-
1217	AO	2-5	-	-	0.60	0.55	-	-	-
	B	5-9	-	-	0.80	0.76	0	12.50	0.05
	BC	50-60	-	-	1.06	1.48	0	15.00	0.11

- = was not determined.

4.1.2. Alpine-tundra mucky-peaty-gley soils

Alpine-tundra mucky-peaty-gley soils *(Geli-Saprihistic Gleysols [Leptic])* are the most typical for humid high mountains in the north of Mongolia (Khubsugul mountain region, northern parts of Khangay and Khentey). They formed on flat tops of mountain ridges, highland terraces, and the upper parts of northern slopes, under shrub-moss and lichen-moss communities that thaw slowly in summer. In some years, ice-rich permafrost was documented at depths of 20 to 25 cm. The processes of cryoturbation, solifluction, frost jacking, and sorting are well pronounced in these soils.

A continuous moss layer overlies a horizon of loose poorly decomposed peat 5 to 10 cm thick, which overlies a dark mucky horizon 11 to 17 cm thick with abundant plant roots in its upper part. The latter is followed by the humus-gley horizon (Ag) up to 25 cm thick and a Bg horizon 25 to 30 cm thick, enriched in organic residues and illuvial humus. Table 2.10.4 gives the analytical characteristics of some Alpine tundra mucky peaty gley soils from the following sites: Profile 647, Tarbagatay, near the Solongot Mountain pass, 2650 m asl; and profile 819, the same locality, at 2620 m asl.

These soils are strongly acidic and have a high content of exchangeable aluminum, while the horizons Ag and Bg are rich in humus, owing to its migration from the upper horizons.

4.1.3. Alpine-tundra destructive soils

Alpine-tundra destructive soils *(Stagni-Gelic Regosols)* occur as mud circles – patches devoid of vegetation. They are composed of a mineral mass strongly mixed by cryoturbation processes. This is a homogeneous bluish gray-brown mass with pale and rusty stains. Tables 2.10.3 and 2.10.4 present some properties of this soil type as derived from the site of Profile 731*c* (Gobian Altai, Ikh-Bogd Mountain, 3620 m asl). Compared to the neighboring soils (profiles 731a and b) under vegetative cover, they have a finer texture in the upper 10 cm and hence are better aggregated, are more skeletal, and have a low humus content.

4.2. High mountain meadows – High-mountain meadow soils (*Lepti-Gelic Phaeozems*)

High-mountain meadow soils *(Lepti-Gelic Phaeozems)* formed under short-herbage high-mountain meadows, where *Cobresia* and *Carex* species predominate, and under alpine forbs in the Khubsugul region. In the humid climate, they occur only on flat mountain tops. In the subhumid climate, they occupy mainly flat mountain tops, high highland terraces, and north-facing slopes only slightly higher than the upper forest limit. In the subarid regions, High-mountain meadow soils occupy similar positions but are more confined to northern slopes. Their distribution shifts entirely onto northern slopes in the most arid regions.

The soil cover of high-mountain meadows is discontinuous because of frost sorting and cryoturbation, though to a lesser extent than on tundras. The fine earth in the deeper part of soil profiles is thixotropic and, by filling voids between boulders, in some cases can act as an aquiclude.

These soils have a clear-cut and rather thick (10 to33 cm) humus horizon of brownish dark gray color, with a grainy structure and a large amount of roots. The A horizon is usually free of coarse fragments and is usually a clay loam texture, while the horizon AB formed in the upper part of loose stony debris enriched in fine earth. The latter is followed by the horizons B and BC in loose stones with a small amount of fine earth, basically in the form of pendants on the upper sides of stones, which indicate clay migration within the profile.

In the lower part of the BC horizon, the fine earth becomes thixotropic and occurs in the form of inclusions of yellow pale loam between boulders. Humus-sesquioxide films occur on the lower sides of stones throughout the profile. In some cases, large boulders have white calcite mottles on their surfaces, which indicate the periodic drying of the upper part of these soils, as the 6-year-long moisture measurements in the Profile 1288 (Tables 2.10.6 through 2.10.8) in the north-eastern Khangay region confirm.

Table 2.10.6. Particle-size and microaggregate analysis of soils of high-mountain meadows.

Pro-file	Hori-zon	Depth (cm)	Course Fragments (>1 mm)	Fractions (mm), fine earth (<1 mm) (%)				Aggregation degree (%) according to Beiver and Roades
				<0.001	<0.01	> 0.05	Aggre-gates >0.05	
621	A	4-14	27	13	36	27	63	57.1
	A	14-24	30	19	39	29	63	54.0
	B1	24-34	33	23	43	24	62	61.3
	B2	34-44	42	17	36	32	61	47.5
	BC	44-54	55	15	33	32	59	45.8
1288	A	0-3	no	10	28	24	-	-
	A	3-13	no	20	44	17	-	-
	A	13-20	no	26	48	14	-	-
	AB	20-30	92	24	50	14	-	-
	B	30-40	77	28	52	13	-	-
	BC	50-60	53	24	42	21	-	-
	BC	70-80	49	20	43	21	-	-

-- = was not determined.

In Mongolian Altai, one also can find High-mountain meadow soils (*Eipti-Gelic Phaeozems)* with a lot of calcite pendants on the lower sides of stones and also with pseudomycellia in the fine earth. These soils are not acidic over the whole profile (profile 1252). Variants of these soils also occur in which the upper part of the A horizon is enriched in humus (profile 1255). Tables 2.10.6 through 2.10.8 give the analytical characteristics of these soils.

The site characteristics of these High-mountain meadow soils include: Profile 621, western part of the Bulnay Ridge, 2300 m asl; Profile 1215, western Khangay, Otgon-Tenger Mountain, 2870 m asl; Profile 1252, Mongolian Altai near Angirt River, 2700 m asl; Profile 1255, in the same locality, 3100 m asl; Profile 1260, Mongolian Altai, southern part of the Dayan-Panchin Mountain pass, 2765 m asl; Profile 1288, northeastern Khangay, Tuvshrulekh, Khairkhan Mountain, 2507 m asl.

These soils have a lot of mobile humus, mainly fulvic acid, though humic acids may predominate near the soil surface. The soils are well aggregated. Moderate mobility is characteristic of aluminium, silicon, and iron.

The soils are mostly acidic and have a cold thermal regime. The temperature of 10°C penetrates to between 17 and 27 cm deep, while that of 20°C isotherm does not reach even 5 cm. The lower part of the soil is always cold with a maximal temperature of only 6.2°C at 100 cm deep. At night, negative temperatures on the soil are possible during any summer month. According to observations carried out in 1975-1980, the maximum amplitude of daily temperature oscillations is 45.5°C. The mean lowest and highest temperatures for the same period were -36.9 and +55.5°C, respectively.

Table 2.10.7. Base saturation, pH, and humus properties in high-mountain meadow soils.

Pro-file	Hor izon	Depth (cm)	pH		Base Satura-tion (%)	Organic Matter Con-tent (%)		C:N	C_{ha}:C_{a}
			H_2O	KCl		Total	Water-soluble**		
621	A	4-14	5.47	5.20	65.1	12.00	0.050 (0.4)	-	-
	A	14-24	6.34	5.00	65.2	6.40	0.030 (0.5)	-	-
	B1	24-34	6.26	5.20	62.5	4.50	0.015 (0.3)	-	-
	B2	34-44	5.98	5.00	61.5	2.90	0.005 (0.2)	-	-
	BC	44-54	5.86	4.80	71.5	1.60	0.005 (0.3)	-	-
1215	A	3-10	5.65	4.70	79.1	7.91	0.017 (0.2)	27	0.71
	B	10-19	5.90	4.65	74.6	5.09	0.002 (0.1)	27	0.28
	BC	50-60	6.20	4.80	86.5	0.74	0.003 (0.4)	-	-
1252	AO	0-3	7.20	6.55	-	25.2*	-	-	-
	A	3-13	7.40	6.55	97.1	10.35	0.026 (0.3)	12	1.01
	B1	16-26	8.40	7.40	-	4.41	-	-	-
	B2	35-45	8.60	7.70	-	1.41	0	-	-
	BC	90-100	8.65	7.70	-	0.80	0.008 (1.0)	-	-
1255	AO	0-4	6.15	5.30	84.7	18.06	-	-	-
	A	5-15	6.18	5.50	81.4	10.56	0.028 (0.2)	11	0.74
1260	AO	0-5	6.35	6.05	87.3	20.62	-	17	0.59
	AB	5-15	6.40	5.50	88.1	5.83	0.021 -(0.4)	16	0.76
1288	AO	0-3	7.00	6.85	-	17.20	-	-	1.27
	A	3-13	6.30	4.38	60.2	11.70	-	-	0.75
	A	13-20	5.90	4.42	60.9	9.50	-	-	-
	AB	20-30	6.15	4.65	54.9	9.20	-	-	0.70
	B	30-40	6.05	4.56	61.7	7.70	-	-	0.56
	BC	50-60	6.40	4.25	72.5	2.00	-	-	-
	BC	70-80	6.28	4.30	75.4	1.70	-	-	-

* Value based on loss on ignition method.

** Figures in parentheses indicate % of total humus.

¶ Ratio of humic to fulvic acid carbon.

-- = was not determined.

In the studied soil, seasonal frost disappeared from the solum (90 cm thick) in late June under the action of rains, and it refroze rapidly after October 20. In the summer the soil is permanently moist, while in autumn the upper part of the soil profile undergoes considerable drying.

Among all the soils of high mountains, only these High-mountain meadow soils have economic uses in Mongolia. They support pastures for cattle in Mongolian and Gobian Altai and partly in Khangay.

Table 2.10.8. Labile compounds in soils of high-mountain meadows.

Profile	Horizon	Depth (cm)	Exchangeable Cations (1M KCl) (meq/100g)		Oxalate Extactable (%)		Water extractable (mg/100g)		
			H^+	Al^{+3}	Fe_2O_3	Al_2O_3	Al_2O_3	SiO_2	Fe
621	A	4-14	-	-	0.61	0.63	19.38	-	-
	A	14-24	-	-	0.56	0.14	22.44	-	-
1215	A	3-10	0.02	0.14	1.06	0.98	0	15.00	0
	B	10-19	0.04	0.08	1.00	0.56	0	-	0.09
	BC	50-60	0.02	0.06	0.92	0.36	0	15.00	0.12
1252	A	3-13	-	-	1.10	1.26	-	-	-
1255	A	5-15	-	-	1.40	2.40	-	7.20	-
1260	AO	0-5	-	-	0.70	0.86	-	-	-
	AB	5-15	-	-	1.00	1.40	-	-	-
1288	A	3-13	0.20	0.70	-	-	-	-	0.50
	BC	50-60	0.10	0.40	-	-	-	-	2.10
	BC	70-80	0.20	0.40	-	-	-	-	-
	BC	80-90	-	-	-	-	-	-	4.00

-- = was not determined.

4.3. Upper-level Taiga – Mountain permafrost taiga peaty-mucky-gley soils *(Turbi-Histic Cryosols [Gleyic])*

Mountain permafrost taiga peaty-mucky-gley soils *(Turbi-Histic Cryosols [Gleyic])* occur under thin forests of taiga and pretundra on the level to depress ional sites on upper parts of northern slopes. They have a surface horizon of slightly compacted, undecomposed moss peat 10 to 20 cm (sometimes up to 30 cm) thick, which in turn usually overlies a black mucky horizon and an A horizon very rich in humus. The thickness of these horizons does not exceed 10 to 12 cm.

Beneath these upper horizons is a gray or brownish gray gleyed horizon, which sometimes is enriched in the illuvial humus. The bottom of the profile is always highly moist, due to the presence of a permafrost table that even in summer does not retreat below 50 cm and sometimes is found at depths of 20 to 30 cm.

The soils have acidic pH values, with iron and aluminium in labile forms. Humus of mineral horizons is fulvic. The upper part of these horizons are clay loams or clay texture, while downward the coarse fragment content increases and the texture may change to gravelly sandy loam.

4.4. Middle Level Taiga – Mountain permafrost taiga soils *(Gelic Leptosols, Turbi-Histic Cryosols)*

Mountain permafrost taiga soils *(Gelic Leptosols, Turbi-Histic Cryosols)* occur under cedar-larch, cedar, and sometimes pine-larch forests with mossy-dwarf shrub vegetation cover. On the surface layer of such soils is a thin (up to 10 cm) organic horizon, composed of forest litter. A very thin (usually 2-3 cm) humus (A) horizon underlies it, followed by a loamy B horizon(s) with a varying amount of stones and rock debris of pale brownish, dark color passing downward into yellowish brown, usually slightly bluish colored.

The fine earth layer is commonly not very thick, ranging from 15 to 40 cm. Beneath is a BC horizon formed in a stony colluvium (former stone streams), with an accumulation of thixotropic loam between stones forming an aquiclude horizon.

Tables 2.10.9 through 2.10.11 give the analytic characteristics of these soils. These are site properties of soils classified as Mountain permafrost taiga soils of the Middle Level Taiga zone: Profile 533, northeastern Khangay, Tuvshrulekh, 1957 m asl, on June 19, 1971, frozen at 60 cm depth; Profile 548, central Khangay, near the Egiyn-daba Mountain pass, June 26, 1971, frozen at 36 cm depth; Profile 600, Hubsugul Mountain region, eastward of Khatgal, 1900 m asl, on July 23, 1971, frozen at 68 cm depth; Profile 1007, northern Mongolian Altai, Turgen Ridge, 2070 m asl, on July 7, 1973, frozen at 85 cm depth; Profile 1160, northeastern Khangay, Tuvshrulekh, 2050 m asl, on July 16, 1980, frozen at 11 cm depth.

The soils are acidic, and saturation ranges from 70-80%. The extent of saturation increases slightly with depth. Silicon, iron, humus, and aluminium are labile. The texture is mostly clayey loams.

The soils are permanently cold. Observations in the proximity of profile 1160 (1975-1980) show that the 10°C temperature penetrates only to a depth of 3-9 cm; i.e., it acts practically within the limits of the litter layer. The maximum temperature at a depth of 100 cm varied in different years from 0.3 to 2.4°C, while the minimum temperature did not decrease below -3.5°C. The thawing of soils continues sometimes until the end of July, but more often it stops in mid-July.

Permafrost is often present within one meter of the surface, and the soils are classified as Cryosols. The rainwater penetrating into frozen soil forms a temporarily frozen ice-bonded aquiclude. On the whole, these soils are always moist; however, their moisture retention capacity is not high, due to the small thickness of the fine earth layer. An appreciable share in their water balance belongs to the moisture from the lateral intrasoil runoff and condensation.

Table 2.10.9. Particle-size and microaggregate analysis of soils of middle-level taiga.

Profile	Hor-izon	Depth (cm)	Course Fragments (>1 mm)	Fractions (mm), fine earth (<1mm) (%)				Aggregation degree (%) according to Beiver and Roades
				).001	<-0.01	>0.05	Aggre-gates >0.05 mm	
533	A	3-10	19	16	42	22	67	67.2
	B	10-15	34	19	43	24	41	41.5
	B	15-23	46	20	44	20	37	45.9
548	A	4-6	no	12	35	18	71	74.6
	B	6-16	40	22	52	7	44	84.1
	BC	20-30	40	20	47	20	40	50.0
600	A	4-9	no	13	40	26	68	61.8
	B1	9-13	10	14	40	25	64	60.9
	B2	13-18	10	16	40	27	57	52.6
653	A	5-9	10	13	32	44	68	43.6
	B1	10-20	50	12	26	51	71	28.2
	EB	21-31	25	13	24	55	80	31.2
	B2	33-43	15	17	32	44	67	34.3
	BC	60-70	85	11	17	61	80	23.8
1007	AO	0-4	no	6	20	43	95	54.7
	A	4-6	no	18	43	27	75	64.0
	B1	6-13	8	15	45	26	60	56.7
	B2	20-30	24	21	42	37	63	41.3
	BC	35-45	17	19	34	36	68	47.1
	BC	70-80	90	6	13	75	88	14.8
1008	AO	4-10	2	12	34	25	73	65.8
	B1	10-20	43	18	39	37	66	43.9
	B2	22-32	19	19	41	32	50	36.0
1081	B	15-25	6	19	42	29	53	45.3
	BC	60-70	60	13	29	43	58	25.9

Table 2.10.10. Base saturation, pH, and humus properties in permafrost-affected soils of middle-level taiga.

Profile	Horizon	Depth (cm)	pH		Base Saturation (%)	Organic Matter Content (%)		C:N	$C_{ha}C_{fa}$
			H_2O	KCl		Total	Water-soluble**		
533	O	0-3	-	5.80	-	66.0*	-	-	-
	A	3-10	5.65	5.40	70.4	14.90	0.160 (1.1)	-	-
	B	10-15	5.50	5.40	76.1	4.70	0.071 (1.5)	-	-
	B	15-23	6.95	5.40	78.4	2.70	0.030 (1.1)	-	-
	BC	45-55	-	5.20	-	-	0.020	-	-
548	A	4-6	-	5.20	77.4	12.80	-	-	-
	B	6-16	-	4.80	67.8	8.90	0.041 (0.5)	-	-
	BC	20-30	-	5.00	77.8	3.70	0.040 (1.1)	-	-
600	O	0-1	-	5.20	-	81.2*	-	-	-
	AO	1-4	-	5.00	74.3	85.0*	-	-	-
	A	4-9	-	4.80	62.0	12.00	0.070 (0.6)	-	-
	B1	9-13	-	4.60	62.4	5.50	0.040 (0.7)	-	-
	B2	13-18	-	4.80	60.4	3.20	0.030 (0.9)	-	-
	BC	22-32	-	4.80	-	2.10	0.020 (1.0)	-	-
	BC	50-60	-	5.00	-	1.90	-	-	-
	BC	70-80	-	5.00		1.80	0.040 (2.2)	-	-
1007	AO	0-4	6.46	5.72	90.0	86.4*	0.065 (0.1)	-	-
	A	4-6	6.15	5.68	86.7	11.80	0.071 (0.6)	-	-
	B1	6-13	6.15	5.45	82.1	4.20	0.052 (1.2)	-	-
	B2	20-30	6.67	5.75	88.9	1.70	0.032 (1.9)	-	-
	BC	35-45	6.56	5.80	89.7	1.40	0.026 (1.9)	-	-
	BC	70-80	6.96	6.17	91.1	0.80	0.019 (2.4)		
1160	O	0-10	4.77	-	-	-	-	-	-
	A	10-12	4.66	-	-	11.95	-	-	1.08
	B1	12-20	4.90	-	-	3.43	-	-	0.77
	B2	20-30	5.39	-	-	3.24	-	-	6.75
	BC	30-40	5.79	-	-	-	-	-	-
	BC	60-70	6.27	-	-	-	-	-	-

* Value based on loss on ignition method.

** Figures in parentheses indicate % of total humus.

¶ Ratio of humic to fulvic acid carbon.

\- = was not determined.

Table 2.10.11. Labile compounds in permafrost-affected soils of middle-level taiga.

Pro-file	Hori-zon	Depth (cm)	Exchange-able Cations (1M KCl) (meq/100g)		Oxalate extact-able (%)		Water extractable (mg/100g)		
			H^+	Al^{+3}	Fe_2O_3	Al_2O_3	Al_2O_3	SiO_2	Fe
533	B	10-15	-	-	0.48	0.22	-	-	-
	B	15-23	-	-	0.51	0.11	-	-	-
1160	O	0-10	-	-	-	-	-	-	0.9
	A	10-12	-	-	-	-	-	-	1.4

4.5. Lower-level taiga – Mountain soddy-taiga deep permafrost soils *(Humi-Gelic Regosols)*

Mountain soddy-taiga deep permafrost soils *(Humi-Gelic Regosols)* occur widely in the lower part of the taiga belt, as well as in the open woodlands between forest and steppe. They occur under larch, pine-larch, and birch-larch vegetation with green moss-herbaceous cover.

The surface layer is a thin forest litter. This is underlain by a brown-gray humus horizon, clearly distinct in the profile, and then by a homogeneous bright gray-brown loamy horizon with rock debris. On the lower and upper sides of rock fragments, one can observe iron-humus films and fine earth coatings, respectively.

The lower solum is divided conditionally into horizons B1 and B2. The thickness of all horizons are as follows (mean and range derived from 16 profiles): O = 4 cm (1-8 cm), A = 7 cm (3-12 cm), AB = 13 cm (0-34 cm), B1 = 19 cm (11-30 cm), B2 = 24 cm (0-50 cm). At depths of 45-75 cm, the B horizons usually are underlain either by deluvial stones with thixotropic loam or with a strongly fissured siltstones or granites.

Tables 2.10.12 through 2.10.14 give the analytical characteristics of these soils. Site characteristics of soils of the Lower-level taiga zone are: Profile 590, left bank of Selenga River, Shabartym-gol, 1180 m asl; Profile 1162, northeastern Khangay, Tuvshrulekh vicinity, 1840 m asl; Profile 1278, the same locality, 1940 m asl.

These soils are composed of clayey loams and loams, well aggregated and rich in humus in their uppermost horizons; the humus is labile, with the ratio of humic and fulvic acids >1 in the upper and <1 in the lower horizons (Table 2.10.12). A fairly wide range of C:N ratio values points to some inhibition of the humification of organic matter. The soils are mostly acidic, being sometimes neutral in the lower part. Migration of iron, aluminium, silicon, and humus can be identified by content of labile, water-extractable forms (Table 2.10.13)

Table 2.10.12. Particle-size and microaggregate analysis of frozen soils of lower-level taiga.

Profile	Horizon	Depth (cm)	Coarse Fragments (>1 mm)	Fractions (mm), fine earth (< 0.01 mm) (%)				Aggregation degree (%) according to Beiver and Roades
				<0.001	<-0.01	>-0.05	Aggregates >0.05	
590	A	3-13	0	21	40	35	81	56.8
	B1	15-25	0	18	34	41	66	37.9
	B2	32-42	9	20	33	34	70	51.4
	BC	60-70	9	14	28	39	74	39.1
1278	A	1-11	0	17	42	20	67	70.1
	B1	13-23	0	18	45	22	44	50.0
	B2	30-40	16	17	44	25	38	34.2

Table 2.10.13. Base saturation, pH, and humus properties in permafrost-affected soils of lower-level taiga.

Profile	Horizon	Depth (cm)	pH		Base Saturation (%)	Organic Matter Content (%)		C:N	$C_{ha}:C_{fa}$ [¶]
			H_2O	KCl		Total	Water-soluble**		
590	O	0-1	6.27	5.63	-	81.7*	-	-	-
	AO	1-3	6.55	5.20	-	61.8*	-	-	-
	A	3-13	6.68	5.34	83.8	9.20	0.045 (0.5)	-	-
	B1	15-25	7.27	5.51	90.5	2.92	0.052 (1.8)	-	-
	B2	32-42	7.34	5.31	92.9	1.20	0.032 (2.7)	-	-
	BC	60-70	7.49	5.37	92.2	0.75	0.039 (5.2)	-	-
1162	A	3-10	6.75	5.77	75.7	13.07	-	-	1.48
	B1	10-20	6.45	4.84	47.5	3.84	-	-	1.76
	B1	20-30	6.38	4.70	48.3	-	-	-	-
	B2	40-50	6.16	4.54	54.3	2.22	-	-	0.65
	BC	90-100	6.74	-	59.7	-	-	-	-
1278	A	1-11	5.90	5.10	73.2	27.69	-	18	1.49
	B1	13-23	5.55	4.20	56.2	5.40	0.041 (0.8)	-	-
	B2	30-40	5.65	4.15	66.2	2.20	0.028 (1.3)	-	-
1279	O	0-4	5.80	5.35	-	61.9*	-	-	-
	A	4-7	6.60	5.85	80.9	15.51	-	25	-
	AB	7-11	6.55	5.40	77.2	7.10	0.024 (0.3)	-	-
	B1	11-21	6.50	5.10	73.9	3.75	0.033 (0.9)	-	-
	B2	30-40	6.25	4.80	-	5.19	-	-	-

* Value based on loss on ignition method.

** Figures in parentheses indicate % of total humus.

¶ Ratio of humic to fulvic acid carbon.

-- = was not determined.

Table 2.10.14. Labile compounds in permafrost-affected soils of lower-level taiga.

Profile	Horizon	Depth (cm)	Exchangeable Cations (1M KCl) (meq/100g)		Oxalate extractable (%)	Water extractable (mg/100g)			
			H^+	Al^{+3}	Fe_2O_3	Al_2O_3	Al_2O_3	SiO_2	Fe
590	A	3-13	0.20	0.15	-	-	-	-	-
	B1	15-25	0.10	No	-	-	-	-	-
	B2	32-42	0.10	No	-	-	-	-	-
	BC	60-70	0.10	No	-	-	-	-	-
1278	B1	13-23	-	-	2.00	2.56	-	40.00	-
	B2	30-40	-	-	0.76	0.52	-	50.00	-
1279	AB	7-11	-	-	1.60	3.28	-	60.00	-
	B1	11-21	-	-	1.36	2.44	-	18.00	-

-- = was not determined.

These soils are cold, though somewhat warmer than the taiga soils of higher elevation. According to the observations in the vicinity of profile 1162, the temperature of 10°C penetrates to a depth of 10-15 cm, i.e., into horizon A, and the maximum and minimum temperatures at a depth of 100 cm are 6.2 and -1.8°C, respectively. At the same time, the Mountain soddy-taiga deep permafrost soils *(Humi-Gelic Regosols)* freeze later than the steppe soils, due to the warming influence of forest and an appreciable moisture content.

These soils are covered with snow in every winter. Their thawing proceeds slowly and summer rains accelerate it, bringing atmospheric heat into the soils. The water regime here is periodically non-percolative; during dry years, no deep percolation of soil water occurs. The slowly thawing permafrost, which retreats below 100 cm deep by mid- to late June, controls the depth of percolation. A considerable share in the hydraulic balance of soils belongs to the lateral intrasoil water runoff that is intense during wet years.

4.6. Subtaiga open woodlands – Mountain meadow-forest deep permafrost soils *(Gelic Phaeozems)*

Mountain meadow-forest deep permafrost soils *(Gelic Phaeozems)* formed mostly on the lower parts of northern slopes, which have angles of 10-15° in forest margins at their boundary with valley meadows or meadow steppe. The moisture influx from the taiga sites occurring at higher altitudes enhances the wetting of these soils. The vegetative cover consists mostly of thin larch and birch woods, with a continuous dense herbaceous cover.

These soils have a rather thick (20-40 cm) and dark humus horizon and sometimes contain buried humus horizons. They also have features of cryogenic dislocations (i.e., cryoturbation) and ice schlieren.

The profile is acidic in the upper part and neutral in the lower horizons. The soils are well humified (10-15%), Cha:Cfa >1. They develop under conditions of humus accumulation and slight leaching; they are well microaggregated and possess an appreciable absorption capacity.

The soils are cold; the temperature of 10°C penetrates to a depth of 25 to 30 cm. Seasonal frost is present in the upper horizons until the end of June.

4.7. Meadows and wetlands in intermontane valleys

4.7.1. Meadow deep permafrost soils (Gelic Phaeozems or Gelic Chernozems)

Meadow deep permafrost soils *(Gelic Phaeozems or Gelic Chernozems)* formed in the belt of exposed forest-steppe in intermontane valleys under rich herbaceous meadows. These soils are used in agriculture for grazing in autumn and growing hay in summer. They have high organic matter content (10 to 17% in 0 to 10 cm) and a deep A horizon (50 to 60 cm).

The humus is labile, Cha:Cfa >1. The upper horizons are clayey loams, while the texture of the lower horizons can be very diverse. The soils are weakly or moderately acid and can be neutral or even slightly alkaline in the lower horizons if carbonates are present there.

In the first half of summer, the presence of slowly thawing seasonal frost in the profile determines the water regime to a considerable extent. During wet years, moisture held above the slowly retreating seasonal frost can cause temporary over saturation of soils. The temperature of 10°C does not penetrate below 30 cm depth.

The thawing of soils begins in the first week of April and proceeds slowly. At 100 cm deep, thawing occurs only in mid- and sometimes late June. The thawing accelerates in July, and the soil thaws to a depth well below the soil control section. The maximum temperature at 100 cm is 6.6°C.

4.7.2. Permafrost-swampy soils (Gelic Gleysols)

Permafrost-swampy soils *(Gelic Gleysols)* occur in the narrow intermontane and river valleys of the forested part of Mongolia. Here, there are no thick peat layers. At their boundary with the steppe zone, these soils are slightly saline at the surface.

4.8. Near springs and saline depressions around drying saline lakes in the steppe belt

Permafrost-affected soils of steppe territories generally occur only on saline soils occurring in depressions in the northern half of Mongolia. They often surround ephemeral saline lakes, under which permafrost is present, often within 200 cm of the surface. These are the areas occupied by the Permafrost solonchaks-solonetzes

(Gelic Solonchaks, Gelic Solonetzes). In some places, frost mounds and thermokarst widely developed.

On September 13, 1975, permafrost was found to occur in the profiles of such soils at depths of between 169 and 238 cm and at 170 cm under the bottom of a dried-up saline lake in North-East Khangay, east and northeast of Tuvshrulekh. These soils have a sodic salinity, which cause high pH values and high humus mobility, resulting in the existence in some places of horizons with illuvial humus at depths of 50-60 cm. The soils are moist, and ground waters are shallow.

4.9. River flood plains

Permafrost-affected soils also occur in river flood plains of Mongolia. Cryogenesis influences the flood plain boggy and meadow soils *(Geli-Saprihistic and Geli-Gleyic Fluvisols),* including saline ones, in the peripheries of the flood plains of the rivers in the northern part of Mongolia.

5. Conclusion

The mountainous relief and strongly continental climate result in the deep winter freezing of soils and the wide distribution of cryogenic soils in Mongolia. Permafrost can occur at high mountainous elevations, under forests on north-facing mountain slopes, and under meadow and wetlands in high elevation intermontane and river valleys. In the northern half of Mongolia, permafrost-affected soils sometimes occur in landscape depressions with additional moistening and accumulations of salts.

Mostly, cryogenic soils of Mongolia correlate as WRB Gelic lower level units, and only imperfectly drained taiga soils meet the criteria of the Cryosol soil order, wherein permafrost must occur within 100 cm of the soil surface. Five soil regions exist in Mongolia, based on the percentage of the landscape occupied by permafrost-affected soils and paleocryogenic features.

References

Bannikova, I.A., N.N. Izmailova, and S.V. Maximovich. 1980. Water balance and productivity of a cobresial meadow in East Khangay. Ekologiya. 5:27-35 (In Russian.)

Gerasimov, I.P., and N.A. Nogina, eds. 1984. The Soil Cover and Soils of Mongolia (Pochvennyi pokrov i pochvy Mongolii). Nauka. Moscow. 191 pp. (In Russian.)

Grubov, V.I., ed. 1980. Vegetation of River Floodplains in People's Republic of Mongolia (Rastitel'nost' rechnykh poim Mongolskoi Narodnoi Respubliki). Nauka. Leningrad. 284 pp. (In Russian.)

ISSS Working Group RB. 1998. World Reference Base for Soil Resources. International Society of Soil Science (ISSS), International Soil Reference and Information Centre and Food and Agriculture Organization of the United Nations. World Soil Resources Report 84. FAO. Rome. 91 pp.

Lavrenko, E.M., and I.A. Bannikova, eds. 1983. Mountain Forest Steppe of Eastern Khangay (PRM) (Gornaya lesostep' vostochnogo Khangaya [MNR]). Nauka. Moscow. 191 pp. (In Russian.)

Lonzhid, N.N. 1959. On the problem of permafrost regionalization of PRM. In: Merzlotnye issledovaniya (Permafrost Studies). Moscow University Press. Moscow. 6:160-165. (In Russian.)

Maximovich, S.V., and V.M. Mirkin. 1980. Possibility of Coordinating of Soil and Plant Classifications on an Ecological Basis. Soviet Soil Science. 12:655-666.

Maximovich, S.V. 1993. Cryogenesis and Paleocryogenesis in Soils of Mongolia. In: D.A.Gilichinsky, ed. Joint Russian-American Seminar on Cryopedology and Global Change. Post-Seminar Proceedings. Russian Academy of Sciences. Pushchino. pp. 75-83.

Melnikov, P.I., ed. 1974. Geocryological Conditions in People's Republic of Mongolia (Geokriologicheskie usloviya Mongolskoi Narodnoi Respubliki). Moscow. Nauka. 200 pp. (In Russian.)

Nogina, N.A., ed. 1978. The Soil Cover of Major Natural Zones of Mongolia. (Pochvennyi pokrov osnovnykh prirodnykh zon Mongolii). Moscow. Nauka, 276 pp. (In Russian.)

Nogina, N.A., ed. 1980. The Soil Map of People's Republic of Mongolia (Pochvennaya karta Mongolskoi Narodnoi Respubliki), scale 1:2 500 000. GUGK. Moscow. (In Russian.)

Ogorodnikov, A.V. 1981 Soils of Mountain Forests in People's Republic of Mongolia (Pochvy gornykh lesov Mongolskoi Narodnoi Respubliki). Nauka. Novosibirsk. 144 pp. (In Russian.)

Tumburbator, D. 1975. Seasonal Freezing and Thawing of Grounds in Mongolia. Ulan-Bator. 107 pp. (In Mongolian.)

Chapter 11. The Periglacial Environment and Distribution of Cryosols in China

Chien-Lu Ping[1], Guoqing Qiu [2], and Lin Zhao[3]

[1]Professor, University of Alaska, Agricultural & Forest Experiment Station, 533 E. Fireweed, Palmer, Alaska 99645, USA; e-mail pfclp@uaa.alaska.edu.
[2]Cold and Arid Environmental and Engineering Research Institute, Chinese Academy of Sciences, Lanzhou, China (deceased).
[3]Cold and Arid Environmental and Engineering Research Institute, Chinese Academy of Sciences, Lanzhou, Gansu 730000, People's Republic of China; e-mail linzhao@ns.lzb.ac.cn.

1. Introduction

Permafrost is defined as a thickness of soil or other superficial deposit, even bedrock, that remains frozen for consecutively two years or more (Muller, 1947). Permafrost, or perennially frozen ground, underlies 2.15×10^6 km^2 or 22.4% of the territory of China (Qui and Cheng, 1995) and can be divided into two broad categories, high-latitude permafrost in northeastern China (0.38×10^6 km^2) and high-altitude permafrost in the Qinghai-Xizang Plateau and some alpine regions in western and eastern China (1.76×10^6 km^2) (Figure 2.11.1). The total area of soils affected by permafrost in China is third in the world, after Russia and Canada, but the area of high-altitude perennially frozen soils is the largest in the world.

2. Periglacial Environment in China

2.1. High-latitude permafrost region

The high-latitude permafrost occurs mainly in the Da Hinggan and Xiao Hinggan Mountains in Northeast China between Lat. 46°10' and 53°30' N., and Long. 115°30' and 126°30' E. (Figure 2.11.1). Moderately low mountains with gentle slopes and terraces characterize the landscape. The area includes three zones (Table 2.11.1) (Gu et al., 1992; Liu et al., 1993).

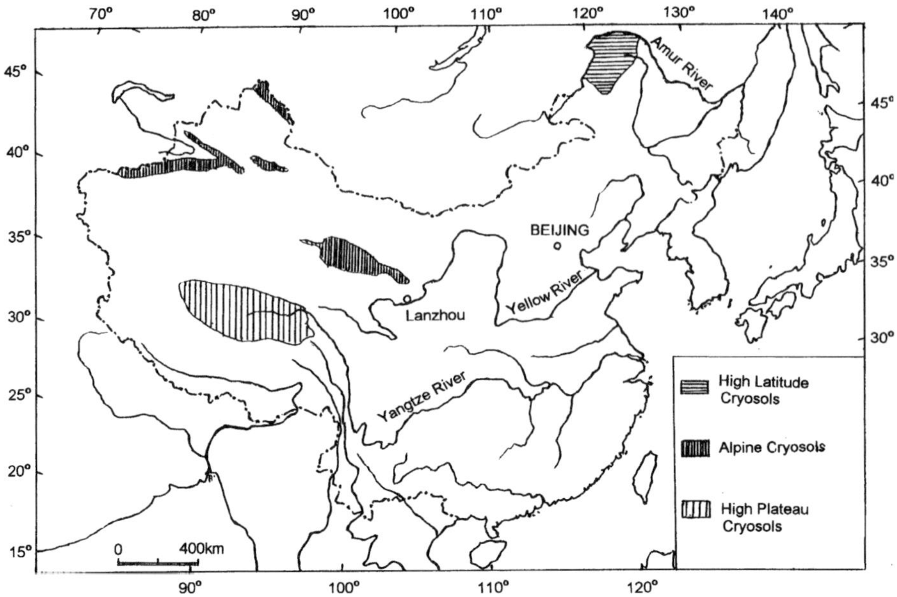

Figure 2.11.1. Schematic map showing types and distribution of Cryosols in China. (The Cryosols are defined as soils having permafrost within 1m or 2m, with cryoturbation features.)

Zone I, commonly called the Amur Area, is the northernmost and is underlain with continuous permafrost. Here, the river valleys are broad with scattered wetlands, and thick tussock and mosses occur in the valley bottoms where permafrost thickness ranges from 60 to 80 m and massive ground ice is common (Shi, 1993). On the south-facing slopes, soils are thin and vegetation is scarce; permafrost is poorly developed, with thickness ranging from 0 to 20 m. The periglacial phenomena include solifluction sediments and terraces, hummocks, sorted circles, asymmetrical valleys, and active and inactive block fields. Some cryoturbated Cryosols are to be expected in this zone.

Table 2.11.1. Soil environment of areas with high-latitude permafrost in northeastern China (Liu et al., 1993; Shi, 1993).

Perma -frost Zone	MAAT, °C	MAP mm	Max. thaw, m	Perma-frost thickness, m	Perma frost %	Ground Temp., °C	FFD days
I	-2.1 to -6.3	440	2.0-4.0 (S) 0.5-1.2 (V) 1.0-1.5 (N)	0-20 50-150 20-50	65-75	0 to 1.0 -2.0 to -4.2 -1.0 to -2.0	<90
II	-2.0 to-4.0	440	1.0-2.0	20-50	50-65	-1.5 to -3.0	90-110
III	-2.0	440	2.0-3.0	5-20	5-20	-1.0 to 0	110-130

S: south-facing slope, V: valley, N: north-facing slope, FFD: frost-free days.

The mean annual air temperature (MAAT) ranges from -2.1to -6.3°C, with a mean of -4.7°C, and the mean annual soil temperature (MAST) ranges from -3 to -1 °C. The mean annual precipitation (MAP) is 440 mm, with most falling between May and September. The active layer in this region ranges from 0.3 to 1 m thick in the lowlands and 1 to 3 m in the uplands.

To the south of Zone I, Zone II occupies the rolling hills and discontinuous permafrost underlies it. The MAAT is approximately -3.0°C, and MAST ranges from -1.5 -1.0°C. Frost-free days range from 90 to 110. The depth of active layers generally ranges from 1 to 2 m.

Zone III lies further south and west of Zone II and has sporadic permafrost. The MAAT is -2.0°C, and the active layer is generally too deep for the soils to be considered Cryosols, except for some isolated permafrost in the wet lowlands.

2.2. High-elevation permafrost regions

2.2.1. Plateau permafrost region

The permafrost region on the Qinghai-Xizang Plateau is the highest and largest plateau in the world, with an area of 1.54×10^6 km^2 and mean elevation above 4000 m (Shi, 1993; Li and Cheng, 1996). It borders the Kunlun Mountain to the north, the Yarlung Zangbo River to the south, the Anyemaqen Mountain and Bayen Har Mountains to the northeast, the Gandes Mountain to the west, and the Hengduan Mountains to the east. The plateau is relatively flat and broad, with high mountains around it, and the Nienqing Tangula Mountains, run east to west in its middle. This high elevation produces a severely cold and dry periglacial environment for the development of permafrost and cryogenic soils. The region has strong solar radiation and wide diurnal temperature fluctuation. Table 2.11.2 describes the physical environment of the plateau permafrost region.

Table 2.11.2. Soil environment of Alpine and Plateau Permafrost Regions in China.

Region	Lat. °N.	Lower limit, m	MAAT °C	MAGT °C	MAP mm	MAEP mm	Aridity
Alpine Permafrost							
Altay Mts.	46-49	2200-2800	-5.4 to -9.4	0 to -5.0	500-800		
Tianshan Mts.	41-43	2700-3300	-5.4 to -9.0	-0.1 to -0.6	600-1000		
Qilian Mts.	36 - 40	3400-5000	0 to -2.0	-0.6 to -2.3	400-800		
Hengduan Mts.	29 - 30	4600-4900	-3.2 to -4.9		500-1000		
Himalaya Mts.	28	5100-5300	-5 to -6	0 to -40	300-800		>3
Kunlun Mts.	34-36	4700-4900	-3 to -7	0 to -30	200-600		>15
Plateau Permafrost							
Continuous permafrost	28-35	3800-5500	-2.5 to -6.5	-1.0 to -3.5	200-500	1800-3000	6-35
Sporadic (isolated)		3700-4500	-0.8 to -2.5	0 to -1.5	400-600	1400-2000	2-15

Qiu and Cheng, 1995; Tong et al., 1985; Zhou et al., 2000.

Since the Pleistocene, the plateau has been rising steadily and the climate has become drier and colder. Periglacial processes are very active, and periglacial phenomena are well developed (Shi, 1993; Gao et al., 1985). The most common periglacial landforms are block fields, stone streams, stone stripes, debris slopes, frost cracks, polygons, hummocks, icings, involutions, thermokarst slumps and settlements, gelifluction lobes, and thermokarst lakes. Some of these features are good indicators of cryoturbated soils and Cryosols.

The vegetation changes from forest in the eastern and southeastern parts of the plateau, to grassland dominated by *Kobresia* spp. interspaced with shrublands in the southeast of Xizang and east Qinghai, to alpine steppe in the northwestern part of the plateau, and to alpine cold desert in the northern fringe and some isolated ridges on the plateau (Gao et al., 1985).

2.2.2. Alpine permafrost regions

The alpine permafrost occurs in the high mountains rather than on high plateaus. Alpine permafrost exists mainly in the Altay, Tienshan, Qilian, Kunlun, Hengduan, and Himalayan Mountains in western China and in isolated localities in eastern China. Altitude is the controlling factor in the formation and development of alpine permafrost; thus there is a clear vertical zone of permafrost and soils (Qui and Cheng, 1993).

Since the latitude and environment of different mountains vary, the characteristics of alpine permafrost and the lower limit of permafrost and Cryosols in different mountains also vary. Table 2.11.2 describes the physical environment of the alpine permafrost regions.

2.2.2.1. Altay Mountains

The Altay Mountains stretch from northwest to southeast along the boundary between China and Russia and between China and Mongolia, with elevation ranges from 1000 to 4300 m. Most of the ridges are composed of metamorphous rocks, gneiss, and sedimentary rock of Paleozoic age. The discontinuous permafrost exists at elevations from approximately 2800 m to the base of glaciers at 3100 m (Tong et al., 1983; Qiu and Cheng, 1993). Permafrost can be up to 400 m thick.

The climate at these elevations is cold and humid. MAAT is -9.4°C, and MAP ranges from 600 mm in the east to 800 mm in the western part of the region. Snow cover is more than 1 m deep and is continuously present through the winter. The area has tundra vegetation.

The sporadic permafrost zone extends to lower elevations from 2200 to 2800 m (Tong et al., 1983). The climate here is also cold and humid. The MAAT ranges from -5.4 to -6.8°C, and MAP is about 500 mm. Snow depth ranges from 1.5 to 2.5 m and stays throughout the winter. The depths of the active layer are <1m under peat, 0.9-1.6 m in loamy soils, and >1.9 m under sandy and gravelly soils.

Palsas with peat >2 m thick are common in this zone. Besides palsas, other periglacial features include earth hummocks, frost cracks, icings, sorted circles, block streams, block slopes, gelifluction lobes, and polygons in glaciofluvial deosits (Tong et al., 1983; Shi, 1993). The vegetation is dominantly herbs, with some shrubs.

2.2.2.2. Tienshan Mountains

The Tienshan Mountains lie across the middle of Xinjiang, stretching 1700 km east-west inside China. Most ridges are composed of metamorphic and igneous rocks. Elevation and aspect strongly affect the ecosystems and thus the soils. Generally, the precipitation increases toward the northwest, and temperature decreases toward the northeast (Qiu et al., 1983). The MAP ranges from 600 mm at the forest-steppe zone at 1800-1500 m, to 1000 mm at snowline, with 90% falling between April and October.

The lowest limit of permafrost is 2700 m on north-facing slopes and 3100 m on south-facing slopes, and this limit lies about 1000 m below snowline. Ground temperature ranges from -0.1 to -0.2°C at the lower limit, where permafrost is <20 m thick and the active layer is 4 to 5 m deep. At the higher limit, where the ground temperature is about -2.0°C , the permafrost is more than 100 m thick and the active layer is <2 m.

Permafrost with ground ice occurs in depressions where segregated ground ice in finely grained sediments forms an ataxitic fabric (a Russian geocryological term referring to an ice-rich mineral layer) in which ice exceeds 65% by volume. In wetlands, ice layers are <1 m and ice contents are >95%. The periglacial features include hummocks, oriented stones, sorting, stone stripes, and sorted circles. In the permafrost-affected zone, meadows occur at the higher elevations and the active layer is usually less than 2 m deep.

2.2.2.3. Qilian Mountains

The Qilian Mountains are in the northern fringe of the Qinghai-Xizang Plateau, with elevation ranging from 1000 to 4700 m. Here it is drier than in the Altay and Tienshan Mountains. The precipitation increases with elevation, from 300 to 400 mm in the alpine steppe, to 500 mm in the forest zone and >500 mm in the alpine shrub. Most of the precipitation falls between June and September (Qui and Cheng, 1995). The annual evaporation is 1000 to 1500 mm and the aridity index is >2.

The alpine shrub zone is above 3500 m and has a strong diurnal temperature fluctuation, up to 40 or 50°C. The vegetation consists mainly of *Caragana jubata* and some junipers.

The forest zone is above 2800 to 3000 m and has a cold and humid climate with diurnal temperature fluctuations ranging from 30 to 35°C. The MAAT ranges from 0 to -2.0°C. The forest consists of *Picea crassifolia* and *Pinus tabulaeforms,* with patches of grassland. The depth of the active layer ranges from 1 to 2 m.

The alpine steppe is below the forest zone and has sporadic permafrost at depths below 2 m.

2.2.2.4. Kunlun Mountains

The Kunlun Mountains lie at the northern fringe of the Plateau. The north slopes and the higher part of the south slopes are cold desert. Permafrost occurs at elevations ranging from 4000 to 5500 m. Periglacial phenomena include block fields, hummocks, and rock glaciers (Shi, 1993).

The climate of this region is cold and arid. The MAP ranges from 200 to 600 mm. The MAAT ranges from -3 to -7°C, and MAST ranges from 0 to -3°C. Vegetation is dominantly scattered shrubs in the alpine desert and grasses in the alpine steppe.

2.2.2.5. Hengduan Mountains

The Hengduan Mountains are on the east- and the north-facing slopes of the Himalayas to the south of the plateau, at elevations ranging from 4600 to 4900 m and 5000 to 5500 m, respectively. Isolated permafrost exists in these regions, and common periglacial features include hummocks, rock glaciers, gelifluction lobes, and sorted circles (Shi, 1993).

This region has a cold and humid climate because it receives seasonal storms from the southeast. The MAAT ranges from -2 to -6°C, and MAST ranges from 0 to -3.0°C. The MAP ranges from 500 to 800 mm. Vegetation consists mainly of *Kobresia* spps and shrubs.

3. Cryosols in the Chinese and other Soil Classification Systems

The term *Cryosol* in this chapter is used synonymously with the Cryosol order in the Canadian soil classification system (Agriculture Canada Expert Committee on Soil Survey, 1987), the Gelisol order in *Soil Taxonomy* (Soil Survey Staff. 1999), the Cryosol units in the World Reference Base for Soil Resources (WRB) system (FAO, 1998), and the permagelic taxons of different soil orders in the Chinese soil taxonomic system (Gong, 1995). The definitions and classification systems of Cryosols all are based on the presence of gelic materials (Bockheim et al., 1997) but differ according to the thickness of the active layers.

It is important to introduce the definitions in each system for cross-reference and correlation purposes. In the following discussion of the distribution of Cryosols in China, *ST* indicates classifications according to *Soil Taxonomy*, *CST*, Chinese soil taxonomy, *CAN*, the Canadian system, and the WRB nomenclature appears after the taxonomic units.

3.1. Chinese system

In the Chinese soil classification system (Gong, 1995), the hierarchy includes order, suborder, great group, and subgroup. Soil order is based on the dominant soil-forming process. The suborders are differentiated based on soil climates. Great groups are differentiated based on soil profile characteristics, and the subgroups are differentiated based on accessory features.

The revised Chinese soil classification system adopted the quantitative measures of *Soil Taxonomy* and thus the definitions are parallel to *Soil Taxonomy's*. In the Chinese soil taxonomic system (Gong, 1995), the formative element *permagelic* is defined as "having permagelic soil temperature or having permafrost layers within 200 cm of the soil surface." The permagelic soil temperature regime (STR) is soil temperature perennially at or lower than 0°C. The gelic STR is the mean annual soil temperature (MAST) $\leq$ 0°C.

In the Chinese system, permagelic STR and permafrost layers within 2 m are recognized at the suborder level of Histosols, great group level of Cambisols, and subgroup level of Entisols. In Gleysols, only the gelic suborder is recognized.

3.2. Canadian system

In the Canadian soil classification system, the Cryosolic Order includes both mineral and organic materials having permafrost either within 1 m of the surface or within 2 m if more than 1/3 of the pedon has been strongly cryoturbated, and they have a MAST <0°C.

In *Soil Taxonomy*, permafrost-affected soils key out in the Gelisol Order, which includes both mineral and organic soils having permafrost within 100 cm of the soil surface or having gelic materials (Bockheim et al., 1997) within 200 cm of the soil surface.

3.3. WRB

In the WRB system (FAO, 1998), Cryosols are mineral soils that contain one or more cryic horizon(s) within 100 cm of the soil surface. Organic soils with permafrost key into Cryic Histosols. Cryic horizons include those that are perennially frozen, ice-cemented or dry, and cryoturbated. Cryosols are differentiated into "soil units" based on characteristics of the soil profile.

4. Descriptions of Cryosols in China

4.1. High-latitude Cryosols

In China, high-latitude Cryosols exist mainly in the northeast area south of the Amur River to the slopes of the Da Hinggan Mts., which run northeast to southwest, and the Xiao Hinggan Mts., which run northwest to southeast (Figure 2.11.1). The distribution of Cryosols in this region shows the effects of both latitude and altitude. The latitude ranges from 46°30' N. to 53°40' N, and the elevation ranges from 600 m in the valleys of the northern part to mountains up to 2000 m in the south part. Slope aspect also plays an important role in the distribution of soil types.

4.1.1. Boreal forest regions

In the zone of predominantly continuous permafrost, Cambisols dominate upland slopes. These soils previously were called Podzolized Brown Conifer Soils (Zhang, 1986), Brown Conifer Forest soils (NSSI, 1976), and Cryic Brown Soils (Zhao et al., 1993). They formed under mixed conifer forest of *Pinus pumila* and *Larix gmelinii* with *Sabina davuvicus, Juniperus* spp*., Arctous alpinus, Empetrum nigrum*, and *Betula exilis* as understory. Generally, the podzolized brown soils are more concentrated on the north-facing slopes in the Amur River region.

A typical profile sequence is Oi-Oe-A-Bw-BC-CR. The organic horizon is generally <15 cm thick, overlying a thin A horizon followed by cambic horizons. The solum ranges from 40 to 60 cm thick. The Bw, BC, and C horizons are either gravelly or very gravelly. Under the conifer forest, the A horizon may turn into a weakly developed E horizon.

Since most of the parent materials are acidic or leached under the conifer forest, the soils are generally acidic with a base saturation <50%. Most pedon data does not include information related to the depth of the active layer and to cryogenic features. Based on geotechnical and meterological data, the upper limit of permafrost is within one meter of the soil's surface (Zhang, 1986; Guo et al., 1993). Thus these soils key out as Permagelic Gelic Cambisols (CST).

In other classification systems, these soils may be classified as Typic Haplorthels or Histic Haplorthels (ST), Orthic Dystric Static Cryosols or Histic Dystric Static Cryosols (CAN), and Haplic Cryosols (WRB).

Under forest cover and continuous organic layers, cryoturbation is not common on the upland slopes, although solifluction occurs in places. These soils have a brownish hue, and the organic carbon content in A horizons ranges from 6 to 9% and in Bw ranges from 2 to 4%. According to the US and the Canadian systems, these soils key out as Umbrorthels and Static Cryosols, respectively.

Saturation caused by a perched water table above the permafrost creates a zone of reduction and results in redoximorphic features, such as mottled or gleyed horizons. If the reduced zone is below 25 cm, then soils key out as Aquic Umbrorthels (ST). But on convex slopes with fractured bedrock less than 50 cm, the soils key out as Lithic Umbrothels (ST).

Zhang (1986) also reported podzol-like soils in this region. Thus, in the WRB system, most soils are Umbric Haplic Cryosols, but some may be in the Spodic Haplic Cryosols units, and some are Lithic and Aquic Haplic Cryosols.

4.1.2. Lowlands

In the lowlands, common vegetation includes *Betula fructicosa, Salix* spp.*, Vaccinium uliginosum, Carex* spp.*, Cassandra* spp., and *Deyeixia* spp (Zhang, 1986).

Poorly drained soils are classified as Gleysols, and those with Gelic and colder STR key into the Gelic suborder (CST). Generally, Gelic Gleysols form on concave slopes or at the footslopes below the Cambisol zone. *A permagelic suborder should be introduced to include those having permafrost within 1 m.*

There are two great groups of Gelic Gleysols, Histic and Entic. In other systems, they are classified as Typic Historthel (ST), Histic Dystric Static Cryosols (CAN), and Gleyic Haplic Cryosols (WRB).

Previously classified as Bog Soils (NSSI, 1976), the Histosols exist mainly in the lowlands of the Da Hinggan Mts. and on terraces with elevations ranging from 600 to 1000 m.

Zhang (1986) described a peaty marsh soil from the Da Hinggan Mountains region. Peat is recorded to 45 cm, and the pH (H_2O) ranges from 4.8 in the surface tier to 5.4 in the bottom tier. These soils key out in the Permagelic Histosols suborder (CST) and further into three great groups, Folic, Fibric, and Hemic. In the other classification system, these soils key into either Hemistels or Fibristels (ST), Fibric or Mesic Organic Cryosols (CAN), and Histic Cryosols (WRB).

4.2. High-altitude Cryosols

The high-altitude Cryosols can be divided into alpine Cryosols and high-plateau Cryosols. The high-plateau Cryosols differ from their high-latitude counterparts in that they form under arid and cold conditions, whereas most alpine and high-latitude Cryosols form under humid or sub-humid conditions.

4.2.1. Alpine Cryosols

Qiu and Cheng (1993) have studied in detail the altitudinal zone of soils in the Altay, Tienshan, and Kunlun Mountains. To the north, Permagelic Histosols and Cambisols and Gelic Gleysols (alpine tundra soils) (CST) occur, starting at 2200 m in the Altay Mountains, 3900 m in the Tienshan Mountains, and 4700 m in the Kunlun Mountains.

The Altay Mountains are more humid; thus, Gelic Gleysols form in the tundra zone, and Permagelic Cambisols form under conifer forest on the north-facing slopes above 1800 m. In contrast, in the Tienshan Mountains, Gelic Cambisols and Gleysols form in the forest below the tundra zone at elevations above 3000 m. However, the thickness of the active layer of these Gelic great groups needs verifying. The Kunlun Mountains lie southeast of the Tienshan Mountain. Based on the study by Harris et al. (1998), Permagelic Cambisols are expected to form in the gelifluction slope deposits.

The permafrost-affected area in the Tienshan Mts. is estimated at 63,000 km^2. The distribution of permafrost in eastern Tienshan closely relates to elevation and aspect; the lower limit of permafrost on the south slope and north slope is at 2700 and 3100 m, respectively, corresponding to the -2.0°C isotherm (Qui et al., 1983; Jin et al., 1993). The MAAT ranges from -2.5 to -5.4°C (Hu et al., 1998).

Depths of the active layer are generally within 2 m, but the maximum thaw depth of some soils may reach 3 m. The MAP ranges from 220 to 470 mm, with three-fourths coming as snow between June and September and melting within days. The average snow depth in winter is about 5 mm. Thus, the ground is exposed to severe conditions in winter, and cryoturbated soils are common, as earth hummocks, patterned ground, polygons, and frost cracks indicate (Qui, 1993).

Hu et al. (1998) studied soil formation along a chronosequence in the Central Tienshan Mts. The modern glacial deposit, at 3730 to 3740 m, is the youngest surface, about 30 years old. Soils form as isolated spots among stone and rocky moraines dominated by freeze-thaw crust, sorted circles, and hummocks. There are no vascular plants, only algae and microorganisms. The biochemical weathering is weak, but physical weathering is strong and results in very stony, sandy textured soils. The soils have an A-C sequence and they key out as Permagelic Gelic Psamment Entisols (CST), Lithic Psammorthels (ST), Regosolic Static Cryosols (CAN), and Lithic Calcic Cryosols (WRB).

The Little Ice Age deposits dated as 100 to 400 years old occur at 3650 to 3730 m. Here the MAAT and MAST are estimated to be -6.9°C and -1.8°C, respectively. Soils on this stony surface are not continuous. Vegetation is limited to lichens and mosses. The soils have a profile sequence of A-C. The organic carbon ranges from 0.03 to 0.05%, and the pH is over 8. Carbonates have accumulated in the subsurface horizon. Periglacial features, including cryoturbation, are common to this area. These soils key out as Calcic Permagelic Gelic Cambisols (CST), Aquic Haploturbels (ST), Gleysolic Turbic Cryosols (CAN), and Turbic Calcic Cryosols (WRB).

The New Ice Age glacial deposits aged 2800 to 6000 years old occur at the elevation of 3400 to 3650 m. The MAAT is estimated -5.9°C, and permafrost temperature is -1.6°C. Alpine meadow vegetation, such as *Kobresia* and *Carex,* has established and formed a patchy sod layer. These soils also have an A-C sequence and key out as Calcic Permagelic Gelic Cambisols (CST), Aquic Haplorthels (ST), Gleysolic Static Cryosols (CAN), and Aquic Calcic Cryosols (WRB).

At lower elevations, from 3000 to 3400 m, soils form in glacial deposits from the last glaciation more than 6000 years ago. Here, *Kobresia* and *Festuca* give continuous ground cover and form alpine meadows. The MAAT is estimated at -2.5 to -3.0°C, with the permafrost temperature at -0.7 to -1.0°C. These soils have a profile sequence of A-Bw-C, and carbonates have accumulated in subsurface horizons. They are classified as those in the 3400- to 3650-m zone. Similar soils also exist in the Altay and Qilian Mts areas (NSSI, 1976).

4.3. High-plateau Cryosols

4.3.1. High plateau cold desert

The high plateau cold desert exists below the snowline on the high slopes, on the mountains rising above the broad plateau in the Qinghai-Xizang Plateau, at elevations ranging from 3800 in central Qinghai to 5300 m in the northern part of Xizang, at about Lat. 34-35°N. The area is within the zone of high mountain discontinuous permafrost (Li and Cheng, 1996).

The soils are very young, as they developed following the most recent deglaciation. The climate is cold and dry. The sparse vegetation consists of *Ceratoides compacta* and some lichens. The MAAT, estimated at 3800 m, is -3 to -4°C, and at 4200 m is -7 to -9°C, and the MAP is 120 mm, with 80% falling between June and August (QARDO, 1988). The annual evaporation is >3000 mm; thus, the climate is extremely arid and the aridity index is 23-24.

Although the MAST is estimated <0°C based on MAAT, Cryosols are not likely to occur because the active layers are generally >2 m, due to strong diurnal temperature fluctuations, dry conditions, and coarsely textured soils. The parent materials consist of glacial deposits or colluvium.

4.3.2. High plateau alpine steppe

This is the most common vegetation association on the northwestern Qinghai-Xizang Plateau. This steppe system is in the northern part (south of Lat. 37° N.) of the Plateau, ranging from 3900 to 4300 m. The landforms are largely the result of alpine glaciation about 12,000 years B.P. Parent materials include glacial drift, outwash, and lacustrine. The parent rocks include granite, sandstone, slate, conglomerate, siltstone, and limestone (QARDO, 1988). This ecosystem is within the continuous permafrost, and the depth of the active layer is generally >2 m (Shi, 1993).

Stipa purpurea, Oxytropis spp., *Astragalus* spp., *and Leontopodium* spp. dominate the vegetation in the area. The coverage ranges from 20 to 50%. This ecosystem is the most extensive in the Plateau, and it supports wildlife's grazing.

Based on an average of 11 meteorological stations in the Qinghai region (QURDO, 1988) and records from Jiangtan Plateau of Xizang (Gao et al., 1985), the MAAT ranges from -2 to -6°C, and MAP range from 100 to 432 mm, with 80% falling between June and August. The annual evaporation exceeds 2000 mm. The aridity index is more than 20. Growing degree days (GDD) >0°C are 94, freezing degree days (FDD), 2012, and the days with minimum temperatures <0°C range from 250 to 320.

This area also experiences strong diurnal and seasonal freeze-thaw cycles that have resulted in granular structures in surface horizons and reticular or platy structures in subsurface horizons. Thus, cryoturbation occurs in these soils.

In most soil descriptions, the depth to permafrost was not recorded. However, based on permafrost maps (Shi, 1993; Li and Cheng, 1996), the active layer ranges from 1 to more than 3 m thick. In the gelifluction deposit in Kunlun Pass, Harris et al. (1998) measured the thickness of active layers ranging from 12 to 30 cm in July at 4780 m and increasing to 1.5 to 2 m at about 4650 m. They also found ice-cemented permafrost in this seemingly arid environment, and the ice content is usually more than 30%.

Generally, the MAAT -2°C isotherm is the lower elevation boundary of sporadic permafrost, and the MAAT -3°C isotherm is the lower boundary of continuous permafrost. Although most of this ecosystem is associated with the zone of continuous permafrost, Cryosols are rare. Based on a recent exploratory investigation of soil vegetation (Ping and Murray, 2000), the soils developed under the alpine steppe generally have permafrost at a depth of >2m. The soils may have a gelic but not permagelic STR; thus, they do not meet the definition of Cryosols in any classification system.

4.3.3. Alpine meadow

The Alpine meadow zone occupies the upland slopes on both sides of the Tangula Mountains, and the lowlands of the northern Qinghai-Xizang Plateau, and is in the zones of both continuous and discontinuous permafrost. The elevation ranges from 4200 to 4500 m in northeastern Qinghai and from 4500 to 4800 m in northeastern Xizang. The soils formed in this zone are commonly called Matti Soils (NSSI, 1976; Gao et al., 1985; QUARDO, 1988). The climate is cold, semiarid to subhumid, with MAAT of -0.2 to -6.0°C, MAP of 300 to 400 mm, 85 to 90% of precipitation falling between June and September, and annual evaporation of >2000 mm; thus, the aridity index is greater than 10.

Parent materials includes residuum and colluvium of granite, gneiss, sandstone, slate and conglomerate, glacial till, and lacustrine. The active layer is generally >1.5m. In addition to seasonal freeze-thaw, the area experiences strong diurnal freeze-thaw cycles.

The dominant vegetation is *Kobresia pyguaca, Kobresia humilis, Carex* spp., *Polygonum,* and *Leontopodium nanum*. The coverage is 70 to 90%. *Kobresia* is only 3 to 5 cm high but forms a dense rootmat of 5 to 10 cm.

The well drained soils have a dense matti epipedon (Bao et al., 1995) that consists of a black A1 and a dark A2 horizon, followed by a strong brown Bw and pale brown BC and C or CR horizons. Organic carbon content ranges from 3 to 15% in the top 30 cm, and the pH value ranges from 5.5 to 7. The Bw horizon shows evidence of a cambic horizon, including the accumulation of carbonates, Fe, Mn oxides, humus to a lesser extent, coatings on soil particles, and well developed reticular and fine subangular blocky structures. The strong freeze-thaw cycle results in frost-heaved grass mounds, frost cracks, and gelifluction lobes.

These soils key out as Permagelic Gelic Cambisols (CST). In areas where permafrost is within 2 m with cryoturbation, these well drained soils key out as Typic Molliturbels (ST), Brunisolic Turbic Cryosols (CAN), and Cambisols in the WRB system.

In lowlands and depressions of the Alpine Meadow zone, where drainage is limited, gleization features are prominent in the subsurface horizons. In the Chinese system, these soils are classified as either Histic or Haplic Gelic Gleysols, depending on the thickness of the histic epipedon. In the US systems, they are Ruptic-Histic Aquiturbels and Typic Aquiturbels, respectively. They are Gleysolic Turbic Cryosols in the Canadian system, and Histic or Gleyic Cryosols in the WRB system. Where the organic horizon is over 40 cm, the soils are classified as

Hemic Permagelic Histosols (CST), Typic Hemistels (ST), Mesic Organic Cryosols (CAN), and Histic Cryosols (WRB).

5. Cryosols and Climate Change

The Qinghai-Xizang Plateau has continuous records of past environment and climate, not only in the glaciers and ice fields but also in soil profiles (QARDO, 1988). Due to the continued rising of the plateau since the Pliocene, the region experienced different climates and developed different soils. Paleosols of the Tertiary and Quaternary were preserved, and the study of these soils and the pollen in them would provide valuable information about the past climate and environmental changes.

In northeastern China, Qiu and Cheng (1995) used ice wedges and sand wedges to trace the southern boundary of permafrost, which extended to Lat. 41 to 42° N during the coldest stage of Last Glaciation. Then it retreated northward during the Holocene Optimum. But ice wedges and permafrost formed in the Late Pleistocene still exist in the Da Hinggan Mts. (Guo and Lee, 1981). On the Qinghai-Xizang Plateau, ice wedges and sand wedges of the Late Pleistocene indicate a severe cold climate about 15000 to 49000 YBP.

The plateau has become warmer over the past 30 years, with an increase of 0.1 to 0.5°C in the mean annual ground temperature (MAGT) (Wang et al., 2000). The northeastern part of the plateau becomes drier, whereas the rest of it becomes wetter (Zhao et al., in review). The matti epipedon, torn by frost heave, compounded by overgrazing, caused increased erosion.

Since the region has areas with subhumid to semiarid environments and vegetation similar to that during the Late Pleistocene in the Beringia, as Goetcheus and Birks (2001) and Höfle et al. (2001) described it, it is an ideal area for studying modern analogs of the Beringia during the Late Pleistocene or earlier.

6. Summary and Conclusion

The vast areas of permafrost-affected soils in China and the abundant information on these soils provide an excellent opportunity for the scientific community to examine the current soil classification systems and to improve our understanding of these soils. Reviewing the current available information and comparing the different classification systems reveal several major issues.

First, most of the pedon data do not include information on the thickness of active layers or depth to the permafrost and evidence of cryoturbation (frost churning). This information is critical to understanding the cryogenic nature of the soils and their correlation with other soil classification systems. In the previous discussion, we inferred most of the active layer depths from geotechnical data; thus they are not site-specific. However, different agencies and some international cooperative projects have undertaken some long-term ecological monitoring, including of soil temperature and soil moisture. The results will help us determine the classification of these soils.

Second, although China has the third largest area of permafrost, its area of Cryosols is considerably smaller. In much of the Chinese literature, the term Cryosol (Dung'to) is synonymous with frozen ground. The depths to permafrost vary in the frozen ground; they must be within 2 m to meet the definition of Cryosols in the Chinese, Canadian, and the U.S. systems, and within 1 m in the WRB system.

Third, with modification, the current soil classification systems can accommodate the high-altitude permafrost-affected soils but not too well. This is because they ignore the most important soil-forming processes and their resultant properties.

The unique nature of pedogenesis in this environment is not just the presence of ice-cemented permafrost at depths, but also the strong diurnal temperature fluctuation. The diurnal fluctuation may reach over 25 to 40°C. It is hot during the day because of the strong solar radiation on the plateau, but the surface soils cool off fast and freeze at night. Thus, the freeze-thaw cycle manifests not only seasonally but also daily. Such daily freeze-thaw cycles move soluble salts, organic acids, and Fe Al-gels, and the combination of all of these in the upper horizons results in unique properties (Gao and Bao, 1995).

Such properties include vesicular crust surface layers (Cao and Lei, 1995), mudflows, salt accumulations in the subsurface horizons, and Fe and/or Al-humus gel coatings on mineral grains. These properties, coupled with those from the seasonal and freeze-thaw cycles, and permafrost characteristics such as frost cracks, hummocks, polygons, and frost churning, present a unique group of soils that deserve a separate category in both the Chinese and US soil classification systems.

References

Agricultura Canada Expert Committee on Soil Survey. 1998. The Canadian system of soil classification, 3rd ed. NRC Research Press. Ottawa.

Bao, X.K., G.M. Cao, and Y.X. Gao. 1995. Formation environment and genetic mechanisms of Mattic Epipedon. ACTA Pedologica Sinica. 32:45-52. (In Chinese.)

Bockheim, J.G., C. Tarnocai, J.M. Kimble, and C.A.S. Smith. 1997. The concept of gelic materials in the new Gelisol order for permafrost-affected soils. Soil Sci. 162:927-939.

Cao, S.G., and W.J. Lei. 1995. Characteristics of Aridic Epipedon and its significance in soil taxonomic classification. ACTA Pedologica Sinica. 32:34-39. (In Chinese.)

FAO. 1998. World Reference Base for Soil Resources. Food & Agriculture Org. of the UN. Rome.

Gao, Y.X., and X.K. Bao. 1995. Effect of soil freezing and thawing on soil properties on the Qinghai-Xizang Plateau and its significance in taxonomic classification. ACTA Pedologica Sinica. 32:168-176. (In Chinese.)

Gao, Y.X., H.Z. Cheng, Z.D. Wu, H.L. Sun, and M.S. Lee. 1985. Soils of Xizang (Tibet). Science Press. Beijing. 318 pp. with 12 plates and 3 maps. (In Chinese.)

Goetcheus, V.G., and H.H. Birks. 2001. Full-glacial upland tundra vegetation preserved under tephra in the Beringia National Park, Seward Peninsula, Alaska. Quat. Sci. Rev. 20:135-147.

Gong, Z.T. 1995. Chinese Soil Taxonomic Classification (revised). Institute of Soil Science, Academia Sinica. China Agricultural Publishing House. Beijing. 218 pp.

Gu, G.A., L.D. Zhang, and Z.B. Li. 1992. Genetic characteristics of Solonetz in the Qinghai-Xizang Plateau. ACTA Pedologica Sinica. 29:442-446.

Guo, D.X., and Z.F. Lee. 1981. Formation and development of permafrost since Late-Pleistocene in northeast China. J. Glaciol. and Cryopedol. 3(4):1-14.

Guo, Z.W., Y.W.Zhou, F.X. Liang, L.H. Liang, and Q.B. Zhang. 1993. Permafrost features and their changes in Amur area, Da Hinggan ling Prefecture. J. Glaciol. and Crypedol. 15:34-40. (In Chinese.)

Harris, S.A., Z.J. Cui, and G.D. Cheng. 1998. Origin of a bouldery diamicton, Kunlun Pass, Qinghai-Xizang Plateau, Peoples Republic of China: gelifluction deposit or rock glaciers? Earth Surf. Process. Landform. 23:943-952.

Höfle, C., M. E. Edwards, D. M. Hopkins, D. H. Mann, and C. L. Ping. 2000. The full-glacial environment of the northern Seward Peninsula, Alaska, reconstructed from the 21,5000-year old Kitluk Paleosol. Quat Res. 53:143-153.

Hu, S.X., J.E. Zhang, Y.C. Ma, N.A. Wang, Y.T. Zhang, and C.H. Liu. 1998. Formation and evolution of alpine soils and the glacial environment changes of Central Tienshan Mountains. ACTA Pedologica Sinica. 35:243-253. (In Chinese.)

Jin, H.J., G.Q. Qiu, L.hao, S.C. Wang, and Z.F. Zhang. 1993. Thermal regime of alpine permafrost in the upper reach of Urumoqi River, Tienshan. J. Glacialo. and Crypedol. 15:110-114. (In Chinese.)

Li, S.D., and G.D. Cheng. 1996. Map of Permafrost Distribution on Qinghai-Xizang Plateau, one sheet, 1:3,000,000 with explanation, 5pp. Lanzhou Institute of Glaciology and Geocryology, Chinese Academy of Sciences. Gansu Culture Press. Lanzhou. (In Chinese.)

Liu, Q.R., Z.K. Sun, and Y.S. Cui. 1993. Study of the distribution of permafrost and vegetation in Da Hinggan Ling. J. Glaciol. and Cryopedol. 15(2):246-251. (In Chinese.)

Muller, S.W. 1947. Permafrost or permamnently frozen ground and related engineering problems. J.W. Edwards. Ann Arbor, MI. 231 pp.

NSSI (Nanjing Soil Scientist Institute). 1976. Soil Atlas of China. Science Press. Shanghai. 1 map with appendix.

Ping, C.L., and D.F. Murray. 2000. Investigation of alpine soils and vegetation of Qinghai-Tibetan Plateau, Preliminary report to the National Geographic Society, Committee for Research and Exploration. Agricultural and Forestry Experiment Station, University of Alaska. Fairbanks.

QARDO (Qinghai Agricultural Resources Development Office). 1988. Soils of Qinghai. China Agricultural Publishing House. Beijing. 414 pp. with 16 plates and one map. (In Chinese.)

Qui, G.Q. 1993. Development conditions of alpine permafrost in the Tianshan Mountain, China. J. Glaciol. and Cryopedol. 15:96-102.

Qiu, G.Q., and G.D. Cheng. 1995. Permafrost in China: Past and present. Permafrost and Periglacial Processes. 6:3-14.

Qiu, G.Q., and G.D. Cheng. 1993. Development condition of Cryosols in alpine and plateau regions of western China.p. 59-75. In: D.A. Gilinchinsky, ed. Post-seminar proceedings, Joint Russian-American Seminar on cryopedology and global change, November 15-16, 1992, Pushchino, Russia. Russian Academy of Sciences. Pushchino, Russia.

Qiu, G.Q., Y.Z. Huang, and Z.F. Li. 1983. Alpine Permafrost in Tien Shan, China. p. 1012-1023. In Proceedings, Permafrost Fourth International Conference, July 17-22, 1983, Fairbanks, Alaska. National Academy Press. Washington, D.C.

Shi, Y.F. 1993. Map of snow, ice, and frozen ground in China, one sheet, 1:4,000,000 with explanation. LIGG, Chinese Academy of Sciences. China Cartographic Publishing House. Beijing. 17pp.

Soil Survey Staff. 1999. Soil Taxonomy: A basic system of soil classification for making and interpreting soil surveys (2nd Ed.). Agriculture Handbook No. 436. USDA-NRCS. Washington, D.C.

Tong. B.L., S.D. Li, T.J. Zhang, and Y.X. He. 1983. Frozen ground in the Altai Mountains of China. p. 1267-1272. In: Proceedings, Permafrost Fourth International Conference, July 17-22, 1983, Fairbanks, Alaska. National Academy Press. Washington, D.C.

Wang, S.L., H.J. Jin, S.X. Li, and L. Zhao. 2000. Permafrost degradation on the Qinghai-Tibet plateau and its environmental impacts. Permafrost periglac. Process. 11:43-53.

Zhang, W.R. 1986. Chinese Forest Soils. Chinese Forest Science Institute. Science Press. Beijing. 1012 pp.

Zhao, L., G.G. Cheng, Y.J. Ding, S.Y. Liu, C.-L. Ping, and D.Q. Yang. 2003. Changes of seasonally frozen ground thickness in the past 30 years in Qinghai-Xizang (Tibetan) Plateau, China. Global and Planetary Change. (In review.)

Zhao, Q.G., H.Q. Wang and G.A. Guo. 1993. Gelisols of China. ACTA Pedologica Sinica. 30(4): 354-364. (In Chinese.)

Zhou, Y.W., D.X. Guo, G.Q. Qiu, G.D. Cheng, and S.D. Li. 2000. Geocryology in China. Science Press. Beijing, China. (In Chinese.) 450 pp. with plates.

Chapter 12. Cryosols of the Arid Antarctic

I.B. Campbell[1] and G.G.C. Claridge[2]

[1]Research Scientist, Land and Soil Consultancy Services, 23 View Mount, Nelson 7001, New Zealand; e-mail iaincampbell@xtra.co.nz.
[2]Research Scientist, 33 Kotari Road, Days Bay, Eastbourne, New Zealand; e-mail graeme.claridge@xtra.co.nz.

1. Introduction

1.1. Area

Antarctica, with an area of approximately 14 million km^2, is the world's largest continent. The Southern Ocean, which extends from about the 40th parallel to the Antarctic Circle at 60° S, surrounds it. The continent is roughly circular, and its surface rises steeply from the coast to a vast interior ice-filled plateau.

Topographically, Antarctica divides easily into three units (Figure 2.12.1). The largest, East Antarctica, lies between 30°W and 150°E and rises to an altitude over 4000 m asl. The smaller part, West Antarctica, rises to about 1500 m asl, while the third part, the Antarctic Peninsula, extends from the continent toward South America.

The ice sheets reach the sea almost everywhere around the perimeter of the continent and merge into very large floating ice shelves in many places. Estimates vary of the area of the continent and the floating ice shelves, of the total ice volume of Antarctica, and of the area of ice-free ground. The latest estimate (Fox and Cooper, 1994), based on accurate mapping and satellite imaging, replaces Bardin and Suvetova's (1967) earlier estimates. It sets the total area of Antarctica, including floating ice shelves, at 13.9 x 10^6 km^2, of which 12.3 x 10^6 km^2 is grounded ice. Only 0.033 x 10^6 km^2, or approximately 0.3% of Antarctica, is ice-free ground on which soils may form. The ice-free areas are widely scattered and occur along the Transantarctic Mountains, around the margins of the continent, and in the Antarctic Peninsula.

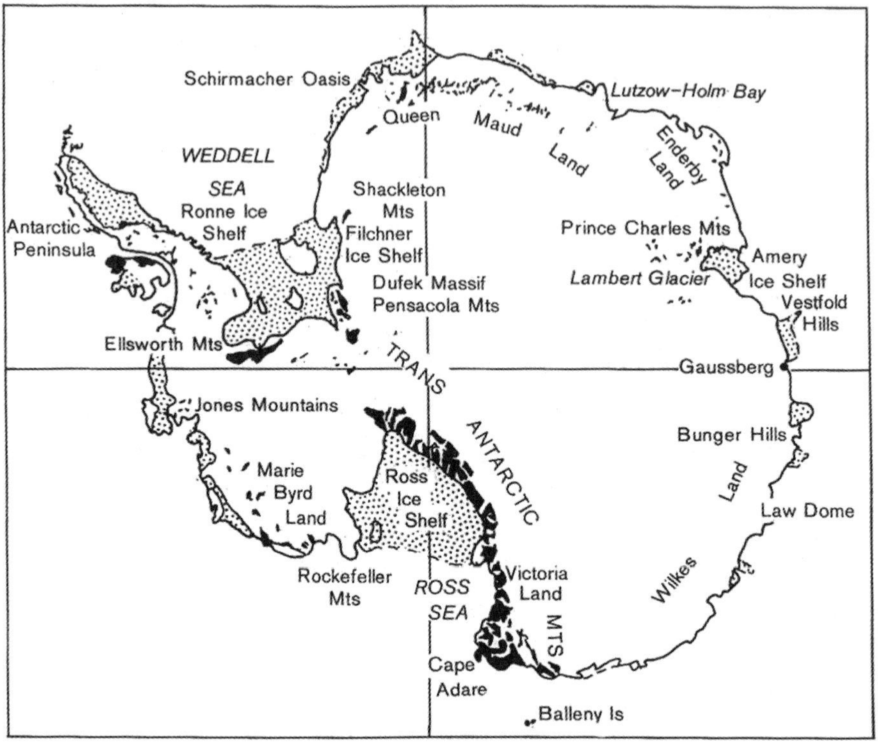

Figure 2.12.1. The Antarctic continent, showing locations the text mentions, including the general location of ice-free areas, although the shading exaggerates the actual extent.

1.2. Temperature

Antarctica is a polar continent surrounded by an ocean, in contrast to the Arctic region, which is an ocean surrounded by continents. This ocean-landmass configuration largely accounts for the climatic difference between the Arctic and Antarctic regions. Antarctica is cold because it receives only a small amount of solar radiation, approximately 16% of that reaching equatorial regions, and also because of the high average surface elevation of the ice sheet. Since outgoing terrestrial radiation exceeds absorbed solar radiation during the year, the Antarctic continent is a major heat sink.

The coldest temperatures have been recorded in the central part of the ice sheet, -89°C at Vostok (Figure 2.12.1) and a mean annual temperature at the South Pole of -49°C. Temperatures increase northward, however, and with proximity to the coast. At Vanda Station, near Lake Vanda in the McMurdo Oasis (or Dry Valley region), for example, the mean annual temperature is 20°C, and it is -18°C at the coastal McMurdo Station (Figure 2.12.2). Farther north, in coastal areas of East

Antarctica, are warmer climates (Burton and Campbell, 1980; McNamara, 1973). At Davis Station in the Vestfold Hills (Figure 2.12.1), for example, the mean annual temperature is -10.2°C with temperatures rising to +13°C in January, and Molodezhnaya and Casey (Figure 2.12.1), experience a similar temperature range.

1.3. Preciptation

Mean annual precipitation over Antarctica is about 50 mm/y, with the least falling inland and the most in coastal regions. In the McMurdo Dry Valley region, precipitation over 20 years averaged 13 mm/y on a valley floor site near Lake Vanda and 100 mm/y in adjacent upland mountains. Precipitation is much higher around the periphery of East Antarctica, with 650 mm recorded at Molodezhnaya, in Enderby Land (MacNamara, 1975). The precipitation nearly always falls as snow.

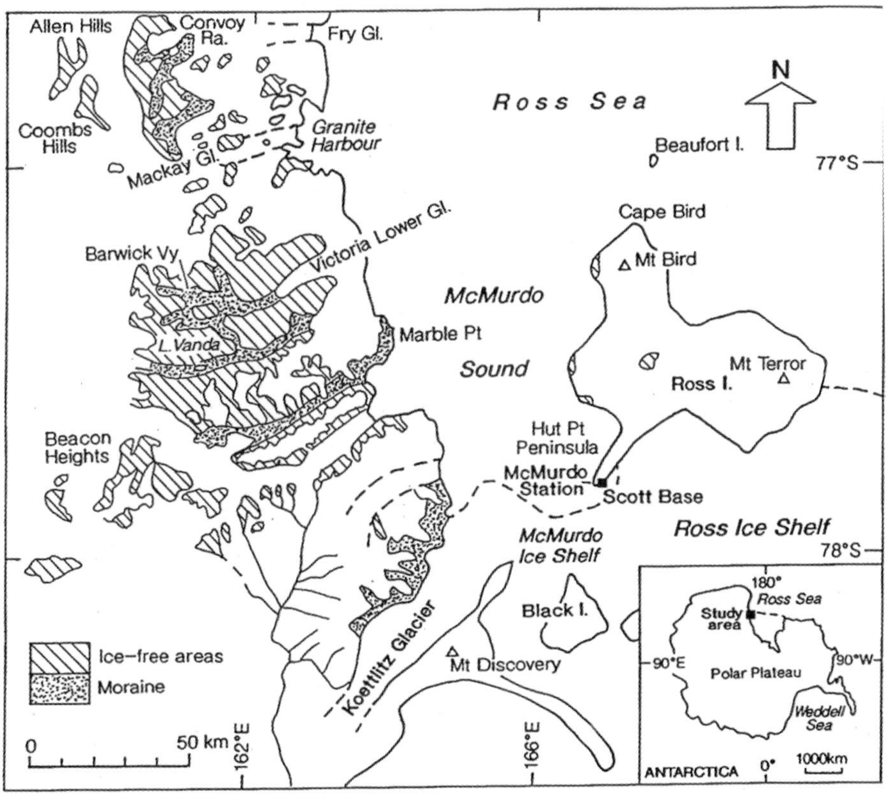

Figure 2.12.2. The ice-free areas of the McMurdo Sound region of Antarctica. The area of ice-free ground in this region is approximately 6000 km^2 and is the largest contiguous area of ice-free ground in Antarctica. The figure also locates the soils in Table 2.12.1.

2. Humid Cryosols

Because the Antarctic Peninsula extends north from the Antarctic Continent into the cyclonic circulation zone, the climate differs much from that of the rest of Antarctica. This provides the basis for distinguishing humid Antarctic Cryosols from arid Antarctic Cryosols, paralleling previous pedological separations of Antarctic soils into cold desert soils, for the soils of East and West Antarctica, and polar desert soils, for soils of the Antarctic Peninsula (Tedrow and Ugolini, 1966; Tedrow, 1977; Campbell and Claridge, 1987).

In the Antarctic Peninsula, mean temperatures range from -11°C at Marguerite Bay, toward the base of the peninsula, to -4°C at Signy Island, at its northern extremity (Figure 2.12.1), with maxima of +9°C and +12°C, respectively. Some stations have recorded over 100 mm annual precipitation, falling at times as either sleet or rain. In the maritime zone of the Antarctic Peninsula and the associated islands where warmer temperatures prevail, the soils are commonly moist, with up to 15% water content in the active layer of soils on King George Island, on the east coast of the Antarctic Peninsula (Zhu et al., 1991).

Under these conditions, small patches of organic soils occur, formed mainly by debris from mosses and algae while weathering is more active. They are humid Antarctic Cryosols, owing to their higher organic matter contents, brunification processes, and greater cryic activity (Beyer et al., 1999; Blume et al., 1997). Some more favored parts of East Antarctica, such as those near Casey Station described by Beyer and Bolter (1999), also may support plant growth and carry humid Cryosols.

3. Arid Cryosols

In contrast, the arid Cryosols of the Transantarctic Mountains and the northern coastal regions (Bockheim, 1997; Campbell and Claridge, 1987) occur under a cooler and more arid climate. They have developed under the most extreme conditions for soil formation anywhere on earth, with the cold temperatures and aridity restricting both weathering and soil biological processes to minimal levels. Their defining features are negligable organic matter contents and cryic movement, very low moisture contents, accumulation of soluble salts in salic horizons, and progressive reddening with increasing age.

3.1. Distribution

Arid Cryosols occur mostly along the Ellsworth Mountains and Transantarctic Mountains and in Victoria Land (Figure 2.12.1) and, to a much lesser extent, in isolated areas around the fringe of eastern and western Antarctica. Since they cover an area of only about 35,000 km^2, their distribution is obviously sporadic.

The largest contiguous area of the soils, about 6,000 km^2, is in the Dry Valleys on the eastern margin of McMurdo Sound (Figure 2.12.2).

Numerous authors have investigated the arid Cryosols since the 1950s (Markov, 1956; Ugolini, 1963; Tedrow and Ugolini, 1966; Claridge, 1965; McCraw, 1967; Campbell and Claridge, 1966, 1969, 1975, 1978, 1987; Claridge and Campbell, 1977, 1984; Everett, 1971; Bockheim, 1977, 1979, 1982). These investigations were aimed primarily at understanding the nature of soil genetic processes in this extremely cold and arid environment and also at determining the significance of pedogenic features in respect to Antarctic landscape formation and glacial history. More recently, attention has focused on environmental and ecosystem relationships and concerns about human impact (Campbell et al., 1994, 1997a, 1997b, 1998a, 1998b, 1998c).

Because of the soils' wide distribution in small areas around the Antarctic continent, no one has tried to show major or significant pedological distinctions on a continental scale through soil maps. The only significant soil map published (Scale 1: 63360) is for a portion of Taylor Valley in the McMurdo Sound region, with units that are essentially lithologic (McCraw, 1967).

3.2. General properties and pedological distinctions

Because of the severe environmental conditions, soil weathering is minimal. The two main pedological processes, oxidation and salinization, are responsible for the main physical differences in the arid Cryosols. Soil moisture levels and the organic regime are also important attributes.

Soil oxidation takes the form of the very slow release of iron oxides from mineral particles and a gradual reddening and progressive increase in oxidation depth over time. Since land surfaces are extremely stable, with some surfaces and soils dating to the Miocene (Campbell and Claridge, 1978; Denton et al., 1993), the soils on the oldest surfaces have the greatest degree of oxidation. Chemical weathering and mineral alteration are at minimal levels and are generally insignificant.

Salinization, or the accumulation of salts in the arid Cryosols, is widespread and results from the high evaporation rates, which typically exceed precipitation. The salts in the soils predominantly come from atmospheric transport. There are clear geographic and climatic differences in soil salt content and age-related differences in salt abundance (Claridge and Campbell, 1977; Campbell and Claridge, 1987). Salt accumulation is essentially linear with time (Bockheim, 1979).

Soil moisture content is another property by which the arid Cryosols can be separated. Balks et al. (1995), Campbell et al. (1997a, 1998a), and MacCulloch (1996) have investigated the thermal regimes and moisture content of soils in the McMurdo Dry Valley region and showed that water contents of the active layer here average from around 5% to 1% during the brief 6-week summer. Ice-cemented permafrost with higher moisture content typically underlies the active layer, but the permafrost is commonly dry-frozen in soils that have a very low active-layer moisture content.

A distinguishing feature of the arid Cryosols is the absence of a significant organic regime. Soil biological occurrences are negligible and generally consist only of discontinuous populations of bacteria, fungi, yeasts, protozoa, or algae. These are more common in warmer and more moist sites. Higher plants are restricted to sporadic lichen and small patches of mosses, usually a few square meters in area alongside summer melt water flows.

3.3. Subdivisions of arid Cryosols

In spite of the extreme conditions under which arid Cryosols form, they do show clearly recognizable regional geographic and pedologic distinctions related, essentially, to differences in landform age and climate.

3.3.1. Weakly developed arid Cryosols

Almost all the landforms are of glacial origin. Thus Antarctic glaciation, with its associated sequences of glacial deposits, has greatly influenced soil distribution.

Antarctic glacial history has been complex and largely involves the interaction of the East and West Antarctic Ice Sheets and also the Alpine Glaciers over a very long time. In most coastal areas and low altitude valley surfaces, soil-forming surfaces have been exposed as a consequence of retreat of ice from a maximum reached at the height of the last glaciation (Wisconsin or Ross Glaciation). Campbell and Claridge (1975) described the soils on these younger surfaces as weathering stage 1 soils (<50,000 yrs), essentially unweathered.

3.3.2. Well developed arid Cryosols

Land surfaces typically increase in age with increasing altitude, and some high altitude glacial surfaces have remained ice-free for a very long time. The oldest soils and those showing the greatest degree of development occur on these high altitude surfaces. They are believed to be of Pliocene to Miocene age (Campbell and Claridge, 1978; Denton et al., 1993; Marchant et al., 1993). Campbell and Claridge (1978) described them as belonging to weathering stage 5.

Well developed arid Cryosols occur in inland parts of the Transantarctic Mountains bordering the East Antarctic Ice Sheet. They are distinguished by their yellowish brown or red (10YR 5/4, 2.5YR 3/6) colors, by soil oxidation which may extend to depths of more than 75 cm, and by the accumulation and segregation of soluble salts into a salt horizon. Paleosols are not uncommon and frequently have the same features of the surface soil, including a salt horizon.

Soils intermediate in development between the weakly developed arid Cryosols and the strongly developed arid Cryosols commonly occur on intermediately aged glacial surfaces.

3.3.3. Arid Cryosols with salt horizons

Climate exerts a strong influence on the properties and distribution of the arid Cryosols through factors which govern both salinity and soil moisture levels. The composition of the salts in the soils depends on their transport pathways; most of them come from the atmosphere

In coastal regions, for example, sodium chloride dominates the soil salts, and the ions present are in the same ratio as in sea water, reflecting an origin from coastal air masses. Farther inland, precipitation contains less chlorides, and the proportion of sulphate increases, while chloride is absent. Ions present include sulphate, nitrate, and iodate, derived via stratospheric transport.

Because the soils inland are generally older, they often have thick salt horizons. In the younger soils near the coast, salt horizons are seldom present. The arid Cryosols therefore can be distinguished geographically by the extent of development of the salt horizon as well as by the composition of the salts present. Figure 2.12.3 shows a typical arid Cryosol.

3.4. Soil moisture distinctions

Soil moisture is an important climate-related feature of the arid Cryosols, and geographic distinctions reflect soil moisture levels (Campbell et al., 1969, 1997a).

In coastal regions, which are at low altitudes, temperatures are warmer and precipitation is greater than elsewhere, moisture contents of the active layer are higher than elsewhere, and ice-cemented permafrost is commonly present around 60-70 cm deep. In inland regions at higher altitudes, temperatures are much lower and the moisture contents of the active layer are significantly lower. The permafrost occurs at much shallower depths, sometimes less than 10 cm, and it is often dry-frozen with insufficient moisture present to cement the soil.

3.5. Organic distinctions

Because of the extremely low levels of biological activity, only in few places may soils contain more than a trace of organic matter. Mosses only grow in small patches of a few square meters or less—typically, salt free, moist situations adjacent to ephemeral streams in coastal regions. Their roots may penetrate 2 to 3 cm into the, soil but accumulations of organic material are rare. Accumulations of algal material into mats sometimes occur, but there is little evidence of organic transformations.

The most significant organic accumulations are those associated with penguin rookeries, estimated to cover around 20 km^2 (Claridge et al., 1999). The organic material is derived from excreta and dead animals and in places has built up to depths of several meters, with evidence of decomposition (Campbell and Claridge, 1966; Spier and Cowling, 1984). Apart from their high organic content, the soils are characterized chemically by their very high nitrate and phosphate concentrations.

Figure 2.12.3. An arid Cryosol from the Asgard Range, Wright Valley, showing well developed desert pavement, gravelly sand texture, and deep staining on ice-cemented permafrost.

Table 2.12.1. Proportion of fine earth, moisture content, pH, and soluble salt content and calcium, sodium, chloride, and nitrate contents of water extracts of three soils from the McMurdo Sound region of Antarctica (classification according to *Soil Taxonomy*).

Depth, cm	Color, Munsell	<2 mm, %	Moisture Wt %	pH	Salts Wt %	Ca	Na	Cl	SO_4
						meq./100g fine earth			
Soil A, Marble Point, altitude 100 m, M. A. T. -18°C (est), Pptn 150 mm (est)									
Glacic Haplorthel, ice-cemented below 65 cm.									
4-0	5Y 6/3	25	0.5	9.8	1.6	0.05	26	24.2	0.10
0-27	5Y 6/3	46	3.5	9.7	0.09	0.05	1.0	1.00	n.d.
27-48	5Y 6/3	44	4.3	9.9	0.11	0.05	1.4	1.00	0.02
48-65	5Y 6/3	55	5.9	9.9	0.17	0.05	2.0	1.30	0.01
65-80	5Y 6/3	54	15.2	9.7	0.28	0.10	1.4	1.00	n.d.
80-95	5Y 6/3	73	67.2	9.6	0.32	0.10	1.1	1.00	n.d.
95-110	5Y 6/3	80	20.9	9.0	0.03	0.20	2.9	3.00	n.d.
Soil B, Wright Valley, Altitude 460 m, M.A.T. -22°C (est), pptn 50 mm (est)									
Salic Anhyorthel, dry frozen permafrost at 40 cm.									
0-5	2.5Y 6/4	57	0.9	8.9	3.0	8.1	1.7	14.5	0.5
5-6	2.5Y 6/4	n.d.	1.3	7.7	45.1	15.6	1090	1500	4.0
6-20	10YR 5/4	64	1.0	7.9	21.6	5.3	278	249	14.0
20-40	2.5Y 5/4	64	1.0	7.9	13.8	4.0	171	303	9.5
40-60	2.5Y 5/4	65	1.0	7.8	6.1	3.3	63	63.5	5.0
60-100	2.5Y 5/4	65	1.2	8.7	1.2	1.8	11	8.3	0.5
Soil C, Quartermain Range, altitude 1200 m, MAT -30°C (est), Pptn. <50mm (est)									
Nitric Anhyorthel, dry frozen permafrost at 20 cm.									
0-4	10YR 5/4	63	4.6	6.8	1.7	14.4	24	10.4	8.5
4-10	10YR 5/4	96	13.0	7.9	9.2	10.7	315	56.3	10.5
10-16	10YR 5/6	70	2.5	7.9	23.3	2.1	1250	45.8	1082
16-27	2.5Y 6/6	73	3.4	7.6	5.9	4.0	188	49.7	148
27-65	2.5Y 6/4	63	2.9	7.7	6.6	30.1	144	53.6	85
65-76	10YR 6/4	82	0.6	8.0	6.8	36.2	125	32.4	117
76-82	2.5Y 7/4	60	4.1	7.8	6.1	24.1	87	29.0	77
82-100	2.5Y 7/4	75	3.4	7.5	1.5	7.5	37	25.7	26

n.d. not detected, MAT mean annual temperature, pptn precipitation.

3.6. Classification

Other chapters of this book discuss the classification of the arid Cryosols. They all fall within the Gelisol order of *Soil Taxonomy* (Soil Survey Staff, 1998). Table 2.12.1 presents representative analyses of three soils, covering the range of arid Cryosols. The warmest and most moist soil, on the McMurdo Sound coast at Marble Point, is a Glacic Haplorthel. At the extreme of aridity and age is a Nitric Anhyorthel, an old soil at high altitude

4. Conclusions

The arid Cryosols occur on most of the ice-free ground in eastern and western Antarctica, but they cover only a tiny portion of the Antarctic continent. Their widely scattered distribution is a function of the glacial history of Antarctica, which extends from Miocene or older times.

Although the arid Cryosols form under a climate dominated by extreme cold and aridity, the very slow rates of weathering that have occurred over extraordinary lengths of time have produced recognizable pedological distinctions. Soils of different regional climates and ages show different characteristics.

Although no systematic mapping exists, research has produced considerable data about the nature and distribution of the arid Cryosols, the pedological processes operating on them, and their significance in relation to the environment and the evolution of the Antarctic landscape.

References

Balks, M.R., D.I. Campbell, I.B. Campbell, and G.G.C. Claridge. 1995. Interim results of 1993/94 soil climate, active layer and permafrost investigations at Scott Base, Vanda and Beacon Heights, Antarctica. University of Waikato, Department of Earth Sciences Special Report. pp. 1-64.

Bardin, V.I., and Y.A. Suvetova. 1967. Basic mathematic characteristics for Antarctica and budget for the Antarctic ice cover. Scientific Reports of the Japanese Antarctic Research Expedition. Special Issue. 1:92-100.

Beyer, L., and M. Bolter. 1999. Formation, ecology and geography of Cryosols of an ice-free oasis in Coastal East Antarctica near Casey Station (Wilkes Land). Aust. J. Soil Res. 37:209-44.

Beyer, L., J.G. Bockheim, I.B. Campbell, and G.G.C. Claridge. 1999. Genesis, properties and sensitivity of Antarctic Gelisols. Antarc. Sci. 11:387-398.

Blume, H-P., L.Beyer, M. Bolter, H. Erlenkeuser, E. Kalk, S. Kneesch, U. Pfisterer, and D. Schneider. 1997. Pedegenic zonation in soils of the southern circumpolar region. Adv. in GeoEcol. 30:69-90.

Bockheim, J.G. 1977. Soil development in the Taylor Valley and McMurdo Sound area, Antarctica. Antarc. J. U.S. 12:105-108.

Bockheim, J,G. 1979. Relative age and origin of soils in eastern Wright Valley, Antarctica. Soil Sci. 128:142-152.

Bockheim, J.G. 1982. Properties of a chronosequence of ultraxerous soils in the Transantarctic Mountains. Geoderma. 28:239-255.

Bockheim, J.G. 1997. Properties and genesis of cold desert soils from Antarctica. Soil Sci. Soc. Amer. J. 61: 224-231.

Burton, H.R.. and P.J. Campbell. 1980. The climate of the Vestfold Hills, Davis Station, Antarctica, with a note on its effects on the hydrology of a deep saline lake. A.N.A.R.E. Scientific Reports, Series D. Meteorological Publication No. 129. Australian Government Publishing Service. Canberra.

Campbell, D.I., R.J.L. MacCulloch, and I.B. Campbell. 1997b. Thermal regimes of some soils in the McMurdo Sound region, Antarctica. In: W.B. Lyons, C. Howard-Williams, and I. Hawes, eds. Ecosystem Processes in Antarctic ice-free Landscapes. Proceedings of an International workshop on Polar Desert Ecosystems, Christchurch, New Zealand, 1-4 July 1996. Balkema. pp. 45-56.

Campbell, I.B., and G.G.C. Claridge. 1966. A sequence of soils from a penguin rookery, Inexpressible Island, Antarctica. N. Z. J. of Sci. 9:361-72.

Campbell, I.B., and G.G.C. Claridge. 1969. A classification of frigic soils - the zonal soils of the Antarctic Continent. Soil Sci. 107:75-85.

Campbell, I.B., and G.G.C. Claridge. 1975. Morphology and age relationships of Antarctic soils. In: R.P. Suggate and M.M. Cresswell, eds. Quaternary Studies. Royal Soc. of N.Z. Bull. 13:83-88.

Campbell, I.B., and G.G.C. Claridge. 1978. Soils and Late Cenozoic history of the upper Wright Valley, Antarctica. N. Z. J. of Geol. and Geophys. 21:635-43.

Campbell, I.B., and G.G.C. Claridge. 1987. Antarctica: soils, weathering processes and environment. Elesvier Science Publishers. Amsterdam. 368 pp.

Campbell, I.B., G.G.C. Claridge, and M.R. Balks. 1994. The effect of human activities on the moisture content of soils and underlying permafrost from the McMurdo Sound region, Antarctica. Antarc. Sci. 6:307-314.

Campbell, I.B., G.G.C. Claridge, M.R. Balks, and D.I. Campbell. 1997a. Moisture content in soils of the McMurdo Sound Dry Valley region of Antarctica. In: W.B. Lyons, C. Howard-Williams, and I. Hawes, eds. Ecosystem Processes in Antarctic ice-free Landscapes. Proceedings of an International workshop on Polar Desert Ecosystems, Christchurch New Zealand, 1-4 July 1996. Balkema. pp. 61-76.

Campbell, I.B., G.G.C. Claridge, D.I. Campbell, and M.R. Balks. 1998a. The Soil Environment. In: J.C. Priscu, ed. Ecosystem Dynamics in a Polar Desert: the McMurdo Dry Valleys, Antarctica. Antarctic Research Series 72. American Geophysical Union. pp. 297-322.

Campbell, I.B., G.G.C. Claridge, M.R. Balks, and D.I. Campbell. 1988b. Permafrost properties in the McMurdo Sound Dry Valley region of Antarctica. Proceedings of the 7th International Conference on Permafrost, Yellowknife, Canada 1998. pp. 121-126.

Campbell, I.B., G.G.C. Claridge, D.I. Campbell, and M.R. Balks. 1998c. Short and long term impacts of human disturbances on snow-free surfaces in Antarctica. Polar Record. 34:17-26.

Claridge, G.G.C. 1965. The clay mineralogy and chemistry of some soils from the Ross Dependency, Antarctica. N. Z. J. of Geol. and Geophys. 8:186-220.

Claridge, G.G.C., and I.B. Campbell. 1977. The salts in Antarctic soils, their distribution and relationship to soil processes. Soil Sci. 123: 377-84.

Claridge, G.G.C., and I.B. Campbell. 1984. Mineral transformation during the weathering of dolerite under cold arid conditions in Antarctica. N. Z. J. of Geol. and Geophys. 27:537-45.

Claridge, G.G.C., I.B. Campbell., and D.S. Sheppard. 1999. Carbon Pools in Antarctica and their significance for global climate change. Adv. in Soil Sci. (In press.)

Denton, G.H., D.E Sugden, D.R. Marchant, B.L. Hall and T.I. Wilch. 1993. East Antarctic Ice Sheet sensitivity to Pliocene climatic change from a Dry Valleys perspective. Geografiska Annaler. 75A:155-204.

Everett, K.R. 1971. Soils of the Meserve Glacier area, Wright Valley, Southern Victoria Land, Antarctica. Soil Sci. 112:425-438.

Fox, A.J., and P.I. Cooper. 1994. Measured properties of the Antarctic Ice Sheet, derived from SCAR digital database. Polar Record. 30:201-204.

MacCulloch, R.J.I. 1996. The microclimatology of Antarctic soils. Unpublished M. Sc. Thesis. University of Waikato. Hamilton, New Zealand.

MacNamara, E.E. 1973. Macro and microclimate of the Antarctic coastal oasis, Molodezhnya. Bull. Periglacjalny. 23:201-236.

McCraw, J.D. 1967. Soils of Taylor Dry Valley, Victoria Land, Antarctica, with notes on soils from other localities in Victoria Land. N. Z. J. of Geol. and Geophys. 10:498-539.

Marchant, D.R., G.H. Denton, D.E. Sugden, and C.C. Swisher. 1993. Miocene glacial stratigraphy and landscape evolution of the western Asgard Range, Antarctica. Geografiska Annaler. 75A:303-330.

Markov, K.K. 1956. Some facts concerning periglacial phenomena in Antarctica (Nekotorye dannye o periglastialnykh yavleniiakh v Antarktide).Vest. Mosk. Univ. (Geograf) 1:139-148. (In Russian.)

Soil Survey Staff. 1998. Keys to Soil Taxonomy. USDA. Washington, D.C. 326 pp.

Speir, T.W., and J.C. Cowling. 1984. Ornithogenic soils of the Cape Bird Adelie Penguin rookeries, Antarctica. Polar Biol. 2:199-205.

Tedrow, J.C.F., and F.C.Ugolini. 1966. Antarctic soils. In: J.C.F. Tedrow, ed. Antarctic soils and soil forming processes. Antarctic Research Series 8. American Geophysical Union. pp. 161-171.

Tedrow, J.C.F. 1977. Soils of the Polar Landscapes. Rutgers University Press. New Brunswick. pp. 1-638.

Ugolini, F.C. 1963. Soil investigations in the lower Wright Valley. Proceedings of the International Permafrost Conference. Nat. Acad. Sci. Nat. Res Council Pub. 1287:55-61.22.

Zhu, C., Z. Ciu, and H. Xiong. 1991. Characteristics of the active layers on Fildes Peninsula of King George Island, Antarctica. Antarc. Res. (Polar Research Institute China). 2:24-37.

Chapter 13. The Soil Cover of Central Siberia

I.A. Sokolov[1], T.V. Ananko[2], and D. Ye. Konyushkov[3]

[1]Professor, V.V. Dokuchaev Soil Science Institute, Pyzhevskii per. 7, Moscow, 109017 Russia; e-mail dkon@agro.geonet.ru.
[2]Senior Researcher, V.V. Dokuchaev Soil Science Institute, Pyzhevskii per. 7, Moscow, 109017 Russia; e-mail dkon@agro.geonet.ru.
[3]Head, Laboratory of Soil Evolution and Ecology, V.V. Dokuchaev Soil Science Institute, Pyzhevskii per. 7, Moscow, 109017 Russia; e-mail dkon@agro.geonet.ru.

1. Introduction

Central Siberia is a vast area (about 4,000,000 sq. km) between the Yenisei and Lena rivers. It stretches from Cape Chelyuskin (77°43'N, 104°18'E) in the north to the foothills of Eastern Sayan, Baikal, and the Stanovoi ridges (52°N) in the south. This chapter also considers the mountainous Trans-Baikal region, to the southeast of Central Siberia, which usually is included in the physiographic province of the Southern Siberian Mountains.

The Russian Tsar Peter the Great initiated the exploration of Central Siberia in the first quarter of the 18th century. Peter Chichagov, who investigated the Yenisei River valley in 1725 to 1730, obtained the first data on the western margins of the Central Siberian Plateau.

In the 1840s, the expedition of the Russian Emperor Academy of Sciences, headed by Alexander von Middendorf, explored the Taimyr Peninsula, in particular, the Byrranga Ridge (1843).

The Department for Peoples' Migration initiated the soil survey in Central Siberia in 1907 to 1914. In the Soviet period, it continued under the aegis of the Commission on the Study of Productive Forces of the USSR. Konstantin D. Glinka (1912, 1927) summarized the results of these expeditions. Along with general data on soil geography, the researchers studied the impact of cryogenic processes and permafrost conditions on the soils (see Section 1).

In spite of new factual data obtained during these expeditions, classification decisions for Siberian soils mostly were extrapolated from the European part of Russia, and the list of soil types was relatively short (Tundra soils in the tundra zone, Podzolic soils in the taiga zone, Gray Forest soils and Chernozems [or Meadow-

Chernozemic soils] in the forest-steppe and steppe zones, respectively, with inclusions of Mucky-Calcareous soils in the areas of calcareous rocks). Because the morphology and properties of Siberian soils do not meet the diagnostic criteria of corresponding soil types, the qualifiers "weakly," "crypto," or "latent" often were applied with soil names, e.g., weakly or cryptopodzolic soils, weakly or cryptogley soils, etc.

In the 1950s, it became clear that the soils of Central Siberia have a lot of specific regional features. O.V. Makeev (1959) proved that Podzolic soils are not typical of taiga environments in the south of Central Siberia. He described the type of Soddy Taiga soils and explained the absence of podzolization in them by the richness of parent rocks in weatherable minerals. V.G. Zol'nikov (1954, 1965) described the type of Palevye (Pale) metamorphic soils without podzolization in the taiga zone.

I.P. Gerasimov (1963) advanced a concept of genetic specificity of Siberian soils and explained it by the continentality of Siberian climate and widespread permafrost. I.A. Sokolov and T.A. Sokolova (1962) suggested that the degree of genetic specificity of soils in permafrost-affected regions depends on soil texture. Well drained, coarsely textured mesomorphic soils with "dry" or low-ice permafrost should be categorized separately from poorly drained, finely textured hydromorphic soils with ice-rich permafrost. Thus, soil science should characterize permafrost-affected territories as a series of zonal soil types.

In the 1960s and 1970s, a lot of new soil names appeared on soil maps of Central Siberia. At present, the soil cover of Central Siberia is well known. The *Soil Map of the Russian Federation*, 1:2.5 M scale, presents its main features. Several monographs describe the soils of particular regions (Gorbachev & Popova, 1992; Elovskaya et al., 1979; Ivanov, 1966; Koposov, 1983; Kuz'min, 1976; Krasyuk, 1927; Makeev, 1959; Martynov, 1965; Mikhailenko, 1967; Nadezhdin, 1961; Nikolaev, 1949; Nogina, 1964; Prasolov, 1927; Vtorushin, 1982; Ufimtseva, 1967; Vasil'evskaya, 1980; Zol'nikov et al., 1962; etc.). A number of publications (Karavaeva et al., 1965; Sokolov et al., 1982, 1993; Sokolov, 1988, 1991) detail the modern concept of soil formation in permafrost-affected areas of Siberia.

2. Physiography

2.1. Relief and geological structure

Geologically, the major part of Central Siberia is confined to the ancient Siberian platform and is a plateau with an average altitude of 500 m a.s.l. The map in Figure 2.13.1 shows the lithological and orographic structures of Central Siberia. Numbers in parentheses in the text, following geographical names, appear on this map.

The Byrranga Ridge (19) (600 to 1000 m a.s.l.) is a Hercynian structure rejuvenated during the Alpine orogenesis that stretches latitudinally across the Taimyr Peninsula in the northern part of Central Siberia.

The North Siberian Lowland (1) separates the Byrranga Ridge from the Central Siberian Plateau. It is confined to the ancient epiplatform trough filled with marine, continental, and lagoon sediments of the Mesozoic age and silty and sandy alluvial and glaciomarine Quaternary sediments. It has a relatively flat relief, complicated by diverse cryogenic form—thermokarst depressions, pattern grounds, pingos, etc.

The Central Siberian Plateau, south of the Lowland, has an uneven surface dictated by the peculiarities of the tectonic structure of the Archean-Early Proterozoic Siberian platform. *The Tungus syneclise* in the west of the platform is filled by mafic volcanic (trappean) rocks of Triassic age and has an inverted relief represented by trappean plateaus: *Putorana (3)* (up to 1200 to 1700 m), *Syverma (9)* (600 to 900 m), and *Tungus (10)* (400 to 600 m). The total area of trappean rocks in Central Siberia is estimated at 1,000,000 sq. km.

The northeastern part of the Siberean platform is confined to *the Ananbar anteclise,* with the Anabar shield in the center. *The Anabar Tableland (4)* (800 to 1000 m a.s.l.) is composed of Archean gneiss rocks. *Kotui (5) and Olenek (6) plateaus* (400 to 600 m a.s.l.) composed of Late Proterozoic and Early Paleozoic calcareous rocks encircle the Anabar shield from the west and east, respectively.

The Vilyui syneclise in the eastern part of the platform is filled with Mesozoic marine and continental deposits overlain by Quaternary alluvial and eolian sands and loams. In the modern relief, it corresponds to the *Central Yakutian Lowland (2)* (150 to 200 m a.s.l.).

Cambrian calcareous rocks outcrop to the surface on the northern slope of the *Aldan anteclise.* This area is distinguished as the *Lena-Aldan Plateau (11)* (400 to 600 m in the north and up to 1000 to 1100 m in the south). The *Aldan Tableland (12)* (1000 m, with some ridges up to 2264 m) is composed of Archean and Early Proterozoic gneiss with intrusions of granite.

The Baikal folding epoch (the Early Cambrian period) gave birth to *mountain chains of the Trans-Baikal region (16)*, rejuvenated in the Cenozoic era. These chains have an altitude of 1500 to 2800 m and stretch southwest to northeast. The *Vitim Plateau (14)* and *Stanovoi Highland (15)* lie between these chains. The *Stanovoi Ridge (13)* (up to 2412 m a.s.l.) to the south of Aldan Plateau divides the Pacific and Arctic ocean basins.

The southwestern part of the platform is composed mainly of Early Paleozoic calcareous rocks with several trappean massifs; the altitude of *Angara (7)* and *Angara-Lena (8)* plateaus averages 350 to 500 m. The *Yenisei Ridge (18)* (1000 to 1100 m, Baikal folding) stretches in the meridional direction along the right bank of Yenisei and borders the platform in the southwest. To the south, within the *Eastern Sayan foredeep,* the elevation is somewhat lower (250 m). *The Eastern Sayan Ridge (17)* (up to 3491 m) originated during the Baikal folding epoch and was rejuvenated in the Alpine orogenic stage. It stretches to the south of Krasnoyarsk in the southeastern direction.

2.2. Soil-forming rocks

The Quaternary glaciers in Central Siberia did not occupy vast areas. Valley glaciers existed in the highest mountains (the Byrranga glacier, the Putorana glacier, and, probably, the Anabar glacier) in the northern part of this region. Soils develop from the residuum of bedrocks or from their loose derivatives of colluvial and solifluctional origin.

The mantles of loamy clay rocks (with considerable pebble content) on the plateaus in the northern part of Central Siberia have relatively low thickness (about 2 to 5 m). The origin of these mantles is uncertain. Usually, their composition resembles the composition of underlying bedrock (thus, the mantles developing on calcareous rocks contain carbonates; those on mafic rocks, typical mafic minerals; etc.). They also contain coarse rock fragments from underlying bedrock and may contain allochthonous fine earth and coarser material. Stony residuum of bedrock, arranged into sorted nets, circles, and stripes, covers the surface of the highest plateaus in the northern part of Central Siberia.

Sandy glaciofluvial and alluvial deposits occur in relief depressions, including the North Siberian Lowland. Loesslike and loess rocks of aerial origin are widespread in the Central Yakutian Lowland and in the southwestern part of Central Siberian Plateau.

In the Trans-Baikal region, sandy and loamy derivatives of granitoid rocks fill intermontane depressions and river valleys.

The soils inherit many features and properties from parent rocks. In similar climatic conditions, the character of pedogenesis differs markedly, depending on the composition and properties of soil-forming rocks (Figure 2.13.1).

2.3. Climate

Central Siberia lies in the center of Eurasia and has an extremely continental climate. The degree of continentality increases from west to east. In Yakutia, the difference between maximum and minimum air temperatures reaches 100°C, and the difference between mean January and July temperatures is about 55 to 65°C.

The distribution of precipitation is very uneven, from more than 1000 mm on western slopes of the Putorana Plateau and Yenisei Ridge to less than 150 to 200 mm in the Central Yakutian Lowland. Mean annual evapotranspiration is about 600 to 800 mm, so the western part of Central Siberia has a humid climate, whereas semiarid and arid conditions prevail in the eastern part (Figure 2.13.2). Orographic effects on the precipitation are strongly pronounced in the Trans-Baikal mountains, where the annual precipitation increases at higher elevations and on the slopes of western and northwestern aspects (in the eastern part, Pacific monsoons condition better moistening of eastern slopes).

Mean annual air temperature varies from -16.2°C in Taimyr tundra, to -6.7°C in the middle taiga zone of Central Yakutia, to -2.5°C in the forest zone of Trans-Baikal region, to -0.5 to -1.0°C in the southwestern part of Central Siberia.

The development of Asian baric maximum causes dry and cold weather in the winter. Atlantic cyclones penetrate into Central Siberia along the northern periphery of the anticyclone, so that winter temperatures and precipitation in the north (Taimyr Peninsula and North Siberian Lowland) are somewhat higher than in the center of the region.

Because of the cooling effect, extremely low winter temperatures (-65 to -68°C, the Essei weather station) characterize relief depressions in the northern part of the Central Siberian Plateau. Temperature inversion occurs in depressions.

Snow cover depth varies from 70 to 90 cm in the west to 20 to 30 cm in the east of Central Siberia. Stable snow cover occurs about 9 months per year (280 days) in the northern areas and six months per year in the southern areas of Central Siberia.

In summer, the atmospheric pressure lowers. Arctic air masses heat and transform into continental air masses with high aboveground temperatures. Maximum summer temperatures may exceed 30°C. Mean July temperature at the latitude of 70°N is about 11 to 12°C; in Yakutsk (62° N), it rises to 18.9°C. The precipitation from Atlantic cyclones is high in the western part of the Central Siberian Plateau (especially on western slopes of the Putorana Plateau and Yenisei Ridge) and rapidly decreases to the east.

In the Trans-Baikal region, summer precipitation mainly connects with Pacific monsoons with the maximum in August through September. The active rise of air temperatures in the late spring and early summer, when the soils are still frozen, causes rapid thawing of snow and snowmelt runoff into the rivers. This water does not do much to moisten the soil. However, summer thawing of ice-rich permafrost provides additional water.

2.4. Vegetation

The vegetative cover of Central Siberia displays distinct latitudinal zonality (from Arctic Tundra to Boreal Steppe) complicated by orographic effects, the diversity of parent rocks, and a decrease in the degree of humidity of the climate from west to east (Figure 2.13.2).

Three subzones of plain tundra vegetation occur in the North Siberian Lowland. **Arctic Tundra,** which includes the northernmost coast areas and islands, is predominantly sedges (*Carex reptobunda, C. subspathacea, C. marina*), some forbs, and grasses (*Papaver radicatum, Ranunculus borealis, Dupontia Fischeri, Atropis phryaganodes, Calamagrostis deschampsioides*) and occupies the northern part of the Taimyr Peninsula and coastal areas to the east.

Typical Tundra, which includes most of the northern Siberian lowland, is predominantly Hypnum mosses, lichens (*Cetraria cuculata, C. crispa, Cladonia rangiferina, Cl. mitis, and Cl. alpestris* in the western part and *Alectoria nigricans* and *Al. ochroleuca* in the eastern part), dryad (*Dryas punctata*), saxifrages (*Saxifraga caespitosa, S. nivalis*), and low shrubs (*Betula exilis, Salix polaris, Cassiope tetragona, Ledum decumbens, Vaccinum uliginosum, Vac. vitis-idaea*) and occupies the area south of the Arctic Tundra.

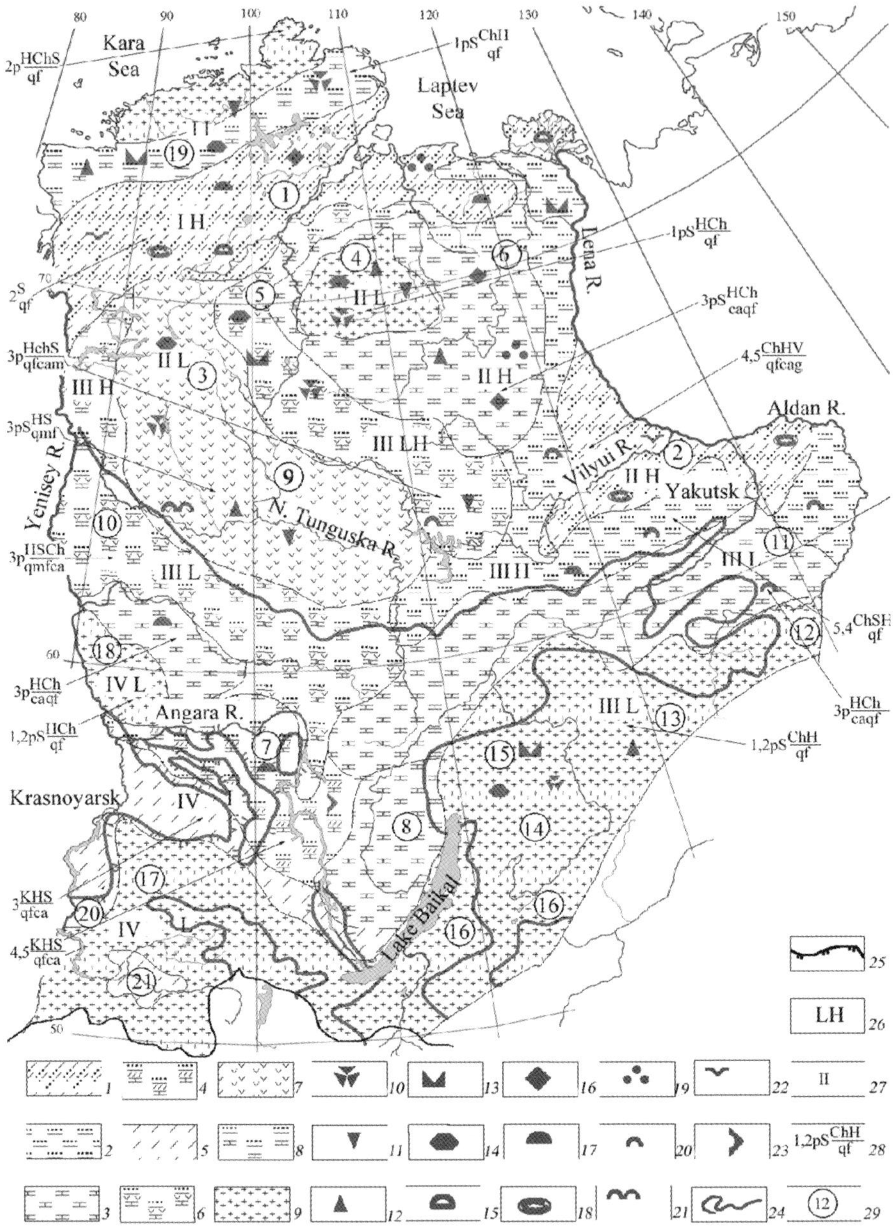

Figure 2.13.1. Lithogeomorphic basis of soil formation in Central Siberia. See the "Legend to Figure 2.13.1" on the next page.

Legend to Figure 2.13.1.

Lithological-geomorphological regions:

(*1*) *aggradational plains and lowlands* (<200 m a.s.l.) composed of loose Quaternary deposits of various geneses;

***Denudation plains and plateaus* (200 to 600 m) composed of sedimentary rocks:**

(*2*) Mesozoic noncalcareous (sandstone, slates, argillite, siltstone),
(*3*) Late Proterozoic and Cambrian calcareous (limestone, dolomite),
(*4*) Early Paleozoic calcareous and noncalcareous (marl, argillite, siltstone, sandstone) overlain by Quaternary covering loams of uncertain genesis,
(*5*) Paleozoic calcareous and nonclacareous overlain by Quaternary loesslike loams of considerable thickness;

***Denudation plateaus* composed of**

(*6*) Paleozoic calcareous and noncalcareous rocks with numerous trappean intrusions (400 to 800 m),
(*7*) Triassic trappean rocks (600 to 1200 m),
(*8*) *denuded neotectonic plateaus and ridges* composed of Paleozoic and Mesozoic calcareous and noncalcareous rocks (600 to 1000 m);
(*9*) *folded ridges and tablelands* (400 to 2000 m) composed of Archean, Proterozoic, and Paleozoic gneiss and igneous rocks of predominantly acid composition.

***Cryogenic landforms*:**

(*11*) kurums (stone fields and streams), (*12*) dells,
(*13*) altiplanation terraces, (*14*) sorted polygons,
(*15*) barren spots, (*16*) polygonal grounds,
(*17*) hummocks, (*18*) thermokarst depressions and lakes,
(*19*) small thermokarst depressions, (*20*) pingos,
(*21*) peat mounds, (*22*) ice-wedged polygons,
(*23*) relic polygonal cryogenic topography.

***Permafrost features*:**

(*24*) southern boundary of continuous permafrost,
(*25*) southern boundary of discontinuous and isolated permafrost;
(*26*) ice content (L, low (<25%); H, high (25 to 60%));
(*27*) summer thawing depth (I, <50 cm; II, 50 to 100 cm; III, 100 to 200 cm; IV, >200 cm).

***Texture and mineralogy* of soil-forming rocks:**

(*28*): texture
(1, sands and loamy sands;
(2, sandy loams and loams;
(3, clay loams and clays;
(4, sandy loams underlain by clay loams;
(5, stratified; index "p" indicates the presence of pebble and coarser material; index "s" denotes the underlying by hard bedrock at a depth <1.0); clay minerals (K, kaolinite; Ch, chlorite; H, hydromica; S, smectite; V, vermiculite); coarse fraction minerals (q, quartz; f, feldspar; m, weatherable minerals of mafic rocks; ca, calcareous; g, gypsum).

(29) *Geographic names* of the main morphostructures:

lowlands:
(1) North Siberian,
(2) Central Yakutian; plateaus:
(3) Putorana,
(4) Anabar,
(5) Kotui,
(6) Olenek,
(7) Angara,
(8) Angara-Lena,
(9) Syverma,
(10) Tungus,
(11) Lena-Aldan;

tablelands and ridges:
(12) Aldan Tableland,
(13) Stanovoi Ridge,
(14) Vitim Plateau,
(15) Stanovoi Highland,
(16) Trans-Baikal Mountains,
(17) Eastern Sayan Ridge,
(18) Yenisei Ridge,
(19) Byrranga Ridge; intermontane depressions:
(20) Minusinsk,
(21) Tuva).

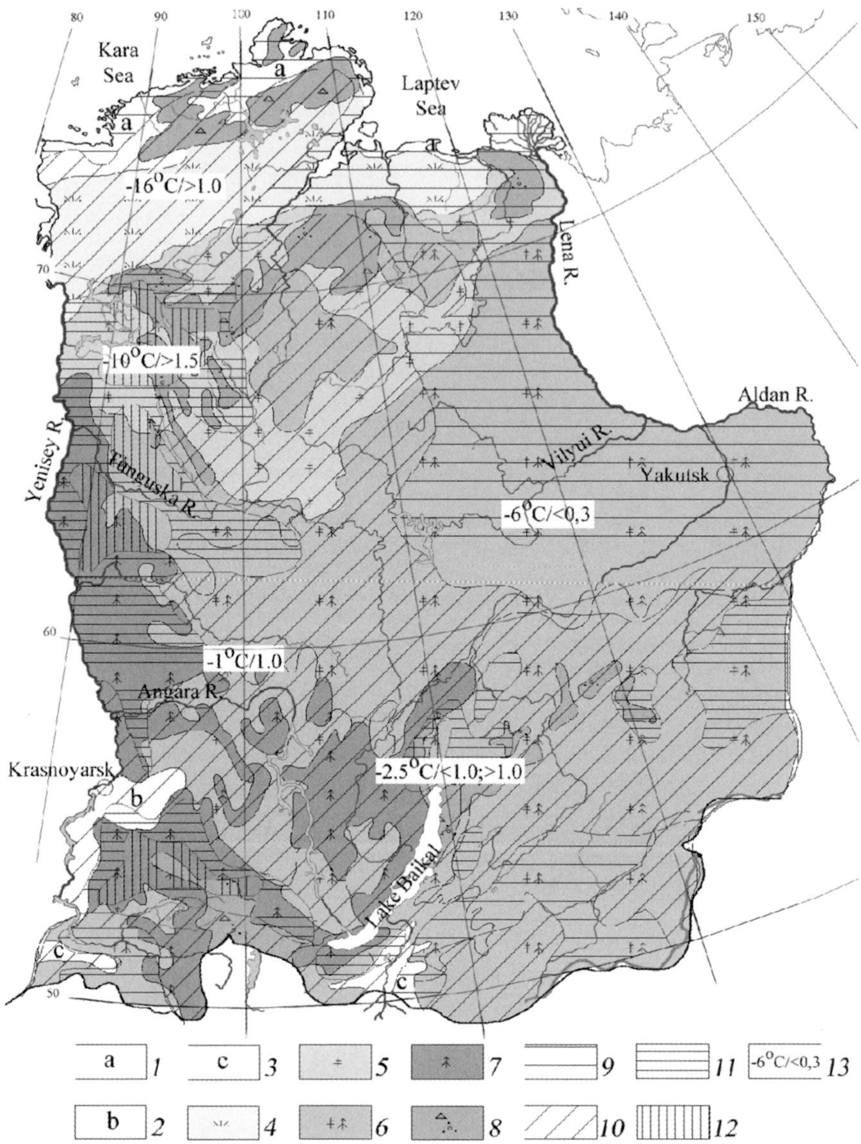

Figure 2.13.2. Schematic map of bioclimatic conditions in Central Siberia. Vegetation: (*1*) dwarf shrubs and herbs of arctic tundra and subpolar deserts, (*2*) grasses and herbs of meadow steppes, (*3*) grasses of dry steppe, (*4*) shrub moss-lichen tundra with polygonal mires, (*5*) open larch forests, (*6*) larch-pine and mixed (dark and light) coniferous forests, (*6*) dark coniferous forests, and (*8*) mountainous stony and moss-lichen tundra with shrubs and dwarf shrubs. Annual precipitation (mm): (*9*) <200 to 400, (*10*) 400 to 600, (*11*) 600 to 800, and (*12*) >800. (*13*) Mean annual temperature (°C) and humidity factor (precipitation/evapotranspiration).

Southern (Shrubby) Tundra, with better manifested and higher shrubs, also appears within the North Siberian Lowland. Cotton grass (*Eriophorum vaginatum, E. angustifolium*) often covers thermokarst depressions.

Thin **Subarctic Forests** alternating with tundra patches (**the Forest-Tundra zone**) occupy the southernmost part of the North Siberian Lowland and the northern part of the Central Siberian Plateau (to the north of the Polar Circle). At high altitudes, **mountainous stony tundra** and virtually barren rock outcrops predominate.

Boreal Forest (Taiga) vegetation occupies the rest of the territory. In the western part, to the south of Nizhnaya Tunguska, along with larch (*Larix sibirica*) forests, dark conifers (*Picea obovata* and *Abies sibirica*), pine trees (*Pinus silvestris*), and Siberian cedar (*Pinus sibirica*) cover considerable areas. Birch (*Betula platyphylla*), alder (*Alnaster fruticosus*), and aspen (*Populus tremula*) form secondary forests in places of forest fires.

In the eastern part of the territory, *Larix dahurica* is the dominant tree species; pine forests often occupy well drained positions. The most contrastive vegetative cover is typical of the Central Yakutian Lowland. Numerous thermokarst depressions, often with thermokarst lakes in their centers, dictate the concentric distribution of vegetation. Grassy and sedge-grassy meadow and meadow-bog communities occur in the center of dried depressions around the lakes. In this zone with arid climate, the soils often contain soluble salts. Alkali grass (*Pucinellia distans*) is a common species.

Meadow-steppe or typical steppe vegetation with meadow grass (*Poa stepposa.*), feather grass (*Stipa capillata*), fescue (*Festuca lenensis*), and wormwood (*Artemisia pubescence*) occupy somewhat higher positions. Larch taiga with grassy cover encircle these areas. On loesslike rocks, larch forests with lingonberry (*Vaccinium vitis-idaea*) predominate on the interfluves. Pine or larch-pine forests occupy better drained sites with sandy deposits (river terraces).

Steppe vegetation also occurs on southern slopes of Trans-Baikal mountains and in the south of Central Siberia, beyond the permafrost zone.

2.5. Cryological conditions

All the soils of Central Siberia undergo long-term (>8 months) freezing. The active layer depth in tundra and taiga areas varies from 0.2 (0.3) to 0.6 (0.8) m in peat soils, to about 1.0 m in loamy clayey mineral soils, and to 1.0 to 1.5 m in coarsely textured sandy loamy and sandy soils. In the south, the depth of thawing increases to 1.0 to 1.5 m in peat soils and heavily textured mineral soils, and to 2.0 to 2.5 m in coarsely textured mineral soils. Plain and mountainous tundra, forest-tundra, and northern taiga lie within the zone of continuous permafrost.

Discontinuous permafrost is typical of the middle taiga subzone. In the southern taiga and steppe regions, permafrost occurs in relatively small isolated areas confined to local depressions with clay loamy substrates and northern slopes under forest vegetation. Permafrost is absent or is deeper than 2.5 to 3.0 m on slopes with a southern aspect and on the interfluves with a relatively thin mantle of loose loamy rocks. The ice content in the upper part of the permafrost varies from 5 to 10% of volume in sandy and pebbly rocks to 50 to 70% of volume in loamy-clayey substrates.

The development of ice wedges, frost cracking, cryogenic sorting, frost heave and cryogenic involutions (cryoturbation), and thermokarst processes mainly dictate the surface topography and soil cover patterns in permafrost-affected areas. Virtually all forms of cryogenic microtopography occur in Central Siberia.

Various sorted nets and 'stone seas' are typical of the highest plateaus with a thin mantle of loose rocks. These are virtually barren areas with rock outcrops and poorly developed soils along cryogenic fissures. Earth hummocks predominate in tundra, forest-tundra, and northern taiga zones on the interfluves. Large low-centered polygons bordered by ice wedges are common on the floodplains and river terraces. Thermokarst depressions (alases) are widespread in Central Yakutia (Figure 2.13.1).

3. Soils

The diversity of soil-forming factors in Central Siberia, especially the sharp climatic differences (from the extremely humid western Putorana Plateau to the semiarid and arid Central Yakutia), and a wide spectrum of parent rocks, together dictate the complexity of the soil cover at the macrolevel. At the microlevel, Cryogenic microtopography causes the specific complexity of soil cover patterns. Recently, we compiled a series of small-scale (1:16 M) soil maps on the territory of Central Siberia. They are based on the 1:2.5 M scale *Soil Map of the Russian Federation* but take into account some recent changes in soil nomenclature and present information in a generalized form.

This series consists of four maps: a lithogenetic map (a detailed characterization of soil forming rocks; some information appears in Figure 2.13.1) a pedogenetic map (Figure 2.13.3), and maps of soil temperature and soil water regimes. The legend to the pedogenetic map groups soils into several classes of soil formation.

3.1. Eluvial-Illuvial Al-Fe-Humus soils (Podzols, Spodosols)

3.1.1. Podzols and Parapodzols

Podzols and Parapodzols are acid soils with distinct chemogenic differentiation of sesquioxides in the soil profile and their accumulation in the form of amorphous Al-Fe-humus compounds in the Bh or Bhs horizon. In pebbly and stony soils, Al-Fe-humus compounds tend to accumulate on lower sides of pebbles, forming very distinct dark brown coatings, whereas the upper sides of pebbles are often bleached and strongly corroded.

If the eluvial E (albic, bleached mineral) horizon is present, these soils are referred to as *Podzols* (Haplic Podzols, WRB). If the Bh horizon lies immediately under the peaty litter or raw-humus (Oi or AOi) horizons, the soils are classified as *Podburs* (Entic Podzols) (Targulian, 1971).

Taking into account the genetic similarity of Podzols and Podburs, we suggest that the latter can be named *Parapodzols*. These soils are widespread in cold humid climate on well drained, coarsely textured (sandy or sandy loams) derivatives of acid silicate rocks under northern taiga and tundra vegetation. They have a shallow profile; as a rule, the permafrost table is found beneath the soil profile.

The further division of Podzols and Parapodzols can be based on the character of surface organic and/or organomineral horizons and on the degree of accumulation and particular composition of Al-Fe-humus compounds in the illuvial horizon.

In extremely humid conditions, the rate of litter accumulation on the surface is much higher than the rate of decomposition of plant remains, and a thick (10 to 30 cm) litter layer forms on the surface. These soils are distinguished as *Dry-Peaty (Folic) Podzols and Parapodzols*. Often, they develop under dwarf pine (*Pinus pumila*) forests.

In conditions of the permafrost zone, the further increase in the thickness of peaty horizons leads to a rapid decrease in the depth of soil's summer thawing and the ascending movement of the permafrost table. As a result, permafrost comes closer to the surface; often it is immediately under or within the peat layer and the soil transforms into a Foli-Cryic Histosol. However, dry-peaty soils are very susceptible to fires that prevent their transformation into Histosols.

The usual thickness of litter horizons in Podzols and Parapodzols is about 5 to 10 cm. The soils with considerable accumulation of amorphous sesquioxides in the Bhs horizon are distinguished as *Dark Podzols and Parapodzols*; these soils fit the definition of Spodosols in *Soil Taxonomy*.

The soils with low content of amorphous sesquioxides in the Bhs horizon fall into the category of *Light Podzols and Parapodzols*; they do not meet quantitative criteria of spodic horizon and can be distinguished at the level of spodic subgroups of Inceptisols and Entisols.

Podzols and parapodzols predominate in the soil cover of mountainous taiga and tundra in the Trans-Baikal region, mainly in the upper and middle parts of slopes of northern aspect, in well drained geomorphic positions. They develop from sandy or moderately coarsely textured loamy derivatives (of colluvial or alluvial origin, or from bedrock residuum) of granites.

Figure 2.13.3. Schematic soil map of Central Siberia. See the "Legend to Figure 2.13.3" on the next page.

Legend to Figure 2.13.3.

Soil sectors:

(1) with the predominance of acid unsaturated eluvial-illuvial and gley soils of humid environments,
(2) transitional, and
(3) with the predominance of neutral to alkaline accumulative and metamorphic (including thixotropic cryozems) soils of semiarid to arid environments;
(4) areas with the predominance of hydromorphic soils;
(5) typical soil combinations (+ sign) and soil microcomplexes (separated by a period).

Soils:

(1) Slightly gleyed humus-accumulative soils of arctic tundra,
(2) humus-illuvial and dry-peat podzols,
(3) parapodzols (including dry-peat parapodzols),
(4) ocherous parapodzols,
(5) granuzems,
(6) raw-humus brown taiga sois,
(7) soddy brown taiga soils,
(8) typical pale soils,
(9) pale calcareous soils,
(10) pale-podzolic soils,
(11) pale solodic soils,
(12) pale-cryozemic soils,
(13) raw-humus and soddy rendzinas,
(14) podbels and soddy podbels,
(15) gray forest soils,
(16) chernozems and meadow-chernozemic soils,
(17) chestnut soils,
(18) peat and raw-humus gley soils,
(19) thixotropic cryozems,
(20) mucky-peaty (histic) cryozems,
(21) peat and peat gley bog soils,
(22) peat soils of cryogenic troughs,
(23) soils of barrens (including surface-calcareous soils),
(24) primitive stony mountainous soils (lithosols) and rock outcrops,
(25) sandy regosols (including sandy alluvial soils).

Example 2.13.1. Pit 126 (Aug. 26, 1966). Vitim Plateau, 1200 m a.s.l., upper part of a 5° slope of northern aspect. Al-Fe-humus Podzol (Sokolov, 1970).

Larch taiga with bog bilberry, mountain cranberry, wild rosemary, and lichens in the ground cover. Large granite boulders on the surface.

Oi, 0 to 8 cm. Loose gray-brown forest litter, densely penetrated by roots, weakly decomposed, charcoal particles, admixture of strongly bleached mineral particles in the low part, gradual transition, wavy boundary.

E, 8 to 18 cm. Whitish, with gray tint, slightly dry, sandy loam, with admixture of granite gravel and pebble, bleached mineral particles, rock fragments are bleached from the top and covered by dark brown coatings from the bottom, few fine live roots, sharp transition, wavy boundary.

Bh, 18 to 35 cm. Bright ocherous-brown (after calcination, reddish ocherous), wet, loamy, with abundant rock fragments, fine earth is aggregated into finc (0.8 mm) granules; all mineral particles are covered by opaque brownish films, single roots, gradual transition, wavy boundary.

BC, 35 to 50 cm. Bright brown, wet, stony, loamy fine earth, silt cappings on rock fragments; the amount of stony material increases downward; no permafrost.

In the same zone, the soils without a distinct bleached horizon (*Parapodzols*) occur. The tendency is toward the prevalence of Podzols with the eluvial horizon in the upper parts of slopes receiving higher precipitation. The accumulation of sesquioxides in the Bh (Bhs) and BC horizons is evident from the results of both total elemental analysis and oxalate extraction (Table 2.13.1).

Amorphous silica (oxalate extraction) also accumulates in the Bh horizon. At the same time, the maximum bulk SiO_2 content occurs in the E horizon, which the residual accumulation of stable quartz minerals explains. Another specific feature of Trans-Baikal Al-Fe-humus soils is a high humus content in the E (3 to 8%) and Bh (1.6 to 4.5%) horizons. The depth of permafrost thawing exceeds the thickness of soil solum.

Podzols and Parapodzols also occur in the northern (the Ananbar Shield) and western parts of the Central Siberian Plateau.

Table 2.13.1. Properties of permafrost-affected soils in Central Siberia.

Horizon	Depth, cm	pH_w	Humus, %	Exchangeable, meq/100 g soil		Particles (mm), %		Total elements, % on ignition			Fe_2O_3 by Tamm, %
				Ca^{2+}+ Mg^{2+}	H^+	< 0.01	< 0.001	SiO_2	Fe_2O_3	Al_2O_3	
Pit 126. Al-Fe-humus Podzol on the residuum of granite. Trans-Baikal region (Sokolov, 1970).											
Oi	0-8	4.8	65.6[(1)]	23.3	5.9	-	-	64.49	4.19	22.52	0.24
E	8-18	4.1	7.5	2.9	7.7	26	8	72.04	2.50	17.53	0.09
Bh	18-40	5.0	4.2	2.3	7.8	37	15	67.29	5.56	18.79	2.91
BC	40-50	5.5	4.1	6.9	5.1	40	18	-	-	-	0.77
Pit 5. Ocherous Parapodzol on the derivatives of trappean rocks. Putorana Plateau (Sokolov & Gradusov, 1978.)											
Oi	0-8	4.3	78.4[(1)]	-	-	-	-	46.2	16.6	17.4	-
AO(E)	8-15	4.3	62.0[(1)]	11.0	4.0	-	-	48.0	15.8	17.6	1.3
Bh	15-20	5.2	11.8	4.2	7.9	18	2	43.3	19.3	20.7	2.6
Bh	20-30	5.6	3.6	4.1	0.9	30	20	46.7	14.4	20.6	1.3
BC	30-40	5.8	0.8	5.0	0.3	10	3	48.1	13.9	17.9	0.6
C(D)	80-85	-	-	-	-	8	4	49.4	13.5	16.8	0.7

(continues)

Table 2.13.1. (cont.)

Horizon	Depth, cm	pH_w	Humus, %	Exchangeable, meq/100 g soil: Ca^{2+}+ Mg^{2+}	Exchangeable, meq/100 g soil: H^+	Particles (mm), %: < 0.01	Particles (mm), %: < 0.001	Total elements, % on ignition: SiO_2	Total elements, % on ignition: Fe_2O_3	Total elements, % on ignition: Al_2O_3	Fe_2O_3 by Tamm, %
Pit 19. Granuzem on loamy clay derivatives of trappean rocks. Putorana Plateau (Sokolov & Gradusov, 1981).											
Oe	0-5	4.9	34.7[(1)]	22.1	13.0	-	-	57.00	10.09	17.35	1.05
ABwh	5-15	5.3	6.4	20.3	12.5	57	28	58.53	10.39	18.26	1.30
Bwh	15-25	5.3	3.1	22.9	3.4	49	38	59.21	10.95	18.77	1.35
A_{tongue}	20-30	5.7	5.3	23.3	2.7	51	19	58.17	10.49	17.71	1.71
Bw	40-50	6.2	1.0	27.9	0.7	65	30	59.19	10.55	18.10	1.22
Bw	55-65	7.4	1.0	41.4	0.0	34	17	61.64	9.45	17.80	0.94
Bg	80-90	7.0	1.2	26.4	0.0	52	30	59.68	10.09	18.16	0.78
Pit 58. Palevaya (Pale) soil on solifluction derivatives of calcareous slates. Syverma Plateau (Sokolov, 1986).											
Oie	0-8	5.4	79.7[(1)]	49.5	6.8	-	-	-	-	-	-
Oe	8-12	5.5	52.2[(1)]	52.8	5.5	-	-	66.82	6.60	15.05	-
AOe	12-20	5.6	5.2	38.3	1.0	52	33	69.10	6.70	15.81	1.02
Bw	20-30	6.3	0.8	35.0	0.2	40	26	70.06	6.53	15.03	0.98
BC	40-50	6.7	0.5	25.2	0.0	28	16	74.49	4.90	12.84	1.00
Ckf	60-70	7.9	0.6	-	-	40	20	69.74	5.73	15.81	0.90
Ckf	70-80	7.9	1.0	-	-	54	26	63.37	7.58	18.69	0.88
Tongu	20-30	6.2	2.0	38.5	0.3	54	32	68.88	6.43	16.72	0.71
e	40-50	6.4	1.5	41.4	0.0	53	32	69.29	6.54	16.38	0.85
Pale-Podzolic loamy-sandy soil (averaged data on six pits). Vilyui basin, Central Yakutia (Sokolov et al., 1974).											
OAE	2-4	5.4	2.8	13.0	0.7	10	5	80.3	1.8	9.5	0.32
AE	4-10	5.6	1.5	6.4	0.6	9	5	82.3	1.7	9.8	0.33
EBt	10-16	5.6	0.8	7.3	0.3	13	7	81.5	2.0	10.1	0.30
Btw	16-32	5.8	0.5	10.6	0.3	16	12	78.3	2.9	12.5	0.30
BC	32-42	6.0	0.3	10.0	0.2	9	7	82.4	2.4	9.9	0.26
C(D)	>42	6.3	0.2	7.6	0.1	6	4	81.4	1.9	10.0	0.23
Pit 1065. Pale Calcareous soil on calcareous loesslike loam. Central Yakutia (Sokolov et al., 1976).											
AO	2-8	6.1[(3)]	26.9[(1)]	49.9	0.4	40	21	68.18	4.34	14.96	0.70
A1Bm	8-17	7.2	3.2	26.8	0.1	40	21	68.80	4.60	17.36	0.40
Bk	17-28	8.7	1.8	-	-	33	19	63.97	4.61	16.94	-
BCk	30-40	8.7	1.1	-	-	34	19	67.48	4.31	16.23	-
Ck	45-60	8.5	1.0	-	-	34	19	67.96	4.00	15.36	-
Pit 1066. Pale Solodic surface-gleyed soil on calcareous loesslike loam. Central Yakutia (Sokolov et al., 1976).											
O	2-6	4.4[(3)]	-	39.9	10.2	-	-	-	-	-	-
AO	6-9	4.5[(3)]	34.5[(1)]	35.4	8.4	40	22	69.77	3.98	15.18	1.10
AEg	9-23	6.2[(3)]	1.3	10.3	0.1	32	10	70.87	4.06	14.79	0.33
Btw	23-34	8.0	2.0	-	-	51	33	65.82	5.79	17.36	0.30
Bk	50-65	8.7	1.2	-	-	38	21	64.34	4.92	16.94	-
Pit 249. Meadow-Chernozemic soil on covering loesslike loam. Trans-Baikal steppe (Sokolov, 1967).											
A	0-10	7.4	5.9	40.0[(2)]	-	58	33				
A	20-30	7.4	4.5	39.1[(2)]		58	36				
A/B	45-55	7.9	2.8	38.0[(2)]		60	39				
A/Bk	55-65	8.2	1.8	26.6[(2)]		61	33				
Bk	80-90	8.2	0.9	25.9[(2)]		57	37				
Bk	110-120	8.2	-	-		-	-				
CDg	150-160	8.1	-	26.2[(2)]		54	33				
Dg	190-200	8.1	-	27.4[(2)]		37	23				
D	290-300	7.8	-	-		26	15				
Tongue	0-7	8.1	4.8	34.6[(2)]		56	36				
Tongue	7-15	8.1	3.2	29.0[(2)]		61	35				
Tongue	15-25	8.0	2.0	23.9[(2)]		57	37				
Tongue	32-42	8.0	1.4	24.4[(2)]		56	36				
Tongue	55-65	8.2	1.2	22.4[(2)]		57	36				

(continues)

Table 2.13.1. (cont.)

Horizon	Depth, cm	pH_w	Humus, %	Exchangeable, meq/100 g soil: Ca^{2+}+ Mg^{2+}	Exchangeable, meq/100 g soil: H^+	Particles (mm), %: < 0.01	Particles (mm), %: < 0.001	Total elements, % on ignition: SiO_2	Total elements, % on ignition: Fe_2O_3	Total elements, % on ignition: Al_2O_3	Fe_2O_3 by Tamm, %
Pit 71. Meadow permafrost-affected soil on colluvial clay loams. Trans-Baikal region (Sokolov, 1967).											
A	0-20	6.8	7.9	45.3	0.2	53	31	68.98	5.26	17.92	0.49
A	22-32	6.4	2.2	27.3	0.1	46	31	69.58	5.59	18.53	0.32
A/B	40-50	6.2	1.1	24.1	0.2	48	31	69.34	5.72	17.72	0.40
B	65-75	5.6	0.9	24.8	0.2	50	31	69.39	5.23	19.27	0.46
B	90-100	6.2	-	26.6	0.2	50	33	-	-	-	0.38
Bg	120-130	6.4	0.7	29.4	0.1	57	35	67.35	4.25	22.15	0.44
BCg	150-160	6.5	0.8	34.5	0.1	56	36	-	-	-	0.37
C	180-190	6.6	0.6	26.9	0.1	48	31	69.22	5.62	18.75	0.32
C	225-235	6.8	0.8	26.1	0.1	52	33	68.83	6.22	19.08	0.39
Pit. 74. Mucky-Calcareous permafrost-affected soil on limestone. Lena-Aldan interfluve (Gerasimov, 1965).											
AOe	2-8	7.5	12.1	52.2[(2)]	-	82	50				0.55
Ak	20-25	7.7	1.9	23.7[(2)]		83	42				0.33
BCkg	45-50	7.7	0.4	15.1[(2)]		79	43				0.31
Pit 253. Tundra Gley soil (trough/hummock with barren surface). Southern Taimyr (Vasil'evskaya, 1980).											
AO	0-10	-	43.6[(1)]	73.8	-	-	-	-	-	-	0.91
Bg	10-20	6.0	2.98	29.6	4.0	-	-	-	-	-	1.57
G	20-37	7.0	2.26	32.2	3.6	-	-	-	-	-	1.44
Bg	0-10	6.7	2.53	22.0	2.2	60	29	69.70	6.97	14.65	1.44
G1	10-30	5.6	2.45	27.0	5.9	60	26	69.96	5.84	14.81	0.74
G2	30-50	6.4	2.19	32.4	4.0	63	29	69.92	6.28	17.30	1.44
Pit 214. Histic Cryozem on the derivatives of trappean rocks. Putorana Plateau (Sokolov, 1980).											
Oi	0-8	5.8	43.0[(1)]	135.1	0.8	-	-	-	-	-	-
BOjj	8-18	6.0	8.7	32.3	0.6	15	9	48.76	12.35	17.58	2.10
BOjj	20-30	5.8	8.9	39.5	0.6	22	13	48.34	12.45	18.56	1.82
Pit 312. Thixotropic Cryozem on the derivatives of calcareous slates and tuff rocks. Central Siberia (Sokolov, 1980b).											
OeA1	10-20	6.6	34.4	-	-	-	-	-	-	-	0.40
BwA1	20-30	6.7	4.9	25.9[(2)]	-	41	24	69.85	5.88	13.40	0.56
Bwtix	30-40	6.3	2.3	17.2[(2)]	-	42	25	70.20	5.83	13.08	0.46
Pit 73. Thixotropic Cryozem on the derivatives of tuff and trappean rocks. Central Siberia (Sokolov, 1980b).											
Oi	0-5	5.8	66.2[(1)]	90.5	2.0	-	-	-	-	-	-
AOe	5-10	6.0	27.0[(1)]	51.6	0.4	44	23	54.66	11.50	19.31	0.89
BA	10-15	6.3	5.2	47.7	0.2	42	24	56.09	11.05	19.30	1.04
Bwtix	45-55	6.9	2.0	43.2	0.04	45	23	55.58	10.67	18.62	0.70
Bwtix	60-70	7.0	2.3	46.8	0.04	45	20	55.90	10.77	18.24	0.89

[(1)] Loss on ignition
[(2)] Cation exchange capacity.
[(3)] pH of salt (KCl) suspension.

3.1.2. Ocherous Podzols and Parapodzols

Ocherous Podzols *and* ***Parapodzols*** (Vitric Podzols and Vitri-Entic Podzols) (including Dry Peaty subtypes) are widespread in the northwestern part of Central Siberia, within the Putorana, Syverma, and Tungus plateaus composed of trappean rocks. This is the only territory in Russia where mafic intrusive and effusive rocks compose very large plateaus; plagioclases (40 to 60%) and monoclinic pyroxenes (20 to 40%) predominate in their mineral composition. The rocks also contain some amount of vitreous material. Clay minerals include chlorites and chlorite-smectites. The contents of quartz and potassium feldspar are very low.

Ocherous Parapodzols and Podzols develop in well drained sites from coarsely textured residuum or colluvium of dolerites under thin larch taiga with small shrubs and moss-lichen ground cover at altitudes up to 450 m a.s.l. They have a shallow (50 to 80 cm) profile underlain by stony material; the depth of permafrost is about 150 cm.

A typical profile of these soils is: Oi (5 to 15 cm), AO(OE) (0 to 5 cm), Bhs (10 to 30 cm), BC(R) (10 to 30 cm). The thickness of mineral horizons distinctly transformed by pedogenetic process is very low (20 to 30 cm). As a rule, the E horizon is indistinct or has very low thickness. Often, this horizon has a rather dark color because of the residual accumulation of pyroxenes.

The accumulation of amorphous sesquioxides and even allophanes in the Bhs horizon is very significant. *In situ* weathering of plagioclases in the Bhs horizon also contributes to the accumulation of amorphous sesquioxides. The latter serve as active structuring agents, which results in perfect microaggregation of fine earth fractions (Sokolov and Gradusov, 1978).

3.2. Illuvial-Metamorphic soils

Granuzems (Eutric Cambisols) are highly specific soils that develop from loamy-clayey derivatives of trappean rocks in conditions of the northern and middle taiga zones. The prepedogenic stage of their hydrothermal transformation can enhance the release of clay material from trappean rocks. The breakage of coarse mineral grains that serve as containers of clay material may lead to production of clay minerals (Sokolov & Gradusov, 1981).

Granuzems develop in drained positions on river terraces (near the brows) and on colluvial fans at the footslopes under larch taiga with mossy-lichen or mossy-herbaceous ground cover. As a rule, the soils developing from finely textured rocks in conditions of cold humid climate are subject to strong gleyzation; in the permafrost zone, they are very susceptible to cryoturbation. However, these processes are weakened in Granuzems.

Example 2.13.2. Pit 19 (July 18, 1968). Southwestern part of the Putorana Plateau, terrace of the Severnaya River, 200 m a.s.l., 25 to 30 m above the water level. Granuzem (Sokolov & Gradusov, 1981).

Spruce-larch taiga with admixture of birch; blueberry, bilberry, geranium, Galium, green mosses, and lichens in the ground cover. A network of cryogenic troughs, forming a polygonal microtopography, dissects the surface. The thickness of a layer of live mosses is about 10 cm.

Oe, 0 to 5 cm. Dark brown loose forest litter, densely penetrated by roots, some admixture of bleached mineral grains; sharp transition.

ABwh, 5 to 15 cm. Gray-brown, loose, fine granular, slightly dry; there are no films on mineral grains; abundant decomposed plant tissues, fine roots, gradual transition.

Bwh, 15 to 35 cm. Brown, loose, granular (the size of aggregates is about 1 mm), mineral grains are colored by brown films.

Bw, 35 to 65 cm. Dark brownish, slightly compact, with angular blocky structure; fine earth material is cemented by clay; microaggregates are about 0.5 mm in size; wet; gradual transition.

Bg, 65 to 100 cm. Grayish brown, with dull brown mottles; compact; indistinct platy structure; wet; contains pebble and coarser fragments of mafic rocks.

In the zone of cryogenic troughs, frost fissures penetrate to the depth about 50 cm. These fissures are filled with humified plant remains in the upper part and a mixture of plant remains and brown fine earth in the bottom.

The most characteristic feature of Granuzems is their perfect aggregation and microaggregation in the Bwh horizon. The physical clay (<0.01 mm) content is about 50%; however, the sum of fractions <0.01 mm during the microaggregate analysis is just about 10%. There is no eluvial-illuvial redistribution of clay; no optically oriented clay is registered in thin sections.

Granuzems are acid to strongly acid in the upper (O, Bwh) horizons and neutral (pH 7.0 to 7.4) in the Bw and BC(g) horizons. The saturation degree changes from 70 to 100%. The abundance of amorphous sesquioxides (2 to 4% Fe_2O_3, 4 to 6% Al_2O_3 [% of ignited soil weight]) dictates their perfect coagulative structuring in the Bwh and Bw horizons; in the Bg horizon, the development of platy structure is due to cryogenic processes.

Soil structuring improves the internal drainage of the soil profile and impedes the development of gleyzation and cryoturbation. The essence of weathering processes in Granuzems is similar to that in Ocherous Parapodzols: intense selective weathering of plagioclases and vitreous minerals, leaching of silica and calcium, active residual accumulation of silica (in pyroxenes), manganese, and titanium. and very high content of oxalate-extractable iron, aluminum, and silica in the Bws horizon (Sokolov & Gradusov, 1981).

In warmer and less humid conditions, Burozems (Brown Forest soils) replace Granuzems; in semihumid to semiarid conditions, Palevye (Pale) metamorphic soils replace them; in colder environments (tundra), Cryozems do (see below).

3.3. Metamorphic soils

3.3.1. Raw-Humus and Soddy Brown Forest soils (Burozems)

Raw-Humus and Soddy Brown Forest soils ***(Burozems)*** develop in conditions of somewhat warmer and less humid climate (middle and southern taiga zones) from loamy or clay-loamy derivatives of silicate rocks, in the basins of Podkamennaya Tunguska and Angara Rivers, and in the mountains of Eastern Sayan Ridge. These are acid to weakly acid, slightly unsaturated soils, with the accumulative pattern of humus distribution and weak features of illuviation of humus and sesquioxides.

Their typical profile is: Oi-AO-(A1)-Bw-BC-C. Predominant *in situ* weathering of minerals results in the development of autochthonous films on the surface of mineral particles in the Bw horizon with typical brown color. Amorphous (oxalate-extractable) sesquioxides predominate in this horizon. However, in contrast to Al-Fe-humus soils, there is no maximum of amorphous sesquioxides in the Bw horizon.

Raw-Humus Burozems (Dystric Cambisols) tend to develop from relatively poor rocks in somewhat cooler climate under mossy taiga. *Soddy Burozems* (Umbric Cambisols) occur in warmer conditions, under richer vegetation with participation of herbs in the ground cover. The enrichment of the substrate in weatherable minerals (if the substrate contains the admixture of mafic and/or calcareous minerals) also contributes to the development of mull-like humus horizons. The main area of these soils lies beyond the permafrost zone, though isolated patches of permafrost occur under peatlands.

In similar conditions, Al-Fe-humus podzols develop from the derivatives of quartz sandstone; and texture-differentiated Luvisols (Podzolic and Soddy-Podzolic soils, according to the legend of the *Soil Map of Russia*) with clay-depleted E horizon and clay-enriched Bt horizon, develop from covering silty clay or loesslike loams.

The development of both Burozems (Cambisols) and Albeluvisols from similar rocks in similar bioclimatic conditions (often, it is impossible to predict the presence or absence of textural differentiation in the soil profile) is hardly probable. The origin of textural differentiation of Luvisols in this case connects with initial heterogeneity of covering loams; i.e., textural differentiation in them has a lithogenic origin (Sokolov, 1988). Our map distinguishes these soils as *Podbels.*

3.3.2. Podbels

Podbels *(Eluvial-Metamorphic soils)* develop in the southwestern part of Central Siberia, in the basin of the Angara River (minor areas, within the low reaches of the Podkamennaya Tunguska River) and at the foothills of Eastern Sayan, under larch, spruce, and secondary birch forests with rich (with considerable share of herbs and grasses) ground cover. In Central Siberia, texturally differentiated acid to slightly acid soils usually have well manifested humus horizons (*Soddy Podbels*): (A-E-IIB(t)-IIBC(k)-IIC(k)).

The hypothesis of the lithogenic origin of textural differentiation in the Podbels (Soddy-Podzolic soils) of Central Siberia has yet to be tested. The name *Podbels* is given to these soils to distinguish them from Podzols proper (Al-Fe-humus soils) more clearly. The transformation of the initially differentiated profile of these soils mainly connects with humus accumulation, some leaching of carbonates (if present), and surface eluvial-gley bleaching of the E horizon (iron segregation in concretions). The main areas of Podbels occur beyond the permafrost zone.

3.3.3. Palevye (Pale) soils

Palevye (Pale) soils are a transitional group between Al-Fe-humus soils of humid climate and humus-accumulative Ca-illuvial ("steppe") soils of semiarid climate. Pale soils may have a raw humus or a typical humus-accumulative horizon underlain by a light brownish (pale) metamorphic Bw horizon with a high content of dehydrated crystallized iron compounds that form thin autochthonous films on the surface of mineral particles.

Pale soils develop from sandy and loamy (often, loesslike) loose rocks, as well as from loamy pebbly derivatives of silicate rocks in conditions of good drainage and deep thawing of permafrost. Initially, they were distinguished as the soils of middle and southern taiga developing from carbonate-containing loesslike loams (Zol'nikov, 1954). Afterward, Ivanova (1971) suggested that Pale soils are typical soils of ultracontinental taiga of Central Yakutia and develop not only from loess rocks but also from the other kinds of loose loamy substrates.

We distinguished several types of Pale soils (Sokolov et. al., 1974, 1976; Sokolov, 1986). Pale soils form intergrades with Al-Fe-humus soils (Unsaturated Pale soils and Pale-Podzolic soils), humus-accumulative Ca-illuvial soils (Pale Calcareous soils on calcareous loamy rocks), Solodic soils (Solodized Pale soils on clayey loams enriched in sodium), and hydromorphic Cryozems (Pale-Cryozemic soils on pebbly clayey rocks with shallow permafrost). Geographically, Pale soils predominate in the central and eastern parts of Central Siberia.

3.3.4. Typical Pale soils

Typical Pale soils (Eutri-Gelic Cambisols, Haplic Cryosols) develop from clay loamy colluvial or solifluction deposits (often, with coarse rock fragments), low-calcareous loesslike loams, and ancient alluvial sandy loamy substrates in tundra and taiga areas, in the middle basins of Nizhnaya and Podkamennaya Tunguska rivers and in the basins of Vilyui and Lena under tundra and taiga vegetation. Very often, the derivatives of mafic intrusive rocks that outcrop to the surface in upper parts of slopes enrich the surface deposits.

Example 2.13.3. Pit 58. Left bank of the Nizhnaya Tunguska River, 3 km to the west of Lake Askita; high flat (terrace-like) surface. Pale soil on solifluction deposits (Sokolov, 1986).

Larch taiga with bilberry, wild rosemary, mountain cranberry, green mosses, and lichens in the ground cover. A polygonal network of cryogenic troughs dissects the surface. Average distance between the troughs is about 1 m; trough width, 20 to 30 cm.

Oie, 0 to 12 cm. Weakly decomposed litter, densely penetrated by roots; admixture of mineral material in the lower part with intermediate decomposition of plant remains. Mineral grains are covered by thin brownish films.

AOe, 12 to 20 cm. Brownish gray, slightly dry, loose; abundant roots; plant remains of intermediate decomposition; silty loamy, loose fine subangular blocky structure, well manifested flocculated microstructure; mineral grains are covered by thin brownish films; abundant charcoal particles; gradual transition.

Bw, 20 to 40 cm. Brown, slightly dry, with loose fine subangular blocky structure; slightly sticky; mineral grains are uniformly covered by brown films; gradual transition.

BC, 40 to 60 cm. Lighter color; silty loamy, with admixture of sand; brown films on mineral particles; structureless; clear transition, even boundary.

Ckf, 60 to 80 cm. Light brown, moist; weak effervescence, no visible carbonate concentrations; ice-rich permafrost at 60 cm (July 20); no water stagnation above permafrost.

In the zone of cryogenic troughs, cryogenic fissures filled with a mixture of mineral and organic material penetrate to the depth of 100 cm. Live roots are very abundant; the degree of decomposition of plant remains in fissures is higher than in the upper AOe horizon; all mineral particles are covered by brown-gray films; abundant coprolites.

Typical Pale soils have a relatively homogeneous profile; there is no textural differentiation. There are no features of illuviation of Al-Fe-humus compounds. Brown films in the Bw horizon uniformly cover the surface of mineral particles. The content of oxalate-extractable iron is low, with a maximum in O and AO horizons. The ratio of dithionite-extractable to oxalate-extractable iron increases downward through the soil profile, which is conditioned by crystallization of amorphous iron compounds.

In spite of a heavy texture of these soils, they characteristically feature an absence of water stagnation above ice-rich permafrost and relatively weak cryogenic phenomena. Typical Pale soils develop from the rocks enriched in weatherable minerals. In conditions of northern taiga and tundra, Pale soils usually have raw-humus horizons. These soils are common in the eastern part of the Putorana plateau.

3.3.5. Pale Podzolic soils

Pale Podzolic soils (Gelic Luvisols) (eluvial-metamorphic soils) occur on low plateaus composed of sandstone and slates of Jurassic and Cretaceous periods. These soils develop from sandy or sandy-loamy residuum and colluvial derivatives of Mesozoic rocks, as well as from ancient sandy alluvial deposits impoverished in weatherable minerals. In the Vilyui and Lena basins, they formed under larch taiga vegetation.

Example 2.13.4. Typical Pale-Podzolic soils, from sveraged data from studied soil pits.

Oe, 0 to 2 cm. Thin loose forest litter.
OAE, 2 to 4 cm. Transitional horizon, very loose; consists of a mixture of undecomposed and humified plant tissues and light-colored mineral material, which is partly aggregated; the aggregates are glued by grayish humic substances; sandy particles are bleached.
AE, 4 to 10 cm. The most light-colored (whitish pale) horizon in the soil profile; indistinct platy structure; very loose; sand particles are bleached.
EBt, 10 to 16 cm. Transitional horizon, often, with distinct tonguing.
Btw, 16 to 32 cm. Illuvial horizon of bright brown color, sometimes, with reddish tint; angular blocky (nutty), with clay films on the surface of peds; the texture is heavier than in the upper and lower-lying horizons.
BC, 32 to 42 cm. Transitional horizon.
C(D), (>42 cm), sandy soil forming (or underlying?) rock. Permafrost table is at a depth of 150 to 250 cm (in the fall). Both low-ice and high-ice permafrost can be found. However, there are no features of water stagnation even above the icy permafrost. Cryogenic movements are not manifested.

Pale Podzolic soils have a neutral or slightly acid reaction (pH 5.2 to 6.2) and do not contain carbonates. The degree of base saturation in upper horizons varies from 50 to 90%. The bleached AE horizon is depleted in sesquioxides and magnesium and enriched in silica. The Btw horizon is enriched in clay and has a high content of dithionite-extractable free crystallized iron oxides.

These are the soils that develop from poor and well drained substrates under taiga vegetation in semiarid climate. Metamorphic weathering and *in situ* accumulation of sesquioxides in the Btw horizon combines in them with some illuviation of clay material and, probably, amorphous Al-Fe-humus compounds (Sokolov et al., 1969, 1974)

3.3.6. Pale Calcareous and Pale Solodic soils

Pale Calcareous and Pale Solodic soils develop within the ancient alluvial plain (terraces of the Lena River) on loesslike clay loamy polymictic aerial deposits. They are typical of alas depressions in Central Yakutia. As a rule, Pale calcareous soils occupy somewhat higher levels around alas depressions. Pale Solodic soils tend to develop in meso- and microdepressions.

These soils are distinguished from Typical Pale soils by the presence of distinct Ca-illuvial horizon at a shallow depth. In Pale Calcareous soils (Calci-Gelic Cambisols), the maximum accumulation of calcium carbonates occurs in the uppermost part of the Bk horizon, which the ascending migration of solutions in hot summer periods (the pulsating exudative water regime) often explains. Pale Solodic soils (Calci-Gelic Luvisols) also may contain gypsum and salt accumulations in deep horizons.

Pale Calcareous and Pale Solodic soils contain ice-rich permafrost within 2 m from the surface; the soil mass is dissected by frost fissures forming a regular network. In Pale Solodic soils these fissures serve as the main channels for the downward migration of colloids. Pale Calcareous and Pale Solodic soils have neutral to alkaline reaction. Pale Calcareous soils have no textural differentiation; they have an accumulative pattern of humus distribution and an eluvial-illuvial pattern of carbonates.

Pale Solodic soils have a maximum clay content in the Bt horizon and eluvial-illuvial distribution patterns of humus, carbonates, and sesquioxides (especially, dithionite-extractable iron). One of the main processes in their E horizon is iron segregation in concretions, which favors the removal of iron from the soil matrix and provides for the downward migration of dispergated clay (Surface eluvial-gley metamorphic soils, Solodic Planosols). A network of cryogenic fissures that serve as the main pathways for solute transport enhances the downward migration.

The role of cryogenic processes in the soil profile morphology increases from Pale Podzolic (minimal influence) to Typical, Calcareous, and Solodic types of Pale soils. As mentioned earlier, Pale soils form intergrades with different types of soil formation.

In particular, ***Pale Cryozemic soils*** develop from heavily textured rocks in conditions of the northern taiga and tundra (in relatively well drained sites). These soils have evident features of cryoturbation in their profile and a distinct polygonal or hummocky microtopography on the surface; at the same time, all soil horizons are clearly distinguished and have a continuous character. This subtype of meso-hydromorphic Pale soils is considered a transition to cryohydromorphic Cryozems.

3.4. Humus-Accumulative Ca-Illuvial soils

Humus-Accumulative Ca-Illuvial soils include the types of *Chernozems, Meadow-Chernozemic, Meadow, Chestnut, Cryoarid Steppe,* and *Tundra-Steppe soils*. Chestnut soils develop beyond the permafrost zone, in intermontane depressions of Eastern Sayan in the southwest of the territory. Cryoarid Steppe and Tundra-Steppe soils develop in the northeastern part of Central Siberian Plateau on the slopes of southern aspect; they do not form vast areas in Central Siberia and are not shown on our map as separate areas.

3.4.1. Chernozems (Chernic Chernozems) and Meadow-Chernozemic soils

Chernozems (Chernic Chernozems) and ***Meadow-Chernozemic*** soils (Haplic Phaeozems) develop under steppes and meadows in the Trans-Baikal region. Steppe soils of the Trans-Baikal region develop in conditions of extremely continental climate with very cold winters that favor deep soil freezing. Often, these soils have permafrost at a depth of less than 2 m. It is reasonable to distinguish Chernozems proper and Chernozem-like permafrost-affected soils (Sokolov, 1967). Chernozems proper develop from sandy or loamy sandy derivatives of silicate rocks with considerable pebble content. Permafrost is absent or occurs at a depth exceeding several meters.

3.4.2. Permafrost-affected Chernozem-like soils

Permafrost-affected Chernozem-like soils include the subtypes of Meadow and Meadow-Steppe soils. Permafrost-affected Meadow (or Chernozemic-Meadow) soils are common in the northern part of the Trans-Baikal region. They develop clayey or loamy-clayey colluvial deposits and are subject to intense cryogenic fissuring; a polygonal network of cryogenic fissures is distinct on the surface.

Along cryogenic fissures, the tongues of the humus horizon penetrate to the depth of 150 to 200 cm; the average depth of the humus layer (A1 + A/B) is about 40 cm. The humus content may reach 12 to 15%.

Though permafrost may occur at a depth exceeding 2 m, its effect on soil properties is very distinct; along with cryogenic fissuring and deep humus tongues, water stagnation takes place in the bottom of the soil profile, with the development of gleyzation and platy cryogenic structure. As a rule, permafrost-affected Chernozemic-Meadow soils have neutral or slightly acid reaction and do not contain Ca carbonates; at the same time, they are enriched in oxalate-soluble sesquioxides.

3.4.2. Meadow-Chernozemic soils

In the southern part of the Trans-Baikal region, the subtype of ***Meadow-Chernozemic*** soils occurs. They have somewhat deeper humus layers (50 to 70 cm), though the humus content is lower (5 to 9%). The reaction is neutral in the upper horizons and slightly alkaline in the BC horizon. In contrast to Chernozems proper, Chernozem-like Meadow soils have rather diverse morphological forms of carbonate neoformations - powdery impregnation of the soil mass (dispersed carbonates), calcareous pseudomycelium, and very soft (earthy) calcareous nodules.

Example 2.13.5. Pit 249 (Sept. 4, 1956). Flat summit of a low hill, 15 km to the north of the village of Kuti. Meadow-Chernozemic soils (Sokolov, 1967).

Meadow steppe; herbs with predominance of tansy.

A1, 0 to 7 cm. Gray with brownish tint, loose, silty clayey, granular structure; uneven irregular boundary; in some places, calcareous material of the Bk horizon reaches the surface.

A1', 7 to 40 cm. Gray with brownish tint, rather compact; angular blocky, blocky peds parting to granular peds; gradual transition (except for the "inverted" tongues of the Bk horizon).

A/B, 40 to 65 cm. Heterogeneous; dark gray humus tongues against the background brown mass; compact; with platy cryogenic structure in the lower part; mottles with calcareous pseudomycelium; gradual transition.

Bk, 65 to 125 cm. Light brown, with whitish calcareous mottles; wet; blocky-platy structure; humus tongues (to the depth of 200 cm); soil texture becomes somewhat coarser downward the profile.

Dg, 125 to 230 cm. Mottley; loamy to sandy loamy with sandy laminae in the lower part; evident features of gleyzation (ocherous mottles, iron-manganic nodules). Permafrost lies deeper.

Thus, ice-containing (impermeable) permafrost causes deep cryogenic fissuring and tonguing of the humus horizon, as well as the development of gleyzation and platy cryogenic structure in deep horizons of Meadow-Chernozemic soils, though the thawing depth often exceeds 2 m and may reach 3 to 5 m. However, the features of above-permafrost water stagnation and gleyzation often occur in Meadow permafrost-affected soils of the Trans-Baikal steppes (pit 71).

It is interesting to note very sharp contrasts in the soil cover of the Trans-Baikal region. Podzols and Parapodzols develop on high summits and on slopes with a northern aspect under taiga vegetation. Chernozems tend to develop from sandy loamy rocks in intermontane depressions with herbaceous and grassy-herbaceous steppes. Meadow-Chernozemic soils occupy similar geomorphic positions but develop from loamy clayey material and are subject to intense cryoturbation.

3.4.4. Chestnut and Dark Chestnut soils (Kastanozems)

Chestnut and ***Dark Chestnut*** soils (Kastanozems) occur in some intermontane basins in the southeastern and southern parts of Trans-Baikal region, predominantly beyond the permafrost zone. They also occur in the intermontane depressions of Eastern Sayan (Minusinsk Depression, Tuva Depression).

Chestnut soils develop from sandy or loamy sandy rocks of alluvial fans under dry steppe sparse vegetation (wormwood, fescue, xerophytic sedges) and are distinguished from Chernozems by the lower humus content (about 3 to 4% in the A1 horizon). Volkovintser (1978) suggested that these soils should be included in a large group of Steppe Cryoarid soils.

Similar soils also develop from pebbly coarsely textured rocks in the northeastern part of Central Siberia and in the northeast of Asia within the permafrost zone (Naumov & Andreeva, 1963; Bystryakov & Kulinskaya, 1980). Usually, they occur only on slopes with a southern aspect with patches of steppe vegetation among taiga or even tundra communities. Though the area of these soils is not large in Central Siberia, they serve as important indicators of extremely continental and relatively dry climatic conditions.

3.5. Accumulative Ca-Humus soils

As shown above, calcareous rocks occupy very large areas in Central Siberia. Proterozoic and Early Paleozoic limestone and marl (in some places, dolomite) outcrop to the surface within the Lena-Aldan, Lena-Angara, Lena-Vilyui, and Angara-Podkamennaya Tunguska interfluves. They also compose the periphery of the Anabar anteclise in the northern part of Central Siberian Plateau.

In the southwestern part of the territory (the Angara-Lena and Angara-Podkamennaya Tunguska interfluves), with humid or semihumid moderately cold climate (in the zone of isolated permafrost), the soils developing from the residuum of calcareous rocks or from their carbonate-rich colluvial derivatives can be classified as typical Rendzinas (Soddy-Calcareous soils) (Rendzic Leptosols), with a thin litter horizon, distinct humus (A1 or A1k) horizon, transitional Bk horizon, and underlying Ck or Rk horizon. V.G. Zol'nikov (1954) and Iv. P. Gerasimov (1963, 1965) described the soils developing from limestone residuum on the plateaus in the Lena-Vilyui and Lena-Aldan interfluves as *Mucky-Calcareous Permafrost-Affected soils* (Calcari-Leptic Cryosols). They described Typical, Leached, and Gleyed variants of these soils.

Example 2.13.6 . Pit 74 (July 27). Lena-Aldan interfluve, 500 m a.s.l. Mucky-Calcareous permafrost-affected soils (Gerasimov, 1965).

Larch taiga with an admixture of pine; birch (Betula Middendorffii) shrubs, bearberry, mountain cranberry, green mosses, and herbs (Limnas Stelleri) in the ground cover.

Oi, 0 to 2 cm. Nondecomposed forest litter.

AOe, 2 to 12 cm. Mucky horizon of dark brown color, slightly dry, silty loamy, densely penetrated by horizontal roots, with distinct semidecomposed plant tissues admixed to the mineral mass.

A1k, 12 to 30 cm. Yellowish brown, with pinkish tint; moist; clayey, with limestone rock fragments; few roots; the fine earth material and rock fragments display effervescence.

BCkg, 30 to 50 cm. Pinkish, with dove-colored gleyed mottles; nonsatiated wet; clayey, with abundant limestone pebble; strongly effervescent.

Ckg, 50 to 70 cm. Similar in morphology; satiated wet; with higher amount of limestone rock fragments having a predominantly vertical orientation; ice-rich permafrost at a depth of 70 cm.

Mucky-Calcareous permafrost-affected soils have the humus content in the A1 horizon of about 2 to 4%. Their reaction is close to neutral. In contrast to Pale Calcareous soils, they do not contain adsorbed sodium and potassium. The content of carbonates increases downward the soil profile.

Considerable content of coarse rock fragments ensures sufficient internal drainage of the upper part of the soil profile. Cryoturbation is not very active. Water stagnation above the permafrost table results in the development of gley features. In conditions of southern Yakutia (within the Lena-Aldan interfluve), ice-rich permafrost is present in the soil profiles under natural (larch taiga) vegetation.

In the northeastern part of the Central Siberian Plateau (the Anabar-Olenek Plateau), with more severe climatic conditions, soil formation on the derivatives of calcareous rocks has a different character. The highest parts of the plateau are subject to active denudation, so that the mantle of loose fine earth material on them is very thin. These surfaces are virtually barren, with well manifested, sorted patterned grounds. Scarce vegetation appears only on the network of cryogenic fissures.

These are virtually barren areas with primitive stony soils, thin humus-accumulative soils of fissures, and surface-calcareous soils of cryogenic circles (soils of spots). A mantle of loamy and clayey material of colluvial and solifluctional origin covers lower parts of the plateau and slopes. Active frost heave results in the development of typical hummocky microtopography.

Peat soils with shallow (20 to 30 cm) permafrost develop in cryogenic troughs between the hummocks; the soils of hummocks are distinguished as Thixotropic Cryozems (see below). Mucky-Calcareous soils develop only in well drained geomorphic positions (near the brows) from the substrates that are enriched in weatherable minerals (the derivatives of mafic intrusions) or nutrients (nesting places).

3.6. Cryohydromorphic soils

Most of mineral heavily textured (loamy and clayey) soils in the permafrost zone display vivid features of cryoturbation and hydromorphism; they have water-saturated horizons during the whole summer and ice-rich permafrost underlies them. These soils can be referred to as *cryohydromorphic soils*. They are much less diverse than mesomorphic soils; i.e., the soils with similar types of soil profile morphology (O-AOjj-Bjj-Cfjj with strongly manifested cryoturbation) develop from different kinds of parent material under different climatic conditions (from extremely cold humid tundra to moderately cold semiarid larch taiga).

At lower taxonomic levels, cryohydromorphic soils can be subdivided into several groups by (a) the presence (or absence) of gley features, (b) the character of organic horizons, (c) soil acidity and base saturation, (d) mineralogical and chemical composition, (e) texture and physical properties, etc. The main groups are Cryohydromorphic Gley soils and Cryohydromorphic Nongley soils that may form intergrades.

3.6.1. Cryohydromorphic Gley soils

Cryohydromorphic Gley soils with Peaty, Mucky-Peaty, or Mucky organic horizons (Gleyi-Turbic Cryosols) predominate in the plain tundra and forest-tundra landscapes in the western part of the North Siberian Lowland. They develop from marine (or glaciomarine) clays and loams, as well as from glaciolacustrine and lacustrine-alluvial sediments with some admixture of pebble material.

Mafic rocks of the Putorana trappean plateau served as the main source of these sediments, as the predominant montmorillonitic composition of clay minerals and the presence of coarse fragments of mafic rocks in them show. Mucky-Peaty Gley soils develop under cotton grass-moss shrub tundra vegetation. V.D. Vasil'evskaya (1980) described these soils in the middle reaches of the Pyasina River. The surface has a hummocky microtopography; barren spots often occur on the tops of hummocks.

Example 2.13.7. Soil profile in the trough zone between hummocks (Tundra Gley soil).

O/AOa, 0 to 10 cm. Wet, dark brown material composed of moss remains, with an admixture of loamy fine earth in the lower part; densely penetrated by fine roots; distinct transition.

Bg, 10 to 20 cm. Wet, brownish dove-colored, with ocherous mottles; compact; sticky; clayey; structureless; densely penetrated by fine roots, especially in the upper part; gradual transition.

G, 20 to 37 cm. Satiated wet; clayey; mottled, with vertical ocherous streaks against the bluish dove background color; compact; sticky; structureless; ice rich permafrost from the depth of 37 cm.

Example 2.13.8. Soil profile of hummocks with a barren surface.

Bg, 0 to 10 cm. Mottled: grayish ocherous-brown with dove-colored mottles; clayey; loose crumbly structure; compact; densely penetrated by roots; gradual transition.

G, 10 to 50 cm. Wet, dove-colored; with bright ocherous streaks and mottles from the depth of 30 cm; clayey; structureless; with few roots.

Cgf, 50 to 110 cm. Grayish, with dove tint; frozen; clayey; with abundant ice lenses and schlieren.

The ocherous-brown color of the soil mass in the Bg horizon attests to active oxidation processes; oxidized amorphous iron compounds disperse in the soil mass and also form relatively soft iron nodules of up to 0.5 mm diameter. In the G horizon, iron compounds are mainly reduced. Relatively weak acidity and high base saturation characterize Tundra gley soils developing from montmorillonitic clayey substrates.

The degree of gleyzation usually increases downward in the soil profile. At the same time, the horizon with the minimum percentage of oxidized iron compounds (manifested by ocherous mottles and streaks) often occurs in the middle part of the profile (Elovskaya et. al., 1979).

Though the main area of Cryohydromorphic Gley soils lies in the northwestern part of Central Siberia, their ecological amplitude is much wider. As some secondary components of the soil cover, these soils occur in virtually all permafrost-affected landscapes. Peat and Mucky-Peat Gley hydromorphic soils with less pronounced cryoturbation or without it at all also occur in taiga landscapes beyond the permafrost zone.

3.6.2. Cryohydromorphic Nongley soils, or Cryozems

Cryohydromorphic Nongley soils, or **Cryozems** (Sokolov, 1980a, b) develop within permafrost-affected territories in conditions of ultracontinental climate in central and eastern Siberia. This group includes two types of soils: Mucky-Peaty (Histic) Cryozems and Thixotropic Cryozems. Vividly manifested cryoturbation and a lack of gleyzation in the mineral soil profile at the background of its saturation with water characterize both types.

3.6.2.1. Mucky-Peaty Cryozems (Oxyaqui-Histic Cryosols)

Hydromorphic soils with evident features of cryoturbation, thick organic horizons, and shallow depth of ice-rich permafrost were first described under mountainous taiga in the Trans-Baikal region under the name of *Taiga frozen (permafrost-affected) soils* (Sokolova & Sokolov, 1963), with the note that gley features are not vividly expressed in them, despite high water saturation.

In the legend of the 1:2.5 M *Soil Map of Russia*, these soils were named "Taiga Hydromorphic Nongley High-Humus Soils." The absence of distinct gley features served as the basis for separating these soils from the group of Gley permafrost-affected soils under the name of Homogeneous Cryozems (the term "homogeneous" designated the process of cryoturbation leading to homogenization of the upper mineral horizons).

Homogeneous Cryozems can develop from different substrates with medium (silty to sandy loamy) texture in autonomous geomorphic positions and on slopes within a wide range of bioclimatic conditions, from Subarctic Tundra to Boreal Taiga (Sokolov, 1980a). They do not form large areas in the soil cover and occur together with different kinds of mesomorphic soils (Al-Fe-humus soils, Pale soil, Mucky Calcareous soils, etc.), Peat soils, and stony strips. Cryohydromorphic Gley soils replace them in relief depressions, as well as on heavily textured clayey rocks.

Field studies of the 1970s proved that Homogeneous Cryozems are more common in the western (humid) part of the permafrost zone in Central Siberia and do not develop in more arid conditions. They should be distinguished as *Histic Cryozems* to stress the importance of peat accumulation on their surface.

Example 2.13.9 . Pit 214 (July 21, 1969). Putorana Plateau, the slope (3°) of southeastern aspect to Lake Ayan. Thin larch forest with shrubs (dwarf birch, willow, and wild rosemary); mosses, lichens, sedges, and dwarf shrubs in the ground cover.

Oi, 0 to 10 cm. Dark brown peaty horizon with weakly decomposed moss tissues.

BOjj, 10 to 30 cm. Gray-brown (10YR 4/2); homogeneous mixture of plant tissues and mineral material; satiated wet; structureless; friable when dry; mineral grains are devoid of cutans; ice lenses from the depth of 20 cm.

C(D)f, 30 to 50 cm. virtually black frozen loamy sand (solifluction derivatives of doleritic rocks); without evident features of pedogenesis.

Histic Cryozems have a thick (10 to 20 cm) organic horizon, underlain by a cryohomogenized mixture of organic and mineral material (BOjj). The degree of decomposition of plant tissues and humification processes depends on the character of parent material. Peaty organic horizons are typical of the derivatives of acid silicate rocks, and mucky horizons can develop on the derivatives of silicate-calcareous rocks.

A high content of semidecomposed plant tissues in the BO horizon explains the low bulk density of these soils, their lack of compaction, and the friable character of the soil mass upon drying. These are the soils with significant water retention capacity and, at the same time, the possibility of lateral water migration toward the zones with perfect drainage in surrounding stone stripes. The humus content is also high, which contributes to the high adsorption capacity (40 to 60 cmol$^{(+)}$/kg soil).

Chemical properties of Histic Cryozems vary, depending on the nature of their parent material. Histic Cryozems on the derivatives of calcareous rocks have neutral pH (7 to 7.5) and are saturated with bases, though the upper part of the peat horizon may be very strongly acid (pH 4 to 4.5). Histic Cryozems developing from nonmafic silicate rocks have a base saturation about 20 to 40% and a strongly acid reaction (pH 5.0 to 5.5).

At the same time, the degree of structuring of the mineral material in Histic Cryozems is very low, which the high water content and frequent cryoturbation explain. The lack of gley (redoximorphic) features in the profile can be due to its aeration in the course of cryoturbation, saturation of soil solutions with oxygen in the network of cryogenic fissures, and low temperatures that suppress microbiological activity and favor the solubility of oxygen in soil solutions.

3.6.2.2. Thixotropic Cryozems (Oxyaquic Cryosols)

Thixotropic Cryozems (Oxyaquic Cryosols) represent another type of cryohydromorphic nongley soils characterized by a water-saturated, heavily textured, cryoturbated, structureless, sticky, rather compact mineral profile with thixotropic properties and the lack of gley features. When dried, the thixotropic mineral horizon becomes very hard. Peat, muck, and thin raw humus and proper humus horizons can develop at the surface; the thickness of organic horizons is usually less than 20 cm.

These soils are widespread in the central and northeastern parts of Central Siberia, within the zone of middle and northern larch (in the western part, with admixture of spruce and Siberian pine) taiga with oligotrophic hydrophilic moss-lichen vegetation in the ground cover. They occupy leveled interfluves, gentle and moderately steep slopes, terrace-like surfaces in river valleys, and colluvial fans.

Thixotropic cryozems develop from heavily textured (loamy clayey and clayey, with relatively low pebble content) colluvial and solifluctional deposits of different mineralogical composition. They are especially typical of the areas composed of calcareous and mafic rocks.

Cryogenic hummocky microtopography is very distinct on the surface. Earthy hummocks have a diameter about 1 to 2 m and relative elevation above the trough zone of about 30 to 50 cm. Sometimes, their tops are barren. Thixotropic cryozems of hummocks alternate with peat soils underlain by shallow (0.2 to 0.5 m) ice-rich permafrost in cryogenic troughs with ice wedges. The maximum thawing depth on hummocks is about 0.4 to 1.0 m, depending on particular climatic conditions and the thickness of the organic horizon.

Example 2.13.10 . Pit 312 (July 15, 1971). Central Siberian Plateau, basin of the Podkamennaya Tunguska River, 3 km to the southeast of the Taiga River mouth. Slope (about 5°) of northeastern aspect, 260 m a.s.l. Thixtropic cryozem developed from colluvial-soliflcutional derivatives of calcareous slates, sandstone, and vitric tuff penetrated (up the slope) by mafic (doleritic) intrusions (Sokolov, 1980b).

Sparse larch taiga with an admixture of spruce and Siberian pine; willow, birch, and alder shrubs; dwarf shrubs (bog bilberry, mountain cranberry, wild rosemary), mosses and lichens in the ground cover. Hummocky microtopography. Live parts of mosses and lichens on the soil surface have a thickness of about 10 cm.

Oi, 0 to 10 cm. Brown mossy peat, weakly decomposed; gradual transition.

OeA1, 10 to 20 cm. Gray, peaty-mucky, densely penetrated by fine roots; with an admixture of humiferous mineral material; sandy grains are bleached; charcoal particles. Clear transition.

BwA1, 20 to 30 cm. Grayish brown (7.5YR 4/3), grayish tint weakens downward. Structureless, loamy clayey, satiated wet; indurated upon drying. Sand-size particles are covered by brown and grayish (humus-rich) films; charcoal particles; few roots. Gradual transition, wavy boundary.

Bwtix, 30 to 40 cm. Brown (7.5YR 5/6), with very homogeneous color. Loamy clay with admixture of gravel and pebbles. Structureless, satiated wet, thixotropic, indurated upon drying. Brown films on sand particles. Ice-rich permafrost at a depth of 40 cm. In thin sections, the material of B horizons has visible porosity about 10%; there are no films on pore walls; stress cutans of optically oriented clay are seen around coarse skeletal particles represented by quartz, feldspars, pyroxenes, hypersthene, dark vitreous material, and fine fragments of mafic rocks.

Thixotropic cryozems are extremely vulnerable soils. Forest fires, vehicle passages, reindeer overgrazing and other natural and humanly induced impacts on the environment cause the activation of cryoturbation and solifluction. On slopes, the latter can completely destroy (remove) the soil layer with thixotropic properties.

3.7. Peat soils

Relatively high elevations, predominantly ascending neotectonic movements, and the general richness of substrates in bases do not favor the development of vast peatlands in Central Siberia. Usually, peat soils develop in the zones of cryogenic troughs forming complexes with different kinds of mesomorphic and hydromorphic soils. They are more common in the western, more humid part of Central Siberia and in the lowlands and flood plains. In the permafrost zone, the depth of the summer thaw in peat soils varies from 20 cm in the north (Cryic Histosols) to about 1 to 1.5 m in the southern part (Gelic Histosols). Sphagnum peat often develops on top of ice wedges.

3.8. Rock outcrops and nonsoil formations

High mountains and plateaus in Central Siberia are virtually devoid of higher vegetation; they are stony barrens with diverse forms of sorted patterned grounds. Vast areas of these formations occur in the Putorana, Anabar, and Olenek plateaus and at high elevations in the Trans-Baikal Mountains.

Given similar climatic conditions, the upper boundary of more or less continuous vegetation depends on the geological background. In conditions of cold semiarid climate (eastern Putorana and Anabar and Olenek plateaus), barren stony surfaces in the areas with acid silicate rocks occur at elevations above 500 to 600 m a.s.l., areas composed of limestone and clayey-calcareous rocks (marl) occur lower, at 300 to 500 m a.s.l. (in the northern part, to 100 to 200 m a.s.l.). Dolomites occupy an intermediate position.

Rock outcrops and talus deposits are also typical of steep slopes in the valleys of main rivers. Glaciers and snow patches existing for several years occupy relatively small areas in the highest mountains. In general, nonsoil formations occupy no less than 20% of the area in the northern part of Central Siberia.

4. Soil-forming Factors and Soil Distribution

Analysis of the soil map (Figure 2.13.3) shows two principally different *soil ecological fields*: (a) the field of acid unsaturated soils with an eluvial-illuvial or gleyed profile (Al-Fe-humus soils and Gley soils) and (b) the field of neutral to alkaline saturated soils with an accumulative-metmorphic profile (Pale soils and Humus-Accumulative Ca-illuvial soils [chernozem-like soils, steppe cryoarid soils, etc.]) combined with specific Hydromorphic Cryoturbated (cryomorphic) Nongley soils, or Cryozems.

Peat (Histic) soils can occur in both fields, though their relative abundance in the latter ("semiarid to arid") field is much lower than in the former ("humid") field. A transitional field also appears. Different soils fill every particular climatic niche, depending on the character of parent rocks. Thus it is reasonable to consider "lithogenic spectrums" of soils developing in given bioclimatic conditions.

The position of climate-controlled boundaries between different types of soil formation depends on the character of parent rocks. Thus, the niche of neutral metamorphic soils is much wider on the rocks enriched in weatherable minerals ("rich rocks"). In contrast, the ecological niche of Al-Fe-humus soils extends into semiarid and even arid climate on "poor rocks" (e.g., quartz sands). In real geographic space, soil-ecological fields correspond to particular typical soil combinations that compose soil sectors (Figure 2.13.3).

In permafrost-affected areas, the degree of internal drainage of the soil profile and the ice content in permafrost predetermine the activity of cryoturbation and frost heave. Well drained mesomorphic or xeromorphic soils on coarsely textured rocks are less susceptible to cryoturbation (except for cryogenic sorting) and have deeper permafrost. Poorly drained hydromorphic soils on heavily textured rocks are subject to active cryoturbation, leading to considerable disturbance of soil horizons. The difference between meso-xeromorphic and cryohydromorphic soils is very distinct.

In general, ecological niches of cryohydromorphic soils are much wider than those of mesomorphic soils. Thus, Peat and Gley soils may occur in virtually all ecological niches on poorly drained rocks. Ecological niches of Histic and Thixotropic Cryozems are more specific; these soils occur in an ultracontinental climate, which favors frost cracking and saturation of soil water with oxygen that prevents the development of gleyzation.

5. Conclusion

Central Siberia can be considered the cradle of Russian geocryology and cryopedology. Starting from the first permafrost temperature measurements by Middendorf (1844) and first descriptions of cryogenic phenomena in soils by the researchers working under the aegis of Siberian expeditions of the Department for Peoples' Migration in the beginning of the 20th century, soil studies in this region enriched pedology with new concepts related to peculiarities of soil formation in permafrost-affected environments.

There were, roughly, three stages of scientific thought. First, pedogenetic concepts worked out in the European territory of Russia were extrapolated for the territory of Central Siberia without due correction for the effect of cryogenesis (though particular manifestations of cryogenic phenomena in Siberian soils were thoroughly described). In the 1960s, scientists widely accepted the concept of genetic specificity of Siberian soils related to the presence of permafrost and ultracontinental climatic conditions. Later came the idea of Siberian soils resulting from intricate interplay between traditional soil-forming factors, permafrost conditions, and cryogenic processes.

Researchers showed that the effect of permafrost and cryogenic processes on soil formation varies considerably, depending on soil texture and climatic conditions. At the same time, soil diversity in Central Siberia results from the great diversity of all other soil-forming factors, including climate (from humid to semiarid, with mean annual temperatures from -16 to about 0°C), parent material (virtually all kinds, from granite residuum to the derivatives of mafic and calcareous rocks, loesses, sands, and even the remains of red-earth formations), vegetation (tundra, taiga, steppe, and cryosteppe), relief (mountain peaks, flat plateaus, steep and gentle slopes, leveled lowlands, alas depressions, etc.), and time (recent soils on fresh solifluction, alluvial and colluvial deposits, full Holocene soils with well developed profiles).

As a result, soil diversity in Central Siberia is much wider than that at similar latitudes of the European part of Russia. In permafrost areas, predominant soils of the humid climate include Typical, Ocherous, and Dry-Peat (Folic) Podzols and Parapodzols, Granuzems, and Raw-Humus Brown Taiga soils (Burozems) in well drained conditions and Peat, Nongley Cryohydromorphic (Histic Cryozems), and

Gley soils in poorly drained conditions. In the semiarid climate, the types of Palevye (Pale Metamorphic), Meadow-Steppe Chernozemlike soils, Chernozems, Kastanozems, and Cryoarid Tundra-Steppe soils predominate in well drained conditions, whereas Peat soils and Nongley Cryohydromorphic soils (Thixotropic Cryozems) are widespread in poorly drained conditions.

Virtually all the soils in the permafrost zone and even beyond it bear evident features of cryogenic processes (frost cracking, tonguing boundary of the humus horizon, ruptic character of surface horizons, frost heave, uneven surface topography, separation of coarse fragments, ice wedging, cryoturbation, development of specific cryogenic structures, burying of organic horizons, cryohomogenization, etc.). The latter gain maximum development in heavily textured soils, especially in Cryozems and some Gley soils.

Along with soils proper, it is reasonable to distinguish cryopedoliths as the bodies with thin peat, muck, and/or raw-humus horizons regularly disturbed by cryoturbation so that the "normal" development of the soil profile and more or less distinct horizonation of the latter is impossible. Cryopedoliths are common on solifluction slopes. The destructive activity of cryogenic processes greatly increases when there are human impacts on the soil cover. Thus special soil care is mandatory in areas of mining and local agriculture (including reindeer pasturing).

References

Bystryakov, G.M., and Ye.V.Kulinskaya. 1980. Soils of Steppe Cryoarid Landscapes in the Upper Reaches of Kolyma and Indigirka Rivers. The Geography and Genesis of Soils in the Magadan Region (Geografiya i genesis pochv Magadanskoi oblasti). DVNTs AN SSSR. Vladivostok. pp. 143-160. (In Russian.)

Elovskaya, L.G., E.I. Petrova, and L.V. Teterina. 1979. Soils of Northern Yakutia (Pochvy Severnoi Yakutii). Nauka. Novosibirsk. 303 pp. (In Russian.)

Gerasimov, I.P. 1963. Genetic Specificity of Siberian Soils (Sibirskii geogr. Sbornik). Material on the Geography of Siberia, Iss. 2, Nauka. Moscow: pp. 7-27. (In Russian.)

Gerasimov, Iv.P. 1963. Soils of the Amur-Aldan Interfluve. About the Soils of Eastern Siberia (O pochvakh Vostochnoi Sibiri). Nauka. Moscow. pp. 153-195. (In Russian.)

Gerasimov, Iv.P. 1965. Mucky-Calcareous Permafrost-Affected Soils of the Aldan-Lena Interfluve. Pochvovedenie. 2:32-42. (In Russian.)

Glinka, K.D. 1912. Geographic Results of Soil Studies in the Asian Russia. Pochvovedenie. 1:43-63. (In Russian.)

Glinka, K.D. 1927. An Essay on Yakutian Soils, Yakutiya. Akad. Nauk SSSR. Leningrad. pp. 131-164. (In Russian.)

Gorbachev, V.N., and E. P. Popova. 1992. Soil Cover of the Southern Taiga of Central Siberia (Pochvennyi pokrov yuzhnoi taigi Srednei Sibiri). Nauka. Novosibirsk. 224 pp. (In Russian.)

Ivanov, A.D. 1966. Eolian Sands in the Western Part of Trans-Baikal Region (Eolovye peski Zapadnogo Zabaikal'ya). Buryat knizhn. izd. Ulan-Ude. 232 pp.

Ivanova, E.N. 1971. The Soils of Central Yakutia. Pochvovedenie. 9:3-17. (In Russian.)

Karavaeva, N.A., I.A. Sokolov, T.A. Sokolova, and V.O. Targulian. 1965. On the peculiarities of soil formation in tundra and taiga areas of Eastern Siberia and the Far East. Pochvovedenie. 7:14-26. (In Russian.)

Koposov, G.F. 1983. The Genesis of Mountainous Soils in the Baikal Region (Genezis pochv gor Pribaikal'ya). Nauka. Novosibirsk. 256 pp. (In Russian.)

Krasyuk, A.A. 1927. Soils of the Lena-Amga Interfluve, Materialy komissii po izucheniyu Yakutskoi ASSR.: Izd. Akad. Nauk SSSR. Leningrad. (In Russian.)

Kuz'min, V.A. 1976. The Soils of Intermontane Basins of the Baikal Region (Pochvy kotlovin Baikal'skogo tipa). Vostochno-Sibirskoe knizhn. izd. Irkutsk. 144 pp. (In Russian.)

Makeev, O.V. 1959. Soddy Taiga Soils in the South of Central Siberia (Dernovye taezhnye pochvy yuga Srednei Sibiri). Ulan-Ude. 345 pp. (In Russian.)

Martynov, V.I. 1965. Mountainous Soils of the Baikal Region (Pochvy gornogo Pribaikal'ya). Buryat knizhn. izd. Ulan-Ude. 165 pp. (In Russian.)

Mikhailenko, M.M. 1967. Soils of the Southern Taiga Zone in the Western Trans-Baikal Region (Pochvy yuzhnoi taigi Zapadnogo Zabaikal'ya). Nauka. Moscow. 160 pp. (In Russian.)

Naumov, Ye.M., and Andreeva, N.A. 1963. Soils of Steppe Slopes in the Yana-Yukagir Plateau. Pochvovedenie. 3:62-70. (In Russian.)

Nadezhdin, B.V. 1961. The Lena-Angara Forest-Steppe Region (Leno-Angarskaya lesostep'). Nauka. Moscow. 328 pp. (In Russian.)

Nikolaev, I.P. 1949. The Soils of Irkutsk Region (Pochvy Irkutskoi oblasti). OGIZ. Irkutsk. 404 pp. (In Russian.)

Nogina, N.A. 1964. The Soils of Trans-Baikal Region (Pochvy Zabaikal'ya). Nauka. Moscow. 314 pp. (In Russian.)

Prasolov, L.I. 1927. Southern Trans-Baikal Region (Yuzhnoe Zabaikal'e). Izd. Akad. Nauk SSSR. Leningrad. 422 pp. (In Russian.)

Sokolov, I.A. 1967. About the Soils of Trans-Baikal Steppes. Pochvovedenie. 3:42-55. (In Russian.)

Sokolov, I.A. 1970. Podzolic Soils of the Trans-Baikal Region. Pochvovedenie. 8:26-37.

Sokolov, I.A. 1980a. Hydromorphic Non-Gley Soils. Pochvovedenie. 1:21-32. (In Russian.)

Sokolov, I.A. 1980b. On the Diversity of the Forms of Hydromorphic Non-Gley Pedogenesis. Pochvovedenie. 2:5-18. (In Russian.)

Sokolov, I.A. 1986. Palevye Soils of the Central Siberian Plateau. Pochvovedenie. 8:5-18. (In Russian.)

Sokolov, I.A. 1988. Zonal Spectrum of Autonomous Soils and Its Ecologic and Genetic Analysis. Pochvovedenie. 3:15-25. (In Russian.)

Sokolov, I.A. 1991. Some Theoretical Results and Problems of the Study of the Soils of Eastern Siberia and the Far East. Pochvovedenie. 5:131-145. (In Russian.)

Sokolov, I.A., and B.P. Gradusov. 1978. Soil Formation and Weathering on Basic Rocks in Conditions of Cold Humid Climate. Pochvovedenie. 2:5-17. (In Russian.)

Sokolov, I.A., and B.P. Gradusov. 1981. About the Exogenesis in the Areas of Basic Rocks (by the example of the Putorana Plateau). History of Great Lakes in the Central Subarctic Region (Istoriya bol'shikh ozer tsentral'noi subarktiki). Nauka. Novosibirsk. pp. 41-68. (In Russian.)

Sokolov, I.A., and T.A. Sokolova. 1962. About the Zonal Type of Soils in Permafrost-Affected Regions. Pochvovedenie. 10:23-32. (In Russian.)

Sokolov, I.A., B.P. Gradusov, T.V. Tursina, I.G. Tsyurupa, and N.A. Tyapkina. 1974. About the Soil Formation on Loose Silicate Substrates in the Permafrost Zone. Pochvovedenie. 5:29-42. (In Russian.)

Sokolov, I.A., E.M. Naumov, B.P. Gradusov, T.V. Tursina, and I.G. Tsyurupa. 1976. Ultracontinental Taiga Pedogenesis on Calcareous Loams in Central Yakutia. Pochvovedenie. 4:11-27. (In Russian.)

Sokolova, T.A., and I.A. Sokolov. 1963. About Mountainous Taiga Soils in the Eastern Part of Trans-Baikal Region. About the Soils of Eastern Siberia. (O pochvakh Vostochnoi Sibiri). Nauka. Moscow. pp. 3-52. (In Russian.)

Sokolov, I.A., T.V. Tursina, and N.I. Belousova. 1969. Recent Podzolization in the Plains of Central Yakutia. Pochvovedenie. 12:22-27. (In Russian.)

Targulian, V.O. 1971. Soil Formation and Weathering in Cold Humid Regions (Pochvoobrazovanie i vyvetrivanie v kholodnykh gumidnykh oblastyakh). Nauka. Moscow. 268 pp. (In Russian.)

Ufimtseva, K.A. 1967. Soils of Intermontane Depressions in the Southern Taiga of Trans-Baikal Region (Pochvy mezhgornykh kotlovin yuzhnoi taigi Zabaikal'ya). Vost. Sibirskoe knizhnoe izdat. Irkutsk-Chita. 102 pp. (In Russian.)

Vasil'evskaya, V.D. 1980. Soil Development in the Tundra of Central Siberia (Pochvoobrazovanie v tundrakh Srednei Sibiri). Nauka. Moscow. 236 pp. (In Russian.)

Volkovintser, V.I. 1978. Steppe Cryoarid Soils (Stepnye krioaridnye pochvy). Novosibirsk. Nauka, 208 pp. (In Russian.)

Vtorushin, V.A. 1982. Automorphic Soils of the Mountainous Taiga in the Southern Trans-Baikal Region (Avtomorfnye pochvy gornoi taigi yuzhnogo Zabaikal'ya). Novosibirsk. Nauka, 176 pp. (In Russian.)

Zol'nikov, V.G. 1954. Soils of the Eastern Part of Central Yakutia and Their Management. Materials on the Environment and Agriculture in Central (Yakutia Materialy o prirodnykh usloviyakh i sel'skom khozyaistve Tsentral'noi Yakutii). Izd. Akad. Nauk SSSR. Moscow. pp. 55-221. (In Russian.)

Zol'nikov, V.G. 1965. Soils. Yakutia. Nauka. Moscow. 1965, pp. 217-243. (In Russian.)

Zol'nikov, V.G., L.G. Elovskaya, L.V. Teterina, and E.I. Chernyak. 1962. Soils of the Vilyui River Basin and Their Management (Pochvy Vilyuiskogo basseina i ikh ispol'zovanie). Izd. Akad. Nauk SSSR. Moscow. 204 pp. (In Russian.)

Section 3. Properties and Processes of Cryosols

Properties and Processes of Cryosols: Introduction

Brigitte Van Vliet-Lanoë[1]

[1]Director of Research, Processus et Bilan des Domaines Sédimentaires, UMR 8110 CNRS & University of Lille 1, 59655 Villeneuve d'Ascq, France; e-mail Brigitte.Van-Vliet-Lanoe@univ-lille1.fr.

In Arctic or alpine soils, frost is an important pedogenic agent. Most of the frost features relate to one dominant process, the segregation of ice in the soil, usually in the form of lenses and the heaving and associated strains resulting from both freezing and thawing. Frost is specifically a thermally oriented desiccation.

Physical weathering related to frost dehydration, hydraulic pressure, and ice crystallization (Claridge & Campbell, S. 3, Ch. 4, 6; Blume et al., S. 3, Ch. 5), in combination with other agents such as chemical weathering or soil faunal activity, can lead to the fragmentation of rocks. The high latitude and/or altitude reduces the biologically active period, limiting the extent of fragmentation due to the low kinetics of the biochemical reaction. But life is present everywhere and biological activity much more widespread than believed.

In the soil, texture (in the pedological sense) controls frost dynamics, while organic matter, amorphous clays, and local drainage conditions play complementary roles. The equilibrium between the thermal deficit at the freezing plane and the thermal input from water (heat of crystallisation and soil heat) (V. Ostroumov, S. 3, Ch. 1) control ice segregation. Optimum conditions for formation of ice lenses occur in loamy textured sediments in the capillary fringe of the water table (pF≤2).

The capability to develop ice lenses is called *frost heave*. In spring, the inversion of the thermal gradient (the frozen soil is colder than the air) promotes the infiltration of melting water in the still frozen substratum, which causes supplementary heave and stresses in the soil just before the thaw.

The size of the aggregates increases progressively with depth and with the thermal deficit's gradual decrease, except in the presence of permafrost. After the ice lenses melt, the arrangement of the soil peds between the residual fissures produces a pattern of soil morphology called *platy to foliated soil fabric* or *microfabrics* (see Van Vliet-Lanoë et al., S. 3, Ch. 2).

The placement of the interpedal fissures affects both hydraulic conductivity and translocation of particles. At the soil surface, ice lensing improves the soil fabric; farmers know this as *frost mulching*. At depth, the frost-formed aggregates become more stable when thawing is slow, coldest winter temperatures are low, and content in colloids (organic or mineral) is high. This can lead to over-consolidation, which is a natural compaction by dehydration and mechanical pressure.

Differential frost heaving, resulting from the accumulation of segregated ice, appears to be the main mechanism of involution in a cold environment, but cryostatic pressures can intensify its impact with differential swelling, exceptional salt heave, and, in poorly drained sites, load cast. Organics or amorphous clays, which increase soil's water retention capability, can emphasize differential frost heaving. Gradients of susceptibility to frost reveal the geometry of the deformation in close relationship to the level of the water table in the active layer (layer freezing and thawing every year).

Permafrost is not the *sine qua non* of cryoturbation. Permafrost has no direct influence on the thermal regime but can modify it with respect to the desiccation of the active layer in winter. In particular, it has impacts on the hydraulic regime (Burn, S. 3, Ch. 3), as it behaves like a rigid bedrock layer. The lowering of the water table in summer and thermal cracking in winter provide polygonal development, in which slow upward injection of unfrozen material occurs at the onset of the frost season, when the water table rises again and cryostatic pressures develop.

Mud liquid expulsion is an exceptional phenomena, occurring only from hydrostatic pressure which topographic slope differences create or in the event of vibrations (i.e., trampling, earthquake, or mechanical traffic) which destroy the soil aggregates. Pedogenic processes can influence the rate of these hydraulic events, can create frost-susceptibility gradients by translocating particles with spring melt, or can facilitate precipitation of pedogenic components, such as amorphous clays or organics.

Deformations can occur simultaneously with frostcreep on a slope. Neither thermal convection of soil or water nor load casting is the only mechanism responsible for these deformations; the original sedimentary fabric, which may be deformed but is still preserved, has an impact. Load cast usually begins with deglaciation or when thermokarst conditions lead to an abrupt melting which induces slumping and vibrations in the water-saturated soils.

Disturbance is associated mainly with deep seasonal frost or occurs within the active layer. It ranges from macroscopic cryoturbation and thrusting to microscopic disturbances. Various granular aggregates formed by mechanical stresses related to differential frost heave (for example, at the contact between an organic and a non-organic horizon) are common features visible under the microscope. In this case, the disturbance disrupts the horizons, allowing the papules (disrupted clay skins) of Bt horizons and granules of spodic B horizons to develop or iron nodules formed in pseudogley during milder paleoclimate conditions to break up, and finally deforming the horizons with differential frost heave, as Chapter 2 in this section discusses. Frost jacking (vertical tilting) of either stones or other pedogenic features, such as nodules, is common.

Climatic conditions have not been constant in the Eurasian Arctic either during the Holocene period or during the last 20000 years. Thermokarst features date to the Late Glacial in northern Yakoutia. There, during the warmest part of the Holocene, 7000 YBP, the climate was probably 2 to 4°C warmer than it is today, restricting permafrost to the northern fringe of NWT Canada and to the northeast of Siberia (Figure 3.0.1A, B, and C, CLIMEX Maps, CCWM). This warming led to a particularly ice-rich permafrost table in the Arctic.

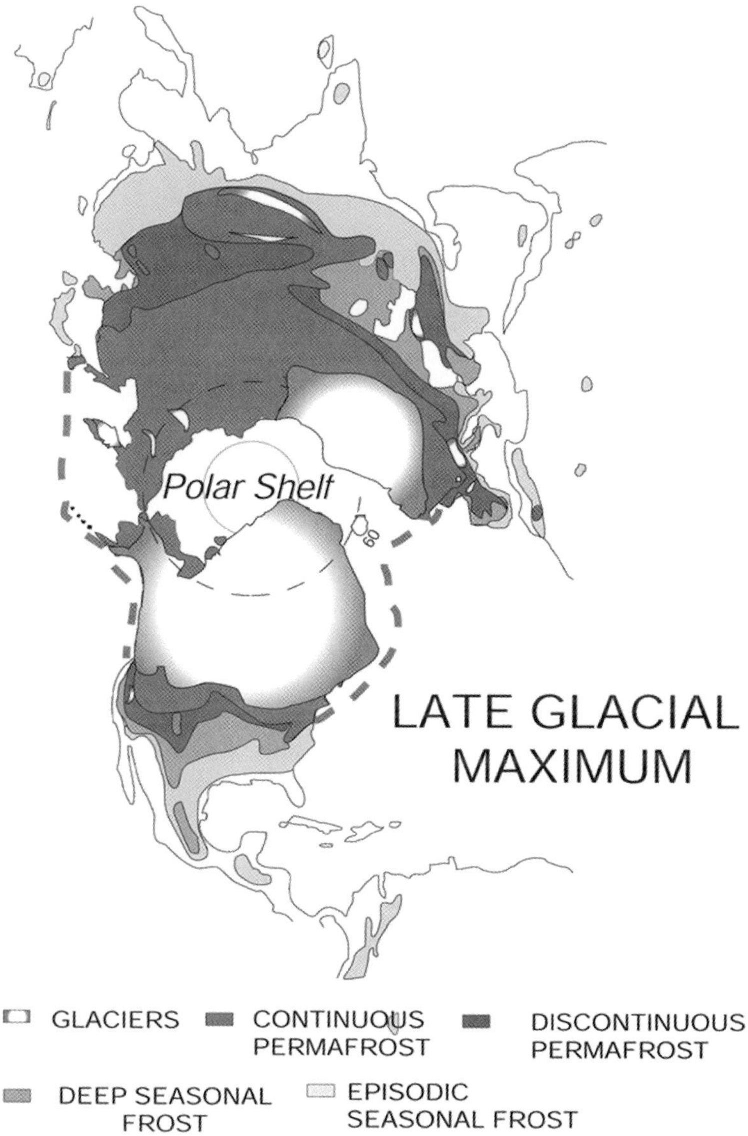

Figure 3.0.1 A. Fluctuation of the extent of permafrost during the last 20,000 years: A. Last Glacial Maximum.

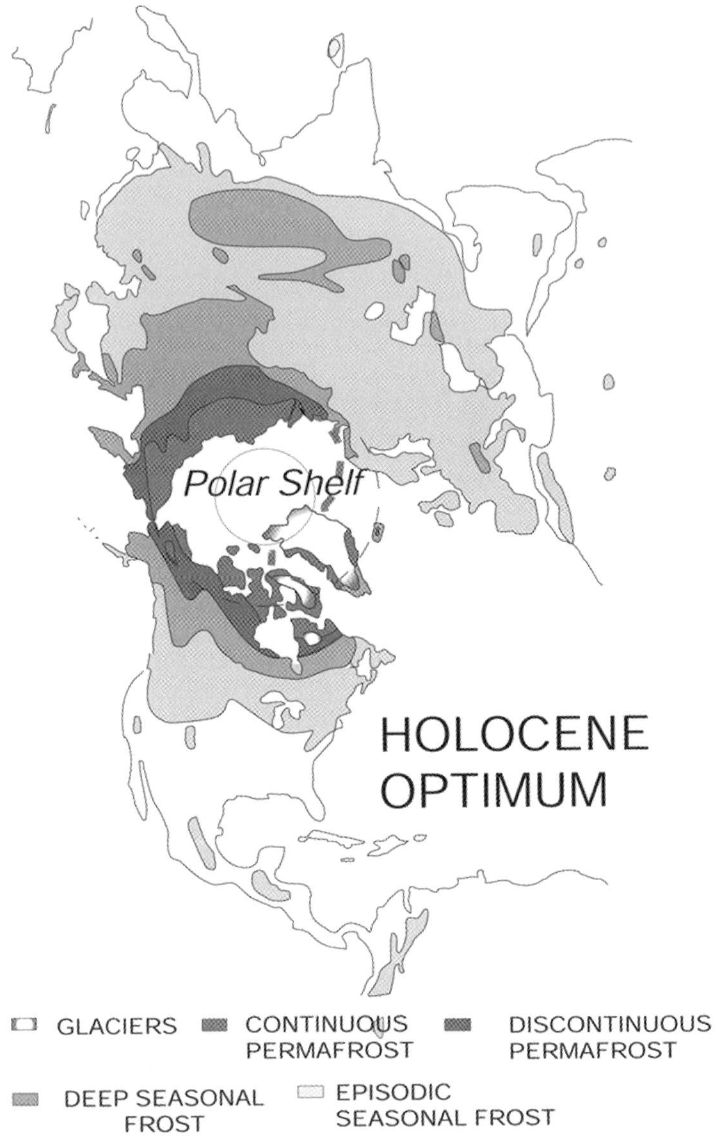

Figure 3.0.1 B. Fluctuation of the extent of permafrost during the last 20,000 years: B. Holocene Optimum.

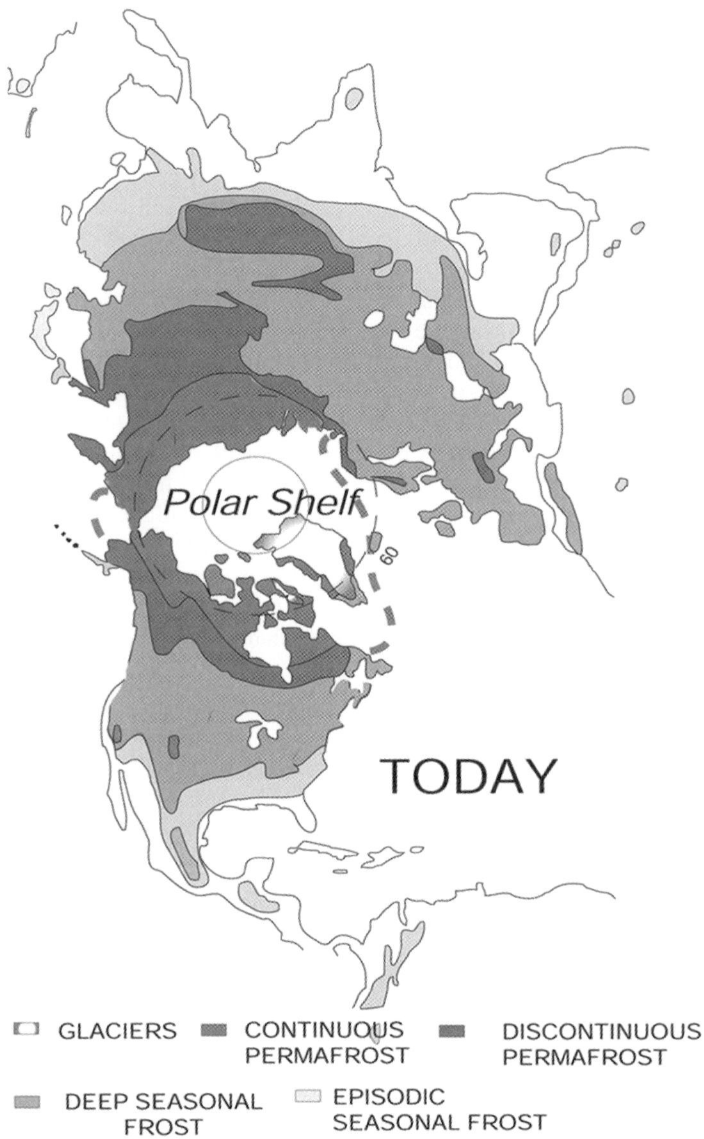

Figure 3.0.1 C. Fluctuation of the extent of permafrost during the last 20,000 years: C. Today.

Since the Subboreal, particularly in last 2000 years, permafrost has re-extended by successive steps to the south. Since the Little Ice Age (Figure 3.0.1A), in the circa-Atlantic provinces it is shallower, poorer in ice, but much colder, as more efficient thermal cracking and ice wedging evidence. An opposite situation to that in Yakoutia exists in Alaska and Western Canada (Burn, S. 3, Ch. 3). The situation is even more complicated in Antarctic regions, where permafrost history extends to the early Tertiary, before the onset of the Quaternary, as Chapters 4-6 in this section stress.

Drainage changes that relate to cooling or warming are recorded in the form of a rise or a drop in the water table within depressions (with or without permafrost) and a change in frost penetration at depth. The consequences for pedogenesis are cryoturbation on wet sites, hummocky microrelief on dryer soils, and solifluction on slopes, all of them disturbing the Hypsithermal soil (Van Vliet-lanoë et al., S. 3, Ch. 2).

Many periglacial features described today as indicators of a cold environment and permafrost in fact developed during the Late Holocene and relate to gradual climate degradation since 6000 YBP. Today, a slight warming leads to the melting of the permafrost, with abrupt failures such as those on the ice-rich permafrost or ice wedges of Yakoutia (C. Burn , S. 3, Ch. 3) or of the Mackenzie, or to progressive deformations with restored pedogenesis in less continental conditions such as those in northern Quebec and Ungava. A warming, as expected from anthropogenic climate modification or increasing deforestation, as in Yakoutia today, leads to intense thermokarst activities, such as those during the Holocene.

Knowledge of the various processes which frost action causes in soils and the interactions of these processes with classical soil-forming processes allows a more accurate reading of soil evolution with time and climate change.

Chapter 1. Physico-Chemical Processes in Cryogenic Soils

V. Ostroumov[1]

[1]Senior Researcher, Cryosol Laboratory, Institute of Physico-Chemical and Biological Problems of Soil Science, Russian Academy of Sciences, 2 Institutskaya St., Puchshino, Moscow region, 142290 Russia; e-mail vostr@issp.psn.ru.

1. Introduction

Processes involved in soil formation (gley, podzolic, and metamorphic processes, leaching, accumulation and retention of humus, and others) are based on "elementary" physico-chemical soil transformations (phase transitions, surface interactions, transfer of mobile phases, etc.). In cryogenic soils, ice, which substantially modifies thermodynamic conditions, complicates physico-chemical processes and induces a characteristic set of specific indices. The soils inherit a thin-platy structure from the cryogenic texture (from ice lensing), the silty fraction predominates, fissuring forms are diverse, and contrasting distribution patterns of soluble components occur (Makeev, 1978).

Understanding the peculiarities of mass transfer, physico-chemical changes and structural modification related to ice segregration in soils helps define the impact of cryogenic processes as soil-forming factors.

When freezing modifies physico-chemical processes in soil, the translocation of solutes and particles is of prime importance. Physico-chemical changes, appearance of stresses and deformations, and structural modification and rearrangements related to ice segregration intimately relate to the transfer of solutes and particles.

2. Transfer of Moisture as Vapor in Freezing Soils

At low moisture content in freezing soils, i.e., when there is free volumetric water and ice, water vapor plays the major role in the moisture transfer and film liquid accounts for less than 1% of the total flow intensity. According to Ershov's experimental data, the diffusion coefficient for vapor (Dv) is $4*10^{-6}$ ... $4.5*10^{-6}$ m^2/s and for liquid unfrozen water $2.5*10^{-10}$ ...4.5...10^{-10} (clay soil, moisture 2.5%, temperature -5°C) (Ershov et al., 1978).

In its pure state, moisture transfers as vapor during freezing in dry sandy soils. Such soils have no mobile liquid film because they do not retain any unfrozen moisture even at close to zero negative temperatures. The permeability of dry sandy soils to gases is sufficient for an effective transfer of vapor (the Dv is about $13*10^{-6}$ ... $15*10^{-6}$ m^2/s under such conditions) (Ershov et al., 1978). In a dry freezing soil, the separate needle-shaped crystals of ice mark the freezing front (Hallet, 1990).

Upon freezing of sandy-loam, loamy, and clayey soils, agglomerates of ablimational ice usually form in large pores and fissures which work as channels for the transfer of vapor (Ershov et al., 1987). Fissuring of soils upon freezing is associated with consolidation (Van Vliet-Lanoë, 1991) during cryogenic dehydration, while, in frozen soils, heterogeneity of the distribution of temperature stresses complicates the situation (Kudrjavtsev et al., 1973). Fissuring of frozen soils favors their aeration, leads to sublimation drying and additional cooling, and enhances heterogeneity of temperature distribution (Miller, 1967).

3. Sublimation of Ice and Transfer of Moisture as Vapor in Frozen Soils

In frozen soils, the structure of convective gas fluxes determines the transfer of moisture as vapor. The distribution of soil temperatures influences substantially these fluxes and the formation or sublimation of ablimational ice. Ablimational ice forms on relatively cold walls of pores and fissures. This directs the resulting transfer of vapor toward the heat flow (Kudrjavtsev et al., 1973). The distribution of ice in soil changes because of the vapor transfer during daily temperature oscillations and throughout longer thermal periodicities. Local ice ablimation (i.e., the formation of ice from water vapor) and sublimation (Ostroumov and Makeev. 1980; Zheleznyak and Kochmenyova, 1992) controls this change.

The heating of soils in the surficial horizon of the layer of daily temperature fluctuations induces sublimation drying, whereas ablimational ice accumulates in its inner portion. Sublimation predominates in the inner part of the layer of daily temperature fluctuations upon soil's cooling from the surface.

Alternating ice ablimation and sublimation leads to the appearance of a horizon of sublimational drying, underlain with a layer enriched in ablimational ice. Such ice-rich layers are typical in regions with an hypercontinental cold climate. The distribution of ice in the permanently frozen soil at a snow-free site in the Dry Valleys region (Antarctica) is a good example of this (Figure 3.1.1). Here, under permanently negative temperatures and low air humidity (Stearns et al., 1993), the moisture transfer as vapor in the soil plays the critical role in its mass exchange with the atmosphere and the underlying layers.

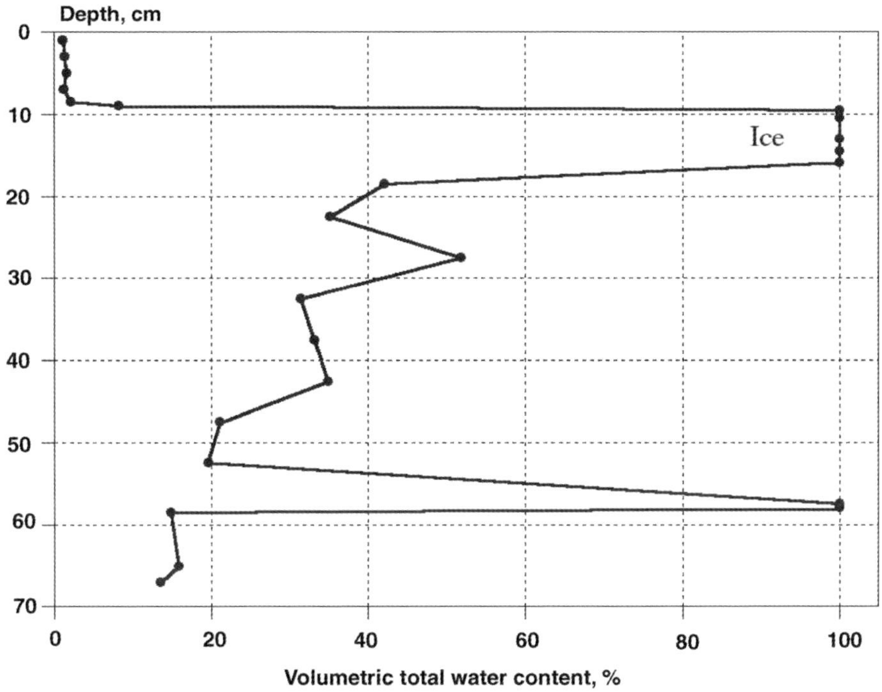

Figure 3.1.1. A layer of sublimational drying (0 to 9 cm) and accumulation of ablimational ice (9 to 17 cm) in a sandy soil (Mount Feather plateau, Dry Valleys region, Antarctica, 2960 m above sea level, mean annual air temperature -26°C).

The relative enrichment of ice with light oxygen isotopes in the accumulation horizon (Table 3.1.1; Dr. C. Siegert, Alfred Wegener Institute of Polar and Marine Researches, Potsdam, Germany, kindly provided the isotope analysis data) shows that the ice largely originated from ablimation and vapor transfer.

Figure 3.1.2 shows data on changes in total moisture content in the vicinity of the wall of a frost fissure in a frozen loamy cryotic boggy soil (northeastern Transbaikal area). The examples in Table 3.1.1 show that the loss of moisture in the soil layer with sublimational drying results in the appearance of a deeper horizon of accumulated ablimational ice (Ostroumov and Makeev, 1980).

Table 3.1.1. Oxygen isotope composition of ice at the Mount Feather Plateau (Figure 3.1.1).

Mass Spectrometer Determination						
	Ice rich horizons (Figure 3.1.1)				Glacier ice	Modern snow
Depth, cm	9.5	10.5	13	57.5		
$\delta^{18}/^{16}$	-34.0	-33.7	-33.8	-33.9	-31.9	-30.6

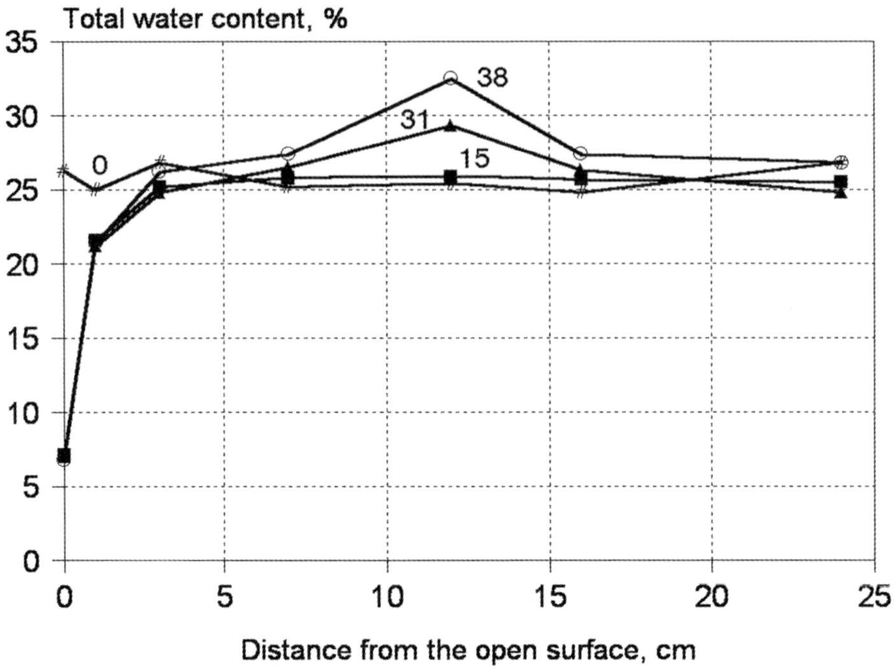

Figure 3.1.2. Dynamics of the total moisture content in a frozen loamy meadow boggy soil in the course of sublimational drying on an open vertical surface (Tungir graben, northeastern part of Transbaikal region, West Siberia). Digits at the curves indicate the time of drying, in days.

The transfer of moisture as vapor also plays an important role in the movement of gas inclusions in ice. Ice sublimates on the warm wall of an inclusion, while crystals of ablimational ice form in its cold wall. This dislocates inclusions in the direction opposite to the heat flow (Lipenkov, 1989).

Frozen soils dehydrate during sublimational drying. Sublimation of ice in a frozen soil leads to its deep drying and dehydration, which the low air humidity and the absence of local moistening in the frozen soil favor. According to I. Zheleznyak (Zheleznyak et al., 1992), in the course of their sublimational drying, the soils lose not only all of the capillary moisture but also a part of adsorbed moisture. This shows in changes of soil characteristics. Studies of properties of the frozen soils in the Transbaikal region have shown that individual soil structured units, from which ice is removed by sublimational drying, become waterproof (Ostroumov and Makeev, 1980). Apparently, in the course of sublimational drying, aggregates become water-resistant, owing to dehydration flocculation of colloids.

Redox processes occurring in the aeration zone of frozen soils favor stabilizing individual soil peds. Figure 3.1.3 shows data on changes in the content of mobile ferrous ions in frozen loamy meadow boggy soil (northeastern part of the Transbaikal region). A network of cryogenic fissures aerate the frozen boggy soil. A thin discontinuous snow cover and the increase in soil permeability to gases because of its sublimational drying facilitate aeration.

In the aeration zone, oxidation gradually decreases ferrous iron. Open fissures in frozen soils favor deep aeration, which induces oxidative processes even in boggy soils. The data in Figure 3.1.3 A, B, and C illustrate the dynamics of the oxidation degree of ferrous iron (extraction with sulfuric acid, reduction with hydroquinone, and complexonometric completion [Kkhudyakov, 1977]) in the frozen meadow boggy soil in the northeastern of Transbaikal region (Ostroumov, 1978).

In a frozen soil of the bog complex, a relative increase in ferric forms occurs during winter, anomalous for the reducing properties of bogs. The relative decrease in the content of ferrous ions occurs in the absence of its noticeable migration, owing to deep aeration of the frozen fissured soil. In the locked filtration transfer of the liquid phase, the redox processes constitute the key mechanisms responsible for the *in situ* metamorphization of podburs and other cryogenic soils (Targulian, 1971).

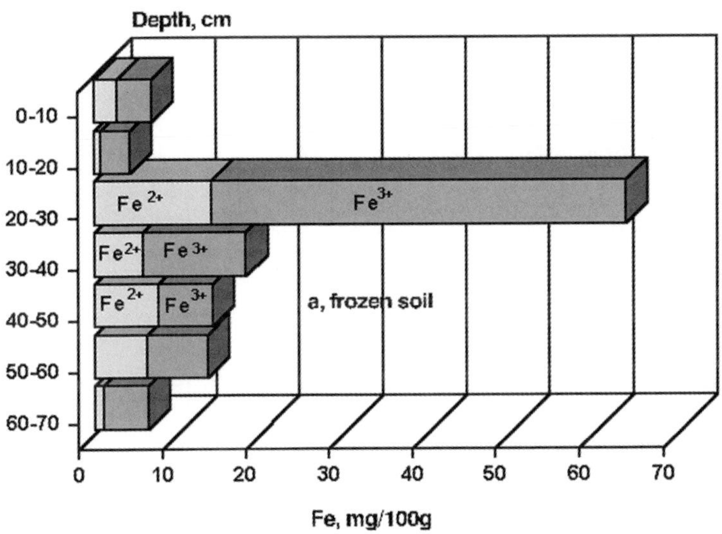

Figure 3.1.3 A. Forms of the mobile iron in frozen loamy meadow boggy soil (winter).

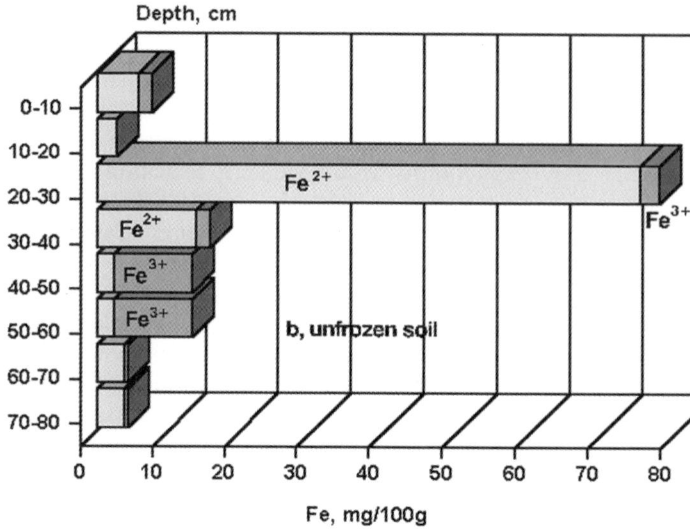

Figure 3.1.3 B. Forms of the mobile iron in unfrozen loamy meadow boggy soil (summer).

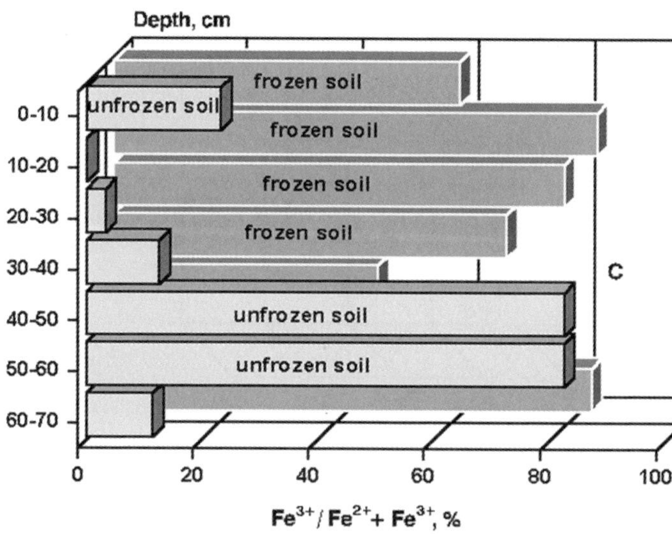

Figure 3.1.3 C. The oxidation degree of iron ($Fe^{3+}/Fe^{2+} + Fe^{3+}$, %) increasing during the sublimational drying in the flacked layer of frozen soil (0 to 40 cm).

4. Transfer of Liquid Moisture in Freezing Soils

4.1. Formation of massive texture in frozen soil

When ice cement forms in sandy soils, practically no migration of moisture occurs, and interstitial ice generally develops during freezing (Hallet, 1978).

4.2. Formation of segregation ice and generation of cryogenic textures

The freezing of loamy and clayey moisture-saturated soils induces the formation of inclusions of pure segregation ice as lenses or layers. Segregation ice is one of the most widespread types of soil ices (Vtyurin, 1975). It grows during soil's freezing as liquid film migrates from the thaw zone and available space allows an increase in the size of streaks (Ershov et al., 1987). According to Ershov's hypothesis (Ershov et al., 1979), the flux of unfrozen water toward the growing ice inclusions also stimulates segregation ice.

As waterlogged soils freeze, the moisture migrates from the active layer toward the freezing front, because the water potential is minimal at the ice's surface (McGaw et al., 1983; Voronin et al., 1989).

An inclusion of segregation ice begins forming as plate-forming crystal nuclei appear at the surface of soil particles. Heat channeled from the surrounding space to the ice crystals induces the influx of pore solution. The ice inclusion grows as long as moisture is supplied onto its surface and as long as the pressure of the growing ice crystals is sufficient for drawing the walls of the ice-filled pore farther apart. The liquid water flux to the growing ice inclusion drains the unfrozen zone close to the freezing front. When the liquid flux ceases, the inclusion growth stops. The phase transition isotherm rapidly shifts through the desiccated zone of soil, where no phase transition takes place, to the area where the moisture content is high enough to form a new ice lens (Zestkova, 1987; Ostroumov and Makeev, 1980).

These conditions ***promote*** the formation of lenses and layers of segregation ice: total saturation of soil with moisture; possible migration of moisture to the freezing front; presence of a colloidal fraction; pore solution of low salinity; presence of minerals susceptible to shrinkage on dehydration (to provide space for the growth of ice crystals); susceptibility of soils to compaction under the pressure of growing crystals; low freezing rate. The factors that ***prevent*** growth of segregation ice inclusions are low moisture content, high concentration of pore solution, and absence of swelling minerals that shrink upon dehydration (Hallet, 1978).

Layers, lenses, and streaks of segregation ice are typical elements of the cryogenic fabric of ice-saturated frozen soils of loamy and clay texture (Vtyurin, 1975). In sandy soils, despite their high permeability, inclusions of segregation ice form only in the presence of colloidal fractions (Popov, 1981). For sandy soils, the presence of segregation ice and layered and streaky textures associated with it is an exception rather than the rule (Popov, 1981).

4.3. Concentration of dissolved components at the freezing front

In the freezing zone, the pore solution is drawn toward the surface of crystals. According to Yazynin's observations performed using cryomicroscopic techniques (Yazynin, 1986), the moisture flow migrating to the freezing front involves not only dissolved substances but also clay particles. He also observed dislocations of larger grains and aggregates. Water molecules complete the buildup of pure ice crystals, causing the growth of a pure ice lens or layer. The dissolved components of pore solution (ions, colloids) remain in the unfrozen liquid, the concentration of which increases with the growth of ice crystals. The growing crystals expel the brine toward the heat source. In this situation, the solution concentration of the unfrozen liquid increases by tens of times in the pore. Thus, owing to the press-off of dissolved components, the interface between the frozen and thawed zones becomes the site of their concentration; the pore solution remains liquid after freezing (Miller, 1967).

Figure 3.1.4 shows an example of the concentration zone of potassium ions (microsampling [Ostroumov, 1997]; acetate-ammonium extracted with atomic emission-based determination [Ostroumov, 1997; Ostroumov, 1998.]) at the interface between the frozen and thawed layers in a freezing sample of loamy soil. The ion concentration zones are preserved after freezing of the soil (Figure 3.1.5). The position of concentration zones relative to the streaks and layers of segregation ice indicate the direction of freezing. Soil aggregates enclosed between segregated ice lenses or streaks lose the greater part of their moisture as they freeze. In the frozen soils with a layered or streaky texture, moisture content in the interlenses approximates the level of molecular water retention (Ershov et al., 1987).

Because of the drop in soil temperature after freezing, the zones of concentrated dissolved components become the sites that accumulate crystalline and amorphous (oxide) precipitates, form crystal hydrates, flocculate colloids, and form solid eutectics (Ershov et al., 1996), which processes occur simultaneously with redox reactions. As a result, individual cryogenic peds prove to be chemically fixed.

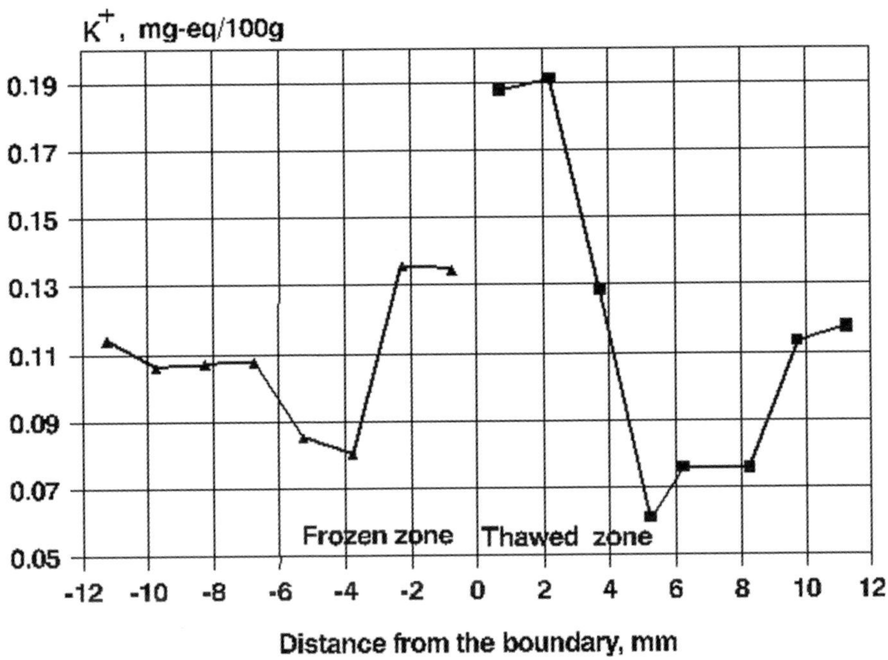

Figure 3.1.4. Distribution pattern of mobile potassium at the frozen/thawed zone interface in a sample of freezing loamy soil (laboratory experimental data).

They become waterproof; their surface is sealed with films and coatings that repeat the zones of concentration of dissolved components (Ostroumov and Makeev, 1980). This favors the preservation, after the soil's thawing, of cryogenic aggregates produced by ice segregation. During the next freezing, the ice segregation renucleates exactly in the soil inter-aggregate porosity which the former lensing of ice left. This further stabilizes the cryogenic peds (Ostroumov and Makeev, 1980).

5. Transfer of Unfrozen Liquid Moisture in Frozen Soils

In frozen soils there is no mobile bulk moisture. For this reason, filtration of the bulk pore solution, which is the major mechanism of mass transfer in thawed soils, is blocked (Murman, 1973). The basic role in the transfer of unfrozen liquid in frozen soils is played by other mechanisms, such as osmotic transfer, diffusion, flowing of films, and translation of liquid inclusions (Cass and Miller. 1959; Frolov, 1998).

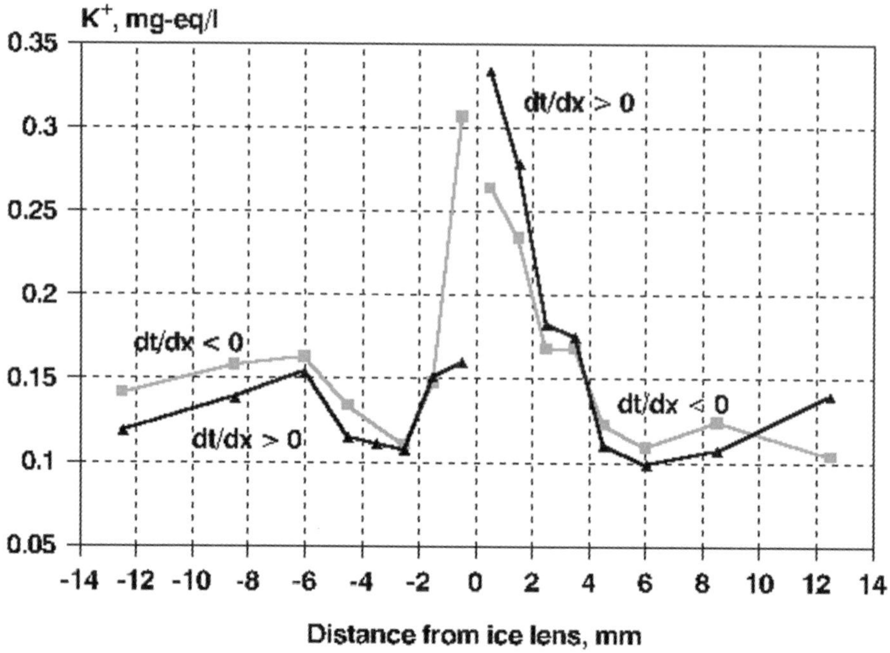

Figure 3.1.5. Distribution pattern of mobile potassium at the interface of dehydrated dispersed material and lens-like inclusion of segregational ice in a sample of frozen loamy soil (laboratory experimental data).

The existing classifications differentiate the following categories of unfrozen moisture in frozen dispersed systems: films at the surface of oxides; silicates and organic matter; films at the surface of ice crystals; and microzones with high concentrations of dissolved components. Of all these types, the liquid films at the surface of ice crystals display the highest mobility (Ershov et al., 1979). Despite its low absolute content, this type of unfrozen moisture largely determines the permeability of frozen soils for dissolved components, in particular for ions.

In frozen soils, the highest content of unfrozen water occurs in the films at the surface of silicates, oxides, and organic matter. This moisture is in the field of surface attraction forces, and its involvement in the transfer of mobile components is not important (Maeno, 1992). The unfrozen moisture of liquid inclusions is localized in closed microvolumes. Migration of such inclusions in soils does not make any appreciable contribution to the transfer of unfrozen moisture (Ershov et al., 1979). However, the material inside these inclusions contains no ice, and it is as highly permeable as the unfrozen soil.

6. Transfer of Dissolved Components in Frozen Soils

Contrary to common current concepts, the frozen ice-saturated soil is not an impermeable substrate. The data reported by Maeno (Maeno, 1992) indicate that the electric conductivity of frozen soils is only several-fold lower than that of their thawed analogs. The electric conductivity of frozen soils is almost exclusively ionic, which indicates rather high permeability of frozen soils to ions (Maeno, 1992). Transfer of ions and other mobile compounds occurs in frozen soils not only under the influence of electromagnetic fields but also from the action of gradients of temperature, pressure, concentration, and other factors affecting the thermodynamic potentials of mobile phases.

Figure 3.1.6 shows the dependence of the effective coefficient of ion diffusion on moisture content in samples of frozen loamy grey forest soil. At both positive and negative temperatures and with the moisture content below hygroscopicity (the left bars in Figure 3.1.6), this soil proves virtually impermeable for ions (the ion diffusion coefficient Di < 10^{-14} m^2/s). At this moisture level, no ice is in the samples, whereas all unfrozen moisture is localized in the non-solubilizing volume (Ananian and Poltev, 1987).

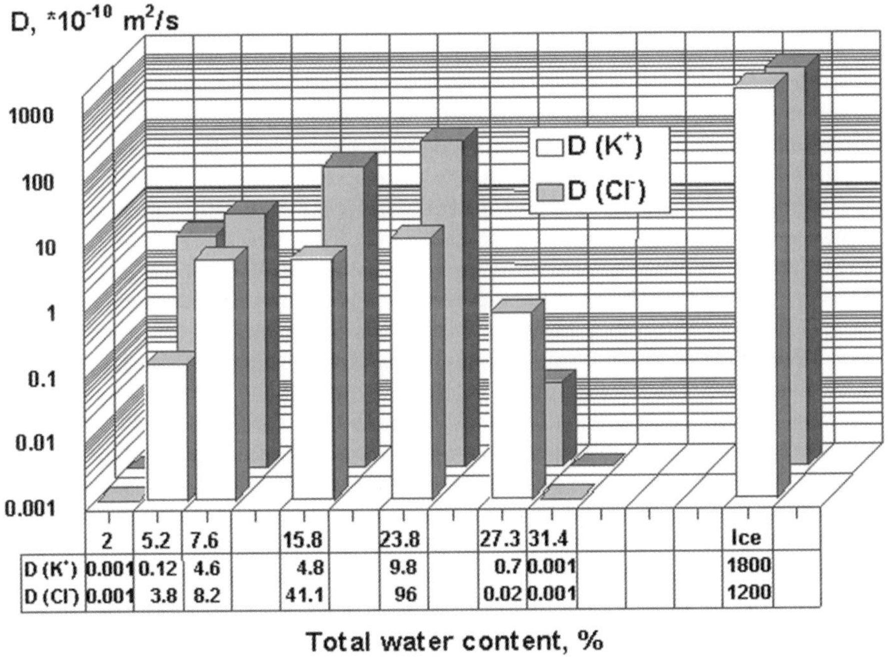

Figure 3.1.6. Effect of the total moisture content (without differentiation of ice and unfrozen moisture) on the effective diffusion coefficient of potassium ions in a frozen loamy soil (middle temperature, -6°C).

As the moisture content increases, ice crystals appear in the samples, and the diffusion coefficient of ions rises dramatically (the higher bars in Figure 3.1.6). The permeability of the frozen loam sample for both cations and ions is maximal ($D_i < 10^{-10} \ldots 10^{-9}$ m^2/s) when the bulk moisture content is somewhat lower than porosity. The surface of ice crystals in the frozen soil is then maximal (Fedoseyeva, 1998). This indicates that the transfer of ions in the frozen loamy soil occurs basically in the unfrozen moisture of quasi-liquid films at the ice/pore air interface. The maximum coefficients of ion diffusion in the frozen soils are only one order of magnitude lower than those in their thawed analogs. R. Murman (Murman, 1973), E. Ershov (Ershov et al., 1996), and others carried out such comparisons.

The high ion permeability of snow and the transfer of ions and complex compounds in snow are associated with films of unfrozen water on the surface of ice crystals (Nechaev et al., 1981). In particular, these processes induce formation of a zone of enrichment of snow in metals in the areas with geochemical anomalies. The basic mechanism underlying the transfer of metals from soil to snow is assumed to be a transfer in films at the surface of ice crystals, induced by the thermal gradient (Nechaev et al., 1981). The distribution pattern of dissolved components in individual crystals of diagenetic ice (firn) provides evidence of the involvement of the surface of ice crystals in the transfer of ions.

Figure 3.1.7 shows the data derived from the slow thawing of a firn sample under the action of the filtration air flow with a temperature of +0.5°C (this experiment was in collaboration with Dr. V.V. Demidov). As the thawing of soil proceeded, the water formed was sampled to determine the content of sodium ions (atomic emission); the size of firn crystals was measured with the aid of a telescopic lens. In the course of thawing, the mean size of crystals diminished linearly, making it possible to interpret the temporal changes in the concentration of ice melt water as the distribution of sodium ions in firn grains. The data of Figure 3.1.7 indicate that the greater part of ions is localized in the surface layers of ice crystals, whereas the inner part of the crystals has a pure composition. The high content of ions in the surface layers of ice crystals in firn and their high concentration in the early melt water confirm the important role ice crystals' surface play in the ion transfer.

In soils, the unfrozen moisture localized in films at the surface of ice crystals performs a similar transport function, which provides exchanges sustaining the metabolism of viable microorganisms, despite frost.

When the frozen soil contains the segregated ice lenses and layers, the ion permeability gradually decreases, ultimately becoming much lower (by 3 to 4 orders of magnitude) than that of pure ice. Thus, frozen soil containing segregation ice has a lower permeability than either pure ice or low ice.

The barrier role of the interface separating two highly permeable media, soil aggregates dehydrated during freezing and inclusion of pure segregation ice, explains the abnormally low ion permeability of ice-saturated frozen soils. The data in Figure 3.1.7 allow comparison of two distribution patterns of ions. The first (curve 1) resulted from a slow freezing of soil when the heat source was placed on the right side. The second came from an analogous slowly freezing sample but after a week-long exposure under an inverted temperature gradient (after freezing, the heat source was placed near the left side of the sample). The inversion of the temperature gradient resulted in a decrease in the content of ions in the concentration zone. A new concentration zone began to form at the opposite face of the soil aggregate. This example shows that changes in the temperature gradient induce a transfer of ions inside the mineral soil enclosed between its interfaces with ice inclusions.

Figures 3.1.8 and 3.1.9 feature the temperature dependencies of dielectric parameters of two frozen samples of a loamy soil. The samples differed in ice content and ice forms. Both samples were prepared from a homogeneous paste, into which electrodes of a dielectric permitivity sensor (Vitel Moisture Probe) were impaled prior to freezing. The first sample (Figure 3.1.8) had an ice content of about 30% and a massive cryogenic texture from rapid freezing. In the second sample (Figure 3.1.9), a slow single-front freezing with the free supply of distilled water formed a thick layered cryogenic texture. After preparation and storage at a constant negative temperature, the samples were slowly heated. In the course of heating, their polarizability and dielectric losses were measured.

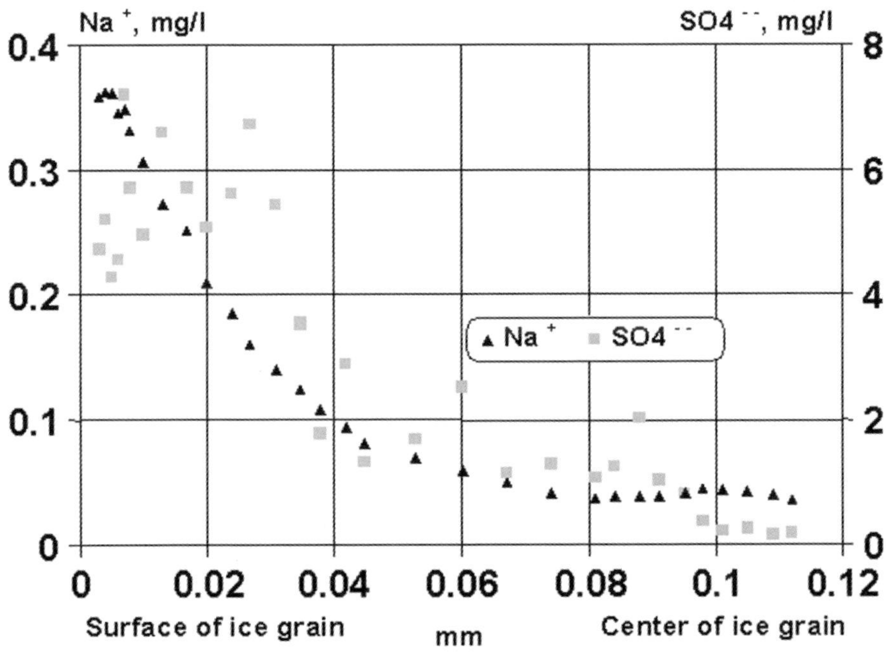

Figure 3.1.7. Distribution of ions in an ice grain (firn).

In the sample with a massive cryogenic texture (Figure 3.1.8), the polarizability and dielectric losses increased monotonously upon heating, whereas the corresponding curves had the same shape as these of unfrozen moisture content. With the increase of temperature, the content of liquid moisture in the frozen loam sample increased gradually and induced a monotonous growth of polarizability (basically owing to the orientation with quasi-elastic rotation of dipolar molecules within the liquid phase) and dielectric losses (basically owing to the forward migration of charged particles in the liquid). It is fairly probable that this occurred due to a gradual increase in the thickness of unfrozen water films.

In the sample with a layered cryogenic texture, both polarizability and dielectric losses also increased during heating. However, at temperatures above -8°C, both parameters successively reached several extreme values, each of which corresponded to a phase transition of a portion of moisture in the zones of ion concentration and in the dehydrated soil aggregates. Comparison of the aforementioned curves confirms that, in the frozen ice-saturated soils containing inclusions of segregation ice, a part of the unfrozen moisture is localized in the zones of concentration of dissolved components and does not form a film but, rather, inclusions of concentrated solution.

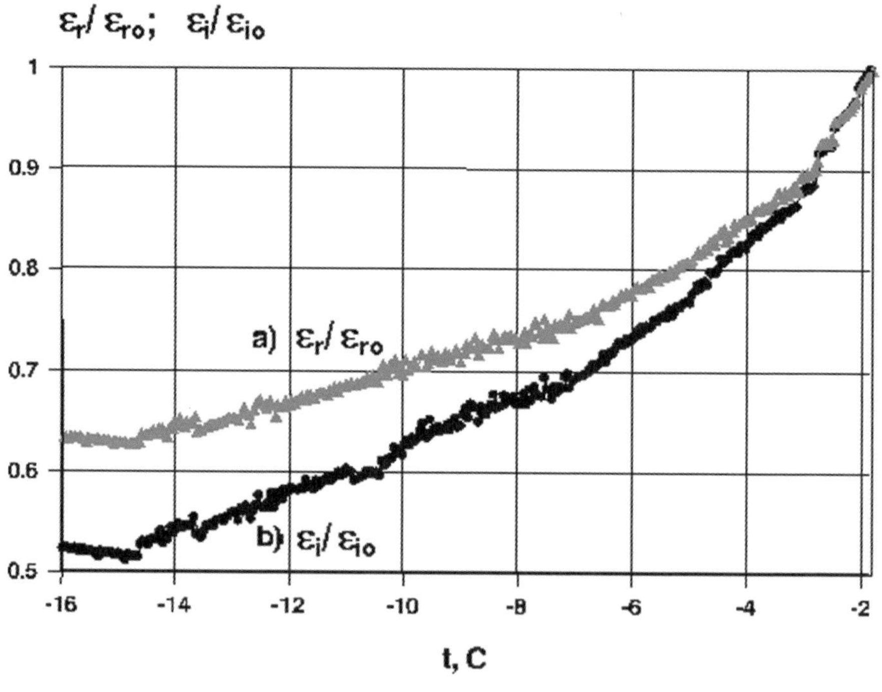

Figure 3.1.8. Temperature dependence of polarizability (a) and dielectric losses (b) in a sample of frozen loamy soil with a massive cryogenic texture. Homogeneous distribution of finely dispersed ice and soluble components.

The frozen soil may be featured as a spatial entity of well permeable domains (soil aggregates and ice inclusions) isolated from each other by the impermeable boundaries. Inside the domains, the medium permeability is fairly high (10^{-10}... 12^{-12} m^2/s in soil aggregates and 10^{-9}...10^{-11} m^2/s in ice against 10^{-7}...10^{-9} m^2/s in soils at positive temperatures).

The permeability of interfaces between pure ice and soil aggregates restricts the overall permeability of a bulk of frozen icy soil and is estimated with a coefficient of ion diffusion of about 10^{-14} m^2/s. For frozen soils, there are descriptions of processes of ion exchange, surface interactions, chemical reactions, and probable microbial metabolism, the limiting factor of which is the ion permeability of the medium. Apparently, these processes can occur inside microorganism's cells where soil has a sufficiently high permeability. The mass exchange does not play the role of the limiting factor for these processes in closed domains. However, transit liquid transfer involving large volumes does not take place in frozen soils. This favors *in situ* metamorphization of soils.

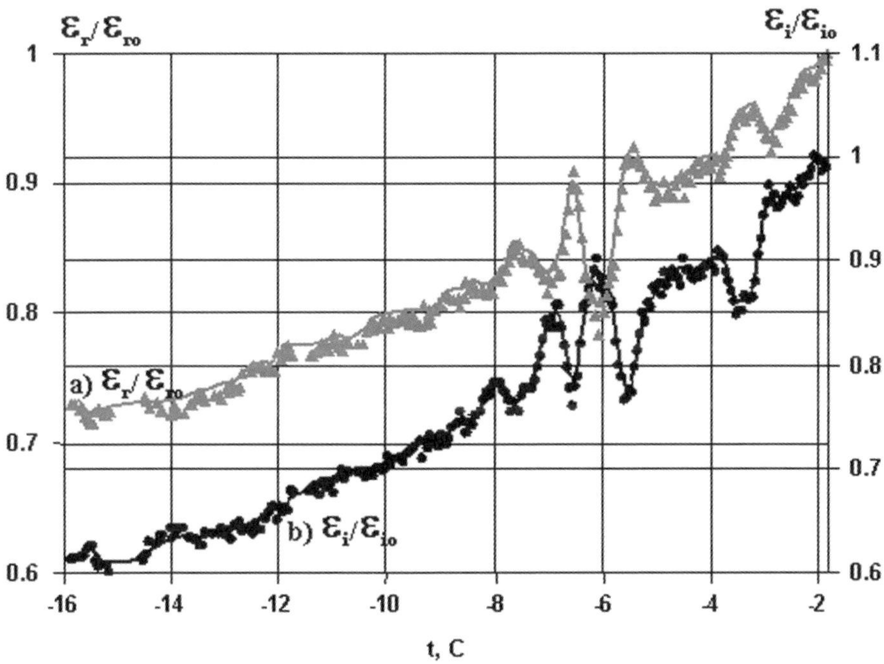

Figure 3.1.9. Temperature dependence of polarizability (a) and dielectric losses (b) in a sample of frozen loamy soil with a layered cryogenic texture. Ice forms segregational layers 20-30 mm thick. The greater part of soluble components is localized in narrow concentration zones inside the layers of dispersed material at the interface with segregational ice.

This model of frozen soil structure corresponds to Frolov's conception of the spatial-skeleton structure of frozen dispersed media (Frolov, 1998). In addition to a network of cryogenic fissures promoting free gas exchange, the spatial entity of permeable cells and impermeable ice/soil aggregates interfaces is an important type of soil heterogeneity of a specific cryogenic origin.

7. Acknowledgments

I gratefully acknowledge Dr. Ch. Siegert, Mr. V.A. Sorokovikov, Mr. A. Vlasenko, Dr. D. Fedorov-Davydov, , Dr. G. Wilson, Dr V. Demidov, Ms. N. Ostroumova, Dr. D. Gilichinsky, Dr. S. Gubin, and Ms L. Pasnitskaya for their help in experiments and field work. Prof. A. Frolov and Prof. B. Van Vliet-Lanoë made important contributions to the work. The Russian Foundation of Basic Researches (projects RFBR-03-05-64455, 03-04-49325, and 03-05-79102) and the International Association for the Promotion of Cooperation with Scientists from the independent states of the Former Soviet Union (project INTAS 01-2329) support these investigations.

References

Ananian, A.A., and N.F. Poltev. 1987. Non-solubilizing volume of water in disperse grounds. Geocryological researches, v. 18. Moscow Univ. Press. Moscow. pp. 193-195.

Cass, L.A., and R.D. Miller. 1959. Role of the electrical double layer in the mechanism of frost heaving. CRREL Research Report. 15 pp.

Ershov, E.D., E.Z. Kuchukhuv, Yu.P. Akimov, I.A. Komarov, and V.V. Gurov. 1978. Ice sublimation as the method of study of ice and frozen grounds. Geocryologial researches, v. 15. Moscow Univ. Press. Moscow. pp. 243-247.

Ershov, E.D., V.G. Cheveriev, Yu.P. Lebedenko, O.M. Yazynin, and A.V.Shashkova. 1979.Water transfer and segregate ice formation in frozen zone of thawing grounds. Geocryological researches, v. 18. Moscow Univ. Press. Moscow. pp. 179-192.

Ershov, E.D., Yu.P. Lebedenko, V. V. Kondakov, E. M. Chuvilin, and M.N Provorova. 1987. Cryogenic texture formation in water rich sediments. Geocryological researches. Moscow Univ. Press. Moscow. pp. 133-140.

Ershov, E.D., E.M.Chuvilin, O.U.Smirnova, and N.S. Naletova. 1996. Transfer of chemical element ions in freezing grounds. 2nd Forum on Heat/Mass Transfer Proc., v. 7. Minsk. pp. 16-20.

Fedoseyeva, V.I. 1998. Experimental study of gold migration in frozen sands. Probs. geocryol. Yakutsk. pp. 105-109.

Frolov, A.D. 1998. Electric and elastic properties of frozen earth materials. Pushchino. 515 pp.

Hallet, B. 1978. Solute redistribution in freezing ground. 3th Internation. Conference on Permafrost Proc., v. 1. Ottawa. pp. 86-91.

Hallet, B. 1990. Self-organization in freezing soils: from microscopic ice lenses to patterned ground. Can. J. Phys. 68(9):842-852.

Khudiakov, O.I., O. Makeev, A. Ksenzova, and A. Buzenko. 1977. Soil resources of the Mari regions, Baikal-Amur Zone, of northeastern Transbaikalien. Pushchino. 24 pp.

Kudrjavtsev, V.A., E.D. Ershov, V.V Gurov., and V.G. Kondrat'ev. 1973. Peculiarities of ice sublimation under the interaction in disperse grounds. Geocryological Researches, v. 13. pp. 183-190.

Lipenkov, V.IA. 1989. Formation and decomposition of air hydrates in glacier ice. Materialy gliatsiologicheskikh issledovanii. Khronika obsuzhdeniia. 65:58-64;

Maeno, M. 1992. Dielectric response of water and ice in frozen soils. Phys. and chem. Hokkaido Univ. Press. pp. 381-386.

Makeev, O.V. 1978. The soils of polar climatic regions their major limitations for food production and he overriding climatic restraints. In: plenary sessions papers, 11th Intern. Congr. Soil Sci. Edmonton. pp. 234-237.

McGaw, R.W.; R.L. Berg, and J.W. Ingersoll. 1983. An investigation of transient processes in an advancing zone of freezing. In: Permafrost : Fourth International Conference. Proc. National Academy Press. Washington, D.C. pp. 821-825,

Miller, R.D. 1967. On the mobility of water molecules in the transition layer between ice and solid surface. J. Colloid Sci. 25(2):166-173.

Murrman, R. 1973. Ionic mobility in permafrost. In: Permafrost: Second International Conference. Proc. 1973, Yakutsk, U.S.S.R. Nat'l Acad. Sci. Washington, D.C. pp. 352-359.

Nechaev, E.A., V.I. Fedoseyeva, and N.F Fedoseyev. 1981. Surface properties of disperse ice. J. Phys. Chem. 55(7):1822-1826.

Ostroumov, V. 1978. Influence of oxidation and reduction of mobile iron on the active layer formation. 5th Conference of young scientists. Moscow Univ. Press. Moscow. pp. 47-54.

Ostroumov, V., and O. Makeev. 1980. Winter drainage of frozen fissuring soils. Soil Sci. 10:80-87.

Ostroumov, V. 1997. Redistribution of ions close to the growing ice inclusions. Ground Freezing. International Symposium of Ground Freezing and Frost Action in Soils Proc. Rotterdam. pp. 397-400.

Ostroumov, V. 1998. Impact of freezing on distribution of ions in soils. Euro. Soil Sci. 5:614-619.

Popov, A.I. 1981. Peculiarities of ice formation in tixotrophic disperse grounds. Problems of Cryolithology, v. 9. Moscow Univ. Press. Moscow. pp. 30-40.

Stearns, C.R., L.M. Keller, G.A. Weidner, and M. Sievers. 1993. Monthly mean climatic data for Antarctic automatic weather stations. Antarc. J.

Targulian, V. O. 1971. Soil formation and weathering in cold humid regions. Nauka Press. Moscow. 267 pp.

Van Vliet-Lanoë, B., and A. Dupas. 1991.Development of soil fabric by freeze-thaw cycles. Its effect on frost heave. In: Ground Freezing. International Symposium of Ground Freezing. Proc. 1991. Balkema. pp. 189-195.

Voronin, A.D., E.M. Savel'eva, and V.E. Ostroumov. 1989. Physical chemical processes in soil cryogenesis. Vsesoiuznyi s'ezd pochvovedov, 8th, Novosibirsk, Aug. 14-18, 1989. pp. 264-268.

Vtyurin, B.I. 1975. Underground ices of Soviet Union. Nauka Press. Moscow. 214 pp.

Yazynin, O. M. 1986. Structure formation in freezing grounds. Moskow Univ. Moscow. 17 pp.

Zestkova, T.N. 1987. Peculiarities of ground freezing in open system. Geocryological researches, v. 18. Moscow Univ. Press. Moscow. pp. 1124-132.

Zheleznyak, I.I., and N.V. Kochmenyova. 1992. Sublimation drainage of frozen soils on Transbaikalien Cryosoils: The effect of cryogenesis on the processes and peculiarities of soil formation. 1 Int. Conf. on Cryopedology Proc. Pushchino. pp. 122-126.

Chapter 2. Micromorphology of Cryosols

Brigitte Van Vliet-Lanoë[1], Catherine A. Fox[2], and Stanislaw V. Gubin[3]

[1]Director of Research, Processus et Bilan des Domaines Sédimentaires, UMR 8110 CNRS & University of Lille 1, 59655 Villeneuve d'Ascq, France; e-mail Brigitte.Van-Vliet-Lanoe@univ-lille1.fr.
[2]Research Scientist, Agriculture and Agri-Food Canada, Southern Crop Protection and Food Research Centre, 1391 Sandford Street, London, Ontario, Canada, N5V 4T3; e-mail foxc@em.agr.ca.
[3]Leading Researcher, Institute of Physico-Chemical and Biological Problems of Soil Science, Russian Academy of Sciences, Institutskaya St., Pushchino, Moscow region, 142290 Russia; e-mail gubin@issp.serpukhov.ru.

1. Introduction

Micromorphological analysis makes it possible to study the interrelationships between the various individual components, particles, and pores that make up sediments and soils.

1.1. Methodology

First, we removed intact samples from the various horizons or layers of material described in the field. Then, in the laboratory, to preserve the integrity of the various components in an undisturbed state, we impregnated the samples with a polyester resin to consolidate the individual particles yet preserve the arrangement of the soil pores, aggregates, and soil material (Murphy, 1986; Fox et al., 1993). We cut the resin-impregnated sample, mounted it onto a glass slide (referred to as a "thin section"), and ground the soil material to approximately 30 µm thick to facilitate viewing its components with a polarizing light microscope. One then could examine the sample for features associated with various soil processes.

Identifying the types of features and then interpreting them in context with the environmental setting provides information about the genesis of the soil. For example, one often can distinguish features resulting from primary sedimentation (original deposition of materials) from features developed from such subsequent processes as redeposition, frost action, and soil formation, and often can arrange modifications to soil morphology in their sequence of formation.

1.2. Research history

Micromorphological investigations of (sub) Arctic and Antarctic soils are few. These scientists undertook some of the pioneering work on the morphology of these soils: Kubiena (1953, 1970), in the Antarctic and Arctic; Fitzpatrick (1956), in Svalbard; Kosheleva (1958), in the Yamal peninsula; Dumanski and St Arnaud (1964), on the boreal soils of Canada; Morozova (1965), in central Yakoutia; Parfenova and Yarilova (1967), Fedorova and Yarilova (1972), and Brewer and Pawluk (1975). They described various types of platy fabric and granular fabric and eluviation-illuviation processes (mud cutans; Kubiena, 1970, and Fitzpatrick, 1956) as common features of these soils.

Micromorphological analysis usually applies to unfrozen material from summer field samples. Tarnocai (1993) describes methods to sample frozen soils and gives suggestions for taking field samples intended for preparation of thin sections. Gubin and Gulyaeva (1997) made a special attempt to compare the microfabric of frozen soils.

1.3. Conceptual background

Rock-inherited features at present dominate cold regions and mountain soils, which also are subject to cryoturbation processes and a reduced rate of soil profile development. Evidence of strong weathering can be present but often derives from Neogene or even Cretaceous pedogenesis, as Kubiena (1970) stressed, or from an early Holocene or late glacial phase of pedogenesis, as in recently deglaciated areas.

Saprolitic materials (that is, residual material formed during another period) often are reworked and incorporated into the till material, as reported for Fennoscandia, the Canadian Shield, and Siberia (goethite clasts, kaolinite microaggregates, vermiculite microaggregates, hematitized grains) (Kubiena, 1970; Van Vliet-Lanoë, 1988a; Tarnocai and Valentine, 1989). Saprolites often are the parent material in the Siberian lowlands, outside the wide fluvial and aeolian quaternary cover (Gubin, 1997).

Consequently, lithological fabric can prevail or strongly influence the microfabric development; obliquely oriented clast fabric related to fluvial deposition, stretching, and the disruption of thinly laminated sediments might exaggerate frost-induced microfabrics. Sedimentary aggradation is also recorded in soil, as well as fluvial, colluvial, or aeolian importation.

In addition, concepts have evolved during the past few years concerning the "absence of biological activity" or the prominent role of cryostatic pressures in permafrost soils. Measurements during experimentation have limited drastically the action of cryostatic pressures by opposition with differential frost heave.

This chapter discusses the patterns and morphologies one can observe in cryosolic soils in terms of the physical processes related to ice formation and freezing and thawing; the biological processes as affecting humus formation; the chemical processes related to organo-chemical weathering; and the large-scale environmental impacts related to cryoturbation and movement of materials due to slope position.

2. Cryogenic Processes

2.1. Basic fabric of ice lensing

Cryosols for some time have included regular platy to angular platy peds resulting from the formation of ice lenses (Figure 3.2.1). McMillan and Mitchel (1953) and Dumanski (1964) and Dumanski and St. Arnaud (1966), who referred to the pattern as an "isoband fabric" (Figure 3.2.1A), provided some of the first micromorphological descriptions from soils and laboratory experiments. This fabric most often occurs in soils with silty textures or at depth in Cryosols as a result of retraction fissures (desiccation).

Smooth planar fissures with unconforming boundaries delimit the platy aggregates. If ice lensing is weak, one can confuse it with desiccation features. Triangular or square convex interpedal pores (Figure 3.2.1D-E) often occur (Van Vliet-Lanoë, 1976). Large pores exist commonly at the base of lithoclasts or compact aggregates, which probably relates to a reduced water supply for ice lens formation and increased retraction. Prismatic (or blocky, compact) fabric (Figure 3.2.1D) develops in either originally water-saturated unconsolidated clays or silty clay (marine or lacustrine) or by abrupt cooling (Van Vliet-Lanoë et al., 1984); it is common in minerogenic palsas and in some fragipan horizons. Compaction of the aggregate relates to frost desiccation and the mechanical stresses, which the growth of ice lenses induce and which supplementary sources of water supply (migration of adsorbed water or of vapor, and refreezing of melting water) may enhance. Various patterns of soil fabric directly relate to ice lens formation.

2.2. Freeze-thaw microfabric

When alternating freeze-thaw cycles affect a soil, the soil morphology that results from ice lensing evolves to lenticular to foliated peds somewhat cooled by internal micro-erosion, particle translocation, and some plastic deformation (Figures 3.2.1, 3.2.2, 3.2.3, 3.2.4, 3.2.5). Dumanski (banded fabric, 1964) and Yarilova and Fedorova (1972) first described these fabrics (Figures 3.2.1F, G, H, and 3.2.4B) in active soils, and Romans and Robertson (1964), Fitzpatrick (1976), and Van Vliet-Lanoë (1976), in fossil material.

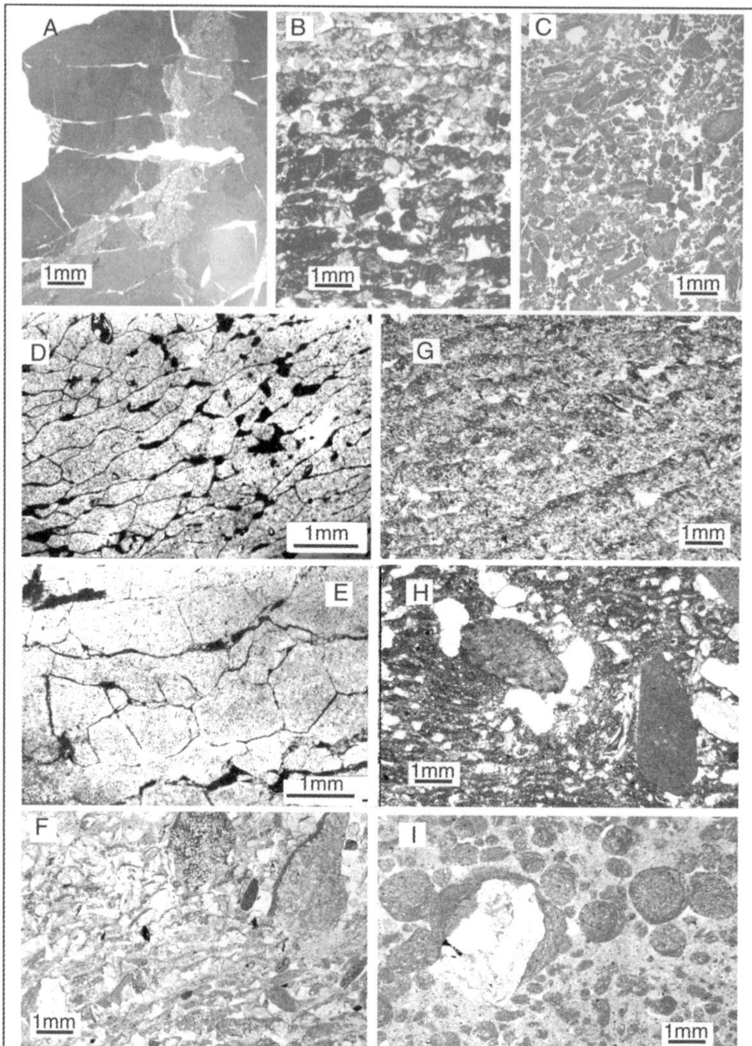

Figure 3.2.1. Cryogenic fabrics. A. Upper permafrost horizon, in glaci-marine silts (George River, Nunavut, Canada) (VV). B. Seasonally frozen saprolite (Mont Jacques Cartier, Québec, Canada) (VV). C. Surface of a recent mudboil in silt clays (Ny Alesund, Svalbard) (VV). D and E. Thin sections in frozen samples, at 30 and 60cm (Siberia) (SG). F. Freeze-thaw (banded) microfabric in a silty sandy paleosoil on till (Gasebu, Svalbard) (VV). G. Banded microfabric in a very fine soliflucted sand (glacial) covered by a cryptogamic crust (Finnish Lapland) (VV). H. Banded microfabric in a dolomitic residual silt (sorted circles, Blomstrand Svalbard) (VV). I. Granular /orbiculic fabric in a former laminated loamy sand, treched by combined solifluction and cryoturbation, C horizon of a Cryorthod, (N.Quebec, Canada) (VV). Photos by Van Vliet-Lanoë (VV) and Gubin (SG).

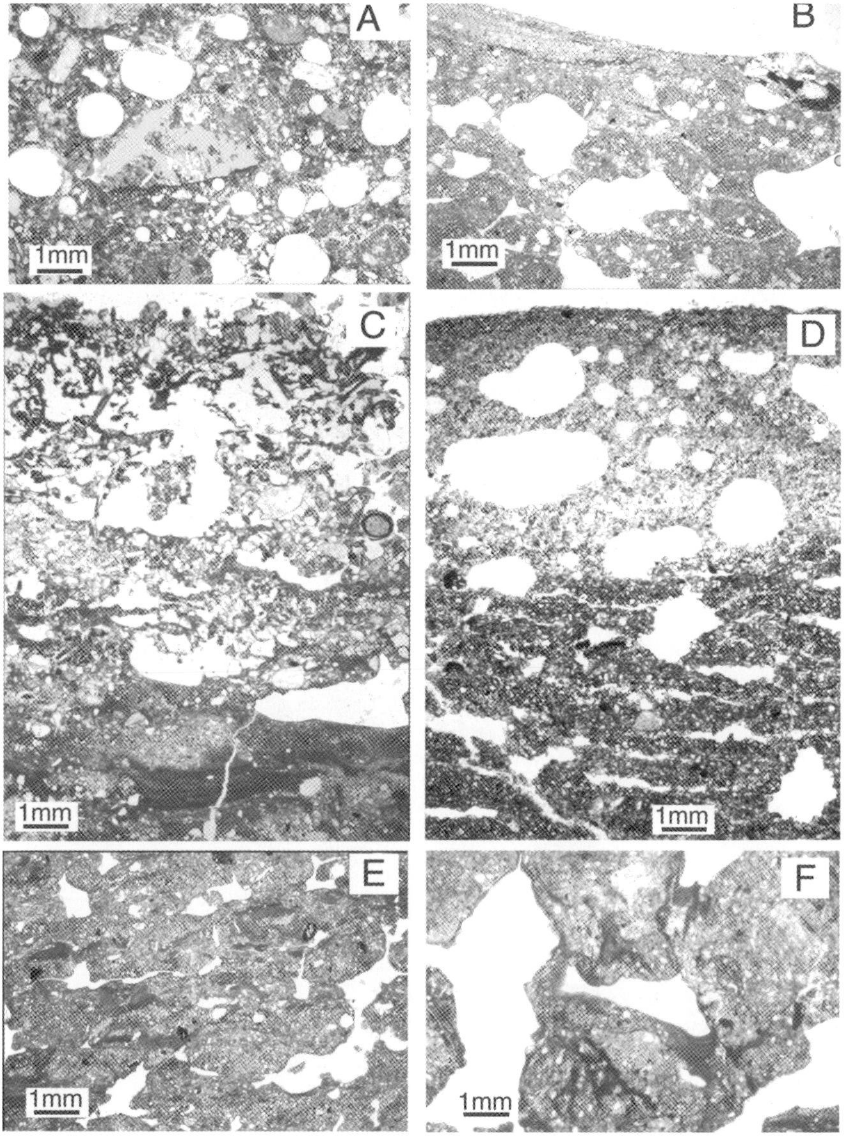

Figure 3.2.2. Vesicules and translocation. A. Rounded vesicles, surface of a mudboil in till (Salluit, Nunavut, Canada). B. Mamillated vesicles and surface splash crusting on clayey silt (Ny Alesund, Svalbard). C. Particle translocation below a cryptogamic crust on a mudboil (N.Quebec, Canada). D. Particle translocation at the surface of a liquefaction susceptible mudboil on till (Salluit, Nunavut, Canada). E. Coarse clay illuviation integrated in a mudboil on a silt cored palsa (northern Quebec, Canada). F. summer clay skin in frost shattered shales (Ny Alesund, Svalbard). Photos by Van Vliet-Lanoë.

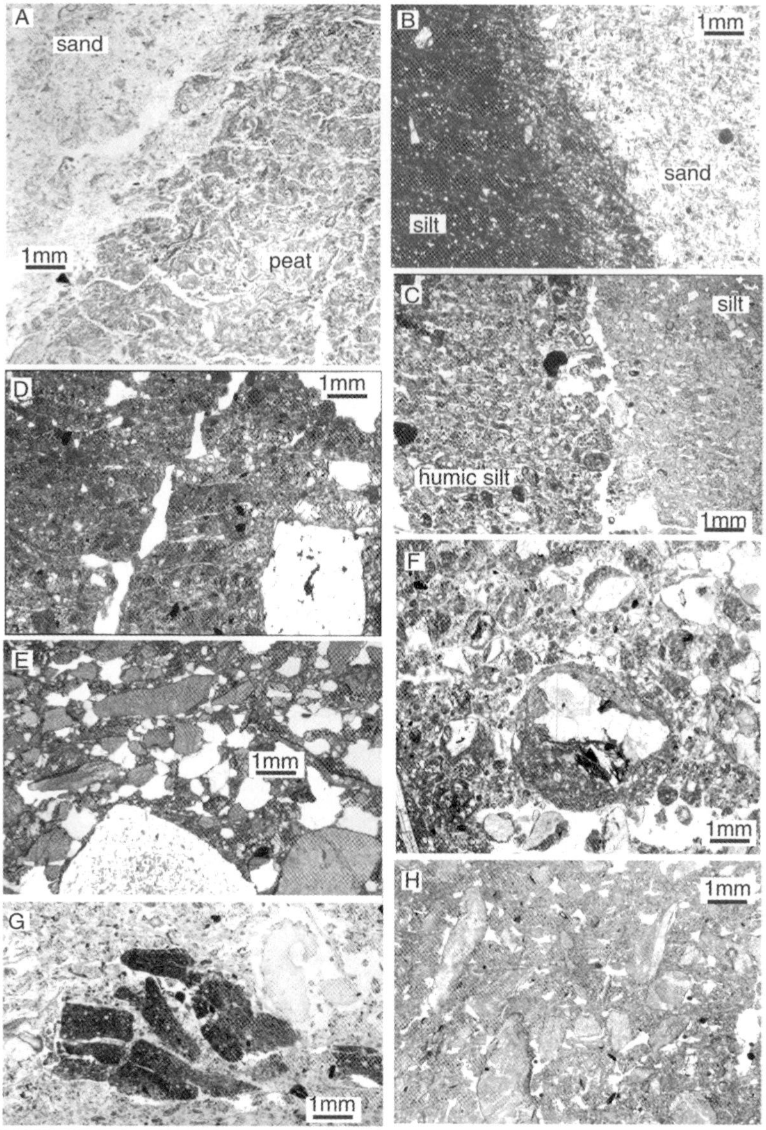

Figure 3.2.3. Cryoturbation and solifluction. A. Cryoturbated contact between a peat (right) and a sand (Kvadehuksletta, Svalbard). B. Cryoturbated contact between a dolomitic silt and an aeolian sand, Gasebu Svalbard). D. Gentle frostcrep in a clayey silt (Brogger moraine, Svalbard); E. Frost creep in frost-shattered dolomitic limestone with some residual silts (Ny Alesund, Svalbard). F. Gelifuction fabric in granitic sand (Finnish Lapland). G. Frost-shattered shale in a loess (A1; Advent, Svalbard); stone frost jacking in a residual dolomitic silt (Kvadehuksletta, Svalbard). Photos by Van Vliet-Lanoë.

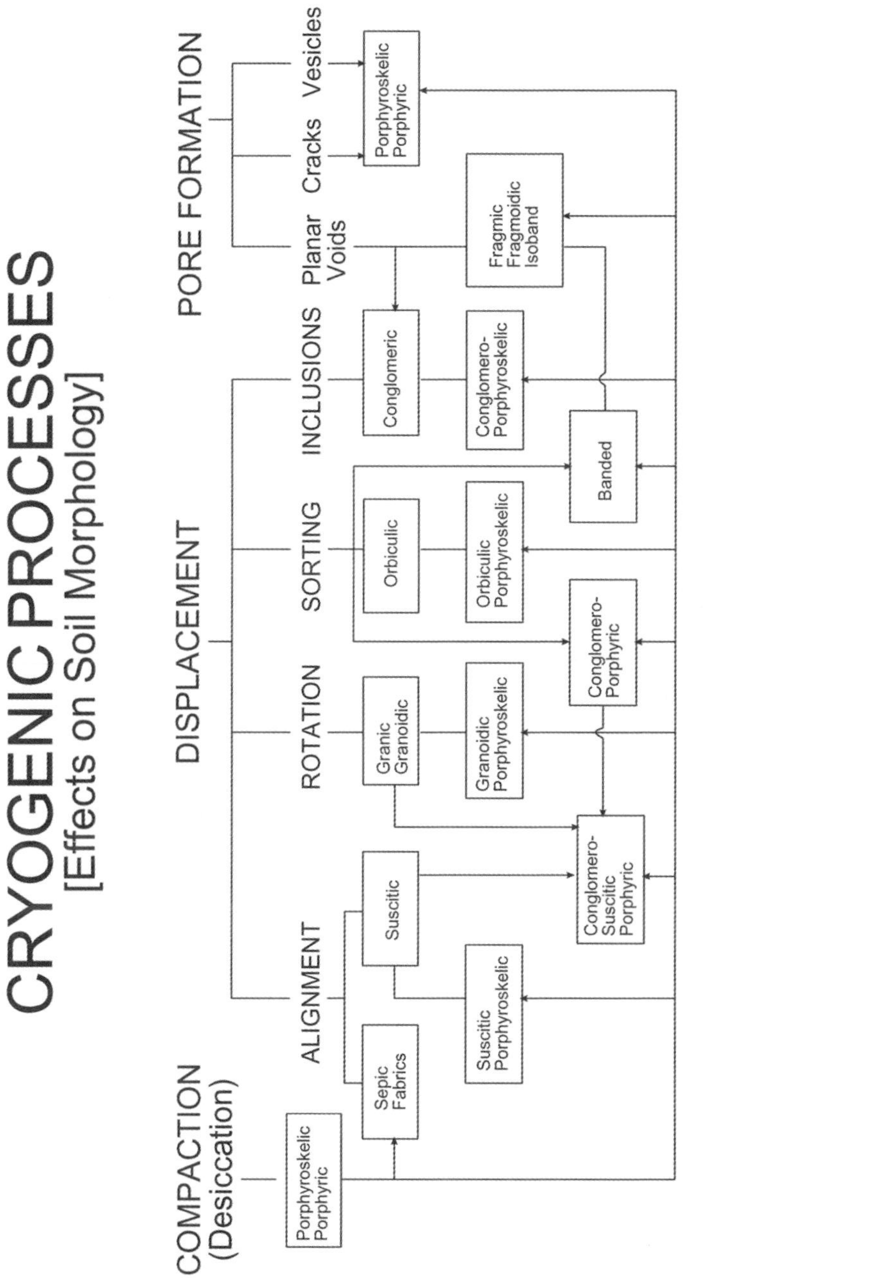

Figure 3.2.4. Classification of cryogenic microfabric (Fox, 1994).

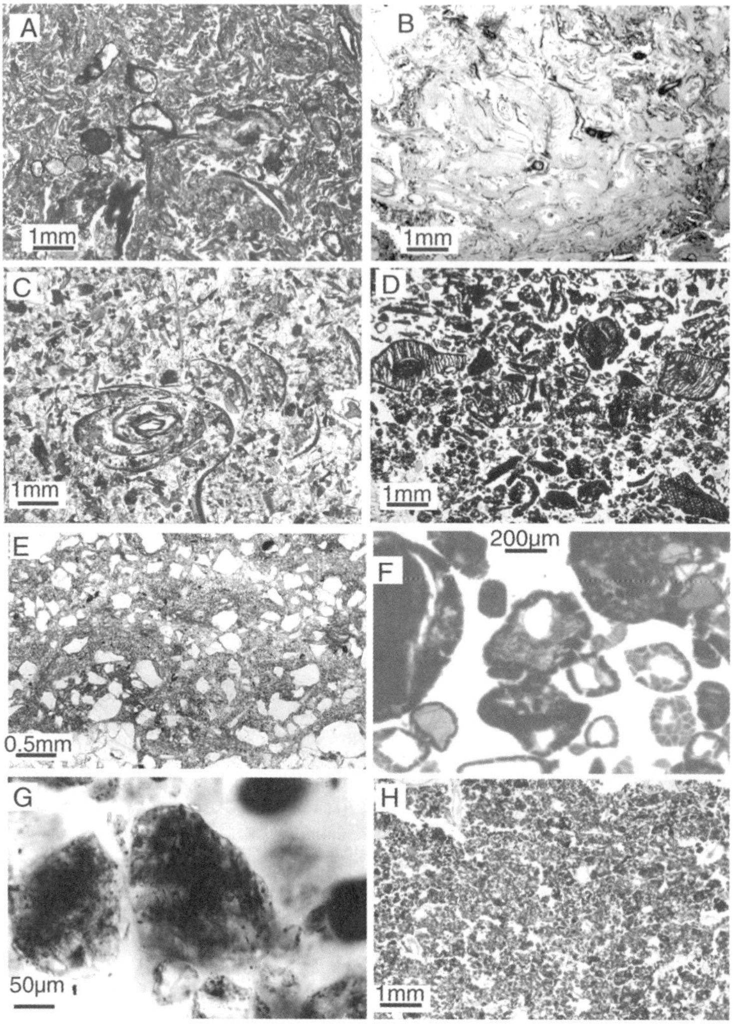

Figure 3.2.5. Humus. A. Incompletely disgested humus of Salix polaris (mor, with acarian droppings) and ectomycorhizo (Ny Alesund Svalbard) (VV). B. Sphagnum peat A0g, unaltered, weakly bioturbated (Hardanger Vidda, Norway) (VV). C. Aeolian fixation on a Polytrychum mat (Salluit, Nunavut, Canada) (VV). D. Burned A0 horizon of a Pinus mariana stand (northern. Quebec, Canada) (VV). E. Orbiculic microfabric in an Orthic Turbic Cryosol, Bmy horizon, 0 to 30 cm (Mackenzie Mountains, Carcajou Range, Northwest Territories, Canada) (CF). F. Microcraked monomorphic organo-aluminic gel in the Bs of a Cryorthod (Finnish Lappland)(VV); G. Retinized organic matter in a Bw (Siberia) (SG). H. Coprogenic moder (collembolla; Blomstrand, Svalbard) (VV). Photos by Van Vliet-Lanoë (VV), Gubin (SG), and Fox (CF).

A capping of segregated fine particles progressively develops on the top face of peds or stones (Figure 3.2.1H, Figure 3.2.3E, F). The cappings consist usually of silt and coarse clay-sized particles, generally thinly stratified, and form up to 2-cm-thick layers on large stones. Coarse inclusions such as mica flakes, quartz, charcoal, or pollen grains or soil fauna fecal material appear in the caps. While the cappings form, sorting and accumulation of skeleton grains may occur, together with some vertical grains and a reverse sorting of particle sizes (Van Vliet-Lanoë, 1976). The specific surface of the minerals making up the skeleton grains and the capability of the grains to be propelled on a thin water film at the front of growing ice crystals control the translocation of particles during thaw periods and the sorting of particles (Corte, 1966; Rowell and Dillon, 1972). For slope situations, this sorting process greatly diminishes because of shearing or rotational movements (Figures 3.2.1I and 3.2.3F).

The accumulation of clean skeleton grains into microfissures occurs rather quickly after 5 to 10 freezing cycles; after a minimum of 30 cycles, distinct capping of mineral grains or particles develops (Dumanski, 1964).Vertical or heaved grains appear after only a few cycles. In alpine or oceanic Arctic environments, several frost cycles occur each year, accelerating the process; abundance of snowmelt water and low clay content can favor their expression at depth. Often each freeze-thaw cycle produces a distinctive layer on the cap or coating. In climatic zones with only one effective cycle of frost each year, one can use the development and thickness of the capping for relative chronological purposes, to determine the rate of formation.

2.3. Formation of granular fabrics

Granular fabric (Figures 3.2.1, 3.2.3, and 3.2.5) is very common in Cryosols and here refers to the rounded aggregates in soils having very fine silt to clayey textures (normal or amorphous colloids). Granular aggregates often occur in contemporary soils that past permafrost has affected; the persistence with time attests to their increased resistance.

Smith et al. (1991) suggested that the alignment of the clay particles near the outer surface of the aggregates would contribute to increased strength and stability. Van Vliet-Lanoe (1985) noted that coatings, formed probably as a result of ultra-desiccation, would make the granular units resistant to cryoturbation effects, mass wasting stresses, and collapse on thawing.

In Cryosols, distinct granular aggregates often occur close to or on the surface of such patterned-ground features as sorted circles and hummocks (Figure 3.2.1C, I). These aggregates frequently occur within the active layer. At greater depths, they are often at the interface with a material less susceptible to frost (cryoturbation) or in solifluction lobes (displacement of materials) (Van Vliet-Lanoë, 1976, 1982; Fox and Protz, 1981; Fox, 1983; Smith et al., 1991).

In addition, these features can form from disruption of constituted clay coatings (papules) or sheets (Van Vliet-Lanoë, 1988a). Capping can develop simultaneously with the mechanical stresses on plastic materials (i.e., clays and organics) which can move or protrude into regions or which enclose and/or move mineral grains. This can lead to distinctive soil morphologies, such as "ovoid" (Morozova, 1965; Bunting and Fedoroff, 1974; Pawluk, 1988), "microcircles" (Konischev et al., 1973), or "conglomeric" and "orbiculic" fabrics (Fox et Protz, 1981; Figure 3.2.5E).

Because of the rounded surfaces of individual aggregates, granular fabric derived solely from cryogenic processes sometimes can be confused with biological mulching (Van Vliet-Lanoë et al., 1984). Granular fabric ("coffee meal fabric," resembling freeze-dried coffee) commonly occurs at the surface (Figure 3.2.1C) and may occur both in non-permafrost and permafrost regions. It usually occurs in the upper portion of the soil most susceptible to freeze-thaw cycles.

Rotation of soil particles may result from differential frost heave within the horizon (Figure 3.2.1I), or at the microtopographical scale, or from lateral displacement by solifluction on slope. Differential thaw consolidation completes the movement with aging, particle migration, and water content at thaw (the internal or external water supply reduces the shear stresses), as in downslope situations (Van Vliet-Lanoë, 1988a). Smith et al. (1991) attribute the development of well rounded granular aggregates with distinctive smoothing of the outer surfaces to highly saturated soil environments, together with water flow.

2.4. Fabric stability and particle translocation

Platy aggregates created solely by desiccation without any compaction are often very unstable upon moistening during thaw periods. This is usually the case in the A horizon and near the permafrost horizon, where the propagation of the freezing front is fast. In contrast, platy aggregates formed at depth by ice lensing under natural conditions are usually extremely firm and resistant, despite not being cemented with amorphous materials as would be the case in an Arctic podsol. The formation of platy aggregates may help to explain the reduction of porosity at depth (Figure 3.2.1C, D, E) that Pettapiece (1974) observed.

The mechanical action of enlarging ice lenses may lead to frost heave of freshly frozen material and platy aggregates. The resulting compaction may contribute to the reduction in internal porosity of the peds observed at very low temperatures (below -10°C) and the resultant increased stability of the aggregates (Van Vliet-Lanoë, 1988a). Exclusion of mineral salts during water freezing also can enhance this process, thereby resulting in a strong flocculation (coagulation, in Russian) of the particles (Rowell and Dillon, 1972).

When melting is slow, air that was entrapped in the desiccated peds now can exist between the lenses and be expelled during moistening, thereby facilitating the preservation of the soil fabric. In contrast, sudden moistening can promote a collapse of the soil morphology. In both instances, vesicles (circular shaped pores, Figure 3.2.2A, B, D) can form (Van Vliet-Lanoë, 1985). In clayey or organic soils, vesicles often are mamillated (Figure 3.2.2B) but usually perfectly rounded in silts or silty-sands (Figure 3.2.2A, D). Vesicles often occur in the upper 5 cm of the soil but also may occur along cracks or in locations where the presence of many rock fragments disturbs the thermal conductivity of the melting soil. Elongated pores occur at depth in silty materials and often may correspond to a partial collapse of the platy fabric (Figure 3.2.2B, C, E).

This surficial collapse (partial or total) of the soil promotes particle translocation and eluviation of the surface horizons (Figure 3.2.2D), as Schmertmann and Taylor (1965) observed, even below a cryptogamic crust (Figure 3.2.2C). A direct consequence of this is the constitution of textural accumulations at depth (B silt and Bt, Fitzpatrick, 1956; Forman and Miller, 1983; Van Vliet-Lanoë, 1985), resulting in a modification of the frost susceptibility of the material within the profile; the same evolution exists between up-slope (depleted) and down-slope positions (Van Vliet-Lanoë, 1987).

Cutanic features (Figure 3.2.2C, F) rarely are preserved, except in drained soils as in northern Quebec or in Yakoutia. They generally form from fine silt and/or coarse clay and are poorly laminated (Van Vliet-Lanoë, 1985). Heave stresses can incorporate particle accumulations as a stress cutan (porostrian) (Figure 3.2.2E). Cutanic features form during the spring and summer. They can be silty or coarse clay skins in the upper horizons or even fine clays in deep melting soils. They are preserved if they accumulate below the lower boundary of ice lensing (the frost limit or desiccated layer). This increase in "plastic" colloidal fraction with depth may help explain the downward development or enclosure of granular material, i.e., "conglomeric" fabric (Figure 3.2.6.) which Fox and Protz (1981) observed (Figure 3.2.6E) in the Bs of podsol (Jakobsen, 1989) (Figure 3.2.6F).

In tills, under high precipitation and recurrent frost action, strong translocation can occur, leading to the development of a B silt layer (Boulton and Dent, 1974; Locke, 1986; Frenot et al., 1995) (Figure 3.2.2D). This process also is called "pervection." The occurrence of a capillary barrier also promotes the accumulation of fines above a coarser, open-work layer (Bybrodi, 1967), leading to exaggerated ice lensing and lateral internal drainage in spring (Van Vliet-Lanoë, 1995). If permafrost exists, the inverted thermal gradient at the onset of thaw allows the temporary inefficiency of the capillary barrier (Van Vliet-Lanoë, 1995), via enhanced suction at depth and a particle translocation to depth.

This process shows the effects of freeze-thaw cycles and the direction of the freezing front on particle size at centimeter scale. Differences in the specific surface of grains, as experiments by Corte (1966) demonstrated, can lead to separation of the finer from coarser materials, resulting in distinct morphological arrangements of the coarser and finer soil materials, i.e., orbiculic fabrics (Figure 3.2.5E) (Fox and Protz, 1981) with a distinct circular pattern which the coarser grains form.

In mud boils and sorted circles, the surface layers often lose fine particles over time and become increasingly susceptible to liquefaction or to forming a structural crusting. In contrast, fine particles accumulate at depth, which enhances the effects of cryoturbation processes. Needle ice action in combination with subsequent summer rainfall promotes the forming of splash crust (Figure 3.2. 2B) or alternatively liberates particles or pseudo-particles available for deflation or soil translocation.

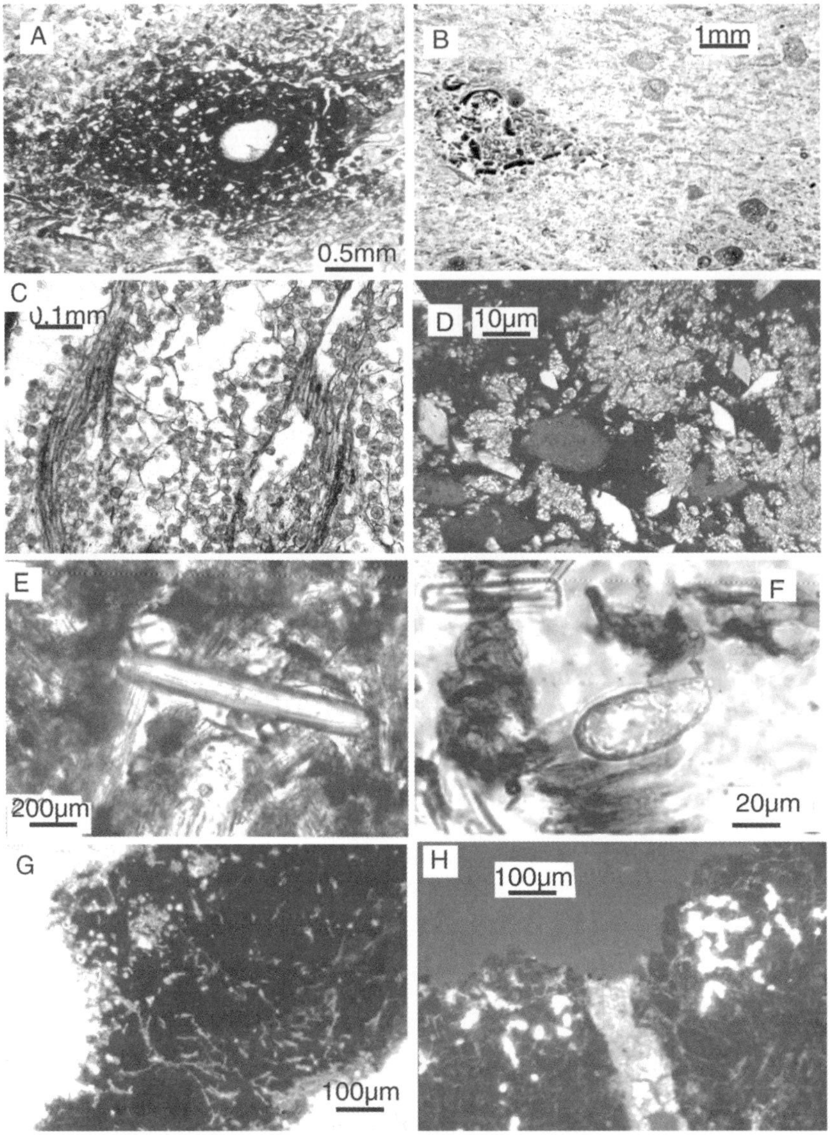

Figure 3.2.6. Biochemistry. A. Bacterian iron hydroxide precipitation along a root pore (Advent, Svalbard). B. Frost-shattered iron coating, Bwg (northern Quebec, Canada). C. Stress calcitic nodules of cyanobacteria, A0 of turf hummock (Ny Alesund, Svalbard). D. Gypsum crystals and oxalates micrite (cross nicol) in a cryptogamic crust (Advent Svalbard). E. Diatoms in a cryoturbated muck (hummocks of Finnish Lappland). F. Tecamoeba: A0/A1 of a hummock (Finnish Lappland). G. Dissolution of limestones by cyanobacteria (inbedding with a fluorescent plastic, reflected UV light) (Blomstrand, Svalbard); cryptoendolithic algae development in a sandstone (natural, reflected fluorescent blue light; Ny Alesund, Svalbard). Photos by Van Vliet-Lanoë.

One can confuse these structural crusts with liquefaction at the surface of pattern ground ("mudboils," if previous ice lensing partly deformed them in winter). "Mud cutans," as Kubiena (1970), Fitzpatrick (1956), and Wilkinson and Bunting (1975) described, can result from this process as well as from percolation of surface runoff.

2.5. Frost shattering and stone jacking

Frost jacking or frost heave displaces individual particles to a vertical position, especially coarsely sized particles. Vertically oriented particles occur at both macroscopic and microscopic scales (Figure 3.2.1H, F and 3.2.3H). The vertical position results from the traction exerted by growing ice lenses on fragments (stones, bone fragments, nodules) included in a matrix susceptible to frost. It leads progressively to rotation and the vertical alignment of the clasts.

Experiments by Kaplar (1965) and Pissart (1969) have demonstrated this. A large open pore at the base of affected fragments is a good indicator that this process is present (Van Vliet-Lanoë, 1976; Fox and Protz, 1981). The fact that the clast sets up a hydraulic barrier enhances this process (Van Vliet-Lanoë and Dupas, 1991). Because the fragments are rotated (Figure 3.2.3H), any silty cap that may have formed on the upper surface now appears at the side or bottom of the fragment, following successive rotations. Fox and Protz (1981) defined a microfabric dominated by vertical clasts as "succitic."

Nevertheless, frost jacking is always in competition with other displacement processes, such as solifluction, that cause the fragments to align in a predominant direction (Harris and Ellis, 1981) (Figure 3.2.3E). Frost shattering, a result of increasing the moisture retention in a fissure with subsequent ice growth, is common in rocks and may occur in combination with dissolution, biological activity, and particle migration (Figure 3.2.3G, H). Frost shattering also may affect pedofeatures, i.e., disrupting soft iron nodules, clay coatings (papules), or carbonate precipitates that may be remnants from milder or dryer pedoclimates, into rounded fragments (Figure 3.2.5B).

2.6. Summary of soil fabric types related to cryogenic processes

Cryogenic processes result in distinctive soil morphology patterns. Displacement of materials realigns coarse and clay-sized materials, rotates aggregates and mineral particles, sorts both coarse and fine fractions of the soil material, and protrudes soil material from one area to another in the soil profile. Compaction produces massive soil material and influences both the alignment of particles, through packing, and the shape of aggregates, due to compression from the formation of ice lenses. Cryogenic processes influence pore shapes and sizes, resulting in planar voids, cracks, and vesicles.

Figure 3.2.4 summarizes the types of fabric (Fox, 1994) observed in soils affected by cryogenic processes. The lines joining the various fabric types indicate the main contributing process or processes that produce a particular arrangement of soil fabric. The main cryogenic processes lead to displacement of particles by vertical alignment, sorting, rotation, and inclusion of materials from other areas in the profile, various types of pore formation, and compaction of soil material through removal of water, due to freezing. The terminology used to identify the different patterns comes from Brewer (1964,1976), Pawluk and Brewer (1975), Fox and Protz (1981), and Brewer and Sleeman (1988).

3. Biological Processes

Weak humification or, alternatively, the absence of humification in wet soils at high latitude or high altitudes, dominates humus formation. Cryogenic deformations may translocate plant residues into middle and lower parts of the soils, in the form of fragments or particles that ice lensing breaks apart. In very cold and permanently waterlogged conditions, as in Western Svalbard, or at high altitude, as on the shore of Tibetan Lake, humification can be nearly absent, so that even the chlorophyll may be partly preserved.

Humification is often reddish brown in waterlogged conditions (peat). In more drained conditions, it evolves to blackish colors (Van Vliet-Lanoë, 1988a), or black (i.e., melanization) induced by frost desiccation and sometimes by the presence of Ca^{++} (Ugolini, 1996; Ugolini and Sletten, 1988). This polymerisation and microdivision led to Karayeva and Targulian's (1960) development of the concept of "cryogenic humus retinization" (Figure 3.2.6G).

Under lichen tundra, the antibiotic properties of lichen acids can reduce bacterial digestion (Hrutfjord and Ugolini, 1984). Bacterial activity and the oxidation-reduction potential control most of the iron precipitation in thin section (Figure 3.2.5A). Released iron hydroxides impregnate various components of the aggregates, charcoal fragments, and plant remains. This leads also to the "black chips" or Svalbard peat and humus, consisting in iron-stained whole leaves of *Salix* or *Dryas*.

A common observation is the high frequency in the A1 horizons of ectomycchorizes (Figure 3.2.6A), in association with *Dryas*, *Salix*, or *Betula* roots (Van Vliet-Lanoë, 1986). Hyphae of various basidiomycetes and actinomycetes are common, along with spores and sclerotia in well to poorly drained humus. Bleached areas in the A1 horizon often relate to mycelium's causing the partial digestion of the amorphous organics, as in Northern Finland and Ungava. Mineral weathering of calcite and quartz grains by dissolution often results in vermicelli patterns or hatching on the grains (Figure 3.2.5G).

These patterns often relate to the prospecting of hyphae or rhizules for mineral nutrients. The indentations result in increased porosity of grains, which in turn increases their frost susceptibility. The precipitation of organic material or accumulation of droppings into the fissures can enhance this susceptibility to frost, thereby increasing the migration potential of water. The presence of diatoms, chrysophycae, and tecamoeba (Figure 3.2.6H) in organic layers are also evidence of water logging during thaw periods.

Different types of organic matter can appear under a light microscope with normally transmitted light or under epifluorescent microscopy. Lignin, cellulose, and sporopollinin (pollen grains) are usually strongly fluorescent (Altemüller and Van Vliet-Lanoë, 1985). Dark mycelium is usually absorbing (black) and chlorophyll (chloroplastes) is reddish in fluorescence. Humified fractions become increasingly black with the degree of polymerisation. Humic acids which consist of large molecules are black, though fluvic acids are small molecules and fluoresce mostly in the brownish-orange range, as occurs in the lower Bh of Spodols.

3.1. Holorganic deposits (peat)

In acidic conditions, peat usually is constituted by reddish, humified, more or less compacted mats of Sphagnum or other mosses (Figure 3.2.5B, C), locally associated with vivianite, some diatoms, and no bioturbation.

In neutral to alkaline conditions, peat is constituted by the accumulation of dark to black humified organic fragments of various species, rich in diatoms and often phytoliths, shells, aquatic snails, postcards, and algae calcite. Fragments are usually well preserved. Exceptionally, chlorophyll may be preserved (epifluorescence microscopy). Bioturbation is present. Bacterial pyritization (framboids) may have occurred.

In some situations, pure algae peat can form in an eutrophic environment, leading to amorphous organic material with a specific oncolithic fabric (i.e., Lapland, northern Quebec [Van Vliet-Lanoë and Sëppälä, 2000]).

In some terrestrial environments, importation of materials by rill wash or aeolian processes occurs; mineral grains, papules, and charcoals appear. Change in humification (darkening) by frost desiccation (Ugolini and Sletten, 1986), oxido-reduction (drainage), and vegetation (root prospecting) indicates climate or drainage modifications. This seems common in Svalbard (Frenot and Van Vliet-Lanoë, 1995), northern Quebec (Van Vliet-Lanoë, 1998), and Siberia (Gubin, 1987).

3.2. Minerorganic deposits

Minerorganic deposits are rich in mineral fraction, usually interstratified with pure organic beds. The deposits represent lacustrine or colluvial landscape positions. Darkening of organic materials usually indicates oxidation-reduction processes or alkaline conditions. Bacterial pyritization (framboids) can exist at shallow depth or in association with a wet cryptogamic crust, especially in brackish conditions (Svalbard).

A complex environmental setting for deposition of fluvial materials exists when floated woods and low diatom content combine. Humus formation can be raw, either synchronous with Sphagnum growth and aeolian deposition (Figure 3.2.5C), or with grassy vegetation (arctic meadow) with an aeolian fraction of silt, 90% mulched by enchytraeids worms and collembolla (Svalbard), which leads to a lightly colored moder (10YR 4/4). Special humus forms occur downslope of snow banks as stratified leaf fragments, combined with faecal material of various sizes from soil fauna and diatoms, such as in northern Ungava (northeastern Canada). Similar deposits occur in the organic materials that accumulate in trough areas between hummocks in the High Arctic.

3.3. Humus and pedoturbation

Organic horizons in Cryosols have both litter and humus components. Leaves, needles, wood fragments, and seeds, together with humification agents such as mycelium, algae, and faecal material from various arthropods and worms, mostly compose the litter. Usually, no mineral material is present except for some particles which aeolian or needle ice processes introduce. Sometimes burned *in situ*, most of the large organic fragments can be carbonized (Bunting, 1973) (Figure 3.2.5D), and cryogenic processes translocate smaller fragments to depth. Large charcoal fragments can further evolve as porous mineral grains. The humus is weakly to strongly humified, weakly to strongly bioturbed, and intermixed with mineral matter (Bunting and Hathout, 1971).

In acidic humus (mor, ranker), humification is low despite a high coprogenic activity (Figure 3.2.5H). Mycelium activity is important and bacterial digestion is less so, especially in tundra or heath soils. Acarina (i.e., Oribatida), collembola, and nematodes mostly represent soil fauna in wetter places (Remmert, 1980). Fly and beetle larvae are also very active, as the common presence of their faecal material in drained sites shows. The integration rate of mineral soil is low.

In slightly acidic soils, such as Arctic meadow soils, enchytraeids (white pot worms) and collembola (springtails) are common. Organic integration into the mineral soil is imperfect. Some other worms, like Dendrobena, are very rare and only active in the upper humus. Flies, mites, and beetle larvae are very efficient. In mull, ranging from boreal to chernozemic soils, as in Yakoutia, on permafrost (Dendrobena), worms are the most efficient, in association with snails. Integration of the organics to the mineral soil is mostly complete. Humification is rather dark, because of the presence of calcium and of desiccation (as also frost). Calcite concretions of salivary glands of earthworms occur.

An incipient form of humus is the cryptogamic crust (mosses, algae, lichens), extremely important in an aridic, cold, or post-fire environment (Figure 3.2.2C). It reduces the erosion susceptibility of the soil and, when the soil is dry, the crust's hydrophobic comportment promotes rill concentration. It is extremely susceptible to the action of needle ice and trampling, but it resists frostcreep and cryoturbation. It is a "biotextile" consisting of hyphae (lichens), mosses, and cyanobacteria. In central Svalbard, this humus is very thin, often associated with microarthropod faecal material, occasional gypsum crystals (Figure 3.2.5D) or cyanobacteria stress cyst (aragonite) (Figure 3.2.5C). In addition, the organic material combines with the aeolian silts and seasonal iron bioprecipitation.

In more continental regions such as Yakoutia, when the available humidity is locally important, true mull, churned by earthworm (Dendrobena), occurs. But the most common form is a mor humus, mostly 50 to 70% coprogenic, related with the seasonal digestion and mulching of microarthropods, collembolla, and nematodes. Insect larvae droppings are common in the litter, as well in discrete vegetative mats, on pattern ground and thick raw humus. As the humification process is weak, dilaceration by ice lensing is common, leading to dusty organic accumulation at depth and micro-cracked accumulation of gel-like products (monomorphic organics).

4. Organo-Chemical Weathering

Two of the main physical factors contributing to the evolution of Cryosols are spring water logging and frost desiccation. Since suction potential at -20°C can extract the adsorbed water, this water film's activity becomes very high (Chaussidon and Pedro, 1979), which promotes weathering at very high pF. This explains the formation of thin iron coatings at the surface of rocks, as reported for the Antarctic (Campbell and Claridge, 1987), and Konishchev and Rogov (1982) observed the solution's hatching on minerals.

Biodegradation of humus is very slow. Thus, small organic molecules like fulvic acids (seen as fluorescent gels in thin sections) accumulate, as well as the nitrates that cyanobacteria produce. Cold temperatures favor the retention of carbon dioxide in water, so that acidity in microsites can become extremely high, leading to rapid decalcification and easy organo-complexing of iron. This process complements the low oxido-reduction potential that water maintains via frost degassing and water logging. Iron, aluminium hydroxide, and silica ions can migrate with the thermally induced water potential, especially to the permafrost table, starting from the onset of thaw (Tyutyunov, 1964), leading to an accumulation of gels in this horizon (Figure 3.2.5F) or close to it, when pH rises (i.e., imogolite) (Ugolini and Sletten, 1986).

Iron hydroxides are mostly of bacterial origin, as they occur either along root tracks in soil, likely formed during summer and broken apart or crushed after becoming firm during winter (Figure 3.2.6A, B), or on open fissures left by ice lensing at the permafrost table, i.e., in palsa. Materials sampled from under surficial water ponds (or the trough portion of ice wedge furrows) also are stained from thick iron accumulation. Thin section shows this accumulation to be cyanobacteria mucigels, stained with ferri-hydrite intermixed with diatoms (Figure 3.2.6E).

The soil matrix characterizing many patterned ground sites which have developed on palaeozoic dolomitic limestone parent materials (i.e., the sorted circles of Western Svalbard; Canadian and Greenland shields; or the Southern Lena valley [Siberia]) has a mineral grain component of authigenic crystals of residual dolomite, with occasional pinkish residual clay, leading to B horizon silt accumulation (Tedrow, 1975). This material is extremely unstable during thaw. Massive mineral material with weak orientation of silt-sized particles, abundant vesicles, and frost-jacked rock fragments with silty caps characterizes the soil microfabric (Figure 3.2.1H). At depth, the clay content increases and the cryogenic fabric is partly preserved. Carbonate cement never occurs, though some organic particles can be translocated at depth.

Carbonate precipitation is rare, except for calcite or dolomite pendants (Sweet, 1974). These consist of stratified fibro-radiated calcite (rarely dolomite), forming microstalagtite in either rock fissures or below blocks in porous soil (beaches gravel, scree). They occur at the base of surficial blocks when lichen deriving from calcium oxalates colonized them, or in shallow temporary ponds that precipitation of cyanobacteria produces (Svalbard, Antarctica). Deeper, in the drained active layer, they relate to cryodesiccation of soil solution, as isotopic analysis shows (Marlin et al.1995), in relation to some biological activity.

These microstalagtite forms do not differ from those of other aridic environments (Freytet and Verrecchia, 1998), despite many affirmations in the literature. Wetter climate conditions can induce a partial solution before a new precipitation or frost shattering detaches them from the support, especially when silts infiltrate the profile.

Their exceptional preservation in cold environments relates to low biological mulching. Other forms of precipitation derive from oxalate accumulated in root tissues of grasses (Parfenova and Yarilova, 1974). Bacterial concretions may form in silty material at the level of the water table (talik) or along root tracks, but this also is not specific to Cryosols (Ivarson, 1975).

5. Cryoturbation and Slope Processes

Most of the soil processes are similar to those of boreal pedogenesis. Cryoturbation and movement of materials on slopes can modify the frost susceptibility of the horizon; with time, these processes cause differential frost heave or translocation of materials between superposed or juxtaposed soil horizons.

5.1. Cryoturbation processes

Cryoturbation involves many mechanisms: aggregate rounding by stress and rotation, in response to differential frost susceptibility of the soil horizon or sediments; slow upward injection of sediment along fissures (thermal or desiccation); translocation of particles and their integration by stress processes into soil aggregates. Micromorphological analysis of materials sampled from active field sites, as well as experimental laboratory analyses, led to several important conclusions (Van Vliet-Lanoë, 1988a, 1988b).

We identified the annual sequence of plastic deformations related to frost and to the translocation of particles during melt. In other words, the preservation of the platy fabric during the summer shows usually that liquefaction does not take place during spring (on a flat surface) and that most of the plastic deformation of layers at the microsite scales occurs just before the frost consolidation.

We showed that the progression of the freezing front could be made from the successive positions of the ice lenses, which often related to textural composition (capillary and adsorbed water content) (Figure 3.2.3A, B, C). This allowed us to define a contrast in frost susceptibility between two sediments and to define a gradient of frost susceptibility that gives the direction of the contrast: it is positive if the upper layer is more susceptible than the underlying one.

We identified the dominant mechanism behind cryoturbation: the stresses resulting from ice lensing and expressed in the field, in the form of differential frost heave and deformation, are recorded in the sediments in the form of platy fabric, elongated vesicles, or cappings in a silty or sandy matrix. We consider the pseudo-liquefaction observed commonly in the field to be, in fact, splash crusting (Figure 3.2.2B).

Cryoturbated materials mainly result from differential heave, related to drainage differences associated with thermal conditions (thermal gradient) and with frost susceptibility (the intensity related to porosity, texture, amorphous content, and mineralogy). Load cast and cryostatic pressures tend to be restricted to wet soils. Most of the cryoturbated features relate also to seasonal frost in imperfectly to poorly drained soils. They do not indicate permafrost except in well drained situations or in the case of flat-bottomed involutions. Deformations and translocation processes are also very similar, in morphology and appearance, to those in swelling clay soils (Van Vliet-Lanoë, 1988a)

5.2. Solifluction processes

Solifluction is usually considered the result of frost heave (and thus the frost susceptibility) of the soil, orthogonal to the topographical surface, and of the soil's thaw consolidation in response to gravity (Figure 3.2.3D, E, F). Harris and Ellis, (1981), Harris (1986), and Van Vliet-Lanoë (1982, 1985, 1988a, 1995) performed microfabric analysis of materials subjected to solifluction. Basically, platy fabric can be deformed plastically up to granular fabric (gelifluction). But micro-erosion, as in the case of repeated freeze-thaw microfabric, is active during the same time. Field data from motion measurements and micromorphological investigations show these coherent processes (Figure 3.2.7.):

- A laminar soil displacement (slipping) occurs plate by plate, as the successive ice lenses melt (Figure 3.2.3D).
- Both the lateral water supply (i.e., the effect of slope or snow bank) and its drainage capability accentuate the displacement of materials. Well aggregated material (granular) drains like a sand, and its mobility is limited, in contrast to that of collapsing silty fabric, which is unable to drain freely. The excess hydraulic pressure results in motion, evolving into a mud flow characterized by massive vesicular fabric and preservation of silt cappings on clasts.
- Slow motion (frost creep: few mm/cycle) gradually changes to high motion (mud flow: few meters).

Many macroscopic forms (i.e., solifluction sheet, lobes, elongated patterned ground) investigated from a micromorphological viewpoint present clear frost-creep fabrics. The granular fabric's gelifluction is often active only superficially. The microfabric approach to periglacial dynamics allows recognition of the main process responsible for the observed morphology, but it does not necessarily record the initial phase of sedimentation. But we easily can combine these observations with observations of cryoturbation processes (i.e., differential frost heave and frost susceptibility contrast and gradient) to understand the macroscopic deformations on slope. Micromorphology usually confirms the existence of both dynamics.

Domain	Speed	Microfabric	Hydraulic pressure	cryogenic fabric	Texture
Normal frostcreep	1-10mm	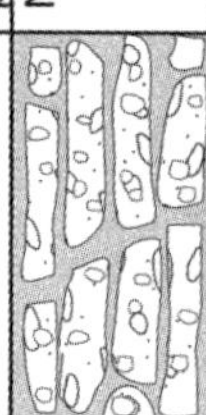	None	without influence	Frost susceptible material
Natural frostcreep	2-3 times the normal frostcreep	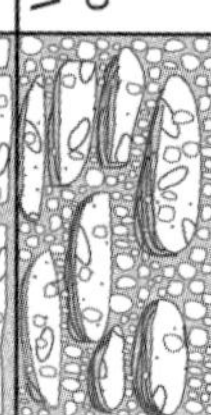	Weak, lateral drainage	weak influence	Frost susceptible material
Accelerated frostcreep	4-5 times normal frostcreep	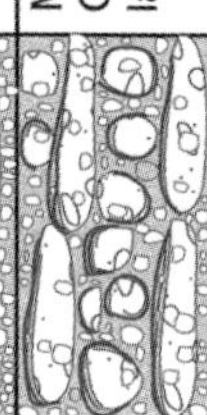	Moderate Cumulative lateral drainage	stable to metastable	Frost susceptible material, excepted sands and silty sands
Gelfluction	10-20 times normal frostcreep	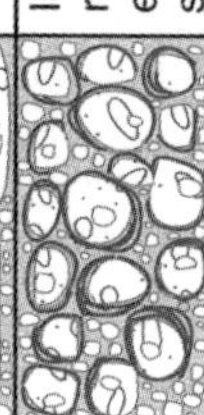	Important, related to an external water source	stable	Coarse sands mixed with clay loam, clay loam, clays
Mudflow	>20 times normal frostcreep		Important, related to an external water source	metastable to unstable	Sands , silty sands, silts

Figure 3.2.7. Relationship between displacement domain (a), and (b) microfabric, slope dynamic, drainage, and aggregate stability (Van Vliet-Lanoë, 1995).

6. Applications for Micromorpological Assessment of Cryogenic Processes

6.1. Geotechnical applications

Van Vliet-Lanoë and Dupas (1991) used micromorphology in Cryosols to understand the soil's ability to develop ice lensing around a chilled gas pipeline. Replacing ice with plastic allowed us to recognize why ice lenses nucleated in relation to a minute difference in soil's hydraulic conductivity. Gubin and Gulyaeva (1997) also comparatively studied the microfabric of frozen soils (Figure 3.2.1D, E).

6.2. Applications to soil classification

Cryosols are soils whose materials are affected by freezing fronts from the surface and from the permafrost zone. Various cryogenic processes result, causing displacement of materials, compaction, and re-organization of the structure which, in turn, cause distinct horizons to develop. Micromorphological assessment provides evidence for these cryogenic processes; identifies the dominant process or combination of processes that have contributed to the formation of the soil fabric; and provides information for interpreting the intensity or extent of the impact of cryogenic processes on soil materials.

The effects of cryogenic processes on soil material is observable at macroscales in the field and at microscales under detailed microscopic analysis of associations between the soil material and soil pores. Field macrostructure strongly relates to micromorphology (Smith et al., 1991). Granular structure in the field appears as granic or granoidic fabrics in thin section; blocky and lenticular structure, as fragmic and fragmoidic; and massive structure, as porphyroskelic or porphyric fabrics.

In some cases, what appears to be massive in the field, when examined in thin section may be closely packed lenticular structural units or granular material, or we may find that the coarser particles have been sorted or aligned so as to create a distinctive morphology, i.e., silt cappings. Such observations from thin section help provide detailed information on the specific effects of cryogenic processes active in these soils now or for an extended time in the past (Van Vliet-Lanoë, 1998).

With respect to the *Canadian Soil Classification System*, micromorphological assessments of the type of fabrics related to specific cryogenic processes will provide supporting evidence for cryoturbation, which is a classification requirement for the Turbic Cryosols. The main processes in these soils include sorting of differently sized particles and mixing of both mineral and organic material from different horizons (Soil Classification Working Group, 1998).

In the study of paleosol materials, we can use to advantage the fact that cryogenic processes result in very distinctive fabrics which reflect freeze-thaw cycles and the presence of permafrost. When more than one of the distinct fabrics occur in soil materials from suspected paleosols, they are evidence of permafrost conditions in past soil environments, and one thus can assign a taxonomic designation for permafrost. Understanding the past history provides data for identifying soil properties affected by climate change.

7. Conclusions

Micromorphological descriptions provide data that helps explain the cryogenic processes which produce particular pedological features at both microtopographical and soil horizon scale. It facilitates defining the successive events recorded in the soil as pedofeatures from previous edafic or climatic conditions. Mechanical processes predominate over weathering or biological activity in Cryosols, in striking contrast with other soil types except swelling clay soils, but they never mask it completely. The presence of two freezing fronts and the predominance of physical displacement of materials through the formation of ice lenses are the main characteristics for defining Cryosols and the main contributing factors for producing distinctive soil morphologies in these soils.

References

Altemüller, H.J., and B. Van Vliet-Lanoë. 1990. Soil thin section fluorescence microscopy. In: J.A. Douglas, ed. Soil micromorphology, a basic and applied science. Elsevier publ. Devel. Soil Sci. 19:565-576.

Brewer, R. 1964. Fabric and Mineral Analysis of Soils. John Wiley and Sons. New York. 470 pp.

Brewer, R. 1976. Fabric and Mineral Analysis of Soils (revised). R. E. Krieger Pub. Co. Huntington, New York. 482 pp.

Brewer R., and S. Pawluk. 1975. Investigations of some soils developed in hummocks of the canadian Sub-Arctic and Southern-Arctic regions. 1. Morphology and Micromorphology. Can. J. Soil Sci. 55:301-319.

Brewer, R., and J.R. Sleeman. 1988. Soil structure and fabric. Division of Soils. CSIRO. Adelaide, Australia.

Bunting, B.T. 1983. The effect of fire on humus profiles in Canadian boreal forest. In: P. Bullock and C. Murphy, eds. Soil Micromorphology. ABA Publ. Co. London. pp. 681-687.

Bunting, B., and N. Fedoroff. 1974. Micromorphological aspects of soil development in the canadian High Arctic. In: G.K. Rutherford, ed. Soil Microscopy. Limestone Press. Kingston, Ontario, Canada. pp. 350-365.

Bunting, B.T., and S.A. Hathout. 1971. Physical characteristics and chemical properties of some High Arctic Organic materials from South West Devon Island, NWT Canada. Soil Sci. 112(2):107-115.

Campbell, I.B., and G.G Claridge. 1987. Antarctica: soils, weathering processes and environment. Elsevier Pub. Devel. Soil Sci. 16. 368 pp.

Chaussidon, J., and G. Pedro. 1979. Rôle de l'état hydrique du système poreux sur l'évolution du milieu. Réalité de l'altération dans les systèmes à faible teneur en eau. Science du Sol, vol. Altération des roches cristalines en milieu superficiel. pp. 223-239.

Corte, A.E. 1966. Particle sorting by repeated freezing and thawing. Biuletyn Peryglacjalny. 15:175-240.

Crampton, C.B. 1977. A study of the dynamics of hummocky microrelief in the Canadian north. Can. J. Earth Sci. 14:639-649.

Dumanski, J. 1964. A micropedological study of eluviated horizons. Master Thesis, Univ. Saskatchewan, 124 pp.

Fedorova, N.N., and E.A. Yarilova. 1972. Morphology and genesis of prolonged seasonally frozen soils in Western Siberia. Geoderma. 7:1-13.

Fitzpatrick, E.A. 1976. Cryons and Isons. Proc. North Engl. Soil Discuss. Group, 1974. 11:31-43.

Forman, S., and G.H. Miller. 1984. Time-dependent soil morphologies and pedogenetic processes on raised beaches, Bröggerhalvöya, Spitsbergen, Svalbard. Arctic and Alpine Res. 16:381-394.

Fox, C. 1983. Micromorphology of an Orthic Turbic Cryosol, a permafrost soil. In: P. Bullock and C.P. Murphy, eds. Soil Micromorphology. ABA Publ. Berkhamsted, UK. pp. 699-705.

Fox, C.A. 1994. Micromorphology of permafrost affected soils. IN: J.M. Kimble and R.J. Ahrens, eds. Proceedings of the Meeting on the Classification, Correlation, and Management of Permafrost-Affected soils, July, 1994. USDA, SCS, NSSC. Lincoln, NE. pp. 51-62.

Fox, C.A., R.K. Guertin, E. Dickson, S. Sweeney, R. Protz, and A.R. Mermut. 1993. Micromorphological methodology for inorganic soils. In: M. Carter, ed. Soil Sampling and Methods of Analysis. Lewis Publishers. Boca Raton, Florida. pp. 683-709.

Fox, C.A., and R. Protz. 1981. Definition of fabric distributions to characterize the rearrangement of soil particles in the Turbic Cryosols. Can. J. Soil Sci. 61:29-34.

Frenot, Y., B. Van Vliet-Lanoë, and J.C. Gloaguen.1995. Particle translocation, and initial soil development on a glacier foreland, Kerguelen Islands, Subantarctic. Arctic and Alpine Res. 27,2:107-115.

Freytet, P., and E. Verreccchia. 1998. Freshwater organisms that build stromatoliths: a synopsis of biocrystallization by prokaryotic and eukarotic alagae. Sedimentology. 45:535-563.

Gubin, S.V., 1987. Paleogeographic aspect of pedogeneis in lowlands of Northern yakoutia. Pushchino. 27 pp.

Gubin, S.V. 1995. Structural evolution of tundra permafrost non-gleyed soils (tundra cryozems). Eurasian Soil Sci. 27:1-13.

Gubin, S.V. 1994. Relic features in recent tundra soil profiles and tundra soil classification. In: J.M. Kimble and R.J. Ahrens, eds. Proc. Meet. Classification, correlation, and management of permafrost affected soils, Alaska. USDA. pp. 63-65.

Gubin, S.V., and L.A.Gulyaeva. 1997. Dynamic of soil micromorphology fabric. In: S. Shoba, M. Gerasimova and R. Miedema, eds. 10th Internat.Working meeting on soil Micromorphology. Moscow-Wageningen. pp. 23-27.

Harris, C. 1986. Mechanisms of mass movement in periglacial environments. In: M. Anderson and K. Richards, eds. Slope Stability. Wiley Publ. London. pp. 531-559.

Harris, C. 1990. Micromorphology and microfabrics of Sorted Circles, front Range, Colorado USA. Proc.5th Canad. Permafrost Conf. Quebec, Col. Nordicana. 54:89-94.
Harris, C., and S. Ellis. 1980. Micromorphology of soils in soliflucted materials, Okstindan, Northern Norway. Geoderma. 23:11-29.
Hrutfjord, B.F., and F. Ugolini. 1984. Mobility of lichen compounds from Cladonia mitis in arctic soils. Soil Sci. 138 (1):48-56.
Ivarson, K.C. 1975. The microbiology of exposed areas of aquatic habitats of Northern Ellesmere island. Arctic. 28,4:295-298.
Jacobsen, B.H. 1989. Evidence of translocation into the B horizon of a subarctic Podzol in Greenland. Geoderma. 45:3-17.
Kaplar, C.W. 1965. Stone migration by freezing of soil. Science, 149:1520-1521.
Kaplina, T.N. 1965. Kriogennyye sklonovyye protessy. Moskva, Izdatel'stvo Literatury po Stroitel'stvu. 95 pp.
Kaplina, T.N., and A.V .Lozhkin. 1987. History of vegetation on the coastal plain of Yakoutia in the Holocene. Development of the USSR nature in Late Pleistocene and Holocene. Nauka publ. Moscow. pp. 207-219.
Kubiena, W. 1970. Micromorphology of polygenetic soils an paleosoils in Polar regions. Annales de Edafologia y Agrobiologia. pp. 845-856.
Lafeber, P. 1966. Soil structure concepts. Engineering Geol. 1(4):261-290.
Marlin, C., and L. Dever. 1995. Interactions eaux-roches en milieu périglaciaire: exemples des carbonates secondaires de la Presqu'île de Brögger (Spitsberg). In: C. Causse et F. Gasse, ed. Hydrologie et géochimie Isotopique, Paris 1998. ORSTOM. pp. 97-116.
McMillan, N.J., and J. Mitchell. 1953. A microscopic study of platy and concretionary structures in certain Saskatchewan soils. Soil Sci. Soc. Am. J. 45:578-586.
Morozova, T.D. 1965. Micromorphological characteristics of pale yellow permafrost soils of Central Yakoutia in relation to cryogenesis. Soviet Soil Sci. 7:1333-1342.
Murphy, C.P. 1986. Thin Section Preparation of Soils and Sediments. A B Academic Pubs. 149 pp.
Nersesova, Z.A., and N.A. Tsytovich. 1966. Unfrozen water in frozen soils. Proc.1st International Permafrost Conference. Washington. Nat. Ac. Sc. Washington, D.C. pp. 230-234.
Pawluk, S. 1988. freeze-thaw effects on granular structure reorganisation for soil materials of varying texture and moisture content. Can. J. Soil Sci. 7:1333-1342.
Parfenova, E.I., and I. Yarilova. 1967. Humus microforms in the soils of the USSR. Geoderma. 1:197-207.
Parmuzina, O.Y. 1978. Cryogenic structure and certain ice segregation phenomena in the active layer. In: A.I. Popov, ed. Problems of cryolithology. Moscow Univ. Press. Moscow. pp. 141-164. (In Russian.)
Pawluk, S., and R. Brewer. 1975. Micromorphological and analytical characteristics of some soils from Devon and King Christian islands, N.W.T. Can. J. Soil Sci. 55:349-361.
Pettapiece, W.W. 1974. A hummocky permafrost soil from the Subarctic of Northwestern Canada and some influences of fire. Can. J. Soil. Sci. 54:343-355.
Pissart, A., 1969. Le méchanisme périglaciaire dressant les pierres dans le sol. Résultats d'expériences. C.R. Acad. Sc. Paris. 268:3015-3017.
Remmert, H. 1980. Arctic Animal ecology. Springer Verlag, Berlin, New York, 250pp.
Rowel D.,and P. Dillon. 1972. Migration and aggregation of Na-ca clays by freezing of dispersed and flocculated suspensions. J. Soil Sci. 23:442-447.

Smith, C.A.S., C.A. Fox, and A.E. Hargrave. 1991. Development of soil structure in some Turbic Cryosols in the Canadian Low Arctic. Can. J. Soil Sci. 71:11-29.

Soil Classification Working Group. 1998. The Canadian System of Soil Classification. Agric. and Agri-Food Can. Publ. 1646 (Revised). 187 pp.

Sweet, K. 1974. Calcrete crusts in an arctic permafrost environment. Am. J. Sci. 274:1059-1063.

Tarnocai, C. 1993. Sampling frozen soils. In: M. Carter, ed. Soil Sampling and Methods of Analysis. Lewis Pubs. Boca Raton, Florida. pp. 755-765.

Tarnocai, C., and R.Valentine. 1989. Relict soil properties of the Arctic and Subarctic regions of Canada. Catena Suppl. 16:9-39.

Tarnocai, C., C.A.S. Smith, and C.A. Fox. 1993. International tour of permafrost Affected Soils – The Yukon and Northwestern territories of Canada. Centre for Land and Biol. Resources Res., Rsrch Branch, Ag. Canada. 197 pp.

Tedrow, J.C.F. 1977. Soils of the polar landscape. Rutgers Univ. Press. New Brunswick, N.J. 638 pp.

Tursina, T.V. 1985. Micromorphological diagnostics of cryogenic features in soils. Problems of soil cryogenesis. Nauka. Moscow. pp. 32-34.

Tursina, T.V. 1994. The peculiar features of microsture and geneisi of cryosols of the permafrost regions of Russia. In: J.M. Kimble and R.J.Ahrens, eds. Proc. Meet. Classification, correlation, and management of permafrost affected soils, Alaska. USDA. pp. 155-159;

Tyutyunov, I.A. 1964. An introduction to the theory of formation of frozen rocks (traduc. I. Muhlhaus), Pergamon Press. Oxford. 125 pp.

Ugolini, F.C. 1986. Pedogenic zonation in the well drained soils of the arctic regions. Quaternary Res. 26(1):100-121.

Ugolini, F.C., and R. Sletten. 1988. Genesis of Arctic brown soils (Pergelic cryochrept) in Svalbard. International Permafrost Conference Trondheim, Norway, August 1988. TAPIR Pub. pp. 478-483.

Van Vliet-Lanoë, B. 1976. Traces de segrégations de glace associées aux sols et phénomènes périglaciaires fossiles. Biuletyn Peryglacjalny. 26:41-54.

Van Vliet-Lanoë, B. 1982. Structures et microstructures associées à la formation de glace de ségrégation. Leurs conséquences. In: H.M. French, ed. Proc. 4th Canadian Permafrost Conf., Calgary, 1981. N.R.C. Ottawa, 116-122.

Van Vliet-Lanoë, B. 1985. Frost effect in soils. In: J. Boardman, ed. Soils and Quaternary Landscape Evolution. Wiley Publ. London. pp. 117-158.

Van Vliet-Lanoë, B. 1986. Interaction entre activité biologique et glace de ségrégation en lentilles; exemples observés en milieu arctique et alpin. In: N. Fedoroff, L.M. Bresson, and M.A. Courty, eds. Micromorphologie des sols - Soil micromorphology. AFES. pp. 337-344.

Van Vliet-Lanoë, B. 1988a. Le rôle de la glace de ségrégation dans les formations superficielles de l' Europe de l' Ouest. Processus et héritages. Thèse d'Etat, Univ. Paris I, Octobre 1987. Editec. Caen. 854 pp.

Van Vliet-Lanoë, B. 1988b. The significance of cryoturbation phenomena in environmental reconstruction. J. Quaternary Sci. 3:1:85-96.

Van Vliet-Lanoe, B. 1995. Solifluxion et transferts illuviaux sur les versants périglaciaires lités. Etat de la question. Géomorphologie, processus et environnement. 2:85-113.

Van Vliet-Lanoë, B. 1998. Frost and soils : implications for paleosols, paleo-climates, and stratigraphy. Catena. 34:157-183.

Van Vliet-Lanoe B., Coutard J.P et Pissart A. 1984. Structures caused by repeated freezing and thawing in various loamy sediments. A comparison of active, fossil and experimental data. Earth Surface Processes and Landforms. 9:553-565.

Van Vliet-Lanoë, B., and A. Dupas. 1991. Development of soil fabric by freeze-thaw cycles. Its effect on frost heave. In: Yu and Wang, eds. Ground Freezing. Balkema. pp. 189-195.

Van Vliet- Lanoë, B., and M. Sëppälä. 2000. Stratigraphy and formation of pounus (frost mounds) in Finish Lapland. Holocene. 12,2:187-799.

Wilkinson, A., and B.T. Bunting. 1975. Overland transport by rill water in a periglacial environment. Geografiska An. 57A:105-116.

Chapter 3. The Thermal Regime of Cryosols

C.R. Burn[1]

[1]Professor, Department of Geography and Environmental Studies, Carleton University, 1125 Colonel By Drive, Ottawa, Ontario, K1S 5B6, Canada; e-mail crburn@ccs.carleton.ca.

1. Introduction

Cryosols are soils whose development is affected by permafrost, ground that remains frozen for two or more years (ACGR, 1988). Cryosols develop both within the active layer, the surface layer of the ground above permafrost that freezes and thaws each year (Muller, 1947), and in the upper portion of the permafrost. For the requirements of classification, the control section for a Cryosol may be 1 or 2 m deep and must contain permafrost (Soil Classification Working Group, 1998; Soil Survey Staff, 1999).

This chapter reviews the thermal regime of the active layer and near-surface permafrost, illustrating various points with field data from the Yukon and western Arctic Canada. This region encompasses a range of permafrost and active-layer conditions, from isolated patches of perennially frozen ground in southern Yukon, to continuous permafrost along the western Arctic coast (Figure 3.3.1; Smith et al., 1998).

The fluctuation of temperature within the soil, or the soil's thermal regime, occurs in response to variations in both air temperature and conditions deeper in the ground. After briefly considering the thermal regime of permafrost, this paper considers: (1) factors that affect the relation between air temperature and ground surface temperature; (2) propagation of the surface temperature wave into the soil and the development of the active layer; (3) mechanisms of heat transfer within the soil; and (4) the overall impact of the active layer's development on annual mean ground temperatures. While the discussion has a theoretical framework, it emphasizes the physical variables that influence field conditions.

Figure 3.3.1. Permafrost map of the Yukon and adjacent Northwest Territories, Canada, indicating locations for which this chapter cites data (after Heginbottom et al., 1995).

2. Thermal Regime of Permafrost

Figure 3.3.2 illustrates the thermal regime of permafrost terrain under equilibrium conditions. The two principal features are an increase in ground temperatures with depth and an envelope of ground temperature near the surface. The envelope describes the range in near-surface temperature that occurs as a result of the seasonal variation in air temperature. The two sides of the envelope represent the annual maximum and minimum temperatures at each depth. The depth at which the difference between the annual maximum and annual minimum temperature is 0.1°C represents, for practical purposes, the depth of zero annual amplitude, z_0. The temperature at this depth is the annual mean ground temperature, T_g, which is distinct from the annual mean soil temperature, T_{sl}, measured 50 cm below the ground surface.

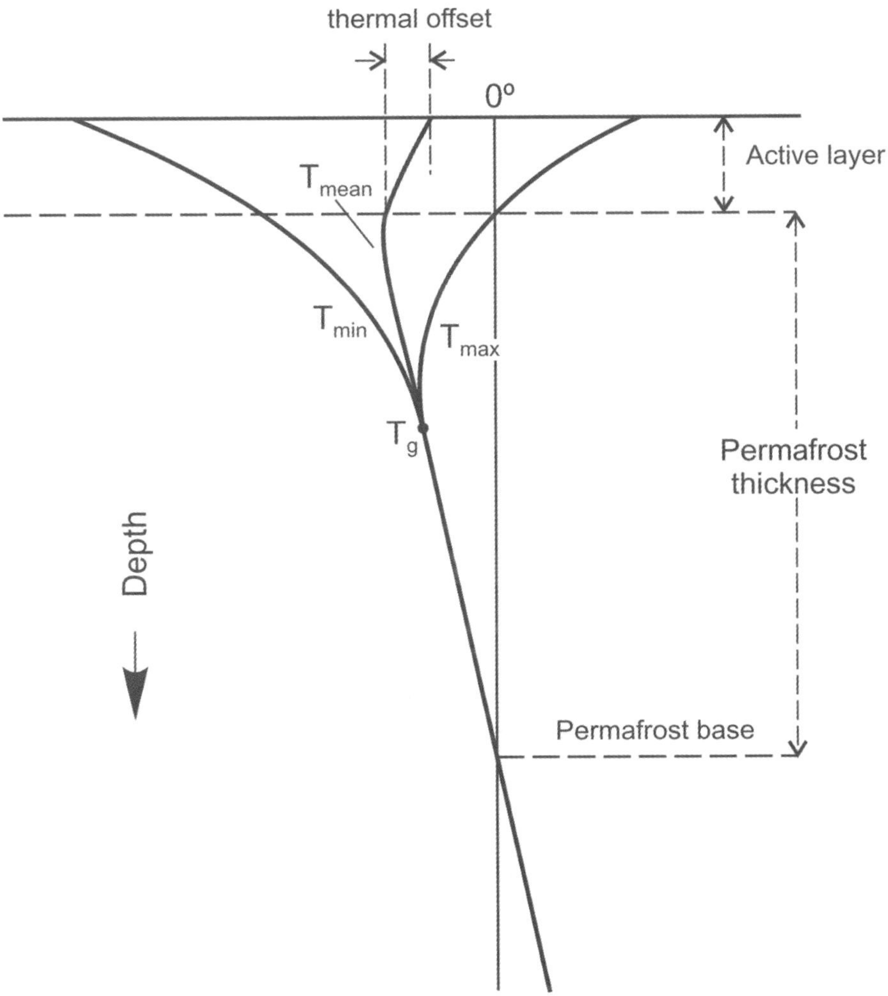

Figure 3.3.2. The thermal regime of permafrost terrain under equilibrium conditions (after Burn and Smith, 1988).

Table 3.3.1. Thermal properties of soil materials.

Material	Thermal conductivity λ, W/m/C	Heat capacity C, J/m^3/C	Thermal diffusivity κ, m^2/s
Air	2.2×10^{-2}	8.6×10^{2}	2.6×10^{-5}
Water	5.6×10^{-1}	4.2×10^{6}	1.3×10^{-7}
Ice	2.2×10^{0}	1.9×10^{6}	1.2×10^{-6}
Quartz	8.0×10^{0}	2.1×10^{6}	3.8×10^{-6}
Mica	3.0×10^{0}	2.4×10^{6}	1.2×10^{-6}
Feldspar	2.0×10^{0}	2.0×10^{6}	1.0×10^{-6}
Granite	2.0×10^{0}	2.1×10^{6}	9.5×10^{-7}
Limestone	2.9×10^{0}	2.5×10^{6}	1.2×10^{-6}
Shale	1.5×10^{0}	1.6×10^{6}	9.4×10^{-7}
Peat (dry)	6.0×10^{-2}	5.8×10^{5}	1.0×10^{-7}
Peat (wet)	5.0×10^{-1}	4.0×10^{6}	1.2×10^{-7}
Peat (wet, frozen)	1.1×10^{0}	1.6×10^{6}	6.8×10^{-7}

Latent heat of fusion of water: 3.33×10^{8} J/m^3.
Latent heat of vaporization of water: 2.45×10^{9} J/m^3.
Data from Williams (1982, Tables 5.1 and 5.2); Williams and Smith (1989, Table 4.1).

Lower in the profile, the temperature increases with depth, due to the geothermal flux (Judge, 1973). The base of permafrost occurs where the profile crosses 0°C. The geothermal flux, Q (~ 0.05 W/m^2), relates to environmental variables by:

$$Q = \lambda\left(\frac{dT}{dz}\right) \tag{3.3.1}$$

where λ is the thermal conductivity of earth materials (W/m^1/°C), T is temperature (°C), and z is depth (m). Table 3.3.1 provides a range of values for λ. Under equilibrium conditions, with a linear temperature profile, the base of permafrost is at depth z_b below z_0, where:

$$Q = \lambda\left(\frac{T_g}{z_b}\right) \tag{3.3.2}$$

Rearrangement of equation 3.3.2 indicates that the depth of permafrost is directly proportional to λ (Williams and Smith, 1989, p.95). Increases in λ occur as a result of an increase in the quartz content of ground materials, and when water or ice replaces air in the pore spaces. One can estimate the thermal conductivity of a composite ground material with (Johansen, 1975):

$$\lambda = \lambda_m^{1-p} \cdot \lambda_w^{p-q-r} \cdot \lambda_a^{q} \cdot \lambda_i^{r} \tag{3.3.3}$$

where λ, λ_m λ_w, λ_a, and λ_i are the thermal conductivities of the ground, mineral materials, water, air, and ice, respectively, p is the porosity, and q and r are the volumetric fractions of air and ice.

Development of Cryosols occurs primarily in the active layer, but features established during previous soil development also are often evident in the upper portion of permafrost (e.g., Burn et al., 1986; Burn, 1997). Although the active layer is defined in terms of temperature (Burn, 1998a), an ice-rich zone at the top of permafrost usually underlies the base of the active layer (Mackay, 1970). The ice-rich zone forms by downward movement of water into permafrost during freezing of the ground at the end of summer (Mackay, 1983; Burn and Michel, 1988; Harris, 1988). This horizon, with its high volumetric latent heat content, impedes further active-layer development once the surface has thawed. A consequence of this ice-rich zone is that the active-layer thickness in most Cryosols varies little from year to year, perhaps by 10%.

3. Ground Surface Temperature

At regional scale, the ground surface temperature (T_s) directly relates to air temperature (T_a), so that, as T_a declines, the proportion of the landscape underlain by permafrost and Cryosolic soils increases (Heginbottom et al., 1995). T_s is a product of the surface energy balance, representing the flux of energy at the ground surface:

$$Q^* = Q_H + Q_E + Q_G \tag{3.3.4}$$

where Q^* is the net radiation, Q_H is the sensible heat transfer between the ground and atmosphere, Q_E is the evaporative flux, and Q_G is the soil heat flux. Each of these fluxes depends on conditions within the ground or at the ground surface. The albedo, which changes seasonally as snow cover accumulates or melts, and changes in vegetation cover, most drastically wrought by forest fire, influence Q^* (Rouse, 1976). Turbulence at the ground surface influences Q_H, which depends on the structure of vegetation, with more developed plants increasing the loss of heat from the surface. Q_E depends on the availability of moisture at the ground surface and the ability of vegetation to transpire once the surface is dry. Q_G relates to the thickness and density of the snow cover and to the nature of soil materials, varying directly with bulk density and moisture content through thermal conductivity.

Any combination of these fluxes, at a specific site, has a unique value of T_s, which numerical methods can estimate (Outcalt, 1972). Variations in these fluxes explain the spatial variation in T_s responsible for discontinuous permafrost beneath much of the boreal forest in North America. At the southerly margins of permafrost in the northern hemisphere, Organic Cryosols occur in peatlands with peat plateau bogs (Zoltai et al., 1988). At these sites, the water table is close to the surface, promoting evaporation, and λ is low when the peat is dry in summer (Brown, 1970). Characteristically the peat becomes wet in autumn from rain and intermittent snowmelt before it freezes, and λ increases (Nicholas and Hinkel, 1996). At such sites, soil thermal properties and moisture conditions combine to sustain permafrost. Throughout much of the discontinuous permafrost zone, drainage determines the location of Cryosols, so that moist sites may support

Cryosols, while drier sites support non-Cryosolic soils, typically Brunisols or Regosols (Swanson, 1996).

In central Yukon, an investigation by Williams and Burn (1996) determined that Cryosols occur where the gravimetric moisture content of mineral soil at a 50-cm depth remains over 25% and the surface organic horizon is over 10 cm thick. These factors override the importance of aspect, forest type, and elevation in the study area near Mayo (Figure 3.3.1). The same ranking of factors is likely appropriate over large parts of the boreal forest, although the critical values for development of permafrost may vary. For conditions representative of Whitehorse, YT (Figure 3.3.1), modelling indicates that the influence on ground temperature of changes in soil moisture content may be greater than variations in air temperature expected during climatic change (Smith and Riseborough, 1983).

Soil moisture content and the thickness of surface organic accumulations are controlling variables associated with summer soil temperatures. In winter, variations in snow depth predominantly influence T_s (Mackay and MacKay, 1974). The snow provides insulation from frigid air temperatures. In the boreal forest, snow depth is broadly uniform, although variable around trees, and its effect on the distribution of Cryosols is secondary (Sturm, 1992), except in peatlands near the southerly limits of permafrost (Camill, 1999). Another exception occurs near rivers and lakes, where accumulation of snow blown off the ice into willow and alder thickets at the water's edge may lead to permafrost eradication (Smith, 1975). However, an abrupt cooling of soil temperatures occurs across the tree line through the transition to tundra, where the snow cover is thinner and packed more densely than in the forest. In 1993 at Inuvik, N.W.T., south of the tree line, Smith et al. (1998) estimated an annual mean surface temperature (AMST) of -1.7°C, while north of the tree line, at Parson's Lake, 60 km from Inuvik, they estimated the AMST at -5.5°C.

The effects of forest fire illustrate the critical importance of surface conditions on T_s. Ground temperatures usually increase after forest fire, not because of the heat of the fire, which is a transient phenomenon, but due to the surface disturbance. Removal of vegetation reduces shade and evaporation, raising T_s; eradication of the canopy leads to increased snow depth in winter. The result is an increase in soil temperatures, active-layer depths, and permafrost degradation (Viereck, 1982; Mackay, 1995; Burn, 1998b).

Ground temperatures collected over several years near Mayo, central Y.T., and in Takhini River valley, near Whitehorse, southern Y.T., illustrate the variation in soil temperatures within small areas, due to differences in surface conditions, soil composition, and moisture content. At Mayo, in the widespread discontinuous permafrost zone (Figure 3.3.1), near-surface ground temperatures have been monitored to a 5-m depth since 1985 at seven sites.

Table 3.3.2 indicates the range in soil temperature at a 50-cm depth (T_{sl}) and in active-layer depths between September 1987 and 1988 at these sites. The mean value for T_{sl} at the seven sites was 0.1°C, and the range in this variable was 1.8°C. At the same sites, the mean and range for T_g were -0.6° and 2.4°C, respectively. In the Takhini River valley, in sporadic discontinuous permafrost (Figure 3.3.1), the range in T_{sl} for 1998-99 (-1.3° to 1.8°C) illustrates the difference between permafrost-free sites, a site with permafrost, and a site burned by a forest fire in 1958, where, in 1999, permafrost was still degrading. At the permafrost-free sites, the data illustrate the effect of snow on annual mean ground temperature, for one site is undisturbed and, at the other, the snow cover is cleared every fortnight.

Figure 3.3.3 presents the annual ground temperature cycle from a 50-cm depth at the permafrost-free sites in 1997-99. The effect of snow's clearing is evident in cooler soil temperatures not only in winter, but also in summer, even though surface conditions are then similar. This indicates the influence of subsurface conditions, i.e., Q_G, on T_{sl}, for the cooler near-surface temperatures are in response to cooler ground at depth.

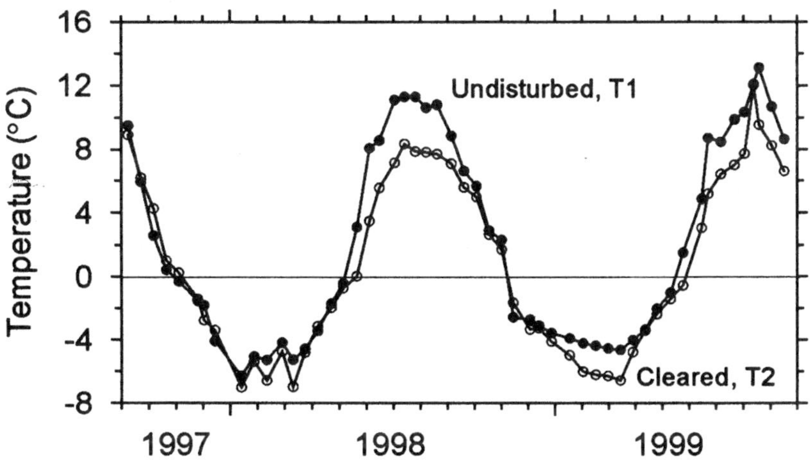

Figure 3.3.3. Soil temperatures at a 50-cm depth over two years, starting in September 1997, from two sites without permafrost in the Takhini River valley (T1 and T2, Table 3.3.2), 50 km west of Whitehorse, YT. The sites are adjacent to each other, but at one the ground surface was cleared of snow every fortnight throughout the winter. The mean temperatures over the two-year period for the undisturbed and cleared sites were, respectively, 2.3° and 1.0°C. The temperature measurements were made with calibrated thermistor cables.

Table 3.3.2. Annual mean near-surface temperatures at Mayo (September 1987-88) and the Takhini Valley (1998) study sites, YT.

Site	Vegetation	Organic layer thickness (cm)	Soil texture	Moisture content at 50 cm (cm^3/cm^3)	Snow depth in January (cm)	T_{sl} (°C)	T_g (°C)	Active-layer thickness (cm)
M1	Closed birch woodland	17	Sandy loam	0.24	30-35	1.1	0.8	-
M2	Closed spruce/birch forest	2	Silty clay loam	0.30	30-40	0.7	-0.4	160
M3	Willows/spruce shrubland	16	Silty clay loam	0.38	35-40	-0.4	-1.6	110
M4	Closed spruce forest	42	Organic/silt loam	0.38	20-25	0.1	-0.9	95
M5	Open birch/spruce wood	23	Silt loam	0.30	30-40	-0.1	-0.3	110
M6	Open spruce forest	18	Silty clay	0.34	30-40	-0.7	-1.6	80
M7	Closed spruce forest	18	Silt loam	0.42	30-45	-0.3	-0.5	75
T1	Grassy meadow	0	Silty clay	na	15-20	1.8	1.0	-
T2	Grassy meadow	0	Silty clay	na	0	0.4	0.0	-
T3	Closed spruce forest	15	Sandy loam	0.18	10-20	-1.3	-1.2	145
T4	Burned willow shrubland	0	Silty clay	0.29	25-30	0.9	-0.1	360

4. The *n*-Factors

The *n*-factor, or ratio of air-to-ground surface temperature, summarizes the relation between air and ground surface temperatures for modelling purposes (Lunardini, 1978). An *n*-factor is determined for the thawing season, n_t, and for the freezing season, n_f. The values of n_f and n_t are usually the ratio of freezing degree-days in the air and at the ground surface and the ratio of thawing degree-days, respectively. The value of n_f normally declines as depth of snow increases (Taylor, 1995). Most *n*-factors available for natural surfaces have been determined empirically, but some theoretical values are available for engineered surfaces (Lunardini, 1978; Jorgenson and Kreig, 1988; Taylor, 1995; Burn, 1998b; Smith et al., 1998). Characteristic values from sites in the boreal forest of northwest Canada are 0.5 to 0.6 for n_t and 0.1 to 0.3 for n_f (Taylor, 1995; Burn, 1998b). Data sets from natural sites in Alaska have provided slightly higher determinations (Jorgenson and Kreig, 1988).

5. Soil Thermal Regime

Soil temperature varies diurnally and annually following changes in T_s. The ground temperature gradient and soil thermal conductivity dominate heat flow upward or downward within the soil, as in equation 3.3.1. At a point where the temperature gradient changes, the difference in heat flux above and below the point indicates the energy available to cool or warm the soil. The specific heat capacity of the soil, C (J m^{-3}°C^{-1}), the energy released or required to change the temperature of 1 m^3 of soil by 1°C, modulates actual temperature changes within the soil. The heat capacity of soil is the sum of the constituent heat capacities:

$$C = (1-p)C_m + (p-q-r)C_w + qC_a + rC_i \tag{3.3.5}$$

where C_m, C_w, C_a, C_i are the specific heat capacities of the soil constituents (Table 3.3.1). C is sensitive to soil wetting and drying, because C_w is 4.18 x 10^6 J/m^3/C, while C_a is 1 J/m^3/C. Changes in bulk density, i.e., porosity, also affect C, particularly in organic soils (Farouki, 1981).

The heat conduction equation (Carslaw and Jaeger, 1959, p. 37):

$$\frac{dT}{dt} = \frac{\lambda}{C}\frac{d^2T}{dz^2} \tag{3.3.6}$$

relates the change in temperature with time, t, to the rate of change in the soil temperature gradient. The ratio of thermal conductivity to specific heat capacity is the thermal diffusivity, κ; which indicates the ability of a medium to propagate a temperature change. For Cryosols, κ increases substantially during freezing, for λ_i is four times λ_w, and C_i is about half C_w. So, if there is little snow cover, cooling of the soil may accelerate once pore water has frozen, and warming of the soil is slower once the ground thaws (Figure 3.3.4).

We can calculate the diurnal and annual range in soil temperatures with equation 3.3.6, if we know the surface temperature regime. Often T_s is considered to follow a sinusoidal pattern, so that:

$$T_s = AMST + A_s \sin(\omega t) \tag{3.3.7}$$

where A_s is the amplitude of the surface temperature oscillation and ω is the angular velocity. The solution of equation 3.3.6 with 3.3.7 as a boundary condition gives (Carslaw and Jaeger, 1959, p.65):

$$T(z,t) = AMST + A_s e^{-z\sqrt{\frac{\omega}{2\kappa}}} \sin(\omega t - z\sqrt{\frac{\omega}{2\kappa}}) \tag{3.3.8}$$

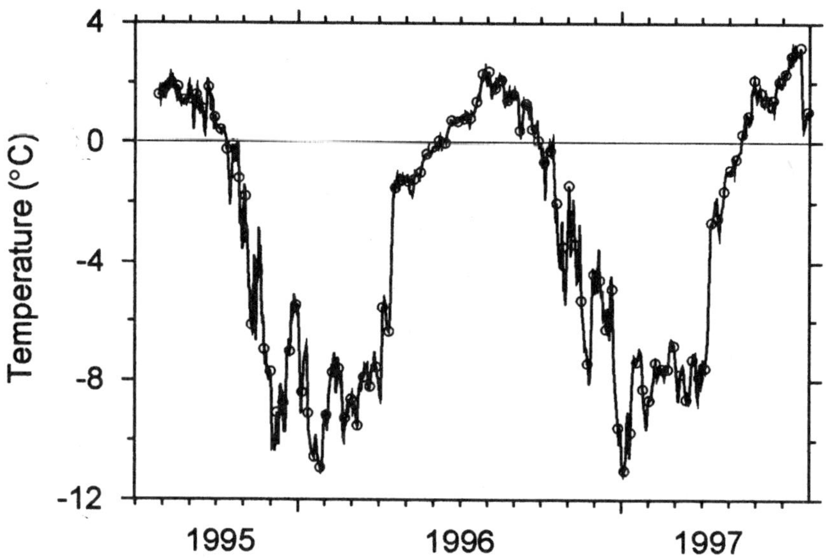

Figure 3.3.4. Soil temperature series from a 50-cm depth in a Turbic Cryosol near Mayo, YT (M6, Table 3.3.2), 28 July, 1995, to 31 July, 1997. The data are daily means of hourly observations. The circles indicate weekly intervals. A Campbell Scientific CR10 data logger connected to a calibrated thermistor cable collected the data. The soil has a gravimetric water content, at a 50-cm depth, greater than 30% throughout summer, and the active layer is less than a meter thick. The mean soil temperature was -3.1°C, and the range in temperature was 15.1°C. Soil temperatures in 1995-97 were cooler than in 1987-88 (Table 3.3.2), mainly due to thinner snow covers in winter during 1995-96 and 1996-97.

Equation 3.3.8 shows that the amplitude of the temperature wave decreases exponentially with depth in the soil, and that the wave is increasingly offset with depth. Figure 3.3.5 illustrates this with near-surface temperature waves from 50 cm and 1.5 m deep, recorded within a soil in the Takhini River valley (site T3). Figure 3.3.6 shows a similar pattern in soil temperatures from 25- and 75-cm depths at Illisarvik, an experimentally drained lake on Richards Island, near the western Arctic coast (Tarnocai, 1994; Mackay, 1997).

The exponent in equation 3.3.8 is inversely proportional to the thermal diffusivity, so that, as soil diffusivity increases, the temperature envelope widens. Dry sands, containing substantial quantities of quartz, have high diffusivities because the heat capacity is low, and hence there is a large range in subsurface temperatures. The rate of damping of the annual temperature wave is the principal determinant of active-layer depth, and its relation to thermal diffusivity illustrates why moist Turbic Cryosols characteristically form with a thin active layer, while dry Static Cryosols are associated with thick active layers. Compare, for instance, the active-layer thicknesses at Mayo, in moist soil, and in the dry Takhini River valley (Table 3.3.2).

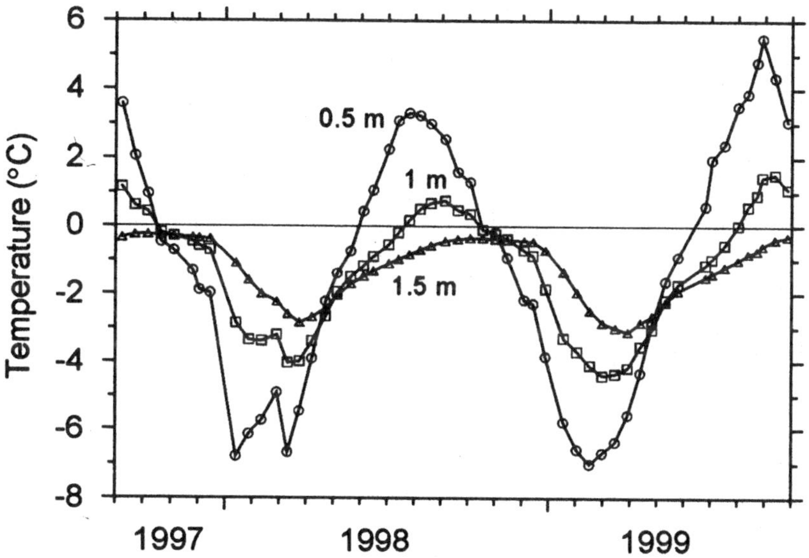

Figure 3.3.5. Near-surface ground temperatures in the Takhini River valley (site T3, Table 3.3.2), September 1997-99, illustrating damping and delay of the temperature wave with depth. The mean temperatures over the two-year period at depths of 50 cm, 1 m, and 1.5 m were, respectively, -0.9°, -1.2°, and -1.3°C. The measurements were made on calibrated thermistor cables. The snow's depth at this site was rarely more than 20 cm during winter.

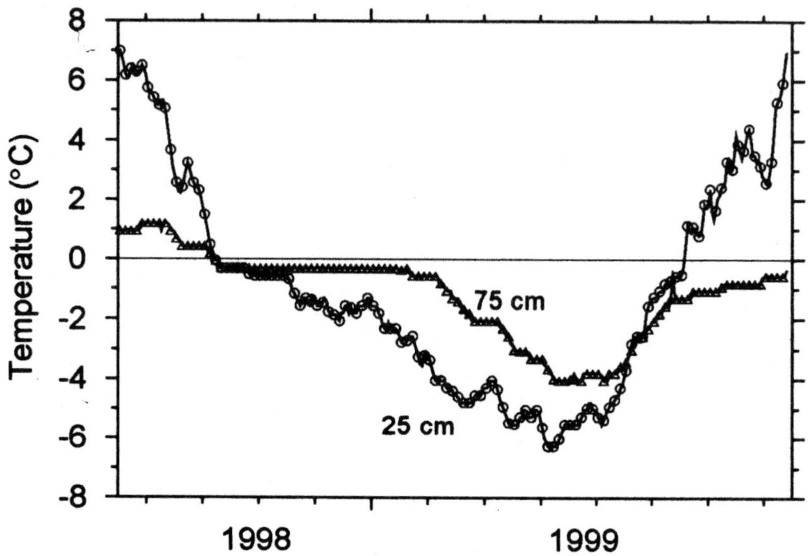

Figure 3.3.6. Soil temperature series from the Illisarvik experimentally drained lake on Richards Island, western Arctic coast, mid-August 1998-99. The mean temperatures at depths of 25 cm and 75 cm were, respectively, -1.1° and -1.2°C. Temperature measurements plotted are the daily means of five readings on an HOBO data logger. Snow accumulates to a depth of 40 to 50 cm in the grasses that have grown up in the lake bottom since drainage. The active layer is up to 1 m thick at this site.

The analysis just presented assumes that heat flow is entirely by conduction and that there is no horizontal transfer of energy. Both assumptions may be violated if groundwater moves within the active layer. In addition, the analysis assumes the surface temperature follows a sinusoidal pattern and that latent heat has no effects. In regions with snow cover and seasonal freezing, the surface temperature wave is asymmetric, and considerable energy may be associated with melting and freezing of soil water (Figure 3.3.6).

6. Latent Heat

In soils with substantial quantities of silt and clay, the freezing and thawing of soil water occurs over a range of temperatures near 0°C, corresponding to the unfrozen water content characteristic for the soil (Koopmans and Miller, 1966; Patterson and Smith, 1981). Soil water is prevented from freezing at 0°C by its confinement in restricted pore spaces, its solute content, and the proximity of adsorbed water to charged particle surfaces, all of which decrease the potential of the water and restrict water molecules from joining the ice crystal lattice (Williams and Smith, 1989, p.7). The unfrozen water content (θ_u) is usually described by an equation of the form:

$$\theta_u = aT^{-b} \tag{3.3.9}$$

but the characteristic curve exhibits hysteresis in the relation between θ_u and T during warming and cooling of soil (Figure 3.3.7).

Unfrozen water distributes latent heat over a range of temperatures, so that saturated freezing and thawing soils exhibit an apparent heat capacity, C_a:

$$C_a = C + L\frac{d\theta_u}{dT} \tag{3.3.10}$$

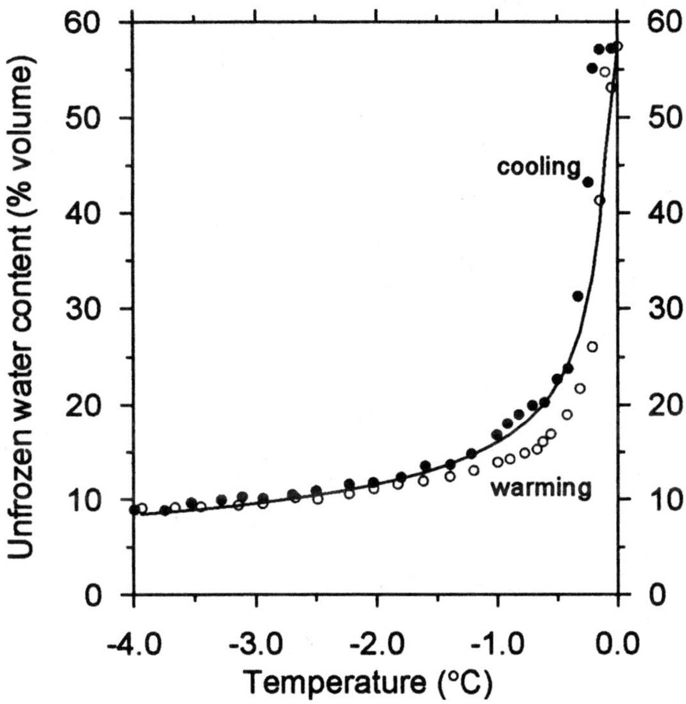

Figure 3.3.7. Characteristic curve of unfrozen water content for glaciolacustrine sediments from Takhini River valley, YT. The data are from Time Domain Reflectometry (Patterson and Smith, 1981). The line indicates the best fit power relation, summarizing data obtained during both warming and cooling, $\theta_u = 16.02T^{-0.46}$. The difference in content of unfrozen water during warming and cooling of soil indicates the hysteresis in this relation, as with the soil-moisture characteristic.

where L is the latent heat of fusion (3.33×10^8 J/m^3). This implies that soils below 0°C have an apparent thermal diffusivity. Soil thermal properties often are calculated incrementally, over a range of temperatures, for instance, -4° to 0°C (e.g., Smith and Riseborough, 1985), and numerical methods are used to determine changes in ground temperatures (Goodrich, 1982).

7. Development of the Active Layer

Cryosols thaw in spring and summer, freeze in autumn, and remain frozen in winter. In winter, ice normally concentrates at the top and bottom of the active layer, where it accumulates during freezing (Mackay, 1980). Thawing of the active layer follows snowmelt and is initially rapid because of the abundant radiant energy and the steep temperature gradients near the ground surface. The temperature gradient declines as depth of thaw increases, reducing the rate of thawing (Figure 3.3.8). Once thaw reaches the ice-rich soil at the base of the active layer, the rate of thawing reduces further, because of the latent heat in this horizon. Thus, the depth of the active layer is relatively constant from year to year, if surface conditions do not change.

Figure 3.3.8 indicates development of the active layer at two sites near Mayo in the summer of 1984, when ground temperatures were measured twice each week. Relatively little deepening, once the thaw reached the ice-rich zone, followed the rapid penetration of thaw at the beginning of summer. The Stefan solution for thawing frozen ground summarizes the curvilinear relation between thaw depth (z_t) and time (e.g., Andersland and Ladanyi, 1994, p.73):

$$z_t = \sqrt{\frac{2\lambda T_s t}{L}} \qquad (3.3.11)$$

Equation 3.3.11 requires calibration for site conditions and is used for first-order estimates of the depth of the active layer (e.g., Hinkel and Nicholas, 1995; Burn, 1997). However, the relation also may be used to estimate inter-annual variability of active-layer depth, because z_t is proportional to cumulative thawing degree days (TDD), if other factors remain constant (Mackay, 1995).

Freezing of the active layer occurs both from the top and bottom during autumn, and the rate of upfreezing depends on the temperature in the permafrost. Above "cold" permafrost, upfreezing may account for 25% or more of the active layer, while above "warm" permafrost, upfreezing may account for less than 10% of the active layer. Once freezing begins, the active layer soon reaches a near-isothermal state, at about 0°C. The progressive freezing means that, in the middle of the active layer, the soil may remain at a constant temperature for a prolonged period. This is the so-called "zero curtain."

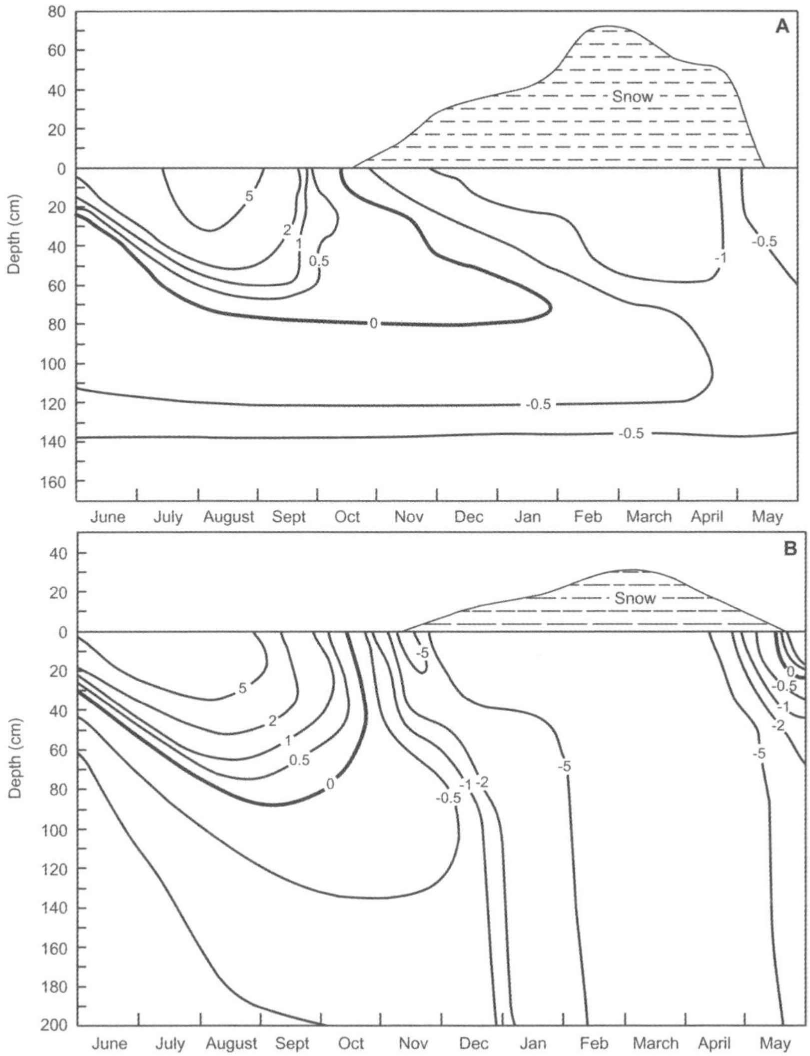

Figure 3.3.8. Annual soil thermal regime at two sites near Mayo, YT, 1 June 1984-1985. Data are from Burn (1986, Figure 4.1). The data were collected manually on calibrated thermistors spaced 15 cm apart on their cables. Readings were taken twice each week in summer and weekly or fortnightly during the rest of the year. Site A was a saturated, swampy site, and the plot exhibits an extended zero curtain during prolonged freezing within the active layer. In contrast, site B was in a relatively dry location, and freezing there was relatively rapid. The influence of the unfrozen water content on freezing of the soil is apparent at the sites in the considerable area between 0° and -0.5°C.

Figure 3.3.6 shows a period of 100 days in the autumn at Illisarvik when the temperature remains close to 0°C as the active layer freezes. Latent heat released during freezing maintains relatively warm temperatures in the ground and at the surface, while the air temperature may be well below 0°C. The duration of the zero curtain depends on air temperature, snow depth, and soil moisture content. In the dry Cryosol at Takhini (Figure 3.3.5), the zero curtain lasts a short time, as it does at Mayo site 6 (Figure 3.3.4). At both Yukon sites, interception in the forest canopy delays accumulation of snow. At Illisarvik, where snow is trapped in tall grasses on the lake bottom early in autumn, the zero curtain is prolonged. A similar zero curtain may occur in spring during snowmelt, for the near-surface ground temperature rises to close to 0°C, due to refreezing of meltwater at the ground surface and within the Cryosol.

These two periods introduce error into calculations of n-factors, for the latent heat effects imply the ground temperature is not responding to air temperature alone. The n-factor then is underestimated in comparison with periods without such pronounced latent heat effects. Burn (1998b) illustrated this point with data from the sites in the Takhini River valley.

8. Non-conductive Heat Transfer

Hinkel and Outcalt (1993) have presented spectral data that indicate soil temperatures vary at diurnal, annual, and other frequencies. While the diurnal and annual cycles are associated with heat conduction following well recognized variations in T_s, the various other frequencies have been attributed to non-conductive processes. In particular, infiltration may lead to heat transfer following snowmelt or summer storms (Hinkel et al., 1997). Wave periods of less than a day have been attributed to redistribution of heat within the soil by evaporation, movement of vapor, and condensation, collectively termed "distillation" (Hinkel and Outcalt, 1994).

Convection may be an important mechanism of heat transfer in organic soils and sands or gravels. Baker and Osterkamp (1988) reported direct observations of convective effects in saturated sand, and Outcalt and Hinkel (1996) have demonstrated the potential for vapor flow in organic soils. However, in finely grained material there may be much less, for Romanovsky and Osterkamp (1995) found little evidence of convection in long-term data from three sites on the North Slope of Alaska, and they concluded that conduction could model the thermal regime at the sites efficiently. Similarly, Burn (2000) found, with a temperature series collected daily over two years at Mayo, only limited evidence for convection in finely grained soil. The data collected at Illisarvik (Figure 3.3.6) show a small spike in early June, 1999, for 75 cm, which may be due to such non-conductive effects.

9. Thermal Offset

When the soil annually is in thermal equilibrium, the heat flow into the ground in summer and the small geothermal flux equal the heat extracted from the surface in winter. The change in thermal properties of Cryosolic soils between thawed and frozen states implies that the soil temperature gradient is, in general, steeper in summer than in winter. The result is that the annual mean soil temperature declines with depth to the base of the active layer, where thermal properties vary little seasonally. The shape of the annual mean ground temperature profile within a Cryosol is, therefore, curvilinear (Goodrich, 1982; Burn and Smith, 1988; Romanovsky and Osterkamp, 1995), so measurements at several depths are necessary for estimating the AMST (Taylor, 1995; see Figure 3.3.2). The difference between AMST and the annual mean temperature at the surface of permafrost is the thermal offset, ΔT. Kudryavstev (1981) presented an expression for ΔT, which Romanovsky and Osterkamp (1995, eq. 12) derived analytically and validated by numerical methods and with field data:

$$\Delta T = \left(\frac{\lambda_t}{\lambda_f} - 1\right)\frac{TDD}{365} \qquad (3.3.12)$$

where TDD are the annual thawing degree days at the ground surface, ΔT depends on the ratio of thawed to frozen conditions and hence is greatest in wet, organic soils. Burn and Smith (1988) measured offsets of up to 1.7°C in the Mayo study area.

The physical importance of the offset is that it enables the presence of permafrost, and hence Cryosols, for AMST > 0°C. At the limit, the offset implies a Cryosol may have a mean annual temperature of over 0°C within the active layer, as is the case for many Cryosols in the discontinuous permafrost zone. In a counterintuitive fashion, equation 3.3.12 implies that the offset is greater in moist turbic Cryosols than in dry static Cryosols, so that turbic Cryosols may be warmer, annually, than static Cryosols.

10. Cryosols, Permafrost, and Climatic Change

Presently, interest in the impact of climate change on the thermal regime of permafrost is considerable (e.g., Lachenbruch and Marshall, 1986; Osterkamp and Romanovsky, 1999). However, to reach permafrost, any air temperature variation must pass through the active layer, where the variation may change (Lachenbruch et al., 1988). Two aspects of changing permafrost conditions are of immediate relevance to the distribution and nature of Cryosols. First, if permafrost disappears from a site, the soil evolves out of its Cryosolic state. Second, changes in the thickness of the active layer directly indicate an altered thermal regime.

Permafrost conditions have changed in North America over several time scales. During the 20th century, near-surface permafrost has warmed from a cooler thermal regime that was established during the Little Ice Age (Lachenbruch and Marshall, 1986). Evidence of such warming is the curvature of ground temperature profiles between the surface and to depths of up to about 100 m. These profiles have been obtained from "cold," continuous permafrost in northern Alaska, and from warmer, discontinuous permafrost in the central Mackenzie River valley (Mackay, 1975). Ground temperatures have continued to rise over much of Alaska through 1996, the last year of data that Osterkamp and Romanovsky present (1999). Similar ground warming was recorded in the Yukon Territory in the 1970s and 1980s (Burn, 1992). However, permafrost has cooled in eastern Alaska and the Yukon during the 1990s, principally in response to a decrease in snow accumulation (Burn, 1998c; Osterkamp and Romanovsky, 1999). Cooling of permafrost has occurred in Quebec, following a general climatic trend (Allard et al., 1995).

The curvilinear nature of the unfrozen water content characteristic (Figure 3.3.7) implies that changes in ground temperature may occur rapidly at low temperatures but more slowly close to 0°C, where latent heat exchanges are important (Riseborough, 1990). Thawing of permafrost follows equation 3.3.11, with the rate of degradation decreasing as the top of the permafrost descends (Burn, 1998b). Ice-rich permafrost less than 5 m thick may thaw over about a century, but greater thicknesses of frozen ground may require considerably longer to thaw. In general, the latent heat in ice-rich ground delivers a considerable temporal persistence to permafrost (Burn, 1998c).

However, ground warming since the Little Ice Age has led to the disappearance of relatively thin permafrost from a broad swath of peatlands in northern Alberta, Saskatchewan, and Manitoba (Kwong and Gan, 1994; Halsey et al., 1995). Permafrost developed in this region during Neoglacial times, when mean annual temperatures were approximately 1°C lower than at present. Forest fires have helped degrade this "warm" permafrost (Zoltai, 1993). Nearly 30,000 km^2 of permafrost peatland is expected to thaw in response to the recent climate change (Vitt et al. 2000). Where permafrost is thawing, it is relict but still, by definition, supports Cryosols. The spatial extent of permafrost similarly increased in Europe and Asia during the Little Ice Age, so Cryosols are currently more widespread than maps of permafrost drawn on the basis of present air temperatures may imply.

Changes in the thickness of the active layer also have been recorded as a result of climate change, either due to general climate warming or to an increase in summer temperature alone. In northwest Canada, the active layer was about twice its present thickness during the warmest period of the Holocene, 9,000 cal. years ago (Burn et al., 1986; Burn, 1997). The thickness of the active layer is a critical component of the thermal regime of Cryosols, because deepening of the active layer normally involves thaw of the ice-rich zone at the top of permafrost and initiation of ground subsidence (Burn, 1998c; Wolfe et al., 2000). Monitoring the thickness of the active layer is one of the principal circumpolar efforts underway to detect the response of permafrost terrain to climate change and climate variation (Nixon and Taylor, 1998; Brown et al., 2000).

11. Conclusion

The principles this chapter outlines form the basis for explaining the transfer of heat between the atmosphere and the ground, with the flux downward in summer and upward, out of the ground, in winter. At site scale, the analysis must involve detailed characterization of thermal properties, ground surface temperatures, and snow conditions, with data collected over small spatial and temporal scales. For application over larger areas at greater time scales, summary relations such as equation 3.3.12 provide an appropriate index of the thermal impact of Cryosols on permafrost.

12. Acknowledgments

The Natural Sciences and Engineering Research Council of Canada, the Polar Continental Shelf Project of Natural Resources Canada, the Aurora Research Institute of Aurora College, and the Northern Research Institute of Yukon College support the field research in northwestern Canada. The Meteorological Service of Canada collaborates with collecting ground temperatures at Mayo. The extent of the research would not be possible without the generous hospitality of Jim and Shann Carmichael of Mayo and encouragement from Joan Ramsay Burn and Andrew Burn. I also thank J.R. Mackay, C.A.S. Smith, and B. Van Vliet-Lanoë for their helpful comments. This chapter is Polar Continental Shelf Project Contribution 02500.

References

ACGR (Associate Committee on Geotechnical Research). 1988. Glossary of permafrost and related ground-ice terms. National Research Council of Canada, Associate Committee on Geotechnical Research. Technical Memorandum 142.

Allard, M., B. Wang, and J.A. Pilon. 1995. Recent cooling along the southern shore of Hudson Strait, Quebec, Canada, documented from permafrost temperature measurements. Arct. Alp. Res. 27:157-166.

Andersland, O.B., and B. Ladanyi. 1994. An introduction to frozen ground engineering. Chapman and Hall. New York, NY.

Baker, G.C., and T.E. Osterkamp. 1988. Implications of salt fingering processes for salt movement in thawed coarse-grained subsea permafrost. Cold Reg. Sci. Tech. 15:45-52.

Brown, J., K.M. Hinkel, and F.E. Nelson. 2000. The circumpolar active layer monitoring (CALM) program: research designs and initial results. Polar Geog. 24:166-258.

Brown, R.J.E. 1970. Permafrost in Canada. Univ. of Toronto Press. Toronto, Ont.

Burn, C.R. 1986. On the origin of aggradational ice in permafrost. Ph.D. thesis. Carleton Univ. Ottawa, Ont.

Burn, C.R. 1992. Recent ground warming inferred from the temperature in permafrost near Mayo, Yukon Territory. In: J.C. Dixon and A.D. Abrahams, eds. Periglacial geomorphology. Chichester. pp. 327-350.

Burn, C.R. 1997. Cryostratigraphy, paleogeography, and climate change during the early Holocene warm interval, western Arctic coast, Canada. Can. J. Earth Sci. 34:912-925.

Burn, C.R. 1998a. The active layer: two contrasting definitions. Permafrost and Periglacial Processes. 9:411-416.

Burn, C.R. 1998b. The response (1958-1997) of permafrost and near-surface ground temperatures to forest fire, Takhini River valley, southern Yukon Territory. Can. J. Earth Sci. 35:184-199.

Burn, C.R. 1998c. Field investigations of permafrost and climatic change in northwest North America. In: A.G. Lewkowicz and M. Allard, eds. Proceedings, 7th International Conference on Permafrost, 23-26 June 1998, Yellowknife, N.W.T. Québec City, Québec. Collection Nordicana 57. pp. 107-120.

Burn, C.R. 2000. The thermal regime of a retrogressive thaw slump near Mayo, Yukon Territory. Can. J. Earth Sci. 37:967-981.

Burn, C.R., and F.A. Michel. 1988. Evidence for recent temperature-induced water migration into permafrost from the tritium content of ground ice near Mayo, Yukon Territory, Canada. Can. J. Earth Sci. 25:909-915.

Burn, C.R., and C.A.S. Smith. 1988. Observations of the "thermal offset" in near-surface annual ground temperatures at several sites near Mayo, Yukon Territory, Canada. Arctic. 41:99-104.

Burn, C.R., F.A. Michel, and M.W. Smith. 1986. Stratigraphic, isotopic, and mineralogical evidence for an early Holocene thaw unconformity at Mayo, Yukon Territory. Can. J. Earth Sci. 23:794-803.

Camill, P. 2000. How much do local factors matter for predicting transient ecosystem dynamics? Suggestions from permafrost formation in boreal peatlands. Global Change Biol. 6:169-182.

Carslaw, H.S., and J.C. Jaeger. 1959. Conduction of heat in solids. 2nd edition. Oxford Univ. Press. Oxford.

Farouki, O.T. 1981. Thermal properties of soils. U.S. Army, Cold Regions Research and Engineering Laboratory. Hanover, NH. Report 82-08.

Goodrich, L.E. 1982. The influence of snow cover on the ground thermal regime. Can. Geotech. J. 19:421-432.

Halsey, L.A., D.H. Vitt, and S.C. Zoltai. 1995. Disequilibrium response of permafrost in boreal continental western Canada to climate change. Climatic Change. 30:57-73.

Harris, S.A. 1988. Observations on the redistribution of moisture in the active layer and permafrost. In: K. Senneset, ed. Proceedings, 5th International Conference on Permafrost, Trondheim, Norway. August 2-5, 1988. Tapir Publishers. Trondheim. Vol. 1, pp. 364-369.

Heginbottom, J.A., M.A. Dubreuil, and P.A. Harker. 1995. Canada-Permafrost. In: National Atlas of Canada, 5th ed. National Atlas Information Service, Natural Resources Canada. Ottawa. Plate 2.1. MCR 4177.

Hinkel, K.M., and J.R.J. Nicholas. 1995. Active layer thaw rate at a boreal forest site in central Alaska, U.S.A. Arct. Alp. Res. 27:72-80.

Hinkel, K.M., and S.I. Outcalt. 1993. Detection of non-conductive heat transport in soils using spectral analysis. Water Resources Res. 29:1017-1023.

Hinkel, K.M., and S.I. Outcalt. 1994. Identification of heat-transfer processes during soil cooling, freezing and thaw in central Alaska. Permafrost and Periglacial Processes. 5:217-235.

Hinkel, K.M., S.I. Outcalt, and A.E. Taylor. 1997. Seasonal patterns of coupled flow in the active layer at three sites in northwest North America. Can. J. Earth Sci. 34:667-678.

Johansen, Ø. 1975. Thermal conductivity of soils. Ph.D. thesis, Norweigan Technical Univ., Trondheim. Translated in U.S. Army, Cold Regions Research and Engineering Laboratory, Translation 637. (In Norweigan.)

Jorgenson, M.T., and R.A. Kreig. 1988. A model for mapping permafrost distribution based on landscape component maps and climatic variables. In: K. Senneset, ed. Proceedings, 5th International Conference on Permafrost, Trondheim, Norway. August 2-5, 1988. Tapir Publishers. Trondheim. Vol. 1, pp. 176-182.

Judge, A.S. 1973. The prediction of permafrost thickness. Can. Geotech. J. 10:1-11.

Koopmans, R.W.R., and R.D. Miller. 1966. Soil freezing and soil water characteristic curves. Soil Sci. Soc. Am. Proc. 30:680-685.

Kudryavstev, V.A., ed. 1981. Permafrost. Moscow State Univ. Press. Moscow. (In Russian.)

Kwong, Y.T.J., and T.Y. Gan. 1994. Northward migration of permafrost along the Mackenzie Highway and climatic warming. Climatic Change. 26:399-419.

Lachenbruch, A.H., and B.V. Marshall. 1986. Changing climate: geothermal evidence from permafrost in the Alaskan arctic. Sci. 234:689-696.

Lachenbruch, A.H., T.T. Cladouhos, and R.W. Saltus. 1988. Permafrost temperatures and the changing climate. In: K. Senneset, ed. Proceedings, 5th International Conference on Permafrost, Trondheim, Norway. August 2-5, 1998. Tapir Publishers. Trondheim. Vol. 3, pp. 9-17.

Lunardini, V. 1978. Theory of n-factors and correlation of data. In: Proceedings, 3rd International Conference on Permafrost, Edmonton, Alberta, July 10-13, 1978. National Research Council of Canada. Ottawa, Ont. Vol. 1, pp. 40-46.

Mackay, J.R. 1970. Disturbances to the tundra and forest tundra environment of the western Arctic. Can. Geotech. J. 7:420-432.

Mackay, J.R. 1975. The stability of permafrost and recent climatic change in the Mackenzie Valley, N.W.T. In: Geological Survey of Canada. Paper 75-1B, pp. 173-176.

Mackay, J.R. 1980. The origin of hummocks, western Arctic coast, Canada. Can. J. Earth Sci. 17:996-1006.

Mackay, J.R. 1983. Downward water movement into frozen ground, western Arctic coast, Canada. Can. J. Earth Sci. 20:120-134.

Mackay, J.R. 1995. Active layer changes (1968-1996) following the forest tundra fire near Inuvik, N.W.T., Canada. Arct. Alp. Rsch, 27:323-336.

Mackay, J.R. 1997. A full-scale field experiment (1978-1995) on the growth of permafrost by means of lake drainage, western Arctic coast: a discussion of the method and some results. Can. J. Earth Sci. 34:17-33.

Mackay, J.R., and D.K. MacKay. 1974. Snow cover and ground temperatures, Garry Island, N.W.T. Arctic. 27:287-296.

Muller, S.W. 1947. Permafrost or permanently frozen ground and related engineering problems. Edwards. Ann Arbor, MI.

Nicholas, J.R.J., and K.M. Hinkel. 1996. Concurrent permafrost aggradation and degradation induced by forest clearing, central Alaska, U.S.A. Arct. Alp. Res. 28:294-299.

Nixon, F.M., and A.E. Taylor. 1998. Regional active layer monitoring across the sporadic, discontinuous and continuous permafrost zones, Mackenzie Valley, northwestern Canada. In: A.G. Lewkowicz and M.Allard, eds. Proceedings, 7th International Conference on Permafrost, 23-26 June 1998, Yellowknife N.W.T. Québec City, Québec. Collection Nordicana 57. pp. 815-820.

Outcalt, S.I. 1972. The development and application of a simple digital surface climate simulator. J. Applied Meteorol. 11:629-36.

Outcalt, S.I., and K.M. Hinkel. 1996. Thermally driven sorption, desorption, and moisture migration in the active layer in central Alaska. Phys. Geog. 17:74-90.

Osterkamp, T.E., and V.E. Romanovsky. 1999. Evidence for warming and thawing of discontinuous permafrost in Alaska. Permafrost and Periglacial Processes. 10:17-37.

Patterson, D.E., and M.W. Smith. 1981. The measurement of unfrozen water content by time domain reflectometry: results from laboratory tests. Can. Geotech. J. 18:131-144.

Riseborough, D.W. 1990. Soil latent heat as a filter of the climate signal in permafrost. In: Proceedings, 5th Canadian Permafrost Conference, 5-9 June 1990, Québec City, Québec. Collection Nordicana 54. pp. 199-205.

Romanovsky, V.E., and T.E. Osterkamp. 1995. Interannual variations of the thermal regime of the active layer and near-surface permafrost in northern Alaska. Permafrost and Periglacial Processes. 6:313-335.

Rouse, W.R. 1976. Microclimatic changes accompanying burning in subarctic lichen woodland. Arct. Alp. Res. 8:357-376.

Smith, C.A.S., C.R. Burn, C. Tarnocai, and B. Sproule. 1998. Air and soil temperature relations along an ecological transect through the permafrost zones of northwestern Canada. In: A.G. Lewkowicz and M. Allard, eds. Proceedings, 7th International Conference on Permafrost, 23-26 June 1998, Yellowknife N.W.T. Québec City, Québec. Collection Nordicana 57. pp. 1009-1015.

Smith, M.W. 1975. Microclimatic influences on ground temperatures and permafrost distribution, Mackenzie Delta, Northwest Territories. Can. J. Earth Sci. 12:1421-1438.

Smith, M.W., and D.W. Riseborough. 1983. Permafrost sensitivity to climatic change. In: Proceedings, 4th International Conference on Permafrost, Fairbanks, Alaska, July 17-22, 1983. National Academy Press. Washington, D.C. Vol. 1, pp. 1178-1183.

Smith, M.W., and D.W. Riseborough. 1985. The sensitivity of thermal predictions to assumptions in soil properties. In: Proceedings, 4th International Symposium on Ground Freezing. Balkema. Rotterdam. pp. 17-23.

Soil Classification Working Group. 1998. The Canadian system of soil classification. Agriculture and Agri-food Canada, Publication 1646 (revised).

Soil Survey Staff. 1999. Soil taxonomy. 2nd ed. USDA, NRCS. Ag. Handbook 436.

Sturm, M. 1992. Snow distribution and heat flow in the taiga. Arct. Alp. Rsch, 24:145-152.

Swanson, D.K. 1996. Soil geomorphology on bedrock and colluvial terrain with permafrost in central Alaska, U.S.A. Geoderma. 71:157-172.

Tarnocai, C. 1994. Genesis of permafrost affected soils. In: J.M. Kimble and R.J. Ahrens, eds. Proceedings of the Meeting on the Classification, Correlation, and Management of permafrost-affected soils, 18-30 July 1993, Yukon and Northwest Territories, Canada, and Alaska, U.S.A. USDA, SCS, NSSC, Lincoln, NE. pp. 143-154.

Taylor, A.E. 1995. Field measurements of n-factors for natural forest areas, Mackenzie Valley, Northwest Territories. In: Current research 1995-B. Geological Survey of Canada. pp. 89-98.

Viereck, L.A. 1982. Effects of fire and firelines on active layer thickness and soil temperatures in interior Alaska. In: H.M. French, ed. Proceedings, 4th Canadian Permafrost Conference, Calgary, Alta., March 2-6, 1980. National Research Council of Canada. Ottawa, Ont. pp. 123-135.

Vitt, D.H., L.A. Halsey, and S.C. Zoltai. 2000. The changing landscape of Canada's western boreal forest: the current dynamics of permafrost. Can. J. Forest Res. 30:283-287.

Williams, D.J., and C.R. Burn. 1996. Surficial characteristics associated with the occurrence of permafrost near Mayo, central Yukon Territory, Canada. Permafrost and Periglacial Processes. 7:193-206.

Williams, P.J. 1982. The surface of the earth: An introduction to geotechnical science. Longman. London.

Williams, P.J., and M.W. Smith. 1989. The frozen earth: Fundamentals of geocryology. Cambridge Univ. Press. Cambridge.

Wolfe, S.A., E. Kotler, and F.M. Nixon. 2000. Recent warming impacts in the Mackenzie Delta, Northwest Territories, and northern Yukon Territory coastal areas. Geological Survey of Canada. Current research 2000-B1, 9 pp. (online: http//www.nrcan.gc.ca/gsc/bookstore).

Zoltai, S.C. 1993. Cyclic development of permafrost in the peatlands of northwestern Alberta, Canada. Arct. Alp. Res. 25:240-246.

Zoltai, S.C., S.Taylor, J.K. Jeglum, G.F. Mills, and J.D. Johnson. 1988. Wetlands of boreal Canada. In: National Wetlands Working Group, ed. Wetlands of Canada. Environment Canada, Ottawa, and Polyscience Publications. Montreal. pp. 97-154.

Chapter 4. Cryosols in the Extremely Arid Transantarctic Mountains Region of Antarctica

I.B. Campbell[1] and G.G.C. Claridge[2]

[1]Research Scientist, Land and Soil Consultancy Services, 23 View Mount, Nelson 7001, New Zealand; e-mail iaincampbell@xtra.co.nz.
[2]Research Scientist, 33 Kotari Road, Days Bay, Eastbourne, New Zealand; e-mail graeme.claridge@xtra.co.nz.

1. Introduction

The Transantarctic Mountains are the dominant physical feature of the Antarctic continent. Some 2500 km long and up to 100 km wide, the Transantarctic Mountains form an escarpment, throughout most of their length, comprised of crystalline basement rocks, chiefly granites, overlain by Devonian to Jurassic, predominantly sandstone sediments, with massive intrusions of Jurassic dolerites (Bradshaw, 1990).

The range separates the East Antarctic Ice Sheet, which reaches an elevation of 4000 m, from the much smaller West Antarctic Ice Sheet, with its average elevation of around 1500 m (Figure 3.4.1). In many places, glaciers flowing from the East Antarctic Ice Sheet have cut valleys through the mountains and, as a result of fluctuations in ice levels, many small scattered areas of ice-free ground with soils occur in the valleys along the length of the mountains. The total area of bare ground and soils in the Transantarctic Mountains is around 23,000 km^2, approximately half of the bare ground and soils in Antarctica (Claridge et al., 2000).

The climatic data, though derived from very few stations, indicate the severity of the general climatic environment for Cryosol formation. At South Pole Station (Lat 90°S) on the Polar Plateau, for example, the mean annual temperature is -49°C, but on the land nearest to this station (Mt. Howe), approximately 300 km to the north and 300 m lower in elevation, the mean annual temperature is around -40°C. Farther north, at Scott Base, which is a coastal site (Lat 77°51'), the mean annual temperature is -20°C, while at Cape Hallett (Lat 72°18'), 600 km farther to the north (Figure 3.4.1), the mean annual temperature is -15°C.

Although mean temperatures distinctly increase northward, the greatest temperature gradient is east to west, from the lower altitude coastal regions to the higher altitude inland mountains. Precipitation follows a similar pattern. Derived mainly from cyclonic storms which move inland from the Southern Ocean, the most precipitation, which always falls as snow, is in northern and coastal regions, while the least falls in inland regions bordering the Polar Plateau. Precipitation is about 50 mm/y over much of the central part of the ice sheet, increasing to about 600 mm/y in some coastal areas (Bull, 1971). Amounts vary greatly, to as little as 13 mm/y in the McMurdo Sound Dry Valley region (Chinn, pers. comm.).

Temperature and precipitation differences, in conjunction with the extreme age of some of the land surfaces, which date to Miocene times (Denton et al., 1993; Campbell and Claridge, 1978), have caused some very distinctive characteristics in the Transantarctic Mountains region's soils. Their main soil features are coarse textures, minimal alteration apart from some clast disintegration and progressive oxidation and reddening since the Miocene, salinization and accumulation of soluble salts in salic horizons, and minimal presence of organic life and absence of a significant biological regime. Environmental differences appear largely through thermal and hydrological properties and soil salt chemistry.

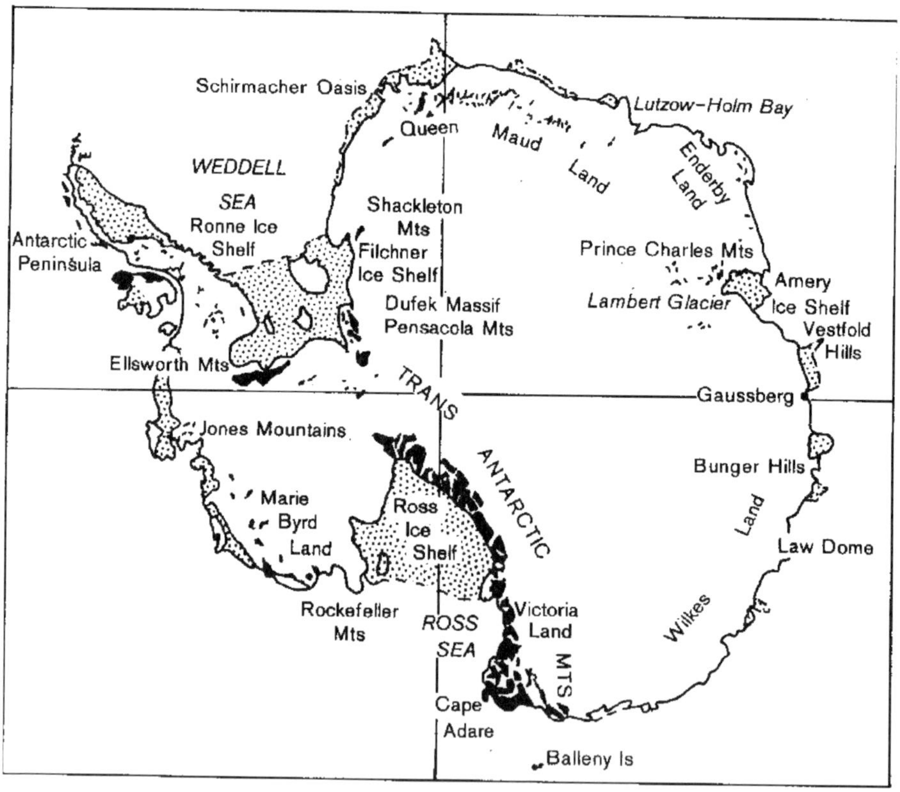

Figure 3.4.1. The Antarctic continent. The figure shows the location of the Transantarctic Mountains. The shaded areas indicate locations of ice-free areas but not their extent.

Table 3.4.1. Maximum, minimum, and mean (MAST) temperatures for 1994 and 1995 at Marble Point, for 1996 and 1997 near Lake Vanda, and for the warmest (MSST) and coldest (MWST) 90-day periods at both sites.

Marble Point		**1994**			**1995**		**90 days**	
Depth	max	min	MAST	max	min	MAST	MSST	MWST
Surface	+13.9	-34.1	-16.3	+12.1	-35.8	-16.9	+0.1	-28.1
16 cms	+9.1	-32.0	-15.6	+7.4	-35.8	-16.3	-0.3	-27.4
25 cms	+6.3	-31.5	-16.2	+4.8	-35.3	-16.8	-1.4	-26.4
33 cms	+5.7	-30.1	-16.1	+2.1	-33.6	-16.8	-2.7	-26.7
40 cms	+3.2	-30.1	-16.1	+2.1	-33.6	-16.8	-2.7	-26.7
55 cms	+2.4	-29.4	-15.8	+1.4	-32.5	-16.4	-2.8	-26.1
Lake Vanda		**1996**			**1997**		**90 days**	
Depth	max	min	MAST	max	min	MAST	MSST	MWST
Surface	+16.4	-45.6	-17.1	+13.1	-47.0	-16.6	+3.9	-29.8
15 cms	+8.0	-39.1	-17.1	+6.4	-43.2	-21.0	-8.5	-27.9
25 cms	+3.5	-36.1	-17.6	+2.8	-40.2	-20.8	-1.8	-27.6
35 cms	+0.4	-32.6	-16.9	+6.6	-35.9	-19.4	-3.3	-26.1
45 cms	-2.9	-30.2	-17.6	-2.9	-32.9	-19.2	-6.9	-25.7
55 cms	-3.3	-28.6	-17.2	-3.6	-31.2	-18.5	-7.0	-24.9

2. Soil Temperature and Properties of Cryosols

Few systematic recordings of soil temperatures exist for the Cryosols of the Transantarctic Mountains; consequently, we have limited knowledge of spatial and temporal soil temperature differences and other important temperature characteristics. However, the geographic trends in soil temperature data follow the climatic trends that the meteorological data indicate.

The climate is relatively warm near the coast but considerably colder (up to 30°C less) on the Polar plateau. The air temperature gradient is essentially a function of increasing elevation, with temperatures decreasing inland by 10°C for every 100 m increase (Mullan and Sinclair, 1990). Mean air temperatures for soil sites along the Transantarctic Mountains follow this trend, with the coldest temperatures inland, bordering the Polar Plateau, and the warmest near the coast.

Single sets of soil temperature observations from scattered description sites along the Transantarctic Mountains, and some short-term, continuously recorded measurements from a few sites, indicate that soil temperatures also decline with increasing distance and elevation from the coast. However, while the mean soil temperatures, including for the warmest and coldest 90-day periods, are important (Table 3.4.1), several other soil temperature attributes are significant for the region's Cryosols.

2.1. Thickness of the active layer and depth to ice-cemented permafrost

In the Cryosols of the Transantarctic Mountains, the thickness of the active layer and depth to the permafrost are essentially a function of the site climate, but soil water content and soil surface features modulate them. Figure 3.4.2 illustrates the general relationship between climate and the thickness of the active layer. The figure reflects single sets measurements of soil temperature and continuously recorded measurements from scattered sites in the Transantarctic Mountains. The depth of the active layer diminishes from around 70 cm in coastal regions to as little as 2 cm at high altitudes inland, to the south (Campbell and Claridge, 2000). In many locations, however, the permafrost is not ice-cemented but is dry frozen and lacks insufficient water for cementation (Balks et al., 1995; Bockheim, 1995; Campbell et al., 1998a, 1998b). In these situations, soil change proceeds at depth within the permafrost, irrespective of the depth of the active layer and position of the 0°C isotherm.

Moisture content and salinity of the soils also strongly influence the thickness of the active layer in the Cryosols of the region (Campbell et al., 1997a). When ice melts, water absorbs energy equivalent to the latent heat of fusion (3.33×10^5 J/kg) with no resulting increase in soil temperature. An equivalent amount of heat is released into the soil when liquid water freezes, resulting in warming. These latent changes in heat buffer change in temperature in the soil during moisture-phase changes (Anderson and Tice, 1989). Some Transantarctic Mountains Cryosols with a high moisture content use the available heat to convert ice into water; consequently, the depth of their active layer is typically shallow in moist sites. Where soil salinity is high, the suppression of freezing delays soil moisture-phase changes and the 0°C isotherm penetrates deeper (Campbell et al., 1997b).

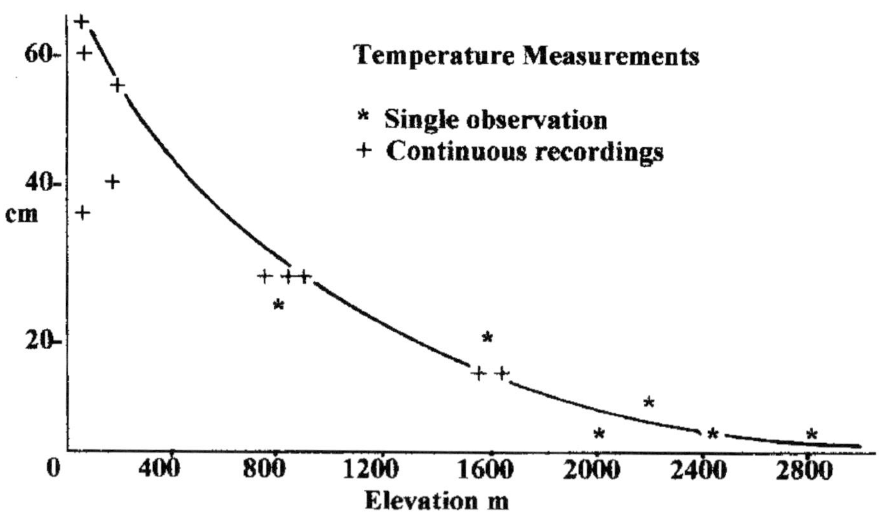

Figure 3.4.2. A plot of the thickness of the active layer against altitude, from measurements of soil temperature at many sites in the Transantarctic Mountains.

The balance of surface radiation is perhaps the most important thermal factor, as it determines the amount of energy available to heat the soil. The ice-free areas of the Transantarctic Mountains receive negligible solar radiation throughout much of the year, but, during the brief summer, they have receive the most solar radiation in the world (Dana et al., 1998). Albedo, or short-wave reflectivity, is the most important property affecting net radiation balance and is a function of the soil's surface color, surface roughness, rock polish, and moisture content (Balks et al., 1995; MacCulloch, 1996). In the McMurdo Sound Dry Valley region, light colored soils from predominantly sandstone rocks have an albedo approximately 4 times greater than that measured on soils from predominantly dolerite rocks with a dark surface; the latter have more energy to heat the soils and cause ablation of subsurface ice.

2.2. Frequency of freeze and thaw cycles

Another important thermal attribute of these soils, not only for the effects of climate on cryogenic processes but also for biological interactions in soil, is the number of freeze/thaw cycles over the period of summer warming and cooling of soil. No continuous soil temperature data are available from the meteorological stations in the region. However, some continuous, although short-term measurements of soil temperature at varying soil depths have been made at a number of sites in the McMurdo Dry Valley area (Balks et al., 1995; Campbell and Claridge, 2000; McKay, 1998; Greenpeace, pers. com.).

Figure 3.4.3 illustrates changes in soil temperature from continuous recordings at various depths at a coastal site at Marble Point during the summer of 1993-94. Although temperature rises and falls daily in response to the angle of the sun, the surface soil only passes through the freezing point 12 times during the summer, from initial thaw to freeze up. Figure 3.4.3 also shows temperature data from an inland site on a valley floor near Lake Vanda, at an altitude of 130 m. At this site, 30 freeze-thaw cycles occurred, probably influenced by shading from adjacent mountains.

At a site in the Asgard Range (Figure 3.4.1), at an altitude of 1400 m, the number of freeze/thaw cycles was approximately 35 (McKay, 1998). At Marble Point, the number of freeze/thaw cycles at 15 cm deep was similar to that experienced at the soil surface, but, at greater depths, soil temperatures were more consistent and fewer freeze/thaw cycles occurred. The Vanda site recorded only 15 freeze/thaw cycles at 15 cm deep and none at 45 cm, while the Asgard Range site showed no freeze/thaw cycles at 17 cm deep. These values are all for unshaded sites. When the observing site is shaded for part of each day, the number of freeze-thaw cycles is considerably greater (Campbell, 2003).

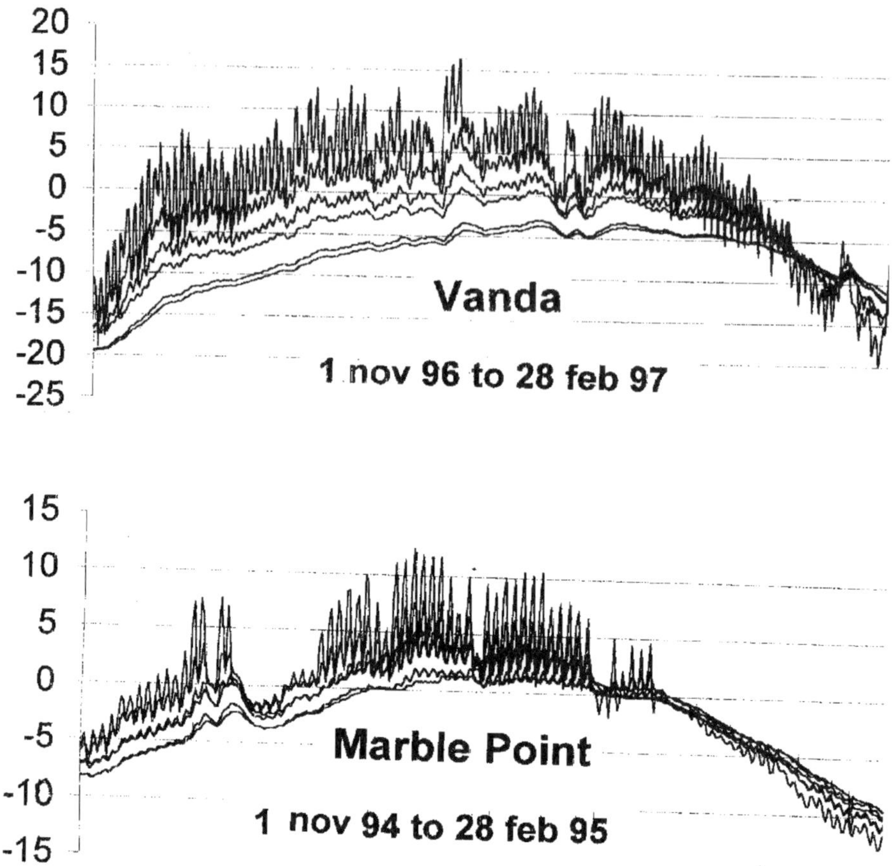

Figure 3.4.3. Soil temperatures for a site at Marble Point during summer, 1994-95, and a site near Lake Vanda in the Wright Valley during summer, 1996-97. The deepest temperature sensor was at 55 cm.

Although no continuously recorded data are available from sites farther inland and at higher altitudes, their frequency of freeze/thaw cycles should fall, relative to those of coastal regions, because of severe cooling from lower air temperatures. Freeze and thaw is commonly cited as a significant factor in Cryosol and cryogenic processes but, in the Transantarctic Mountains where soil moisture levels are very low, the influence of freeze and thaw has clear limits.

Two other thermal characteristics of soil which vary considerably in relation to climate are the length of time from first summer thaw to the start of freezing and the maximum summer diurnal temperature variation. Both of these properties are especially important for biological interactions in soil. At the coastal Cape Evans site, the surface soil remained unfrozen for 87 days. At the Marble Point site, adjacent to a small seasonal snow pack, the surface soil was unfrozen for 40 days in 1994 (Balks et al., 1995). At the Vanda site, which is snow-free, it was unfrozen for 94 days in 1996. About 50 days were recorded for the Asgard Range site in 1994. These variations depend on seasonal snow cover and other site factors which affect the quantity of solar radiation.

With increasing depth of soil, the duration of the thaw period becomes progressively less. The maximum diurnal temperature range for surface soil recorded at Marble Point was 12°C, but a 28°C variation occurred at an inland mountain site in the Convoy Range, to the north of the Dry Valley region (MacCulloch, 1996) at an altitude of 1500 m. All sites have the capacity for surface soil temperatures to rise well above 0°C when radiation inputs are large. The difference between high and low altitude sites is that soil temperatures drop dramatically in high altitude sites because of very cold air when cloud cover, topographic shading, or snow restrict radiation.

3. Precipitation, Soil Moisture, and Properties of Cryosols

Soil science has long recognized the lack of available moisture as a key feature of the genesis of cold desert soil and previously used it as a basis for subdividing the Transantarctic Mountains Cryosols into climate-related soil classes (Campbell and Claridge, 1969). These classes ranged from soils on the fringe of the Polar Plateau where liquid water was never visibly present (ultraxerous), to soils in coastal locations where visible moisture was present over much or part of the brief Antarctic summer (subxerous). The availability of soil moisture also has been recognized as a major factor influencing the ecology of arid Cryosols (Cameron, 1969; Wyn-Williams et al., 1997; Freckman and Virginia, 1998).

3.1. Soil moisture sources

Precipitation in the Transantarctic Mountains region derives solely from snow, which is greatest in coastal regions and least in the inland colder regions with very low humidities. Precipitation on valley floors may be very low, however, because of föhn effects from westerly winds (Bromley, 1985). The accumulation of drifting snow may increase effective precipitation at many sites, but the removal of snow by wind may reduce it at others. Because of the very cold conditions and low humidity throughout most of the year, little of the snow finds its way into the soils as liquid moisture, as most either blows away or sublimes. During the summer, some melting around the edges of accumulations of snow may occur as the soils warm, but measurements have shown that the soil may lose this moisture in about 10 days (Campbell et al., 1997a; Campbell (in prep). Some moisture finds its way into the soil in summer when snow falls on warmed soil surfaces, but this moisture generally evaporates within a day or so (Campbell and Claridge, 2000). Evaporation rates of 3 mm per day have been measured at a coastal site (Balks et al., 1995).

At some sites, mainly in the warmer coastal regions, thawing of snow packs around the margins of some glaciers results in significant transient stream flows. The soils commonly are saturated in the zone of the water flow and also are moistened up to several meters away by capillary flow.

3.2. Soil moisture properties and geographic relationships

Balks et al. (1995) and Campbell et al. (1997a, 1998a, 1998b) have studied the relationships between climate differences and the moisture status of the active layer of permafrost in the Transantarctic Mountains. The soil moisture data they gathered over a period of 8 years from 240 sites within the McMurdo Sound Dry Valley region provide a good indication of relationships between soil moisture differences and the regional climate (Campbell and Claridge, 2000).

The highest moisture contents in the active layer occur in soils of the coastal regions. At a typical well drained site (Figure 3.4.4a), the moisture content generally ranges from around 0.1% at the soil surface to around 10% near the surface of the underlying ice-cemented permafrost. Within the permafrost, the moisture content rises abruptly, with average values of around 40% within the first meter (Campbell et al., 1998b).

In inland mountain areas, values of soil moisture content within the active zone are lower, ranging from <0.5% at the soil surface to 3.5% at depth, with values decreasing farther inland. The presence of ice-cemented soil, however, commonly does not coincide with the permafrost table but may occur some depth below it, with a zone of dry-frozen soil between the permafrost table and the ice-cemented soil beneath. When present, ice-cemented permafrost in the inland mountains typically has a lower moisture content than the ice-cemented permafrost of the coastal regions.

Soils with the lowest measured moisture contents in both the active layer and the permafrost occur in some of the dry valleys, with values of 0.4% at the soil surface to 0.65% in the dry frozen soil below the permafrost table (Figure 3.4.4b). Although no data are available for soils from some of the colder parts of the Transantarctic Mountains, similar or even lower values should occur.

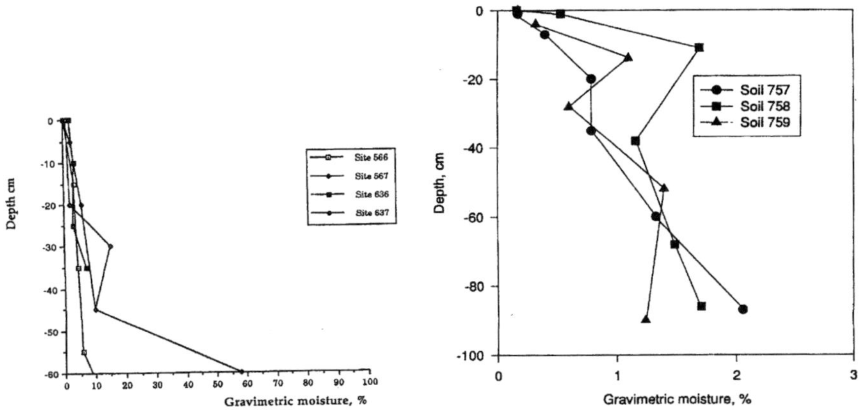

Figure 3.4.4. Moisture contents of (a) relatively moist coastal sites (Marble Point) with icy permafrost at 60 cm and (b) inland sites (near Lake Vanda in the Wright Valley) which lack icy permafrost, even though the soil is permanently frozen below 35 cm.

4. Soil Climate and Soil Salinity

Because of the prevailing aridity, potential evaporation exceeds precipitation almost everywhere in the Transantarctic Mountains, and soluble salts accumulate within the soils. These salts come mainly from atmospheric transport, with lesser amounts of the salts from soil weathering. Claridge and Campbell (1977), Campbell and Claridge (1987), and Bockheim (1982) have described the origin and distribution of the salts in these Cryosols.

Salt accumulations take differing forms, depending on their composition, the soil's age, and the soil's moisture content. They may occur as surface efflorescences and encrustations, as diffuse or concentrated horizons usually within 5 to 15 cm of the surface, as scattered flecks throughout the soil, or as thick concentrations beneath surface stones or stones within the soil profile. Where the salts strongly concentrate, the salt horizon often is cemented. With increasing soil age, the concentration and thickness of the salt horizon usually increases.

In coastal regions, the salts in the soils consist largely of sodium chloride and sodium sulphate, derived from sea water and transported atmospherically by way of coastal storms before being deposited by snow fall. In the soils of inland regions, chloride is absent, while sulphates of sodium and magnesium and nitrate are present, derived via a different atmospheric pathway (Claridge and Campbell, 1977). The nitrate salts accumulate in the inland very arid soils because moisture is insufficient to leach them or for organisms to use them.

5. Conclusions

The Cryosols of the Transantarctic Mountains occur under the coldest and driest conditions on earth. Extremely low temperatures and effective precipitation are the main environmental parameters influencing soil development, but receipt of solar energy is the principal factor governing environmental inputs and relationships among soil properties.

While the environmental conditions for soil formation are extreme, local and regional differences in conditions are responsible for significant differences in soil properties. Depth of the active layer, length of the thaw period, number of freeze and thaw cycles, regional differences in the active layer and permafrost moisture contents, and regional differences in soil salinity and other chemical characteristics are among the soil properties closely related to differences in environmental conditions in Transantarctic Mountains Cryosols.

References

Anderson, D.M., and A.R. Tice. 1989. Unfrozen water content of six Antarctic soil materials. Cold Regions Engineering. Proceedings of the Fifth International Conference. American Society of Civil Engineers, New York. pp. 355-366.

Balks, M.R., D.I. Campbell, I.B. Campbell, and G.G.C. Claridge. 1995. Interim results of 1993/94 soil climate, active layer and permafrost investigations at Scott Base, Vanda and Beacon Heights, Antarctica. Univ. Waikato, Dept. Earth Sciences. Special report. 64 pp.

Bockheim, J.G. 1979. Relative age and origin of soils in eastern Wright Valley, Antarctica. Soil Sci. 128:142-152.

Bockheim, J.G. 1982. Properties of a chronosequence of Ultraxerous soils in the Transantarctic Mountains. Geoderma. 28:239-255.

Bockheim, J.G. 1995. Permafrost distribution in the Southern Circumpolar region and its relation to the environment: a review and recommendations for further research. Permafrost and Periglacial Processes. 6:27-45.

Bradshaw, M.A. 1990. Geological History. In T. Hatherton ed., Antarctica, the Ross Sea Region. 42-63. DSIR Publishing. Wellington.

Bromley, A.M. 1985. Weather observations Wright Valley, Antarctica. New Zealand Meteorological Service. Information Publication. 11. 37 pp.

Bull, C. 1971. Snow accumulation in Antarctica. In: L.O.Quam, ed. Research in the Antarctic. Am. Assoc. Advan. Sci. Pub. 93. pp. 367-424.

Cameron, R.E. 1969. Cold Desert characteristics and problems related to other arid lands. In: W.G. McGinnies and B.J. Goldman, eds. Arid lands in perspective. Univ. of Arizona Press. pp. 167-205.

Campbell, I.B. 2003. Soil characteristics at a long-term ecological research site in Taylor Valley, Antarctica. Aust. J. Soil Res. (In press.)

Campbell, I.B., and G.G.C. Claridge. 1969. A classification of frigic soils - the zonal soils of the Antarctic Continent. Soil Sci. 107:75-85.

Campbell, I.B., and G.G.C. Claridge. 1978. Soils and Late Cenozoic history of the upper Wright Valley, Antarctica. N. Z. J. Geol. Geophys. 21:635-43.

Campbell, I.B., and G.G.C. Claridge. 1987. Antarctica: soils, weathering processes and environment. Elsevier Science Publishers. Amsterdam. 368 pp.

Campbell, I.B., G.G.C. Claridge, M.R. Balks, and D.I. Campbell. 1997a. Moisture content in soils of the McMurdo Sound Dry Valley region of Antarctica. In: W.B. Lyons, C. Howard-Williams, and I. Hawes, eds. Ecosystem Processes in Antarctic Ice-free Landscapes. Proceedings of an international workshop on Polar desert Ecosystems, Christchurch, New Zealand, 1-4 July 1996. Balkema. pp. 61-76.

Campbell, I.B., G.G.C. Claridge, D.I. Campbell, and M.R. Balks. 1998a. The Soil Environment. In: J.C. Priscu, ed. Ecosystem Dynamics in a Polar Desert: the McMurdo Dry Valleys, Antarctica. Antarctic Research Series Volume 72. American Geophysical Union. pp. 297-322.

Campbell, I.B., G.G.C. Claridge, M.R. Balks, and D.I. Campbell. 1998b. Permafrost properties in the McMurdo Sound-Dry Valley region of Antarctica. Proceedings of the 7th International Conference on Permafrost, Yellowknife, Canada 1998. pp. 121-126.

Campbell, I.B., and G.G.C. Claridge. 2000. Soil temperature, moisture and salinity patterns in Transantarctic Mountains Cold Desert ecosystems. In: W. Davidson, C. Howard-Williams, and P. Broady, eds. Antarctic ecosystems, models for wider ecological understanding. Caxton Press. Christchurch, NZ. pp. 233-240.

Campbell, D.I., R.J.L. MacCulloch, and I.B. Campbell. 1997b. Thermal regimes of some soils in the McMurdo Sound region, Antarctica. In: W.B. Lyons, C. Howard-Williams, and I. Hawes, eds. Ecosystem Processes in Antarctic Ice-free Landscapes. Proceedings of an International workshop on Polar desert Ecosystems, Christchurch, New Zealand. 1-4 July 1996. Balkema.

Claridge, G.G.C., and I.B. Campbell. 1977. The salts in Antarctic soils, their distribution and relationship to soil processes. Soil Sci. 123:377-384.

Claridge, G.G.C., I.B.Campbell, and D.S. Sheppard. 2000. Carbon pools in Antarctica and their significance for global climate change. In: R. Lal, J.M. Kimble, and B.A. Stewart, eds. Global Climate Change and Cold Desert Ecosystems. Lewis Publishers. Boca Raton, FL. pp. 59-75.

Dana, G.L., R.A. Wharton, and R. Dubayah. 1988. Solar radiation in the McMurdo Dry Valleys, Antarctica. In: J.C. Priscu, ed. Ecosystem Dynamics in a Polar Desert: the McMurdo Dry Valleys, Antarctica. Antarctic Research Series Volume 72. American Geophysical Union. pp. 39-64.Denton, G.H., D.E Sugden, D.R. Marchant, B.L. Hall, and T.I. Wilch. 1993. East Antarctic Ice Sheet sensitivity to Pliocene climatic change from a Dry Valleys perspective. Geografiska Annaler. 75A:155-204.

Freckman, D.W., and R.A. Virginia. 1997. Soil biodiversity and community structure in the McMurdo Dry Valleys, Antarctica. In: J.C. Priscu, ed. Ecosystem Dynamics in a Polar Desert: the McMurdo Dry Valleys, Antarctica. Antarctic Research Series Volume 72. American Geophysical Union. pp. 323-335.

MacCulloch, R.J.L. 1966. The microclimatology of Antarctic soils. Unpublished M.Sc. Thesis. Univ. Waikato. Hamilton, New Zealand.

McKay, C.P., M.T. Mellon, and I.E. Friedmann. 1998. Soil temperature and stability of ice-cemented ground in the McMurdo Dry Valleys, Antarctica. Antarc. Sci.10:31-38.

Mullan, A.B., and M.R. Sinclair. 1990. Climate and weather. In: T. Hatherton, ed. Antarctica, the Ross Sea Region. DSIR Publishing. Wellington. pp. 98-117.

Wyn-Williams, D.D., N.C. Russell, and H.G.M. Edwards. 1997. Moisture and habitat structure as regulators for microalgalcolonists in diverse Antarctic terrestrial habitats. In: W.B. Lyons, C. Howard-Williams, and I. Hawes, eds. Ecosystem Processes in Antarctic Ice-free Landscapes. Proceedings of an International workshop on Polar desert Ecosystems, Christchurch, New Zealand. 1-4 July 1996. Balkema. pp. 61-76.

Chapter 5. Mineralogy and Weathering of Antarctic Cryosols

H.-P. Blume[1], J. Chen[2], E. Kalk[3], and D. Kuhn[3]

[1]Professor, Institute for Plant Nutrition and Soil Science, Christian-Albrechts-University Kiel, Olshausenstr, 40, 24098 Kiel, Germany; e-mail hblume@soils.uni-kiel.de.
[2]Professor, Institute of Soil Science, Chinese Academy of Sciences, East Beijing Road 71, P.O. Box 821, Nanjing 210008, People's Republic of China; e-mail jchen@issas.ac.com.
[4]Dr., Institute for Plant Nutrition and Soil Science, Christian-Albrechts-University Kiel, Olshausenstr, 40, 24098 Kiel, Germany; e-mail hblume@soils.uni-kiel.de.1.

Introduction

Many soil researchers of the Antarctic (e.g., O'Brien, 1979; Campbell and Claridge, 1987; Bockheim and Ugolini, 1990) hold that no considerable chemical weathering and no new formation of minerals has taken place in soils of the Antarctic Desert and Tundra. They describe intensive changes in soil condition from cryoturbation and cryoclastic weathering but do not recognize a stronger influence of chemical weathering. However, pedologists from Central Europe assume that intensive cryoclastic weathering facilitates easier chemical weathering at low temperatures and thus allows a new formation of minerals (Kopp and Kowalkowski, 1990). Some even assume that, in many soils of Central Europe, brownification and clay formation mainly occurred under periglacial conditions (Kowalkowski and Borzyskowiski, 1973; Kopp et al., 1982).

To support these different hypotheses and to understand the mineralogy of Antarctic soils, we sampled representative soils from Wilkes Land, Continental Antarctica, and King George Island, Maritime Antarctic. We also studied the influence of lichen-induced weathering of rocks and stones.

2. Methods

Methods included **particle size analysis** (sieve and pipette method following H_2O_2 and dithionite/citrate [necessary because aggregates were stabilized by iron oxides in many cases] treatment and $[Na\ PO_3]_6$ dispersion); **organic matter** (org.m.: dry oxidation at 1200°C and colorimetric CO_2 measurement, org. C. multiplied by 2); **carbonates** (H_3PO_4 extraction at 80°C and colorimetric CO_2 measurement); **pedogenic oxides and humates** (Fe_o and ODOE [= optical density of oxalate extract] for NH_4-oxalate extraction, Fe_d for dithionite-citrate extraction; Fe_p and ODPE [= optical density of pyrophosphate extract] for pyrophosphate extraction); **total and available P** (P_t extraction with HCl/HNO_3 at 105°C, P_a extraction with lactate, colorimetric measurement); **mineralogy** (X-ray diffraction of clay and fine silt [2-20 μm], microscopic analysis of coarse silt [20-63 μm], fine (63-200 μm) and medium [200-630 μm] sand fraction); **soil classification** (after ISSS/ISRIC/FAO, 1998). For further details of methodology, see Schlichting et al. (1995).

3. Site Locations and Conditions

The **first site** is in the Continental East Antarctic (-9.3° mean annual temp.; 180 mm precipitation, mainly snow) close to Casey Station, Windmill Islands, Wilkes Land (66°17′S, 110°32′E). The rocks of the region consist of a layered sequence of gneiss with schist and migmatite, the so-called "Windmill Metamorphics" (Blight and Oliver, 1977), and are covered partly by Neoglacial solifluction deposits and moraines. The region lost its ice cover 5500 years ago and rose during that time (Goodwin, 1993). The studied soils of this region include a Hapli-turbic Cryosol on moraine deposits, a Stagni-turbic Cryosol on basic gneiss to schist, and a Skeleti-gelic Podzol of a former penguin rookery.

The three soils studied from **Casey** are frozen nearly year-round, and permafrost occurs at 50 to 80 cm deep. The Hapli-turbic Cryosol in a downslope position is free of snow in warm years only but not for longer than one month, so it supports no vegetative cover. The two other profiles are in upper slope position and therefore free of snow one to two months every year.

The **second site** is in the Maritime Antarctic (-1.8° C, 510 mm), close to Arctowski Station, Admiralty Bay, King George Island (62°10′S, 58°28′W). The rocks of the region are mainly Tertiary tholeiite basalt (Birkenmayer, 1980) covered by Neoglacial (decades to centuries old) and older moraines and by solifluction deposits. Volcanic ash from neighboring islands (e.g., Deception Island) also were deposited during the Quarternary. The studied soils here induced a Calcari-regic Cryosol on a blocky moraine, a Skeleti-gelic Podzol on basalt, and a Humi-gelic Umbrisol of a former penguin rookery.

The three soils studied from **Arctowski** formed under somewhat warmer conditions, so that dense vegetation covers two of them. Only the Regic Cryosol is sparsely covered by plants, due to its position on a very young moraine (perhaps not more than one hundred years).

4. Soils and their Mineralogy

The Hapli-turbic Cryosol lies in the center of a stone garland. This, and discontinuous depth functions of texture fractions, as well as organic matter (Table 3.5.1a), demonstrate a strong influence of cryoturbation. The mineralogy of the soil (Table 3.5.2) features a high quartz content, feldspars and mica thus reflecting the geology of many leuco gneiss. Primary minerals dominate in the clay fraction also, reflecting a strong influence of cryoclastic weathering. Low pH values, some mineral grains with weathered surface, pedogenic iron, and very little smectite demonstrate the beginning of chemical weathering.

The parent rock of the **Stagni-turbic Cryosol** is a long but only 0.5 to 2 meters wide stripe of vertically standing schist and basic gneiss, from former tectonic influences. This produced a high clay content, especially in the subsoil (Table 3.5.1b), whereas leuco gneiss in the surrounding areas influenced the topsoil. Strong cryoturbation also formed this soil, so that stones and sands are enriched in the topsoil; the function of organic C is discontinuous with depth.

Soil and snow algae (together with bird excrements; see high concentration of available P in Table 3.5.1b) accumulated the organic C, because cryoturbation prevents the growth of lichens. Strong cryoclastic weathering formed clay. This also promotes chemical weathering. We found pedogenic iron, together with a higher amount of partly weathered minerals, e.g. feldspars (Table 3.5.2). Smectite also formed. In addition to acidic conditions, changing redox conditions (strong mottling of subsoil; see Munsell colors in Table 3.5.1b) also may have influenced this process. Most pedogenic Fe seems to exist as organic complexes (see Fe_p together with ODPE = pyrophosphate extractable C) in relation to Fe_{d-p} (= mostly crystallized iron oxides). We have yet to understand the higher concentrations of Fe_p in relation to Fe_o.

The third profile, a **Podzol**, formed on a former penguin rookery. The birds gathered medium to coarse gravels (Table 3.5.1c) together with excrements, bones, and feathers (R. Smith, UK, personal communication). This resulted in extremely high concentrations of different phosphate minerals, mainly apatite, but also strengite and vivianite (the latter shows that reduced conditions do exist in the deeper subsoil) in the clay fraction (Table 3.5.4). Due to the presence of the clear horizontation and a strong enrichment of sea borne sponge spicules (transported by wind) only in the topsoil, we do not believe that cryoturbation influenced the soil over the last centuries. Instead, a strong chemical weathering of special mineral species (e.g. chlorites, feldspars, micas) occurred (perhaps due to a formation of nitric and sulfuric acids from the proteins of birds excrements). Most of the pedogenetic Fe seems to be organically bound (see Fe_p and ODPE due to Fe_{d-p} in Table 3.5.1c). But some iron should be bound by phosphates (see strengite and vivianite contents of the clay and the fine silt fraction in Table 3.5.4). Fe- and Al-phosphates are extracted by pyrophosphate but nearly not by oxalic acid.

The first soil, a **Calcari-regic Cryosol**, formed on a gravelly to blocky moraine of mainly tholeiite basalt origin (see Table 3.5.6; plagioclases, especially andesine, zeolites, and pyroxenes are abundant; volcanic glasses and ashes and some carbonates are also present). Kuhn (1997) observed an enrichment of stones on the soil surface together with a weak stone circle formation. We classified the soil therefore as a Cryosol, although it seems that soil ice melts more than 2 meters during warm summers. Weathered minerals (feldspars, pyroxenes) exist, together with many smectites (see Table 3.5.6) and relatively high contents of pedogenic iron (Table 3.5.5a). We believe that strong physical weathering facilitates chemical weathering, also. We also found chalcedony, which indicates that chemical weathering and new formation of minerals could have occurred before glacier movement, under formerly warmer climate conditions.

The second soil, a **Skeleti-gelic Podzol**, formed on tholeiite basalt. The mineralogy of this soil reflects this (Table 3.5.7). Permafrost was present to at least two meters deep. We found no influence of cryoturbation in the field. Rising contents of stones and sand, together with decreasing contents of silt and clay with soil depth, show the influence of weathering. Any cryoturbation would have produced higher stone and sand contents, together with lower clay contents in the topsoil (Table 3.5.5b). The soil is strongly acidified and enriched with pedogenetic iron: crystallized oxides, phosphates, and organically bound compounds, also. We found many partly weathered grains, especially in the topsoil: feldspars, mica, and pyroxenes (Table 3.5.7). Smectites also formed as mean clay minerals.

Highly weatherable volcanic glasses and olivine without any weathering phenomena occurred in the topsoil on the other side. We believe that this reflects an influence of a very recent volcanism; the last eruption of a volcano on Deception Island (around 160 km away) took place in the 1960s. A noticeable influence of birds was evident from the high contents of total phosphorus (Table 3.5.5b). The soil is podzolized, because the B horizons fulfill the definition of spodic horizons; they are enriched with pedogenic oxides and with pyrophosphate extractable C; metal organic compounds cement the Bms horizon to ortstein. We therefore classified the soil as Podzol. Also criteria of andic properties (beside volcanic glasses) may exist, like $Al_o+{}^1/_2Fe_o$ =1.2 % (ISSS/ISRIC/FAO, 1998: andic should have > 2), which are similar to spodic conditions of Podzols on the other side.

The third soil, an **Humi-gelic Umbrisol,** formed out of a former penguin rookery. Some layers between 20 and 80 cm deep contained penguin bones (Kuhn, 1997) and extremely high P contents (up to 12 %) and high gravel contents (Table 3.5.5c). We found variscite, vivianite, and apatite and other phosphate minerals, which we could not identify by x-ray (Table 3.5.8). Tatur (1985) identified and described Arktowskite $((Al_{8.4}Fe_{0.6})(OH)_3(PO_4)_8)$ and taranakite $((K, NH_4)_3Al_5H_6(PO_4)_8)$, and Myrcha and Tatur (1991) found amorphous Al-phosphates at similar sites.

Table 3.5.1. Soils from Casey, Wilkes Land*.

Horizon	Depth cm	Skel. %	Color moist	pH $CaCl_2$	sa %	si %	cl %	C_{org} mg/g	Fe_p mg/g	Fe_o mg/g	OD PE	Fe_{d-p} mg/g	P_a mg/g
a) Hapli-turbic Cryosol with gelundic and salic phase on moraine with some algae.													
Az	0-1	54	2.5Y4/2	4.6	66	33	1.4	1.8	.54		.02	2.4	.13
AC1	-5	17	2.5Y4/2	4.5	54	42	4.0	1.4	.58		.02	3.0	.14
AC2	-20	41	2.5Y4/2	4.4	67	28	4.7	1.8	.76		.04	2.5	.21
Cw1	-50	66	2.5Y5/2	4.4	70	25	5.4	3.9	.82		.04	1.5	.15
Cw2	-70	62	2.5Y5/2	4.0	83	11	6.3	3.0	.92		.04	1.9	.17
b) Stagni-turbic Cryosol, on basic gneiss to schist, with gelundic phase, with algae.													
Ah	0-0.5	33	2.5Y5/2	4.4	63	22	15	7.2	1.4	0.6	.02	2.1	.06
AB	-5	20	2.5Y6/4	4.4	36	31	34	4.9	6.2	1.5	.03	2.1	.10
Bg1	-20	13	7.5YR5/8	4.3	33	29	38	6.6	5.5	1.9	.04	1.3	.26
Bg2	-30	10	a. GY6/1	4.4	31	24	45	2.6	7.0	1.8	.03	7.0	.17
c) Skeleti-gelic Podzol of a former penguin rookery on leuco gneiss, under lichens.													
AE	6-8	99	10YR6/2	4.2	88	7	5	39	1.5	0.6	.37	0.9	0.33
Bh	-11	85	10YR3/2	4.3	83	12	5	41	5.9	2.5	1.5	1.0	0.47
Bsh1	-18	85	10YR3/4	4.4	67	19	13	24	9.6	4.7	1.1	1.4	0.86
Bsh2	-30	88	8YR4.5/5	4.6	"	"	"	29	6.5	3.7	.52	0.3	n.d.
Bsh3	-40	85	10YR4/5	4.8	78	17	5	30	4.5	3.4	.41	1.9	1.36
Cgw	-50	70	2.5Y5.5/4	5.3	"	"	"	3	3.1	2.2	.06	1.6	n.d.

* a) and b) after Blume (1994); c) after Blume et al. (1997); **skelet.**: stones + gravel; **sa** sand, **si** silt, **cl** clay.

Table 3.5.2. Mineralogy of a loamy Hapli-turbic Cryosol[1].

a) mean of 20-630 μm Ø in %.									
Horizon	**Depth, cm**	**Chlorite, unw.**	**Chlorite, w.**	**Pyroxenes**	**Amphiboles**	**Mica unw.**	**Mica w.**	**Weath., min.**	**Opaque min.[2]**
Az	0-1	4.2	0	0.9	0.9	16.6	0	0	2.0
AC	5-20	4.2	0	0	3.3	14.9	0	0.2	1.7
Cw2	50-70	5.3	0	0.7	2.0	13.8	0	0.2	1.6
Horizon	**Depth, cm**	**Plagiocl.**	**Alkali, feldsp.**	**Weath. feldsp.**	**Epidote**	**Quartz**	**Garnet**	**Rock frag.**	**Others**
Az	0-1	22.6	16.9	0.9	0.7	28.6	0.2	3.8	1.6
AC2	5-20	22.0	16.3	1.3	0.4	29.5	0.7	4.4	1.1
Cw2	50-70	21.1	16.9	1.8	1.8	30.0	1.3	2.4	1.1
b) clay and fine silt fraction									
Horizon	**Depth, cm**	**Quartz**	**Feld-spars**	**Pyroxene**	**Olivine serpent**	**Chlorite**	**Illite /mica**	**Smectite**	
clay fraction (< 2 μm Ø) in %, semiquantitative									
Az	0-1	20	43	0	7	<1	30	<1	
Cw2	50-70	14	56	0	0	<1	30	<1	
silt fraction ((2-20 μm Ø) in %, semiquantitative									
Az	0-1	27	43	11	0	9	10	<1	
Cw2	50-70	29	20	0	10	3	36	<1	

[1]After Blume et al. (1997); unw.=unweathered, w.=medium weathered.
[2]Anatase and rutile also present.

Table 3.5.3. Mineralogy of a loamy Stagni-turbic Cryosol, developed on basic gneiss to schist[1].

a) mean of 20-630 µm Ø in % (unw. unweathered, w. medium weathered).

Horizon	Depth cm	Chlorite unw.	Chlorite w.	Pyroxenes	Amphiboles	Mica unw.	Mica w.	Weath., min.	Opaque, min.*
Ah	0-1	2	0	0	2	10	1	0	0
Bg2	20-30	3	0	2	3	16	0	0	1

Horizon	Depth cm	Plagiocl.	Alkali-feldsp.	Weath feldsp	Epidote	Quartz	Garnet	Rock fragm.	Others
Ah	0-1	21	8	4	0	35	0	13	4
Bg2	20-30	19	10	5	2	33	2	4	2

b) clay and fine silt fraction.

Horizon	Depth cm	Magnetite	Quartz	Feldspars	Chlorite	Illite/ mica	Smectite	Mixed layers	Kaolinite
clay fraction (<2µm Ø) in %, semiquantitative									
Ah	0-1	<1	20	58	<1	22	<1	0	0
AB	1-5	<1	17	16	4	43	20	<1	0
Bg2	20-30	<1	11	58	0	31	0	0	<1
silt fraction (2-20µm Ø) in %, semiquantitative									
Ah	0-1	6	22	34	5	33	0	0	0
AB	1-5	14	20	30	6	30	<1	0	0
Bg2	20-30	24	17	13	11	35	<1	0	0

1) Blume et al. (1997).

*Anatase and rutile also present.

Table 3.5.4. Mineralogy of a Skeleti-gelic Podzol (former penguin rookery) on gneiss.

a) mean of 20-630 µm Ø in %[1].

Horizon	Depth, cm	Volc. Glas	Apatite	Chlorite, unw.	Chlorite, w.	Pyr.+ Amp., unw.	Pyr.+ Amp., w	Micas, unw.	Micas, w.	Strongly w. min.
AE	6-8	1.1	0.7	9.8	0	1.5	0.4	12.6	0	1.3
Bh	8-11	0	0	9.4	4.2	3.3	0	12.5	0.4	1.1
Bsh	11-30	0	0	16.7	29.8	0.4	0	17.5	0.4	3.8
Cgw	40-50	0	0	6.9	0	3.1	0	16.9	0.9	1.6

Horizon	Depth, cm	Plagioc.	Alkali-feldsp	Weath. feldsp.	Quartz	Chalcedone	Opaque mineral[2)]	Rock-fragm.	Others[3]	Biota[4)]
AE	6-8	18.6	14.2	1.8	30.8	0.9	2.1	1.6	1.9	many
Bh	8-11	19.3	17.7	2.2	25.1	0	0.6	1.8	1.9	0
Bsh	11-30	8.2	6.0	3.8	7.5	0.2	2.7	1.1	1.9	0
Cgw	40-50	21.5	13.3	2.7	18.9	1.3	1.3	7.3	4.3	0

b) clay and fine silt fraction.

Horizon	Depth, cm	Quartz	Feldspar	Pyroxene	Mica illite	Mixed layers	Apatite	Strengite	Vivianite	Kaolinite	Chlorite
clay fraction (< 2µm Ø) in %, semiquantitative.											
AE	6-8	41	33	8	9	0	4	5	0	<1	0
Bh	8-11	26	36	17	6	0	15	0	0	0	0
Bsh	11-30	0	16	0	<1	0	80	4	0	0	0
Cgw	40-50	24	32	0	31	<1	6	0	7	0	0
silt fraction (2-20 µm Ø) in %, semiquantitative.											
Bh	8-11	40	13	5	15	0	27	0	0	<1	0
Bsh	11-30	10	10	0	23	0	45	10	0	0	2

[1]unw.=unweathered; w.=medium weathered; pyr.=pyroxenes, amp.=Amphiboles.
[2]Anatase, ilmenite, rutile, zircon.
[3]Garnet, epidote, silimanite.
[4]Sponge spicules.

Table 3.5.5. Soils from Arctowski, King George Island[1].

Horizon	Depth cm	sto. %	color moist	pH $CaCl_2$	sa %	si %	cl %	C org mg/g	Fe_p mg/g	Fe_{o-p} mg/g	OD OE	Fe_{d-o} mg/g	P_t mg/g	car mg/g
a) Calcari-regic Cryosol on stony moraine without vegetation.														
Aw	0-4	33	2.5Y3/2	6.4	53	29	18	1.0	0	7.1	.02	5.5	1.39	0
Cw1	-10	55	2.5Y3/2	7.1	52	26	22	0.5	0	7.4	.03	4.7	1.72	2.6
Cw2	-30	58	2.5Y3/2	7.2	48	30	22	0.5	0.1	4.5	.04	8.0	2.13	1.3
Cw4	50-80	40	5Y2.5/2	7.1	74	18	9	0.5	0.1	1.6	.03	11.2	2.89	1.3
b) Skeleti-gelic Podzol on basalt.														
Ah	0-2	9	9YR3/3	4.3	48	18	34	31.4	3.0	4.0	.19	12.6	5.7	0
AE	-13	12	6YR3/3	4.2	64	15	21	37.2	3.8	3.1	.22	13.5	6.3	0
Bh	-15	30	9YR5/4	4.1	64	15	21	31.8	7.1	1.3	.37	15.1	18.0	0
Bhs	-18	37	8YR3/3	3.9	76	11	13	44.1	7.0	2.2	.47	14.5	19.5	0
Bms	-25	60	5YR4/3	4.0	76	11	13	9.6	3.0	6.3	.11	10.7	15.0	0
CBw	-34	70	8YR3/4	3.9	83	6	11	2.1	2.1	4.6	.08	6.7	6.0	0
c) Humi-gelic Umbrisol of former penguin rookery on basalt.														
Ah1	0-19	37	7YR3/3	4.2	89	9	2	18.0	4.6	10.2	.20	10.5	17	0
Ah2	-36	31	9YR5/3	4.2	75	16	9	19.1	4.1	6.5	.42	11.6	64	0
Ah3	43-48	5	9YR6/4	4.1	48	36	16	26.5	7.6	6.6	.11	14.1	120	0
Ah4	64-86	62	8YR5/6	4.1	91	5	4	25.8	1.6	3.8	.08	3.5	76	0

[1]After Kuhn, 1997; **sa** sand, **si** silt, **cl** clay; **car** $CaCO_3$.

Table 3.5.6. Mineralogy of a Calcari-regic Cryosol on stony moraine without vegetation, Arctowski.

a) mean of 63-630 µm Ø in %[1].

Horizon	Depth, cm	Volc. glass	Olivine	Serpentine	Chlorite	Mica		Pyroxene		Strongly w. min.
						unw.	w.	unw.	w.	
Aw	0-4	0.3	0	4.7	0	0	0	5.4	0	15.6
Cw2	10-30	0	0.3	2.0	0	1.0	1.7	5.2	0	25.1
Cw4	50-80	0.7	o.7	1.3	0.7	0.6	0	5.3	0	29.0

Horizon	Andesine	Plagioclase	Alkali feldsp.	Weath. feldsp.	Zeolites[3], chalcedony		Opaque min.	Rock fragm.	Apatite	Others
Aw	22.4	10.8	2.	4.3	4.4	0.3	1.7	25.6	0.3	0.8
Cw2	16.7	6.5	1.7	10.0	2.4	0.7	0.3	21.4	0	0
Cw4	12.0	12.4	1.0	12.0	2.7	0	2.3	17.1	0	2.1

b) clay fraction[2].

Horizon	Depth, cm	Serpentine	Feldspars	Chlorite	Illite/mica	Olivine	Smectite	Pyroxene
Aw	0-4	15	26	13	9	0	24	13
Cw1	-10	5	12	4	11	0	54	14
Cw2	-30	10	27	10	9	12	15	17
Cw4	50-80	20	19	31	12	9	0	9

[1]unw.=unweathered, w.=medium weathered.
[2]Fraction (<2µm Ø) in %, clay semiquantitative.
[3]Chabazite, skolezite.

Table 3.5.7. Mineralogy of a loamy Skeleti-gelic Podzol with andic phase on moraine of basalt, Arctowski.

a) mean of 20-630 μm ∅ in %[1]

Horizon	Depth cm	Volc. glass	Olivine	Serpentine	Chlorite	Mica unw.	Mica w.	Pyroxene unw.	Pyroxene w.	Strongly w. min.
Ah	0-2	1.6	6.8	2.0	1.2	1.6	2.8	6.0	0	36.8
Bhs	15-18	1.7	1.4	2.4	0.3	1.7	0	7.9	0	9.6
CBw	25-34	0	0	2.3	0	0	0	5.3	2.3	5.7

Horizon	Andesine	Plagioclas	Alkali feldspar	Weath. feldspar	Zeolites[2], chalcedony		Opaque min.	Rock fragm	Others
Ah	2.0	6.8	5.6	9.2	0.4	3.2	2.0	10.8	1.6
Bhs	11.6	1.7	2.7	8.9	0.2	3.8	4.1	41.6	0.3
CBw	19.7	1.0	1.7	9.0	0	0.7	1.3	50.2	0.3

b) clay fraction

fraction (<2μm ∅) in %, clay semiquantitative

Horizon	Depth, cm	Quartz	Feldspars	Chlorite	Illite/ mica	Olivine	Smectite	Pyroxene
Ah	0-2	<10	30	<5	10	<5	35	5
Bhs	15-18	<5	30	25	<5	<5	25	10
CBw	25-34	<5	10	30	<5	<5	40	<5

[1]unw.=unweathered, w.=medium weathered.
[2]Skolezite, thomsonite.

Table 3.5.8. Mineralogy of the clay fraction of a Humi-gelic Umbrisol (former penguin rookery) on basalt, Arctowski.

Depth, cm	Smectite	Illite	Chlorite	Variscite	Vivianite	Apatite	Other phosphates
43-48	<10	<10	<10	<10	10-20	<10	>20
64-86	<10	<10	<10	<10	10-20	<10	>20

5. Rock Weathering by Lichens

The lichens occupying rocks or stones as substrate, generally known as saxicolous species, can be divided into three distinct groups: crustose, foliose, and fructicose species (Jones and Wilson, 1985). In terms of the modes of attachment to substrate rocks, lichens can be grouped into epilithic (living on the surface of rock) and endolithic types (living in the interior of rock) (Golubic et al., 1981).

In general, lichens induce the **physical weathering** of rocks by these mechanisms: (1) the penetration of hyphae through intergranular voids and mineral cleavage planes, (2) expansion and contraction of thallus by microclimatic wetting and drying, (3) freezing and thawing of lichen thallus and associated microenvironment, (4) swelling action of organic and inorganic salts originating from lichen activity, and (5) incorporation of mineral fragments into thallus (Figure 3.5.1). In some cases, if not many, lichens' mechanical action on their mineral substrate dominates the **bio-weathering process**, especially in the early stages of colonization. The main possible **chemical processes** by which lichens can dissolve minerals are (1) generation of respiratory CO_2, (2) the excretion of oxalic acid, and (3) the production of biochemical compounds with complexing ability (Chen et al., 2000).

We chemically analyzed weathered rock samples with lichens and investigated thin sections with a light microscope and scanning electron microscope. We found chemical changes in the surface layer in relation to the unweathered interior (Table 3.5.9). There was an enrichment of Fe, P, K, and partly Mn in the first millimeters (like in desert varnish), whereas other elements were lost. There may be an influence of dust and/or sea spray, too.

In Antarctica, as far as we are aware, the lichen communities of most investigations are the endolithic communities inhabiting the Beacon Sandstone in the Antarctic cold desert (e.g., the Ross Desert) (Vestal, 1988). However, observable effects of epilithic lichens on the weathering process of various rocks also have been reported from regions including the maritime Antarctic (Campbell and Claridge, 1987; Ascaso et al., 1990; Chen and Gong, 1995).

Figure 3.5.1 shows an influence of lichens on rock weathering and new formation of minerals, especially iron oxides and oxalates.

Table 3.5.9. Chemical changes in surface layer of rocks by weathering and soil-forming processes in relation to the interior; Fildes Peninsula, Antarctica[1].

Sample[4]	Position	Loss[2] %	SiO_2 %	Fe_2O_3 %	Al_2O_3 %	CaO %	MgO %	TiO_2 %	MnO %	K_2O %	Na_2O %	P_2O_5 %
5	interior	5.9	53.3	7.6	16.1	0	10.5	0.7	.13	.32	2.5	.14
	surface	5.9	54.6	8.2	16.9	0	8.9	0.7	.14	.42	2.9	.14
	%[3]	+1.2	+2.4	+ 8.7	+ 5.0	0	-15.5	+7.4	+6.1	+31	+17.1	0
13	interior	5.6	49.2	8.7	17.7	4.6	7.7	0.8	.14	.35	3.1	.16
	surface	7.1	50.1	9.1	17.3	4.0	7.6	0.7	.13	.36	2.8	.18
	%	+ 25.9	+1.8	+4.7	-2.3	-12.9	-1.3	-12.7	-11.4	+2.9	-8.4	+9.9

[1]After Chen and Gong (1995).
[2]Ignition loss.
[3]Change in %.
[4]5 basaltic andesite w. *Usnea Antarctica*; 13 gabbro w. *Xantorua elegans*.

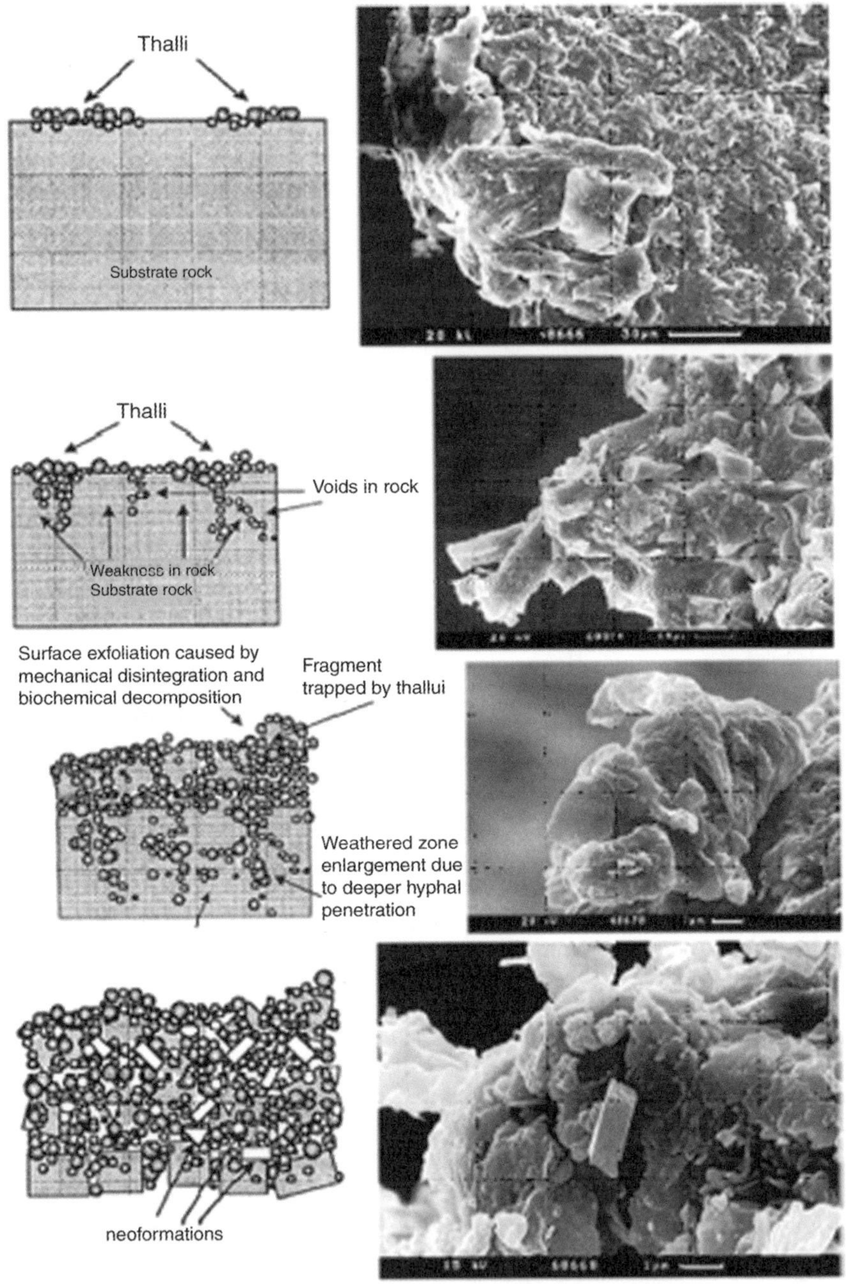

Figure 3.5.1. Suggested sequence of the weathering effects on substrate rocks of the Fildes Peninsula, Antarctica. (Photo [of an electron microscope] 68668, andesite with *Haematomma erythromma*, 68666, like 68668, 68534, andesite porphyry with *Usnea fasciata,* 68678, hyperstene-olivine gabbro with *Usnea fasciata*. Chen et al., 2000.)

6. Weathering, Acidification, and Mineral Formation

A strong, **physical weathering** in the form of frost weathering takes place together with high cryoturbation (Bockheim, 1995) in areas with permafrost. This weathering primarily is from the freezing and thawing of the soil water. The extent to which the rock fragments and minerals of the topsoil are crushed depends on the frequency of freezing and thawing. The weathering therefore occurs primarily during the warm summer, with no snow on the soil (Campbell and Claridge, 1987).

These conditions exist for one to two months on hill backs in Wilkes Land and on down-slopes for not more than one month. Depressions remain snow-covered all year. The weathering conditions exist for 3 to 4 months on King George Island, but soils here do not show frost changes throughout because of higher average temperatures and lower temperature amplitudes. In addition, a thick plant cover here (below 60 m above sea level) muffles the frost change. Only the soils of very young moraines like the Regic Cryosol are missing an isolating vegetative blanket.

Initially the frost weathering is restricted to the topsoil: a few cm for loamy soils like the Stagnic Cryosol, but several dm for gravel-rich soils like the Podzol at Casey, due to low water contents and therefore low heat capacity. The weathering of a loamy soil nevertheless will be more intense as intensive cryoturbation also occurs during the summer.

Cryoturbation translocates the soil particle permanently and forms new targets. The higher clay content of the Stagnic Cryosol in the subsoil is based on an upward movement of rougher particles, which, in turn, favors their disintegration.

Sandy to gravelly soils have a well developed lichen cover which may insulate them from freeze/thaw cycles (Melick et al., 1994). The extremely high gravel content of the Podzol prevents frost weathering due to dry conditions, so that only temperature weathering occurs.

Soils from loose sediments are favored for weathering, e.g., the Haplic Cryosols at Casey and Arctowski. The latter also contains volcanic glass and porous slake which disintegrate particularly easily. In the Stagnic Cryosol, the gneiss and schist layers stretch vertically and favor penetration and freezing of water in fissures. High contents of primary minerals in the clay fraction demonstrate that the physical weathering was intense; easily fissionable minerals like feldspar and mica occur, but so does quartz. These findings agree with those of Bockheim and Ugolini (1990) and Beyer et al. (1999).

Nearly all examined soils of the humid Maritime and East Continental Antarctic are strongly **acidified** (Blume et al., 1997; Beyer et al., 1999). Only soils on very young moraines with carbonates were pH neutral to alkaline, such as soils of the dry Antarctic Cold Desert that Campbell and Claridge (1987) and Beyer et al. (1999) described. Organic acids are present, but primarily nitric acid and sulfuric acid.

Snow and soil algae cause the soils to be free of vegetation, like the two Haplic Cryosols. Microorganisms later mineralize the biomass. Nitric and sulfuric acid (Blume and Boelter, 1996) then arise from nitrification and sulfurication. Blue-green algae also promote these processes through binding air nitrogen. The algae of many lichens also have this ability.

The relatively strong acidification of soils from leuco gneiss results from its low neutralization capacity. Soils from basalt and schist have a higher neutralization capacity in principle, but low temperatures seem to retard the neutralization.

Birds also intensify soil acidification. Feces of birds are alkaline at first but promote later acidification of the soils (Tatur, 1989). The excrements of the birds are enriched with nutrients so that a luxuriant vegetation follows the birds' rookeries. Primarily nitric and sulfuric acid arise from the proteins of the feces, however, due to a strong microbial activity in the soil. Many soils of the Antarctic close to the sea reflect this. Penguins lay out their rookeries up to 50 m high. Large areas, now up to 100 m, rose during the holocene due to the melting of ice (Myrcha and Tatur, 1991) and were contaminated by bird excrements. Phosphate analyses of many soils around Arctowski have shown that this applies to almost all sites there below 60 to 70 m above sea level; only the soils of very young moraines were affected little (Blume et al., 2000).

Chemical weathering also took place. Clear surface corrosions show some mineral species like feldspar, pyroxene, and mica. The appearance of pedogenic iron oxides (see also Beyer and Bölter, 1998), as well as illites and smectites in the clay fraction, indicates solution-chemical weathering. Despite low temperatures, strong physical weathering and cryoturbation promote chemical weathering, because a permanent formation of new surfaces allows acids to react intensively (Kopp, 1970; Kopp and Kowalkowski, 1990). The water is available for solution-chemical weathering in the warm summer months with humid climate conditions. In addition, relatively many formers of organic complexes appear in the soils with plant cover, despite a small biomass production, because of the very low microbial activity in the soil. These complex-formers bind extracted weathering products and promote further solutions. They facilitate podzolization (Blume et al., 1997; Beyer and Bölter, 1998). For the Antarctic Peninsula region, Myrcha and Tatur (1991) postulated that strong action of guano solutions entirely leached iron from the soil.

These findings of acidification, brownification, and sometimes podzolization under cold but humid periglacial conditions agree with those of Kowalkowski and Borzyskowiski (1973), Kopp et al. (1982), and others who believe the same holds true for Central Europe during the late Pleistocene.

How is it possible that others think that chemical weathering does not play a remarkable role in polar desert soils? Some made observations similar to ours, but their interpretation was different. For example, O´Brien et al. (1979) analyzed a soil on an old moraine from metamorphic rocks of Elephant Island. This soil is extremely acidic, and most of the minerals of the sand fraction are intensively weathered and thus could not be identified. O´Brien et al. (1979) interpreted a low intensity of weathering because they did not find differences between topsoil and subsoil. In our opinion such a conclusion is not possible, as cryoturbation greatly mixed the horizons.

The **mineral formation** also proved meaningful. Both the Stagnic Cryosol formed from schist and the Gelic Podzol formed from basalt showed illite and smectite but no kaolinite in the clay fraction. Illite formed from mica, due to cryoclastic weathering and a loss of potassium. Conceivably, however, smectites formed from basalt from a solution of pyrocene products, because the mica contents are low.

In addition, iron oxides have formed. Goethite was present and brownification of the soils with vegetative cover occurred. Ferrihydrite was suspected in relation to the reddish brown colors of the B horizons (e.g. 7.5 YR 5/8 in Table 3.5.1b). But the contents of Fe_o were generally low, whereas the contents of Fe_p were often high, together with high ODPE values (and phosphate contents). This indicates organically bound iron (together with iron phosphates: e.g. the B horizons of the Podzol in Table 3.5.4).

Bockheim's and Ugolini's (1990) "rubification" seems to be mainly a formation of organic iron complexes, together with phosphates, at least in our soils (Beyer et al., 1997). On rocks as well as stones, and in the first millimeters of soil, iron oxides also formed under the influence of lichens (Chen and Gong, 1995).

The activity of birds, particularly of penguins, led to the **formation of phosphate minerals** in many soils of the continental and maritime Antarctic. Also, new compounds resulting from reactions with weathering products, primarily Al and Fe, formed in addition to the apatite (the phosphate of the bird bones), including arktowskite, strengite, taranakite, variscite, vivianite, and others (see also Myrcha and Tatur, 1991). Strengite and vivianite will develop only under reduced conditions. At the Umbrisol, a former penguin rookery, we analyzed up to 12% P_t in the fine soil (Table 3.5.5c). This would be 61% variscite or 65% apatite. The contents in the clay fraction should be even higher; however, with X-ray, we found only 20-30% (Table 3.5.8): A great part of the phosphate in the clay fraction is probably amorphous, perhaps as iron phosphate to organic complexes. This could explain the relatively high Fe_p content of these soils. They are only partially oxalate extractable, and these high contents correspond with high contents of mobile humus (Table 3.5.1c).

7. Summary

We studied the mineralogy of representative soils near Casey, Wilkes Land, Continental Antarctic, and near Arctowski, King George Island, Maritime Antarctic, using a microscope, an electronic microscope, X-ray, and chemical analysis. The mineralogy of the sand and silt fractions, and parts of the clay fraction, reflect the geology of the sites: leuco gneiss, basic gneiss, and schist near Casey; tholeiite basalt and volcanic ashes near Arktowski. Cryoclastic weathering therefore was the main soil-forming process (beside cryoturbation).

But most of the studied soils were medium to highly acidified, down to pH 4. Chemical weathering also took place, especially for soils with vegetative cover. Clear surface corrosions show some mineral species like feldspar, pyroxene, and mica. The appearance of pedogenic iron oxides, as well as illites and smectites, in the clay fraction also indicate solution-chemical weathering. We believe that physical weathering and cryoturbation promote this. A specialty of many Antarctic soils are phosphate minerals like apatite, arktowskite, strengite, varicite, and vivianite, due to a large influence of birds, especially penguins.

References

Ascaso, A., L. Sancho, and C. Rodrigues-Pasquale. 1990. The weathering action of saxicolous lichens in Maritime Antarctica. Polar Biol. 11:33-39.

Beyer, L., H.-P. Blume, H. Knicker, and M. Bölter. 1997. Soil organic matter of suggested spodic horizons in relic ornithogenic soils of Coastal Continental Antarctica in comparison to this of spodic horizons in Germany. Soil Sci. 162:518-527.

Beyer, L., and M. Bölter. 1998. Formation, ecology, and geography of Cryosols of an ice-free oasis in Coastal East Antarctica near Casey Station, (Wilkes Land). Austr. J. Soil Res. 37:209-244.

Beyer, L., J. Bockheim, I. Campbell, and G. Claridge. 1999. Review - Genesis, properties and sensitivity of Antarctic Gelisols. Antarc. Sci. 11:387-398.

Birkenmeyer, K. 1980. Geology of Admiralty Bay, King George Island. Polish Polar Res. 1:29-54.

Blight, D., and R. Oliver. 1977. The metamorphic geology of the Windmill Islands, Antarctica. J. Geol. Soc. Australia. 24:239-262.

Blume, H.-P. 1994. Soils of Casey, Antartica. 15th World Congr. of Soil Sci. Acapulco, Transact. 6b:382-383.

Blume, H.-P., L. Beyer, M. Bölter, H. Erlenkeuser, E. Kalk, S. Kneesch, U. Pfisterer, and D. Schneider. 1997. Pedogenic zonation in soils of the Southern Circum-Polar Region. Adv. in GeoEcol. 30, 69-90 (Reiskirchen).

Blume, H.-P., and M. Bölter. 1993. Soils of Casey Station, Wilkes Land, Antarctica. In: Post-Sem. Proc. of 1st Intern. Conf. on Cryopedology. Russian Academy of Science. Pushchino, Russia. pp. 96-103

Blume, H.-P., and M. Bölter. 1996. Wechselwirkngen zwischen Boden- und Vegetationsentwicklung in der Kontinentalen Antarktis (Interactions between soil formation and plant development of the Continental East Antarctic). Verhandl. Gesell. für Ökologie.

Blume, H.-P., D. Kuhn, and M. Bölter. 2002. Soils and soilscapes. In: M. Bölter and L. Beyer, eds. GeoEcology of Antarctic ice-free coastal landscapes. In: Ecological Studies 154. Springer. New York, Berlin. Ch. 3.2.

Bockheim, J. 1995. Permafrost distribution in the Southern circumpolar region and its relation to the environment. Permafrost and Periglac. Proces. 6:27-45.

Bockheim, I., and F. Ugolini. 1990. A review of pedogenic zonation in well-drained soils of the Southern Circumpolar region. Quarternary Res. 34:47-66.

Campbell, I.B., and G.G. Claridge. 1987. Antarctica: Soils, weathering processes and environment. Elsevier. Amsterdam.

Chen, J., H.-P. Blume, and L. Beyer. 2000. Weathering of rocks induced by lichen colonization – a review. Catena. 39:121-146.

Chen, J., and Z. T. Gong. 1995. Role of lichens in weathering and soil-forming processes in Fildes Peninsula, Antarctic. Pedosphere. 5:305-314.

Golubic, S., E. Friedmann, and J. Schneider. 1981. The lithobiontic ecological niche, with special reference to microorganisms. J. Sedim. Petrology. 51:475-478.

Goodwin, J. 1993. Holocene deglaciation sea level change and the emergence of the Windmill Islands, Budd Coast, Antarctia. Quarternary Res. 37:10-18.

ISSS/ISRIC/FAO. 1998. World reference base of soil resources. World Resources Reports 84. FAO. Rome.

Jones, D., and M. Wilson. 1985. Chemical activity of lichens on mineral surfaces – a review. Int. Biodeterior. 21:99-104.

Kopp, D. 1972. Periglaziäre Umlagerungs-(Perstruktions-) zonen im nordmitteleuropäischen Tiefland und ihre bodengenetische Bedeutung (Periglacial perstruction zones in the Northern Central- European Lowland and their role in soil genesis). Tag.-Ber. Deutsche Landwirtsch.-Wiss. Berlin 102:55-81.

Kopp, D., K.-D. Jäger, and M. Succow. 1982. Naturräumliche Grundlagen der Landnutzung (Natural bases of land use). Akademie-Verlag, Berlin.

Kopp, D,. and A. Kowalkowski. 1990. Cryogenic and pedogenic perstruction in Tertiary and Quarternary deposits, as exemplified in the outcrop of Sternebeck. Quarternary Stud. in Poland. 9:51-71.

Kowalkowski, A., and J. Borzyskowiski. 1973. The role of periglacial and extraperiglacial perstructions in the formation of soil profile in Central Europe. Folia Quarternaria. 49:37-45.

Kuhn, D. 1997. Genese, Ökologie und Soziologie einer Bodengesellschaft in einem Periglazialgebiet der König-Georg-Insel (West-Antarktis) (Genesis, ecology and sociology of a soilscape in a perglacial area of the King George Island, West Antarctic). Schriftenr. Inst. für Pflanzener. Bodenk. Univ. Kiel. 40:1-173.

Melick, D., M. Hovenden, and R. Seppelt. 1994. Phytogeography of bryophyte and lichen vegetation in the Windmill Islands, Continental Antarctica. Vegetatio. 111:71-87.

Myrcha, A., and A. Tatur. 1991. Ecological role of the current and abandoned penguin rookeries in the land environment of the Maritime Antarctic. Polish Polar Res. 12:3-24.

O´Brien, R., G. Romans, and L. Robertson, 1979. Three soil profiles from Elephant Island, South Shetland Islands. Brit. Antarc. Survey Bull. 47:1-12.

Schlichting, E., H.-.P. Blume, and K. Stahr. 1995. Bodenkundliches Praktikum (Pedological practical work). Blackwell. Berlin.

Tatur, A. 1989. Ornithogenic soils of Maritime Antarctica. Polish Polar Res. 10:481-532.

Vestal, J. 1988. Biomass of the cryptoendolithic microbota from the Antarctic Desert. Appl. Environm. Microbiol. 54:957-959.

Chapter 6. Weathering Processes in Arid Cryosols

G.G.C. Claridge[1] and I.B. Campbell[2]

[1]Research Scientist, 33 Kotari Road, Days Bay, Eastbourne, New Zealand; e-mail graeme.claridge@xtra.co.nz.
[2]Research Scientist, Land and Soil Consultancy Services, 23 View Mount, Nelson 7001, New Zealand; e-mail iaincampbell@xtra.co.nz.

1. Introduction

The arid Cryosols form in a very low-energy environment. The prevailing temperatures are very low, mean annual temperatures being everywhere below freezing, and the active layer of the soil is only above freezing for a very short time (Campbell and Claridge, S. 3, Ch. 4)). Moisture availability is also very low, because for most of the time any moisture present is in the form of ice, while the absolute amount of water is also extremely low because of very low precipitation.

Nevertheless, weathering processes operate, leading to alteration of soil materials, the formation of soil horizons, and obvious although minimal soil development. Both physical and chemical weathering processes operate, although the degree of chemical weathering is relatively small and the regolith is very largely a product of physical or mechanical weathering processes.

2. Physical Weathering

The major processes of physical weathering and rock disintegration in Antarctica are glacial action; the action of water in various forms, as liquid, ice, and vapor; salt weathering; insolation; and wind action. The regolith forms through the combination of these processes.

2.1. Glacial action and rock decay

Undoubtedly, direct glacial erosion has been a major agent in the destruction of rock and the ultimate formation of landscape and regolith in Antarctica. Although soils do not occur exclusively on till, the main product of glacier erosion, most soils, and particularly those showing the greatest profile development, formed on these deposits. Features such as cirques, "roches moutonnées," glacial benches, mamillated surfaces, etc., formed as a result of abrasion and ice plucking, perhaps best provide evidence of glacial erosion.

The main product of glacial erosion of course is till, which glacial retreat deposits. Subglacial till, derived largely through glacial abrasion, is poorly sorted, although it generally has relatively high amounts of sand and silt.

However, much of the till from which soils formed in Antarctica appears to be supraglacial (ablation till) rather than subglacial. It formed from debris that fell onto the glacier after rock disintegrated on the valley walls and subsequently was deposited at the glacier surface as the ice stagnated and ablated.

2.2. Rock disintegration

The nature of the rock strongly influences much of the rock's disintegration. For example, dolerites initially split along cleavage planes, at times forming huge prismatic blocks, but subsequent weathering of the surfaces commonly occurs by small-scale exfoliation and crumbling, giving rounded rock surfaces.

Granite is particularly susceptible to granular disintegration. Argillite shows a strong tendency to split along cleavage planes, while sandstone and orthoquartzite crumble into individual grains. Surfaces cemented by the formation of a strong weathering skin tend to flake off.

2.3. Water and rock decay

Water, in its various forms, helps rock to decay in Antarctica, but its effectiveness varies according to local conditions. In areas of high elevation and very low temperatures on the inland edge of the Transantarctic Mountains, free water is rarely present and ablation removes snow and ice. Under these conditions, little water is available for rock decay.

It seems likely that rocks in the coldest and most arid areas may split because of temperature changes or through water vapor's freezing within fine rock cracks, rather than through the freezing of liquid water. Hairline cracks in some rocks, on splitting apart, have been found to contain thin films of clay-like material, suggesting that moisture may be present at times, possibly as vapor or as strongly saline solutions with depressed freezing points (see below).

At lower elevations in glacial valleys, in some northern and coastal situations and in the dry valleys, northerly winds and accompanying low clouds give warmer and moister conditions. Ablation and evaporation are still high, but, generally, there are more signs of free water, which forms occasional, small melt pools in hollows in rocks and collects as icicles. Notwithstanding the visible presence of some water, there is little sign that physical disintegration of rocks is substantially more active or effective than in the very arid regions at the edge of the central plateau.

2.4. Salt weathering

Crystallization of salts from solutions confined within pores and fine cracks exert pressures that can fragment rocks. Wellman and Wilson (1965) and Cooke and Smalley (1968), among others, have shown that the process is similar to frost weathering, with crystal growth exerting sufficient pressure to separate mineral grains or rock fragments. Wellman and Wilson (1965) consider that the necessary conditions for salt fretting are a supply of salts, sites protected from wind and rain in which salt can accumulate, and cyclic changes in humidity and/or temperature that pass through the crystallization point of at least one of the salts present. For erosion to occur, as distinct from rock breakdown, an agent such as wind must remove the disaggregated material. Salt fretting is similar to most other erosive processes, in that there is a well defined water or permafrost base level below which it does not act, but it differs from other erosive processes in acting most rapidly on the undersides of rock surfaces.

2.5. The effect of insolation

Although air temperatures are very low, solar radiant heating of soil and rock surfaces can raise them to temperatures in excess of 25°C. Black (1973) reported over 100 freeze-thaw cycles annually in some areas and considered that these would cause granular disintegration and splitting of small stones.

We have found (Balks et al., 1995) that, at Marble Point, on the coast of McMurdo Sound, and beside Lake Vanda, in the Wright Valley, relatively few freeze-thaw cycles occur each year, at the beginning and end of the short summer period, although daily temperatures fluctuate. However, insolation does not appear significant in causing particle-size reduction.

In some places, however, some evidence for the process can be seen. Some sandstones or coarsely grained granites show distinct flaking, whereas similar rocks close by do not, indicating that slight differences in porosity or cracking pattern may render some rocks more susceptible. Likewise, heaps of shattered darkly colored rocks suggest that insolation may have been an important mechanism in the breakdown of the rock, possibly due to a greater number of freeze-thaw cycles as a consequence of the dark color of the rock.

2.6. Wind action

Wind long has been recognized as an important agent in rock decay in Antarctica; it may, in many situations, be more important in the modification of landscapes and regolith formation than ice or water. Wind not only removes and transports rock particles effectively; it also contributes to rock disintegration. Strong winds carrying particles of snow, ice, or sand abrade the rocks, break them down by repeated bombardment, and remove the disintegrated rock materials. This gradually reduces the size of the rocks protruding above ground and redistributes fragmental material. The abrasion generally forms faceted rather than rounded rocks.

Where boulders protrude above an open ground surface, the influence of wind in transporting, sorting, and abrading is clear. The wind redeposits fine material from the sides and the upwind end of the boulders in the lee as a "tail" of fine gravel. Close to the boulder side or, in some cases, beneath the boulders, where wind velocities are greatest, the removal of finer particles leaves the coarser material concentrated as a lag pavement; at the same time, the lag stones are abraded to a greater degree than stones farther away from the boulders. This creates the desert pavement characteristic of arid Cryosols. Wind also moves the products of rock disintegration from exposed surfaces, where lag gravels do not protect them.

The ultimate destination of particles that move on moraine and valley floor surfaces is not clear because few extensive deposits of windblown material occur. In the ice-free valleys of South Victoria Land, aeolian deposits are rare and generally comprised of poorly sorted gravel ripples, sand sheets, and some dunes. In many places, small accumulations of sand occur in patterned ground cracks. These are obvious sites for the accumulation of saltating or drifting material because of the sheltered environment the depressions create. It is likely that most of the wind-transported sand becomes incorporated into the regolith through patterned-ground processes, although some may be blown out to sea.

2.7. Frost action and stirring within the regolith

Comparison of soils from young, intermediate, and old land surfaces shows that, with increasing age, the soils contain increasing amounts of material of fine particle size. The proportion of clay in the soils distinctly increases with time, mainly because of physical weathering, although the increase is small. Processes resulting in physical breakdown of rock material include frost action, contraction and expansion with associated ice-wedge phenomena, and thawing and ablation processes.

Frost action occurs mainly within the top few centimeters of the soil and is a result of freeze and thaw action after periodic wetting by snow meltwater. It often forms a crusty surface horizon and weakly cohesive layers, just below the surface, that are slightly vesicular through freezing of water droplets. In hollows in low-lying depressions, freeze and thaw is more effective and soils have a higher proportion of silt- and clay-sized material than in drier soils. Freeze-thaw processes can produce materials less than 0.5 μm (McDowall, 1960). Even dry soils give evidence of some particle-size reduction in the regolith, judged by particle-size changes down the soil profiles. Water vapor within the soil may cause this, as occasionally shown by precipitation of soluble salts on the faces of profile pits that have been exposed to warming by the sun.

3. Chemical Weathering

Although physical weathering processes are dominant in Antarctica, clear evidence exists of chemical weathering, although its extent may be difficult to gauge. The most noticeable indication is a red or brown staining on rocks; dolerite is stained a very dark reddish brown; granite and metamorphic rocks a lighter rust color; while pale yellow quartzite of the lower Beacon Supergroup is flecked with the remnants of a rusty colored coating which appears to be disintegrating under present conditions.

Rocks containing traces of copper minerals sometimes show pronounced green surface staining. Only the younger volcanic rocks show no evidence of chemical weathering; they remain starkly black.

However, when examined more closely, the effects of chemical weathering are not readily apparent in analytical results. Some workers consider physical weathering predominant and chemical weathering negligible. However, they do observe many of the features characteristic of soils of the McMurdo Sound region: high pH, the presence of water-soluble salts, and slightly expanded, dioctahedral-micaceous, clay minerals.

3.1. Staining

The most obvious effect of chemical weathering is staining on rock surfaces, commonly although mistakenly termed *desert varnish*. Staining occurs on the rocks of most older glacial land surfaces and cliff faces, particularly in sheltered situations. On hard rocks, such as dolerite, granite, or schist, the staining may completely coat the rock surfaces, although the stain may be blotchy rather than continuous, as is generally the case on granite or schist in coastal situations. Coatings on granite are thinner and paler in color than those on dolerite, due to the lower content of ferromagnesian minerals, largely biotite, in granite.

The staining is a thin cover on the rock surface, produced by the weathering of ferromagnesian minerals, which are almost completely weathered to a depth of about 1 mm from the surface. The iron contained in these minerals either filled the voids or migrated to the surface to form a surface coating, enriched in iron and silica, on quartz and feldspars, which resist further weathering (Glasby et al., 1981).

Coarsely grained sandstones often show a rusty brown surface stain, forming an iron oxide crust that cements the grains together and gives the surface a somewhat greater strength than the bulk of the rock. However, the fretting processes tend to remove the crust rather readily, and small pits of relatively unweathered and palely colored rock may pockmark an apparently smooth rock surface, contrasting with the more coherent, iron-oxide stained and cemented coatings. In general, coatings are more extensive on the surfaces of finely grained rocks, since coarsely grained rocks tend to crumble under the influence of intergranular weathering.

Specific minerals, such as goethite or hematite, have not been identified positively in the coatings on rock surfaces. The iron oxides probably only have short-range order, but the various colors indicate they may range from hydrous iron oxides with a goethite-like structure to anhydrous hematite-like material. The hematite-like minerals in coatings on rock surfaces suggest, therefore, that goethite-like structures can dehydrate, even under low temperature and humidity, given sufficient time. It is significant that the reddest colors occur in the most arid regions and on surfaces that have been exposed to weathering for very long periods of time, on the order of millions of years.

3.2. Pitting

On older weathering surfaces, strongly stained surface stones, particularly finely grained dolerite, develop prominent pits. The development of pits clearly relates to rock texture, just as surface staining and polish relate to rock texture. Very finely grained dolerite shows a micropitting pattern of closely spaced pits as small as 5 mm in depth and in diameter. More coarsely grained dolerite shows a much coarser pitting pattern, with pits up to 40 mm across and 20 to 100 mm deep. The bottom surfaces of the pits are often brighter in color and much rougher than the polished upper surfaces of the rock and the pit openings.

Accumulation of moisture formed the pits, probably salt solutions with a lowered freezing point in irregularities on the rock surface, selective chemical weathering of the ferromagnesian minerals, and removal of unweathered grains by salt-wedging. As the irregularities enlarge and deepen, further moisture accumulates and weathering continues. Wind removes loosened fragments and abrades the upper surfaces of the pits, polishing them and leaving a smooth surface on which the oxide coating can accumulate.

3.3. Soil colors

The more finely textured material within tills with age becomes progressively stained more strongly and to a greater depth. Young soils are almost completely lithochromic, with very pale colors, while stongly weathered soils range from yellowish brown to yellowish red. The color reflects the presence of amorphous iron and aluminium oxides on the surface of mineral grains.

In the laboratory, reducing and complexing agents, such as citrate-bicarbonate-dithionite, can remove these coatings and allow one to measure the amount of iron in the coatings (Mehra and Jackson, 1960). In general, the older or more weathered soils contain larger amounts of extractable oxides. Soils developed on dolerite, with a higher content of weatherable minerals containing iron, generally have higher contents of extractable Fe_2O_3, and hence stronger colors, than soils from granite or sandstone.

3.4. Influence of salts on soil weathering

During chemical weathering, ions are released from rock minerals but, because of the arid climate and the absence of leaching, they remain in the soil as accumulations of water soluble salts or encrustations on rocks. Study of their distribution and composition provides clear evidence for chemical weathering processes.

The presence of water soluble salts within the soil is almost essential to chemical weathering processes, which cannot proceed in the absence of moisture. Although very little moisture is available in the arid Cryosols, concentrated salt solutions may remain unfrozen and, thus, chemical weathering and movement of ions may occur within the crevices of rocks and on the surfaces of apparently dry rocks and may lead to mineral transformations.

The composition of the salts in arid Cryosols clearly indicates chemical weathering processes. The salts consist largely of the chlorides, nitrates, and sulphates of sodium, potassium, calcium, and magnesium. Combinations of these anions may form almost all of the possible crystalline phases, including double salts, and cations have been identified in soils (Keys and Williams, 1981; Claridge and Campbell, 1977).

The salts may occur in the soil as coatings or efflorescences on the soil surface, as accumulations on the undersides of stones where the evaporation or sublimation of saline solutions moving to the soil surface leave them, or as discrete horizons, sometimes up to 15 cm thick, of crystalline salt, with very little mineral matter. Where there is a large proportion of calcium, especially where marble is present in the parent material, calcite encrustations may form on the undersides of surface stones.

Calcium and magnesium derive largely from weathering of ferromagnesian minerals (augite, hornblende, etc.) in dolerite and biotite, etc., in granites. Potassium derives largely from the weathering of muscovite and to a small extent from potash feldspar. Sodium may be derived from the weathering of sodium feldspars, etc., but is largely of marine origin. Smaller amounts of calcium, magnesium, and potassium also may be of marine origin.

Soils formed largely from dolerite have a high proportion of calcium and magnesium ions, while soils formed from granite contain more potassium. Sodium is almost everywhere the dominant cation, indicating the greater contribution of salts of marine origin to the soils. The anions all are derived atmospherically, rather than being a consequence of chemical weathering, and indicate the course of atmospheric circulation.

The nitrate and sulphate ions which dominate the anions in soils of the inland regions of the Transantarctic Mountains originate as protein compounds in the foam blown from tropical oceans and transported through the upper atmosphere as part of global atmospheric circulation (Claridge and Campbell, 1968). During the transport process, chloride is removed preferentially, while nitrate and sulphate, in the form of nitric and sulphuric acids, remain. Chloride, on the other hand, derives from precipitation from moist air masses moving inland from the polar oceans and carrying salts with the composition of sea water.

3.5. Clay mineral formation

The presence of authigenic clay minerals in rocks and soils is clear evidence of chemical weathering processes in some arid Cryosols. In soils formed from transported material and, possibly, also in soils formed apparently from weathering *in situ*, the dominant minerals found were illite and hydrous micas, derived largely from finely grained micas contained within the Beacon Supergroup sediments (Claridge, 1965; Claridge and Campbell, 1968a).

The only evidence for chemical weathering in these minerals is the increase in hydration, shown by the transformation of mica though illite to interlayered hydrous micas and vermiculite. However, where weathering occurs in the presence of abundant magnesium ions and high pH, such as in undrained hollows and depressions where moisture and salts are abundant, or in soil formed directly on weathering dolerite, smectite can be detected, sometimes in quite appreciable amounts (Claridge, 1965; Campbell and Claridge, 1967, 1982).

Weathering of mica usually proceeds as far as vermiculite, but further weathering may either transform it to aluminium chlorite, by precipitation of aluminium hydroxide in the interlayer, or degrade it to kaolin. Although kaolin minerals, apart from halloysite, do not occur in the arid Cryosols of the Transantarctic Mountains, aluminium chlorites occur in some of the more strongly weathered soils in Antarctica.

Weathering of dolerite leads to a smectite mineral, characterized as an iron-rich nontronite-montmorillonite intergrade, in some cases very well crystallized (Claridge and Campbell, 1984). Smectite formation depends on moisture availability. In the most arid situations, on the inland edge of the Transantarctic Mountains, smectites do not form, and the clay fraction contains instead almost unweathered fragments of glassy dolerite.

Kaolin minerals do not occur in arid Cryosols, as the weathering intensity is not great enough, even though the time available for weathering can be very great. However, traces of halloysite, a tubular kaolin-type mineral, occur in some soils of the McMurdo Sound region, where it is associated with the weathering of widely distributed volcanic ash, which may have been deposited up to 15 million years ago and, on glaciological evidence, was deposited probably more than 3.5 million years ago (Claridge and Campbell, 1974b; Marchant et al., 1993).

Where salt concentrations are high and water is present, as in undrained basins and surrounding saline lakes, the soils may be damp or even saturated with concentrated saline solutions of low freezing point. In these situations, weathering processes can proceed farther than in the drier soils. Soils can have very much higher clay contents than the dry soils that surround them, sometimes up to 30%, with a high proportion of smectite (Claridge and Campbell, 1967). Soils in saline situations usually have an olive color.

Where no obvious signs of saline groundwater appear, thin films of saline solution may be present in the interstices between grains of soil or in cracks in rocks, thus permitting ionic migration and mineral transformations to occur. Ugolini and Anderson (1973) showed that, even at -5°C, an appreciable amount (0.003g H_2O per g soil) of liquid water was present within the soil interstices and between grains. This small amount of highly saline and reactive liquid allows chemical weathering processes to continue, even at very low temperatures.

The reactivity of water generally is enhanced in these films only a few molecular diameters thick (Fripiat et al., 1965; Chaussidon and Pedro, 1979), and the effective acidity can be high, perhaps of the order of 0.5M). Thus the removal of potassium from the interlayers of muscovite occurs, leading to the formation of illite and vermiculites and the neoformation of smectites. At the same time, mineral grains are attacked and the released cations added to the soil solution. Liquid films can persist in soils at temperatures as low as -51°C, the freezing point of a saturated calcium chloride brine.

In some cases, zeolites may form in soils. Traces of chabazite occur in a soil in the floor of the Wright Valley (Gibson et al., 1983), consistent with the finding of chabazites in other arid environments under alkaline, saline conditions. Philipsite occurs in some soils of the Taylor Valley (Linkletter, 1974), formed by the alteration of ash that had fallen into saline lakes with a high pH and sodium content and had weathered to phillipsite, along with some hydrous micas. The zeolites occur in old soils formed after the saline lakes had dried up. In the same area, we found the halloysite which has already been described and which probably was formed from the same ashy material. Philipsite also occurs in an old lake deposit at Marble Point.

4. Biological Weathering

Soil organisms have little or no observable influence on weathering processes in arid Cryosols, because of the very limited distribution of plants and animals (Ugolini and Perdue, 1968; Ugolini and Grier, 1969; Ugolini, 1977). Mosses, which occur in small isolated patches on very young soils near ephemeral streams, tend to concentrate salts preferentially with respect to the underlying mineral soil, even to the point of toxicity and eventual destruction of the plant cover (Ugolini, 1977).

However, cryptoendolithic organisms cause some degree of rock breakdown, which may contribute to a characteristic weathering pattern, exfoliating the surface skins of sandstone and other porous rocks (Friedmann, 1982). By this process, exudates from the lichen hyphae dissolve the cementing substance between the sandstone grains, and the upper crust peels off, exposing the lichen, which then penetrates farther into the sandstone. Lichen exudates also may mobilize iron compounds within the sandstone so that they move to the surface where they are precipitated. This produces bleaching within the zone of lichen growth and darkens the surface layer. Some iron may be moved deeper into the rock.

This type of rock breakdown may be more widespread than previously thought and many rock surfaces may be affected, although this process is very slow, with weathering cycles possibly on the order of ten thousand years. Crustose lichens can be observed breaking up the surface of weathered granite boulders and lifting the exfoliating surface of weathering dolerite. Whether these lichens are responsible for much of the chemical weathering and exfoliation of such rocks, or whether a weathered rock surface provides a porous medium for endolithic organisms to invade and occupy, remains to be determined. The scarcity of biological occurrences means that the overall effect of any biological weathering must be slight.

5. Conclusion

Due to the relative insignificance of chemical weathering, physical processes are most important for regolith formation and soil development in Antarctica, but the very arid environment restricts the effectiveness of processes such as freeze and thaw, while the intense cold diminishes the influence of glacial erosion. The combination of extreme cold and aridity acting over long periods of time has produced many unusual features, which need to be taken into account when considering the origin of the regolith and subsequent soil formation. Differences in mode of ablation, windiness, available moisture, salt deposits, aspect, rock type, etc., all may exert an influence on regolith formation. Within the regolith itself, physical processes including patterned ground formation and granular disintegration also influence the expression of soil morphology.

While chemical weathering is not the dominant process causing rock disintegration and formation of soil parent material in arid Cryosols, it is significant and its presence can be demonstrated in a number of ways. It is not sufficiently important to cause major changes in bulk composition of rocks or soils, but there is some weathering and decompositon of mineral grains, largely of ferromagnesians but also, in some cases, feldspars. Secondary weathering products are present as iron oxides coating grain and rock surfaces, as clay minerals formed within soil profiles and sedimentary deposits, as deposits of insoluble salts such as calcite or gypsum, as water soluble salts contained within the soil or as discrete salt horizons or encrustations, and in the formation of zeolites in certain special environments such as lacustrine sediments.

The rate of chemical weathering is extremely slow, and most surfaces show little trace of its presence. However, the time scale in Antarctica is extremely long. Consequently, on old surfaces, which may have been exposed to weathering processes like those of the present day for many millions of years, weathering may be relatively well advanced.

References

Balks, M.R., D.I. Campbell, I.B. Campbell, and G.G.C. Claridge. 1995. Interim results of 1993/94 soil climate, active layer and permafrost investigations at Scott Base, Vanda and Beacon Heights, Antarctica. Univ. of Waikato, Dept. of Earth Sciences. Special rpt. 1. 64 pp.

Black, R.F., 1973. Cryomorphic processes and micro-relief features, Victoria Land, Antarctica. In: B.D. Fahey and R.D. Thompson, eds. Research on polar and alpine geomorphology. Proceedings of the third Guelph Symposium in Geomorphology, Norwich. pp. 11-24.

Campbell, I.B., and G.G.C. Claridge. 1967. Site and soil differences in the Brown Hills region of the Darwin Glacier, Antarctica. N. Z. J. Sci. 10:563-577.

Campbell, I.B., and G.G.C. Claridge. 1982. The influence of moisture on the development of soils of the cold deserts of Antarctica. Geoderma 28:221-238.

Chaussidon, J., and G. Pedro. 1979. Rôle de l'état hydrique du système poreux sur l'évolution du milieu. Réalité de l'altération dans les systèmes à faible teneur en eau. Science du Sol. pp. 223-239.

Claridge, G.G.C. 1965. The clay mineralogy and chemistry of some soils from the Ross Dependency, Antarctica. N. Z. J. Geol. Geophys. 8:186-220.

Claridge, G.G.C., and I.B. Campbell. 1968a. Soils of the Shackleton Glacier region, Queen Maud Range, Antarctica. N. Z. J. Sci. 11:171-218.

Claridge, G.G.C., and I.B. Campbell. 1968b. Origin of nitrate deposits. Nature. 217:428-430.

Claridge, G.G.C., and I.B. Campbell. 1974b. Presence of halloysite as an indication of volcanic ash in the McMurdo Sound region, Antarctica. N. Z. J. Geol. Geophys. 17:579-585.

Claridge, G.G.C., and I.B. Campbell. 1977. The salts in Antarctic soils, their distribution and relationship to soil processes. Soil Sci. 123:377-384.

Claridge, G.G.C., and I.B. Campbell. 1984. Mineral transformation during the weathering of dolerite under cold arid conditions. N. Z. J. Geol. Geophys. 27:537-545.

Cooke, R.U., and I.J. Smalley. 1968. Salt weathering in deserts. Nature. 220:1226-1227.

Friedmann, I. 1982. Endolithic microorganisms in the Antarctic cold desert. Sci. 215:1045-1053.

Fripiat, J.J., A. Jelli, G. Poncelet, and G. Andre. 1965. Thermodynamic properties of absorbed water molecules and electrical conductivities in montmorillonite and silica. J. Phys. Chem. 69:2185-2197.

Gibson, E.K., S.T. Wentworth, and D.S. McKay. 1983. Chemical weathering and diagenesis of a cold desert soil from Wright Valley, Antarctica: an analog of Martian weathering processes. J. Geophys. Res. 88, suppl. A912-A918.

Glasby, G.P., J.G. McPherson, B.P. Kohn, J.H. Johnston, J.R. Keys, A.G. Freeman, and M.J. Tricker. 1981. Desert varnish in southern Victoria Land, Antarctica. N. Z. J. Geol. Geophys. 24:389-397.

Keys, J.R., and K. Williams. 1981. Origin of crystalline, cold desert salts in the McMurdo Region, Antarctica. Geochim. Cosmochim. Acta. 45:2299-2309.

Linkletter, G.O. 1974. Authigenic phillipsite in lacustrine sediments, Taylor Valley, Antarctica. N. Z. J. Geol. Geophys. 17:603-609.

McDowall, I.C. 1960. Particle size reduction of clay minerals by freezing and thawing. N. Z. J. Geol. Geophys. 3:337-343.

Marchant, D.R., G.H. Denton, and C.C. Swisher. 1993. Miocene-Pliocene glacial history of Arena Valley, Quartermain Mountains, Antarctica. Geografisker Annaler. 75A:269-302.

Mehra, O.P., and M.L. Jackson. 1960. Iron oxide removal from soils and clays by a dithionite-citrate system, buffered with sodium bicarbonate. Clays Clay Mins. 7:317-327.

Ugolini, F.C. 1977. The protoranker soils and the evolution of an ecosystem at Kar Plateau, Antarctica. in G. A. Llano (ed.) Adaptions within Antarctic ecosystems. Proceedings of the third SCAR symposium on Antarctic Biology. pp. 1091-1110.

Ugolini, F.C., and D.M. Anderson. 1973. Ionic migration and weathering in frozen Antarctic soils. Soil Sci. 115:461-470.

Ugolini, F.C., and C.C. Grier. 1969. Biological weathering in Antarctica. Antarc. J. U.S. 4:110-111.

Ugolini, F.C., and M.J. Perdue. 1968. Biological weathering in Antarctica. Antarc. J. U.S. 3:166.

Wellman, H.W., and A.T. Wilson. 1965. Salt weathering, a neglected geological erosive agent in coastal and arid environments. Nature. 205:1097-1098.

Section 4. Ecological Processes of Cryosols

Ecological Processes of Cryosols: Introduction

Lothar Beyer[1]

[1]Professor, Institute for Polar Ecology, Christian-Albrechts-University Kiel, Wischhofstr. 1-3, Bldg. 12, 24148 Kiel, Germany; e-mail soil_science@web.de.

By definition, ecology is heavily oriented to organisms; its concern is all forms of life in relation to their environment (Coleman et al., 1992). Soil ecology thus concerns soil organisms and their interactions with chemical and physical soil compounds (Coleman and Crossley, 1996; Gisi et al., 1997).

In both polar regions, low temperatures and the impact of permafrost restrict soil's formation (Goryachkin et al., Section 2, Chapter 1). If aridity is not the driving force in a polar ecosystem (Campbell and Claridge, Section 3, Chapter 6 and Section 4, Chapter 6) accumulation of organic matter is the most striking soil-forming process because of the considerably high reservoirs of carbon in both circumpolar regions (Batjes, 1996; Blume et al., 1996; Beyer et al., 1999). The bioavailability of soil organic matter (SOM) determines microbiological processes (Killham, 1994), and the release of nutrients does not inhibit plants' adaptation to the microclimate (Woodin and Marquiss, 1997). Thus, organic matter, decomposition, and biological activity are this section's main topics.

Michaelson et al. (Section 4, Chapter 1) focus somewhat more on the soil profile level in Arctic Alaska. Cryosols to 1 meter there can contain large amounts of SOM, which vary with geomorphic region, character of the tundra cover, soils, and depth. In a pattern consistent with cold temperature preservation, decomposable fractions occur in largest proportions in the SOM of the permafrost, along with high concentrations of soluble C and cellulose, which easily could decompose under warmer conditions.

Zamolodchikov and Federov-Davydov (Section 4, Chapter 2) focus more on the landscape level and summarize the current data available on biological productivity and decomposition and their impact on soil formation in the Russian Arctic. More than half the cited references are in Russian, so the chapter is a large source of information for non-Russian terrestrial polar scientists.

Soil ecological information is less avalaible from the southern circumpolar regions (Beyer et al., 1999). In addition, most pedogenic knowledge derives from the highly arid Dry Valley Region (Campbell and Claridge, 1987; Bockheim and Ugolini, 1990), but assumptions have changed remarkably since Blume and Bölter's 1993 publication (see Goryachkin et al., Section 2, Chapter 1; Tarnocai et al., Section 5, Chapter 1).

This section's third and fourth chapters focus on the SOM of Antarctic soils, which has been overlooked for decades (e.g., Campbell and Claridge, 1990; Bockheim and Ugolini, 1990). Both of these chapters support Bölter and Kandeler's point (Section 4, Chapter 5) that we need to discover the origin of Antarctic SOM. This should be a research priority.

Bölter and Kandeler (Section 4, Chapter 5) describe the microbial community and summarize the impact of temperature, moisture, and aeration on microbes and microbial activity. They also focus on the interactions between particulate and dissolved organic matter (together with SOM) and microbes and their activity and conclude that SOM quality and availability are major concerns.

Vassiljevskaja et al.'s Chapter 7 focuses on the biology of arid Cryosols of the Transantarctic Mountains. These extremely arid soils are generally unfavorable environments for plant and animal life because of low temperatures and lack of available moisture. However, organisms have managed to colonize these soils, and simple ecosystems have developed. Current data about soil biology from the Dry Valley region suggests that more information on the interactive processes in soil is required to characterize the soil's ecology. However, because of the extreme fragility of these highly aged soils, the installation of technical equipment (e.g., litter bags) and/or frequent sampling should be minimized. Scientific curiosity is no justification for the destruction of soil.

References

Batjes, N. H. 1996. Total Carbon and Nitrogen in the Soils of the World. Eur. J. Soil Sci. 47:151-163.

Beyer, L., J. G. Bockheim, I. B. Campbell, and G. G. C. Claridge. 1999. Review – Genesis, Properties, and Sensitivity of Antarctic Gelisols. Antarc. Sci.11:387-398.

Blume, H.-P., and M. Bölter. 1993. Soil of Casey Station. In: D. Gilichinski, ed. Proceedings of the 1st International Conference of Cryopedology, November 1992. Russian Academy of Science. Pushchino. pp. 96-104.

Blume, H.-P., D. Schneider, and M. Bölter. 1996. Organic Matter Accumulation in and Podzolization of Antarctic Soils. J. Plant Nutrition and Soil Sci. 159:411-412.

Bockheim, J. G., and F. C. Ugolini. 1993. A Review of Pedogenic Zonation in Well drained Soils of the Southern Circumpolar Region. Quaternary Res. 34:47-66.

Campbell, I. B., and G. G. C. Claridge. 1987. Antarctica: Soils, Weathering Processes and Environment. Elsevier. Amsterdam.

Coleman, D. C., E. P. Odum, and D. A. Crossley. 1992. Soil Biology, Soil Ecology, and Global Change. Biol. and Fertil. of Soils. 14:104-111.

Colemann, D. C., and D. A. Crossley, Jr. 1996. Fundamentals of Soil Ecology. Academic Press. San Diego.

Gisi, U., R. Schenker, R. Schullin, F. X. Stadelmann, and H. Sticher. 1997. Bodenökologie Thieme, 2nd Ed. Stuttgart. (In German).

Killham, K. 1995. Soil Ecology. Cambridge Univ. Press. Cambridge.

Woodin, S. J., and M. Marquiss, eds. 1997. Ecology of Arctic Environments. Blackwell Science. Osney Mead.

Chapter 1. Organic Matter and Bioactivity in Cryosols of Arctic Alaska

G.J. Michaelson[1], X.Y. Dai[2], and C.-L. Ping[3]

[1]Research Associate, University of Alaska Fairbanks, Agricultural and Forestry Experiment Station, Palmer Research Center, 533 E. Fireweed, Palmer, Alaska 99645, USA; e-mail pngjm@uaa.alaska.edu.
[2]Senior Research Associate, Dept. of Rangeland Ecology and Management, Texas A&M University, 2126 TAMU, College Station, Texas 77843, USA ; e-mail xdai@ag.tamu.edu.
[3]Professor, University of Alaska, Agricultural & Forest Experiment Station, 533 E. Fireweed, Palmer, Alaska 99645, USA; e-mail pfclp@uaa.alaska.edu.

1. Introduction

Soils of northern regions contain about 26% of the terrestrial C-stocks; the Arctic and boreal zones each hold about 12 to 13% (Post et al., 1982). Under the climatic conditions of the recent geological past, these regions have served as a sink for atmospheric carbon. Recent research efforts in Alaska have shown that tundra which has exhibited sink activity up to the 1970s has now, in the past decade of 1980 to the 1990s, reversed net C-flux activity and become a source of atmospheric carbon (Oechel et al., 1993).
This reversal has occurred along with a period of warming climatic conditions for the region. Models of the earth's climate identify the northern regions as having the potential for the greatest warming as recent climate warming trends continue (Kahl et al., 1993). With soil respiration being the largest component affecting the flux of C from the tundra to the atmosphere, understanding the Cryosols, their biological activity and available soil substrates, will be necessary to predict the effects of climate change on the ecosystems of the northern regions.

Temperature and moisture conditions control soil profile development, microbial respiration in soil, and, hence, the character of soil organic matter. Temperature and moisture affect the accumulation and distribution of C-stocks in the soil profile through controls on vegetation, cryogenic soil processes (such as cryoturbation), and the presence of or depth to permafrost.

The following discussion is in two parts: first, C-storage amounts and the distribution of C-stores; second, the character of soil organic matter (SOM) and its influence on bioactivity of C-stores under cold conditions. This information is necessary to more fully revealing the function that soils of cold regions play in northern ecosystems and how they will respond to changing environments.

2. Storage and Distribution

2.1. C storage

Several studies have reported C-stock estimates for tundra Cryosols (Table 4.1.1). Values for tundra soils vary from 3 to 110 kgC/m^2. The estimates vary with characteristics such as soils selected (under differing tundra types at different locations) and the depth of measurement. Many estimates include only the soil's organic layer and/or the upper mineral horizon. These surface soil layers were thought to be most active in nutrient cycles and in their exchange of C with the atmosphere, and, for most soils, these layers contain virtually all of the C-stocks.

Table 4.1.1. C-stores in tundra Cryosols, from selected literature.

Location	Land Cover	Stores Measured	C-Stores KgC/m^2	Reference
Barrow, AK Coastal plain	Tundra	0-20 cm	20	Chapin et al., 1980
Circumpolar sites	Moist tundra		11	Post et al., 1982
	Wet tundra		22	
	Rain tundra		37	
Coastal plain	Wet tundra	0-100 cm	110	Schell and Ziemann, 1983
Alaskan Arctic	Tussock tundra	O, A	29	Oechel and Billings, 1992
	Wet sedge	O, A	13	
	Overall	0-20 cm	20	
High Arctic to Boreal			4-63	Tarnocai and Smith, 1992
		0-100 cm	3-101	Kimble et al., 1993
Central AK	All major tundra			Michaelson et al., 1996
Arctic		0-100 cm	16-94	
Coastal plain		0-100 cm	62	
Arctic foothills		0-100 cm	44	
Active layer		0-100 cm	27	
Permafrost		0-100 cm	26	

However, we have known for some time that many Cryosols typically contain significant C-stores at depth (Douglas and Tedrow, 1959; Tarnocai and Smith, 1992; Michaelson et al., 1996). More recently, we have recognized that the C-stocks of the subsurface active layer and upper permafrost are important to the function of Cryosols in Arctic ecology and to the functioning of the Arctic on a global scale. Subsurface C-stocks are thought to be active in cold-season respiration of C to the atmosphere, through continued biological active at low temperatures (Zimov et al., 1993; Oechel et al., 1997). Recognition of the significance of subsurface C-stocks in Cryosols points to the importance of knowing more about their quantity and quality.

Ping et al. (1998) reported detailed characteristics of Alaskan Cryosols, and Michaelson et al. (1996) and Bockeim et al. (1998) made detailed studies of Cryosol C-stocks. Soils in these studies represent the Gelisol order of the U.S. classification system (Soil Survey Staff, 1998), as Gelisols support the major tundra types of Arctic Alaska.

The Arctic region of Alaska is divided into two physiographic areas: first, the coastal plain that extends from the Arctic Ocean to the northern edge of the Arctic foothills; second, the Arctic foothills south of the coastal plain, extending to the northern foothills of the Brooks Range.

The studies of Cryosols mentioned above were in the Kuparuk River basin that occupies the central part of the Alaskan Arctic and spans the two physiographic areas. Auerbach and Walker (1995) mapped the Basin's vegetative cover classes, and the basin contains the major vegetative land cover classes found across Arctic Alaska. Walker et al. (1998) recognized that surface soil reaction (as measured by pH) exerts a fundamental influence on the composition of the tundra plant community and thus provides the major division for the classification of vegetative cover (acidic versus nonacidic tundra), with soil moisture as the secondary classifier (wet, moist, and dry).

Ping et al. (1998) and Michaelson et al. (1996) measured soil characteristics and C-stocks in the soils of the major tundra cover classes and found the same factors to be important in both the formation of Cryosols and accumulation of C-stocks. These important factors include low temperature, frost action (ice wedge polygon formation and cryoturbation), and soil moisture or redox condition.

The cooler temperature regime of the Cryosols favors accumulation of SOM and increasing C-stores. In general, average C-stocks were higher, at 62 kg C/m^2, for soils of the northern coastal plain region than at 44 kg C/m^2 for those of the more southern foothills region. The coastal plain region's MAAT is cooler at -12.8 to -10.3°C and MAST at 50 cm -7 to -9°C than the foothills', where MAAT and MAST temperatures are 2 to 4 °C warmer (Ping and Moore, 1993).

Cryoturbation or frost mixing is a major factor that serves to increase soil C-stocks (Bockheim et al., 1998; Ping et al., 1998). Soil C fixed by plants in the organic surface horizon is mixed into the lower mineral horizons. Through cryoturbation, SOM is exposed to mineral interactions, lower temperatures, often more reducing moisture conditions, and possible encasement in the permafrost. All of these conditions of exposure result in the preservation or protection of SOM and buildup of soil C-stocks.

The amounts of soil C-stocks under each tundra vegetative cover class vary considerably, as do the presence and degree of the conditions that promote soil development and the build-up of SOM, i.e., temperature, cryoturbation, and moisture. A rather broad range of C-stocks occurs for both tundra cover and soil class. In general, however, a broad picture of C-stocks appears if we allow for exceptions due to variability in local conditions that promote soil and SOM development (Figure 4.1.1).

Michaelson et al. (1996) found that C-stocks were highest in the wet acidic tundra, followed by moist acidic tundra dominated by shrub and moss, nonacidic tundra with large polygons, nonacidic tundra with nonsorted circles or frost boils, and moist acidic tundra with Eriophorum tussocks dominating. The data of Bockheim et al. (1998) showed a similar trend and presented a conceptual model for the effects of cryoturbation on soil C-stock enhancement.

From this sequence (Figure 4.1.1) we can see that moisture and substrate reactions, which reflect soil development, are influential in preserving C-stocks, and cryoturbation is important in translocating C to lower horizons. The form in which frost action is expressed on the soil surface (either as large polygons or as active frost boils) also influences C-stocks through its effects on vegetation and soil temperature. Frost action not only mixes SOM but also exerts influence through its affect on surface vegetation, soil temperature, and depth to permafrost.

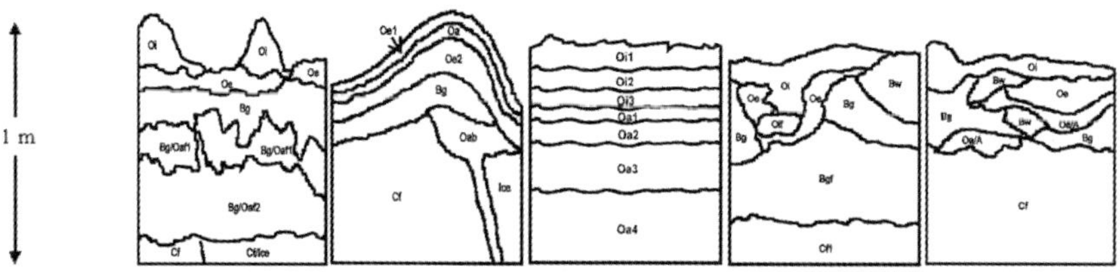

USDA Soil Classification[1]	Ruptic-Histic Aquiturbel		Sapric Glacitel		Typic Fibristel		Mollic Aquiturbel		Typic Aquiturbel	
C-Stocks (1m)[2]	TOC	HI-C	TOC	HI-C	TOC	HI-C	TOC	HI-C	TOC	HI-C
0-20 cm	20	0.72	14	1.20	4	0.11	9	0.17	9	0.41
20-permafrost	22	0.74	17	0.73	35	0.36	11	0.33	7	0.32
Permafrost (kg OC m^{-2})	46	1.35	30	1.02	32	0.01	29	0.37	17	0.83
General Tundra Cover Class[3]	Acidic – moist Shrub / moss		Nonacidic Polygons (rim shown)		Acidic Wet		Nonacidic Nonsorted circles		Acidic - moist Eriophorum tussocks	
General Range of C-Stores[4] To 1 m (kg OC/m^2)	61-94 n=3		42-82 n=6		34-82 n=5		40-48 n=2		16-58 n=4	
Mechanisms for SOM Preservation	Cryoturbation Acidic Conditions Slope Movement		Cryoturbation High Water Content		High Water Table Acidic Conditions		Cryoturbation		Cryoturbation Slope Movement	

Figure 4.1.1. Examples of some major tundra cover classes, their soil classification (Ping et al., 1998), range of C-stores or TOC (Michaelson et al., 1996), carbon in active hydrophilic organic matter fractions (HI-C), and important mechanisms for SOM preservation.

2.2. C distribution

Contribution of Cryosol C-stocks to terrestrial C are underestimated if based on older tundra C-stock estimates (Table 4.1.1) that account for only near-surface C-stocks. The C distribution in Cryosols follows a different pattern from that of soils in more temperate areas, due to the mixing action of frost processes. The ice-forming and -mixing processes of cryoturbation occur in many different ways, including the formation of ice wedge polygons and ground ice (Bockheim and Tarnocai, 1998), diapiric and frost boil movement of soil layers (Swanson et al., 1999; Shilts, 1978), and mixing due to slope movement. These processes heavily affect soil formation and C distribution in Cryosols.

For Cryosols of the central Alaskan Arctic, including C-stores from the entire active layer and upper permafrost to 1 meter increases terrestrial C-stock estimates by as much as 2.5 times over those made using only C-stocks of the upper soil horizons (Michaelson et al., 1996). Analysis of C distribution over a range of soils supporting the major tundra classes showed that the average distribution of C-stocks is in the ratio of about 1:1:2 for the active-layer organic horizons, active-layer mineral horizons, and permafrost (to 1 m), respectively.

The genetic horizons of Cryosols typically contain different levels of C-stocks (Figure 4.1.2). The surface organic horizons, especially the Oa, are high in C (average 29% OC) and contain a large portion of C-stocks, as one would expect for any soil. The mineral B horizons, although lower in C content (average 4.4% OC), typically are thick and contain nearly the same amounts of C as the Oe horizons. The mixed mineral/organic (B/O) horizons result from cryoturbation and are intermediate in C content (average 10% OC), but they occur in thick zones with higher bulk densities and thus, on average, contribute to C-stocks in amounts similar to those of the Oa horizons. The upper permafrost (to 1 m) is often a Cf horizon with lower C content (average 7% OC), but this zone is often 40-60 cm thick and thus contains large portions of the 1 m C-stocks.

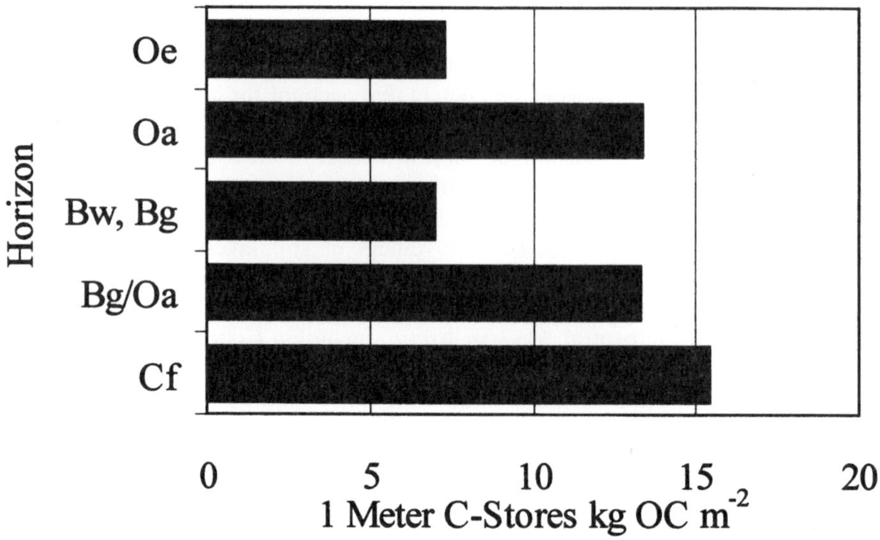

Figure 4.1.2. Average Carbon stocks contained in some commonly found horizons of Arctic Alaskan Cryosols. Data from Michaelson et al. (1996).

3. Character and Bioactivity

3.1. Character

Cryosols can contain relatively large proportions of SOM substrates that, under warmer conditions, quite easily decompose (Ping et al., 1998; Michaelson and Ping, unpublished data). Sequential digestion of SOM indicates a high proportion in the fractions that include cell wall and cell wall-like components or reminates, the hemicellulose (Hcel) and neutral detergent soluble substances (NDS).

Cold soil conditions are likely responsible for the persistence of higher proportions of these components that are derived from detritial plant materials. These components account for 50 to 57% of OC in organic horizons, 35 to 49% of OC in B horizons, and 49% to 77% of OC in the upper permafrost (Cf horizons). The Hcel and NDS normally occur in the surface organic horizons, as expected with annual production of vascular plants, but in Cryosols the highest proportions commonly occur in the SOM of permafrost.

Mixing of SOM, resulting from cryoturbation, coupled with the much lower temperatures and higher moisture contents of lower horizons, retards decomposition of SOM. Decomposition of cellulose and related compounds is retarded or stopped both with low temperature, through enzyme inhibition, and with high moisture levels, due to the limit they impose on physical processes (Lukins et al., 1984).

Detailed analysis of SOM from permafrost Cf horizons (C.L Ping, unpublished data; Michaelson and Ping, 2003) shows that they also contain elevated levels of organic fractions of low molecular weight (35-75% of soluble OC) that include easily decomposed polysaccharides. In contrast, soils of the warmer active layer contain a lower range of the organic fractions of low molecular weight (9-30% of soluble OC).

Michaelson and Ping (2003; unpublished data) examined SOM of the active layer and permafrost from acidic and nonacidic tundra soils. They found differences in SOM character between the types of soil-vegetation substrates, with differences in soil moisture-temperature conditions for the two major tundra types. Acidic tundra contains higher proportions of humified SOM components in the active layer (23% of OC) when compared to nonacidic tundra (12% of OC). Soils associated with acidic tundra also contain larger stocks of labile water-soluble SOM components than do adjacent soils of non-acidic tundra. The greater buildup and preservation of SOM in acidic tundra is consistent with its greater moisture content and C-sink activity (Walker et al., 1998).

Permafrost-encased SOM contains larger proportions of the fibrous cellulose component than does active-layer SOM. Cellulose in the active-layer SOM averages 5% of OC, compared to 8% in permafrost. Low temperature and moisture control both the accumulation and character of SOM in Cryosols. For example, cryoturbated soil horizons contain, on average, 30% more water-soluble C than do surface organic layers, and these stocks average 66% higher labile-C content than those of the surface organic layers (Michaelson and Ping, 2003). The character of C-stocks, especially as it relates to decomposition, could be important to the function of the Arctic system in the sequestration and flux of C under a changing climate.

3.2. Bioactivity

Climate change affecting the temperature and/or the moisture regime of Cryosols could result in shifts in the balance of decomposition of SOM. Soil systems with sufficient nutrients increase respiration and the rate of decomposition of SOM to a new optimum level with an increase in temperature and moisture conditions. Studies of soil respiration of Alaskan Cryosols show this response (Douglas and Tedrow, 1959; Flanagan and Veum, 1974; Nadelhoffer et al., 1991).

C-flux studies on the landscape level also indicate the importance of the temperature-moisture dependency of respiration in the Arctic (Vourlitis et al., 1993; Christensen et al., 1998). The optimum moisture level for respiratory activity in organic soils occurs at a water content of about 200% of dry weight (Flanagan and Veum, 1974) and, for mineral soils, at a water content of about 60% of pore volume (Douglas and Tedrow, 1959).

3.2.1. Temperature

Bioactivity in Cryosols must occur under a range of temperature and moisture conditions. Relatively small differences in soil depth result in very different temperature-moisture environments. During the warm season, all depths of the active layer begin warming in April, with temperatures rising to a maximum that occurs in July for the surface and in September for the bottom of the active layer. Cooling begins after the maximum is reached, and cooling takes place as the active layer loses heat from both the surface to the air and from the bottom to the permafrost table. The soil surface cools to below 0°C in September, but the subsurface active layer remains at 0°C or unfrozen for a period of weeks to months, depending on losses of surface heat. Losses of surface heat depend on air temperature and the insulative properties and thickness of the surface organic layers and the snow cover.

During the first part of the cold season, most of the subsurface active layer becomes isothermic at or near 0°C for a period of weeks, beginning in September, before cooling differentially with depth. Soils near the surface 20 cm experience warm-season temperature maximums from about ambient air temperature to 5°C. Soil layers from 20 cm deep to permafrost (subsurface active layer) experience temperature maximums that range from about 0 to 5°C, and in upper permafrost maximums are from -5 to 0°C (Romanovsky and Osterkamp, 1997).

Bioactivity of Arctic Alaskan Cryosols has been studied particularly for the range of temperatures for surface soils during the warm season. Nadlehoffer et al. (1991), working with surface organic and subsurface mineral soils, observed an approximate doubling of soil respiration as soil temperatures increase from 9 to 15°C. Dai et al. (2000) found a similar doubling of respiration over a 4 to 25°C range of temperature (Figure 4.1.3, middle). Although respiration rates are lower at subzero temperatures, respiration rate increases are greater, with over 100-fold increases possible as temperatures rise from -5°C to 0°C (Figure 4.1.3., right).

3.2.2. Labile SOM fractions

Activity studies of base-insoluble and soluble fractions found the hydrophilic (HI) fraction most active. Respiration among active-layer and permafrost soils at 4 and 25°C correlates positively to their content of the HI fraction (Figure 4.1.4a). This fraction includes low molecular weight acids and neutrals such as polysaccharides and simple sugars that are highly available substrates for microbes.

Soil stocks of this bioactive HI fraction are distributed throughout the active layer and upper permafrost of Cryosols. In the active layer, the surface 20 cm of the soil profile can contain from 20% to 40% of the total HI stocks to 1 meter (Figure 4.1.1). The subsurface active layer of the mineral soils (from 20 cm down to permafrost) can contain 21% to 38% of stocks to 1 meter and 75% for organic soils, such as the Fibristel example of Figure 4.1.1. Permafrost stocks can range from 35% to 53% of the total HI but only 2% for the Fibristel.

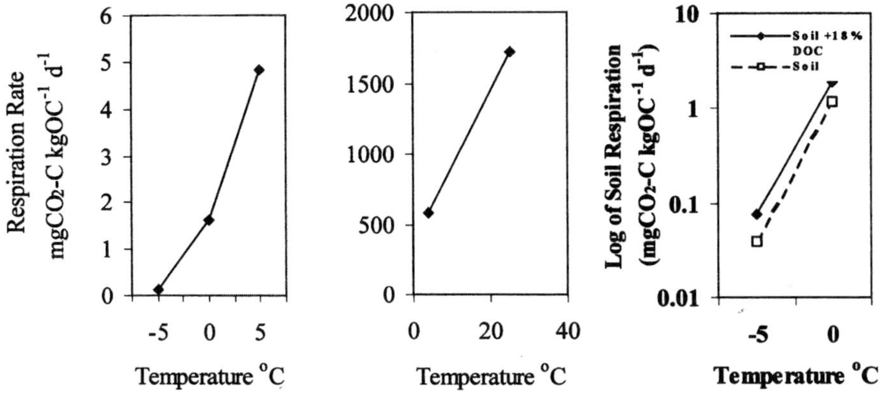

Figure 4.1.3. Respiration response for Arctic soils to temperature for the (a) low range (left data from Michaelson et al., 1998a) and (b) high range (middle data from Dai et al., 2000) and with (c) DOC enhancement (right data from C.L. Ping, unpublished data).

Unpublished data of C.L. Ping shows that active fractions are distributed throughout soil profiles, with differing patterns for various Cryosols. Distribution and partitioning of SOM in the profile appear to vary and are linked to soil processes. Substantial portions of HI stocks occur associated with organic accumulations, such as in the surface layers of the Sapric Glacistel or subsurface active layer of the Fibristel (Figure 4.1.1). But, as cryoturbation distributes SOM, large portions of the HI stocks occur in the subsurface active layer and permafrost for many Cryosols, such as the Aquiturbels (Figure 4.1.1).

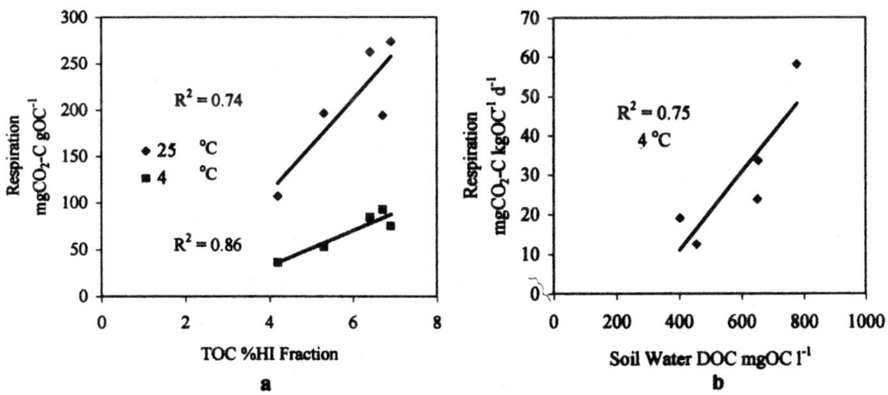

Figure 4.1.4. Respiration of Arctic soils (A) as related to the active hydrophilic (HI) fraction of total organic carbon (TOC) (data from Dai et al., 2000) and (B) as related to water-soluble OC (data from Michaelson and Ping, unpublished).

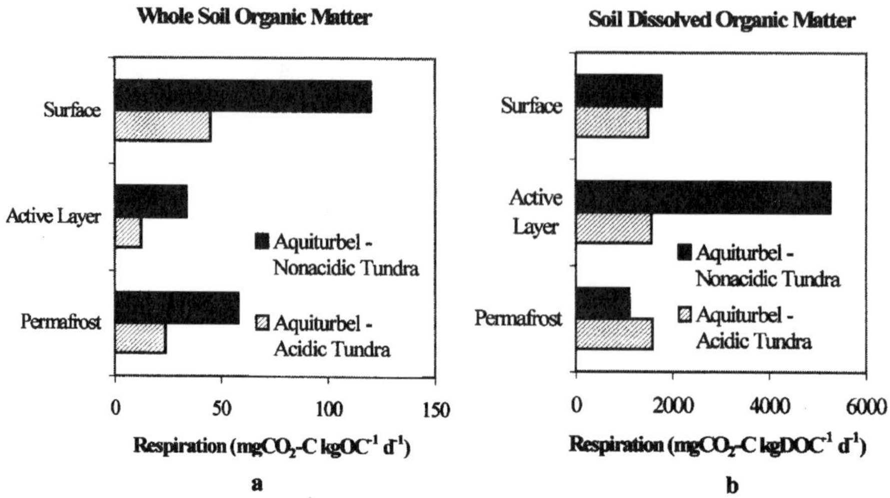

Figure 4.1.5. Respiratory activity (at 4°C) of soil OC from the top 1 meter of the surface (0 to 20 cm), subsurface active layer (20 cm to permafrost), and upper permafrost of Arctic Alaskan soils (data from Michaelson et al., 1998b; Michaelson and Ping, unpublished data).

3.2.3. Respiration

Soil respiration occurs at various depths that are under different temperature-moisture regimes and varies with the SOM amounts and qualities present. Soil incubation studies of both acidic and nonacidic Aquiturbels (Michaelson and Ping, unpublished data) indicate that the SOM present in the upper permafrost respires at a rate that is about twice as high as that of SOM in the active layer just above it but at only about half the rate of the SOM of the surface-layer soils (Figure 4.1.5).

Respiration rates for dissolved organic C (DOC) (Figure 4.1.5b) are about two orders of magnitude higher than for the SOM as a whole. Aquiturbels of acidic and nonacidic tundra also exhibit different patterns of relative respiration with depth in the profile. Differences in respiration rates among soils and soil layers are small, with the exception of DOC from the nonacidic active layer. Nonacidic active-layer DOC respires at a rate at least three times higher than that of any other soil or soil layer. If we compare respiration of soils in the subsurface active layer and in the permafrost, the amount of dissolved organic carbon present in soil water correlates to respiration rate (Figure 4.1.4b), and respiration of the water-soluble C correlates to its HI fraction content ($r = 0.97$).

3.2.3.1. Cold season changes

Significant amounts of C-flux from Arctic soils occur during the cold season (Zimov et al., 1993). Oechel et al. (1997) estimated that C fluxes could be up to an additional 60% of those that occur during the warm season. Although low soil temperature and low availability of moisture reduce respiration, microbes are known to function well below 0°C, and unfrozen water contents of Arctic soils remain sufficient to support activity.

Flanagan and Veum (1974) report microbial activity in tundra detrital materials at -7°C. Nadlehoffer et al. (1991) report a very reduced respiratory sensitivity to temperature change between 2 and 10°C. At these lower temperatures, respiration rates are orders of magnitude lower, but response to temperature and respiration continues. Work with three subsurface active-layer soils from the Alaskan Arctic indicates that respiration reduces about 4-fold as temperature reduces from 5 to 0°C and then 10-fold as temperatures reduce from 0 to -5°C (Figure 4.1.3).

Reduced response of respiration at low temperatures has been attributed to enzyme-substrate limitations, which would affect the release of the higher energy, simple carbohydrates from cellulose and cellulose-like SOM fractions. Given that low temperatures impose energy and enzyme limitations upon the breakdown of the more complex substrates, it follows that use of readily available high-energy carbohydrates in soils will become more important at lower temperatures.

3.2.3.2. DOC activity

Carbohydrates become available as soil dissolved organic carbon (DOC) when soils are exposed to freeze-thaw cycles, and respiratory bursts follow this carbohydrate release (Schimel and Clien, 1996; Skogland et al., 1988). This freezing-release phenomenon provides highly available substrates that could drive respiration in the low ranges during the cold season when other conditions are not limiting.

Soluble substrates from soil DOC could be quite significant in light of the long cold season, high subsurface C-stocks, and low temperature bioactivity that occur in Cryosols. The DOC activity differences that Figure 4.1.5 shows indicate that the type of soil and tundra cover also may play an important role in the soil's potential for respiration at low temperatures, through quality differences for soil DOC. Biological activity requires unfrozen water with available substrates and, during the cold season, these conditions are limited to the unfrozen water films that occur around soil particles (Romanovsky and Osterkamp, 1997).

Impacts of soil DOC substrates in unfrozen water films are difficult to study, but respiration qualities of DOC suggest they are highly available and active relative to whole SOM. For example, soil DOC from surface soil averages 100 times higher respiration than SOM carbon as a whole, and DOC from subsurface and permafrost soil is 40-50 times more active in respiration than corresponding whole SOM (Michaelson and Ping, unpublished data; Michaelson et al., 1998b).

As mentioned above, respiration at 4°C in the subsurface active layer and permafrost soil correlates highly with the amounts of soil DOC present, and the respiration of DOC correlates with its hydrophilic substrate content. Respiration rates at low temperatures are higher with higher DOC levels. Soil respiration at 0 and -5°C averages higher in samples where DOC was enhanced by 18%. As temperatures increased from -5°C to 0°C, soils with 18% enhanced DOC respired at rates 90% and 60% higher than the same control soils receiving only water with no DOC (Figure 4.1.3, right side).

Michaelson and Ping (2003) investigated low-temperature (-2°C) respiration rates using 88 samples from Cryosols in Arctic Alaska. They found a good correlation between water-soluble organic carbon stocks and respiration rates. There was no correlation between water-soluble organic carbon stocks and total soil carbon stocks. This indicates that a complex set of factors, which may include vegetative composition, mineralogy, and environmental conditions of soil, may control cold-season respiration substrate levels.

4. Conclusion

Cryosols of the Alaskan Arctic contain large amounts of SOM to 1 meter deep. The amounts of SOM or C-stores vary with the geomorphic region, character of the tundra cover, soils, and depth to which stocks are considered. On the average, soils of the northern coastal plain contain larger amounts of carbon (62 kgOC/m^2) than those of the southern Arctic foothills (44 kgOC/m^2). The ranges of C-stores of the tundra cover classes overlap but, in general, stocks are in the order of acidic wet ≥ acidic moist shrub/moss ≥ nonacidic w/polygons ≥ nonacidic w/nonsorted circles ≥ acidic moist w/Eriophorum tussocks. Midrange C-store values for the classes are 77, 73, 57, 44, and 35 kgOC m^{-2}, respectively.

The C-stocks vary greatly with depth, largely as a result of cryogenic processes. On average, Cryosol C-stocks to 1 meter are evenly divided between the active layer and permafrost and, in the active layer, between the surface organic layer and the mineral horizons. Large portions of C-stocks commonly occur in the Cf horizon of the uppermost permafrost layer, the Oa horizons of the surface, and the combination B/O horizons of the active layer. The stocks in the B/O and Cf often result from cryogenic processes, such as cryoturbation of patterned-ground soils and slope movement in foothill areas.

Cryogenic processes in the environment, cryoturbation, SOM distribution, and, in turn, SOM preservation influence the character of SOM in Cryosols. Organic fractions containing cell-wall and related constituents, such as hemicellulose and soluble fibers, dominate the SOM of Cryosols. In a pattern consistent with a process of cold-temperature preservation, these decomposable fractions occur in largest proportions in the SOM of the permafrost, along with high concentrations of soluble OC and cellulose, which could decompose easily under warmer conditions.

Bioactivity of Cryosol SOM varies, as in any soil, with temperature and moisture conditions. However, soils experience a significantly different range of seasonal temperatures with relatively small changes in the depth of the soil profile. Surface soil temperatures may rise to near 25°C, subsurface active layer temperatures up to about 5°C, and permafrost up to 0°C. Soil respiratory response to temperature is quite different within this range of -5°C to 25°C. Laboratory incubations indicate that respiratory response is on the order of 10-fold with temperature increases from -5°C to 0°C, 4-fold with increases from 0°C to 5°C, and 2-fold with continued increases up to 25°C. This is consistent with a shift in substrate use from the general SOM above 0°C to water-soluble DOC substrates as the soil matrix water freezes.

In the temperature range above zero, proportions of active hydrophilic components correlate with soil and DOC respiration. DOC-enhanced soil can respire at a higher rate in the subzero temperature range and DOC may be the primary source of substrates in the subzero temperature range which much of the soil profile experiences for most of the year. Assuming that unfrozen water is necessary for respiration and bioactivity, dissolved substrates of high quality become more important in the cold season.

Cryogenic processes serve to both distribute and preserve SOM and result in the concentration of bioactive SOM in the upper permafrost, and respiration rates of the permafrost thus can be double those of the subsurface active layer. Surface SOM, with its fresh inputs seasonally, has respiration rates double those of the permafrost. With the temperature differential down the soil profile, substrates in the different depths will be most active at different times during the year.

Accessing SOM and bioactivity in Cryosols is complicated, and much remains unknown. But the potential is high for very significant impacts resulting from relatively small temperature changes. As an example, one might consider the impact of a 10-fold respiration increase in the permafrost or lower active layer during the cold season. This increase could occur as temperatures change from -5°C to O°C during an extended winter warm period, keeping in mind that, under current conditions, respiration during this period could account for up to 60% of the yearly net flux of CO_2-C.

5. References

Auerbach, N.A., and D.A. Walker. 1995. Preliminary vegetation map, Kuparuk Basin, Alaska: a Landsat-derived classification. Institute of Arctic and Alpine Research. Univ. of Colorado, Boulder.

Bockheim, J.G., and C. Tarnocai. 1998. Recognition of cryoturbation for classifying permafrost-affected soils. Geoderma. 81:281-293.

Bockheim, J.G., D.A. Walker, and L.R. Everett. 1998. Soil carbon distribution in nonacidic and acidic tundra of Arctic Alaska. In: R. Lal, J.M. Kimble, E. Levine, R.F. Follet, and B.A. Stewart, eds. Advances in Soil Science — Soil Processes and the Carbon Cycle. CRC Press, Boca Raton, FL. pp. 143-155.

Chapin, F.S., III, P.C. Miller, W.D. Billings, and P.I. Coyne. 1980. Carbon and nutrient budgets and their control in coastal tundra. In: J. Brown, P.C. Miller, L.L. Tieszen, and F.L. Bunnell, eds. An Arctic Ecosystem: The Coastal Tundra at Barrow, Alaska. Stroudsburg, Penn. Dowden, Hutchison and Ross. pp. 458-486.

Christensen, T.R., A. Jonasson, A. Michelson, T.V. Callaghan, and M. Hastrom. 1998. Environmental controls on soil respiration in the Eurasian and Greenlandic Arctic. J. of Geophy. Res. 103(D22), 29,015-29,022.

Dai, X.Y., C.L. Ping, and G.J. Michaelson. 2000. Bioavailability of soil organic matter in tundra soils. In: R. Lal, J.M. Kimble, and B.A. Stewart, eds. Global Climate Change and Cold Region Ecosystems. Adv. in Soil Sci. CRC Press. Boca Raton, Fl.

Douglas, L.A., and J.C.F. Tedrow. 1959. Organic matter decomposition rates in Arctic soils. Soil Sci. 88:305-312.

Flanagan, P.W., and A.K. Veum. 1974. Relationships between respiration, weight loss, temperature and moisture in organic residues in tundra. In: A.J. Holding, O.W. Heal, S.F. MacLean, Jr., and P.W. Flanagan. eds. Soil Organisms and Decomposition in Tundra. Swedish IBP Com. Stockholm, Sweden. pp. 249-278.

Kahl, J.D., D.J. Charlevoix, N.A. Zaitseva, R.C. Schnell, and M.C. Serreze. 1993. Absence of evidence for greenhouse warming over the Arctic Ocean in the past 40 years. Nature. 361:335-337.

Kimble, J.M., C. Tarnocai, C.L. Ping, R. Ahrens, C.A.S. Smith, J. Moore, and W. Lynn. 1993. Determination of the amount of carbon in highly cryoturbated soils. In: D.A. Gilichinsky, ed. Post Seminar Proceedings of the Joint Russian-American Seminar on Cryopedology and Global Change, Nov. 15-16, 1992. Pushchino, Russia. Russian Academy of Sciences, Inst. of Soil Sci. and Photosynthesis. Pushchino, Russia. pp. 277-291.

Lukins, A.E., J.M. Melillo, and R.L. Sinsabaugh. 1984. Factors affecting cellulose activity in terrestrial and aquatic ecosystems. In: Current perspectives in microbial ecology. Amer. Soc. of Microbiol. Washington, DC. pp. 572-579.

Michaelson, G.J., and C.L. Ping. 2003. Soil organic carbon and CO_2 respiration at subzero temperature in soils of the Arctic. Alaska J. of Geophy. Res. 108(D2). (In press.)

Michaelson, G.J., C.L. Ping, and J.M. Kimble. 1996. Carbon storage and distribution in tundra soils of Arctic Alaska, U.S.A. Arctic and Alp. Res. 28(4) 414-424.

Michaelson, G.J., C.L. Ping, and X.Y. Dai. 1998a. Arctic soils and cold-season respiration, In Proc. of 1999 Science Workshop, Seattle March 11-13, 1999. Science Management Office, ARCSS-LAII. Fairbanks, AK 99775-7740. pp. 65.

Michaelson, G.J., C.L. Ping, G.W. Kling, and J.E. Hobbie. 1998b. The character and bioactivity of dissolved organic matter at thaw and in the spring runoff waters of the Arctic tundra north slope, Alaska. J. Geophy. Res. 103(D22):28,939-28,946.

Nadelhoffer, K.J., A.E. Giblin, G.R. Shaver, and J.A. Laundre. 1991. Effects of temperature and substrate quality on element mineralization in six Arctic soils. Ecol. 72(1):242-253.

Oechel, W.C., S.J. Hastings, G. Vourlitis, M. Jenkins, G. Riechers, and N. Grulke. 1993. Recent change of Arctic tundra ecosystems from a net carbon dioxide sink to a source. Nature. 361:520-523.

Oechel, W.C., and W.D. Billings. 1992. Effects of global change on carbon balance of Arctic plants and ecosystems. In: F.S. Chapin, III, R.L. Jefferies, J.F. Reynolds, G.R. Shaver, and J. Svoboda, eds. Arctic Ecosystems in a Changing Climate. Academic Press. San Diego. pp. 139-168.

Oechel, W.C., G. Vourlitis, and S.J. Hastings. 1997. Cold-season CO_2 emission from Arctic soils. Global Biogeoch. Cycles 11:163-172.

Ping, C.L., J.G. Bockheim, J.M. Kimble, and G.J. Michaelson. 1998. Characteristics of cryogenic soils along a latitudinal transect in Arctic Alaska. J. of Geophy. Res. 103(D22):28,917-28,928.

Ping, C.L., and J.P. Moore. 1993. Soil classification and climatic zones of Alaska. In Proceedings of the Sixth International Conference on Permafrost, July 5-9, 1993, vol. I. South China Univ. of Tech. Press. Beijing. pp. 517-522.

Post, W.M., W.R. Emanuel, P.J. Zinke, and A.G. Stangenberger. 1982. Soil carbon pools in the world life zones. Nature. 298:156-159.

Romanovsky, V.E., and T.E. Osterkamp. 1997. Thawing of the active layer on the coastal plain of the Alaskan Arctic. Permafrost and Periglac. Process. 8, 1-22.

Schell, D.M., and P.J. Ziemann. 1983. Accumulation of peat carbon in the Alaska Arctic coastal plain and its role in biological productivity. In: Permafrost: Fourth International Conference Proceedings, Fairbanks, Alaska. Washington, D.C. Nat'l Acad. Sci. pp. 1105-1110.

Schimel, J.P., and J.S. Clein. 1996. Microbial response to freeze-thaw cycles in tundra and taiga soils. Soil Biol. Biochem. 28(8):1016-1066.

Shilt, W.W. 1978. Nature and genesis of mudboils, central Keewatin, Canada. Can. J. Earth Sci. 15(7), 1053-1068.

Skogland, T., S. Lomeland, and J. Goksoyr. 1988. Respiratory burst after freezing and thawing of soil: experiments with soil bacteria. Soil Biol. Biochem. 20(6):851-856.

Soil Survey Staff. 1998. Keys to Soil Taxonomy. 8th edition. USDA, NRCS. Washington, DC.

Swanson, D.K., C.L. Ping, and G.J. Michaelson. 1999. Diapiric cryoturbation of soils due to thaw of ice-rich material in the upper permafrost. Permafrost and Periglacial Proc. 10:349-367.

Tarnocai, C., and C.A.S. Smith. 1992. The formation and properties of soils in the permafrost regions of Canada. In: D.A. Gilichinsky, ed. Cryosols: The effects of cryogenesis on the processes and peculiarities of soil formation. Proceedings of the 1st International Conference on Cryopedology, Nov. 10-16, Pushchino, Russia. Russian Academy of Sciences, Inst. of Soil Sci. and Photosynthesis. Pushchino, Russia. pp. 21-42.

Vourlitis, G .L., W.C. Oechel, S.J. Hastings, and M.A. Jenkins. 1993. The effect of soil moisture and thaw depth on CH_4 flux from wet coastal tundra ecosystems on the North Slope of Alaska. Chemosphere. 1-4:329-337.

Walker, D.A., J.G. Bockheim, F.S. Chapin, III, W. Eugster, J.Y. King, J.P. McFadden, G.J. Michaelson, F.E. Nelson, W.C. Oechel, C.L. Ping, W.S. Reeburg, S. Regli, N.I. Shiklomanov, and G.L. Vourlitis. 1998. Energy and trace-gas fluxes across a major Arctic soil-pH boundary. Nature. 394:469-472.

Zimov, S.A., G.M. Zimova, S.P. Daviodav, A.I. Daviodava, Y.V. Voropaev, Z.V. Voropaeva, S.F. Prosiannikov, O.V. Prosiannikova, I.V. Semiletova, and I.P. Semiletov. 1993. Winter biotic activity and production of CO_2 in Siberian soils: A factor in the greenhouse effect. J. Geophys. Res. 98:5017-5023.

Chapter 2. The Biological Cycle in Terrestrial Polar Ecosystems and Its Influence on Soil Formation

D.G. Zamolodchikov[1] and D.G. Fedorov-Davydov[2]

[1]Leading Researcher, Forest Ecology and Production Center of Russian Academy of Sciences, Profsouznaya ul., 84/32, Moscow, 117810 Russia; e-mail dzamolod@cepl.rssi.ru.
[2]Senior Researcher, Institute of Physico-Chemical and Biological Problems of Soil Science, Russian Academy of Sciences, Institutskaya St., Pushchino, Moscow region, 142290 Russia; e-mail gilichin@issp.serpukhov.ru.

1. Introduction

Basic pools of organic matter in terrestrial ecosystems are phytomass, mortmass, zoomass, microbiomass, and soil organic matter (SOM). The phytomass includes the living organic matter of plants in the surface and ground spheres, whereas the mortmass is organic matter of dead standing and fallen trees and of the brushwood of lignified parts of plants, dead grass, litter, dead parts of mosses, and dead underground plant organs (Bazilevich, 1993). The phytomass or mortmass contains 2 to 4 times more organic matter than the zoomass and microbiomass, which are composed of the biomass of animals and microorganisms, respectively (Kovda, 1985). SOM is composed of humus and peat (Orlov et al., 1996). The SOM reservoir is the largest in Arctic and boreal ecosystems.

The biological cycle may be defined as "a set of biogenic fluxes of matter and energy which connect the pools of organic matter in ecosystems to each other and with the atmo-, hydro- and lithospheres." Conveniently, the terminology for the carbon cycle also can characterize biological cycling. Carbon is the major structure-forming element of living matter, and its interpool fluxes are almost stoichiometric to the fluxes of organic matter. The cycles of other biogenic elements could differ substantially from those of the carbon cycle.

The flux going from the atmosphere to the phytomass is called the *gross primary production* (GPP); the reverse flux to the atmosphere is due to autotrophic respiration (R_A). The sum of these fluxes, which are oppositely directed and, consequently, have different signs, gives the *net primary production* (NPP).

The flux from the phytomass to the mortmass reflects the processes of necrosis of whole plants (falling off) or their parts (e.g., leaf fall). Peat formation and humification generate fluxes directed from the mortmass to the SOM. Heterotrophic respiration (R_H) characterizes the destruction of organic matter in the mortmass and soils (in its relevance to humus, this process is called *mineralization*). Soil microorganisms (fungi and bacteria) play the leading role in the heterotrophic respiration; the role of the soil zoobiota also can be substantial.

Many types of terrestrial ecosystems also include fluxes associated with vital activities of herbivorous animals, which result in the direct emission of the greater part of carbon to the atmosphere and make up a component of R_H, while the smaller part of carbon passes through the stage of excreta and either is released to the atmosphere or joins the SOM pool. The sum of GPP, R_A, and R_H is called *net ecosystem's production* (NEP) or, in its application to carbon, it is regarded as its *net flux* (NF). This magnitude characterizes changes in the total amount of carbon in all the pools of an ecosystem.

This chapter characterizes a number of components of the biological cycle as relevant to soils of the polar zone. Strictly speaking, the soil proper includes components of this cycle in the underground sphere and the boundary layer separating the under- and above-ground spheres of an ecosystem. However, since soil's transformations of the mortmass closely depend on the mortmass' amount and structure, we must characterize pools and components of the cycle, such as phytomass and net primary production. In addition, biological productivity long has been used to estimate soil fertility and therefore is one of the objects of soil science (Kovda, 1985).

2. Biological Productivity

The peak of studies into biological productivity, from the late 1960s till the early 1980s, related to the implementation of the International Biological Program (IBP). N.I. Bazilevich generalized results relevant to Northern Eurasia (Bazilevich, Rodin, 1971; Bazilevich et al., 1986; Bazilevich, 1993).

This chapter is based on the data numerous investigators reported, characterizing the biological productivity of tundras in the Russian Arctic (Aleksandrova, 1958, 1971; Andreev, Galaktionova, 1978; Andreyashkina, 1972, 1974; Andreyashkina, Gorchakovsky, 1972; Bazilevich et al., 1986; Bogatyrev, 1973; Chepurko, 1971; Chertovsky et al., 1978; Dyakonov, Reteyum, 1971; Gasheva, 1974; Gorchakovsky, Andreyashkina, 1975; Ignatenko et al., 1972, 1973; Ignatenko, Khakimzyanova, 1971; Ignatenko, Pugachev, 1980; Khantimer, 1970; Khlynovskaya et al., 1988; Khodachek, 1969; Manakov, 1970, 1972, 1974; Manakov, Nikonov, 1979; Pavlova, 1969; Pospelova, Orlov, 1984; Pospelova, Zharkova, 1972; Pugachev, 1984; Rakhmanina, 1971; Shamurin, 1970; Shamurin et al., 1972; Smorodnikova, 1983; Stenina et al., 1979; Trotsenko, 1974; Vikhreva-Vasil'kova et al., 1964; Vil'chek, 1984, 1986, 1987; Zyryanov, Kolpashchikov, 1981).

In addition, we used the results of our studies in southern tundras of the European part of Russia (vicinity of Vorkuta city and Khal'mer-Yu township) and various types of tundras in West Taimir (Aya-Turku Lake, Tareya and Kresty stations) and on the Chukotka Peninsula (Chaplino township, Ioni Lake).

To obtain averaged estimates of phytomass and net primary production, we divided the territory of the Russian extreme north into a series of zonal-regional blocks (Tables 4.2.1, 4.2.2, 4.2.3, 4.2.6). When differentiating these blocks, we tried primarily to reflect zonal-latitudinal variability and, when the body of initial data allowed it, also the regional variability. All absolute magnitudes are in units of mass of dry organic matter.

2.1. Above-ground phytomass in Russian polar zone

The values of above-ground phytomass (Table 4.2.1) demonstrate a rather clear-cut zonal variability. For plain ecosystems, the above-ground phytomass is minimal in polar deserts (1.29 t/ha), increases in subarctic typical (6.51 to 7.79 t/ha) and southern (8.41 to 8.76 t/ha) tundras, and reaches its maximum in forest tundras (25.82 to 31.11 t/ha). This trend no longer occurs in subarctic southern tundras of the Pacific-adjacent region, where the above-ground phytomass is even smaller than that in subarctic typical tundras of eastern Europe and Taimir. Alpine polar deserts have the lowest values of above-ground phytomass (0.49 t/ha), while alpine tundras are close in values to subarctic tundras.

For subarctic tundras and forest tundras, one observes a west-east trend toward decrease in the pools of above-ground phytomass. For example, among subarctic southern tundras, the phytomass is maximum in eastern European tundras (8.76 t/ha), somewhat smaller in Taimir tundras (8.41 t/ha), and minimum in the tundras of the Pacific-adjacent region (5.03 t/ha).

Zonal-latitudinal changes in the fractional composition of the above-ground phytomass are also obvious. In polar deserts, lichens make a greater part of above-ground phytomass, whereas mosses dominate Arctic and subarctic typical tundras (47 to 56% [w/w]). In subarctic southern tundras, lignified parts of shrub and dwarf shrub contribute most to the above-ground phytomass (40 to 51%); nonetheless, the percentage of mosses remains appreciable (27 to 42%). In forest tundras, the predominance of lignified parts is enhanced (50 to 81%).

In alpine polar deserts, the above-ground phytomass consists exclusively of lichens. Alpine tundras also show a noticeable participation of lichens. The maximum percentage of herbaceous plants (71%) occurs in meadow ecosystems of subarctic tundras, exceeding by far the percentage of grasses in other ecosystems.

The regional-longitudinal variability of the fractional composition of the above-ground phytomass is not as a rule synchronous in different types of tundra ecosystems. Thus, west to east, mosses increase their representation in subarctic typical tundras and forest tundras and decrease it in subarctic southern tundras. The only regional-longitudinal trend common for all subzones is the west-to-east increase in the ratio of the phytomass of shrub (for forest tundras of shrub and trees) to that of dwarf shrub. In typical and southeastern European tundras, the shrub layer has a larger phytomass than that of dwarf shrub; the ratio of phytomasses of shrub and dwarf shrub is 1.2 and 4.0, respectively. In Asian tundras, the situation is the reverse; the ratio is 0.73 in typical Taimir tundras, 0.92 in southern tundras of western Siberia and Taimir, and 0.63, its minimum, in southern tundras of the Pacific-adjacent region.

Of particular interest is the variability of the absolute mass value of the assimilating organs of vascular plants, i.e., of the overall mass of herbaceous plants, leaves of trees, and shrub and dwarf shrub. These fractions of phytomass basically determine the production capacity of ecosystems. Minimum mass values of assimilating organs occur in polar deserts (0.06 t/ha), whereas, in Arctic tundras, this parameter is almost one order of magnitude higher and reaches 0.55 t/ha. In the zone of subarctic tundras, the mass of assimilating organs of vascular plants is 1 t/ha, on the average, and varies only from 0.80 to 1.18 t/ha within the borders of subzones and regions. In forest tundras, this phytomass fraction reaches 2.48 to 2.52 t/ha. Meadows of subarctic tundra have a similar phytomass value (2.43 t/ha). The magnitude of this parameter for the alpine tundras of eastern Europe is intermediate between those for the subarctic and forest tundras, and the magnitude for alpine tundras of the Pacific-adjacent region is similar to that for subarctic tundras.

2.2. Net above-ground production in Russian polar zone

Like the magnitudes of above-ground phytomass, those of net above-ground production (Table 4.2.2) display a pronounced zonal variability. In plains of polar deserts, the above-ground production is minimum (0.16 t/ha). It increases to 4.5 times that amount in Arctic tundras, while the above-ground phytomasses of these zones differ slightly more than 200% (Table 4.2.1). The above-ground production of subarctic typical tundras of eastern Europe and Siberia amounts to 1.7 to 1.8 t/ha and is higher than that of subarctic southern tundras of these regions (1.4 to 1.5 t/ha). The above-ground production of southern tundras of the Pacific-adjacent region is closer to the production of Arctic tundras than to that of subarctic tundras of other regions. The primary production is maximum in forest tundras (2.3 to 3.2 t/ha), where it is 1.5 to 2 times higher than in subarctic tundras.

Table 4.2.1. Above-ground phytomass and its fractional composition in ecosystems of the Russian polar zone.

Zones (subzones) and regions	Above-ground phytomass, t/ha (mean±stand. err.)	Fractional composition of above-ground phytomass, %							No. sampling sites (n)
		Shrub*		Dwarf shrub		Grasses	Mosses	Lichens	
		Leaves	Woody parts	Leaves	Woody parts				
Polar deserts	1.29	0.0	0.0	0.0	0.0	4.7	31.0	64.3	1
Arctic tundras	2.78±0.51	0.9	3.4	7.9	10.5	11.1	46.8	19.5	9
Subarctic typical tundras of eastern Europe	7.79±1.26	3.4	19.5	7.7	11.4	0.5	48.7	8.8	4
Subarctic typical tundras of Taimir	6.51±0.53	1.6	10.4	6.6	10.0	10.1	56.3	4.9	36
Subarctic southern tundras of eastern Europe	8.76±1.40	4.6	33.5	2.9	6.7	1.6	41.8	8.9	13
Subarctic southern tundras of western Siberia and Taimir	8.41±1.58	4.5	26.2	7.5	25.7	2.1	31.3	2.8	17
Subarctic southern tundras of Pacific-adjacent region	5.03±0.55	3.4	17.9	9.2	23.9	8.2	26.6	10.9	6
Meadows of subarctic tundra	3.21±0.39	2.7	12.8	2.4	7.3	70.7	4.1	0.0	14
Forest tundras of eastern Europe	31.11±4.92	5.3	77.3	2.4	3.6	0.4	8.2	2.8	11
Forest tundras of western Siberia and Taimir	25.82±4.94	3.8	42.5	5.6	7.8	0.2	36.7	3.4	5
Alpine polar deserts	0.49±0.25	0.0	0.0	0.0	0.0	0.0	0.0	100.0	2
Alpine tundras of eastern Europe	7.16±0.73	4.7	33.5	16.8	27.8	0.7	0.7	15.8	15
Alpine tundras of Pacific-adjacent region	10.89±1.91	2.8	12.0	6.6	12.6	1.1	2.0	62.8	9

* Trees and shrub for forest tundras.

Table 4.2.2. Above-ground net primary production and its fractional composition in ecosystems of the Russian polar zone.

Zones (subzones) and regions	Above-ground net primary production, t/ha (mean±stand. err.)	Fractional composition of above-ground production, %							No. sampling sites (n)
		Shrub*		Dwarf shrub		Grasses	Mosses	Lichens	
		Leaves	Woody parts	Leaves	Woody parts				
Polar deserts	0.16	0.0	0.0	0.0	0.0	37.2	37.2	25.7	1
Arctic tundras	0.73±0.15	0.8	0.6	17.7	3.5	51.5	21.5	4.4	7
Subarctic typical tundras of eastern Europe	1.84±0.03	29.1	8.4	0.3	0.0	3.5	58.4	0.2	2
Subarctic typical tundras of Taimir	1.74±0.13	6.6	2.4	13.8	3.3	39.3	33.7	1.0	33
Subarctic southern tundras of eastern Europe	1.42±0.32	25.5	9.0	12.2	3.5	13.1	34.8	1.9	9
Subarctic southern tundras of western Siberia and Taimir	1.45±0.22	29.2	10.7	14.3	11.0	13.3	20.7	0.9	15
Subarctic southern tundras of Pacific-adjacent region	0.94±0.32	10.9	4.5	19.3	10.2	41.6	10.3	2.7	4
Meadows of subarctic tundra	2.49±0.29	3.4	1.0	3.1	0.3	91.0	1.1	0.0	14
Forest tundras of eastern Europe	2.32±0.24	35.5	21.8	9.7	4.3	4.9	22.2	2.0	11
Forest tundras of western Siberia and Taimir	3.18±0.51	30.9	9.5	18.1	6.3	1.7	32.6	0.8	5
Alpine polar deserts	0.03±0.02	0.0	0.0	0.0	0.0	0.0	0.0	100.0	2
Alpine tundras of eastern Europe	1.13±0.17	30.1	13.1	32.0	11.9	4.1	0.8	7.9	15
Alpine tundras of Pacific-adjacent region	1.38±0.23	11.2	5.2	31.0	7.3	9.9	4.7	30.8	7

* For forest tundras, trees and shrub.

Alpine polar deserts have minimal above-ground productivity, not exceeding 0.03 t/ha. Alpine tundras occupy an intermediate position between Arctic and subarctic tundras. In their production potential, the meadows of subarctic tundras come close to that of forest tundras.

Vascular plants dominate the fractional composition of above-ground production. Only in polar deserts (both plain and alpine ones) and subarctic typical tundras of Europe does the lichen or moss-lichen cover account for more than 50% of above-ground production. In Arctic tundras, herbaceous plants make the major contribution to above-ground production (51%). Shrub plants play a more noticeable role in the production in eastern European subarctic typical and southern tundras (35 to 37%), whereas dwarf shrub and herbaceous plants are most important contributors in tundras in Asia (38 to 61%). In forest tundras, the arboreal-shrub layer accounts for 40 to 58% of production.

The most homogeneous above-ground production occurs in alpine polar deserts (lichens account for 100% of production) and meadows of subarctic tundra (herbaceous plants make up 91% of production). Shrubs play a more appreciable role in alpine tundras of Eastern Europe than in the Pacific-adjacent region (43 and 16%, respectively).

Plants with lignified parts make the greatest contribution to above-ground production with their foliage. Thus, on the average, in all types of tundra, the ratio of mass from foliage-producing components to that from lignified parts of plants is 2.6 for shrub and 3.9 for dwarf shrub. The opposite is true of the phytomass (the mass ratio of foliage to lignified parts is 0.17 and 0.54, respectively).

The net primary above-ground production of ecosystems which are in the close-to-climax state is approximately equal to the annual above-ground litter. For this reason, the values in Table 4.2.2 are highly characteristic of the amount and structure of organic matter supplied annually to the litter. We consider the rates and character of further transformations of individual litter fractions in detail below.

2.3. Total phytomass and net primary production in the Russian polar zone

Underground components substantially influence the structure of phytomass and net primary production of ecosystems in the polar zone (Table 4.2.3). Unfortunately, the data on the underground parameters are scarcer than the data on above-ground parameters, due to the laborious nature of studies of the former. When calculating the values of above-ground phytomass and production for Table 4.2.3, we used only those sampling sites for which estimates of underground components were available. Therefore, the values of above-ground components for a number of zones and regions differ from those in Tables 4.2.1 and 4.2.2.

Table 4.2.3. Total phytomass and net primary production of ecosystems of the Russian polar zone.

Zones (subzones) and regions	Phytomass, t/ha (mean±standard error)					Production, t/ha/year (mean±standard error)				
	Above-ground	Underground	Total	Ratio under-ground/ total	n	Above-ground	Under-ground	Total	Ratio under-ground/ total	n
Polar deserts	1.29	0.29	1.58	0.18	1	0.16	0.09	0.25	0.36	1
Arctic tundras	2.73±0.49	5.40±1.33	8.13±1.71	0.66	8	0.79±0.21	1.52±0.57	2.31±0.75	0.66	5
Subarctic typical tundras of eastern Europe	7.79±1.26	12.24±3.58	20.03±4.34	0.61	4	–	–	–	–	0
Subarctic typical tundras of Taimir	6.18±0.71	12.21±1.52	18.39±2.06	0.66	10	1.66±0.22	2.08±0.34	3.74±0.51	0.56	9
Subarctic southern tundras of eastern Europe	7.85±1.80	14.24±2.89	22.09±4.53	0.64	10	1.48±0.21	0.85±0.43	2.33±0.64	0.37	4
Subarctic southern tundras of western Siberia and Taimir	10.96±4.18	16.65±7.03	27.61±11.17	0.60	6	1.72±0.50	0.87±0.26	2.59±0.76	0.33	6
Subarctic southern tundras of Pacific-adjacent region	4.68±0.73	14.61±2.85	19.29±3.08	0.76	4	0.95±0.45	1.35±0.38	2.30±0.74	0.59	3
Meadows of subarctic tundra	2.79±0.46	11.58±2.12	14.73±2.42	0.81	8	2.38±0.34	3.19±0.51	5.56±0.75	0.57	9
Forest tundras of eastern Europe	31.11±4.92	18.25±3.89	49.37±8.58	0.37	11	2.33±0.35	0.94±0.14	3.27±0.48	0.29	7
Forest tundras of western Siberia and Taimir	25.82±4.94	21.95±2.75	47.77±7.32	0.46	5	4.12±0.70	1.36±0.08	5.48±0.78	0.25	2
Alpine polar deserts	0.49±0.25	0.00	0.49±0.25	0.00	2	0.03±0.02	0.00	0.03±0.02	0.00	2
Alpine tundras of eastern Europe	6.82±0.77	4.16±0.74	10.98±1.12	0.38	16	1.07±0.17	0.57±0.10	1.64±0.24	0.35	16
Alpine tundras of Pacific-adjacent region	9.74±1.71	3.08±0.49	12.82±1.98	0.24	11	1.11±0.30	0.40±0.21	1.50±0.42	0.26	7

The total phytomass of polar deserts makes up not more than 1.6 t/ha, while in Arctic tundras it is 5 times as large and amounts to 8.1 t/ha. The total phytomass increases to 18 to 20 t/ha in subarctic typical tundras and 19 to 27 t/ha in subarctic southern tundras. The maximum level of phytomass occurs in forest tundras (48 to 49 t/ha). Little regional-longitudinal variability of underground phytomass appears, while the revealed interregional differences are within the limits of standard errors. The alpine tundras are closer to Arctic tundras in total phytomass, although their above-ground component more resembles that of subarctic tundras.

The contribution of the underground component to the total production in polar deserts is 18%, and in Arctic and subarctic tundras is 60 to 66%, except in subarctic tundras of the Pacific-adjacent region, where it amounts to 75%. In forest tundras, the share of underground phytomass decreases to 37 to 46%. The maximum percentage of underground phytomass occurs in meadows of subarctic tundra (80%). In alpine tundras, the underground component of phytomass is relatively small, 24 to 37%.

In some cases, the geographic variability of total net primary production is paradoxical. While the production of polar deserts is substantially lower than that of other plain zones, the production of Arctic tundras (2.3 t t/ha /year) proves close to that of subarctic southern tundras (2.3 to 2.6 t/ha/year). The production of subarctic typical tundras of Taimir is inferior only to that of forest tundras of West Siberia and Taimir (5.5 t/ha/year) and meadows of subarctic tundra (5.6 t/ha/year) and exceeds the production of eastern European forest tundra (3.27 t/ha /year). The production of alpine tundras (1.5 to 1.6 t/ha /year) is smaller than that of Arctic tundras.

All subzones and regions of the polar zones (except for alpine polar deserts) may be divided into two large groups in relation to the ratio of underground to total production. The first group, in which this ratio ranges from 25 to 37%, includes polar deserts, subarctic southern tundras (except those of Pacific-adjacent region), alpine tundras, and forest tundras. The second group (56 to 66%) includes Arctic tundras, subarctic tundras of Taimir, meadows of subarctic tundra, and southern subarctic tundras of the Pacific-adjacent region.

Unfortunately, the number of sampling sites for calculating averaged estimates of underground production is minimal compared to those for monitoring other components of biological productivity. Determining underground production is also most problematic, methodologically. Thus, estimates of total production are appreciably less reliable than estimates for phytomass and above-ground production.

2.4. Geographical regularities

The basic trends in geographic variability of phytomass and production of ecosystems of the Russian polar zone are:

1. the sharply manifest increment of the above-ground phytomass and a more gradual increase in the total phytomass upon transition from polar deserts to forest tundras
2. maximum underground percentage of phytomass in subarctic tundras

3. clear-cut geographic changes in primary production only at the level of zones, and no such changes at the level of subzones
4. maximum underground component of production in Arctic and subarctic typical tundras
5. longitudinal maxima of total primary production in middle Siberia
6. lower percentage of underground components of phytomass and production in alpine ecosystems.

These trends in the variability of indicators of biological productivity of tundras in the Russian polar zone may relate to geographic variability of climatic conditions. We selected mean July and January temperatures, duration of the period of temperatures above 0°C (the warm period), and the mean wind speed in January as the basic climatic parameters. We derived estimates of these parameters on the basis of a body of climatic data, described in detail in our report on carbon cycle modeling in tundras of Russia (Zamolodchikov et al., 2001).

Minimum levels of phytomass and production in polar deserts are due to the short warm period of the year, which does not exceed 70 days, and to low temperatures (2°C in July, on the average). In Arctic tundras, the duration of the warm period increases to 100 days, while the mean July temperature goes up to 6.5°C. These conditions allow the Arctic tundras to have a higher level of primary production than polar deserts. However, low winter temperatures (from -26 to -31°C in January) and high speeds of winter winds (about 7 m/s) prevent formation of considerable pools of the above-ground phytomass.

In subarctic tundras, the warm period extends to 110 days, while the mean July temperature ranges from 8.3 to 11.9°C. Winter temperatures are higher in European subarctic tundras (-18°C in January) than in Siberian ones (from -28 to -30°C), which larger values of above-ground phytomass reflect. However, the summer temperatures are higher in Siberian tundras and reach their maximum level in typical tundras of Taimir (11.9°C in July), which well agrees with high values of net primary production in this region.

The same trend also occurs at the level of forest tundras. The January temperature is higher in European (-16°C) than in Siberian (-29.6°C) tundras; the situation with the July temperatures is opposite (11.8 and 12.3°C). As a result, the above-ground phytomass and the total phytomass are larger in European forest tundras, while the production is higher in Siberian forest tundras. The temperatures in forest tundras are only slightly milder than those in subarctic tundras of the corresponding regions, and hence the considerable increase in the above-ground phytomass in forest tundras may compare with the noticeable decrease in the speed of winter winds (5.1 m/s in forest tundras against 5.7 to 6.7 m/s in subarctic tundras).

On the whole, the climatic conditions in alpine tundras correspond to those in plain subarctic tundras. The specificity of production parameters of alpine tundras is due to the stony character of soils that prevents formation of appreciable pools of underground phytomass and inhibits underground production. The climatic parameters of alpine polar deserts are similar to those of Arctic tundras; however, when combined with the rocky character of grounds, this leads to the least values of phytomass and production.

Statistics confirm the above reasoning about the interrelation of production and climatic parameters. The total primary production exhibits a statistically significant correlation with the mean July temperature ($R = 0.76$, $p < 0.05$, $n = 12$), while the above-ground phytomass negatively couples to the January wind speed ($R = 0.60$, $p < 0.05$, $n = 13$). Of certain interest is the close correlation between the overall production and the phytomass of assimilative organs of vascular plants ($R = 0.84$, $p < 0.05$, $n = 12$).

The production parameters of ecosystems in the Russian polar zone (Tables 4.2.1, 4.2.2, 4.2.3) agree well with analogous values for other Arctic regions. The phytomass of polar deserts in the Canadian Archipelago (Bliss, 1975; Bliss et al., 1984) is on the average 1.55±0.96 ($n = 6$) t/ha with a net production of 0.09±0.06 t/ha/year. The mean value of phytomass of Arctic tundras in the north of Alaska (Shaver, Chapin, 1991; Webber, 1974) is 11.5±4.4 ($n = 4$) t/ha, while the mean net primary production is 2.3±1.0 t/ha, which is very close to the corresponding values for Russian Arctic tundras (Table 4.2.3). The phytomass and net primary production of forest tundras in southwestern Greenland (Elkington, Jones, 1974) are 54.9 and 2.1 t/ha/year, respectively.

3. Decomposition

A fundamental peculiarity of polar ecosystems is the predominance of dead SOM, represented by non-decomposed or partially decayed plant residues (mortmass) and substances of soil humus, over the living phytomass and the prevalent concentration of mineral nutrition elements in it. According to Bazilevich (1994), this property can be considered adaptive, since it ensures a relative stability of the biota under extreme climatic conditions of Arctic and subarctic ecosystems, which are fragile, vulnerable, and susceptible to various oscillations (Bliss, 1962, 1970). The main cause of the predominance of dead over living organic matter is the low rate of plant litter decomposition in polar biogeocenoses, due to the low biological activity of Arctic and subarctic soils.

3.1. Cellulose degradation

A traditional technique for evaluating biological activity of soils determines the degradation rate of cellulose exposed on the surface of soil or placed inside it (Richard, 1945; Vostrov, Petrova, 1961, Mishustin, Petrova, 1963). The degradable cellulose-containing material may be filter paper, cellophane film, linen or cotton fabric, plates of special standard cellulose (Borregaard A/S Sarpsborg) (Berg et al., 1975), never-dried porous cellulose, sulfite cellulose which is closest to plant cellulose (Lahde, 1974), etc. An appreciable body of data exists to date on the actual cellulosolythic activity of soils in various tundra and forest tundra areas. Some data are expressed as a percentage of the decrease in cellulose mass after a year-long exposure under natural conditions.

In the upper 20 cm of Arctic tundra soils of Spitsbergen, almost 6% of the initial cellulose mass decomposed, and the destruction rate in the 0- to 10-cm layer was nearly twice as high as that in the 10- to 20-cm layer (Zlotin et al., 1989). In North Fennoscandia, the rate of cellulose degradation on the soil surface was from 2.5 to 3.5% in tundras and tundra marshes of Stordalen (Abisko) (Sweden, Tornetrask Lake area) (Rosswall, 1974; Berg et al., 1975) to 6 to 18% in the forest tundra of the Kevo region (Finland) (Berg et al., 1975). In different types of alpine tundra of Hardangervidda (Norway), the actual cellulosolythic activity varied within the range of 0.5 to 6% in one year (Berg et al., 1975; Heal et al., 1974).

A fairly high rate of cellulose decomposition occurred on the surface of an illuvial-humic sandy podzol in the lichen tundra of Khibiny mountains (altitude 560 m). After a one-year experiment, the decrement of the initial mass was over 17% (Parinkina et al., 1994). In the upper layer of surface-gley soils of Bolshezemelskaya tundra, 2 to 6% of cellulose decomposed during the first half of the warm season; therefore, about 4 to 12% of cellulose decomposed during the entire season (Stenina, 1978).

In the willow-moss and lichen-willow-moss Arctic tundras of western Taimir (M. Pronchishcheva Bay station), the rate of cellulose degradation on the soil surface was 4 to 7% per year. In the northern subarctic dryad-sedge-moss spotty and hummocky tundras (Tareya station), the values of cellulose degradation on the surface were the same, whereas inside the profile they could increase up to 18% on some elements of nanorelief. In the middle subarctic dryad-moss and shrub-sedge-moss tundras (Agapa station), values were still lower: 2 to 4.5% on the surface and 6 to 8% at the 5-cm depth. In the southern subarctic shrub-sedge-moss spotty, alder and birch tundras (Kresty Taimirskie station), the cellulosolythic activity was 6 to 7% per year on the surface and about 10% inside the soil profile (Vassiljevskaja et al., 1975: Bogatyrev, 1976; Parinkina, 1980, 1989).

According to our data, in the northeastern part of Kolyma Lowland, on the surface of a podzolized sandy podbur corresponding to the green moss-dwarf shrub-lichen association, 2–3% of cellulose degraded after one year; the annual cellulose decrement was 4 to 5% in the layer of peat horizon (2 cm), 2 to 5% in the upper mineral horizon (10 to 12 cm), and 0.0 to 0.6% at a depth of 40 cm. In intrazonal soils corresponding to different stages of overgrowth with vegetation and transformation of polygonal bogs, the mean-weighted values of cellulosolythic activity for organogenic horizons varied from 1.5 to 5%.

In the hummocky sedge-moss Arctic meadow tundra in Canada (Devon Island), the cellulosolythic activity on the soil surface manifested not at all after a one-year experiment, whereas the cellulose mass decrement was 8.5% at the 5-cm depth and 16% at the 10-cm depth (Bliss, 1975). In the wet polygonal Arctic meadow tundra of North Alaska (Barrow), this annual parameter was about 3.5% in the litter layer and 1.5% directly in soil (Rosswall, 1974; Bunnel et al., 1975,

1977). In alpine tundras of Alaska (Eagle Summit), the decomposition rate on the soil surface was about 3.5% (Rosswall, 1974).

The above data indicate that, in relation to actual cellulosolythic activity, regional differences for soils belonging to different sectors of the northern polar zone are not large. Only some territories of northern Europe stand out against the background of low values. These are primarily Finland and the Kola Peninsula and, to a lesser extent, Bolshezemelskaya tundra. As one can judge from the scarce available data, within the boundaries of each region, no zonal and subzonal differences manifest.

3.2. Litter degradation (litter bag experiments)

Although cellulose is one of the predominant components of plant tissues, the regularities of the litter decomposition differ substantially from those for pure cellulose. Here, the pattern of geographic heterogeneity is much more complicated. In addition to climatic and landscape conditions, characteristics of "primary substrate" (i.e., soil, tundra litter, or other medium) in which the destruction process takes place, it is also necessary to consider the characteristics of the "secondary substrate" (Van Cleve, 1974), i.e., the properties of the decaying litter itself.

On the basis of such criteria as plant taxonomic position, morphological peculiarities, chemical composition, and physical properties, Heal and French (1974) proposed grouping various types of litter into 6 large classes: (1) mosses; (2) lichens; (3) woody residues; (4) shoots of shrub and dwarf shrub; (5) coniferous needles ("hard leaves"); and (6) leaves of shrub, dwarf shrub, and grasses ("soft leaves").

Studies of the plant litter decomposition took place mostly in the framework of IBP implementation in the late 1960s through early 1970s. Some experiments also were performed at later dates. The intensity of destruction processes was estimated from the decrement of the mass of plant residues exposed on the soil surface or below the layer of loose litter in special porous caprone bags (litter bags) with a pore diameter of 1 mm (Bocock and Gilbert, 1957). These experiments investigated the decomposition of both the mixed litter characterizing the community on the whole and various kinds of litter, i.e., leaves, stems, or roots of an individual plant or a group of constituent plants of the community. Table 4.2.4. presents the data from available literature in a highly generalized and rough form. The listed values characterize the limits of variation of the rate of mass loss for a given class of litter during the first year of exposure. (Note that in this chapter, a "rate of mass loss and decomposition" means the specific rate — the part of loss in one year.)

The highest rate of the natural mixed litter loss was characteristic of alpine tundras of Hardangervidda (variants of Wet Meadow, Dry Meadow, and Snow Bad) (33 to 38% per year), while it dropped for the Lichen Heath tundra variant to 12% per year. Such a pattern corresponded to the increase in the C/N ratio to 17 to 22 at the first three sites and to 48 to 65 at the fourth site (Dowding, 1974; Van

Cleve, 1974). In other regions (Barrow and Kolyma Lowland), the decomposition of litter occurred at an annual rate of 6 to 12%.

Against the background of the predominant low rate of loss of moss mass, one should note high rates for the Dry Meadow site at the Hardangervidda station, Pechora station, and Signy Island in the subantarctic zone, where the upper limit of this parameter variation reached 17 to 28%. In the latter case, such a pattern appeared only for *Drepanocladus uncinetus* and only under humid conditions (Collins et al., 1975), whereas the major peat-forming moss *Chorisodontium aciphyllum* degraded at an annual rate of 2% (Baker, 1972; Heal and French, 1974). Apparently, this is due to a higher or lower eutrophicity of the above-listed habitats, compared to other tundras and tundra bogs.

In general, the low rate of decomposition of moss is due in part to the fact that the dead parts of plants in mosses mix with the living plants and have no contact with soil. In addition, their litter is depleted in nitrogen and other nutrient elements and has high contents of degradation-resistant inhibitory substances. For example, the specific substance sphagnol in its tissues inhibits the decomposition of sphagnum (Rosswall et al., 1975).

The rate of decrement of mass in lichens was still lower than in mosses and characterized by prevalent annual values of 5 to 8%. Against this background, only alpine tundras of the Kola Peninsula stood out, where a mixture of fruticose lichens (*Cladina rangiferina, C. stellaris, Cetraria islandica*) lost annually up to 20% of their mass. The decelerated decomposition of lichens probably is due to the increased content of lignin and cellulose (Grishina, 1986) and to the presence of specific inhibitors of decomposition (Kuprevich, Moiseeva, 1957; Berman *et al.*, 1979).

The high lignin content also was responsible for the stable low rate of loss of wood mass in trees and shrubs: 4 to 8% in tundras and forest tundras of Europe and 12 to 13% in those of North America.

The decomposition rate of needles of various pine species was in the range of 11 to 18%. The low destruction of the coniferous litter ("hard" leaves), compared to that of the leaf litter of anthophytes ("soft" leaves), relates to the higher content of resins, waxes, lignin, tannins, and flavonoids (Van Cleve, 1974).

As to the "soft" leaves, the rate of loss of mass in anthophytes showed strong variation. At most geographic sites, the litter of trees and shrubs lost its mass more rapidly than the litter of dwarf shrub, including numerous evergreen species, whereas the decomposition of the litter of herbaceous plants occurred faster than that of trees and shrubs. The decelerated decomposition of the leaves of evergreen dwarf shrubs is due both to their anatomic-morphological peculiarities and to the high content of polyphenols that play an inhibitory role (Heal et al. 1981).

As a rule, the slowest decomposition is characteristic of the dryad, which is particularly clear in the example of low losses of mass in the case of Dry Meadow at Hardangervidda, where the "soft" leaves contained solely the dryad leaves. In dry alpine tundras of the Okhotsk Sea-adjacent region, low values of loss of mass also characterize heath dwarf shrub, which decompose relatively fast in other regions.

The highest annual rates of the dwarf shrub leaf loss of mass occurred at the swamped Sivaya Maska site (19 to 45%) and in biogcocenoses of Taimir (29 to 36%). The decomposition rate of the summergreen dwarf semishrub *Rubus chamaemorus* was maximum on the forest tundra Gusinoe bog (Pechora station) (57%) and minimum in mountain Lichen Heath of Hardangervidda (10%).

Of all European and western Siberian circumpolar territories, the eutrophic bog systems at the Pechora and Sivaya Maska sites had the largest losses of mass of shrub leaf litter. A rather low loss occurred only for the Kevo area, in the example of decomposition of birch (*Betula tortuosa*) leaves (Rosswall et al., 1975), while the absolute maximum for this litter group was in Taimir, in the vicinity of Tareya station. In southern tundras of the Taimir Peninsula (Kresty Taimirskie station), losses of litter mass differed between the shrub-covered and spotty tundra varieties. In the Okhotsk Sea-adjacent area, the rate of loss of the mass of shrub leaves was approximately at the same level as in Europe.

Among different types of herbaceous litter, the leaves of legumes loose their mass faster than the sedge species, while the latter in turn loose their leaf mass faster than cereals, which is believed due to the growth of the C/N ratio in this sequence (Heal and French, 1974; Bunnel et al., 1975, 1977; Parinkina, 1989). On the whole, the dicotyledonous plants usually have higher decomposition rates than those of monocotyledonous plants.

The maximum specific rates of the loss of herbaceous litter mass are confined to alpine tundras of Khibiny mountains and to the Pechora site; at the latter, the high upper limit is due to rapid decomposition of *Menyanthes trifoliata*. In Fennoscandia, such losses have lower values. High decomposition rates again occurred in Taimir (20 to 50%). In the Okhotsk Sea-adjacent area (Aborigen station), the rate of sedge decomposition decreased to 31%.

The rate of destruction of herbaceous litter in the reed grass communities on the Kolyma Lowland was approximately a third that of the sedge-cereal vegetation at the Harp station and half that of the cereal vegetation in South Taimir. The maximum rates in the case of Taimir meadows were determined by fast degradation of the nitrogen-rich *Astragalus* litter (54%).

Table 4.2.4. Specific decomposition rates of various litter types in ecosystems of the polar zone.

Region	Geographic location	Cenoses characterization	Specific decomposition rate of various fall types, share of mass decrement during the first year of exposure, %									Reference
			Mixed litter	Mosses	Lichens	Tree and shrub wood	Needles	Dwarf shrub leaves	Rubus chamaemorus leaves	Tree and shrub leaves	Herbaceous plants	
Fennoscandia	Stordalen	Tundra bog	–	0-10	–	6-8	–	15-29	16-32	21-30	37-42	Heal et al., 1981; Rosswall, 1974; Rosswall et al., 1975; Sonesson, 1973
	Kevo	Low alpine heath	–	–	–	–	–	–	–	5	–	Heal and French, 1974; Rosswall et al., 1975
		Forest tundra, pine area	–	–	6	4-5	15-18	–	–	–	–	–"–
		Forest tundra, birch area	–	–	–	5	–	–	–	24-27	–	–"–
	Hardangervidda	Lichen heath	12	–	5-8	–	–	24	10	–	7-20	–"–
		Dry meadow	38	11-17	–	–	–	5-7	–	–	–	–"–
		Wet meadow	33	0-11	–	–	–	–	–	28	41-42	–"–
		Snow bad	38	–	–	–	–	–	–	–	–	–"–
Kola Peninsula	Khibiny	Alpine tundras	–	–	20	–	–	31	–	–	62-67	Parinkina et al., 1994; Pereverzev, 1987, 1988
Eastern European forest tundra	Sivaya Maska	Eutrophic bog	–	8	–	6	–	19-45	36	32-36	–	Heal and French, 1974
	Pechora	Eutrophic bog	–	7-28	–	5	–	19	57	31	25-76	Heal and French, 1974; Heal et al., 1981
Western Siberian forest tundra	Harp	Tundra areas	–	–	–	–	–	16-34	–	21-26	–	Andreyashkina, 1974; Gorchakovsky, Andreyashkina, 1975
		Swamped meadow	34	–	–	–	–	–	–	–	–	–"–

(continues)

Table 4.2.4. (cont.)

Region	Geographic location	Cenoses characterization	Specific decomposition rate of various fall types, share of mass decrement during the first year of exposure, %									Reference
			Mixed litter	Mosses	Lichens	Tree and shrub wood	Needles	Dwarf shrub leaves	Rubus chamaemorus leaves	Tree and shrub leaves	Herbaceous plants	
Taimir	Tareya	Tundra	–	–	–	–	–	29-35	–	34-48	43-49	Parinkina, 1980, 1989
		Polygonal bog	–	–	–	–	–	31-36	–	39-49	40-50	– " –
	Agapa	Tundra	–	–	–	–	–	–	–	17-30	20-27	Bogatyrev, 1976; Vassiljevskaja et al., 1975
	Kresty Taimirskie	Spotty tundra	–	–	–	–	–	–	–	19	–	Parinkina, 1980, 1989
		Shrub tundra	–	–	–	–	–	–	–	31-34	32	– " –
		Flood-plain meadow	–	–	–	–	–	–	–	–	23-54	– " –
		Tundra bog	–	–	–	–	–	–	31	–	–	– " –
Kolyma Lowland	Akhmelo Lake	Tundra	10-11	–	–	–	–	–	–	–	–	Fedorov-Davydov, 1997
	Malchikovskaya channel	Flood-plain meadow	10-11	–	–	–	–	–	–	–	–	– " –
Okhotsk Sea-adjacent region	Snow Valley	Alpine tundra	–	–	6	–	11	13-18	–	16	–	Berman et al., 1979
	Aborigen	Alpine tundra-steppe	–	–	–	–	14	–	–	–	31	Grishkan, 1995, 1997
Canadian Arctic Archipelago	Devon Island	Meadow	–	–	–	–	–	–	–	–	18	Bliss, 1975
Northern Alaska	Barrow Cap	Meadow	6-10	–	–	–	–	–	–	–	15-28	Bunnel et al., 1975, 1977; Bunnel, Scoullar, 1981; Flanagan, Veum, 1974; Heal, French, 1974
	Eagle Summit	Alpine tundra	–	–	–	–	–	–	–	–	5-6	Heal et al., 1981

(continues)

Table 4.2.4. (cont.)

Region	Geographic location	Cenoses characterization	Specific decomposition rate of various fall types, share of mass decrement during the first year of exposure, %									Reference
			Mixed litter	Mosses	Lichens	Tree and shrub wood	Needles	Dwarf shrub leaves	Rubus chamaemorus leaves	Tree and shrub leaves	Herbaceous plants	
North American forest tundra	Eagle Creek	Forest tundra	–	–	–	12	–	–	–	–	28	Heal and French, 1974
	Elliott	Forest tundra	–	–	–	13	–	–	–	–	21	– " –
	Dempster Highway	Forest tundra	–	–	–	13	–	–	–	–	19	– " –
Colorado	Niwot Ridge	Alpine tundra	–	–	–	–	–	–	–	22	30-51	Heal and French, 1974; O'Lear, Seastedt, 1994
Subantarctic	Signy Island	Tundra	–	2-25	–	–	–	–	–	–	–	Baker, 1972; Collins et al., 1975; Heal and French, 1974

The specific decomposition rates of herbaceous plants in North American tundras were somewhat lower than those in Eurasian tundras. In particular, unfavorable edaphic conditions—low pH, anaerobiosis, and deficit of accessible nitrogen and phosphorus— influence Barrow (Bunnel et al., 1975). The low decomposition rate in the Eagle Summit area seems due not only to harsh climatic conditions in high mountains but also to the high C/N ratio (~60) (Van Cleve, 1974). In tundras and forest tundras of North America, the decomposition rate of herbaceous vegetation increases from north to south, but the alternation of plain and alpine tundras complicates this. *Acomastylis rossii* in alpine tundras of Colorado had the maximum decomposition rates.

Subzonal distinctions do not occur within the tundra boundaries. However, the decomposition rate of plant residues increases upon transition to forest communities, but not for all litter types and not always appreciably. Thus, in the Hardangervidda area, substantial differences appeared where the experimental test litter was barley straw, the decomposition rate of which reaches 33 to 40% in a birch forest at an altitude of 780 m, compared to 10 to 25% in various types of alpine tundra (1250 to 1320 m) (Rosswall et al., 1975). A substantial difference between tundra and forest in the Hardangervidda area also showed in the decomposition of *Rubus chamaemorus* leaves (Heal and French, 1974) and in the Khibiny mountains in the decomposition of fruticose lichens (Parinkina et al., 1994). The decomposition rate of mosses in boreal forests of Central Sweden was also appreciably higher than in the Stordalen area (Berg, 1984).

At the Aborigen station in the Okhotsk Sea-adjacent region, the decomposition rate of vetch (the experimental test litter) was appreciably higher in the forest areas (65 to 89%) than in the tundra-steppe (57%), but it was not much higher than in the Siberian dwarf-pine elfin wood (66 to 77%) and was approximately the same as in the dryad-sedge and steppificated groups (57 to 64%). In forests, in shrub formations, and, in part, at steppe sites, the decomposition of Siberian dwarf-pine needles was appreciably more intense. An appreciable advantage of warmer ecosystems in decomposing grasses appeared only in aspen woods; at steppificated sites, the decomposition rate was even lower than at a tundra-steppe site (Grishkan, 1995, 1997).

When interpreting field data on the decomposition rate of both above- and underground litters, one should bear in mind that their appreciable part is lost not because of microbial destruction but because of abiogenic leaching, which is maximum during the first year of exposure (Witkamp, 1966; Rosswall, 1973; Rosswall et al., 1975; Widden, 1977; Pereverzev, Alekseeva, 1980; Parinkina, 1989; Grishkan, 1997).

During the cold season (September to May, inclusively), up to 50% (Rosswall et al., 1975a) or more (Andreyashkina, 1972, 1974) of the total first-year mass loss often is leached. This happens because soils over moisten in autumn, plant residues physically fragment under the action of freeze-thaw cycles, and snow melts in spring, which favors the runoff of organic matter from the litter (Rosswall, 1973; Bogatyrev, 1989).

The potentially leachable fraction (readily soluble carbohydrates and amino acids) makes up 5 to 30%; it was minimal in wood and maximal in grass leaves (Heal and French, 1974). However, comparison of the carbon loss during respiration with the decrement of exposed litter mass showed that the leaching proper is responsible for the loss of 8 to 35%, and in some species even less than 5%, of the total first-year losses (Bunnel, Scoullar, 1981).

At the same time, the soil microflora rather rapidly consumes an appreciable, and possibly greater, part of substances in the potentially leachable fraction. When a relevant experiment starts at the end of summer, psychrotrophic microorganisms can use these substances in autumn, before the onset of strong frosts, or in spring when the soil temperature rises. At Barrow, for example, the respiratory activity of some soil microorganisms occurred at temperatures of -7.5°C and above (Flanagan, Veum, 1974).

3.3. Mortmass turnover and soil formation

The proportion of plant residues lost because of the uptake by invertebrate saprophages, microbial destruction, and abiogenic leaching is almost always maximum during the first year after entire plants or their individual organs fall and die. In contrast to pure cellulose, the mass of which decreases linearly or, according to some investigators (Rosswall, 1974), even in an accelerated mode, the natural litter shows decelerated degradation upon exposure under natural conditions. In this situation, one can differentiate rapid, decelerated, and slow decomposition stages (Naumov and Naumova, 1993).

Inhibition of the transition to the decelerated stage begins usually during the second to third years of experiment (Heal and French, 1974; Andreyashkina, 1974; Rosswall et al., 1975; Bliss, 1975; Pereverzev, 1987; Parinkina, 1989). In addition, model laboratory experiments on the respiration of different ages of litters showed that inhibition of the destruction of the *Eriophorum angustifolium* and *Carex aquatilis* leaves from Barrow occurs in the second year (Flanagan and Veum, 1974). Thus, one of the basic characteristics of the litter is its age, i.e., the duration of its stay on the soil surface (Heal and French, 1974).

Leaching, or the uptake of soluble fractions, and the relative accumulation of lignin and other substances that are difficult to access, after a certain time cause changes in the chemical composition of the material under degradation. Physical properties of the litter also change, and this decelerates the destruction rate of the same polymers in the old litter. For example, cellulose decomposes in freshly dead residues almost twice as fast as in the litter layer (Naplekova, 1974). This provides ground for presuming that its decomposition is still slower in peat.

At the Harp station in the Trans-Urals, in experiments on the litter decomposition, peat formation from the exposed material, i.e., the loss of the initial morphological structure of leaves and their transformation into a powdery mass, occurred in 5 to 6 years after the beginning of experiments (Andreyashkina, 1974). According to our data, peat formation on the surface of podburs in Kolyma Lowland occurred as early as during the second year, while, in six years after the beginning of the experiment, the greater part of the initial litter changed into a peat-like material. Soil horizons thus form differently enriched in dead plant residues (peat, peaty, duff, etc.). The rapid and decelerated decomposition stages occur in a loose, surface-based litter layer, while the slow stage takes place in all these horizons.

It is more difficult to estimate the time required for the slow-stage occurrence than to determine the rate of fresh litter decomposition (Heal and French, 1974). Heal (1972) presumed that most (95%) of the tested kinds of litters are totally decomposed after less than 50 years. According to Grishina (1986), the total mineralization of cellulose from plant residues requires 30 to 50 years in the spotty tundra soils, 20 to 30 years in the tundra gley typical soil, and about 200 years in the boggy tundra soil. In alpine tundras and Siberian dwarf-pine elfin wood cenoses of the Okhotsk Sea-adjacent region, the total decomposition of an annual lichen litter occurs within less than 26 years, while the total decomposition of the needles of Siberian dwarf-pine elfin wood requires 12 years (Ignatenko, Pugachev, 1979, 1981).

The results of experiments on measurement of respiration determined by decomposition of detritus in various soils of Barrow allow one to assess the rates of SOM's degradation under natural conditions (Douglas and Tedrow, 1959). In the light Arctic brown soils, close to podburs in their taxonomic position, the rate of peaty horizon decomposition was 2.3%, while that of humus and peatified detritus in the upper mineral horizon did not exceed 0.31% per year. At the watershed, in the upland soil, similar to the tundra gley soils, the rate of organic matter's decomposition in the upper horizon was 2.8%. The annual transformation rate of the upper layer of bog peat at the Wet Meadow was lower than in mineral varieties (0.89%), while the decomposition rate of the material of organogenic soil in a frost peat mound was still lower because of a moisture deficit and did not exceed 0.42%.

To assess the intensity of the biological cycle, i.e., the rate of mortmass renewal in its relevance to the forests of California, Costa-Rica, and Columbia, Jenny et al. (1949) proposed 50 years ago to use a coefficient that characterizes a given landscape facies or a given ecosystem and is a function of five Dokuchaev's factors of soil formation (climate, living organisms, relief, parent material, and time):

$$k = L/M + L = f(cl, o, r, p, t)$$

where L is the mean perennial litter fall mass and M is the mean perennial mortmass.

Since that time, this index has been used widely for various natural zones, including polar ecosystems. Heal et al. (1981), by comparing parameters of Jenny's equation for various tundra and bog ecosystems, put forward an idea about their diversity with regard to the rate of organic matter's turnover. Thus, for the tundras and tundra marshes of Stordalen and Barrow (wet meadow), the k values range from 0.001 to 0.01 (i.e., the rate of mortmass turnover varies from 0.1 to 1% per year). In the case of wet tundra of Devon Island (Arctic tundra) and moss (*Cho-*

risodontium aciphyllum) communities of Signy Island (subantarctic tundra), *k* roughly equals 0.01 (1% per year); the *k* values from 0.01 to 0.1 (the mortmass turnover rate from 1 to 10% per year) characterize different tundras of Hardangervidda; for the grassland of Signy Island, *k* equals 0.1 (10% per year); and, finally, for the drepanoclade (*Drepanocladus uncinetus*) communities of Signy Island, the *k* value is higher than 0.1 (> 10% per year).

Based on the averaged values of biological productivity (Table 4.2.3) and the data on the pools of mortmass from the same sources, one can calculate in analogous manner the rates of dead organic matter's turnover for basic zonal-regional blocks of the Russian extreme north (Table 4.2.5). To this end, we used the averaged data on net primary production, assuming that, in the climax ecosystem with a zero carbon balance, the production approximately equals the litter fall, while decomposition of the equivalent amount of dead organic matter compensates for the influx of fresh organic matter.

The rates of organic matter's turnover calculated in this way characterize the mortmass renewal on the whole. Therefore, in magnitude, they occupy an intermediate position between the relatively high rate of fresh litter decomposition and the low decomposition rate of old, often peatified, plant residues in the upper horizons of the soil profile.

It follows from Table 4.2.5 that the turnover rates are similar for plains, are rather high (12.5 to 13.3) in polar deserts and Arctic tundras, decrease somewhat upon transition to subarctic typical tundras (7.1), and are minimal (2.2 to 5.3) in southern subarctic tundras. This trend may relate to changes in the composition of plants and to the increased percentage of lignified parts in the litter. The meadow cenoses of subarctic tundra have much higher values for this parameter than those of tundra communities proper that populate terraces and watersheds. For the subarctic southern tundra, the minimum occurs in the Pacific-adjacent region, while the maximum occurs in Taimir. Southern tundras of eastern Europe have intermediate values. Upon transition to the forest tundra zone, the turnover rate somewhat increases, being higher in Siberian than in eastern European thin forests.

3.4. Geographical regularities

The decomposition rate calculated in this manner is much lower in alpine deserts than in plains. This is due to the different composition of phytomass, with the predominance of lichens in the former case and of herbaceous plants in the latter. For alpine tundras, the decomposition rates vary approximately within the same limits as in plains subarctic southern tundras. Compared to the European tundras, the tundras of the Pacific-adjacent mountains have higher (2.2-fold) turnover rates, which correlates with the decrease of lignified parts in their litter.

The empirically derived data on the decomposition rates of fresh plant litter, and the calculated rates of the dead organic matter's turnover on the whole, allow several geographic conclusions. First, one should note a specific pattern of geographic zonality in the biological cycle.

Table 4.2.5. Mean values of litter (net primary production), pools of mortmass, and estimates of the organic matter turnover in various ecosystems of the polar zone.

Zones (subzones) and regions	Litter fall mass (NPP) L, t/ha/year	Mortmass M, t/ha	k = (L/(L + M)) *100%
Polar deserts	0.25	1.64	13.3
Arctic tundras	2.31	16.11	12.5
Subarctic typical tundras of Taimir	3.74	49.06	7.1
Subarctic southern tundras of eastern Europe	2.33	56.41	4.0
Subarctic southern tundras western Siberia and Taimir	2.59	45.84	5.3
Subarctic southern tundras of the Pacific-adjacent region	2.30	100.99	2.2
Meadows of subarctic tundra	5.56	35.30	13.6
Forest tundras of eastern Europe	3.27	52.19*	5.9
Forest tundras of western Siberia and Taimir	5.48	57.15*	8.7
Alpine polar deserts	0.03	2.71	1.1
Alpine tundras of eastern Europe	1.64	60.74	2.6
Alpine tundras of the Pacific-adjacent region	1.50	24.55	5.8

*The values were approximated from data on neighboring regions.

In this case, the climatic factor acts both directly and indirectly. Directly, it increases the mortmass turnover in the direction of the summary positive temperature growth upon the transition from southern subarctic tundras to forest tundras. Indirectly, it determines the composition of vegetation and plant litter by decreasing the mortmass turnover rate upon the transition from Arctic to subarctic tundras.

In contrast to the cellulosolythic activity, which reached maximum values in alpine and plain eastern European tundras and decreased noticeably in Siberia, the maximum decomposition rates of fresh litter and mortmass turnover on the whole occurred in biogeocenoses of Taimir Peninsula. These phenomena appeared in the northern (Tareya), middle (Agapa, Ary-Mas), and southern (Kresty Taimirskie) subarctic tundra as well as in the forest tundra.

Such a pattern depends on a favorable combination of climatic, landscape-geochemical, and geobotanical peculiarities of Taimir, which is mostly a lowland separated from the northern part of the peninsula by the Byrranga mountain system, which prevents the penetration of cold air masses from the north. The soil-forming deposits are marine loamy sediments with montmorillonite predominant in the clay fraction and minerals of trappic formation, enriched by bases and iron, in coarse fractions (Bogatyrev, 1976; Pospelova, Vasilievskaya, 1985; Bazilevich, 1993a). The high ash content in the litter and the absence of lichens among the dominants of plant communities also have a positive effect on the decomposition processes.

Eastern Europe has lower decomposition rates of litter and mortmass. This is largely due to the common water-logging of plains territories and to the dry character of tundras on elevations and in mountains of Fennoscandia (Heal et al.,

1981), as well as to the wide occurrence of lichen and heath communities. The role of elements of mineral nutrition in the soil dramatically increases at the regional level. In particular, this shows in the rate of the mortmass turnover in alpine tundras of Hardangervidda being appreciably higher than at the Stordalen site, as well as in the high decomposition rates of various litter types on forest tundra eutrophic marshes of eastern Europe compared to those in the oligotrophic ecosystems of northern Fennoscandia.

The inhibited character of the mortmass turnover in subarctic tundras of the Pacific-adjacent region reflects the dryness of litters and small numbers of soil microorganisms and invertebrates. As a result, dry-peat horizons often form in soils, despite good drainage (Berman et al., 1979).

Wet Arctic tundras of North America have rather high decomposition rates of dominant herbaceous plants. However, the still higher magnitudes of net primary production and litter, as well as decelerated transformation of organic matter in thick peat horizons, determine the rather low rate of renewal of mortmass and the stagnant character of the biological cycle.

References

Aleksandrova, V.D. 1958. Experience in the determination of the above- and underground mass of plants in the Arctic tundra. Botan. zhurn. 12:1748-1761. (In Russian.)

Aleksandrova, V.D. 1971. Experience in the determination of the above- and underground phytomass in a polar desert of the Franz Josef Land. In: Biological Productivity and Cycles of Chemical Elements in Plant Communities (Biologicheskaya produktivnost' i krugovorot khimicheskikh elementov v rastitel'nykh soobshchestvakh). Nauka. Leningrad. pp. 33-37. (In Russian.)

Andreev, V.N., and T.F. Galaktionova. 1978. Formation of pools of the above-ground phytomass in communities of subarctic tundra. In: Seasonal and Annual Dynamics of the Phytomass in the Subarctic Tundra (Sezonnaya i pogodnaya dinamika fitomassy v subarkticheskoi tundre). Novosibirsk. Nauka. pp. 164-177. (In Russian.)

Andreyashkina, N.I. 1972. Productivity of communities of hypoarctic shrub and dwarf shrub in the forest tundra of Trans-Urals. Abstract of Ph.D. Thesis (Biol.). 24 pp. (In Russian.)

Andreyashkina, N.I. 1974. Decay and accumulation of plant residues in the forest tundra of Trans-Urals. In: Biomass and Dynamics of the Plant Cover and Animal Population in the Forest Tundra (Biomassa i dinamika pastitel'nogo pokrova i zhivotnogo naseleniya v lesotundre). Sverdlovsk. pp. 129-134. (In Russian.)

Andreyashkina, N.I., and P.L. Gorchakovsky. 1972. Productivity of dwarf shrub and grass communities of the forest tundra and methodology of its estimation. Ekologiya. 3:5-12. (In Russian.)

Baker, J.H. 1972. The rate of production and decomposition of Chorisodontium aciphyllum (Hook. F. & Wills) broth. Brit. Antarc. Survey Bull. 27:123-129.

Bazilevich, N.I. 1993. Biological Productivity of Ecosystems in Northern Eurasia (Biologicheskaya produktivnost' ekosistem severnoi Evrazii). Moscow. 295 pp. (In Russian.)

Bazilevich, N.I. 1993a. Geographic regularities of the indices of biological productivity of soil-plant communities in northern Eurasia, Pochvovedenie. 10:10-18. (In Russian.)

Bazilevich, N.I. 1994. Biological productivity of ecosystems in Siberia. Pochvovedenie. 12:51-56. (In Russian.)

Bazilevich, N.I., O.S. Grebenshchikov, and A.A. Tishkov. 1986. Geographic Regularities of the Structure and Functioning of Ecosystems (Geograficheskie zakonomernosti struktury i funktsionirovaniya ekosistem). Moscow. Nauka. 297 pp. (In Russian.)

Bazilevich N.I., and L.E. Rodin. 1971. Productivity and cycling of elements in natural and cultural phytocenoses. In: Biological Productivity and Cycling of Elements in Plant Communities (Biologicheskaya produktivnost' i krugovorot elementov v rastitel'nykh soobshchestvakh). Leningrad. Nauka. pp. 5-32. (In Russian.)

Berg, B. 1984. Decomposition of moss litter in a mature scots pine forest. Pedobiologi. pp. 301-308.

Berg, B., L. Karenlampi, and A.K. Veum. 1975. Comparisons of decomposition rates measured by means of cellulose. In: F.E. Wielgolaski, ed. Fennoscandian Tundra Ecosystems. Part 1. Plants and Microorganisms. Springer-Ferlag (Ecol. Studies No. 16). Berlin-Heidelberg-New York. pp. 261-276.

Berman, D.I., I.V. Ignatenko, and A.A. Pugachev. 1979. On peat accumulation in intensively drained soils of extreme north-east of the USSR. In: The Biological Cycle in Tundra-Forests of the South of Magadan Region (Biologichesky krugovorot v tundrolesyakh yuga Magadanskoi oblasti). Vladivostok. pp. 134-145. (In Russian.)

Bliss, L.C. 1962. H. Lieth, ed. Die Stoffproduktion der Pflanzendecke. Fischer-Ferlag. Stuttgart. pp. 35-46.

Bliss, L.C. 1970. Primary production within Arctic tundra ecosystems. In: W.A. Fuller and P.G. Kevan, eds. Proc. of the Conf. on Productivity Conservation in Northern Circumpolar Lands, Edmonton, 1969. IUCN pub., new series No. 16. Morges, Switzerland. pp. 75-85.

Bliss, L.C. 1975. Devon Island, Canada. In: T. Rosswall and O.W. Heal, eds. Structure and Function of Tundra Ecosystems. Ecol. Bull. No. 20. Stockholm. pp. 17-60.

Bliss, L.C., J. Svoboda, and D.I. Bliss. 1984. Polar deserts, their plant cover and plant production in the Canadian high Arctic. Holarctic Ecol. 7:304-324.

Bocock, K.L., and O.J.W. Gilbert. 1957. The disappearance of leaf litter under different woodland conditions. Plant and Soil. 9:179-185.

Bogatyrev, L.G. 1973. Biological cycling of ash elements in the dwarf shrub-sedge-moss tundra. In: Soil and Vegetation in Permafrost Regions of the USSR (Pochvy i rastitel'nost' merzlotnykh raionov SSSR). Magadan. DVNTs AN SSSR. pp. 351-356. (In Russian.)

Bogatyrev, L.G. 1976. The biological cycle in tundras of West Taimir. Abstract of Ph.D. Thesis (Biol.). 19 pp. (In Russian.)

Bogatyrev, L.G 1989. On the influence of the freeze-thaw on the leaching of elements from plant litter. In: Interactions of Organisms in Tundra Ecosystems (Vzaimodeistviya organizmov v tundrovykh ekosistemakh). Abstracts of Rpts. Syktyvkar. pp. 156-158. (In Russian.)

Bunnel, F.L., S.F. Maclean, and J. Brown. 1975. Barrow, Alaska, USA. In: T. Rosswall and O.W. Heal, eds. Structure and Function of Tundra Ecosystems. Ecol. Bull. No. 20. Stockholm. pp. 73-124.

Bunnel, F.L., and K.A. Scoullar. 1981. Between-site comparisons of carbon flux in tundra by using simulation models. In: Tundra Ecosystems: a Comparative Analysis. Cambridge Univ. Press. V. 25:685-715.

Bunnel, F.L., D.E.N. Tait, and P.W. Flanagan. 1977. Microbial respiration and substrate weight loss. II. A model of the influences of chemical composition. Soil Biol. and Biochem. 9:41-47.

Chepurko, N.L. 1971. Biological productivity and cycling of chemical elements in forest and tundra communities. In: Biological Productivity and Cycling of Chemical Elements in Plant Communities (Biologicheskaya produktivnost' i krugovorot khimicheskikh elementov v rastitel'nykh soobshchestvakh). Nauka. Leningrad. pp. 213-219. (In Russian.)

Chertovsky, V.G., F.P. Elizarov, B.A. Selinov, and V.S. Kornyak. 1978. Forest growth conditions and productivity of pretundra forests. In: Ecology of Taiga Forests (Ekologiya taezhnykh lesov). Arkhangelsk. pp. 32-43. (In Russian.)

Collins, N.J., J.H. Baker, and P.J. Tilbrock. 1975. Signy Island, maritime Antarctic. In: T. Rosswall and O.W. Heal, eds. Structure and Function of Tundra Ecosystems. Ecol. Bull. No. 20. Stockholm. pp. 345-374.

Dyakonov, K.M., and A.Yu. Reteyum. 1971. Materials on the productivity of tundra and forest tundra on the lower reaches of Ob river. In: Biological Productivity and Cycling of Chemical Elements in Plant Communities (Biologicheskaya produktivnost' i krugovorot khimicheskikh elementov v rastitel'nykh soobshchestvakh). Nauka. Leningrad. pp. 43-46 (In Russian.)

Douglas, L.A., and J.C.F. Tedrow. 1959. Organic matter decomposition rates in Arctic soils. Soil Sci. 88:305-312.

Dowding, P. 1974. Nutrient losses from litter on IBP tundra sites. In: A.J. Holding et al., eds. Soil Organisms and Decomposition in Tundra. Tundra Biome Steering Committee. Stockholm. pp. 363-373.

Elkington, J.J., and B.M.G. Jones. 1974. Biomass and primary productivity of birch (Betula pubescense s. Lat.) in south-west Greenland. J. Ecol. 62:821-830.

Fedorov-Davydov, D.G. 1997. Peculiarities of the biological cycle in the tundra and flood plain landscapes in the lower reaches of Kolyma river. In: The II Int. Conf. on Cryopedology. Abstracts. Syktyvkar. 96.

Flanagan, P.W., and A.K. Veum. 1974. Relationship between respiration, weight loss, temperature and moisture in organic residues in tundra. In: A.J. Holding et al., eds. Soil Organisms and Decomposition in Tundra. Tundra Biome Steering Committee. Stockholm. pp. 249-277.

Gasheva, A.F. 1974. Phytomass pools in some communities of the Harp station. In: Transactions of the Institute of Plant and Animal Ecology, Ural Research Center of the USSR Academy of Sciences. Sverdlovsk. 88:106-107. (In Russian.)

Gorchakovsky, P.A., and N.I. Andreyashkina. 1975. Study of the primary production of forest tundra communities at the Harp station. In: Resources of the Biosphere (Resursy biosfery). Leningrad. Nauka. 1:25-33. (In Russian.)

Grishina, L.A. 1986. Humus Formation and Humus Status of Soils (Gumusoobrazovanie i gumusnoe sostoyanie pochv). Izd. MGU. Moscow. 243 pp. (In Russian.)

Grishkan, I.B. 1995. Decay of plant litter in major communities of the upper Kolyma basin, Ekologiya. 1:7-10. (In Russian.)

Grishkan, I.B. 1997. Mycobiota and Biological Activity of Soils on the Upper Reaches of Kolyma River (Mikobiota i biologicheskaya aktivnost' pochv Verkhnei Kolymy). Vladivostok. 135 pp. (In Russian.)

Heal, O.W. 1972. Decomposition studies in tundra. In: F.E. Wielgolaski and T. Rosswall, eds. Proc. Int. Meeting on Biological Productivity of Tundra. Leningrad, October, 1971. pp. 93-97.

Heal, O.W., and D.D. French. 1974. Decomposition of organic matter in tundra. In: A.J. Holding et al., eds. Soil Organisms and Decomposition in Tundra. Tundra Biome Steering Committee. Stockholm. pp. 279-309.

Heal, O.W., G. Howson, D.D. French, and J.N.B. Jeffers. 1974. Decomposition of cotton strips in tundra. In: A.J. Holding et al., eds. Soil Organisms and Decomposition in Tundra. Tundra Biome Steering Committee. Stockholm. pp. 341-362.

Heal, O.W., P.W. Flanagan, D.D. French, and Ir.S.F. MacLean. 1981. Decomposition and accumulation of organic matter in tundra. Tundra Ecosystems. A Comparative Analysis, V. 25. Cambridge Univ. Press. pp. 587-633.

Ignatenko, I.V, and F.I. Khakimzyanova. 1971. Soils and total pools of phytomass in birch-dryad and willow tundras of the East-European North. Ekologiya. 4:17-24. (In Russian.)

Ignatenko, I.V., A.V. Knorre, and B.N. Norin, B.N. 1972. Standing corp in plant communities at the station Ary-Mas. In: Tundra Biome. Proc. of the IV Int. Meeting on Biological Productivity of Tundra (Leningrad). Stockholm. pp. 140-149.

Ignatenko, I.V., B.N. Norin, and A.T. Rakhmanina.1973. Cycling of ash elements and nitrogen in some biogeocenoses of the East-European forest tundra. In: Soils and vegetation in permafrost regions of the USSR (Pochvy i rasitetel'nost'merzlotnykh raionov SSSR). DVNTs AN SSSR. Magadan. pp. 335-350. (In Russian.)

Ignatenko, I.V, and A.A. Pugachev. 1979. Dynamics of the plant mass and the biological cycle in alpine tundra and Siberian dwarf-pine landscapes of the northern coast of the Okhotsk Sea. In: The Biological Cycle in Tundra Forests in the South of Magadan Region (Biologichesky krugovorot v tundrolesyakh yuga Magadanskoi oblasti). Vladivostok. pp. 92-124. (In Russian.)

Ignatenko, I.V., and A.A. Pugachev. 1980. The biological cycle in biogeocenoses of the Okhotsk Sea mountain province. Pochvovedenie. 8:43-54 (In Russian.)

Ignatenko, I.V., and A.A. Pugachev. 1981. The biological cycle in alpine tundras of the northern coast of the Okhotsk Sea. Ekologiya. 6:82-85 (In Russian.)

Jenny, H, S.P. Gessel, and F.T. Bibngham. 1949. Comparative study of decomposition rate of organic matter in temperature and tropic regions. Soil Sci. 68:419-432.

Khantimer, I.S. 1970. Increment of the above-ground phytomass in virgin and economically disturbed tundra. In: Productivity of Biogeocenoses of Subarctic (Produktivnost' biogeotsenozov Subarktiki). Ural. fil. AN SSSR. Sverdlovsk. pp. 84-86. (In Russian.)

Khlynovskaya, N.I., A.I. Nesterenko, and N.E. Podkovyrkina. 1988. Pools of total phytomass of major vegetation types in landscapes of the northeastern USSR. In: Problems of Studies and Protection of Landscapes in the Northeastern USSR (Problemy izucheniya i okhrany landshaftov Severo-Vostoka SSSR). DVNTs AN SSSR. Vladivostok. pp. 8-27 (In Russian.)

Khodachek, E.A. 1969. Plant mass of phytocenoses in western Taimir. Botan. zhurn. 54:1059-1073. (In Russian.)

Kovda, V.A. 1985. Biogeochemistry of the Soil Cover (Biogeokhimiya pochvennogo pokrova). Nauka. Moscow. 264 pp. (In Russian.)

Lahde, E. 1974. Rate of decomposition of cellulose in forest soils in various parts of the Nordic countries. In: Reports from the Kevo Subarctic Station. 11:72-78.

Manakov, K.N. 1970. Elements of the Biological Cycle in the Polar North (Elementy biologicheskogo krugovorota na polyarnom Severe). Nauka. Leningrad. 160 pp. (In Russian.)

Manakov, K.N. 1972. Productivity and the Biological Cycle in Tundra Biogeocenoses of the Kola Peninsula (Produktivnost' i biologichesky krugovorot v tundrovykh biogeotsenozakh Kol'skogo poluostrova). Nauka. Leningrad. 148 pp. (In Russian.)

Manakov, K.N. 1974. Cycling of mineral elements in thin birch forests of forest tundra of the Kola peninsula. In: Biological Processes and Mineral Exchange in Soils of the Kola Peninsula (Biologicheskie protsessy i mineral'ny obmen v pochvakh Kol'skogo poluostrova). Kol. fil. AN SSSR. Apatity. pp. 110-117 (In Russian.)

Manakov, K.N., and V.V. Nikonov. 1979. Regularities of the biological cycle of mineral elements and soil formation in biogeocenoses of three alpine vegetation zones. In: Soil Formation in Biogeocenoses of the Khibiny Mountains (Pochvoobrazovanie v biogeotsenozakh Khibinskikh gor). Kol. fil. AN SSSR. Apatity. pp. 65-94 (In Russian.)

Mishustin, E.N., and A.N. Petrova. 1963. Determination of biological activity of soil, Mikrobiologiya. 32,3:479-483. (In Russian.)

Naplekova, N.N. 1974. Aerobic Degradation of Cellulose by Microorganisms in Soils of West Siberia (Aerobnoe razlozhenie tsellulosy mikroorganizmami v pochvakh Zapadnoi Sibiri). Nauka. Moscow. 250 pp. (In Russian.)

Naumov, A.V., and E.N. Naumova 1993. Decomposition of plant root mass in "young" soils of the Kansk-Achinsk Thermal Energy Complex (KATEC). Pochvovedenie. 5:47-55. (In Russian.)

O'Lear, H.A., and T.R. Seasteadt. 1994. Landscape patterns of litter decomposition in alpine tundra. Oecologia. 1-2:95-101.

Orlov, D.S., O.N. Biryukova, and N.I. Sukhanova. 1996. Organic Matter in Soils of the Russian Federation (Organicheskoe veshchestvo pochv Possiyskoi Federatsii). Nauka. Moscow. 256 pp. (In Russian.)

Pavlova, E.B. 1969. On the plant mass in tundras of western Taimir. Vestn. MGU. Ser. Biol. 5:62-67. (In Russian.)

Parinkina, O.M. 1980. The rate of organic matter decomposition in tundras of the Taimir Peninsula. In: Biogeocenoses of the Taimir Tundra (Biogeotsenozy Taimyrskoi tundry). Leningrad. pp. 234-249. (In Russian.)

Parinkina, O.M. 1989. The Microflora of Tundra Soils (Mikroflora tundrovykh pochv). Nauka. Leningrad. 159 pp. (In Russian.)

Parinkina, O.M., V.N. Pereverzev, and T.Kh. Piyn.1994. Peculiarities of decomposition of the abovesoil lichens under conditions of alpine tundra, northern and southern taiga. Pochvovedenie. 5:42-48 (In Russian.)

Pereverzev, V.N. 1987. Biochemistry of Humus and Nitrogen in Soils of the Kola Peninsula (Biokhimiya gumusa i azota pochv Kol'skogo poluostrova). 303 pp. (In Russian.)

Pereverzev, V.N. 1988. The decomposition rate of plant residues and humus formation in soils of the Khibiny mountains, Pochvovedenie. 2:68-77. (In Russian.)

Pereverzev, V.N., and N.S. Alekseeva. 1980. Organic Matter in Soils of the Kola Peninsula (Organicheskoe veshchestvo v pochvakh Kol'skogo poluostrova). Nauka. Leningrad. 227 pp. (In Russian.)

Pospelova, E.B., and V.D. Vasiljevskaja. 1985. Primary production of Taimir tundras. Botan. Zhurn. 2:188-198. (In Russian.)

Pospelova, E.B., and M.V. Orlov. 1984. Pools and structure of the phytomass of spotty tundras on the eastern coast of Taimir Lake. Ekologiya. 1:14-21. (In Russian.)

Pospelova, E.B., and Yu.G.Zharkova. 1984. The vegetation cover and phytomass of plant communities at the Agapa station. In: Soils and Productivity of Plant Communities (Pochvy i produktivnost' rastitel'nykh soobshchestv). Izd. MGU. Moscow. 1:14-21 (In Russian.)

Pugachev, A.A. 1984. Pools of biomass and elements of the biological cycle in communities of the alpine tundra zone. In: The Soil Layer of Alpine Tundra Ecosystems of the Annagach Ridge (the upper reaches of Kolyma river (Pochvennyi yarus ekosistem gornykh tundr khrebta Bol'shoi Annagach (verkhovya Kolymy)). DVNTs AN SSSR. Vladivostok. pp. 134-159 (In Russian.)

Rakhmanina, A.T. 1971. Above- and underground phytomass of some communities in the East-European forest tundra. In: Biological Productivity and Cycling of Chemical Elements in Plant Communities (Biologicheskaya produktivnost' i krugovorot khimicheskikh elementov v rastitel'nykh soobshchestvakh). Nauka. Leningrad. pp. 7-42 (In Russian.)

Richard, F. 1945. Der Biologische Abbau von Zellulose- und Eiweiss-Testschnuren im Boden für das Forstliche Versuchswesen. 24:297-397.

Rosswall, T. 1973. Plant litter decomposition studies at the Swedish tundra site. In: Swedish Tundra Biome Technical Report. 14:129-133.

Rosswall, T. 1974. Cellulose decomposition studies on the tundra. In: A.J. Holding et al., eds. Soil Organisms and Decomposition in Tundra. Tundra Biome Steering Committee. Stockholm. pp. 325-340.

Rosswall, T., J.G.K. Flower-Ellis, L.G. Johansson, S. Jonsson, B.E. Ryden, and M. Sonesson. 1975. Stordalen (Abisko), Sweden. In: T. Rosswall and O.W. Heal, eds. Structure and Function of Tundra Ecosystems. Ecol. Bull. No. 20. Stockholm. pp. 265-294.

Rosswall, T., A.K Veum, and L. Karenlampi. 1975a. Plant litter decomposition at Fennoscandian tundra sites. In: F.E. Wielgolaski, ed. Fennoscandian Tundra Ecosystems. Part 1. Plants and Microorganisms. (Ecol. Studies, No. 16). Springer-Ferlag. Berlin-Heidelberg-New York. pp. 268-278.

Shamurin, V.F. 1970. Pools of phytomass in some tundra communities of Vorkuta region. In: Biological Bases for the Use of Natural Resources in the North (Biologicheskie osnovy ispol'zovaniya prirody Severa). Kn. Izd. Komi SSR. Syktyvkar. pp. 25-29. (In Russian.)

Shamurin, V.F., T.G.Polosova, and E.A. Khodachek. 1972. Plant biomass of main plant communities at the Tareya station (Taimir). In: Tundra Biome. Proc. of IV Int. Meeting on Biological Productivity of Tundra (Leningrad). Stockholm. pp. 163-182.

Shaver, G.R., and S.F. Chapin III. 1991. Production: biomass relationships and element cycling in contrasting Arctic vegetation types. Ecol. Monographs. 61:1-31.

Sonesson, M. 1973. Studies in production and turnover of bryophytes at Stordalen 1972. Swedish Tundra Biome Tech. Rpt. 14:66-75.

Smorodnikova, N. Yu. 1983. Pools and structure of plant mass in the Arctic tundra of Vrangel Island. In: Biological Problems of the North (Biologicheskie problemy Severa). Abstracts of X All-Union Symp. Magadan, Part 1. pp. 210-211 (In Russian.)

Stenina, T.A., 1978. Biological activity of soils in the Vorkuta tundra. Pochvovedenie. 10:59-64. (In Russian.)

Stenina, T.A., A.V.Kononenko, and A.N.Tsypanova. 1979. Phytomass of tundra communities in the Vorkuta region. In: Biological Productivity of Soils and Its Increase in the

Interests of People's Economy (Biologicheskaya produktivnost' pochvi ee uvelichenie v interesakh narodnogo khozyaistva). Abstracts of All-Union Meeting, Dec. 18-20, 1979. Izd. MGU. Moscow. pp. 150-153. (In Russian.)

Trotsenko, G.V. 1974. On pools of phytomass in major plant communities of forest tundra in the north of the Ob river basin. In: Biological Problems of the North (Biologicheskie problemy Severa). Abstracts of VI Symp., Yakutsk, No. 3: Botany and Plant Resources. pp. 188-192. (In Russian.)

Van Cleve, K. 1974. Organic matter quality in relation to decomposition. In: A.J. Holding et al., eds. Soil Organisms and Decomposition in Tundra. Tundra Biome Steering Committee. Stockholm. pp. 311-324.

Vassiljevskaja, V.D., V.V. Ivanov, L.G. Bogatyrev, E.V. Pospelova, N.M. Shalaeva, and L.A. Grishina. 1975. Agapa, USSR. In: T. Rosswall and O.W. Heal, eds. Structure and Function of Tundra Ecosystems. Ecol. Bull. No. 20. Stockholm. pp. 141-158.

Vikhreva-Vasil'kova, V.V., V.A. Gavrilyuk, and V.F. Shamurin. 1964. The above- and underground mass of some dwarf shrub communities in the Koryak Land. In: Problems of the North (Problemy Severa). Moscow-Leningrad. pp. 137-147. (In Russian.)

Vil'chek, G.E. 1984. On the productivity of tundra biogeocenoses in the vicinity of Kharasway Cap. Biol. nauki, Botanika. 7:67-71. (In Russian.)

Vil'chek, G.E. 1986. Productivity of some biocenoses in Vorkuta tundras. Ekologiya. 2:8-13. (In Russian.)

Vil'chek, G.E. 1987. Productivity of typical Taimir tundras. Ekologiya. 5:38-43. (In Russian.)

Vostrov, I.S., and A.N. Petrova. 1961. Determination of biological activity of soils by various methods. Mikrobiologiya. 30,4:665-672. (In Russian.)

Webber, P.J. 1974. Tundra primary productivity. In: J.D. Ives and R.G. Barry, eds. Arctic and Alpine Environment. Methuen. London. pp. 445-473.

Widden, P. 1977. Microbiology and decomposition on the Truelove Lowland. In: L.C. Bliss, ed. Trulove Lowland, Devon Island, Canada: a High Arctic Ecosystem. Univ. Alberta Press. Edmonton. pp. 505-530.

Witkamp, M. 1966. Rates of carbon dioxide evolution from the forest floor. Ecol. 47:492-493.

Zamolodchikov, D.G., D.V. Karelin, and N.V. Zukert. 12001. An empirical model of carbon fluxes in Russian tundra. Global Change Biol. 7:147-162.

Zlotin, R.I., A.A. Tishkov, and O.N. Shubnikov. 1989. Biocenological relationships in the Arctic tundra ecosystem (Spitsbergen). In: Interactions of Organisms in Tundra Ecosystems (Vzaimodeistviya organizmov v tundrovykh ekosystemakh). Abstracts. Syktyvkar. pp. 214-217. (In Russian.)

Zyryanov, V.A., and L.A. Kolpashchikov. 1981. Plant resources and their use by the wild reindeer in western Taimir. In: Ecology and Economic Use of the Terrestrial Fauna of the Yenisei North (Ekologiya i khozyaistvennoe ispol'zovanie nazemnoi fauny Yeniseiskogo Severa). SO VASKhNIL. Novosibirsk. pp. 27-38. (In Russian.)

Chapter 3. Soil Organic Matter Storage in Cold Soils of Coastal Eastern Antarctica (Casey Station, Wilkes Land)

Lothar Beyer[1], Kristina Pingpank[2], Manfred Bölter[3], and Rodney D. Seppelt[4]

[1]Professor, Institute for Polar Ecology, Christian-Albrechts-University Kiel, Wischhofstr. 1-3, Bldg. 12, 24148 Kiel, Germany; e-mail soil_science@web.de.
[2]Ph.D., Institute for Plant Nutrition and Soil Science, Christian-Albrechts-University Kiel, Olshausenstr. 40, 24098 Kiel, Germany.
[3]Professor, Institute for Polar Ecology, Christian-Albrechts-University Kiel, Wischhofstr. 1-3, Bldg. 12, 24148 Kiel, Germany; e-mail mboelter@ipoe.uni-kiel.de.
[4]Professor Dr., Institute of Arctic Biology, University of Alaska, Fairbanks, and Australian Antarctic Division, Channel Highway, Kingston 7050, Tasmania, Australia; e-mail rod.seppelt@antdiv.gov.au.

1. Introduction

Recent literature regards organic matter and its accumulation as of minor importance to soil's formation in Antarctica (Bockheim and Ugolini, 1990). This is probably why the global map of distribution of soil organic carbon leaves Antarctica completely white (Batjes, 1995). However, most of the recent data came from soils of dry valleys with a very low humus content (Campbell and Claridge, 1987; Claridge and Campbell, Section 4, Chapter 6). In contrast, Blume et al. (1997) showed that the carbon and nitrogen content of the top soils of the southern circumpolar region is much greater than expected (see also Blume and Bölter, 1993; Blume et al., 1996).

This chapter presents data about soils close to the Australian Casey Station, to specify the carbon and nitrogen storage in the soils of coastal Antarctica and discuss the relationship between the accumulation of soil organic matter (SOM) and soil development.

2. Sites

Casey Station is on the coast of eastern Antarctica. The mean annual precipitation is 170 mm, mainly snow, and the average temperature is -9°C. Spodic Haplic or Leptic and Gleyic Turbic Cryosols and Gelic Histosols are typical in this area (Blume and Bölter, 1993; Blume et al.; 1997, Beyer et al., 2000). Four representative sites (10 x 10 m^2) for collecting soil material were installed near the station.

3. Methods

Soil samples were collected at the four sites at every square meter and from representative profiles in January and February 1996. Analytic methods were in accordance with Schlichting et al. (1995). Carbon was determined after dry combustion with a Ströhlein apparatus. Nitrogen was digested with the Kjeldahl procedure and measured in a flow injection analyzer. The pH values were determined in a 10 mM $CaCl_2$ solution. The total mass of carbon and nitrogen were calculated using the carbon and nitrogen contents in the fine earth, the bulk density, the horizon depth, and the stone content.

4. Geomorphology and Soils

Bare rock or rock fragments from gneiss and schist covered 25 to 56% of the surface (Table 4.3.1) (Paul et al., 1995). The weathered parent material accumulated in small valleys and depressions and initially formed the soil, forming A-C soils (Beyer et al., 2000). In addition, loamy fine earth material separated from sand and gravel, due to intensive cryoturbation (Blume et al., 1997). This was visible at all sites as stone rings and solifluction tongues. The permanent movement of the soil material inhibited plant colonization (Washburn, 1980). In this case, loose soil material reflecting recent cryoturbation covered the surface (Table 4.3.1).

Table 4.3.1. Summarized geomorphological surface features of the investigated sites.

Site	Bare rock and rock fragments	Gravel pavement	Loose soil material without vegetation	Loose soil material with vegetation
	% of total surface			
A	47	NE	27	26
B	56	NE	26	18
C	28	NE	12	48
D	25	27	NE	48

NE: no occurrence.

Table 4.3.2. Soil geography of the investigated sites A-D (100 m^2 plots) and the total landscape (overall) and percentage standard deviation ($\sigma_{\%}$).

Soil Unit Soil Survey Staff , 1998; *(WRB-FAO, 1998)*	**Site** A	 B	 C	 D	**Over all**	$\sigma_{\%}$
	----------- % of covered area ----------					
Orthels & Turbels *(Leptic Cryosols)*	47.0	26.4	12.8	47.9	33.5	44
Loamy-skeletal Lithic Haplorthel or Haploturbel *(Haplic or Turbic Leptic Cryosol)*	27.0	9.0	12.8	47.9	24.2	63
Sandy-skeletal, Lithic Haplorthel *(Haplic Leptic Cryosol)*	20.0	17.4	NE	NE	9.4	100
Orthels (all podzolic) *(Spodic Leptic Cryosols)*	1.5	NE	51.2	26.8	19.9	106
Sandy-skeletal, Lithic Spodorthel[1] *(Spodic Leptic Cryosol)*	NE	NE	51.2	NE	22.2	173
Loamy-skeletal, Typic Spodorthel[1] *(Spodic Haplic Cryosol)*	NE	NE	NE	26.8	6.7	173
Sandy, Lithic Spodorthel[1] *(Spodic Leptic Cryosol)*	1.5	NE	NE	NE	0.4	173
Aquaturbels *(Gleyic Turbic Cryosols)*	4.3	17.5	7.8	NE	7.4	87
Loamy-skeletal, Typic Aquaturbel *(Gleyic Turbic Cryosol)*	NE	16.3	NE	NE	4.1	173
Loamy-skeletal, Lithic Aquaturbel *(Gleyic Turbic Leptic Cryosol)*	4.3	1.2	NE	NE	1.4	128
Fine-loamy, Lithic Aquaturbel *(Gleyic Turbic Leptic Cryosol)*	NE	NE	7.8	NE	2.0	173
Bare rock/rock fragments	47.2	56.1	28.2	25.2	39.2	33

NE: no occurrence.
[1]Taxa proposed by Beyer et al. (1999b).

Geomorphological features remained visible when cryogenic processes were relic, but lower plants (lichens [e.g., *Usnea sphacelata, Usnea antarctica*] and mosses [e.g., *Ceratodon purpurea*]) covered the soil surface (Seppelt and Broady, 1988; Smith, 1990; Melick et al., 1994a). At site D, the gravel pavement reveals an abandoned penguin rookery, which is typical for coastal Antarctica (Campbell and Claridge, 1987; Smith, 1990). Obviously the material was too coarse and the time since the recent colonization by the penguins was too short for plants to have adapted.

Table 4.3.2 summarizes the results of the detailed soil survey at sites A-D. The solum was, in general, <25 cm (Beyer et al., 2000) and therefore, in most cases, the soils were classified as Leptic Cryosols. These predominant soil units had a sandy and loamy skeletal texture.

Very often, the loamy fine earth material separated from sand and gravel due to intensive cryoturbation (Blume et al., 1997). Soils having such recent cryogenic features were classified as Turbic Leptic Cryosols, in contrast to the Haplic Leptic Cryosols with no cryogenic features (WRB-FAO, 1998). Additionally in the loamy parent materials, frequently aquic conditions (Soil Survey Staff, 1998) induced redoximorphic properties, due to stagnic water. In such cases, soil formation produced Gleyic Turbic (Leptic) Cryosols.

Spodic Haplic or Leptic Cryosols occurred at sites A and C and especially at the abandoned penguin rookery at site D. At foot slopes, in small depressions, or along some surface meltwater channels, frequently small patches of Gelic Histosols occur (Beyer et al., Section 4, Chapter 4; Beyer, 2000). However, they did not appear at plots A-D.

Blume et al. (1997), Beyer et al. (1997a), Beyer and Bölter (2000), and Beyer et al. (2000) discuss soil formation processes and soil properties in detail. In summary, the soil survey indicates a great variability of soil formation and an extraordinary small-distance variation in soil geography (see Table 4.3.2, $\sigma_{\%}$). This suggests a possibly wide range in soil ecological properties and carbon and nitrogen stocks which might affect the adaptation of soil organisms, the nutrient supply, and plant colonization and might reflect microbial turnover rates.

Surprisingly, Table 4.3.3 shows no significant differences in the pH values of the top soils. The mean was between 4.0 and 4.5 in all soils. Regardless of the site, the frequency distribution was very similar. In the Gelic Histosols, the variation was somewhat higher due to the formation of different organic layers: the fibric Oi had a mean pH of 4.5, the hemic Oe, 4.2. and the weakly sapric Oa, 3.7 (Beyer, 2000). The latter occasionally dropped below 3.6. This is why most of the soils belong taxonomically to the dysic nutrient regime (Soil Survey Staff, 1998; Beyer et al., 2000).

5. Carbon and Nitrogen Storage

The content of total organic carbon (TOC) in the top soil ranged from <5 mg/g to >50 mg/g (Table 4.3.4). Occasionally, in small depressions, the TOC content was close to the content of peaty horizons (especially at sites B and C; compare data in Table 4.3.5). However, the mean of the Histosol layers ranged between 200 and 320 mg/g, depending on the occurrence of fibric (Oi), hemic (Oe), or weakly sapric (Oa) materials (Beyer, 2000).

In the loamy soils without vegetation, at the inner center of the stone rings and solifluction tongues (site A), TOC was lower. A certain mineral soil unit (Cryosol subgroup) did not indicate a specific TOC content. Especially at sites B and C, we could distinguish Haplic Leptic Cryosols and SOM-enriched Leptic Cryosols (Beyer et al., 2000).

Table 4.3.3. Mean (∅) pH value ($CaCl_2$) in top soils (0 to 5 cm) and frequency distribution at different soil units and landscape sites (A-D).

Soil Unit (WRB-FAO, 1998)	**Site**	**N**	**∅**	**3.6-4.0**	**4.1-4.5**	**4.6-5.0**
Leptic Cryosol	all	203	4.3	16	74	10
Haplic or Turbic Leptic Cryosol	A	27	4.4	0	78	22
(loamy-skeletal)	B	16	4.1	50	50	0
	C	21	4.3	19	71	10
	D	75	4.3	5	86	9
Mean (site A-D)	∅	139	4.3	12	76	12
Haplic Leptic Cryosol (sandy-skeletal)	A	31	4.2	19	78	3
	B	33	4.2	30	58	12
Mean (site A-B)	∅	64	4.2	25	67	8
Spodic Cryosol	all	99	4.3	18	59	23
Spodic Leptic Cryosol (sandy-skeletal)	C	70	4.4	6	61	33
Spodic Haplic Cryosol (loamy-skeletal)	D	24	4.1	46	54	0
Spodic Leptic Cryosol (sandy)	A	5	4.0	60	40	0
Gleyic Turbic Cryosol	all	42	4.2	12	76	12
Gleyic Turbic Cryosol (loamy-skeletal)	B	23	4.3	4	83	13
Gleyic Turbic Leptic Cryosol (loamy-skeletal)	A	9	4.4	13	63	24
	B	2	4.3	0	100	0
Gleyic Turbic Leptic Cryosol (fine-loamy)	C	8	4.2	25	75	0
Gelic Histosol	NE	6[A]	4.1	48[B]	19	33

NE: no occurrence.
[A]Six sites with 21 organic layers.
[B]Incl. 10% <pH 3.6.

In the sandy textured soils, we found no Leptic Cryosols with a TOC content exceeding 50 mg/g, but we did with the loamy soils (Table 4.3.4). The Spodic Cryosols showed also a great variation in TOC content but did not significantly differ from the mean of the non-spodic Cryosols. In contrast, except for the fine loamy Gleyic Cryosols with a low gravel content and a thick moss cover, TOC in the top soils of the Gleyic Cryosols was significantly lower, as in most other mineral soils; and the surface of other Gleyic Cryosols showed no vegetation at all (Beyer et al., 2000).

We observed a positive, well correlated pattern of TOC in relation to nitrogen contents (N_t) (Beyer et al., 2000). For this reason, the mean C/N ratio was much similar in most of the top soils, close to 10, with little variation (Table 4.3.6). Only in the upper organic layers of the Gelic Histosols was the C/N ratio wider; it reached the level of the mineral soils already in the sapric materials of the Oa horizons (Table 4.3.7). Some maxima exceeded 35 in the mineral soils (Table 4.3.6), as well as in the Oi layers of the Gelic Histosols.

Table 4.3.4. Mean (∅) organic carbon content (TOC) in top soils (0 to 5 cm) and frequency distribution at different soil units and landscape sites (A-D).

Soil Unit (WRB-FAO, 1998)	**Site**	**N**	**∅** mg/g	**Percentage** (mg carbon/g fine earth) <5	-10	-20	-30	-50	-100	-150
Leptic Cryosol	all	203	19	4	33	36	13	8	4	1
Haplic or Turbic Leptic Cryosol	A	27	8	15	75	7	4	0	0	0
(loamy-skeletal)	B	16	37	0	0	38	19	25	19	0
	C	21	45	5	24	24	0	10	24	14
	D	75	14	3	21	43	7	1	0	0
Mean (site A-D)	∅	139	20	6	33	40	8	5	6	2
Haplic Leptic Cryosol	A	31	14	0	48	16	35	0	0	0
(sandy-skeletal)	B	33	20	0	18	39	25	27	0	0
Mean (site A-B)	∅	64	17	0	33	28	15	14	0	0
Spodic Cryosol	all	99	22	7	17	29	26	15	5	0
Spodic Leptic Cryosol (sandy-skeletal)	C	70	19	10	24	36	17	6	7	0
Spodic Haplic Cryosol (loamy-skeletal)	D	24	27	0	0	17	58	25	0	0
Spodic Leptic Cryosol (sandy)	A	5	45	0	0	0	0	100	0	0
Gleyic Turbic Cryosol	all	42	9	28	42	19	7	2	2	0
Gleyic Turbic Cryosol (loamy-skeletal)	B	23	5	35	48	17	0	0	0	0
Gleyic Turbic Leptic Cryosol	A	9	6	44	44	12	0	0	0	0
(loamy-skeletal)	B	2	5	0	50	50	0	0	0	0
Gleyic Turbic Leptic Cryosol (fine-loamy)	C	8	24	0	25	25	37	0	13	0

Table 4.3.5. Mean (∅) organic carbon content (TOC) and its standard deviation in typical layers of Gelic Histosols, and the percentage of TOC frequency distribution.

Horizon	N	∅	σ	Percentage				
				<150	-200	-300	-400	>400
				mg carbon/g fine earth				
Oi	6	324	89	0	17	33	17	33
Oe	9	274	73	0	22	44	33	0
Oa	6	203	46	17	50	33	0	0

Tables 4.3.8 and 9 reflect the wide range of the TOC and N_t storage in the soil solum of whole profiles at the very small scale of 1 m^2. The carbon storage in the mineral soils ranged from 0.3 to 8.2 kg/m^2, and the nitrogen storage, from 21 to 736 g/m^2. In the Histosols the stocks were much higher: TOC, from 5.2 to 45.6 kg/m^2; N_t, from 0.2 to 2.6 kg/m^2 (Beyer et al., 1995; Blume et al., 1996; Blume et al., 1997; Beyer, 2000). The storage data of Tables 4.3.8 and 4.3.9 confirm the great variation within the non-gleyic Cryosols (haplic, leptic, spodic, turbic) and within one subgroup and/or at the same site. Among the four sites, the mean TOC storage ranged from 1.8 to 4.0 kg/m^2 and the mean N_t storage, from 180 to 360 g/m^2, a deviation of more than 60%.

Overall, the C-to-N ratio ranged from 5 to 26 (Table 4.3.10). In the mineral soils, carbon and nitrogen storage strongly correlated ($r = 0.786$, $n = 21$). However, in cases of high accumulation of organic matter, the C/N ratio was extraordinarily high (Beyer et al., 1995, 1997b) or extremely low, with an evident nitrogen input by penguins (Blume et al., 1997; Beyer et al., 1997a).

We found no correlation between C and N storage and depth of the solum in the Cryosols ($r = -0.1269$), which ranged from 6 to 50 cm (Beyer et al., 2000), or in the Gelic Histosols (Beyer, 2000). For the mineral soils, this means the Leptic Cryosols do not have lower amounts of TOC and N_t as do their haplic counterparts. In addition, there was no correlation to the stone content in the soils ($r = -0.201$), which ranged from 16 to 83% (Beyer et al., 2000), indicating no lower amounts with a high gravel content.

Table 4.3.6. Mean C/N ratio in top soils (0 to 5 cm) and frequency distribution at different soil units and landscape sites (A-D).

Soil Unit (WRB-FAO, 1998)	Site	N	∅	Percentage ≤7	8-10	11-13	14-19	20-25	26-35	>35
Leptic Cryosol	all	203	12	10	60	20	4	<1	1	4
H. or T. Leptic Cryosol	A	27	12	48	22	15	0	0	8	7
(loamy-skeletal)	B	16	16	0	50	31	0	0	0	19
	C	21	14	5	24	33	29	5	0	5
	D	75	10	4	67	27	4	0	0	0
Mean (site A-D)	∅	139	12	12	50	26	6	1	1	4
Haplic Leptic Cryosol	A	31	9	6	88	6	0	0	0	0
(sandy-skeletal)	B	33	12	3	79	0	0	0	0	9
Mean (site A-B)	∅	64	11	5	82	8	0	0	0	5
Spodic Cryosol	all	99	10	9	71	19	1	0	<1	<1
Spodic Leptic Cryosol (sandy-skeletal)	C	70	9	14	72	10	3	0	0	1
Spodic Haplic Cryosol (loamy-skeletal)	D	24	10	0	58	42	0	0	0	0
Spodic Leptic Cryosol (sandy)	A	5	14	0	40	40	0	0	20	0
Gleyic Turbic Cryosol	all	42	9	17	58	12	2	0	0	2
Gleyic Turbic Cryosol (loamy-skeletal)	B	23	8	9	82	9	0	0	0	0
Gleyic T. Leptic Cryosol	A	9	11	44	44	0	0	0	0	12
(loamy-skeletal)	B	2	9	0	10	0	0	0	0	0
Gleyic T. Leptic Cryosol(fine-loamy)	C	8	10	13	37	37	0	0	0	0

H.: Haplic; T.: Turbic.

Table 4.3.7. Mean (∅) C/N ratio and its standard deviation in typical layers of Gelic Histosols, and the percentage of C/N frequency distribution.

Horizon	N	∅	σ	Percentage <10	10-13	14-19	20-25	26-35	>35
Oi	6	29	10	0	0	33	33	17	17
Oe	9	19	5	11	0	44	33	11	0
Oa	6	14	3	17	50	33	0	0	0

Table 4.3.8. Total organic carbon (TOC in kg per m^2) storage in the soil solum (range 7 to 50 cm) of the investigated profiles sites, means overall soils and sites, and the percentage of standard deviation ($\sigma_{\%}$).

Soil unit (WRB-FAO, 1998)	Site A	Site B	Site C	Site D	**Mean** Over all	$\sigma_{\%}$
H. or T. Leptic Cryosol					3.2	74
loamy-skeletal	1.4	3.4, 3.6, 8.2	6.1, 0.3	1.4	3.4	76
sandy-skeletal	1.6	3.3	NE	NE	2.4	36
Spodic Cryosol					4.6	55
Leptic, sandy-skeletal	NE	NE	3.5	NE	NC	NC
Haplic, loamy-skeletal	NE	NE	NE	2.2	NC	NC
Leptic, sandy	8.2	NE	NE	NE	NC	NC
Gleyic Turbic Cryosol					2.8	57
loamy-skeletal	NE	0.8	NE	NE	NC	NC
Leptic, loamy-skeletal	4.4	4.4	NE	NE	4.4	<1
Leptic, fine-loamy	NE	NE	1.8	NE	NC	NC
Mean overall Cryosols (∅)	3.9	4.0	2.9	1.8	3.4	69
Standard deviation ($\sigma_{\%}$)	71	57	74	23	56	NC
Gelic Histosols[A]	NE	NE	NE	NE	25.9	49

NE: no occurrence; NC: not calculated; T.: Turbic; H.: Haplic.
[A]Six sites outside the plots A-D.

With respect to the soil units, the correlation with carbon and nitrogen was weak; however, we observed a sequence tendency of the mean storage: Gleyic Turbic (Leptic) Cryosol = Haplic or Leptic Cryosol < Spodic (Haplic or Leptic) Cryosol < SOM-enriched (Haplic or Leptic) Cryosol << Gelic Histosol.

Tables 4.3.11 and 4.3.12 compare the summarized TOC and N_t stocks in the soils of the 100 m^2 A-D plots to soils as carbon and nitrogen sinks at a landscape level. Site C showed the highest TOC and N_t stock of the four sites. Its extreme value, caused by the extraordinary occurrence of Spodic and SOM-enriched Cryosols, especially affected the TOC storage variability of 42% (Table 4.3.11) (Beyer et al., 2000). However, comparing landscape level with square meter level of a single profile decreases the variability from 69% to 42% for carbon (compare Table 4.3.8 and 4.3.11, $\sigma_{\%}$) and from 62% to 31% for nitrogen (compare Table 4.3.9 and 4.3.12, $\sigma_{\%}$). Excluding site C, the variability dropped for carbon storage to below 10% (Table 4.3.11) and for nitrogen to 25% (Table 4.3.12). These observations indicate that a detailed soil survey enables a more accurate estimation of the carbon and nitrogen stocks. The results suggest that, at 75% of the landscape sites, the carbon and nitrogen stocks are very similar.

Table 4.3.9. Total nitrogen (N_t in g/m^2) storage in the soil solum (range 7 to 50 cm) of the investigated profiles sites, means overall soils and sites, and the percentage of standard deviation ($\sigma_{\%}$).

Soil Unit (WRB-FAO, 1998)	Site A	Site B	Site C	Site D	Mean Over all	$\sigma_{\%}$
H. or T. Leptic Cryosol					283	72
loamy-skeletal	82	347, 288, 736	285, 21	82	263	76
sandy-skeletal	169	342	NE	NE	256	34
Spodic Cryosol					324	11
Leptic, sandy-skeletal	NE	NE	285	NE	NC	NC
Haplic, loamy-skeletal	NE	NE	NE	372	NC	NC
Leptic, sandy	314	NE	NE	NE	NC	NC
Gleyic Turbic Cryosol					211	41
loamy-skeletal	NE	156	NE	NE	NC	NC
Leptic, loamy-skeletal	309	307	NE	NE	308	<1
Leptic, fine-loamy	NE	NE	115	NE	NC	NC
Mean overall soils (∅)	218	362	177	227	267	62
Standard deviation (σ)	45	49	64	64	62	NC
Gelic Histosols[A]	NE	NE	NE	NE	1561	50%

NE: no occurrence; NC: not calculated; T.: Turbic; H.: Haplic.
[A]Six sites outside the plots A-D.

Table 4.3.10. C/N ratio of total soil profiles, means overall soils and sites, and the percentage of standard deviation ($\sigma_{\%}$).

Soil Unit (WRB-FAO, 1998)	Site A	Site B	Site C	Site D	Mean Over all	$\sigma_{\%}$
H. or T. Leptic Cryosol	12.2	11.1	21.3	16.8	15.4	26
loamy-skeletal	16.8	14.3	21.3	16.8	17.3	15
sandy-skeletal	9.3	9.7	NE	NE	9.5	2
Spodic Cryosol	25.9	NE	12.4	5.9	14.7	57
Leptic, sandy-skeletal	NE	NE	12.4	NE	NC	NC
Haplic, loamy-skeletal	NE	NE	NE	5.9	NC	NC
Leptic, sandy	25.9	NE	NE	NE	NC	NC
Gleyic Turbic Cryosol	14.3	6.1	15.4	NE	13.3	51
loamy-skeletal	NE	4.9	NE	NE	NC	NC
Leptic, loamy-skeletal	14.3	14.4	NE	NE	14.4	<1
Leptic, fine-loamy	NE	NE	15.4	NE	NC	NC
Mean overall soils (∅)	16.7	10.8	16.4	11.4	14.0	NC
Standard deviation ($\sigma_{\%}$)	36	36	23	48	40	-
Gelic Histosols[A]	NE	NE	NE	NE	17.8	19

NE: no occurrence; NC: not calculated; T.: Turbic; H.: Haplic.
[A]Six sites outside the plots A-D.

Table 4.3.11. Total organic carbon (TOC in kg per 100 m^2) in the soil solum (range 7 to 50 cm) of the investigated sites, the mean overall sites and soil units, and the percentage of standard deviation ($\sigma_{\%}$).

Soil Unit (WRB-FAO, 1998)	Site A	Site B	Site C	Site D	Mean Over all	$\sigma_{\%}$
H. or T. Leptic Cryosol	68.4	93.7	59.2	65.8	71.8	18
loamy-skeletal	37.1	36.2	59.2	65.8	49.6	27
sandy-skeletal	31.3	57.5	NE	NE	44.4	30
Spodic Cryosol	12.2	NE	181.5	59.0	84.2	85
Leptic, sandy-skeletal	NE	NE	181.5	NE	NC	NC
Haplic, loamy-skeletal	NE	NE	NE	59.0	NC	NC
Leptic, sandy	12.2	NE	NE	NE	NC	NC
Gleyic Turbic Cryosol	18.9	17.8	13.8	NE	16.8	13
loamy-skeletal	NE	12.5	NE	NE	NC	NC
Leptic, loamy-skeletal	18.9	5.3	NE	NE	12.1	56
Leptic, fine-loamy	NE	NE	13.8	NE	NC	NC
Sum per 100 m^2	99.5	111.5	254.5	124.8	147.8	42
excluding site C					111.9	9

NE: no occurrence; NC: not calculated.

Table 4.3.12. Total nitrogen stock (N_t in g per 100 m^2) in the soil solum (range 7 to 50 cm) of the investigated sites, the mean overall sites and soil units, and the percentage of standard deviation ($\sigma_{\%}$).

Soil Unit (WRB-FAO, 1998)	Site A	Site B	Site C	Site D	Mean Over all	$\sigma_{\%}$
H. or T. Leptic Cryosol	5594	8478	2777	3928	5194	41
loamy-skeletal	2214	2527	2777	3928	2862	23
sandy-skeletal	3380	5951	NE	NE	4651	27
Spodic Cryosol	471	NE	14592	9970	8344	70
Leptic, sandy-skeletal	NE	NE	14592	NE	NC	NC
Haplic, loamy-skeletal	NE	NE	NE	9970	NC	NC
Leptic, sandy	471	NE	NE	NE	NC	NC
Gleyic Turbic Cryosol	1320	2911	897	NE	1709	51
loamy-skeletal	NE	2543	NE	NE	NC	NC
Leptic, loamy-skeletal	1320	368	NE	NE	844	56
Leptic, fine-loamy	NE	NE	897	NE	NC	NC
Sum per 100 m^2	7385	11389	18266	13898	12735	31
excluding site C					10890	25

NE: no occurrence; NC: not calculated.

However, widespread podzolization and/or, as Beyer (2000) demonstrated, an extraordinary accumulation of organic matter greatly increases these stocks. The SOM-enriched Cryosols are transitional segments to the Gelic Histosols and, despite the patchiness in the occurrence of the organic soils, they are probably of predominant geoecological significance with respect to carbon and nitrogen storage (Beyer, 2000).

Table 4.3.13 summarizes the results of carbon and nitrogen stock investigations of a 10 m x 10 m square plot C. Recently the humic Leptic Cryosols represent 13% of the soil geography but 23% of the carbon stock (Table 4.3.13). Between the areas of humic mineral soils, patches of organic soils occurred, mostly too small for a normally scaled soil survey (Beyer et al., 2000). A rough calculation substituting 1% of a humic mineral soil with an organic soil (due to the occurrence of Gelic Histosol) yields between 5 and 6% greater carbon and nitrogen stock (Table 4.3.13). This has a direct impact on colonization with microbes and retention of nutrients and moisture (Beyer, 2000).

Table 4.3.14 summarizes the data for carbon and nitrogen stock of the four sites in order to show the significance of single soil units as carbon and nitrogen sinks. The Spodic Cryosols are probably an important soil for carbon and nitrogen storage in the Casey area. For example, the sandy Spodic Cryosols covered only 13% of the area, but the C and N storage was about 30%. This is because these soils had a high carbon and nitrogen content at a depth of 30 to 40 cm (Beyer et al., 1997a; Blume et al., 1997).

In addition, the loamy Spodic Cryosols, which are not leptic, store a huge amount of nitrogen (Table 4.3.3), because these soils are mostly relic ornithogenic soils with a great influence of penguin guano (Beyer et al., 1997a), which is rich in nitrogen (Croxall, 1987). These soils showed the lowest C/N ratio of all soils in the landscape.

The next important soil unit is the Leptic or Haplic Cryosols, which cover about 33% of the area and contributed 49% of the carbon but only 42% of the nitrogen. The source of this organic matter is not known (Blume et al., 1997). We think such high amounts cannot derive only from the recent and often scarce vegetative cover and might be leftovers from a microbial activity and translocation into the subsoil (Melick and Seppelt, 1992; Melick et al., 1994b; Bölter, 1993).

Table 4.3.13. Recent carbon and nitrogen storage (kg) in a 10 m x 10 m square plot and its calculated increase with the occurrence of Gelic Histosols instead of humic mineral soils (adapted from Beyer, 2000; slightly modified).

Code	Soil Type	Area[A]	Area	Total organic carbon			Total nitrogen		
		%	%	g/m^2	kg[B]	%	g/m^2	kg[B]	%
C4	Gleyic T. Leptic Cryosol	7.8	11	1775	14	5	115	0.9	5
C11	Spodic Leptic Cryosol	51.2	72	3544	181	71	285	14.6	80
C3	Haplic Leptic Cryosol	3.3	5	296	1	<1	21	<0.1	<1
C5	Humic H. Leptic Cryosol	9.5	13	6126	58	23	285	2.7	15
	-1%		+1[C]	30000[D]	82	29	1900	4.3	22
	-2%		+2[C]	30000[D]	106	35	1900	5.9	28
	-3%		+3[C]	30000[D]	130	40	1900	7.5	33

[A]Remainder to 100% is bare rock.
[B]Per 100 m^2.
[C]Increase: % of Gelic Histosols.
T.: Turbic; H.: Haplic.

Table 4.3.14. Significance of mineral Cryosols in TOC and N_t storage (evaluated from 4x100 m^2 plots) in the coastal area of eastern Antarctica near Casey Station (Wilkes Land).

Soil Unit (WRB-FAO, 1998)	Area %[A]	-----TOC----- kg	%	------N_t------ kg	%	TOC/N_t
H. or T. Leptic Cryosol	33	287	49	21.5	42	13.3
loamy-skeletal	24	198	34	12.2	24	16.2
sandy-skeletal	9	89	15	9.3	18	9.6
Spodic Cryosol	21	252	43	25.1	48	10.0
Leptic, sandy-skeletal	13	181	31	14.6	28	12.4
Haplic, loamy-skeletal	7	59	10	10.0	19	5.9
Leptic, sandy	<1	12	2	0.5	1	24.0
Gleyic Turbic Cryosol	7	51	8	5.1	10	10.0
loamy-skeletal	4	13	2	2.5	5	5.2
Leptic, loamy-skeletal	1.4	24	4	1.7	3	14.1
Leptic, fine-loamy	2	14	2	0.9	2	15.6

[A]Percentage of the covered area, excluding bare rock.

The parent material of the local Gleyic Turbic Cryosols was mainly of loamy texture. These soils are still active in cryoturbation and the permanent movement of the soil material obviously inhibits plant colonization (Washburn, 1980). In addition, the aquic conditions suggest a temporary lack of oxygen during the unfrozen season (Beyer et al., 2000). This might affect a plant succession negatively.

In summary, the Gleyic Turbic Cryosols cover only small dots in the coastal Antarctic Casey area and were of minor importance with respect to carbon and nitrogen storage.

6. Conclusions

Antarctic Gelic Histosols and most Cryosols have high C and N concentrations. The accumulation of organic matter in soils of the coastal Antarctic is similar to that of comparable Arctic regions (Batjes, 1995; Jacobsen, 1997). The ice-free areas of coastal eastern Antarctica show a tremendous humus accumulation. The source and ecological function of this organic matter is not known, and unraveling this puzzle should be a future research task.

The very narrow C-to-N ratios indicate a higher accumulation of nitrogen than in the northern polar regions (Beyer et al., 1995; Beyer et al., 1997a, 1997b). This suggests a potentially high availability of organic matter. The humus is probably preserved only because of recent cold conditions (Rosswall and Heal, 1975; Smith, 1990), but this might change with global climate changes (Adamson and Adamson, 1992; Selkirk, 1992).

The small-distance variation of the topographical, geomorphological, geological, and pedogenic patterns induces a great variation in C and N concentration and storage. This complicates calculation of soil total C and N storage. However, the survey at a landscape level suggests that the carbon and nitrogen stock is very similar at 75% of the landscape sites, but widespread podzolization and/or extraordinary accumulation of organic matter greatly increases these stocks. A proper soil survey would improve storage estimates and thus help with modeling levels of carbon and nitrogen release with rising temperatures (Lal et al., 1995; Scharpenseel et al., 1990) and/or anthropogenic disturbances (e.g. Adamson and Seppelt, 1990; Vanhala et al., 1996; Tarnocai, 1997; Zamolodchikov et al., 1997).

7. Acknowledgments

The Australian Antarctic Divison, Kingston, Tasmania, and the Deutsche Forschungsgemeinschaft, Bonn-Bad Godesberg (Be1259/4-1), financially supported this investigation. The Australian Antarctic Research Expeditions and especially the Casey crew in the summer of 1995/1996 supported the field and lab work. We gratefully acknowledge all of them. Special thanks to Prof. H.-P. Blume, who made it possible for the senior author to participate in the field expedition.

References

Adamson, E., and R.D. Seppelt. 1990. A comparison of airborne alkaline pollution damage in selected lichens and mosses at Casey Station, Wilkes Land. In: K. R. Kerry and G. Hempel, eds. Antarctic ecosystems. Ecological change and conservation. Springer. Berlin. pp. 347-353.

Adamson, H., and E. Adamson. 1992. Possible effects of global climate change on Antarctic terrestrial vegetation. In: Department of Arts, Sport, the Environment and Territories. Impact of Climate Change on Antarctica - Australia. Austral. Government Publ. Service. Canberra. pp. 52-62.

Batjes, N.H. 1995. World soil carbon stock. ISRIC. Wageningen.

Beyer, L. 2000. Properties, formation and geo-ecological significance of organic soils in the coastal region of eastern Antarctica. Catena. 39(2):940-120.

Beyer, L., and M. Bölter 2000. Ecology and geography of Spodic Cryosols in terrestrial ecosystems of eastern Antarctica. Catena. 39(2):79-93.

Beyer, L., H.-P. Blume, C. Sorge, H.-R. Schulten, H. Erlenkeuser, and D. Schneider. 1997b. Humus composition and transformation in a pergelic Cryohemist of coastal Antarctica. Arct. and Alpine Res. 29:358-365.

Beyer, L., H. Knicker, H.-P. Blume, B. Bölter, B. Vogt, and D. Schneider. 1997a. Soil organic matter of suggested spodic horizons in relic ornithogenic soils of coastal continental Antarctica (Casey Station, Wilkes Land) in comparison with that of spodic soil horizons in Germany. Soil Sci. 162:518-527.

Beyer, L., K. Pingpank, G. Wriedt, and M. Bölter. 2000. Soil formation on coastal continental Antarctica. Geoderma. 95:283-304.

Beyer, L., C. Sorge, H.-P. Blume, and H.-R. Schulten. 1995. Soil organic matter composition and transformation in gelic Histosols of coastal continental Antarctica. Soil Biol. Biochem. 27:1279-1288.

Blume, H.-P., and M. Bölter. 1993. Soil of Casey Station (Wilkes Land, Antarctica). Proceedings 1st. Int. Conf. Cryopedology. Russ. Acad. of Sciences. Pushchino. pp. 96-103.

Blume, H.-P., D. Schneider, and M. Bölter. 1996. Organic matter accumulation in and podzolization of Antarctic soils. Z. Pflanzenernähr., Bodenk. 159:411-412.

Blume, H.-P., L. Beyer, M. Bölter, H. Erlenkeuser, E. Kalk, S. Kneesch, U. Pfisterer, and D. Schneider. 1997. Pedogenic zonation in the Southern circumpolar region. Advances in GeoEcol. 30:69-90.

Bockheim, J.G., and F.C. Ugolini. 1990. A review of pedogenic zonation in well-drained soils of the southern circumpolar region. Quat. Res. 34:47-66.

Bölter, M. 1993. Effects of carbohydrates and leucine on growth of bacteria from Antarctic soils (Casey Station, Wilkes Land). Polar Biol. 13:297-306.

Campbell, I.B., and G.G.C. Claridge. 1987. Antarctica: Soils, Weathering Processes and Environment. Elsevier. Amsterdam.

Croxall, J.P., ed. 1987. Seabirds, feeding ecology and role in marine ecosystems. Univ. Press. Cambridge.

Jacobsen, B.H. 1997. A soil map of Greenland. Abstracts Cryopedology 1997, August 1997, Syktyvkar, Russia. p. 43.

Lal, R., J.M. Kimble, E. Levine, and B.A. Stewart. 1995. Soil and global change. Lewis Publishers. Boca Raton, FL.

Melick, D.R., and R.D. Seppelt. 1992. Loss of soluble carbohydrates and changes in freezing point of Antarctic bryophytes after leaching and repeated freeze-thaw cycle. Antarc. Sci. 4:399-404.

Melick, D.R., M.J. Hovenden, and R.D. Seppelt. 1994a. Phytogeography of bryophyte and lichen vegetation in the Windmill Islands, Wilkes Land, Continental Antarctica. Vegetation. 111:71-87.

Melick, D.R., M. Bölter, and R. Müller. 1994b. Rates of soluble carbohydrate utilization in soils from Windmill Islands Oasis, Wilkes Land, Continental Antarctica. Polar Biol. 14:59-64.

Paul, E., K. Stüwe, J. Teasdale, and B. Worley. 1995. Structural and metamorphic geology of the Windmill Islands, eastern Antarctica: field evidence for repeated tectonothermal activity. Aust. J. Earth Sci. 42:453-469.

Roswall, T., and O.W. Heal. 1975. Structure and Functions of tundra ecosystems Ecol. Bul. 20. Swedish Natural Research Council. Stockholm.

Scharpenseel, H.W., M. Schomaker, and A. Ayoub. 1990. Soil on a warmer earth. Elsevier. Amsterdam.

Schlichting, E., H.-P. Blume, and K. Stahr. 1995. Bodenkundliches Praktikum Blackwell, 2nd ed.. Berlin.

Selkirk, P.M. 1992. Climate Change and the Subantarctic. In: Department of Arts, Sport, the Environment and Territories (ed.). Impact of Climate Change on Antarctica -Australia. Australian Government Publishing Service. Canberra. pp. 43-51.

Seppelt, R.D., and P.A. Broady. 1988. Antarctic terrestrial ecosystems: The Vestfold Hills in context. Hydrobiol. 165:177-184.

Smith, R.I.L. 1990. Plant community dynamics in Wilkes Land, Antarctica. Proc. NIPR Symp. Polar Biol. 3:229-244.

Soil Survey Staff. 1998. Keys to Soil Taxonomy. USDA-NRCS. Washington D.C.

Tarnocai, C. 1997. The Canadian GIS organic carbon database and its application to Cryosols. Abstracts Cryopedology 1997, August 1997, Syktykar, Russia. p. 59.

Vanhala, P., H. Fritze, and S. Neuvonen. 1996. Prolonged simulated acid rain in the subarctic: effect on the respiration rate and microbial biomass. Biol. Fertil. Soils 23:7-14.

Washburn, A.L. 1980. Geocryology: A survey of periglacial processes and environments. Wiley. New York.

WRB-FAO. 1998. World Reference Base for Soil Resources. World Soil Resources Reports 84. FAO. Rome.

Zamolodchikov, D.G., D.V. Karelin, and A.I. Ivaschenko. 1997. Geoinformation system based approach to studying carbon fluxes in tundra landscapes. Abstract International Conference "The Problems of Earth Cryosphere." Pushchino. April 1997. p. 304.

Chapter 4. Composition and Transformation of Soil Organic Matter in Cryosols and Gelic Histosols in Coastal Eastern Antarctica (Casey Station, Wilkes Land)

Lothar Beyer[1], Daniel M. White[2], Kristina Pingpank[3], and Manfred Bölter[4]

[1]Professor, Institute for Polar Ecology, Christian-Albrechts-University Kiel, Wischhofstr. 1-3, Bldg. 12, 24148 Kiel, Germany; e-mail soil_science@web.de.
[2]Associate Professor, Department of Civil and Environmental Engineering, University of Alaska, Fairbanks, 451 Duckering Bldg., P.O. Box 755900, Fairbanks, Alaska 99775-5900, USA; e-mail ffdmw@uaf.edu.
[3]Dr., Institute for Plant Nutrition and Soil Science, Christian-Albrechts-University Kiel, Olshausenstr. 40, 24098 Kiel, Germany.
[4]Professor, Institute for Polar Ecology, Christian-Albrechts-University Kiel, Wischhofstr. 1-3, Bldg. 12, 24148 Kiel, Germany; e-mail mboelter@ipoe.uni-kiel.de.

1. Introduction

Despite the live-hostile climate in the Antarctic, soil formation processes, due to the accumulation of organic matter, are evident (Beyer et al., 1999). In the maritime Antarctic climate region, organic soils, mainly formed by debris of mosses and algae, are widespread (Campbell and Claridge, 1987) and mineral soils also contain huge amounts of soil organic matter (SOM) (Blume et al., 1996). In contrast, in the coastal regions of eastern Antarctica, only local patches of Gelic Histosols are associated with Leptic, Haplic, and Spodic Cryosols (Beyer et al., 1999).

In every case, only scarce information is available on SOM in Antarctic terrestrial ecosystems. Usually, analyses of SOM in Antarctica mostly are confined to loss-on-ignition (LOI) and the determination of organic carbon and nitrogen (Campbell and Claridge, 1987). Only a few reports are available about the contents of dissolved organic matter (Tearle, 1987; Bölter 1990; Melick and Seppelt 1992; 1994; Roser et al. 1992a, b; 1994). Wilson et al. (1986) and Wilson (1990) investigated Antarctic mosses and soil humic acids with carbon-13 NMR spectroscopy. However, soil ecological and biological approaches require more specific information on SOM in order to quantify its geo-ecological impact on plant adaptation (Broady and Weinstein, 1998; Seppelt and Broady, 1988), microbial colonization (Bölter et al., 1994), and soil and/or environmental fragility (Chen and Blume, 1998; Campbell et al., 1998).

To reach a better understanding of soil formation processes in Antarctica, we investigated SOM of two Gelic Histosols (permafrost-affected organic soils) and three Cryosols (permafrost-affected mineral soils) in the coastal region of eastern Antarctica near Casey Station (Wilkes Land). We used an integrated approach to characterize SOM composition and transformation: wet-chemical analyses, CPMAS carbon-13 NMR spectroscopy, and modern pyrolysis tools (Py-FIMS and Py-GC/MS). We compared the results to well described soil formation processes under temperate climate conditions.

2. Sites and Soils

The sites and soils are south of the Australian Casey Station on Bailey Peninsula (Latitude 66°18'S, Longitude 110°32'E) at Wilkes Land in coastal eastern Antarctica, about 700 m from the shore. The mineral parent materials of the Cryosols are weathered gneiss and schist, moraine deposits, outwash sediments (Paul et al., 1995), and marine deposits from seabirds (Campbell and Claridge, 1987). The coastal landscape was partly deglaciated between 5,000 and 6,000 years B.P. (Goodwin, 1993) and terraces elevated then.

Annual precipitation (180 mm) is mostly snow. Because of strong drift due to persistent high winds (Pickard, 1986), the real input of water in soil is very heterogeneous (Beyer et al., 2000a). The mean annual temperature is -9.3°C. From November to February, sunshine lasts for 5 to 7 h/d. During the Antarctic summer of nearly six weeks, temperatures are above freezing (mean in January, +0.2°C) and plant communities of mosses (*Bryum algens, Ceratodon purpurea, Schistidium antarcticum*), lichens (*Usnea sphacelata, Pseudephebe minuscula, Umbilicaria decussata*), and soil algae can establish and grow very slowly (Seppelt and Broady, 1988).

Soil sampling occurred in January 1991 (organic soils) and February 1996 (mineral soils). The soils were classified according to the *World Reference Base of Soil Resources* (WRB-FAO, 1998) and, for comparison, *US Soil Taxonomy* (Soil Survey Staff, 1998).

The most common organic soil in the Casey area is the (Terric) Leptic Gelic Histosol (US: Lithic Hemistel) (Beyer et al., 2000b). The sampled soil example was at the end of a slope in a small, moist depression. The top layer (LH) consisted of dead mosses (*Ceratodon purpurea*), colonized by the lichen *Usnea antarctica*. In deeper layers, the mosses were more humified and more brownish to black (Table 4.4.1). However, plant fibers were still visible among the sapric organic materials (Table 4.4.1, Post et al., 1988).

The second investigated organic soil was a Fibric Leptic Gelic Histosol (US: Lithic Fibristel) at the edge of a meltwater lake. This soil occurred only locally at small spots near the liquid surface water (Blume et al., 1997), and the complete soils consisted of dead algal materials (*Phormidium* sp.). The peat was not decomposed or only weakly so and was greenish to gray (Table 4.4.1). Most of the time, water or ice covers this soil. During the recent summer, the first 10 to 20 cm were above the water level. Under such conditions, the redox potential was lower than that of the drier Terric Leptic Gelic Histosol (Table 4.4.1).

All three Cryosols (mineral soils) had an extremely high content of coarse materials (380 to 900 g/kg; Beyer et al., 2000b). The fine material was sand to sandy loam. The Spodic Leptic Cryosol with the code C11 (US: Sandy-skeletal Lithic Haplorthel) was in a small depression between single bare rock fragments under a moss and lichen carpet with 100% surface covering. The parent material is weathered gneiss bare rock.

The Spodic Cryosol with the code WP5 (US: Sandy-skeletal Typic Haplorthel) was at the edge of a relic and dried melt water under a thick cushion of mosses. The parent materials are glacial outwash sands and gravels.

The Spodic Cryosol with the code D3 (US: Loamy-skeletal Typic Haplorthel) was at an upper hill slope under a thick gravel pavement, which is typical for relic penguin rookery sites. At this site was no vegetation.

All investigated Cryosols are typical shallow Podzols. They represent 20 to 30% of the mineral soils in the Casey area (Beyer and Bölter, 2000).

3. Methods

3.1. Field measurements

We estimated the degree of humification according to the Post method, by squeezing the moist soil sample between the fingers (Post et al., 1988; Schlichting et al., 1995). Degree 1 indicates no humification (pure fibric materials), and degree 10 means complete humification (totally sapric). We measured the redox potentials (E_h) with platinum electrodes (Pfisterer and Gribbohm, 1989) after 6 to 12 h installation in the field (Schlichting et al., 1995).

Table 4.4.1. Selected properties of peaty horizons in Gelic Histosols in eastern Antarctica.

WRB[A]	US[B]	Depth cm	Color (Munsell)	Post degree[C]	Moisture content[D]	$EC_{2.5}$ mS	E_{h7} mV	pH $CaCl_2$
Terric Leptic Gelic Histosol (US: Lithic Hemistel) from mosses								
LH	Oi	0-3	ND	1-2	dry	0.72	ND	4.8
H1	Oe	-8	5YR2.5/1.5	3-5	dry	0.61	530	3.9
H2	Oa1	-16	5YR2.5/1.5	8-9	moist	0.20	640	3.8
H3	Oa2	-28	5YR2.5/1.5	8-9	moist	0.13	ND	3.8
R	R	>28	bare rock					
Fibric Leptic Gelic Histosol (US: Lithic Fibristel) from algae								
H1	Oi	0-1	5GY5/1	1-2	very wet	1.8	620	4.8
H2	Oe1	-8	5GY5/1	1-2	very wet	4.9	250	4.8
H3	Oe2	-18	5GY5/1	1-3	very wet	0.65	260	4.8
R	R	>18	bare rock					

[A]Taxonomy according to the *World Reference Base for Soil Resources* (WRB-FAO, 1998).
[B]According to *US Soil Taxonomy* (Soil Survey Staff, 1998).
[C]Degree of humification (Schlichting et al., 1995): 1 means not humified (fibric), 10 means completely humified (sapric).
[D]Moisture status estimated in the field.
[E]Actual lake water level 15 to 20 cm below soil surface.
$EC_{2.5}$: electrical conductivity in 1:2.5 water extract; E_{h7}: redox potential measured in the field and converted on pH7 level (Schlichting et al., 1995).

3.2. General and wet chemistry

Most soil property determinations were carried out in accordance with Schlichting et al. (1995) on air-dried soil samples. The pH value was measured in 10 *mM* $CaCl_2$ with a commercial glass electrode, and electrical conductivity (EC), in a 1:2.5 water extract. Loss-on-ignition (LOI) was determined gravimetrically after combustion at 650°C in a furnace. Total organic carbon (TOC) was calculated after dry combustion in a Coulomat 702 (Ströhlein Instruments, Germany).

The samples were heated (600°C) in an induction furnace under oxygen; CO_2 was trapped in $Ba(OH)_2$ and the remaining $Ba(OH)_2$ was neutralized by titration. Total nitrogen (N_t) was digested by the classical Kjeldahl method and determined as nitrate in a flow injection analyzer (Jones, 1991). Pedogenic iron oxides were determined after extraction with dithionate-citrate (Fe_d) and oxalate (Fe_o). The organic bonded iron was determined in the alkaline NaOH extract (Fe_{org}).

The extinction of the oxalate extract was measured at 472 nm (optical density of oxalate extract = ODOE) in accordance with Daly (1982). The potential cation exchange capacity (CEC_p) was calculated by summarizing the single capacities of Na, K, Mg, Ca, H, and Al, in accordance with Schlichting et al. (1995). The percentage of Na, K, Mg, and Ca from CEC_p was defined as base saturation (BS).

The immobile P fraction (P_c) was determined to indicate the penguin impact and was extracted with 2% citric acid (modified, original 1%), in accordance with Dyer (1894). Methoxyl-C (OCH_3-C) was determined in accordance with the Zeisl-Pregl method (Hauptmann, 1992). The OCH_3-C in soils was demethylized into methyl iodide by boiling each sample with hydroiodic acid. Methyl iodide was hydrolized into methanol and silver iodide, which was determined gravimetrically.

Lignin was estimated as 7 x OCH_3 content. The α-NH_2-N was determined after extraction with 10 *M* HCl and 1 *M* formic acid (10:1), in accordance with Stevenson and Cheng (1970), by using the ninhydrin reaction (triketo-hydrindine hydrate).

Nitrate-N (NO_3^--N) and Ammonium-N (NH_4^+-N) were extracted with 10 m *M* $CaCl_2$, in accordance with Schlichting et al. (1995). Concentrations of the inorganic N fractions in the extracts were determined by flow injection analysis (Jones, 1991), using a Tecator flow injection analyzer (Perstorp Analytical, Germany). The organic, non-α-NH_2 nitrogen (N_{org}) was calculated as N_t - (α-NH_2-N + inorganic N fractions).

3.3. CPMAS

The CPMAS carbon-13 NMR spectra were recorded from milled bulk soil samples (<200 μm). (*CPMAS* is cross polarization magic angle spinning [special NMR technique]. *NMR* is nuclear magnetic resonance.). (The spectra were obtained at 2.3 Tesla (25.2 MHz) in a Bruker MSL 100 spectrometer equipped with a commercial 7 mm CPMAS probe at a rotation frequency of 4 kHz. A contact time of 1 ms was used. Due to short T_{1H} in SOM (Fründ and Lüdemann, 1989), we chose a recycle delay of 0.3 s. We report the ^{13}C chemical shifts here relative to tetramethylsilane (TMS = 0 ppm).

We subdivided and integrated the spectra into these ranges for carbon species in SOM: 0 to 46 ppm (alkyl-C), 46 to 60 ppm (methoxyl-C), 60 to 80 ppm (hydroxyl-C), 80 to 110 ppm (acetal-C), 110 to 140 ppm (aromat-C), 140 to 160 ppm (phenol-C), and 160 to 210 ppm (carboxyl-C) (compare Wilson, 1987). Under these experimental conditions, one can quantitatively relate the relative signal area to the C concentration (Fründ and Lüdemann, 1989).

3.4. Py-FIMS

The Py-FIMS mass spectra were recorded using a double-focusing Finnigan MAT 731 mass spectrometer. (*Py-FIMS is* pyrolysis field-ionization mass spectrometry.) The heatable or coolable direct introduction system with an electronic temperature-programming was used (IGT, Much, Germany), adjusted at the +8kV potential of the ion source.

The sample weights were increased with decreasing TOC concentrations to produce strong, approximately equal total ion intensities (TII). Thus, the sample weights ranged from about 0.5 to 1.3 mg air-dry, milled soil (<200 μm).

To determine the percentage of volatilized matter, all samples were weighed before and after each Py-FIMS analysis (error ±0.01 mg). The samples were heated in high vacuum from 50 to 650°C at a heating rate of approximately

0.5°C/s. About 60 magnetic scans were recorded for the mass range 16 to 1000 Dalton during the heating. These single spectra were integrated by the Finnigan MAT SS200 data system, resulting in a summed spectrum for each measurement.

Three laboratory replicates were carried out for every sample. The obtained summed spectra were averaged to one survey spectrum. The TII of the replicated measurements were normalized to 1 mg sample, averaged and plotted versus the pyrolysis temperature, resulting in thermograms. Schulten (1987, 1993) has published detailed descriptions of Py-FIMS for analyses of terrestrial and aquatic humic fractions and whole soils. We tentatively assigned selected nominal masses to selected classes of biomarkers: carbohydrates, amino acids and peptides, N-heterocycles, alkylaromatics, lipids, free n-C_{16} to n-C_{34} fatty acids, and sterols.

3.5. Py-GC/MS

Subsamples with a mass of 25 mg were placed in quartz tubes (2 cm x 2mm ID) and quantified with a Mettler Microbalance. The samples were held in place by a plug of quartz wool at each end of the tube. Py-GC/MS was conducted on each sample and used to identify as many compounds as possible. (*Py-GC/MS is* pyrolysis gas chromatography/mass spectrometry.)

Py-GC/MS was conducted with a CDS Model 1000 pyrolyzer and a Model 1500 GC interface. The interface temperature was set at 280°C. During pyrolysis, the sample was heated from 280°C to 700°C in 0.1 s and held at 700°C for 9.9 s. The pyrolysis reactor was mounted on a GC, HP 5890 Series II with a Hewlett Packard HP-1 column (Cross-linked methyl-siloxane) 25m x 0.2m x 0.33 μm film thickness. The GC temperature program was 35°C for 15 min., 2°C/min ramp to 250°C, and hold for 10 min. The GC was plumbed directly to an HP 5971A Series Mass Selective Detector in electron impact mode.

The MS scanned mass units were 45 to 650. All mass spectra were compared to the Wiley 138 mass spectral library. Helium was the carrier gas at 0.5 cc/min. The sample was injected with a split ratio of 1:50.

For each sample, the mass spectral signals for all compounds were normalized to the compound in the greatest abundance. This study considered only those compounds with a normalized intensity of ≥20%.

Each compound was assigned with a probable parent (compare Table 4.4.8). The parent is a soil biopolymer that was present in the sample before pyrolysis. The soil biopolymer classifications we used were protein, phenol precursors, carbohydrates, aromatic hydrocarbon precursors (AHP), lipids, and amino carbohydrates (Bracewell et al., 1989). We classified all long-chain hydrocarbons, alcohols, aldehyds, and ketones as having lipid parents and considered any compound with no assignment to a probable parent *unresolved.* The relative intensities for all soil biopolymers in each samples were added. We present each soil biopolymer class as a percentage of the total of the assigned compounds.

4. Composition of Soil Organic Matter

4.1. Gelic Histosols

Table 4.4.1 presents some selected properties of the investigated Leptic Gelic Histosols, with the site and soil descriptions and some other parameters. Table 4.4.2 summarizes some SOM properties of the organic soils. Total organic carbon (TOC) concentrations and losses on ignition, as criterion for SOM content, decreased with increasing profile depth. However, TOC increasingly was enriched in the SOM of deeper horizons, as an increase in the TOC/SOM ratio indicates. The OCH_3-concentrations as indicator for lignins (Beyer et al., 1993) were very low over the whole soil profile, in comparison to Histosols under temperate climate conditions (Beyer et al., 1992).

In the Terric Gelic Histosol, between 3 and 16 cm, the total nitrogen (N_t) concentrations were higher, as in the top and lowest horizons. The higher N_t concentrations in these horizons were due to higher proportions of the organic non-α-NH_2 nitrogen (N_{HC}-N). The α-NH_2-N concentrations continuously decreased with increasing soil depth. In the whole profile, soluble, inorganic NO_3^--N and NH_4^+-N fractions were of minor importance.

In the Fibric Gelic Histosol, the TOC concentrations decreased with increasing profile depth, as found for the Terri-Gelic Histosol (Table 4.4.2). In addition, the OCH_3-C concentrations were very low in the whole profile. The N_t concentrations decreased in deeper soil horizons. The nitrogen was mainly α-NH_2-N. Other N compounds were of minor importance. The higher NH_4^+ concentrations, in comparison to those of the Terric Gelic Histosol, correspond to the low redox potentials restricting nitrification (Beyer et al., 1990).

In the Terric Gelic Histosol, peaks dominated the carbon-13 NMR spectra of all soil horizons (Figure 4.4.1) at 73.6 ppm (glucose units) and 173 ppm (carboxylic C units), respectively. The shoulder at 55 ppm arose from methoxyl or amino carbon. Since 29 to 57% of N_t in the soil horizons is α-NH_2-N and the concentration of OCH_3-C is very low (Table 4.4.2), the shoulders at 55 ppm derived probably from amino carbon.

Table 4.4.3 shows the integrated areas below the carbon-13 NMR spectra. In all soil horizons, methoxyl, aromatic, and phenolic structures were of lesser importance than acetal, hydroxyl, and alkylic structures. Carboxylic carbon compounds increased, especially in the H1 and H2 horizons. Likewise, alkylic carbon increased with soil depth, whereas *O*-alkylic carbon (glucose units) decreased. This indicated a decomposition of carbohydrates from the moss debris and an enrichment of carboxylic and alkylic carbon during soil-forming in the Antarctic Terric Gelic Histosol.

Table 4.4.2. Selected SOM properties of peaty horizons in Gelic Histosols in eastern Antarctica.

Hor.	TOC	Nt	TOC/N_t	LOI	TOC/LOI	OCH_3 Mg/g	α-NH_2-N mg[A]	%[B]	NO_3^--N --- ng/g ---	NH_4^+-N	N_{org} mg[A]	%[B]
Terric Leptic Gelic Histosol (US: Lithic Hemistel) from mosses												
LH	287	15.4	19	553	0.52	1.48	8.7	57	4.5	79	6.7	43
H1	195	21.8	9	378	0.52	0.63	8.0	47	105.3	67	13.9	53
H2	188	20.7	9	318	0.59	0.86	5.9	29	31.2	28	14.7	71
H3	159	13.1	12	259	0.61	1.09	5.0	38	7.3	32	8.0	62
Fibric Leptic Gelic Histosol (US: Lithic Fibristel) from algae												
H1	266	13.2	20	729	0.37	1.00	11.1	85	3.4	167	2.0	15
H2	264	10.2	26	550	0.48	1.36	11.0	100	1.2	201	tr.	<<1
H3	164	9.0	18	329	0.50	0.65	7.7	86	0.9	132	1.3	14

[A] mg/g soil.
[B] in % of N_t.
N_{org}: Nt- (α-NH_2-N + NO_3^-N + NH_4^+-N); tr.: traces.

Of special interest was the detection of aromatic carbon, although mosses as plant precursors for SOM do not contain lignins (Hurst and Burges, 1967). The carbon-13 NMR spectra of the horizons from the Fibric Gelic Histosol (Figure 4.4.1) were similar to those of the Terric Gelic Histosol (Figure 4.4.2). However, the aromatic and alkylic peaks were less intensive. The peak at 21 ppm derived from the methyl group of carbohydrates, and the peak at 31 ppm was typical for alkyl-CH_2 units (Breitmaier and Voelter, 1990). The profile of the Fibric Gelic Histosol showed no depth functions of characteristic carbon-13 NMR subunits (Table 4.4.3).

For the Terric Leptic Gelic Histosol, which is the most representative organic soil of the Casey sites (Beyer, 2000), the Py-FIMS approach detected some more detailed information on the molecular-chemical composition of SOM. Figure 4.4.3a-d shows the spectra. Decreasing percentages of volatilized matter and total ion intensities reflect decreasing TOC concentrations in deeper soil horizons.

According to the thermograms (Figures 4.4.3a-d, upper right), the organic matter accumulated in the soil horizons was in each case pyrolyzed over a wide temperature range, reaching two marima around 300°C and 500°C, respectively. The temperature-resolved evaluation of mass spectra gave, for all horizons above 350°C, a signal pattern similar to that of humified plant materials. The mass distribution and intensities resemble the results obtained by Py-FIMS of peat.

In addition, with increasing pyrolysis temperature, thermally induced changes in the structure of the pyrolyzed organic matter have to be considered. Therefore, in Figures 4.4.4a-d, only the summed Py-FI survey mass spectra for the temperature range between 100°C and 350°C appear. For all soil horizons, they are qualitatively similar and point to slight decomposition and transformation of organic matter within the soil profile.

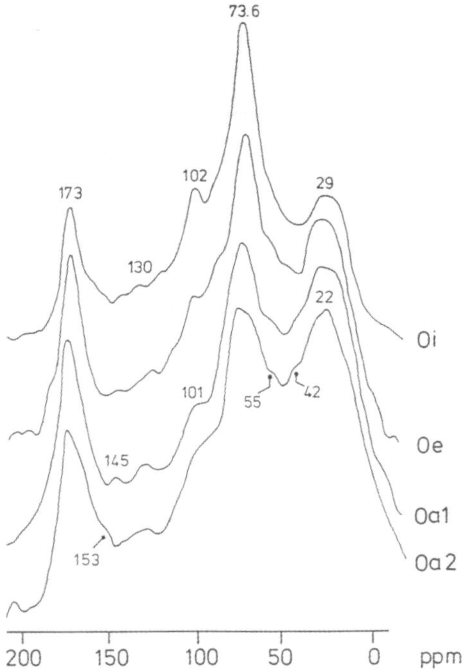

Figure 4.4.1. CPMAS carbon-13 NMR spectra of horizons of a Terric Leptic Gelic Histosol in eastern Antarctica, formed by mosses (Oi = LH, Oe = H1, Oa1 = H2, Oa2 = H3).

Table 4.4.3. Carbon-13 NMR-experiment subunits in peaty horizons in Gelic Histosols in eastern Antarctica.

hori-zon	alkylic C 0-46	methoxyl C 46-60	hydroxyl C 60-80	acetal C 80-110	aromatic C 110-140	phenolic C 140-160	carboxylic C 160-210
Terric Leptic Gelic Histosol (US: Lithic Hemistel) from mosses							
LH	23.1	8.1	23.9	20.0	8.6	4.7	10.6
H1	29.6	8.6	19.2	17.5	8.0	4.4	12.6
H2	34.9	8.1	16.3	14.3	7.4	5.0	14.1
H3	36.4	8.5	16.9	14.8	7.2	5.1	11.0
Fibric Leptic Gelic Histosol (US: Lithic Fibristel) from algae							
H1	19.6	8.7	32.3	22.8	6.0	2.3	8.4
H2	20.2	8.8	34.3	21.9	5.1	2.1	7.6
h3	22.6	9.0	32.3	20.3	5.2	2.2	8.4

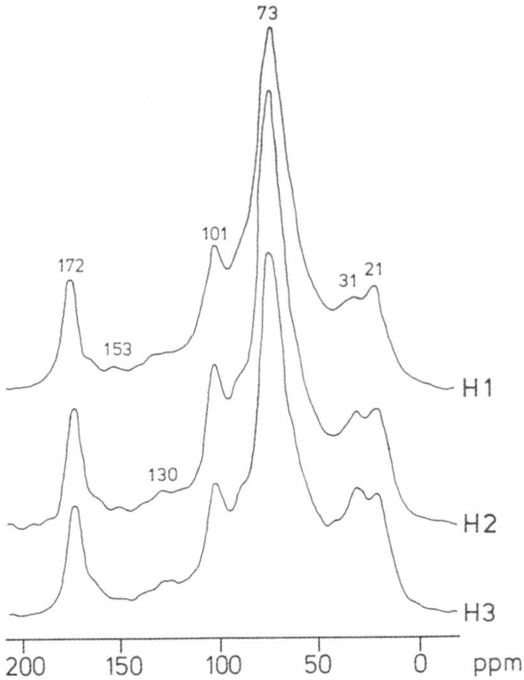

Figure 4.4.2. CPMAS carbon-13 NMR spectra of horizons of a Fibric Leptic Gelic Histosol in eastern Antarctica, formed by algae.

Signals at 84, 96, 98, 110, 112, 126, and 144 m/z were typical pyrolysis products of carbohydrates. N-containing compounds were discernible as amino acids/peptides (57, 70, 73, 74, 75, 84, 87, 91, 97, 99, 115, 120, 129, and 135 m/z) and heterocyclic N-containing pyrolysis products (67, 109, 111, 137 m/z). In contrast to higher developed plants, mosses do not contain lignins (Hurst and Burges, 1967).

In agreement with this observation, the signals in the characteristic mass range for lignin subunits between 94 and 356 m/z unusually compared to Py-FIMS spectra of forest and agricultural soils formed under temperate climatic conditions (Schulten, 1993). However, this mass range was not depleted. This refers to alkylaromatics, e.g., methyl- to docosyl-benzenes (starting with 92 m/z), methyl- to pentamethyl-naphthalenes (starting with 142 m/z), and methyl- to tetramethyl-phenanthrenes (starting with 192 m/z).

Recently, these alkylaromatics have been assigned to significant structural building blocks of humic substances (Schulten et al., 1991; Schulten and Schnitzer, 1993). Homologous series of n-C_{10} to n-C_{20} alkyl diesters (starting with 202 m/z), n-C_{16} to n-C_{34} fatty acids (starting with 256 m/z), and n-C_{20} to n-C_{30} alkenes and alkanes (starting at 280 and 282 m/z, respectively) indicated lipids.

Some relatively intensive signals overlapped these homologous series. These signals at 394, 396, 410, 412, and 414 m/z referred to sterols already observed in

Py-FI mass spectra of various plant materials (Schnitzer et al., 1990; Hempfling and Schulten, 1991). In addition, n-C_{44} to n-C_{46} alkyl-mononesters (648, 662, 676 m/z) reflected lipid-derived structures.

In the temperature range from 100°C to 350°C, between 43% and 56% of the TII/mg sample could be assigned to seven significant biomarkers of SOM: carbohydrates, amino acids, peptides, N-containing pyrolysis products, alkylaromatics, lipids supplemented by free n-C_{16} to n-C_{34} fatty acids and steroles (Schnitzer and Schulten, 1992; Sorge et al., 1993). With increasing profile depth, the absolute ion intensities per mg soil of carbohydrates, alkylaromatics, and lipids, inclusive of free fatty acids and sterols, decreased. In the same trend, the absolute intensity of signals of amino acids, peptides, and heterocyclic N-containing pyrolysis products decreased.

The summed relative ion intensity (in %TII/mg soil) of each biomarker illustrates its contribution to the composition of organic matter (Figure 4.4.4). Carbohydrates and lipids dominate the molecular-chemical composition of SOM in all soil horizons. Proportions of carbohydrates and alkylaromatics changed only insignificantly with profile depth. The proportion of amino acids peptides signals on the TII/mg soil slightly increased with profile depth, whereas that of heterocyclic N-containing pyrolysis products is nearly constant. Lipids increased in H1 and H2 horizons, compared to the top and lowest horizons. The separate recording of signals for free fatty acids and sterols showed that the significant increase of lipids (alkanes, alkenes, n-alkyl ester, bound fatty acids) in the H1 horizon can occur for free fatty acids and sterols.

4.2. Cryosols

All three Cryosols showed the horizon sequence of AE-Bh(Bhs,Bs)-C, with strong acidification and maximum values of aluminum (Beyer et al., 2000b) and iron fractions, optical density of the oxalate extract (ODOE), loss-on-ignition (LOI), and total organic carbon (TOC) and total nitrogen (N_t) in the spodic Bh/Bhs horizons (Table 4.4.4). This is characteristic of podzolic soils (Soil Survey Staff, 1998). According to McKeague et al. (1983), such characteristic depth functions of these parameters suggest a translocation of SOM into the subsoil.

A high amount of phosphorus in soil D3 indicated the influence of guano input into the soils (Blume et al., 1997). The increase of P (Table 4.4.4, P_c) in the subsoil of the Podzol from solid rock (no. C11 in Table 4.4.1) also might indicate guano input in earlier stages of soil formation. The C/N ratios in Antarctic Spodic Cryosols were much narrower, due to extremely high N_t contents, than their temperate counterparts. This was mostly striking in the Spodic Cryosol at an abandoned penguin rookery (no. D3 in Table 4.4.3).

Table 4.4.5 shows the SOM composition according to the carbon-13 NMR experiments. The increase of the summarized alkyl carbon and simultaneous decrease of O-alkyles compares with that observed in Podzols in temperate climate regimes (Post et al., 1988; Sorge et al., 1994; Beyer, 1996).

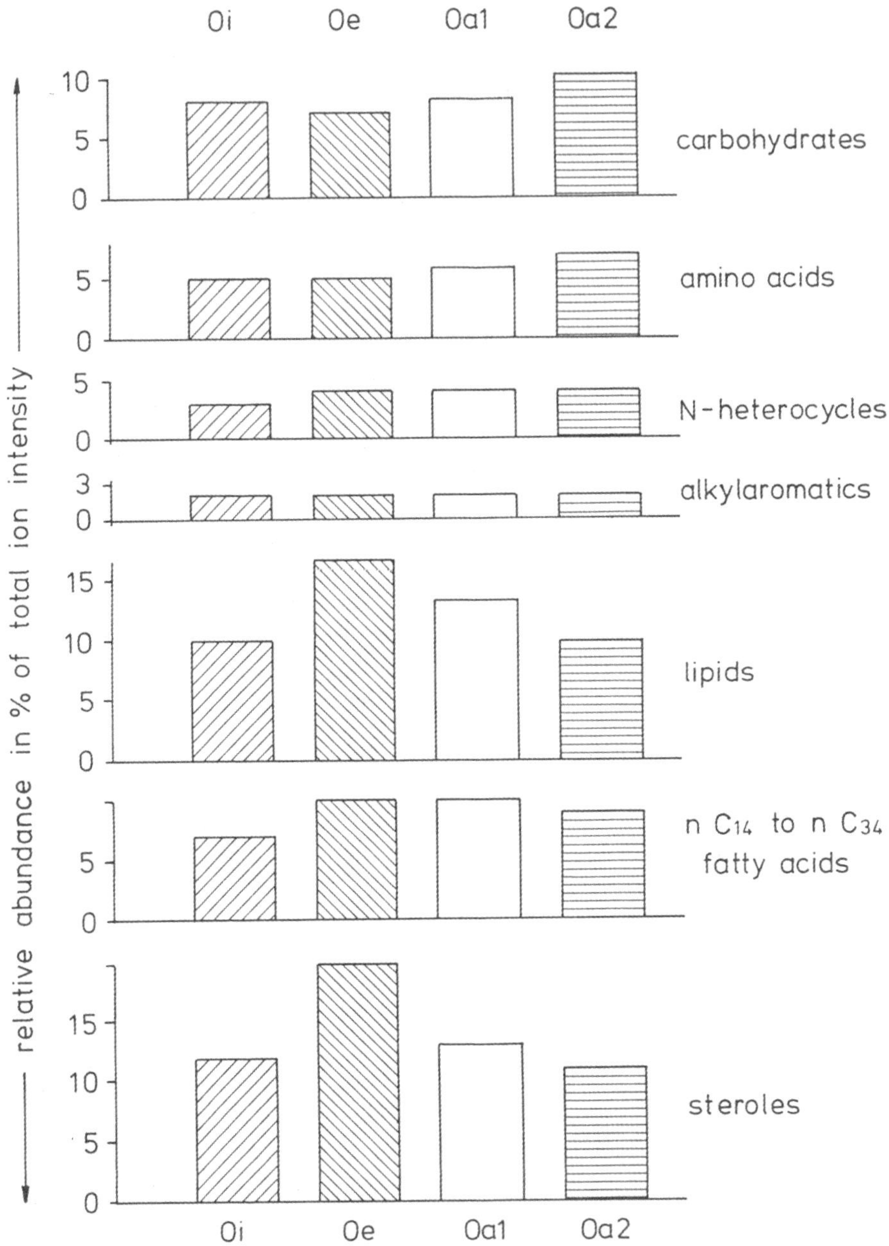

Figure 4.4.3. Summed and averaged Py-FIMS spectra for the temperature range 100°C - 350°C of the LH-horizon (a), H1-horizon (b), H2-horizon (c), and H3-horizon (d) of a Terric Leptic Gelic Histosol. The total ion intensities (TII) in the thermograms (upper right) were normalized in counts*10^6/mg per sample.

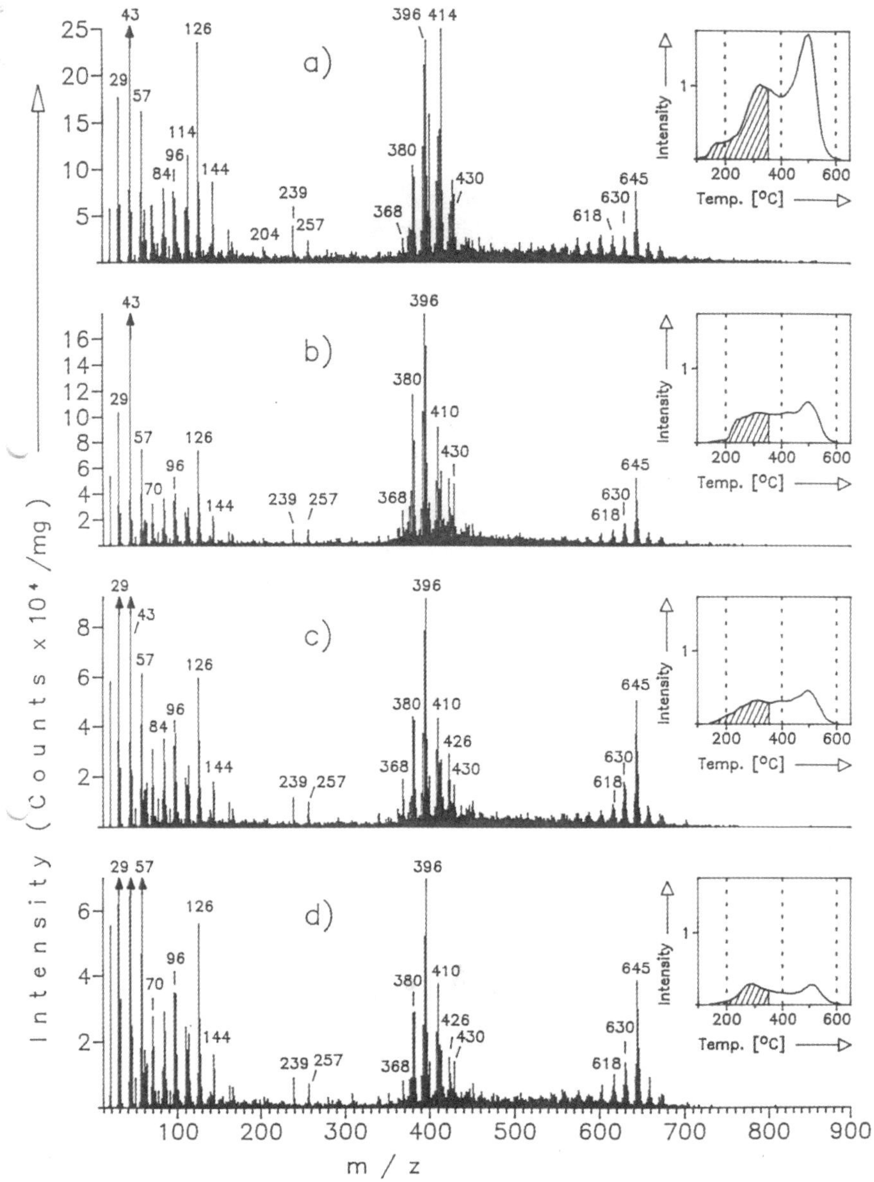

Figure 4.4.4. Summed relative abundance (percentage of TII) biomarkers of SOM after Py-FIMS in horizons (Oi = Lh, Oe = H1, Oa1 = H2, Oa2 = H3) of a Terric Leptic Gelic Histosol.

However, the high level of carboxyl carbon in all soil horizons is very uncommon. This might be the reason for the somewhat higher CEC_p in the soil at the abandoned penguin rookeries (Table 4.4.3). The abrupt change of the SOM composition between the Bhs, the IIBs, and the IIICw horizon of the Podzol on the penguin rookery (D3) is probably due to the layers of penguin droppings and deposits from different materials. Beyer et al. (1997) suggest for the spodic horizons that, in contrast to those in Germany, the SOM had a high percentage of amino derivates from proteins, polysaccharides, urates, and chitin, resulting in a mean C-to-N ratio of 10 and a high content of carboxylic carbon units, which probably derived from amino and other organic acids.

The podzolic soil from solid rock under mosses and lichens (C11) showed an aromatic carbon pattern comparable to that of Podzols in temperate climate regions. However, the aromatic content in the AE was very high, and residual preservation from the mosses cannot explained it. We do not know if there was another source of SOM in the area as relic organic matter or an input by eolian transport. An input by seabirds could be neglected, because the P contents were low (Table 4.4.4).

The greatest similarity in SOM composition to that of Podzols from temperate climates was in the podzolic soil from outwash sediments under mosses (WP5). However, the comparison of both AE horizons under mosses (WP5 and C11) confirmed the assumption that, in the shallow Spodic Cryosol from gneiss and schist (C11), another carbon source than recent vegetation was probable. Comparing the SOM composition of the soil horizons and selected organic parent materials, the correlation coefficient r (Table 4.4.6) suggests that, in soil WP5, the algal and mossy carbon sources influenced SOM to a great extent, whereas, in soil D3 and C11, they did so much less.

Table 4.4.7 shows the major ions we found in the Antarctic soil samples, using the Py-GC/MS technique. In general, no samples gave any indications of lignin. A variety of methylphenols were present, which indicate either protein or lignin. If the methylphenols were derived from lignin, however, there also should have been methoxylphenols (Hempfling and Schulten, 1990), which we did not observe. This means that the phenols are mainly from proteins.

The observed pattern variety of long-chain alcohols, acids, and ketones were not typical for soil (Hempfling and Schulten, 1990; Schulten and Schnitzer, 1992). Usually there are homologous series of alkanes and *1*-alkenes (Schulten and Schnitzer, 1992; 1997), neither of which were present in any of the Antarctic samples. For that matter, there were a few homologous series of alkylated aromatics (alkylated benzenes, naphthalenes, etc.) which usually occur in soil (Schulten, 1996a; 1996b). Most of the long chains in the Antarctic soils samples appeared to contain multiple oxygen- and nitrogen-containing functional groups, cyclic ionized and heterocyclic structures, and alkylations. However, the complex and easily ionized compounds and the low carbon content made mass spectral identification very difficult.

Table 4.4.4. Selected properties of Spodic Cryosols in eastern Antarctica.

Horizon	Depth	Color	pH	CE	BS	P_c	TOC	LOI	TOC	N_t	TOC	OD	Fe_d	Fe_{ox}	Fe_{org}
	cm	(moist)		C_p	%	------ mg/g ------			LOI		N_t	OE	----- mg/g -----		
Sandy-skeletal Spodic Leptic Cryosol (C11) from weathered gneiss rock under mosses and lichens															
AE	0-2.5	10YR4/1	3.83	99	18	0.3	23.7	53.8	0.44	2.21	10.7	246	2.16	1.09	0.61
Bh	-10	10YR2/1	3.73	140	15	0.4	38.5	75.7	0.51	2.52	15.3	377	3.08	1.74	0.90
BsC	-20	10YR4/3	3.88	57	11	0.7	9.7	19.9	0.49	0.99	9.8	246	2.76	1.16	0.34
IIC	-25	2.5Y5/3	4.01	54	9	1.3	5.4	12.1	0.44	0.53	10.3	164	4.14	1.51	0.38
R	>25	bare rock													
Sandy-skeletal Spodic Cryosol (WP5) from glacial outwash sediments under mosses															
AE	0-4	10YR4/1	3.83	163	19	0.2	39.5	92.5	0.43	2.96	13.4	164	1.50	0.84	0.70
Bh	-7	10YR2/1	3.83	222	12	0.4	64.7	133.9	0.44	4.99	13.0	414	2.23	1.43	1.33
C	-20	10YR4/3	4.09	84	10	ND	19.0	36.5	0.52	1.95	10.9	165	1.30	0.71	0.51
Cf	>20	permafrost													
Loamy-skeletal Spodic Cryosol (D3) at an abandoned penguin rookery with nil vegetation															
D	0-6	gravel pavement													
AE	-9	10YR4/1	3.47	137	21	1.6	27.6	61.7	0.45	0.96	28.9	211	3.06	2.15	1.15
Bh	-11	7.5YR2.5/1	3.65	229	20	5.5	36.5	86.2	0.42	3.41	10.7	641	6.58	6.11	6.00
Bhs	-16	7.5YR4/3	4.00	241	27	5.9	30.6	66.3	0.46	3.95	7.8	418	7.86	6.28	1.48
IIBs	-35	7.5YR5/4	4.70	364	57	6.2	25.7	53.2	0.48	3.51	7.3	202	4.83	4.74	0.24
IIICw	-71	10YR5/4	4.90	118	48	5.7	10.5	18.3	0.57	4.57	2.3	108	4.40	2.90	0.24

hor.: horizon; pH: measured in 10 *mM* $CaCl_2$; CEC_p: potential cation exchange capacity in $mmol_c$ /kg; BS: base saturation in % of CEC_p; P_c: citrate soluble phosphorus; TOC: total organic carbon; LOI: loss-on-ignition; N_t: total nitrogen; ODOE: optical density of the oxalate extract (Soil Survey Staff, 1998); Fe_d: dithionate extractable iron; Fe_{ox}: oxalate extractable iron; Fe_{org}: NaOH extractable iron.

Table 4.4.5. Soil organic matter composition according to CPMAS carbon-13 spectroscopy of Spodic Cryosols in eastern Antarctica (signal range in ppm shift to the reference TMS).

Horizon	Alkylic carbon			O-alkylic carbon				Aromatic carbon			Carboxylic carbon		
	0-25	-45	Σ	45-60	-90	110	Σ	110-140	-160	Σ	160-190	-220	Σ
Soil C11 from weathered gneiss rock under mosses and lichens in a small depression													
AE	6	10	16	5	15	12	33	19	11	30	14	7	21
Bh	11	13	24	6	15	8	29	11	11	22	15	10	25
Soil WP5 from glacial outwash sediments under mosses at a dried-up melt water lake side													
AE	10	15	25	8	25	12	45	10	6	16	11	3	14
Bh	11	17	28	8	25	11	14	9	5	14	10	4	14
C	10	15	25	7	20	10	37	14	9	23	12	3	15
Soil D3 under gravel pavement with nil vegetation at an abandoned penguin rookery													
AE	8	11	19	5	19	12	36	14	10	24	15	6	21
Bh	11	14	25	5	16	9	32	13	10	23	15	5	20
Bhs	9	14	23	5	12	10	27	14	11	25	18	7	25
IIBs	15	18	34	6	15	7	28	11	7	18	15	5	20
IIICw	7	19	12	5	11	6	22	22	6	36	18	5	23
Organic parent materials													
moss	ND	ND	23	8	24	20	52	9	5	13	ND	ND	12
algae	ND	ND	20	9	32	23	64	6	2	8	ND	ND	8
guano	ND	ND	29	ND	ND	ND	37	ND	ND	20	ND	ND	14

ND not determined.

Table 4.4.6. Linear correlation coefficients (r) between NMR subunits of the SOM in Spodic Cryosols of eastern Antarctica and organic matter of mosses and algae (for data, see Table 4.4.5, n = 9).

Soil	Horizon	r-value$_{algae}$	r-value$_{mosses}$
C1	AE	0.488	0.589
	Bh	0.480	0.610
WP5	AE	0.931***	0.966***
	Bh	0.911***	0.950***
	C	0.800**	0.856**
D3	AE	0.789**	0.807**
	Bh	0.613	0.730*
	Bhs	0.356	0.500
	IIBs	0.498	0.624
	IIICw	0.041	0.149

* $p < 0.05$.
** $P < 0.01$.
*** $p < 0.001$.

In all three Cryosols, the signal pattern was more detailed in the AE and Bh horizons than in the deeper layers, which indicated the direct influence of local vegetation or microbial colonization (Figures 4.4.5, 4.4.6, 4.4.7), as was found with the NMR data (see above). In soil C11 from weathered gneiss rock under mosses and lichens, only the albic AE and spodic Bh horizon contained smaller molecules, such as phenols and single ring aromatics (Figure 4.4.5, peaks no. 28, 37), whereas, in the deeper parental BsC and IIC horizons, long chain acids, alcohols, and ketones were present (Figure 4.4.5, peak no. 87, 93, 101). Only in the AE and the Bh did we find protein precursors (e. g., peak no. 9, 75), which was in line with a higher occurrence of bacterial microorganisms (Table 4.4.4). However, at the same time, the high peak no. 109 suggests the significance of unresolved compounds, which might have some aromatic structure, as both peak 79 (Figure 4.4.5, Table 4.4.7) and the NMR data suggest (Table 4.4.6).

A similar pattern appeared in soil WP5 from glacial outwash sediments under mosses, with respect to protein (Figure 4.4.6). Protein peaks were present only in the bleached albic AE and the dark-colored spodic Bh horizon, whereas lipids were in all horizons. Especially in the spodic horizon (Bh), the protein peak no. 9 was extraordinarily high. Peak no. 36 in the Bh and peak no. 88.4 in the C material indicated the predominance of amino carbohydrates and carbohydrates in this soil. The occurrence of the latter agrees with the extinct dominance of alkyl carbon units, according to the NMR experiments (Table 4.4.6).

Surprisingly, in soil D3 with nil vegetation at an abandoned penguin rookery, the uppermost AE horizon showed a similar pattern of the chromatogram as the two lower located ones (IIBs and IIICw), with the occurrence of long chain acids, alcohols, and ketones molecules. Phenols and single-ring aromatic carbon units dominated both spodic horizons of soil D3 (Bh and Bhs) (Figure 4.4.7). The highest occurrence of peak no. 98 in the three deeper subsoil horizons suggest a predominance of carbohydrates compared to all other soil layers that this study investigated.

Table 4.4.8 shows the summarized semi-quantitative data as a percentage of the assigned signals. In the Spodic Cryosol (C11) from weathered gneiss rock under mosses and lichens, the SOM of the albic AE and spodic Bh horizons held a large percentage of unresolved compounds but also contained some carbohydrates, protein and/or phenol precursors, and aromatic hydrocarbon precursors (AHP).

For all samples, most of the compounds composing the unresolved fraction appeared to be heterocyclic or cyclic molecules (Table 4.4.8). In addition, NMR suggests 20 to 30% aromatic carbon nature in soil C11 (Table 4.4.6). These molecules probably derived from larger humic molecules. Unresolved compounds (BsC) and lipids (IIC) dominated the deeper horizons.

The decrease of phenol precursors and AHP agree with the decrease of aromatic carbon units within the NMR experiments (Table 4.4.6). The increase of alkyl carbon from the AE to the Bh seen with NMR was not detectable in the lipid fraction of Py-GC/MS. This suggests that the unresolved fraction and/or the undetected part of SOM was more important.

In the Spodic Cryosol (WP5) from glacial outwash sediments under mosses in all samples, lipids were abundant (Table 4.4.8). As observed in the Cryosol from solid rock (C11) in soil WP5, the SOM pattern of the upper AE and Bh horizons was similar, consisting of protein, phenol precursors, carbohydrates, trace amounts of amino carbohydrates, and low amounts of unresolved compounds, whereas the SOM in the parental C horizon had abundances of higher amounts of unresolved carbon units, lipids, and, in contrast to the C horizon of the Podzol C11, amino carbohydrates (Table 4.4.8). As observed in soil C11, proteins occurred only in the two top soils, probably as a consequence of microbial colonization. The higher protein content correlates with the greater occurrence of algal colonization (Table 4.4.4), the main precursor of chlorophyll (Bölter, 1995; 1997).

In the Spodic Cryosol (D3) with nil vegetation at an abandoned penguin rookery was a high abundance of amino carbohydrates, except in the IIBs layer. None of the other Cryosols contained nearly those high concentrations of amino carbohydrates. This might be due to some degraded products of uric acids (White and Beyer, *unpublished*).

Uric acid itself might hide in the unresolved fraction, especially given heterocyclic moieties (Table 4.4.7, peaks 78, 79), which occurred in most of the horizons of soil D3 (Figure 4.4.7). This probably reflects the influence of the organic input from guano. In contrast to the other Cryosols, soil D3 has abundant miscellaneous carbohydrates in all five horizons, without having the clear differentiation from the upper soil horizons and the subsoil, whereas it has abundant lipids only in three horizons (AE, Bh and IIBs). Proteins occurred in the upper soil layers, as in the other Cryosols.

Table 4.4.7. Major ions detected with Py-GC/MS in Spodic Cryosols in eastern Antarctica and their assignment to probable parent compounds.

peak no.	**major ion** (ion in highest abundance)	**most probable chemical compound** *(probable parent compound)*
9	95, 96	dimethylpyrole: *protein*
15	50, 78, 103, 104, 105	styrene: *aromatic hydrocarbon precursors*
23	51, 61, 109, 110	furancarboxaldehyde: *carbohydrate*
28	66, 94, 95	phenol: *phenol precursors*
30	40, 41, 47, 91, 92	toluene: *aromatic hydrocarbon percursors*
31	63, 69, 83, 84, 112	fluorophenol: *carbohydrate*
34	43, 65, 82, 107, 108, 137	benzeneacetonitrile: *amino carbohydrate*
35	51, 90, 97, 117	2-methylphenol: *phenol precursor*
36	42, 53, 68, 69, 70, 96, 97, 98	heptanol: *carbohydrate*
37	42, 51, 59, 77, 79, 90, 107, 108	4-methylphenol: *phenol precursor*
38	42, 47, 55, 62, 97, 126	carbohydrate: *carbohydrate*
42	42, 43, 53, 57, 58, 60, 69, 70, 85, 95, 116	polysaccharide: *carbohydrate*
45	42, 43, 51, 57, 60, 69, 71, 97, 98, 10	*unresolved*
46	42, 59, 69, 91, 131	alkylbezene aromatic: *hydrocarbon pre-cursors*
48	45, 51, 83, 91, 111, 115	*unresolved*
50	43, 44, 56, 57, 90, 117	indole: *protein*
56	41, 49, 59, 65, 75, 114, 122	*unresolved*
60	42, 71, 87, 98	long chain `arrangement´: *lipid*
65	43, 54, 60, 69, 116	*unresolved*
67	42, 55, 60, 63, 71, 73, 78	long chain carboxylic acid: *lipid*
75	53, 92, 93, 186	alkylpyridine: *protein*
77	44, 51, 54, 60, 82, 96, 97, 124	long chain: *lipid*
78	41, 50, 63, 77, 78, 91, 143, 221, 236	heterocyclic: *unresolved*
79	42, 67, 77, 91, 115, 128, 205, 220	alkylnaphthalene: *unresolved*
82	55, 70, 101, 115, 128, 203, 205	long chain `arrangement´: *lipid*
83	42, 48, 52, 91, 119	phenylalkyl `arrangement´: *unresolved*
85	43, 205, 220	*unresolved*
87	44, 58, 81, 91, 117, 195	long chain `arrangement´: *lipid*
88.2	41, 42, 58, 91, 92, 221	*unresolved*
88.4	42, 100	alkylated acetamide: *aminocarbohydrate*
88.6	43, 55, 69, 83, 96, 97, 110, 124, 138	alkenylcyanide: *aminocarbohydrate*
89	43, 57, 73, 81, 82, 91, 94, 95, 110, 141	*unresolved*
93	42, 55, 69, 85, 97, 115, 129, 256	hexadecanoic acid: *lipid*
97	55, 56, 60, 68, 72, 73, 81	*carbohydrate*
98	41, 57, 70, 83, 96, 97, 110, 124, 180, 208	alkenylcyanide: *carbohydrate*
99.2	48, 57, 62, 73, 85, 99, 141	cycloalcohol: *unresolved*
99.4	41, 55, 69, 83, 91, 109	long chain ketone: *lipd*
101	89, 103	long chain alcohol: *lipid*
103	41, 42, 56, 83, 122	long chain acid, alcohol or ketone: *lipid*
108	43, 5, 71, 83, 97, 111, 123, 310	thiaphenyl `arrangement´: *unresolved*
109	43, 77, 83, 280	*unresolved*

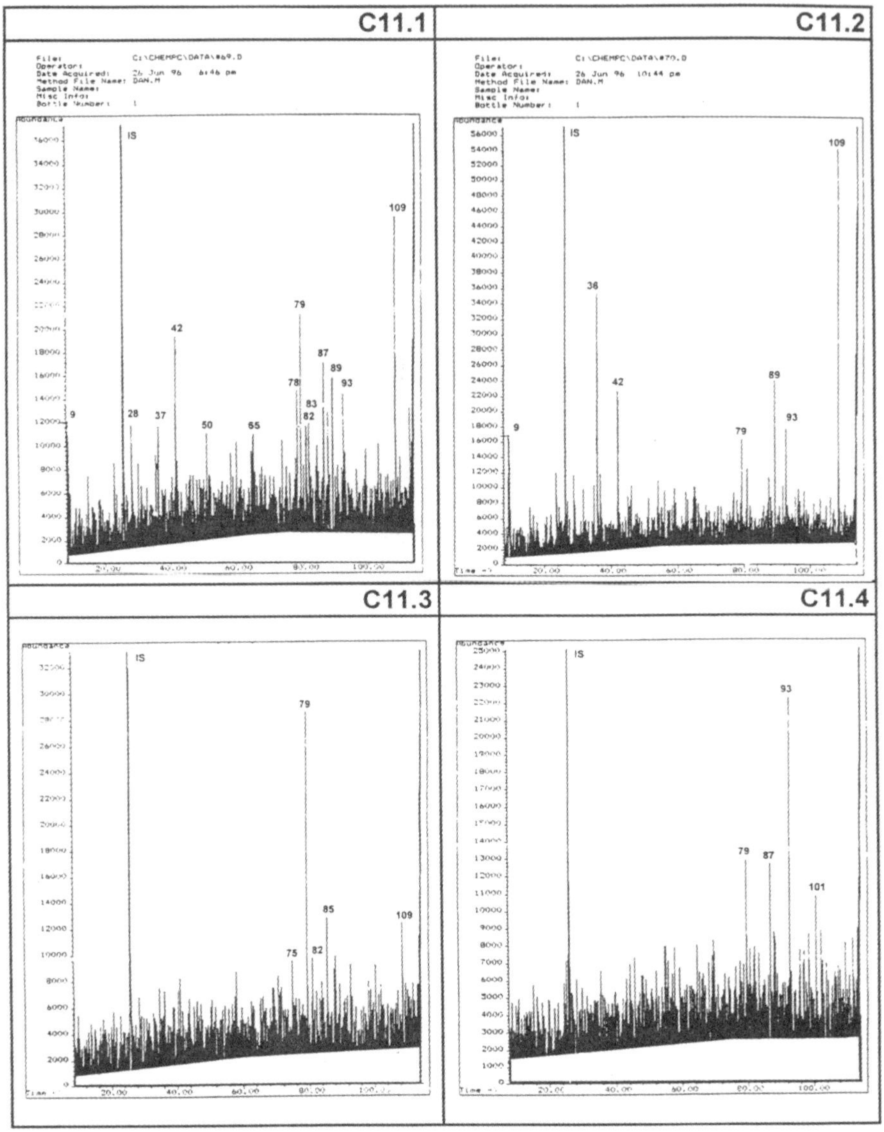

Figure 4.4.5. Ion chromatograms of pyrolysis GC/MS of a Spodic Leptic Cryosol (C11) from weathered gneiss rock under mosses and lichens.

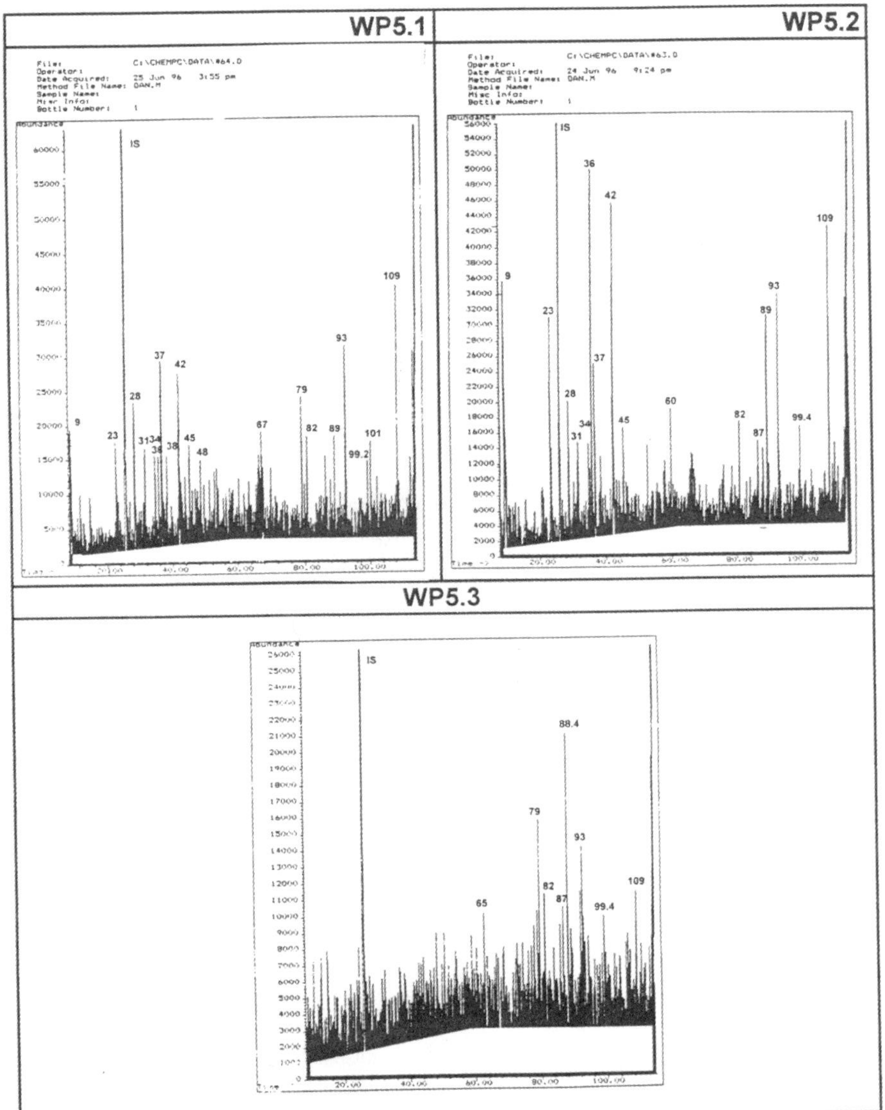

Figure 4.4.6. Ion chromatograms of pyrolysis GC/MS of a Spodic Cryosol (WP5) from glacial outwash sediments under mosses.

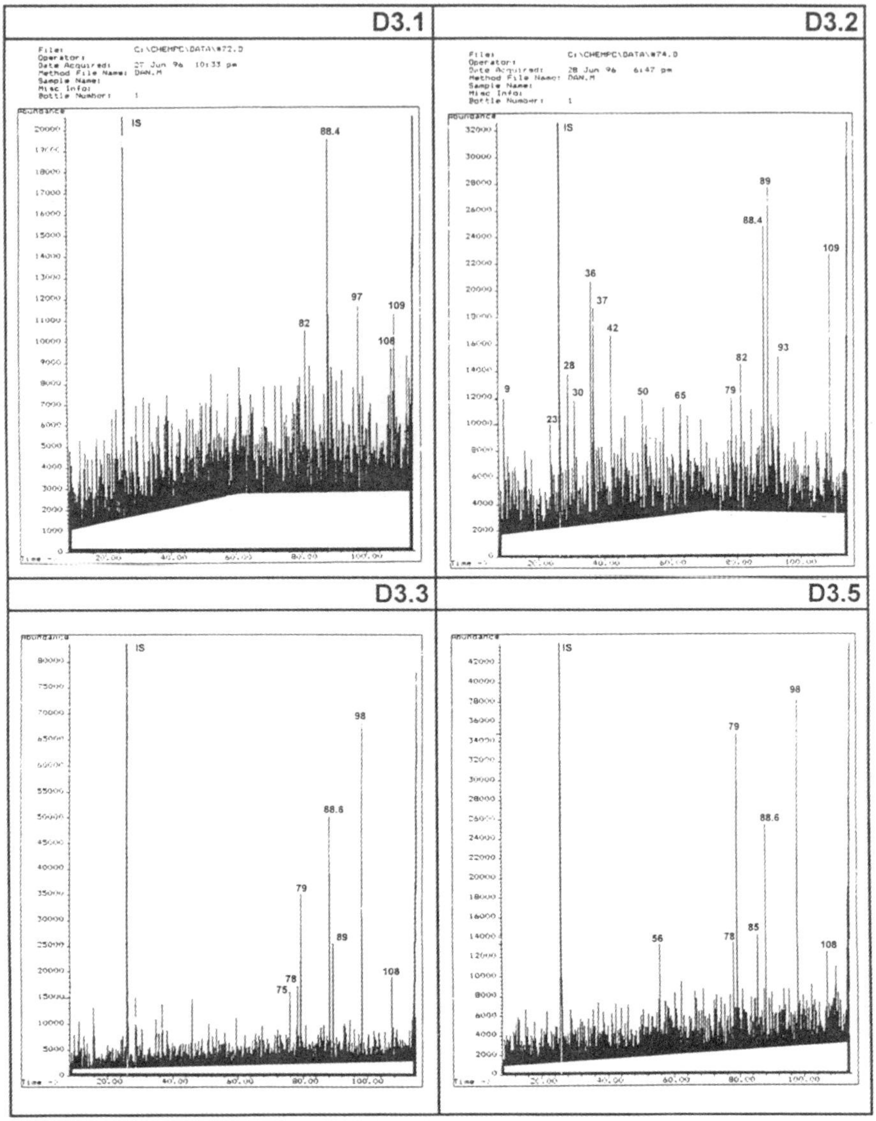

Figure 4.4.7. Ion chromatograms of pyrolysis GC/MS of a Spodic Cryosol (D3) with nil vegetation at an abandoned penguin rookery.

Table 4.4.8. SOM compound classes according to the pyrolysis GC/MS experiments in Spodic Cryosols in eastern Antarctica.

Horizon	Protein	Phenol precursors	Carbo hydrates	Aromatic hydro- pre- cursors	Lip- ids	Amino- carbo- hydrates	Unresolved
	---------- % of assigned signals ----------						
Soil C11 from weathered gneiss rock under mosses and lichens in a small depression							
AE	9	9	14	3	20	ND	46
Bh	8	ND	30	ND	15	ND	48
BsC	11	ND	ND	ND	11	ND	78
IIC	ND	ND	ND	ND	79	ND	21
Soil WP5 from glacial outwash sediments under mosses at a dried-up melt water lake side							
AE	5	14	23	ND	22	4	32
Bh	8	10	33	ND	25	3	21
C	ND	ND	ND	ND	39	20	42
Soil D3 under gravel pavement with nil vegetation at an abandoned penguin rookery							
AE	ND	ND	19	ND	16	33	32
Bh	9	13	24	4	10	10	29
Bhs	7	ND	30	ND	ND	22	41
IIBs	ND	ND	43	ND	34	ND	23
IIICw	ND	ND	27	ND	ND	17	56

ND: none detected.

NMR and Py-GC/MS suggest a very specific SOM pattern in the IIBs horizon. Both techniques indicated the abundance of such carbon moieties to a greater extent. The strong decrease of *O*-alkylic carbon units within the soil profile observed with NMR was not found with Py-GC/MS. The minimum of aromatic carbon in the IIBs horizon might correlate to the occurrence of carbohydrates. However, the observed maximum in carbohydrates was in contrast to the NMR data.

This suggests that pyrolysis data do not reflect total SOM, due to the occurrence of huge amounts of unresolved compounds (Figure 4.4.5, 4.4.6, 4.4.7) and the unconsidered SOM fractions (>20%, compare methods). In any case, the data combination suggests that a huge amount of aromatic carbon hides in the unresolved fraction of Py-GC/MS. The NMR data suggest that a decrease of alkylic carbon and increase of carboxylic carbon units modified SOM from the parent organic matter material, guano, to the soil horizons.

5. Discussion

Dead moss is the initial organic matter for soil formation in the Terric Leptic Gelic Histosols (US: Lithic Hemistel). According to the NMR data, aliphatic carbon units (acetal, hydroxyl, alkyl) dominate in the SOM. Tentative assignments by Py-FIMS indicate that the alkyl compounds consist of lipids, fatty acids, and sterols. *O*-alkylic compounds were partly identified as carbohydrates.

Obviously, the non-identifiable part within Py-FI mass signals (about 50% of TII) is higher within the carbohydrates than within the lipid fraction. This observation is similar to that of Nip et al. (1987) in a comparison of flash pyrolysis and NMR. The aromatic compounds are not from lignin, because the typical signals in Py-FIMS are missing.

Py-FIMS data suggest that alkylaromatics are present in the necrotic mosses. Under temperate conditions, these compounds are typical structural building blocks of humic substances (Schnitzer and Schulten, 1992). Beyond this, in the LH-horizon, only scarce humification is visible (Table 4.4.1). Therefore, the high amount of carboxyl groups probably derive from fatty acids und proteins (Wilson, 1987). With regard to the other peaks in the NMR spectra, the aromatic structures possibly derive from aromatic amino acids like histidine ($C_{carboxyl}$: 174 ppm, C_{β}: 28 ppm, $C_{aromatic}$: 117 to 136) or tyrosine ($C_{carboxyl}$ 172 to 175 ppm, C_{β}: 27 to 28 ppm, $C_{aromatic}$ 114 to 138 ppm) (Breitmaier and Voelter, 1989).

The NMR data show that carbohydrates decomposed in the Terric Leptic Gelic Histosol during the soil's formation from mosses, but the pyrolysis data do not reflect this trend. However, Py-FIMS did not yield evidence for a significant decomposition or transformation of carbohydrates because the NMR and the pyrolysis data are not directly comparable. Py-FIMS probably underestimates carbohydrates in relation to total SOM.

Our results confirm Nip et al.'s investigations (1987) showing that, despite high abundances of polysaccharides in biomacromolecules, flash pyrolysis reflects only a limited and specific part of polysaccharides. The drastic change in the C/N ratio from LH to H1 (Table 4.4.1) confirms the degradation of the nitrogen-free polysaccharides in the soil profile.

The most striking feature of the carbon-13 NMR data is the drastic increase of alkyl carbon. As the methods section here describes, the lipids detected by Py-FIMS reflect only selected signals for alkyl-diesters, fatty acids, alkenes, alkanes, alkyl-monoesters, and sterols in the mass range from 202 to 732 m/z. Alkylics of lower molecular weight or other chemical structure are not considered. Thus, the lipids represent only a special part of the alkylic carbon that carbon-13 NMR detects.

The relative intensities (percentage of total ion intensity) of the specific Py-FI mass signals for lipids increased in the H1 and H2 horizons, compared to the top and lowest horizons. Complementary to carbon-13,NMR spectroscopy's determination of the general enrichment of alkyl carbon with increasing soil depth, Py-FIMS' detection of specific, biologically important lipid-derived structures enabled a deeper insight into the behavior of alkyl structures in the profile of the Terric Gelic Histosol.

The decomposition of organic matter is not as intensive as in soils under temperate climate (Kögel-Knabner, 1992, 1993). The governing influence is the extreme cold, which strongly retards the transformation of fresh organic residues (Wilson, 1990). Therefore, in this Antarctic Histosol, carbohydrates dominate the SOM composition also in deeper horizons, together with recalcitrant, alkylic compounds.

Schnitzer et al. (1990) observed similar effects in the organic layer of Canadian subarctic soils. Mosses as plant precursors of the organic matter in the Terric

Gelic Histosol do not contain lignin (Hurst and Burges, 1967). Thus, the aromatic structures that carbon-13 NMR and Py-FIMS identify demonstrate that lignin input is not necessary for forming aromatic humic structures. These compounds might derive from aromatic amino acids (Breitmaier and Voelter, 1990). On the other hand, the synthesis of these compounds is possible from ketonic and carboxylic structures (Ziechmann, 1980). The occurrence of aromatic constituents in SOM of an Antarctic Histosol is in line with results of Wilson et al. (1986), who observed aromatics in humic acid fractions from Antarctic soils.

The greenish gray soil color in all horizons of the Fibric Leptic Gelic Histosol (US: Lithic Fibristel) corresponds to the wet-chemical and spectroscopic data indicating only very weak transformation processes in this soil. The organic nitrogen is mostly bound in amino compounds. These results suggest that proteins are of major importance.

Obviously, the carboxyl carbon was derived from fatty acids and proteins (Breitmaier and Voelter, 1989; Wilson, 1990), given the lack of significant humification. The alkyl units derive probably from ether-linked *n*-alkyl chains (Gelin et al., 1993). These highly non-hydrolyzable biomacromolecules are located in the outer cell wall of several species of algae (Largeau et al., 1990) and have a high potential for preservation in soils and sediments (Gelin et al., 1993). However, the very slight increase of alkyl carbon units in the Fibric Leptic Gelic Histosol does not reflect their selective preservation (Gelin et al., 1993; Largeau et al., 1990; Nip et al., 1986).

Due to the proximity of a meltwater lake, the whole soil profile is very moist and, consequently, the thermal capacity of this Histosol is high. The solar energy input into the Antarctic is not sufficient to increase soil temperature in such wet soils significantly. Therefore, transformation processes of organic matter are not detectable.

In contrast, in water-saturated soils under temperate climate conditions, the initial SOM can slowly transform, because soil temperatures in summer are high enough to stimulate microbial decomposition processes (Blume et al., 1991). The Terric Gelic Histosol has a lower moisture level, and soil temperatures during the Antarctic summer are high enough to enable transformation processes of organic matter. Bölter (1992) measured soil temperatures up to 12°C in neighboring soils during the Antarctic summer. Microbial activity is much higher under these conditions, and transformation processes of organic matter lead to the brownish to black color of the deeper horizons (Smith, 1990; Blume et al., 1997).

In the mineral Cryosols, the great difference in SOM composition and SOM depth function in the profiles suggests that parent materials and different carbon sources influence the mechanisms of humification and translocation in the soil. Probably not only a special vegetation, such as heather in temperate climate regimes but also more microclimate and soil microbial consumption and/or production of organic compounds are responsible for the occurrence and intensity of podzolization and transformation of organic matter.

For example, Beyer et al. (2001) found that the total bacterial colonization (TBC) of vegetated soils correlates to selected organic compounds. TBC and terminal methyl groups (0 to 25 ppm) showed a weak correlation of $r = 0.674^*$. The same authors found that the alkyl-C/O-alkyl-C ratio correlated well to TBC ($r = 0.709^{**}$), whereas TBC negatively correlated between the signals of olefinic and alkyl aromatics (110 to 140 ppm), with an *r*-value of -0.820**, suggesting negative interactions between soil microbes and aromatic SOM compounds. These correlations also may suggest that bacterial cell carbon affects the SOM composition, or vice versa, since in temperate climates microbial activity and SOM composition well correlate and interact (Capriel et al., 1995).

The soil-forming processes in the Spodic Cryosol from solid gneiss rock under mosses indicate a strong decrease of *O*-alkylic carbon and an increase of aromatic and carboxylic carbon units from the recent moss vegetation and probably some soil algae during humification. This is why White and Beyer (1999), using the Py-GC/FID technique, found a strong correlation between the pyrogram of recent vegetation and the underlying soil horizons. (*Py-GC/FID* is pyrolysis gas chromatography/flame ionization detection.)

NMR and some Py-GC/MS data confirmed a different SOM composition in the bleached AE horizon and the dark spodic Bh horizon, as the classification as a Podzol suggests (WRB-FAO, 1998). Proteins are mainly in the top soils, probably due to a favorite colonization with microbes (Beyer et al., 2000c). In the Antarctic podzolic soil from weathered gneiss rock under mosses, alkylic carbon units have moved from the AE to the Bh and probably into deeper subsurface horizons, as the high abundance of lipids in the subsoil suggests. Aromatic compounds have been preserved in the uppermost topsoil layer.

In the Spodic Cryosol from glacial sediments under mosses, the NMR data indicate a decrease of O-alkyl C but only a weak increase of aromatic and carboxylic carbon units from the recent soil algae colonization (Bölter, unpublished). Surprisingly, neither NMR nor Py-GC/MS data indicated a different SOM composition in the bleached AE horizon and the dark spodic Bh horizon, as a Podzol classification suggests (Soil Survey Staff, 1998; WRB-FAO, 1998). However, both methods suggest a change in SOM quality in the C horizon.

The higher carbohydrate content that Py-GC/MS found agrees with the somewhat higher amount of hydroxyl carbon units of NMR (60-90 ppm). However, the increase of the aromatic carbon from the spodic Bh to the parental C horizon was not detected as phenol precursors or AHP. As described for the soil C11, these moieties are probably in the unresolved fraction or in the amino carbohydrate fraction, as is true for soil D3, where the disappearance of amino carbohydrates in the fourth horizon correlates well to a significant decrease of the aromatic carbon units.

In addition, since we are looking at the largest amount of components, some moieties occurring less could be omitted. In contrast to soil C11, NMR data did not show the increase of the lipid fraction from the albic AE and the spodic Bh to the parental C horizon. In summary, in the Spodic Cryosol from glacial outwash sediments under mosses, the SOM chemistry does not reflect the soil morphology. However, the Py-GC/MS data suggest a translocation of alkylic carbon moieties with podzolization.

The Spodic Cryosol with nil vegetation at an abandoned penguin rookery (D3) had a high abundance of amino carbohydrates, except in the IIBs horizon. None of the other Cryosols contained nearly this concentration. The NMR data suggest SOM was modified from the parent organic matter, guano, to the soil horizons by decreasing alkylic carbon and increasing carboxylic carbon units. However, the impact of layering, due to the possible deposition of different materials by seabirds, complicates interpreting the SOM data with respect to soil-forming processes (Beyer et al., 2000a).

The NMR data indicate a strong downward movement of alkylic carbon, with a simultaneous enrichment of *O*-alkylic carbon, whereas aromatic and carboxylic carbon moieties are stable in the profile, and an influence of layering with a different organic matter source in the deepest horizon is possible. Py-GC/MS showed, in most of the five soil horizons, a completely different SOM pattern, which indicates a much stronger influence of seabird-induced layering than of *in-situ* soil formation. In contrast to soil WP5, in the podzolic soil D3 with nil vegetation at an abandoned penguin rookery, the SOM chemistry reflects the soil morphology with the assignment of a bleached AE and the dark-colored spodic Bh horizon. But the Py-GC/MS data suggest no translocation of lipid moieties.

6. Conclusions

The integrated approach, including morphological observations, wet-chemical analyses, CPMAS carbon-13 NMR, and pyrolysis experiments, allowed new insights into soil organic matter (SOM) composition and transformation of the investigated Antarctic permafrost-affected soils. The SOM of the Gelic Histosols consisted mainly of carbohydrates and lipid-derived compounds. In the water-saturated Gelic Histosol, transformation processes are not evident, but slight decomposition and transformation processes of SOM occurred under drier conditions. Especially carbohydrates decompose and alkylic compounds are enriched. Of special interest is the detection of aromatics in the SOM of Antarctic Histosols. Thus, our data show that aromatic lignins are not necessary for the formation of aromatic humic substances.

In Cryosols on the coast of eastern Antarctica, SOM composition has a high level of carboxyl carbon and lipidic moieties in all soil horizons, which is very unlike that of temperate regions. The correlation of vegetated soils' selected SOM compounds and bacteria suggests that algal and moss carbon sources influence SOM greatly. The great variety in SOM composition and SOM depth suggests that parent materials and different carbon sources influence humification and translocation mechanisms. We think that not only vegetation but predominantly microclimate and soil microbial consumption and/or production are responsible for the complicated geochemistry of these soil processes.

The observed variety in the pattern of long-chain alcohols, acids, and ketones which the Py-GC/MS technique showed are not typical for soil. Most of the long chains in the Antarctic soil samples appear to contain multiple oxygen- and nitrogen-containing functional groups, cyclic ionized and heterocyclic structures, and alkylations, which the occurrence of 20 to 40% aromatic NMR carbon subunits confirm. However, the complex and easily ionized compounds and the low carbon content made Py-GC/MS mass spectral identification very difficult, which results in a high amount of unresolved fractions.

With respect to soil formation of a Spodic Cryosol from weathered gneiss rock under mosses, alkyl carbon units have moved from the AE to the Bh and probably into deeper located horizons, whereas aromatic compounds have been preserved in the uppermost topsoil layer. In a Spodic Cryosol from glacial outwash sediments under mosses, the SOM chemistry does not reflect the soil morphology. However, the pyrolysis data indicate also a translocation of alkyl carbon moieties during podzolization.

In contrast, in a Spodic Cryosol with nil vegetation at an abandoned penguin rookery, the SOM chemistry reflects the soil morphology, with the assignment of a bleached AE and dark spodic Bh horizon. However, pyrolysis data does not suggest any translocation of lipid moieties, and so we think that podzolization is chemically an ill-defined soil-forming process in Antarctic soils. Parent materials, physical and mechanical properties, microclimate, local moisture regime, and microbial colonization affect the processes of formation and translocation of SOM.

Finally, finding the origin of Antarctic SOM should be a future research emphasis of Antarctic soil science. Our results suggest that SOM transfer may play an important role in nutrient cycles in Antarctic soil ecosystems. A final conclusion requires verifying water and soluble SOM movement in the landscapes, which is more a technical than a mythological problem, due to the extreme stoniness and shallowness of most Antarctic soils. Beyond this, investigations of soil-forming processes in Antarctica might be useful for understanding the more complex pedogenic processes under temperate climate conditions.

7. Acknowledgments

The Australian Antarctic Divison, Kingston, Tasmania, and the German Research Council (DFG), Bonn - Bad Godesberg, financially supported these investigations. The Australian Antarctic Research Expeditions, and especially the Casey crew in the summers of 1990-1991 and 1995-1996, supported the field and lab work. We gratefully acknowledge all of them.

References

Beyer, L. 1996. Soil organic matter composition in Podzols of the Northwest German Lower Plain. The Sci. of the Total Environ. 181:167-180.

Beyer, L. 2000. Properties, formation and geo-ecological significance of organic soils in the coastal region of East Antarctica (Wilkes Land). Catena. 39:79-93.

Beyer, L., and Bölter, M. 2000. Ecology and geography of Spodic Cryosols in terrestrial ecosystems of East Antarctica. Catena. 39:95-119.

Beyer, L., H.-R. Schulten, and R. Fründ. 1992. Properties and composition of soil organic matter in forest and arable soils in Schleswig-Holstein. 1. Comparison of morphology and results of wet chemistry, CPMAS ^{13}C-NMR spectroscopy and pyrolysis mass spectrometry. Zeitschrift für Pflanzenernährung und Bodenkunde. 155:345-354.

Beyer, L., C. Wachendorf, and C. Koebbemann. 1993. A simple wet chemical retraction procedure to characterize soil organic matter. 1. Application and recovery rate. Comm. in Soil Sci. and Plant Anal. 24:1645-1663.

Beyer, L., H.-P. Blume, H. Knicker, and M. Bölter. 1997. Soil organic matter of suggested spodic horizons in relic ornithogenic soils of coastal continental Antarctica (Casey Station, Wilkes Land) and spodic horizons in soils of Germany. Soil Sci. 162:518-527.

Beyer, L., J.G. Bockheim, I.B. Campbell, and G.G.C. Claridge. 1999. Review - Genesis, properties and sensitivity of Antarctic Gelisols. Antarc. Sci. 11(4):387-398.

Beyer, L., M. Bölter, and R.D. Seppelt. 2000a. Nutrient and thermal regime and microbial biomass and vegetation pattern of soils in the coastal region of Antarctica (Wilkes Land). Arct., Antarc.. Alp. Res. 32:30-39.

Beyer, L., K. Pingpank, G. Wriedt, and M. Bölter. 2000b. Soil formation in coastal continental Antarctica. Geoderma. 95:283-304.

Beyer, L., D.M. White, and M. Bölter. 2001. Organic matter composition, and transformation, and microbial colonisation of Gelic Podsols in the coastal region of East Antarctica. Austral. J. Soil Res. 39:543-563.

Blume, H.-P., L. Beyer, and F. Friedrich. 1991. Correlations between the microbial activity and the water, air, temperature and nutrient status of different soils under different land use. In: G. Esser and D. Overdieck, eds. Modern Ecology: Basic and Applied Aspects. Elsevier. Amsterdam. pp. 321-346.

Blume H.-P., D. Schneider, and M. Bölter. 1996. Organic matter accumulation in and podzolization of Antarctic soils. Zeitschrift für Pflanzenernährung und Bodenkunde. 159:411-412.

Blume, H.-P., L. Beyer, M. Bölter, H. Erlenheuser, E. Kalk, S. Kneesch, U. Pfisterer, and D. Schneider. 1997. Pedogenic zonation in soils of Southern circumpolar region. Adv. in GeoEcol. 30:69-90.

Bölter, M. 1990. Microbial ecology of soils from Wilkes Land, Antarctica. I. The bacterial population and its activity in relation to dissolved organic matter. Proc. NIPR Symposium on Polar Biology. 3:104-109.

Bölter, M. 1992. Environmental conditions and microbiological properties from soils and lichens from Antarctica (Casey station, Wilkes Land). Polar Biol. 11: 591-599.

Bölter, M. 1995. Distribution of bacterial numbers and biomass in soils and on plants from King-George Island (Arctowski Station, Maritime Antarctica). Polar Biol. 15:115-124.

Bölter, M. 1997. Microbial communities in soils and on plants from King-George Island (Arctowski Station, Maritime Antarctica). In: B. Battaglia, J. Valencia, and D. W. H. Walton, eds. Antarctic communities: species, structure and survival. University Press. Cambridge. pp. 162-169.

Bölter M., H.-P. Blume, and H. Erlenkeuser. 1994. Pedogenic, isotopic and microbiological properties of Antarctic soils. Polarforschung. 64:1-7.

Bracewell, J.M., J. Haider, S.R. Larter, and H.-R. Schulten. 1989. Thermal degradation relevant to structural studies of humic substances. In: M.H.B. Hayes. P. MacCarthy,

R.L. Malcolm, and R.S. Swift, eds. Humic Substances II. John Wiley and Sons. New York. pp. 181-222.

Breitmaier, E., and W. Voelter. 1990. Carbon-13 NMR Spectroscopy. Verlag Chemie. Weinheim.

Broady, P.A., and R.N. Weinstein. 1998. Algae, lichens and fungi in La Gorce Mountains, Antarctica. Antarc. Sci. 10:376-385.

Campbell, I. B., and G.G.C. Claridge. 1987. Antarctica: soils, weathering processes and environment. Elsevier. Amsterdam.

Campbell, I.B., G.G.C. Claridge, D.I. Campbell, and M.R. Balks. 1998. Short and long term impacts of human disturbances on snow free surfaces in Antarctica. Polar Records. 34:15-24.

Capriel, P., T. Beck, H. Borchert, H. Grönholz,and G. Zachmann. 1995. Hydrophobicity of the organic matter in arable soils. Soil Biol. and Biochem. 27:1453-1458.

Chen, J., and H.-P. Blume. 1998. Impact of human activates on the terrestrial ecosystems of Antarctica - a review. Polarforschung. 65:83-92.

Daly, B.K. 1982. Identification of Podzols and podzolized soils in New Zealand by relative absorbance of oxalate extracts of A and B horizons. Geoderma. 28:29-38.

Dyer, B. 1894. On the analytical determination of probable mineral plant food in soils. Transacta Chem. Soc. 65:115-67.

Fründ, R. and H.-D. Lüdemann. 1989. The quantitative analysis of the solution- and CPMAS ^{13}C-NMR spectra of humic material. THe Sci. of the Total Environ. 81/82:157-168.

Gelin, F., J.-P. L.A. Gatellier, J.S. Sinninghe Damsté, P. Metzger, S. Derenne, C. Largeau, and J.W. de Leeuw. 1993. Mechanisms of flash pyrolysis of ether lipids isolated from the green microalga Botryococcus braunii race A. J. Anal. and App. Pyrolysis.27:155-168.

Goodwin, I.D. 1993. Holocene deglaciation, sea-level change, and the emergence of the Windmill Islands, Budd Coast, Antarctica. Quat. Res. 40:70-80.

Hauptmann, S. 1992. Einführung in die organische Chemie.Deutscher Verlag für Grundstoffindustrie. Leipzig.

Hempfling, R., and H.-R. Schulten. 1990. Chemical characterization of the organic matter in forest soils by Curie point pyrolysis-GC/MS and pyrolysis-field ionization mass spectrometry. Org. Geochem. 15:131-145.

Hempfling, R,. and H.-R. Schulten. 1991. Pyrolysis-(gas chromatography/) mass spectrometry of agricultural soils and their humic fractions. Zeitschrift für Planzenernährung und Bodenkunde. 154:425-430.

Hurst, H.M., and N.A. Burges, 1967. Lignin and humic acids. In: A.D. McLaren and G.H. Peterson, eds. Soil Biochemistry Vol. 1. Marcel Dekker. New York. pp. 260-286.

Jones, J.B. 1991. Kjeldahl method for nitrogen determination. Micro-Macro Publishing. Athens.

Kögel-Knabner, I. 1992. Forest soil organic matter: Structure and formation. Bayreuther Bodenkundliche Berichte. 14:1-102.

Kögel-Knabner, I. 1993. Biodegradation and humification processes in forest soils. In: J.M. Bollag and G. Stotzky, eds. Soil Biochemistry, Vol. 8. Marcel Dekker. New York. pp. 101-135.

Largeau, C., S. Derenne, E. Casadevall, C. Berkaloff, M. Corolleur, B. Lugardon, J.F. Paynaud, and J. Connan, 1990. Occurrence and origin of "ultra laminar" structures in "amorphous" cerogens of various source rocks and oil shales. Org. Geochem. 16:889-895.

McKeague, J. A., F. DeConnick, and D.P. Franzmeier. 1983. Spodosols. In: L.P. Wildung, N.E. Smeck, and G.F. Hall, eds. Pedogenesis and Soil Taxonomy. II. The Soil Orders. Elsevier. Amsterdam. pp. 217-252.

Melick, D.R., and R.D. Seppelt. 1992. Loss of soluble cabohydrates and changes in freezing point of Antactic bryophytes after leaching and repeated freeze-thaw cycles. Antarc. Sci. 4:399-404.

Melick, D.R., and R.D. Seppelt. 1994. Seasonal investigation of soluble carbohydrates and pigment levels in Antarctic byophytes and lichens. Bryologist. 97:98-105.

Nip, M., E.W Tegelaar, J.W. de Leeuw, and E.W. Schenck. 1986. A new non-saponifiable highly aliphatic and resistant biopolymer in plant cuticulas. Naturwissenschaften. 73:579-585.

Nip, M., J.W. de Leeuw, P.J. Holloway, J.P.T. Jensen, J.C.M. Sprenkels, M. de Pooter, and J.J.M. Sleeckx. 1987. Comparison of flash pyrolysis differential scanning calorimetry, ^{13}C NMR and IR spectroscopy in the analysis of a highly aliphatic biopolymer from plant cuticles. J. Anal. and App. Pyrolysis.11:287-295.

Paul, E., K. Stüwe, J. Teasdale, and B. Worley. 1995. Structural and metamorphic geology of the Windmill Islands, east Antarctica: field evidence for repeated tectonothermal activity. Austral. J. Earth Sci. 42:453-469.

Pickard, J. 1986. Antarctic oasis. Terrestrial environment and history of the Vestfold Hills. Academic Press. London.

Pfisterer, U,. and S. Gribbohm. 1989. Zur Herstellung von Platinelektroden für Redoxmessungen. Zeitschrift für Pflanzenernährung und Bodenkunde. 152:455-456.

Post, B., H. Hempfling, H. Klamberg, and H.-R. Schulten. 1988. Zur Charakterisierung von Boden-Huminstoffen. Fresenius Zeitschrift für Analytische Chemie. 331:273-281.

Roser, D.J., D.R. Melick, and R.D. Seppelt. 1992a. Reductions in the polyhydric alcohol content of lichens as an indicator of environmental pollution. Antarc. Sci. 4:185-189.

Roser, D.J., D.R. Melick, H.U. Ling, and R.D. Seppelt. 1992b. Polyol and sugar content of terrestrial plants from continental Antarctica. Antarc. Sci. 4:413-420.

Roser, D.J., R.D. Seppelt, and O. Nordstrom. 1994. Soluble carbohydrate and organic acid content of soils associated microbiota from the Windmill Islands, Budd Coast, Antarctica. Antarc. Sci. 6:53-59

Schlichting, E., H.-P. Blume, and K. Stahr. 1995. Bodenkundliches Praktikum. Blackwell. Berlin.

Schnitzer, M., and H.-R. Schulten. 1992. The analysis of soil organic matter by pyrolysis-field ionization mass spectrometry. Soil Sci. Soc. Am. J. 54:98-105.

Schnitzer, M., C. Tarnocai, P. Schuppli, and H.-R. Schulten. 1990. Nature of the organic matter in Tertiary Paleosols in the Canadian Arctic. Soil Sci. 149:257-267.

Schulten, H.-R. 1987. Pyrolysis and soft ionization mass spectrometry of aquatic/terrestrial humic substances and soils. J. Anal. and App. Pyrolysis.12:149-186.

Schulten, H.-R. 1993. Analytical pyrolysis of humic substances and soils: Geochemical, agricultural and ecological consequences. J. Anal. and App. Pyrolysis.25:97-122.

Schulten, H.-R. 1996a. Direct pyrolysis-mass spectrometry of soils: a novel tool in agriculture, ecology, forestry, and soil science. In: T. W. Boutton and S.-I. Yamasaki, eds. Mass spectrometry of soils. Marcel Dekker. New York. pp. 373-436.

Schulten, H.-R. 1996b. A new approach to the structural analysis of humic substances in water and soils. In: J.S. Gaffney, N.A. Marley, and S.B. Clark, eds. Humic and fulvic acids. Isolation, structure and environmental role. ACS. Washington. pp. 42-56.

Schulten, H.-R., and M. Schnitzer. 1992. Structural studies on soil humic acids by curie-point pyrolysis-gas chromatography/mass spectrometry. Soil Sci. 153:205-224.

Schulten, H.-R., and M. Schnitzer. 1993. A state of the art structural concept for humic substances. Naturwissenschaften. 80:29-30.

Schulten, H.-R., and M. Schnitzer. 1997. Chemical model structures for soil organic matter and soils. Soil Sci. 162:115-130.

Schulten, H.-R., B. Plage, and M. Schnitzer. 1991. A chemical structure for humic substances. Naturwissenschaften. 78:311-312.

Seppelt, R.D., and Broady, P.A. 1988. Antarctic terrestrial ecosystems: The Vestfold Hills in context. In: J. M. Ferris, H. R. Burton and I. A. E. Bayly, eds. Biology of the Vestfold Hills, Antarctica. Kluwer Academic Publisher. Dordrecht. pp. 177-184.

Smith, R.I.L. 1990. Plant community dynamics in Wilkes Land, Antarctica. Proceedings NIPR Symposium. Polar Biol. 3:229-44.

Soil Survey Staff. 1998. Keys to Soil Taxonomy. USDA-NRCS. Washington, D.C.

Sorge, C., M. Schnitzer, and H.-R. Schulten. 1993. In-source pyrolysis field-ionization mass spectrometry and Curie-point pyrolysis-gas chromatography/mass spectrometry of amino acids in humic substances and soils. Biol. and Fertil. of Soils. 16:100-120.

Sorge, C., M. Schnitzer, P. Leinweber, and H.-R. Schulten. 1994. Molecular-chemical characterization of organic matter in whole soils and particle-size fractions of a Spodosol by pyrolysis-field ionization mass spectrometry. Soil Sci. 158:189-203.

Stevenson, F.J., and Cheng, C.-N. 1970. Amino acids in sediments: recovery by acid hydrolysis and quantitative estimation by a colorimetric procedure. Geochemica et Cosmochimica Acta. 34:77-88.

Tearle, P.V. 1987. Cryptogamic carbohydrate release and microbial response durin spring freeze-thaw cycles in Antarctic fellfield soils. Soil Biol. and Biochem. 19:381-390.

White, D.M., and L. Beyer. 1999. Pyrolysis-GC/MS and GC/FID of three Antarctic soils. J. Analyti. and App. Pyrolysis. 50:63-76.

Wilson, M.A. 1987. N.M.R. Techniques and Applications in Geochemistry and Soil Chemistry. Pergamon Press. Oxford.

Wilson, M.A. 1990. Application of nuclear magnetic resonance spectroscopy to whole soil. In: P. MacCarthy, C.E. Clapp, R L. Malcolm and P.R. Blomm, eds. Humic Substances in Soil and Crop Sciences. American Soil Sci. Soc. Madison, Wisconsin. pp. 221- 260.

Wilson, M.A., K.M. Goh, P.J. Collin, and L.G Greenfield, 1986. Origins of humus variations. Org. Geochem. 9:225-231.

WRB-FAO. 1998. World Soil Resources Reports 84. FAO, ISRIC, ISSS. Rome.

Ziechmann, W. 1980. Huminstoffe. Verlag Chemie. Weinheim.

Chapter 5. Microorganisms and Microbial Processes in Antarctic Soils

Manfred Bölter[1] and Ellen Kandeler[2]

[1]Professor, Institute for Polar Ecology, Christian-Albrechts-University Kiel, Wischhofstr. 1-3, Bldg. 12, 24148 Kiel, Germany; e-mail mboelter@ipoe.uni-kiel.de.
[2]Professor, Institute of Soil Science, University of Hohenheim, 70599 Stuttgart, Germany; e-mail kandeler@uni-hohenheim.de.

1. Introduction

Antarctica is a large, ice-covered continent, where soils are the exception. Less than 3% of the region is ice-free, and most of that is extremely cold and arid (Lewis Smith 1993a, b) most of the year. The environment severely limits soil organisms' survival strategies and adaptive capacities. Nevertheless, several groups of soil-dwelling animals, plants, and microorganisms have settled in various niches of this continent.

The isolated nature of ice-free areas — nunataks, deserts, and oases along the coast line and islands in the surrounding ocean (Figure 4.5.1) — complements small-scale changes within these landscapes. In contrast to the vast tundras of the northern hemisphere, the Antarctic is a mosaic of different environments. This holds true for geomorphological, pedological, botanical, and microbiological features. Gradients between individual habitats within a region can be as large as between different regions (Bölter et al., 1999).

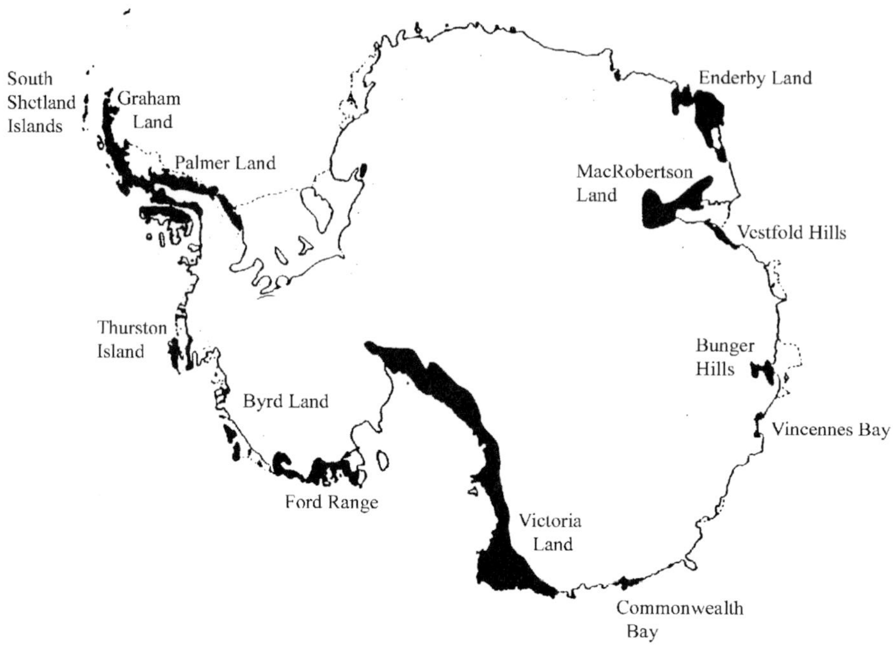

Figure 4.5.1. Locations of exposed rock and snow-free ground in coastal areas of the Antarctic continent (after Green et al., 1967).

Most soils here can be classified as permafrost-affected soils (Gelisols or Cryosols; Beyer et al., 1999) which show low chemical weathering processes but strong mechanical forces on rocks and surface layers of soils, via cryoturbation. Large landscapes thus show features of patterned ground in various grades. Repeated movements of surface layers by erosion or cryoturbation directly influence the colonization of slow-growing plant communities (e.g., lichens) or other fellfield stabilizing processes (Wynn-Williams, 1988, 1993) and thus slow microbially mediated changes in soil properties, such as decreases in pH (e.g., from nitrification), accumulation of soil aggregates, and increases in buffering capacity.

These factors make it difficult to uncover general evidence and create pictures of microbial activity in Antarctic soils on a broad scale. But recent research results can help describe facets of microbial processes related to different kinds of interactions, feedback, and regulation mechanisms. Although the feedback to changing environmental conditions cannot be as large as those in northern landscapes, the processes under investigation can be modeled and compared with those of other systems.

This overview mainly considers the coastal regions, with emphasis on environments of the maritime Antarctic and eastern Antarctic oases.

2. Environmental Constraints within Microbial Habitats

2.1. Temperature

Temperature can be regarded as a synonym of energy input into the system. Temperature conditions have impacts on soil properties (e.g., soil volume, pressure, redox potential, diffusion, viscosity, surface tension, and water aggregate status) and soil organisms. Temperature directly affects the physiological reaction rates of cells. Organisms that grow at or near 0°C have established special ways to cope these conditions. They contain specific lipids (unsaturated fatty acids) or other compatible solutes (e.g., polyols, amino acids) that enable their cellular membranes to retain fluidity when cold (Wynn-Williams, 1990; Paul and Clark, 1996).

Soil microorganisms adapted to these extreme conditions are called *psychrophiles*. Bacteria from most Antarctic habitats are considered *psychrotrophs* (Vishniac, 1993). Their prominent ability to survive in this environment is an eurythermal property which allows them to grow and profit from wide environmental temperature ranges. Although microbial growth and respiration are very low over the long periods of cold and xeric conditions in polar ecosystems, microbial carbon mineralization during these times may account for a substantial component of total annual C loss in these ecosystems, comparable to that observed from Arctic winter conditions (Zimov et al., 1993; Clein and Schimel, 1995; Oechel et al., 1997).

In addition, episodes of freezing and thawing that occur during spring, summer, and fall may increase microbial respiration due to the enhanced supply of substrates for soil microorganisms, which derives from killed and lysed cells during the period of freezing (Wynn-Williams, 1980). Such leaching of dissolved matter holds especially true for the dominating cryptogamic vegetation which shows accelerated release of dissolved carbohydrates, mainly polyols (Tearle, 1987; Melick and Seppelt, 1992; Roser et al., 1994).

Previous studies do not fully clarify whether these freeze-thaws substantially affect the composition and function of microbial communities and therefore the overall functioning of these ecosystems (Schimel and Clein, 1996). Many microbes react very sensitively to such stress factors, and the effect on their survival capacity relates to the range of the temperature change (Cameron, 1974). Survival strategies for such stress factors are especially necessary for communities living in surface horizons of soils or in cryptogamic crusts. Several yeasts and some bacteria cope successfully with such situations (Vishniac and Hempfling, 1979; Wynn-Williams, 1983; Lyatch et al., 1984).

The rise of the temperature above 0°C is only for short times and only in the surface horizons; temperature gradients relate strongly to plant cover (Bölter, 1992a). Such conditions influence strongly the time spans which allow microbial activity at different stages of their metabolism or chemical processes (Bölter et al., 1995).

2.2. Moisture

Most Antarctic regions suffer from low precipitation. A basic feature of the driving variable temperature is its control of the activity and availability of water. Temperature controls soil moisture, and this has consequences on changes in hydrology and chemical and biological processes. Temperatures below freezing prevent the existence of fluid water; high temperatures force evapotransportation and thus produce aridic conditions, which also stress microorganisms within the small window of conditions favorable for metabolism and reproduction.

Rapid changes in temperature are followed by rapid changes in soil moisture, which affect primarily the transport of solutes to osmotrophic organisms but also increase the concentrations of remaining ions or organic substances of low molecular weight and thus pose a strong osmotic force to organisms, especially those which cannot escape such conditions by rapid movement (e.g., bacteria, fungi, algae). They need the ability to produce special survival forms (e.g., resting cells) or to produce and store protective substances, processes which can be very energy-consuming.

Many bacteria show the benefit of such changes in life forms. They can minimize their cell size in order to decrease internal cell's energy needs and their expenses to maintain high membrane potentials (Schink, 1999). Hence, the requirements for general maintenance are much higher than in temperate environments, especially when changes occur at a critical level for life processes, such as at the freezing point.

Only in a few regions in the Antarctic is moisture not permanently limiting. Such regions are in the maritime Antarctic and in a few spots on the continent where geomorphological peculiarities (e.g., terraces) prevent quick drainage. One such region is on the Windmill Islands, where a well developed community of different mosses occurs, even producing histosols of different stages and deepness with special habitats for microorganisms and developing cryofibrists from algal mats (Beyer et al., 1995; Beyer and Bölter, 1999).

2.3. Aeration

Well drained soils are well aerated, and anaerobic processes occur only to a limited extent. Thus the Antarctic is not as important to evaluating CH_4 or N_2O as are the Arctic tundra environments (Christensen et al., 1994; Virzo de Santo et al., 1999). The presence of well aerated soils affects the degradation of soil organic matter; strong accumulations of organic matter, however, can occur only in regions with high plant productivity in combination with water-logging features. In such areas, anaerobic microbial activity and small amounts of CH_4 production have been monitored (Wynn-Williams, 1980, 1985a; Yarrington and Wynn-Williams, 1985). Lewis Smith (1985) showed that methanogenesis accounted for only 0.5 to 3% of total C-decomposition there.

Soils with a high degree of porosity can profit from convection in surface layers. Fine material with high contents of silt or clay provides very different habitats, with anaerobic zones hampered by slowed diffusion in wet conditions. However, these soils mostly lack sufficient organic material to sustain an appropriate microbial aerobic community.

3. Living resources: Organic Matter

3.1. Particulate organic matter

Antarctic environments which bear significant amounts of phytomass mostly are dominated by cryptogams (i.e., mosses and lichens), and the annual primary production and subsequent input of litter is very low (Lewis-Smith, 1985). Lichens are extremely slowly growing plants which produce a limited amount of residue. In addition, mosses change the substrate so that it is not a favorable food source for many decomposing organisms; they create an acidic and anaerobic environment. Further, mosses produce phenolic compounds (Prins, 1982) which need special constraints for effective microbial degradation (e.g., a well aerated environment).

Only small areas in the maritime Antarctic have sufficient climatic amplitudes to allow the growth of the only endemic vascular plants, *Deschampsia antarctica* and *Colobanthus quitensis,* and thus a substrate of higher quality. Some invaders have been registered recently (e.g., *Poa annua*, Olech pers. comm.) but the spread of vascular plants is rather limited and cannot be considered a significant input to the system.

The lack of annual litter input from vascular plants, in combination with hampered decomposition processes, has substantial consequences on soil organisms and thus on soil development (Blume and Bölter, 1996; Blume et al., 1996; Beyer and Bölter, 1999; White and Beyer, 1999). The organic matter in some spodic horizons differs significantly from that in temperate environments (Beyer et al., 1997). Further, the rare abundance of rooting plants and soil-dwelling animals reduces the transport of matter, although plants and animals, where abundant, affect soil development (Block, 1984; Bölter et al., 1997).

Levels of organic matter of high molecular weight nevertheless occur over wide ranges. Table 4.5.1 presents some examples, measured as loss on ignition. The prominent effect on the amount of organic matter is the plant cover. Strong gradients, according to its depth distribution, occur. The controlling factor for the decomposition is the accessibility of promoting substances (e.g., nitrogen and phosphorous), which are at limiting concentrations in most non-coastal environments. These compounds are necessary for the synthesis of new cell material. The general threshold for a ratio which determines N-remineralization vs. N-immobilization of about 25 (Paul and Clark, 1989) also can be accepted for this environment.

Table 4.5.1. Contents of organic matter and C/N ratios of Antarctic soils.

Location	Soil/sample description	Soil depth (cm)	Loss on ignition (LOI) (% of d.wt.)	C/N	Reference
Windmill Islands	surface soil	0-3	11.0	12.7	Bölter, 1990
		3-12	2.2	7.6	
Windmill Islands	Dystri-gelic Regosol	0-70		6-7	Blume et al., 1997
	Lepti-gelic Podzol	0-26		5-7.6	
	Dystri-gelic Leptosol	0-30		9-12	
	Terri-gelic Hitosol	0-28		9-19	
	Fibri-gelic Histosol	0-18		18-21	
Windmill Islands	Lithic Cryorthent	0-15	0.4-0.7	8.5-13.2	Beyer and Bölter, 1999
	Lithic Cryorthent	0-36	0.6-2.2	10.9-22.7	
	Lithic Cryorthent	0-15	3.6-6.5	11.9-20.2	
	Lithic Cryorthent	0-10	14.0-17.3	10.1-15.7	
	Lithic Cryohemist	0-22	6.3-60.5	12.0-35.8	
	Pergelic Haplocryod	0-71	1.8-8.6	2.3-28.9	
	Pergelic Haplocryod	0-20	3.7-13.4	10.9-13.4	
	Lithic Haplocryod	0-25	1.2-7.5	9.8-15.3	
	Lithic Cryaquept	0-35	2.2-2.7	6.3-29.7	
Windmill Islands	surface soils	0-4	0.2-40.2	2.9-13.9	Heatwole et al., 1989
Signy Island	Moss cover	0-2	44.0	15	French, 1981
	Grass meadow	0-2	15.9	13	
	Moraine soil	0-2	1.8	5.7	
Deception Island	Gelic-dystri Andosol	0-30	0.7-4.5	10-19	Bölter et al., 1999
Elephant Island	Eutri-gelic Regosol	0-60	0.7-2.2	7.9-13	
Paradise Bay	Fibri-gelic Histosol	0-25	76.7-98.1	10-41	
Peterman Island	Fibri-gelic Histosol	0-18	8.9-78.9	7.0-17	
Paulett Island	Ornitho-gelic Regosol	0-40	6.0-56.9	4.9-7.8	
Coronation Island	Fibri-gelic Histosol	0-28	70.7-97.7	19-56	
Coronation Island	Humi-gelic Leptosol	0-30	5.1-83.9	7.9-12	

Organic matter from plant production has extremely high C/N-ratios. Kappen (1985) found such ratios for that of *Usnea sulphurea* and *U. fasciata* between 51 and 116; Christie (1987a) reports values up to 60 from moss banks. Soils in the vicinity of penguin rookeries or other concentrations of animal life show extremely low C/N-ratios, with values down to 5 (Bölter et al. 1999). Most reports of soils, however, describe C/N-ratios between 10 and 20 (Table 4.5.1).

A strong patchiness below m-scales has been monitored, due to plant cover and small-scale relief (Beyer et al., 1998). Most of the soil nitrogen, however, has to be considered inorganic N derived from excretes; thus it does not describe the quality of litter or other particulate organic material. It becomes evident from most of these analyses that elevated C/N ratios are typical for surface layers; deeper horizons generally show lower C/N ratios, due to lower contents of refractive material.

Table 4.5.2. Dissolved organic matter in Antarctic soils.

Location	Sample/Plant species	Compounds	Amount (mg/g)	Reference
Singy Island	*Usnea sp.*	Polyols	184.7	Tearle, 1987
	Himantormia sp.	(Arabitol, Ribitol, Mannitol)	159.9	
	Andreae sp.		12.2	
Windmill Islands	surface soils (0-2 cm)	free amino acids	0.14	Bölter, 1990
		monocarbohydrates	0.03	
	deep layer soils (2-12 cm)	free amino acids	0.02	
		monocarbohydrates	0.01	
Windmill Islands	*Grimmia antarctici*	sugars (fructose, glucose, sucrose)	65.3	Melick and Seppelt, 1992
	Bryum pseudotriquetum		52.9	
	Ceratodon purpureus		21.8	
	Cephaloziella exilifora		5.84	
Windmill Islands	*Bryum pseudotriquetum*	soluble carbohydrates	273	Melick and Seppelt, 1994
	Grimmia antarctici		159	
	Cephaloziella exilifora		54	
	Umbilicaria decussata		52	
	Usnea sphacelata		71	
Windmill Islands	*Umbilicaria decussata*	polyols + sugars	30-36	Roser et al., 1992a
	Usnea sphacelata		17.2	
Windmill Islands	var. lichens	polyols + sugars	12-63	Roser et al., 1992b
	var. mosses		16-60	
	Cephaloziella exiliflora		8.7-17	
	var. algae		19-48	

3.2. Dissolved organic matter

Generally, particulate and dissolved organic matter can correlate closely, a fact which shows the importance of primary producers' sustaining the food chain. However, cells burst from freeze-thaw or leaching, overriding this effect. Methodological obstacles and practical problems make studies of such phenomena difficult in Antarctica.

Dissolved carbohydrates, amino acids, and other low-weight organic molecules are substantial for sustaining osmotrophic organisms. The access to these substances may be energy consuming (e.g., via excretion of exoenzymes and specific uptake proteins). Despite the high demand of these molecules, tremendously high concentrations of individual compounds occurred (Table 4.5.2), but polyols and pentoses are not preferred metabolic substrates. Low-weight organic matter often occurs in real excess in Antarctic soils. The question arises about their fate, i.e, their role in nutrition, their persistence, and their possible role in sequestration.

4. The Microbial Community

Living soil organic matter (SOM) – bacteria, cyanobacteria, fungi, soil algae, protozoans and micro-fauna – shows a very distinct live pattern in Antarctic soils. Many studies have tried to describe the quantity, quality, and role of organisms in nutrient cycles. Many studies describe microbial communities, individual taxonomic or physiological groups of these communities, or overall processes (e.g., CO_2-evolution), and they recalculate microbial standing stock (i.e., biomass or biomass equivalents) from measures of activity. But it is difficult to assemble these studies to obtain a valid overall picture of processes and acting organisms. Methodological peculiarities often do not allow direct comparisons of data from individual projects.

Either indirect (= biochemical, physiological) or direct (= microscopic, cultural techniques) methods can determine microbial biomass, mostly in carbon units. Indirect methods mostly measure concentrations of cell constituents (e.g., ATP, or activity parameters, such as substrate-induced respiration). Data from several studies are available for such approaches to Antarctic soils or other terrestrial habitats (Roser et al., 1993a; Bölter, 1994). However, problems arise when converting such data, e.g., of ATP, to carbon equivalents or when applying the appropriate method for extracting ATP. Direct comparisons of such "converted" data thus may lead to long but not fruitful discussions. Studies also have used activity parameters to obtain data of microbial biomass; some approaches are presented below.

Bacteria are doubtless the main organisms responsible for the turnover of organic matter. Their number and biomass relate directly to total organic matter, notwithstanding the methods applied. Hence, high concentrations of organic matter generally are followed by high numbers of bacteria. This can be validated by transects along gradients across different landscapes or by depth profiles at high resolutions in soil profiles (Bölter et al., 1995, 1999).

Obviously, not only total number and total biomass show such combined gradients with organic material. Individual properties of the population, such as mean cell volume; shares of parts of the community, such as those identified by size classes of length or biovolume; and shifts in morphological properties also are involved in particular relationships with SOM. But not all of these parameters show similar strengths of correlations with global or individual compounds. This indicates actions of different bacterial communities and their special requirements for various substrates. It also can be interpreted as individual answers for various kinds of stress on the community.

Data on cell numbers and biomass have been described over wide ranges up to 10^{10} cells per g in ornithogenic soils. For example, from Ross Island and Windmill Islands (Ramsay and Stannard, 1986; Roser et al., 1993b), organic soils contain up to 10^9 cells/g (Wynn-Williams, 1985b; Bölter et al., 1999). The crucial point in estimating bacterial biomass relates to the exact measurements of their geometrical properties of size and shape (i.e., the establishment of a reliable mean cell volume).

Due to the high importance of small cells in nutrient-poor soils, the application of standard figures is not appropriate. Ramsay and Stannard (1986) calculate, for ornithogenic soils, mean volumes between 0.22 and 0.88 μm^3. Data from other places show mean cell volumes at much lower figures, between 0.11 and 0.61 μm^3 in surface samples and between 0.02 and 0.29 μm^3 in deeper layers from King

George Island (Bölter, 1995; Bölter et al., 1997) or between 0.06 and 0.14 μm^3 in organic soils from the Windmill Islands (Bölter, 1992a).

The role of filamentous fungi in Antarctic soils seems limited to organically rich soils, whereas yeasts can occur in many places, from nutrient-poor to nutrient-rich soils (Vishniac, 1993). Their contribution to the overall microbial community seems low (Roser et al., 1993b; Bölter, 1995, 1997; Beyer et al., 2000). On the other hand, they show fast reactions on a fresh supply of dissolved organics, with bursts to high biomass levels (Wynn-Williams, 1982).

A large variety of different fungal species have been found and described taxonomically, which derive from various habitats (Vishniac, 1993; Azmi and Seppelt, 1998). Some studies have introduced eco-physiological investigations (Vishniac and Hempfling, 1979; Pugh and Alsopp, 1982; Azmi and Seppelt, 1999), but the role of these microorganisms in the food web is still not fully understood. A main reason for this is the problem in determining their state of activity in the field. Cultural techniques indicate that several species are present; however, their ability to form resting stages (e.g., spores) doesn't hint about their actual contribution to the degradation or turnover of organic matter. Laboratory experiments show a broad capability to use various kinds of organic matter, indicating that they have a great potential to act in the system very efficiently.

Phototrophs (i.e., cyanobacteria and soil algae) have been given much attention in descriptions of processes of colonization and exudation (Wynn-Williams, 1990). Further, their input of organic matter via snow melting from colonized snow fields to soil surfaces has to be considered. They are present in top soils which contain no or little organic matter, filling their role as colonizers, on plants such as Aufwuchs, or at places with limnetic influence, at banks or just at the moving boundaries of flooded areas with rapidly changing water levels. Most frequent phototrophs in the terrestrial environment are several diatoms (e.g., *Pinnularia sp.*), as well as *Prasiola crispa* and *Nostoc commune*.

Besides their role in the detrital food chains of Antarctic soils, algae and cyanobacteria have two main functions: first, the stabilization of soil via production of slimes, thus producing primary aggregates; second (only *cyanobacteria*), the fixation of nitrogen and thus enrichment of the N-deficient soils with an important nutrient. Studies on interactions between phototrophs and heterotrophs are few, and only some have been carried out to determine relationships between these groups at the community level (Tearle, 1987; Bölter, 1996, 1997; Beyer et al., 2000).

Wynn-Williams (1993) discussed the role of soil stabilizing processes from results of investigations in Signy Island fellfield soils. He found that slime-producing algae force the production of primary particles. Thus they prevent, to some extent, erosion by wind and flowing water and, on the other hand, they provide primary areas for other seeds and thus build up first niches for heterotrophs. This first step in forming small environments of microorganisms is a basic feature also for other low-nutrient, desert-like areas.

The input of nitrogen via autotrophic N-fixation of cyanobacteria can be of great importance for the system (Lewis-Smith, 1985; Bölter et al., 1995). Christie

(1987b) showed a considerable amount of N can be attributed to cyanobacteria's activity. This input, however, is only of local importance, where sufficient conditions for cyanobacteria exist (e.g., moss beds).

There is still a big gap in balancing the N-contents of soils and its possible sources on a broad scale. This uncertainty is mainly due to the lack of data which provide insights into the seasonal cycle of N-fixation in relation to N-precipitation and the input of tremendous bird colonies. Here, it is not only the actual input via feces and thus high amounts of gaseous ammonium in the vicinity of rookeries, but also the transport of marine N-salts via sea spray. Input to soils via weathering of native rocks is insubstantial.

5. Microbial Activity

Methodologies for descriptions of microbial and/or bacterial activity are broad. Approaches for various habitats used individual or complex substrates (e.g., carbohydrates, amino acids, soil extracts), reactions of enzymes (e.g., hydrolytic activity), measurements of end products (e.g., carbon dioxide, methane), microcalorimetry, and microscopic growth measurements.

All these methods provide distinct insights into different processes at different levels of the food web and/or metabolic pathways. Their aim is not only to estimate the fate and turnover of substances but also to describe the role of microorganisms as constituents of the total soil biomass. Organisms must be considered for the decomposition of organic material, and released and sorbed enzymes may show significant activity.

5.1. Enzymes

Enzymes released by microorganisms and sorbed onto organic and mineral particles in soils can degrade and break down substances of high molecular weight. Thus, they have an important function as mediators between producers and osmotrophic consumers. Enzyme activities in soils at subzero temperatures is attributed to enzyme-substrate interaction in unfrozen water at the soil particles (Bremner and Zantua, 1975).

Urease, phosphatase, and arylsulfatase activities have been detected in soils even at -10°C and -20°C. In polar ecosystems, the enzymes may be less limited by the effect of temperature on enzyme activity than by ice formation, which limits diffusion, or the accumulation of end products, which can inhibit enzyme activity (McClaugherty and Links, 1990).

Since microbial growth is reduced to a great extent at low temperature, continued enzyme activity at nearly freezing temperatures could result in an accumulation of products. Enzyme-mediated decomposition may also explain why litter mass occurred in winter under snow pack at near 0°C (McBrayer and Cromak.

1980). Orchard and Corderoy (1983), however, discuss the fact that the time span for an activity might be too short for a significant affect.

Enzymatic activities for different hydrolytic processes have been found in Antarctic soils. Horowitz et al. (1972) explain high turnover rates of glucose in soils of the Dry Valleys by assuming active extra cellular enzymes. Ramsay and Stannard (1986) also relate high metabolic rates in Antarctic soils to free enzymes.

Studies using fluorescein diacetate (FDA) give an overall view of hydrolases and have shown quite high activity in surface soils from the Windmill Islands (Bölter, 1992). Extremes were monitored in samples of a silty soil surface covered with a thin layer of algae and cyanobacteria. Extracellular enzymes probably catalyzed the decay of high amounts of slimes. Similarly, the surface of a lichen showed high values of FDA activity, accompanied by elevated bacterial counts and biomass.

Another study conducted with methylumbelliferyl substrates showed a diverse pattern of derived activity data (Bölter, 1989, 1992b). Surface samples had 5 to 10 times higher activity levels.

5.2. Metabolic Processes and Respiration

Dominant questions while assessing methods for obtaining data on microbial activity are "How many organisms are active?" or "Which part of the microbial community is active?" or "What is the preferred substrate?" or "What are dominant environmental factors?" Several attempts have been made to answer these questions.

Microautoradiography is a basic method for determining the number of organisms involved in the metabolic process. Ramsay (1983) found that the proportion of bacteria from ornithogenic soils taking up ^{3}H-glucose was less than 8%. These results from Ross Island could be confirmed by very low measurements of CO_2 evolution (Orchard and Corderoy, 1983). These data, because of the specificity of this environment, cannot be transposed to other environments.

High concentrations of sugars and polyols at other places indicate a more active community using these substances for metabolism. Given measurements of colony growth and substrate use, carbohydrates, especially glucose, are preferred substrates for growth (Bölter, 1993; Melick et al., 1994). These data sustain results from Wynn-Williams (1982), who showed that peat samples supplemented with sugars increased their activity, a fact which was used as an indication that the dominance of microbial respiration is due to dissolved organic carbon. Similarly, micro-colony growth was at comparable levels when growth was stimulated with either glucose or soil extracts (Bölter, 1993).

Gas exchange measurements can use the consumption rates of oxygen or the evolution of carbon dioxide. Wynn-Williams (1982, 1984, 1985c, d) conducted several experiments on oxygen consumption and found it closely related to CO_2 output (Wynn-Williams, 1982). Calculations of respiratory quotients of data from moss peats from a wider area, the Scotia Arc and Antarctic Peninsula, were between 0.5 and 2 (Wynn-Williams, 1984). Direct correlations between these data, however, were not found (Wynn-Williams, 1985a).

Similar findings resulted from comparing glucose remineralization rates, oxygen consumption, and CO_2 evolution from soils and plant samples from Arctowski and Casey (Bölter, 1991). This indicates a great differentiation in those processes. The share of respired ^{14}C-labelled glucose has high variability and can fluctuate between 40 and 90% (Bölter, 1992b). Thus, a significant proportion of the metabolic process is used to maintain life rather than produce new biomass.

Data on bacterial biomass production in soils from maritime and continental Antarctica have been derived from rates of incorporation of ^{14}C-labeled glucose (Bölter, 1990, 1992b) and show, for organic surface horizons, values up to 0.94 µg C/g soil/h. The lowest values, in deeper horizons with low levels of organic matter, showed figures below 0.1 µg C/g soil/h. Extrapolating such data to a yearly figure yields a maximal bacterial production of up to 20 mg C/m^2.

In the future, estimates of different ecophysiological parameters that are frequent measures in temperate and tropical soil could help describe the energetic efficiency of the microbial community and the maintenance requirement of the microbial biomass in polar ecosystems. For example, the Cmic/Corg ratio could give information about the carbon dynamics of Antarctic soils, whereas the metabolic quotient (qCO_2) calculated from the basal respiration and microbial biomass would show whether soil microorganisms use efficiently available substrates.

References

Azmi, O.R., and R.D. Seppelt. 1998. The broad-scale distribution of microfungi in the Windmill Islands region, continental Antarctica. Polar Biol. 19:92-100.

Beyer, L., B. Vogt, H-P. Blume, and H. Erlenkeuser. 1995. Soil organic matter composition and transformation in a swamp from algae of coastal continental Antarctica. Polar Res. 65:117-122.

Beyer, L., H. Knicker, H-P. Blume, M. Bölter, B. Vogt, and D. Schneider. 1997. Soil organic matter of suggested spodic horizons in relic ornithogenic soils of coastal continental Antarctica (Casey Station, Wilkes Land) in comparison with that of spodic soil horizons in Germany. Soil Sci. 162:518-527.

Beyer, L., K. Pingpank, M. Bölter, and R.D. Seppelt. 1998. Small distance variation of carbon and nitrogen storage in mineral Antarctic Cryosols near Casey Station (Wilkes Land). Z Pflanzenernähr Bodenk. 161:211-220.

Beyer, L., and M. Bölter. 1999. Formation, ecology, and geography of Cryosols of ice-free oasis in coastal East Antarctica near Casey Station (Wilkes Land). Aus. J. Soil Res. 37:209-244.

Beyer, L., M. Bölter, and R.D. Seppelt. 2000. Nutrient and thermal regime, microbial biomass, and vegetation of Antarctic soils in the Windmill Islands Region of east Antarctica (Wilkes Land) Arct. Alp. Res. 32:30-39.

Block, W. 1984. Terrestrial microorganisms and invertebrates. In: R.M. Laws, ed. Antarctic ecology, Vol. I. Academic Press. London. pp. 163-236.

Blume, H-P, and M. Bölter. 1996. Wechselbeziehungen zwischen Boden- und Vegetationsentwicklung in der kontinentalen Antarktis. Verh. Ges. Ökol. 25:25-34.

Blume, H-P, D. Schneider, and M. Bölter. 1996. Organic matter accumulation in and podzolization of Antarctic soils. Z. Pflanzenernähr Bodenk. 159:411-412.

Blume, H-P., L. Beyer, M. Bölter, H. Erlenkeuser, E. Kalk, S. Kneesch, U. Pfisterer, and D. Schneider. 1997. Pedogenic zonation in soils of the southern circum-polar region. Adv. GeoEcol. 30:69-90.

Bölter, M. 1989. Microbial activity in soils from Antarctica (Casey Station, Wilkes Land). Proc. NIPR Symp. Polar Biol. 2:146-153.

Bölter, M. 1990. Microbial ecology of soils from Wilkes Land, Antarctica: I. The bacterial population and its activity in relation to dissolved organic matter. Proc. NIPR Symp. Polar Biol. 3:104-119.

Bölter, M. 1991. Microbial mineralization in soils and plant material from Antarctica. In: G. Weller, C.L. Wilson, and B.A.B. Severin, eds. International Conference on the Role of the Polar Regions in Global Change: Proc. Conf. June 11-15, 1990 Univ. Alaska, Fairbanks. Geophys. Inst. Univ. Alaska. Fairbanks. pp. 418-422.

Bölter, M. 1992a. Environmental conditions and microbiological properties from soils and lichens from Antarctica (Casey Station, Wilkes Land). Polar Biol. 11:591-599.

Bölter, M. 1992b. Vergleichende Untersuchungen zur mikrobiellen Aktivität in Böden und an Kryptogamen aus der kontinentalen und maritimen Antarktis (Casey, Wilkes Land, und Arctowski, King George Island). Habilschr Univ. Kiel, 1-201.

Bölter, M. 1993. Effects of carbohydrates and leucine on growth of bacteria from Antarctic soils (Casey Station, Wilkes Land). Polar Biol. 13:297-306.

Bölter, M. 1994. Estimations of microbial biomass by direct and indirect methods with special respect to monitoring programs. Proc. NIPR Symp. Polar Biol. 7:198-298.

Bölter, M. 1995. Distribution of bacterial numbers and biomass in soils and on plants from King George Island (Arctowski Station, Maritime Antarctica). Polar Biol. 15:115-124.

Bölter, M. 1996. Analysis of soil microbial communities (autotrophs and heterotrophs) from King George Island (Arctowski Station). Proc. NIPR Symp. Polar Biol. 9:283-298.

Bölter, M. 1997. Microbial communities in soils and on plants from King George Island (Arctowski Station, Maritime Antarctica). In: Battaglia B, Valencia J, and D.W.H. Walton, eds. Antarctic communities: species, structure and survival. Cambridge Univ. Press. Cambridge. pp. 162-169.

Bölter, M., H-P. Blume, and L. Kappen. 1995. Bodenbiologische Untersuchungen in der maritimen und kontinentalen Antarktis (King George Island und Windmill Islands). Teil 1. Umweltparameter und anorganische Nährstoffe. Polar Res. 65:41-61.

Bölter, M., H-P. Blume, D. Schneider, and L. Beyer. 1997. Soil properties and distributions of invertebrates and bacteria from King George Island (Arctowski Station), maritime Antarctic. Polar Biol. 18:295-304.

Bölter, M., H-P. Blume, and D. Kuhn. 1999. Soils and their microbiological properties from a transect from Cape Horn to the Antarctic Peninsula. Polar Biosci. 12:54-67.

Bremner, J.M., and M.I. Zantua. 1975. Enzyme activity in soils at subzero temperatures. Soil Biol. Biochem. 7:383-387.

Cameron, R.E. 1974. Application of low latitude microbial ecology to high latitude deserts. In: T.L. Smiley and J.H. Zumberge, eds. Polar deserts and modern man, Univ. Arizona Press. Tucson. 71-90.

Christensen, T.R., S. Jonasson, T.V. Callaghan, H. Mavström, and F.R. Livens. 1999. Carbon cycling and methane exchange in Eurasian tundra ecosystems. Ambio. 28:239-244.

Christie, P. 1987a. C:N ratios in two contrasting Antarctic peat profiles. Soil Biol. Biochem. 19:777-778.

Christie, P. 1987b. Nitrogen in two contrasting Antarctic bryophyte communities. J. Ecol. 75:73-93.

Clein, J.S., and J.P. Schimel. 1995. Microbial activity of tundra and taiga soils at sub-zero temperatures. Soil Biol. Biochem. 27:1231-1234.

French, D.D. 1981. Multivariate comparisons of IBP tundra biome site characteristics. In: L.C Bliss, O.W. Heal, and J.J. Moore, eds. Tundra ecosystems: a comparable analysis. Cambridge Univ. Press. Cambridge. pp. 47-75.

Greene, S.W., J.L. Gressitt, D. Koob, G.A. Llano, E.D. Rudolf, R. Singer, W.C. Sreere, and F.C. Ugolini. 1967. Terrestrial life of Antarctica. Antarctic Map Folio Series 5. Geographical Society. New York.

Heatwole, H., P. Saenger, A. Spain, E. Kerry, and J. Donelan. 1989. Biotic and chemical characteristics of some soils from Wilkes Land, Antarctica. Antarct. Sci. 1:225-234.

Horowitz, N.H., R.E. Cameron, and J.S. Hubbard. 1972. Microbiology of the Dry Valleys of Antarctica. Sci. 176:242-245.

Kappen, L. 1985. Water relations and net photosynthesis of Usnea. A comparison between *Usnea fasciata* (Maritime Antarctic) and *Usnea sulphurea* (Continental Antarctic). In: D.H. Brown, ed. Lichen physiology and cell biology. Plenum Publ Corp. New York. pp. 41-56.

Lewis-Smith, R.I. 1985. Nutrient cycling in relation to biological productivity in Antarctic and sub-Antarctic terrestrial and freshwater ecosystems. In: W.R. Siegfried, P.R. Condy, and R.M. Laws, eds. Antarctic nutrient cycles and food webs. Springer-Verlag. Berlin. pp. 138-155.

Lewis-Smith, R.I. 1993a. Dry coastal ecosystems of Antarctica. In: Evd Maarel, ed. Ecosystems of the world, Vol. 2A. Elsevier. Amsterdam. pp. 51-71.

Lewis-Smith, R.I. 1993b. Dry coastal ecosystems on sub-antarctic islands. In: Maarel Evd, ed. Ecosystems of the world, Vol. 2A. Elsevier. Amsterdam. pp. 73-93.

Lyakh, S.P., T.M. Kozlova, and S.M. Salivonik. 1984. Effect of periodic freezing and thawing of cells of the Antarctic black yeast *Nadsoniella nigra* var. *hesuelica*. Microbiol. 52:486-491.

McBrayer, J.F., and K. Cromck, Jr. 1980. Effect of snow-pack on oak-litter breakdown and nutrient release in a Minesota forest. Pedobiol. 20:47-54.

McClaugherty, C.A., and A.E. Linkins. 1990. Temperature responses of enzymes in two forest soils. Soil Biol. Biochem. 22:29-33.

Melick, D.R., and R.D. Seppelt. 1992. Loss of soluble carbohydrates and changes in freezing point of Antarctic bryophytes after leaching and repeated freeze-thaw cycles. Antarct. Sci. 4:399-404.

Melick, D.R., and R.D. Seppelt. 1994. Seasonal investigations of soluble carbohydrates and pigment levels in Antarctic bryophytes and lichens. The Bryologist. 97:13-19.

Melick, D.R., M. Bölter, and R. Möller. 1994. Rates of soluble carbohydrate utilization in soils from the Windmill Island Osasis, Wilkes Land, continental Antarctica. Polar Biol. 14:59-64.

Oechel, W.C., G. Vourlitis, and S.J. Hastings. 1997. Cold season CO_2 emission from Arctic soils. Global Biogeochem. Cycles. 11:163-172.

Orchard, V.A., and D.M. Corderoy. 1983. Influence of environmental factors on the decomposition of penguin guano in Antarctica. Polar Biol. 1:199-204.

Paul, E.A., and F.E. Clark. 1996. Soil microbiology and biochemistry, 2nd Ed. Academic Press. San Diego, London.

Prins, H.H.T. 1982. Why are mosses in cold environments only? Oikos. 38:374-380.

Pugh, G.H.F., and D. Alsopp. 1982. Microfunig from Signy Island, South Orkney Islands. Br. Antarct. Surv. Bull. 57:56-67.

Ramsay, A.J. 1983. Bacterial bomass in ornithogenic soils of Antarctica. Polar Biol. 1:221-225.

Ramsay, A.J., and R.E. Stannard. 1986. Numbers and viability of bacteria in ornithogenic soils of Antarctica. Polar Biol. 5:195-198.

Roser, D.J., D.R. Melick, and R.D. Seppelt. 1992a. Reductions in the polyhydric alcohol content of lichens as an indicator of environmental pollution. Antarct. Sci. 4:185-189.

Roser, D.J., D.R. Melick, H.U. Ling, and R.D. Seppelt. 1992b. Polyol and sugar content of terrestrial plants from continental Antarctica. Antarct. Sci. 4:413-420.

Roser, D.J., R.D. Seppelt, and N. Ashbolt. 1993a. Microbiology of ornithogenic soils from the Windmill Islands, Budd Coast, continental Antarctica: some observations on methods for measuring soil biomass in ornithogenic soils. Soil Biol. Biochem. 25:177-183.

Roser, D.J., R.D. Seppelt, and N. Ashbolt. 1993b. Microbiology of ornithogenic soils from the Windmill Islands, Budd Coast, continental Antarctica: microbial biomass distribution. Soil Biol. Biochem. 25:165-175.

Roser, D.J., R.D. Seppelt, and O. Nordstrom. 1994. Soluble carbohydrate and organic acid content of soils and associated microbiota from the Windmill Islands, Budd Coast, Antarctica. Antarct. Sci. 6:53-59.

Schimel, J.P., and J.S. Clein. 1996. Microbial response to freez-thaw cycles in tundra and taiga soils. Soil Biol. Biochem. 28:1061-1066.

Schink, B. 1999. Ecophysiology and ecological niches of prokaryotes. In: J.W. Lengeler, G. Drews, and H.G. Schlegel, eds. Biology of the prokryotes. Thieme. Stuttgart. pp. 723-762.

Tearle, P.V. 1987. Cryptogamic carbohydrate release and microbial response during spring freeze-thaw cycles in Antarctic fellfield fines. Soil Biol. Biochem. 19:381-390.

Virzo de Santo, A., P. Ineson, and K. Smith. 1999. Soils as source and sinks of trace gases. In: A Farina, ed. Perspectives in ecology. Blackhuys Publ. Leiden. pp. 107-113.

Vishniac, H. 1993. The microbiology of Antarctic soils. In: EI Friedman, ed. Antarctic microbiology. Wiley-Liss Inc. New York. pp. 297-341.

Vishniac, H.S., and W.P. Hempfling. 1979. Evidence of indigenous microbiota (yeast) in the dry valleys of Antarctica. J. Gen. Microbiol. 112:301-314.

White, D., and L. Beyer. 1999. Pyrolysis-gas chromatography/mass spectrometry and pyrolysis-gas chormatography/flame ionization detection analysis of three Antarctic soils. J. Anal. Appl. Pyrol. 50:63-76.

Wynn-Williams, D.D. 1980. Seasonal fluctuations in microbial activity in Antarctic moss peat. Biol. J. Linn. Soc. 14:11-28.

Wynn-Williams, D.D. 1982. Simulation of seasonal changes in microbial activity of maritime Antarctic peat. Soil Biol. Biochem. 14:1-12.

Wynn-Williams, D.D. 1983. Distribution and characteristics of *Chromobacterium* in the maritime and sub-Antarctic. Polar Biol. 2:101-108.

Wynn-Williams, D.D. 1984. Comparative respirometry of peat decomposition on a latitudinal transect in the maritime Antarctic. Polar Biol. 3:173-181.

Wynn-Williams, D.D. 1985a. Comparative microbiology of moss-peat decomposition on the Scotia Arc and Antarctic Peninsula. In: W.R. Siegfried, P.R. Condy, R.M. Laws, eds. Antarctic nutrient cycles and food webs. Springer-Verlag. Berlin. pp. 204-210.

Wynn-Williams, D.D. 1985b. Photofading retardent for epifluorescence microscopy in soil micro-ecological studies. Soil Biol. Biochem. 17:739-746.

Wynn-Williams, D.D. 1985c. The Signy Island terrestrial reference sites: XVI. Peat O_2-uptake in a moss turf relative to edaphic and microbial factors. Br. Antarct. Surv. Bull. 68:47-59.

Wynn-Williams, D.D. 1985d. The Signy Island terrestrial reference sites: XVII. Peat O_2-uptake in a moss carpet relative to edaphic and microbial factors. Br. Antarct. Surv. Bull. 68:61-69.

Wynn-Williams, D.D. 1988. Microbial colonization of Antarctic fellfield soils. Proc. 4th Intern Symp. Microb. Ecol. Ljubljana, Yugoslavia. pp. 191-200.

Wynn-Williams, D.D. 1990. Ecological aspects of Antarctic microbiology. In: K.C. Marshall, ed. Advances in microbial ecology, Vol. 2. Plenum Publ Corp. London. pp. 71-146.

Wynn-Williams, D.D. 1993. Microbial processes and initial stabilization of fellfield soil. In: J. Miles and D.W.H. Walton, eds. Primary succession on land. Blackwell Scientific Publ. Oxford. pp. 17-32.

Yarrington, M.R., and D.D. Wynn-Williams. 1985. Methonogenesis and the anaerobic micro-biology of a wet moss community at Signy Island. In: W.R. Siegfried, P.R. Condy, and R.M. Laws, eds. Antarctic nutrient cycles and food webs. Springer-Verlag. Berlin. pp. 229-233.

Zimov, S.A., I.P. Semiletov, S.P. Daviodov, Yu.V. Voropaev, S.F. Prosyannikov, C.S. Wong, and Y-H. Chan. 1993. Wintertime CO_2 emission from soils of northeastern Siberia. Arctic. 46:197-204.

Chapter 6. The Biology of Arid Cryosols

G.G.C. Claridge[1] and I.B. Campbell[2]

[1]Research Scientist, 33 Kotari Road, Days Bay, Eastbourne, New Zealand; e-mail graeme.claridge@xtra.co.nz.
[2]Research Scientist, Land and Soil Consultancy Services, 23 View Mount, Nelson 7001, New Zealand; e-mail iaincampbell@xtra.co.nz.

1. Introduction

Because of the prevailing low temperatures, low humidities, frequent freeze-thaw cycles, and salinity of the soil, the Arid Cryosols of the Transantarctic Mountains are generally unfavorable environments for plant and animal life. Nevertheless, careful examination of the soil shows that habitable niches do exist, and organisms have colonized the soil and radiated to fill most of them. Food chains have formed, and a relatively large number of organisms occur in a relatively simple ecosystem.

2. Primary Producers

Within the soil, microorganisms can and do survive. The primary producers in Arid Cryosols are cyanobacteria and eukaryotic algae, fixing carbon and providing nutrients for other heterotrophic organisms (Wynn-Williams et al., 1997) Traces of bacteria occur in most soils, even the coldest and most arid (Cameron et al., 1971; Broady and Weinstein, 1998), while algae are distributed less widely and restricted to more favorable situations (Vincent, 1988; Broady, 1996).

Yeasts and actinomycetes are also present. Together, these organisms transform sunlight into organic matter which can support traces of animal life, and thus an ecosystem can develop. Cameron (1972), for example, found at least 10 genera of bacteria, two species of yeast, and four of algae in soils of the Victoria Valley that contained no visible plant life. However, the distribution of organisms within the soils was highly variable.

On the surface of the soil, a few plants can survive in suitable locations. Mosses grow in small patches where conditions are favorable, generally close to ephemeral streams in coastal areas where the soils are moist and salt-free (Schwartz et al., 1992). The roots of the mosses can penetrate 2 to 3 cm into the soil and contribute traces of organic matter or humus. Away from the areas of flowing water, visible plant life is absent.

Lichens are more widespread and occur far inland at altitudes of more than 2000 m (Claridge et al., 1971) but generally occur on rock surfaces rather than soil, as they are slow-growing and require a stable substrate.

Some organisms have exploited a favorable microenvironment within rock cracks and the pores between grains of sandstone (Friedmann and Ocampo, 1976). These largely occur in vertical exposures of sandstone on the valley walls and generally are lichens, growing in a thin zone, a few mm thick, about 1 or 2 mm below the surface. The algal component of the symbiont lichen generally is separated from the fungal component, in order to make maximum use of the incoming solar radiation and moisture.

These cryptoendolithic lichens frequently occur associated with unidentified colorless bacteria presumed to act as decomposers, so that a small ecosystem develops in this microenvironment. Even on the valley floors, where moisture may be somewhat more available, green algal material grows in cracks in translucent rocks such as massive quartz aggregates or marbles, where it survives in a kind of miniature greenhouse.

3. Factors Governing Organic Life

There are no detectable weathering differences between soils that contained relatively abundant life and those that did not, nor are there any other differences observable that could be attributed to the presence of life. Yet soil properties clearly affect the organisms themselves. The soils that contained organisms are younger and more moist and contained much less salts than those that did not contain life (Cameron and Conrow, 1969; Janetschek, 1970; Ugolini, 1977).

The availability of moisture is considered the major regulator for primary terrestrial colonists in soils of the cold desert (Kennedy, 1993; Wynn-Williams et al., 1997), but since salt concentration in soils is a function both of moisture and age, soil salinity probably is the dominant factor. Temperature is also important. The intensity of life decreases with increasing latitude and altitude, and both factors govern temperature which, in turn, is one of the factors governing available moisture.

The most widespread organisms are microorganisms; higher plants, mosses, lichens, and algae occur only in favorable situations where groundwater from melting snow is available during the short summer. These soils are not saline, as water flow leaches out any salt, and they may appear quite productive in comparison with surrounding areas. Mosses and filamentous cyanobacteria occur in areas of flowing water or where capillary movement moistens soils, alongside flowing water. Away from the area of flowing water, plant life is absent.

4. Animal Life

The bacteria and algae provide a substrate on which grazing animals can survive. In Antarctic soils, the most important grazers are nematodes, although tardigrades and rotifers are also present at some sites (Freckmann and Virginia, 1998). The distribution of nematodes is related to soil factors such as pH and salt concentration and also to the presence of soil organic matter, which itself is related to food sources and soil moisture.

In no other known systems do nematodes represent the top of the food chain or are the relationships so simple. Only very few species of nematode appear in Antarctica, and of these only 3 are widespread in the cold desert soils of the Ross Sea region. The unusually low diversity of the nematode population suggests that the ecosystem will be disrupted greatly by the loss or decline of even a single species that is sensitive to environmental change.

Especially favored situations support additional predators even higher up the food chain than nematodes. These are arthropods, represented by free-living mites, as distinct from mites that are parasitic on birds or marine mammals, and collembola (Gressitt and Shoup, 1967). The diversity of species of these animals is much greater than has been reported for nematodes; for example, in the Hallett Spit area at least 9 species of mite and 3 species of collembola occur. These animals inhabit dry depressions on the underside of stones or in cracks in stones, generally associated with moss patches. Although little is known of their food sources, they probably prey on the nematodes and other microscopic organisms, as well as on plant material.

Thus, in the ice-free areas, the soils of the cold desert do carry a small and simple ecosystem, consisting of primary producers, bacteria, yeasts, and algae, grazed by nematodes, which in turn form the food source for primitive arthropods. This ecosystem has little effect on the soils and is distributed patchily but widely. Usually its presence can be detected only microscopically, and even the largest animals, the collembola, barely can be detected with the naked eye.

The density of the biological community relates to a number of soil factors, of which the most important are moisture availability and salinity, but it largely depends on latitude and altitude (Figure 4.6.1). The most biologically rich areas are in the extremely rare exposures of bare ground in North Victoria Land, in particular, Hallett Spit, in Edisto Inlet (Rudolph, 1966; Boyd et al., 1966), but some kind of biological activity is detectable as far south as bare ground occurs.

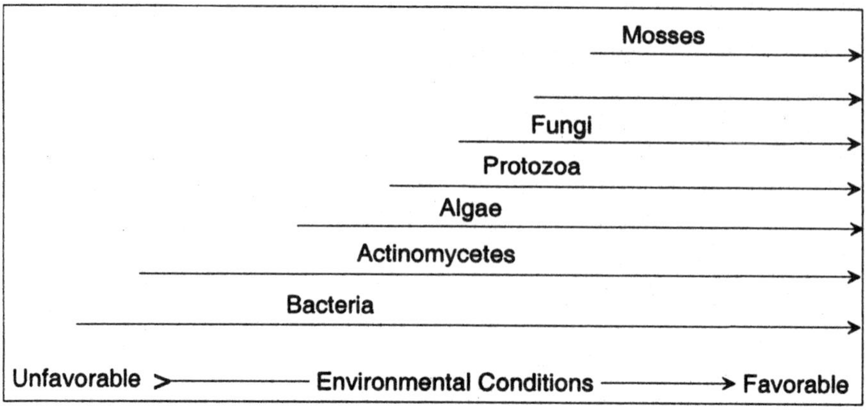

Figure 4.6.1. The relationship between abundance or diversity of species and environmental factors.

5. Ornithogenic Soils and thermally heated Soils

Included within the Arid Cryosols are small areas of very different soils, those formed on penguin excreta (ornithogenic soils) and those formed on thermally heated ground where temperature is not a limiting factor and moisture is available from melting snow.

Ornithogenic soils occupy no more than 20 km^2 of Antarctica (Claridge et al., 1999) but are distinguished by a very high carbon content (up to 25%), together with lesser but still large amounts of nitrogen and phosphorus (Ugolini, 1972; Speir and Cowling, 1984). In these soils, bacterial populations are very high (Ramsay, 1983; Ramsay and Stannard, 1986), but little is known of the other microscopic soil organisms. The guano itself probably provides a very inhospitable environment for many organisms, but the nitrogen-rich runoff from the rookeries encourages considerable algal growth in the streams and is a very rich food source for other plants and animals. Outside the rookery areas, there is very little effect on the soils.

In a very few places, notably near the summits of the few active volcanoes in Antarctica, the ground is warmed from geothermal surfaces, moisture is available from condensing steam emissions, and weathering processes proceed much farther than in the Arid Cryosols. In continental Antarctica, thermally heated ground has been found near the summit of Mt. Erebus (Ugolini, 1967), on Mt. Melbourne (Lyon and Giggenbach, 1974), and on Mt. Rittmann (Bonaccorso et al., 1991). Other hydrothermal vents may occur in Marie Byrd Land (LeMasurier and Wade, 1968).

These sites cover no more than a few hectares but support a unique assemblage of organisms (Ugolini and Starkey, 1966; Broady, 1984; Broady et al., 1987; Bargagli et al., 1996), because of the combination of satisfactory moisture status, suitable temperatures, pH values approaching neutral, and nutrient supply from bird excreta. On the surface of the ground, thin crusts of algae and bryophytes grow at temperatures between 10 and 50°C. Within the soil, a rich population of bacteria, fungi, actinomycetes, and yeasts occurs. Most of the organisms are similar to those in other Antarctic soils, but some of the bacteria are thermophilic, capable of surviving at temperatures up to 70°C.

Thermally heated ground provides a habitat for organisms normally unable to survive in the rigorous Antarctic environment. The sites provide niches in which organisms from outside Antarctica can colonize and spread. However, the transitory nature of patches of heated ground means that communities have little time to establish themselves before the activity moves to another site. The greatest danger here is that thermally heated ground could provide a suitable habitat for exotic species introduced by visitors from outside the Antarctic.

6. Organic Matter Content of the Soils

Because of the very low biomass and metabolic activity of the Arid Cryosols of the Transantarctic Mountains, the organic carbon contents are extremely low. For the range of soils extending from the southernmost occurrence of soil in the world (Cameron, 1971) to the McMurdo Sound region, carbon contents range from 0.02 to 0.08%, except where carbonaceous material occurs in the parent material (Claridge et al., 2000). The nature of the organic matter has not been characterized, as quantities are too small. Much of the carbon measured is expected to be living matter, such as algae or bacteria. It is impractical to measure organic nitrogen contents, because of the large amounts of inorganic nitrogen present in the soils as nitrate salts

Even where organic matter accumulates, such as in algal peats around shallow ponds and in depressions, there are no recorded attempts to characterize the nature of the material, such as have been made for some soils of coastal Antarctica (Beyer et al., 1997). The total mass of carbon in arid Cryosols has been estimated at 15 megatonnes, an insignificant figure in comparison with the world figure or even with the 100 megatonnes estimated for the soils of coastal Antarctica or the 600 megatonnes in the soils of the Antarctic Peninsula (Claridge et al., 2000).

More work has been done on the composition of penguin guano in ornithogenic soils, which cover about 0.04% of the ice-free area of Antarctica. They contain up to 25% carbon, 16% nitrogen, and 3% phosphorus, very different from other Antarctic soils, or indeed from soils of temperate regions, although they resemble guano deposits in other parts of the world (Speir and Cowling, 1984; Orchard and Corderoy, 1983). Despite their limited extent, these soils contain in total about 3.5 megatonnes of carbon.

In the event of an amelioration in the Antarctic climate due to climatic warming, the pattern of vegetation in coastal regions of Antarctica should spread farther south. Unless the temperature warms to a very great extent, moss and lichen patches, at present of very limited extent on the arid Cryosols, would only expand and thicken to a limited extent, allowing the soils under them to thicken and organic matter content to increase. Thus a climatic warming can be expected to increase the amount of carbon contained in the soils and thus provide a sink, although very small, for greenhouse gases.

7. Conclusions

In the arid Cryosols, the influence of the organic cycle and of the organisms that make up that cycle is extremely small. Most of the soils, although not devoid of living matter, contain such a relatively small number of organisms and such a limited range of species that their presence can barely be detected by the carbon content of the soil. The organisms exert almost no influence on soil-forming processes, apart from some weathering of rock surfaces by lichens.

However, as conditions for life become more suitable, in scattered and isolated spots with sufficient moisture, small plant communities can form and the organic content of the soil increases. Although the number of plant and animal species increases in such communities, there appears to be little or no accumulation of humus, in contrast to that of some of the soils of the warmer coastal regions of the Antarctic continent which Beyer and Bölter (1999) described.

While the Arid Cryosols are not devoid of life, they can be regarded as soils in which the influence of living matter on soil development is almost completely negligible.

References

Bargagli, R, P.A. Broady, and D.W.H. Walton. 1996. Preliminary investigation of the thermal biosystem of Mt Rittmann fumaroles (northern Victoria Land, Antarctica). Antarc. Sci. 8:121-126.

Beyer, L., and M. Bölter. 1999. Formation, ecology and geography of Cryosols of an ice-free oasis in coastal East Antarctica near Casey Station (Wilkes Land). Austral. J. Sci. 37:209-44.

Beyer, L., H-P. Blume, C. Sorge, H-R. Schulten, H. Erlenkeuser, and D. Schneider. 1997. Humus composition and transformations in a pergelic cryohemist of coastal Antarctica. Arc. & Alp. Rsch. 29:358-365.

Bonaccorso, A., M. Maione, P.C. Pertusati, E. Privitera, and C.A. Ricci. 1991. Fumarolic activity at Mt. Rittmann volcano (northern Victoria Land, Antarctica). Memorie della Società Geologica Italiana. 46:453-456.

Boyd, W.L., J.T. Staley, and J.W. Boyd. 1966. Ecology of microorganisms in Antarctica. In: J.C.F. Tedrow, ed. Antarctic soils and soil-forming processes. Amer. Geophys. U. Antarc. Ser. 8:125-159.

Broady, P.A. 1984. taxonomic and ecological investigations of algae on steam-warmed soils on Mt. Erebus, Ross Island, Antarctica. Phycologia. 23:257-271.

Broady, P.A. 1996. Diversity, distribution and dispersal of Antarctic terrestrial algae. Biodivers. & Cons. 5:1307-1355.

Broady, P.A., D. Given, L. Greenfield, and K. Thompson. 1987. The biota and environment of fumaroles on Mt. Melbourne, northern Victoria Land. Polar Biol. 7:97-113.

Broady, P.A., and R.N. Weinstein. 1998. Algae, lichens and fungi in La Gorce Mountains, Antarctica. Antarc. Sci. 10:376-385.

Cameron, R.E., and H.P. Conrow. 1969. Soil moisture, relative humidity and microbial abundance in dry valleys of southern Victoria Land. Antarc. J. U.S. 4:23-28.

Cameron, R.E., G.H. Lacy, F.H. Morelli, and J.B. Marsh. 1971. Farthest south microbial and ecological investigations. Antarc. J. U.S. 6:105-110.

Cameron, R.E. 1972. Microbial and ecological investigations in Victoria Valley, southern Victoria Land, Antarctica. In: G.A. Llano, ed. Antarctic Terrestrial Biology. Amer. Geophy. U. Antarc. Rsch. Ser. 20:195-220.

Claridge, G.G.C., I.B. Campbell, and D.S. Sheppard. 2000. Carbon pools in Antarctica and their significance for global climate change. R. Lal, J.M. Kimble, and B.H. Stewart, eds. Global climate change in cold regions ecosystems. Lewis Publishers. Boca Raton, FL. pp. 59-78.

Freckmann, D.W., and R.A. Virginia. 1998. Soil biodiversity and community structure in the McMurdo Dry Valleys, Antarctica. In: J.C. Priscu, ed. Ecosystem dynamics in a Polar Desert. Amer. Geophy. U. Antarc. Rsch. Ser. 72:323-335.

Friedmann, E.I., and R. Ocampo. 1976. Endolithic blue-green algae in the Dry Valleys: primary producers in the Antarctic desert ecosystem. Sci. 193:1247-1249.

Gressitt, J., and J. Shoup. 1967. Ecological notes on free-living mites in North Victoria Land. In: J.L. Gressitt, ed. Entomology of Antarctica. Amer. Geophy. U. Antarc. Rsch. Ser. 10:307-320.

Janetschek, H. 1970. Environments and ecology of terrestrial arthropods in the High Antarctic. In: M.W. Holdgate, ed. Antarctic Ecology. Academic Press, London 2:871-885.

Kennedy, A.D. 1993. Water as a limiting factor in the Antarctic terrestrial environment: biogeographical synthesis. Arc. & Alp. Rsch. 4:308-315.

LeMasurier, W.E., and F.A. Wade. 1968. Fumarolic activity in Marie Byrd land, Antarctica. Sci. 162:352.

Lyon, G.L., and W.F. Giggenbach. 1974. Geothermal activity in Victoria Land, Antarctica. N.Z. J. Geol. and Geophys. 17:511-522.

Orchard, V.A., and D.M. Corderoy. 1983. Influence of environmental factors on the decomposition of penguin guano in Antarctica. Polar Biol. 1:199-204.

Ramsay, A.J. 1983. Bacterial biomass in ornithogenic soils of Antarctica. Polar Biol. 1:221-225.

Ramsay, A.J., and R.E. Stannard. 1986. Numbers and viability of bacteria in ornithogenic soils of Antarctica. Polar Biol. 5:195-198.

Rudolph, E.D. 1966. Vegetation of Hallett Station area, Victoria Land, Antarctica. Ecol. 44:585-586.

Speir, T.W., and J.C. Cowling. 1984. Ornithogenic soils of the Cape Bird Adelie penguin rookeries, Antarctica. 1. Chemical properties. Polar Biol. 2:199-205.

Ugolini, F.C. 1967. Soils of Mt. Erebus, Antarctica. N.Z. J. Geol. and Geophys. 10: 431-442.

Ugolini, F.C. 1972. Ornithogenic soils of Antarctica. In; G.A. Llano, ed. Antarctic Terrestrial Biology. Amer. Geophy. U. Antarc. Rsch. Ser. 20:181-193.

Ugolini, F.C.1977.The protoranker soils and the evolution of an ecosystem at Kar Plateau, Antarctica. In: G.A. Llano, ed. Proceedings of the 3rd Symposium on Antarctic Biology. Smithsonian Institute. Washington D.C. 1091-1110.

Ugolini, F.C., and R.L. Starkey. 1966. Soils and micro-organisms from Mt Erebus. Nature. 211:440-441.

Wynn-Williams, D.D., N.C. Russell, and H.G.M. Edwards. 1997. Moisture and habitat structure as regulators for microalgal colonists in diverse Antarctic terrestrial habitats. In: W.B. Lyons, C. Howard-Williams, and I. Hawes, eds. Ecosystem processes in Antarctic ice-free landscapes. Balkema. Rotterdam. pp. 77-88.

Chapter 7. Biodiversity, primary productivity, and the seasonal dynamic of soil processes in Taimyr soil-permafrost complexes

V. D. Vassiljevskaja[1], B. Pospelova[2], and V. Telesnina[3]

[1]Professor, Faculty of Soil Science, Moscow State University, Vorobievy gory, Moscow, 11899 Russia; e-mail vasilev@soil.msu.ru.
[2]Senior Research Assistant, Geographical faculty, Moscow State University, Vorobievy gory, Moscow, 119899 Russia; e-mail taimyr@orc.ru.
[3]Junior Research Assistant, Faculty of Soil Science, Moscow State University, Vorobievy gory, Moscow, 11899 Russia; e-mail taimyr@orc.ru.

1. Introduction

This chapter discusses:

1. the microclimatic differences and seasonal thawing dynamic of soils
2. the dynamics of redox conditions and of the chemical composition of soil solutions
3. the primary productivity of plant communities in separate elements of simplexes
4. the connection between biodiversity and soil-vegetative cover in complex structures
5. the possible influence of global climate changes on tundra soil cover and on primary productivity.

Our research focused on Cryosols, illustrated by spotty and hummocky tundra (nano-relief) and complex bogs (micro-relief).

2. Materials and Methods

Among Arctic sectors, the Taimyr territory's climate is the most continental, with an average annual temperature of -13. 4°C, average summer temperature of 10 to 13°C, and annual precipitation of 297 to 344 mm. The entire territory is within the perennial permafrost area. Regional and stationary soil investigations (along with botanical and zoological research) began in West Taimyr in 1966, as part of the International Biological Program. In East Taimyr, investigations began in 1978.

The West Taimyr investigations have been carried out mainly in the Pyasina river basin (71°25′ n.l. – Agapa river mouth – the north border of the south tundra; Tareya river mouth – 73°14′ n.l. – typical tundra). The East Taimyr investigations have centered on the Bikada monitoring station (the east coast of Taimyr lake, typical tundra, 74°50′n.l., 105°30′e.l.).

Cryosols are weakly acid; gleyic horizons, especially in spot soils, are sometimes neutral. These soils show a high degree of base saturation (80 to 90%) and high humus content (2 to 5%) within all the seasonal thawing layer. Cryosols of complex bogs have a lower degree of base saturation (40 to 60%) and a more acid reaction (pH_{H2O} is lower than 5. 2 to 5. 6). Organic matter content is higher than in Cryosols (the loss after ignition is 30 to 70%). In large degree, the mineralogical composition of the parent material, in which montmorillonite predominates in the clay fraction, defines the soil chemical features (Vassiljevskaya, 1981; Everett at al., 1981).

All of the seasonal dynamic investigations were carried out with soils of main tundra types at monitoring stations only in the summer. The investigation took into account the structure of the soil complex.

The authors studied the soil temperature regime at Agapa monitoring station during the summers of 1967 through 1970 (Vassiljevskaya et al., 1974; Vassiljevskaya, 1980). The observations covered the parts of the second floodplain terrace and floodplain terrace slopes of different exposition, and they used mercurial thermometers. We made similar observations at the Tareya monitoring station of the Komarov Botanical SA USSR Institute (Romanova, 1978). We measured the thickness of the active layer (AL) with a sharp metallic pivot for all elements of permafrost complexes, making 50 replications.

For studying gley process dynamics and the intensity of gleyzation by some quantitative indexes, we determined mobile forms of iron and redox indexes. On every element of relief soil samples, we collected from two depths, 0 to 10 and 20 to 40 cm., with 5 replications. We extracted the mobile forms of iron by shaking freshly collected soil samples with 0.1n H_2SO_4 (1:10) for 5 minutes.

We determined protoxide forms of iron colorimetrically by a, a' dipiridil and determined trivalent iron by titration with trilon B in the presence of sulphursalicile acid. We measured redox potential (ORP) by potentiometer, with platinum and combined chlorine-silver electrodes. We brought the small, freshly collected monoliths from soil horizons into the laboratory in polyethylene packets for measuring with a constant temperature of 18°C.

We collected the samples for wringing out soil solutions in July through August of 1968 at Agapa monitoring station territory and from May to August of 1969, June to August of 1970, and June to August of 1971 to 1973 at Tareya monitoring station. We collected samples for ten days. These samples were combined of sample parts, equal in size, collected from all depths of the horizon, all five parts being homogeneous. We wrung out soil solutions under 150 ATM, with an hydraulic press.

The soils investigated were from spotty and hummocky tundra and complex bogs. Information about the soil solution composition of the Agapa monitoring station was published first by V. V. Ivanov (1970) and, about the Tareya station, by V. D. Vassiljevskaya and G. A. Kolpashikova (1978).

We calculated above-ground plant biomass for the 50X50 cm plots for above-moss store and 25X20 cm plots for the living moss layer. We replicated tests 20 to 30 times for every tundra type. We determined a preliminary ratio between coeno-elements preliminarily. We collected hay samples separately for every coeno-element, after calculating the average weighted value. We considered the lower border of living moss tissues to be the border between above- and below-ground plant biomass. We sorted the living above-ground parts by species and calculated the dead above-ground residues separately. We defined below-ground plant biomass for every 10-cm layer, using the special bore; we cleaned the samples on a 1-mm soil sieve.

3. Results and Discussion

3.1. Temperature regime

The temperature of soil and of air near the soil are the most important ecological factors in the Arctic. The temperature regime affects the floristic composition of plant communities, the biology, ecology, and productivity of tundra plants, and the composition and spatial distribution of plant biomass. Temperature also plays an important role in soil formation. The thickness of the active layer, the intensity of mineralization from fallen leaves, the rate and intensity of chemical reactions, and other processes within soil also closely connect with soil climate. Our investigation has shown that air temperature at 1.5 m above different tundra types within the geomorphological profile is rather uniform: the changes are not over .01^{0}C. The highest temperature in July (among observation years) was about 10 to 15^{0}C and in August the average monthly air temperature did not exceed 7 to 8^{0}C.

The most rectilinear connection between air and soil temperatures occurred at a 5-cm depth under bare soils (r =0. 98). Under plant cover, the correlation between soil and air temperatures is lower even at 5 cm deep (r = 0. 71 to 0. 79). The influence of the constantly frozen layer on the temperature regime of Cryosols AL increases away from the surface. The correlation coefficient decreases most clearly on plantless plots; the permafrost's influence on the daily temperature dynamic is lower for soils with an organic horizon.

American scientists calculated the correlation coefficient (0. 90) between soil surface (Y) and air (X) temperatures for tundra of the Prudho Bay monitoring station (Brown at al., 1975). The regression equation is y=0. 755x +3. 6.

No sharp differences occur between temperatures of plantless spots and moss-covered spots, although it is important to note the higher heating later. A very rapid decrease in soil temperature occurs with a decrease in air temperature. Commonly, for all seasons, permafrost thus plays a predominant role in the formation of the soil AL temperature regime. When estimating the course of the average daily soil temperature for different tundra types, one must note its correlation with air temperature and that soil temperature drops behind regularly (at the 5- to 10-cm depth), to 2 to 3^{0}C under bare spots and to 2 to 5^{0}C under moss.

Nacano and Brown (Nacano and Brown, 1972) constructed the mathematical patterns for thermal regime of tundra soils of the Barrow monitoring station (Alaska). These authors determined a close connection between the temperature regime of Cryosols and organic horizon depth, and between temperature regime and organic matter content and the intensity of heat diffusion by organic and mineral horizons. They found a good coincidence of results between soil temperatures observed in the field and those calculated for different depths and fixed AL thicknesses.

Romanova (1978) investigated the daily temperature dynamics of soil and air temperature changing within nano-slopes in complexes. The differences in temperature regime between slopes of different ascent of northern and southern exposures are present very strongly in nano-relief (Table 4.7.1).

Comparisons between soil temperatures of bare spots and parts with vegetation reveal that moss cover plays a heat-isolating role. In spite of relatively higher heating, soil temperature under the moss cover (at all depths) is essentially lower. These parts thaw later and weakly. Generally, the temperature regime of plantless spots in comparison with adjacent plant-covered parts is more favorable (during the summer), and bare spots thaw earlier. This explains why bare-spot soil is rich in living roots that penetrate from adjacent plant-covered parts and why soil fauna is relatively various here. In loamy Cryosols of spotty dwarfshrub-moss tundra at the Agapa monitoring station July 15 to August 30 (1968 through 1970), the soil temperature under the plant cover was 5^{0}C at 5 cm and 2 to 6^{0}C at 35 cm, and, under the bare spot, was, respectively, 8.9^{0}C and 3.8^{0}C.

Thawing permafrost or changing AL thickness during the growth period plays a leading role in creating conditions for soil-forming processes and for vital activity of soil invertebrates, microflora, and lower and higher plants. Within micro- and nano-complexes, the differences in AL thickness regulate the migration processes of compounds. During the summers 1968 through 1970, we investigated a number of chemical and physico-chemical soil features to study the dynamics of thawing.

Loamy soils of spotty shrub and dwarf shrub-sedge-moss tundra, watersheds, and terraces thaw on average for 70 to 80 cm. Soils of hummocky dwarf-shrub-sedge-moss tundra on watershed slopes have a coarser texture and thaw for 60 to 70 cm. Soils of hummocky cottongrass-sedge-moss boggy tundra thaw for 50 to 60 cm. In complex bogs, thawing doesn't exceed 40 to 50 cm. Sandy soils of floodplains have the deepest seasonal thawing (100 cm and more).

Table 4.7.1. Differences in soil surface temperature (^{0}C) between northern and southern nano-relief slopes (July, Romanova, 1978).

Time	Ascent of nano-slope, Degrees							
	5	10	15	20	25	30	35	40
Day	4-5	7-8	11-12	14-16	18-20	22-24	26-28	32-34
Night	-1	-1.5-2. 0	-2.0-2.5	-3.0-3.5	-4	-5	-6	-7

We calculated the correlation coefficient between the seasonal thawing level for cracks and for plantless spots with extensive data from Agapa station spotty tundra. The connection between these indexes is rather close (r = 0,95). The regression equation is:

$$y = 1.11x - 16.30.$$

During the initial soil-forming stages in tundra rivers and floodplains formed with lightly textured alluvium and covered with thinned vegetation (grass-forbs meadows), the thawing level is homogeneous. When polygonal bogs develop with sedge-moss vegetative cover and permafrost microrelief (higher floodplain), the thawing level in the polygon-hummock system begins to vary markedly, forming soil complexes, which differ in thawing depth. On the terraces and flat surfaces with spotty and hummocky tundra, contrasts in seasonal thawing of separate nano-complex elements become greater, due to the irregularity of their organic horizons and vegetative cover distribution. This has a pronounced effect on the seasonal dynamics of soil processes.

3.2. Redox regime

Among tundra types (especially for organogenic horizons), regular moisture distribution increases from elluvial landscapes with crack-polygonal and hummocky tundra to accumulative landscapes (polygonal or flat palsa bogs). Within the complexes such as crack-border, crack-palsa, and polygon-border, redistribution of moisture occurs, and moisture content is lower in nano-ridge soils and higher in nano-depressions. Such distribution is typical mainly for organic humus-accumulating horizons; in mineral horizons, these differences are less expressed. Moisture changed comparably little over the years of our observations. The peat horizons of polygonal bogs have the greatest moisture content (120 to 800%). Gleyic horizons of bog soils have moisture content of about 40 to 80%. Gley horizons of spotty and hummocky tundra are the driest among clay and sandy loam soils (20 to 30%).

While completing the morphological description of tundra loamy soils pits, we observed mineral horizons with different visual signs of intensity of gleyzation — dark grayish brown (2. 5Y4 /2), dark bluish gray (5B3 /1), olive-brown (2. 5Y 4 /4), dark greenish gray (10G 4 /1), bright blue, light olive-brown (2. 5Y 5 /4), and other colors of horizons. Among the pits, the profile of p. 251, Stagni-Turbic Cryosol (Humic), is least gleyic. In the spotty tundra soil profile under the bare spot, gleyzation signs are less expressed, probably due to improved aeration due to less organic matter content and less moisture. In spotty polygonal tundra (p. 258, Turbi-Saprihistic Cryosol [Stagnic]), the signs of gleyzation are clearer (olive color, bluish-gray and ochric spots). In most gley soil of p. 253 – Turbi Fibrihistic Cryosol (Gleyic), we observed, from a 20-cm depth, a bluish-gray clear gley horizon (G), both on hummocks and in depressions (this isn't the B_g horizon described for other profiles).

In summer, the content of forms of iron protoxide is lowest in the soils of watershed spotty nano-polygonal tundra — less than 10 mg/100 g of dry soil, on average (Table 4.7.2). The content of that in soils of the second over-floodplain terrace spotty tundra (p. 258) is somewhat higher, 15 to 25 mg/100g (Table 4.7.2). The gley loamy horizons of spotty-hummocky tundra corresponding to slopes of watershed ridges is richest in protoxide forms of iron (80 to 130 mg/100g of soil). Redox potential changes in this sequence, accordingly, 500 to 600, 400 to 500, and 280 to 450 mV. By the end of the vegetative period, the content of forms of protoxide iron increases. This likely connects with the intensification of microbiological reduction processes, due to temperature increases and increasing AL thickness.

The high soil moisture content in p. 258 does not show strong gleyzation, in contrast to p. 253. In this case, probably, the character of pores plays a part. In the heavily textured gley horizons of p. 253, small pores predominate, hampering water exchange and pore aeration, whereas in the thixotropic, especially surface, horizons of p. 258, macro-pores predominate, which distinctly shows on dried soil monoliths. Redox potential (ORP) values confirm better aeration conditions, too.

It is worthwhile to estimate tundra soils' gley process expression in the middle of the vegetative period (mid-July through early August). At the end of that period, iron protoxide and ORP indexes smooth out. During the winter, when the soil is frozen, mobile iron forms also form. Consequently, directly after the thawing of soil, we observe, as a rule, the highest content indexes of forms of mobile iron. In the case of the Agapa monitoring station's soils, iron contents directly after the thawing of soil were from 14 to 96 mg/100g (for different soils and horizons) and, in the middle of July, the range widened to 136 mg, demonstrating the possibility of dividing soils more exactly in terms of their degree of expression of gley process. Elaborating the row of quantity characteristics will let us divide Gleyzems (Gleyi-Turbic Cryosols) and Cryozems (Oxyaqui-Turbi Cryosols) more certainly.

3.3. Migration of substances

We investigated the migration of substances in tundra soils by analyzing soil solutions. The cryogenic processes of changes of mass and water play the essential role in forming the morphology and chemistry of tundra gley permafrost soils, but not enough research has been done on the quantitative character of these factors. Little information exists about the chemical composition of the liquid phase of soil and its change during the vegetative period and for different elements of permafrost complex. In general, solution chemistry for tundra soil shows essentially more mineralization of mineral horizons than organic horizons (Figure 4.7.1). This is probably because of the poor mineralization of organic matter under the strong northern climatic conditions, the increasing role that mineral soil plays in the formation of soil solution, and the richer water supply of organic horizons.

The averaged data for the composition of soil solutions in organic and mineral horizons show that organic horizons contain 23 mg/l Ca^{2+}, 10 mg/l Mg^{2+}, 74 mg/l HCO_3^-, and mineral horizons contain, respectively, 39, 19, and 121 mg/l.

Table 4.7.2. Dynamic of content of mobile iron forms (mg/100g), ORP (mV), and moisture (%) in the soils of Agapa monitoring station (Ivanov and Bogatiryov, 1970).

NN of cross sections	Element of nano-relief	Samples collecting depth	8.06.1969r.			
			Fe0	Fe_2O_3	ORP	Moisture
251	Spot	0-20	19.3	51.5	536	36.4
		20-40	17.5	56.5	524	30.4
	Crack	0-10	41.0	58.8	565	149.3
		20-40	22.8	27.6	625	30.6
253	Spot	0-20	32.9	122.9	516	31.3
		20-40	96.4	23.4	282	39.2
	Crack	0-10	13.6	52.6	580	350.9
		20-40	37.8	59.6	568	36.7
258	Spot	0-20	19.4	31.9	482	24.1
		20-40	17.5	49.3	571	28.7
	Crack	0-10	20.9	63.8	577	145.6
		20-40	19.4	42.4	578	32.5
				14.07.1969r.		
251	Spot	0-20	2.1	46.1	504	16.6
		20-40	2.8	51.9	516	18.4
	Crack	0-10	7.8	23.3	546	58.4
		20-40	3.9	29.2	584	27.1
253	Spot	0-20	10.4	97.8	546	27.6
		20-40	104.1	0.8	288	32.1
	Crack	0-10	60.3	76.9	263	105.6
		20-40	135.0	88.4	333	43.2
258	Spot	0-20	16.6	31.6	518	19.5
		20-40	19.5	32.5	486	21.9
	Crack	0-10	23.7	42.8	472	84.1
		20-40	11.9	30.9	564	233
				10.08.1969r.		
251	Spot	0-20	11.4	55.4	521	18.4
		20-40	9.7	58.3	513	18.8
	Crack	0-10	37.7	27.9	568	33.9
		20-40	41.0	249	579	25.3
253	Spot	0-20	24.3	80.6	480	28.6
		20-40	83.9	62.9	262	31.9
	Crack	0-10	52.9	77.3	466	39.4
		20-40	72.4	64.7	422	34.3
258	Spot	0-20	28.4	47.3	572	18.8
		20-40	34.0	41.7	566	18.6
	Crack	0-10	59.8	86.1	574	100.9
		20-40	42.3	56.7	574	209

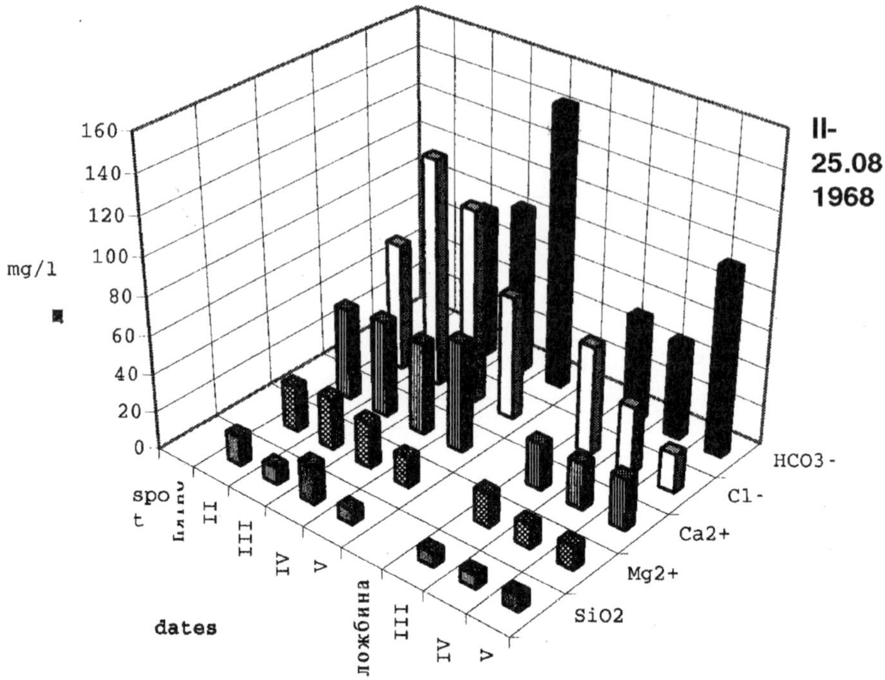

Figure 4.7.1. Dynamic of the composition of soil solutions (mg/l, 0- to 10-cm depth). Spotty nano-polygonal tundra. Monitoring station "Agapa" (p. 251).

Organic horizons of tundra gley soils have a more porous, finely grained texture. During the most summer, a frost gley horizon underlies them; consequently, inter-soil lateral drainage (horizontal above-permafrost elluvation) goes mainly along these organic horizons and causes their leaching. These processes are also evident from the bleached fine earth in the mucky horizons of cracks and from the similar composition of soil solutions of organic horizons and of water springs and surface runoff (Vassiljevskaya, 1980). Lateral runoff affects the mineral horizons to a low degree.

The upper horizons of spotty tundra at both Tareya and Agapa accumulated soluble compounds in uncovered spots during the winter, due to ascending migration toward the freezing front (water migration by temperature gradient). By the middle or end of August, the degree of mineralization of soil solutions also regularly increased, because decomposition of organic matter increased during this period, simultaneously with increasing AL thickness due to thawing. The Barrow monitoring station on the Arctic slope of Alaska (Bilgin and Douglas, 1972) showed more active migration of soluble compounds into the surface runoff with increasing AL.

Comparing soil solutions in different tundra types shows that solutions are essentially more mineralized in spotty and hummocky tundra soils than in polygonal bogs soils.

Soil solutions for Taimyr Cryosols change from poorly acid to alkaline. The SiO_2 concentration varies from some mg/l to 20 to 30 mg/l without some regularities during the vegetative period. Organic carbon content in soil solutions changed from 10 mg/l in mineral horizons to 15 to 36 mg/l in organic horizons in July and, respectively, from 20 to 100 mg/l in August.

The seasonal dynamic investigations not only let us see the differences in the character of processes in tundra gleyzems and bare-spot soils, but also let us confirm the presence of close soil-geochemical conjugation in the migration of compounds within soils composing the soil-permafrost complex. Once freezing has begun, during all winter the compounds migrate toward the coldest surface (surface of spot soil) in frozen soil. This process includes also the soil under vegetation, though cooling of surface and upper horizons is essentially less here.

Due to spring thawing, the permafrost level under the bare spot takes on a cup form. The water migrates from the thawed part of the spot soil profile into frozen horizons, including still frozen soils of cracks. At the same time, intensive flow occurs along the soil surface in hollows. Maximum soil thawing and equalization of thawing levels cause alternate processes of migration of moisture and soluble compounds in the system of adjacent soils, up to the dry crust of the spot (during the hot weather) and down to the freezing front (when precipitation occurs). At this point, a horizontal over-permafrost elluvial process occurs along the organic-accumulative horizons and leveled plots, alternating the migration of compounds from the bare spot to a complex element with plant cover (Figure 4.7.2).

3.4. General biodiversity, plant biodiversity, and primary productivity

General biodiversity, plant biodiversity, and primary productivity represent the intensity of ecosystems functioning in strong Arctic conditions. We observed the connection between plant diversity and definite soil type for Taimyr tundra. The highest vascular plant diversity occurred in locations characterized by prevalent Himi-Gelic Leptosols. Vascular plant diversity decreased with intensified cryogenic processes, increased moistening, and development of gley process. Thus, 54% of 427 vascular plants species, constituting the East Taimyr flora, occur mainly on the drained, deeply thawing soils of mountains, hill tops, and river banks, 32% occur on Gley-Turbic Cryosols of flat watersheds, and only 13% on Humi-Gelic Cryosols of waterlogged locations. The converse tendency applies for mosses.

Many authors (Pospelova, 1974, 1989; Vassiljevskaja et al., 1975; Pospelova, Vassiljevskaja, 1985) have investigated tundra's primary productivity. Data are available about the reserve and structure of phytomass for 15 plant communities of West Taimyr (south tundra) and for about 50 of East Taimyr (typical tundra). Of these, we investigated 35 for structure of underground and above-ground phytomass and the rest for underground phytomass. Associations are given by Matveeva's classification (Matveeva, 1998).

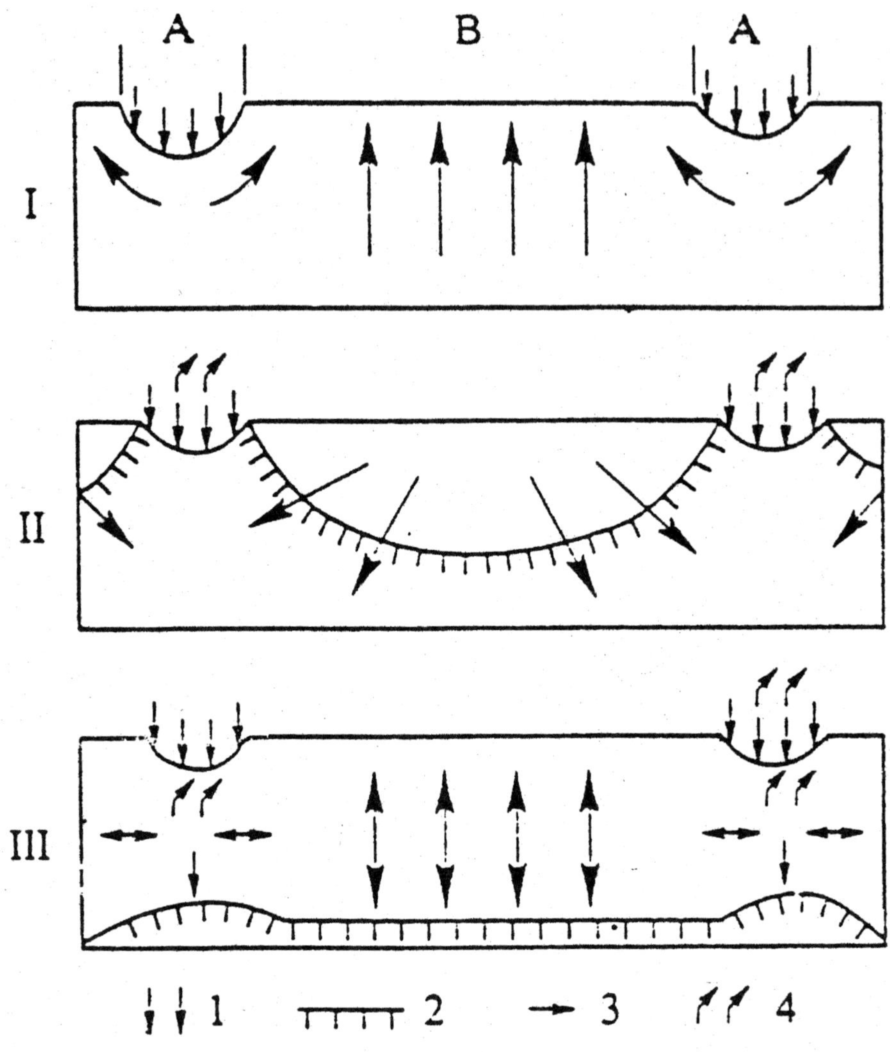

Figure 4.7.2. Seasonal mass and moisture exchange in a polygonal-crack complex of a slightly gleyed Tundra gley Earth (A) and soil of a spot (B). I. Frozen soil (late autumn, winter). II. Spring thaw. III. Maximum thaw (end of summer). 1. Shrub-moss plant cover. 2. Thawing depth. 3. Direction of movement of moisture and water-soluble components. 4. Surface runoff or horizontal suprapermafrost elluviation.

The values of general reserve (GR), which includes leavings and dead parts, vary from 3200 to 4800 g/m² for zone communities of spotty dwarfshrub-sedge-moss tundra (ass. *Carici arctisibiricae – Hylocomietum alascani;* Matveeva, 1994) on weakly acid tundra gley soils of typical and south tundra (the least value, 2000 to 2500 g/m², is typical for hilltop communities, where bare spots cover about 30% of the area). Above-ground plant biomass (APB) is 7 to 10% of GR here. With increasing shrub activity, increasing linear thermokarst, and increasing gley process on slopes, GR reaches 5000 to 7000 g/m^2, and APB, respectively, reaches 10 to 15%. For ass. *Sphagno-Eriophoretum vaginati* subass. *typicum* (Walker et al., 1994) on the same, but more gleyed and more heavily textured soils (only in the south tundra of West Taimyr), GR is 8211 to 9354 g/m^2, including APB of 6 to 7%. The lesser values of GR are typical for communities of intrazonal ass. *Rhytidio rugosi-Dryadetum punctatae* (Matveyeva, 1998), which is distributed on drained neutral soddy soils of gravel and sandy outcrops, 3200 to 4500 g/m^2, but the APV part is highest here, 13 to 18%.

In the intrazonal complexes of polygonal bogs communities (ass. *Meesio triquetris – Caricetum stantis*; Matveyeva, 1994; *Poo arcticae-Dupontietum fisheri*; Matveyeva, 1994) on acid and over-wetted Humi-Gelic Cryosols, GR is from 7000 to 14000 g/m^2, depending on the ratio of complex elements areas, which is connected with the development stage. The part of APB in typical tundra subzone is 4 to 5% and, on more shrub-covered bogs of south tundra, 10 to 15%. For azonal vegetation of flood-plains on neutral drained alluvial soils, GR is 3000 to 5000 g/m^2 (meadows) and 10000 to 12000 g/m^2 (willows), and APB is 3 to 5% in both cases.

Underground biomass sharply predominates in GR for all phytocoenoses. The main root mass concentrates in organic horizons (3000 to 6000 g/m^2) or in the upper mineral horizons of bare spots (800 to 1000 g/m^2); in gley horizons of all tundra types, it doesn't exceed 100 to 200 g/m^2.

These data indicate a rich geochemical background which provides high phytomass values in comparison with other Arctic and subarctic regions (Wielgolaski, 1970; 1981). It probably results from solids flowing from Byrranga, where weathered limestones are distributed widely, as well as from rocks, represented by the montmorillonite clays of Holocene transgressions.

We also recognized some general regularities, such as above-ground phytomass of main plant dominants depending on micro-ecotops features, in particular on soil chemical features, which characterize definite micro- and nano- relief elements.

We observed the highest biomass values for *Salix arctica* (more than 100 g/m^2), for legumes (30 to 50 g/m^2), and for mosses *Rhytidium rugosum*, *Rhacomitrium lanuginosum* on neutral soils that were becoming weakly alkaline in lower horizons (pH is accordingly 6 to 7 and 7 to 8), and on non-gleyzated and poorly gleyzated soils, which have a high degree of base saturation (V is 80 to 90% in organic horizons and 95 to 99% in mineral horizons). These soils also have high K_2O content for all profiles (20 to 40 mg/100 g) and high content of P_2O_5 — 20 to 30 mg/100g in lower (sometimes also in upper) horizons. These are Cryozems (Oxyaqui-Turbic Cryosols) of high bare spots, borders, and rims, located in drained watersheds and slope borders. The same conditions are optimal for *Dryas punctata* and *Hylocomium splendens moss*; although these dominating species are characterized by a wide ecological amplitude, the tendency of prevailing mass accumulation in the described conditions is rather clear.

The main Taimyr tundra dominant, *Carex arctisibirica*, appears almost everywhere, but the highest biomass (APB more than 20 g/m^2) occurs for more acidic (pH 6 to 6.5 for al profile) and less base-saturated soils (V is 70 to 85% in organic horizons and 80 to 90% in mineral horizons) with lower and inconstant phosphorous and potassium content (3 to 13 and 6 to 15 mg/100g accordingly) – the soils of flat watersheds spotty tundra.

The optimum *Salix pulchra* biomass accumulation (APB is more than 50 r/m^2) corresponds with bog elevations and gentle damp slopes. Soil cover is Gleyzems (Gleyi-Turbic Cryosols) or Peaty Gleyzems (Turbi-Histic Cryosols). These are more acid (pH is 4 to 5.5 for all profiles), and the degree of base saturation is low (20 to 50% in organic horizons, changing in mineral horizons). P_2O_5 content is also low, not higher than 14 mg/100 g in the mucky horizon; K_2O content in organic horizons changes from 10 to 20 mg/100g (the bogs) to 30 to 40 mg/100g (dell complexes); in mineral horizons, it is not higher than 10 mg/100g. *Salix reptans*, *Tomentypnum nitens* moss, and *Ptilidium ciliare* moss appear in the same conditions, but their amplitude is rather wider.

Carex stans and *Limprichtia revolvens and Meesia triquetra* mosses are drawn to Cryic Histosols of wet bog polygons, thermokarst depressions, and deep dells. APB values of *C. stans* attain 30 to 50 g/m^2 here. The chemistry of these soils is similar to that of the latter (pH is 4 to 5, V is 20 to 40% and 5 to 30% respectively, and lower phosphorous and potassium content), but they are wetter and the peat horizon thickness is higher (18 to 20 cm in contrast with 7 to 12 cm on elevations).

3.5. Probable changes due to a globally warming climate

A steady average annual temperature increase, especially because of winter warming and an increase in snow thickness, probably will reduce the area of polygonal-crack complexes, because more favorable conditions for plant existence can cause intensive overgrowing on plantless spots. Degradation of ice-wedges can cause increasing thermocarst and its movement toward the north, but an increase in plant biomass reserve can, to some degree, hamper this process.

Biological turnover due to warming will change the character of the biogeochemical cycles of some elements. Additional carbon will be emitted into the atmosphere. More intensive elements will be removed from autonomic landscapes to accumulate in the increasing AL thickness and elluvate laterally and horizontally.

If the predicted average annual air temperature increase of 5^{0}C occurs, the increasing percentage of elluvial-illuvially differentiated soils (Gelic Gleysols) in soil combinations, where Cryosols and Histosols predominate now, will be quite probable. Currently existing relicts of the mid-Holocene climatic optimum confirm this. Podzols, characterized by white bleached horizons with a thickness of about 20 cm (Ivanov, Sopina, 1980), are distributed in Arctic regions on sandy substrata.

4. Conclusion

The hydrothermal regime of spotty and hummocky tundra soils for different micro-combinations is formed rather stereotypically, in spite of concrete differences in separate forms of soil-permafrost complexes. This fact causes the similarity of horizontal and vertical migration processes within complexes in seasons; i.e., causes their similar functional pattern.

For example, in hummocky complexes, the salient element of micro-relief (hummock, tussock, border), the crack-nanopolygonal or the center of the polygon (or bare spot), results from rapid spring-summer warming and autumn-winter cooling. Most inert in the hydrothermal regime are the spaces between hummocks in hummocky complexes and the cracks in polygonal and crack-polygonal complexes. Consequently, there are temperature and moisture gradients within soil, and different rates and depths of thawing. And these influence the direction and intensity of the migration of water and compounds between complex elements (and to the outside of the complex).

Therefore, during the spring thaw, lateral and descending migrations are most expressed in the most heated part of the complex (the center of the polygon, hillock, border) and in locations towards adjacent soils (the outside of the polygon and the crack, the middle of the micro-bog, and the crack in a polygonal complex). At the same time, there are inner migration processes in crack polygonal complexes, mainly vertical migrations.

During the autumn-winter freezing, the flows of moisture and compounds are directed conversely, towards salient elements of relief and the nano-polygon center. However, in spite of the similarity of functional patterns, different complexes differ in period, intensity, and length of migration, as well as in quantitative and qualitative composition of transported compounds and moisture (Vassiljevskaya et al., 1993). Because of increasing temperature during the vegetative period and increasing AL thickness, the processes of migration of water-soluble compounds become more intensive, and the mineralization of soil solutions increases. Climate fluctuations probably will influence most such dynamic short-term processes as elluvial-gleyzation migration, migration and mobilization of organic matter, iron, and aluminum, and processes of descending, lateral, and ascending migration, including biogenic migration (Vassiljevskaya and Pervova, 1995).

References

Bilgin, A., and L. A. Douglas. 1972. Relationship of soils and hydrological nutrient cycling in a tundra ecosystem. Proceedings Tundra Biome Symposium. Univ. of Washington.

Brown J., R. K. Haugen, and S. Parrish. 1975. Selected climatic and soil thermal characteristics of the Prudhoe Bay region. In: Ecological investigation of the tundra in the Prudhoe Bay region, Alaska. Biol. Pap. N2. Univ. Alaska.

Everett, K. R., V. D. Vassiljevskaya, J. Brown, and B. D. Walker. 1981. Tundra and analogous soils. In: Tundra Ecosystems: A comparative analysis. Int. Biol. Progr. 25. Cambridge Univ. Press. Cambridge.

Ivanov, V. V., and L. G. Bogatiryev. 1970. In addition to gley process characteristic in tundra soils (West Taimyr). In: Vestnik MGU. Ser. 6. Biology, Pedology. 6. (In Russian.)

Ivanov, V. V., and E. P. Sopina. 1980. Soils of Lower Ob forest tundra near the Urengoy deposit. In: Vestnik MGU. Ser. 17. Pedology. 1. (In Russian.)

Matveeva, N. V. 1998. Zonality in Arctic plant cover. St. Petersburg. (In Russian.)

Nakano, J., and J. Brown. 1972. Mathematical modeling and validation of the thermal regimes in tundra soils, Barrow, Alaska. Arct. and Alp. Rsch. 14:1.

Pospelova, E. B. 1974. The structure and spatial distribution of plant biomass in main plant communities of "Agapa" monitoring station. In: Soils and productivity of plant communities. MSU. 2. Moscow. (In Russian.)

Pospelova, E. B., and V. D. Vassiljevskaya. 1985. The primary productivity of Taimyr tundra. In: Bot. J. 70:2. (In Russian.)

Pospelova, E. B. 1989. Muscox pastures in Bicada river basin and their productivity. In: Mammals and birds of Middle Siberia North (Mlekopitayushie i ptitsi severa Sredney Sibiri). Novosibirsk. (In Russian.)

Romanova, E. N. 1978. Micro- and meso-climates of Taimyr. In: The structure and functions of Taimyr tundra biogeocenoses. Nauka. Leningrad. (In Russian.).

Vassiljevskaya, V. D., E. B. Pospelova, V. V. Ivanov, and L. G. Bogatiryev. 1974. The soils temperature regime for different tundra types of "Agapa" monitoring station during the vegetative period. In: Soils and productivity of plant communities. MSU. Moscow. (In Russian.)

Vassiljevskaja, V. D., V. V. Ivanov, L. G. Bogatyrev, E. B. Pospelova, N. M. Schalaeva, and L. A. Grischina. 1975. Agapa, USSR. In: Structure and function of tundra ecosystems. Stocholm. Ecol. Bull. 20.

Vassiljevskaya, V. D., and G. A. Kolpaschikova. 1978. The changing of soil solutions composition of 'Tareya" monitoring station during the vegetation period In: The structure and functions of Taimir tundra biogeocenoses. Nauka. Leningrad. (In Russian.)

Vassiljevskaya, V. D. 1980. Soil forming in tundra of Middle Siberia. Nauka. Moscow. (In Russian.)

Vassiljevskaya, V. D. 1981. Agapa. USSR. In: Tundra Ecosystems: A comparative analysis. Int. Biol. Progr. 25. Cambridge Univ. Press. Cambridge.

Vassiljevskaya, V. D., N. A. Karavayeva, and Ye M. Naumov. 1993. Formation of the strusture of the Soil Mantle in Polar Regions. Pedology. 7. (In Russian.)

Vassilyevskaya, V. D., and N. Ye Pervova. 1995. The change of migration intensity of water-soluble compounds in soils of tundra and taiga landscapes due to global changing climate. Vestnik MSU. ser. 17. Pedology. 1. (In Russian.)

Wielgolaski, F. E. 1970. Primary production. In: O. W Heal, ed. Proceedings on the Tundra Biome Working meeting on analysis of ecosystems, Finland, September, 1970.

Wielgolaski, F. E. 1981. Primary production of tundra. In: J. P. Cooper, ed. Photosynthesis and productivity in different environments. Cambridge University Press. U. K.

Section 5. Classification of Cryosols

Classification of Cryosols: Introduction

G. Broll[1] and D. Konyushkov[2]

[1]Professor, Division of Geology and Agroecolgy, ISPA - University of Vechta, P.O. Box 1553, 49364 Vechta, Germany; e-mail gbroll@ispa.uni-vechta.de.
[2]Head, Laboratory of Soil Evolution and Ecology, V.V. Dokuchaev Soil Science Institute, Pyzhevskii per. 7, Moscow, 109017 Russia; e-mail dkon@agro.geonet.ru.

Soil classification is the basis for communication not only among soil scientists, but also among all those people who are involved with soils for various purposes. Soil classification is helpful for a common understanding of soil properties, soil processes, and soil conservation. In the 1880s, the Russian geologist, V.V. Dokuchaev, made it possible to classify soils separately from rocks, vegetation, and climate, on the basis of the soils' own genetic features. Currently, a number of national classification systems are used in various countries.

In temperate regions, a major purpose of soil classification is land use and land management. In polar regions, however, land use is of minor interest and, thus, in most countries except Russia and Canada, soils of these regions have been neglected for many years. Russia, which has the world's largest area of permafrost, included permafrost-affected soils in their soil classification schemes at different taxonomic levels. Several revisions of these schemes have been developed (see this section's Chapter 2, by Mazhitova). In Canada, the Cryosolic Order was developed in the early 1970s for classifying soils in permafrost regions and it has been revised several times (see this section's Chapter 1, by Tarnocai).

Since the early 1990s, polar regions and, therefore, permafrost-affected soils have become of greater interest not only to the scientific community but also to the public, because of global change. Predictions concerning changes in carbon cycling and carbon storage, for example, can be made only with knowledge of these soils, since Cryosols can act as either carbon sinks or carbon sources, depending on the type of soil. The increasing importance of permafrost-affected soils led to a wider interest in their classification, with one result being that soil scientists in the USA wanted to include those soils in their soil taxonomy. After establishment of the International Committee on Permafrost-affected Soils (ICOMPAS) in 1993 and intense international collaboration between cryopedologists, the new Gelisol Order was created (see this section's Chapter 3, by Ahrens et al.).

Parallel to the activities in North America, European soil scientists working on the *World Reference Base for Soil Resources* (WRB), which is intended to be the international reference base for classification of soils, included the Cryosols as a major soil group in this classification system. Revisions are still under discussion (see this section's Chapter 4, by Tarnocai et al.), but acceptance of permafrost-affected soils within the WRB would be a big step toward correlating permafrost-affected soils among national classification systems.

Chapter 1. Classification of Cryosols in Canada

Charles Tarnocai[1]

[1]Research Scientist, Agriculture and Agri-Food Canada, Research Branch (ECORC), K.W. Neatby Building, Rm. 1135, 960 Carling Avenue, Ottawa, Canada, K1A 0C6; e-mail tarnocaict@agr.gc.ca.

1. Introduction

The *Canadian System of Soil Classification* (Soil Classification Working Group, 1998), Canada's national system, classifies soils according to their measurable properties. The classification is hierarchical, having soil orders at the highest level and great groups and subgroups at lower levels. Families and series are also part of the classification.

The current Canadian soil classification includes ten soil orders, one of which is the Cryosolic order. This order was developed in the early 1970s for classifying soils in the permafrost regions of Canada (Tarnocai et al., 1973a and 1973b) and soon was accepted and incorporated into the *Canadian System of Soil Classification* (Canada Soil Survey Committee, 1978). This group of soils, whose genesis is dominated by cryogenic processes, has formed in mineral or organic materials that have permafrost within one meter of the surface, or within two meters if the active layer of the pedon has been strongly cryoturbated, as indicated by disrupted, mixed, or broken horizons. Their mean annual soil temperatures are ≤ 0°C. Differentiating Cryosols from soils of other orders involves either determining or estimating the depth to permafrost.

Soils of the Cryosolic order cover approximately 35% (2.5×10^6 km^2) of the land area of Canada (Tarnocai, 1998) and occupy much of the northern third of the country, where permafrost exists close to the surface in both mineral and organic deposits. They predominate in the Arctic, are dominant in the Subarctic and northern boreal forest areas, and extend into some organic materials in the southern boreal forest and into some alpine areas of mountainous regions. Cryoturbation of these soils is common and may be indicated by such patterned ground as earth hummocks and sorted and nonsorted nets, circles, polygons, and stripes.

This chapter describes the use of the *Canadian System of Soil Classification* to classify these permafrost-affected soils (Cryosols). In addition, it describes the pedon, the basic soil unit used for the classification, and its use for describing, sampling, and classifying Cryosolic soils.

2. The Pedon (the Basic Soil Unit)

2.1. Definition

The pedon, the basic soil unit, is the three-dimensional unit on which the classification of all soils is based. Its relevance is most apparent in Cryosols, since patterned ground generally is associated with the permafrost terrain in which many of these soils occur. Patterned ground is a collection of repeating individual landscape units, about one to two meters in width, having the same general internal and external (microrelief) morphology. These landscape units therefore provide naturally occurring units, or pedons, that are used for classification. The vertical extent of this pedon (also called the control section) is one meter for noncryoturbated (Static and Organic) Cryosols and two meters for cryoturbated (Turbic) Cryosols.

The pedon concept is particularly important for Turbic Cryosols, which generally exhibit cyclic variation. For example, in the case of circles, the intercircle area that forms the perimeter of the circle delimits the patterned ground unit. Even though the profiles beneath the circle and the intercircle area differ markedly, this variation is cyclic and occurs repeatedly over the landscape.

The pedon of a nonsorted circle in Figure 5.1.1 shows this. The diameter of the circle is measured from the midpoint of the intercircle area on one side of the circle to its midpoint on the other side. If this diameter is no greater than two meters, the full cycle forms the pedon (Figure 5.1.1). If the circles are farther apart, so that a full cycle is two to seven meters in diameter, the pedon includes half a cycle, extending from the midpoint of the circle to the midpoint of the intercircle area. If the circles are still farther apart, so that the lateral dimension of the cycle is greater than seven meters, two pedons are identified, one associated with the circle and the other with the intercircle area.

This is also common with low- and high-centered lowland polygons and large ice-wedge polygons. In these types of patterned ground, one pedon characterizes the polygon trench area, including the ice wedge, while the second characterizes the central part of the polygon (Figure 5.1.2). Figure 5.1.3 shows examples of four cryoturbated pedons of nonsorted circles and earth hummocks, two very common types of patterned ground. Figure 5.1.4. shows examples of two noncryoturbated pedons. If no patterned ground exists, the pedon is established by arbitrarily choosing a lateral distance of one to two meters.

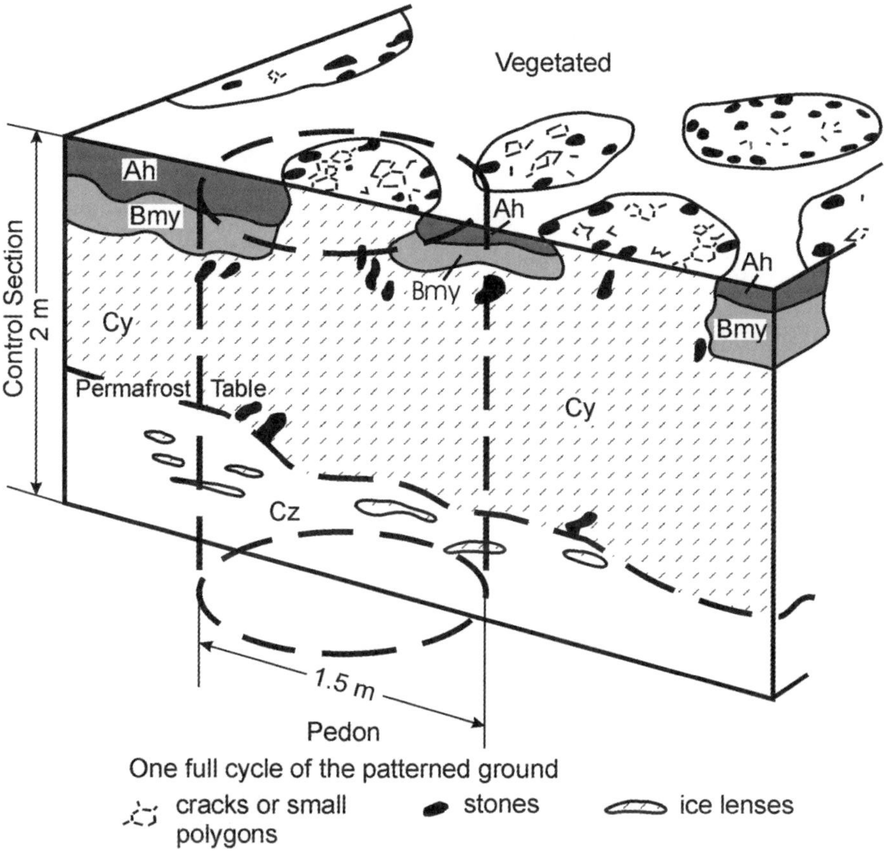

Figure 5.1.1. Diagram showing the location of the pedon of an Orthic Turbic Cryosol in a nonsorted circle type of patterned ground.

2.2. Using the pedon to describe, sample, and classify soils

The type and size of the patterned ground predetermine the location of the soil trench used for describing and sampling permafrost soils. A soil trench is dug to expose the full lateral extent of the pedon. The perennially frozen part of the pedon is excavated using a power hammer or a power-driven corer drill (Tarnocai, 1993).

As soon as the soil profile is exposed, a grid is placed on the profile and an accurate cross-section drawing is prepared (Figures 5.1.3 and 5.1.4). This cross-section includes all morphological features, such as the soil surface, permafrost table, soil horizon boundaries and designations, intrusion of materials, oriented stones, and other macromorphological features. Photographs (Polaroid, digital, and/or color slides) are taken at this time.

The soil then is described according to standard soil description procedures. This description generally differs from soil descriptions prepared for temperate soils because the soil horizons are not arranged in a continuous sequence of depth from top to bottom. Instead, the minimum and maximum horizon thicknesses are recorded for all soil horizons. In order to understand a cryogenic pedon with its complex soil horizons, the pedon description must always be used in conjunction with the cross-section diagram.

The soil then is classified and sampled for the full extent of the pedon. Bulk density and thin section samples are taken from the identified soil horizons, and the locations of these samples are marked accurately on the cross-section diagram. Similarly, composite samples are taken from all of the identified soil horizons. These samples are used for physical and chemical analyses.

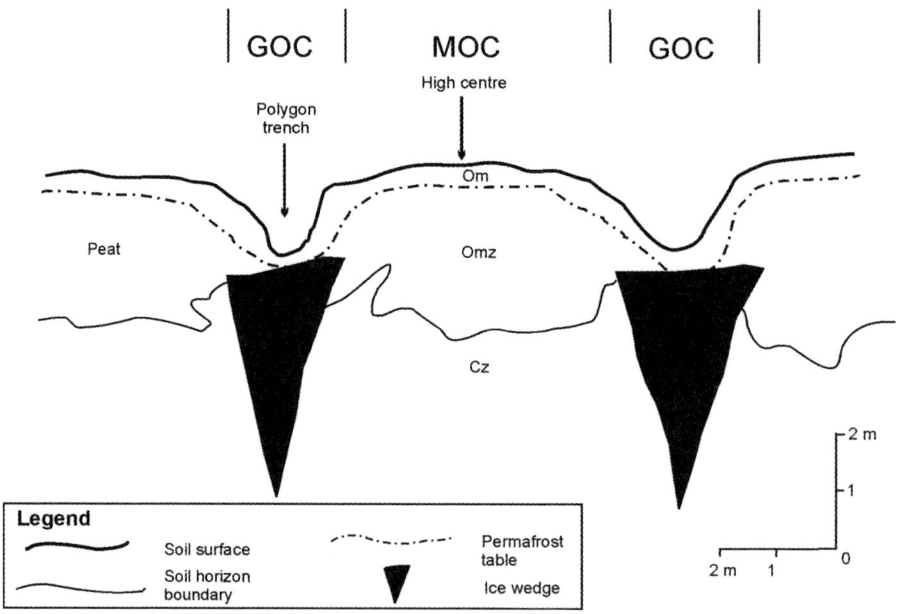

Figure 5.1.2. Schematic diagram of a high-centered lowland polygon showing the locations of the Glacic Organic Cryosol (GOC) in the ice wedge area and the Mesic Organic Cryosol (MOC) in the remaining portion of the polygon.

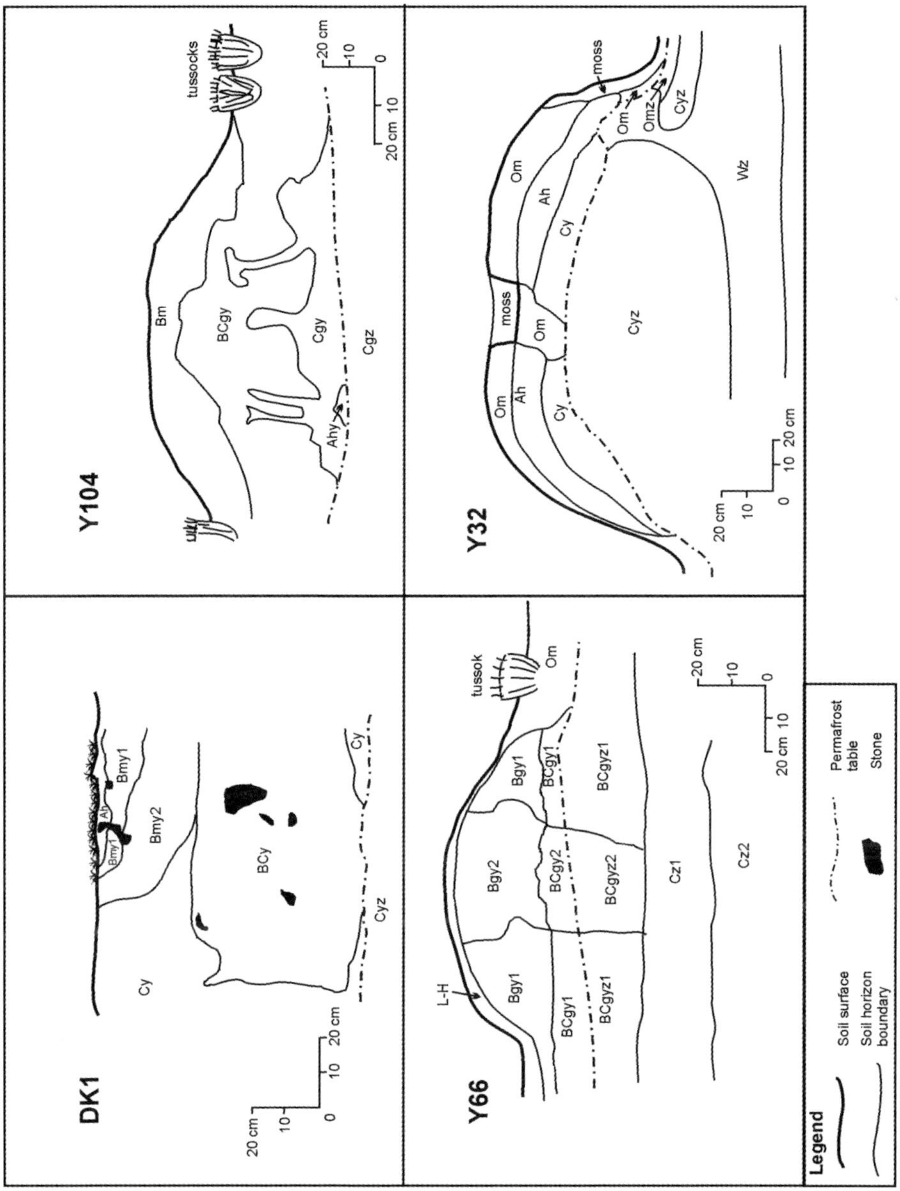

Figure 5.1.3. Cross-sections of cryoturbated soil profiles. Profiles Y32 (Histic Regosolic Turbic Cryosol; lat. 64° 36' N, long. 126° 23' W), Y66 (Gleysolic Turbic Cryosol; lat. 68° 55' N, long. 137° 50' W) and Y104 (Brunisolic Eutric Turbic Cryosol; lat. 68° 57' N, long. 133° 48' W) are associated with earth hummocks, while profile DK1 (Orthic Eutric Turbic Cryosol; lat. 66° 06' N, long. 94° 21' W) is associated with nonsorted circles. Horizon designators follow the *Canadian System of Soil Classification* (Soil Classification Working Group, 1998), where "y" indicates a cryoturbated and "z" a perennially frozen horizon.

3. Soil Horizons and other Layers

Mineral horizons contain 17% or less organic carbon while organic horizons contain more than 17% organic carbon. For Cryosols, the major characteristics of these horizons and layers are listed below.

Mineral horizons and layers

A – mineral horizon generally forming at or near the surface in the zone of accumulation of organic matter (Ah) or of leaching or eluviation (Ae).
B – mineral horizon characterized by the development of soil structure or by a change of color. Color changes include browning due to oxidation of iron (Bm) or mottling and gleying due to reduction (Bg).
C – mineral parent material unaffected by pedogenic processes except for gleying and cryoturbation.
R – consolidated bedrock layer.
W – a layer of water that, in Cryosols, commonly occurs as segregated ice.

Most common lower-case suffixes associated with mineral Cryosols

e – a horizon characterized by eluviation.
g – a horizon characterized by gray colors, prominent mottling, or both, indicating permanent or periodic intense reduction.
h – a mineral horizon enriched with organic matter.
m – a mineral horizon slightly altered by hydrolysis, oxidation, solution, or all three, to give a change in color, structure, or both.
t – an illuvial horizon enriched with silicate clay.
y – a cryoturbated horizon as manifested by broken horizons, incorporation of materials from other horizons, and mechanical sorting.
z – a perennially frozen layer.

Organic horizons

O – organic horizon developed mainly from mosses, sedges, and woody materials and normally associated with wet conditions
L, F, and H – organic horizons developed primarily from the accumulation of leaves, twigs, and woody materials. They are normally associated with upland forests.

Lower-case suffixes associated withthe O horizon

f – an organic horizon consisting largely of undecomposed fibric materials.
m – an organic horizon consisting largely of moderately decomposed mesic materials.
h – an organic horizon consisting largely of well decomposed humic materials.
y – same definition as for mineral horizons.

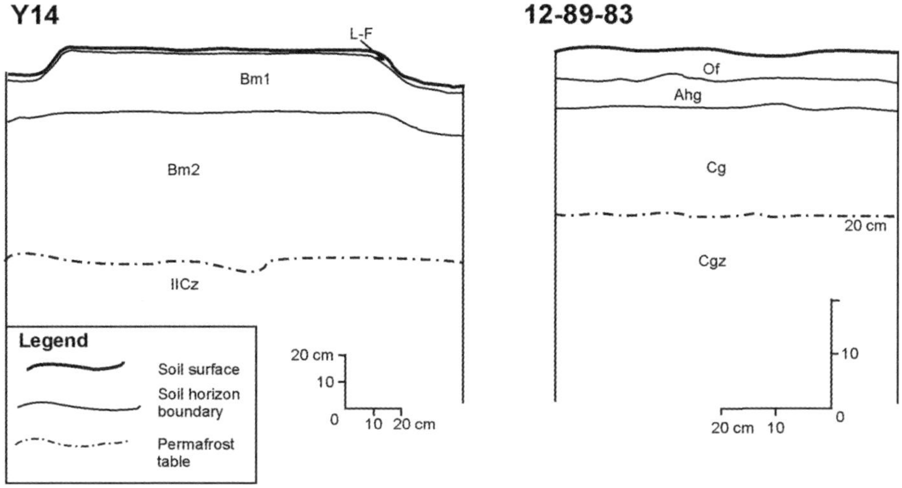

Figure 5.1.4. Cross-sections of noncryoturbated soil profiles. Profile Y14 (Brunisolic Eutric Static Cryosol; lat. 61° 25' N, long. 122° 46' W) is associated with earth hummocks, but 12-89-83 (Gleysolic Static Cryosol; lat. 81° 28' 50" N, long. 76° 30' 30" W) is not associated with patterned ground. Horizon designators follow the *Canadian System of Soil Classification* (Soil Classification Working Group, 1998), where "z" indicates a perennially frozen horizon.

4. Classification of Cryosolic Soils

The Cryosolic order is divided into three great groups, Turbic Cryosol, Static Cryosol, and Organic Cryosol. This division is based on the degree of cryoturbation and the nature of the soil material, mineral or organic, as Table 5.1.1 indicates. Figure 5.1.2 shows a cross-section of an Organic Cryosol while Figure 5.1.3 shows cross-sections of four Turbic Cryosols and Figure 5.1.4 shows those of two Static Cryosols. Brief descriptions of the great groups and subgroups, based on the *Canadian System of Soil Classification* (Soil Classification Working Group, 1998), appear below.

Table 5.1.1. Nature of materials, degree of cryoturbation, and depth to permafrost in various Cryosol great groups.

Parameter	Great Groups		
	Turbic Cryosol	**Static Cryosol**	**Organic Cryosol**
Soil material	Mineral	mineral	organic
Cryoturbation	marked; usually patterned ground	None	uncommon*
Permafrost	within 2 m of surface	within 1 m of surface	within 1 m of surface

*Some shallow Organic Cryosols could be cryoturbated.

4.1. Turbic Cryosols

Turbic Cryosols are mineral soils that have permafrost within two meters of the surface and show marked evidence of cryoturbation within the active layer, as indicated by disrupted, mixed, or broken horizons, or displaced material, or a combination of both. They generally are associated with patterned ground, which includes such cryogenic forms as sorted and nonsorted circles, nets, polygons, stripes, and steps in stony or coarse-textured material and nonsorted units such as earth hummocks in medium- and fine-textured materials.

The pedon includes all elements of the microtopography in cycles less than seven meters wide. Pedogenic processes include sorting of particles of various sizes and mixing of both mineral and organic material from different horizons. Organic horizons (Oy) or organic-rich mineral horizons (Ahy) are characteristically present near the permafrost table, and there is generally a buildup of ice in the upper part of the permafrost layer.

The subgroups within the Turbic Cryosol great group are differentiated on the basis of the type and sequence of horizons. Table 5.1.2 briefly describes these subgroups.

4.2. Static Cryosols

Static Cryosols have developed primarily in coarse-textured mineral parent materials, or in a wide textural range of recently deposited or disturbed sediments where evidence of cryoturbation is still largely absent, or in both. They may have organic surface horizons less than 40 cm thick. Static Cryosols have permafrost within one meter of the surface but show little or no evidence of cryoturbation or features that indicate cryoturbation. They may be associated with patterned ground features such as polygons.

The subgroups within the Static Cryosol great group are differentiated on the basis of the type and sequence of horizons. Table 5.1.3 briefly describes these subgroups.

4.3. Organic Cryosols

Organic Cryosols have developed primarily from organic material containing more than 17% organic carbon by weight and have permafrost within one meter of the surface. They are greater than 40 cm thick or are greater than 10 cm thick over either a lithic contact or an ice layer that is at least 30 cm thick. Common landforms associated with Organic Cryosols include palsas, peat polygons, and high-centered polygons. The subgroups within the Organic Cryosol great group are differentiated on the basis of the depth and degree of decomposition of the organic material. Table 5.1.4 briefly describes these subgroups.

Table 5.1.2. Brief description and classification of Turbic Cryosols (for more details, see Soil Classification Working Group, 1998).

Subgroup and common horizon sequence*	Description
Orthic Eutric Turbic Cryosol Om, Bmy, BCy, Cgy, Omy, Cz	These soils have a relatively high degree of base saturation, as indicated by a pH of 5.5 or greater in some or all of the B horizons.
Orthic Dystric Turbic Cryosol Om, Bmy, BCy, Cgy, Omy, Cz	These soils have a low degree of base saturation, as indicated by a pH of less than 5.5 throughout the B horizons.
Brunisolic Eutric Turbic Cryosol Om, Bm, Bmy or BCy, Cgy, Omy, Cz	These soils have a Bm horizon ≥10 cm thick and a relatively high degree of base saturation, as indicated by a pH of 5.5 or greater in some or all of the B horizons.
Brunisolic Dystric Turbic Cryosol Om, Bm, Bmy or BCy, Cgy, Omy, Cz	These soils have a Bm horizon ≥10 cm thick and a low degree of base saturation, as indicated by a pH of less than 5.5 throughout the B horizons.
Gleysolic Turbic Cryosol Om, Bgy or Cgy (or both), Cz	These soils have evidence of gleying in the form of low chromas or mottling to the mineral surface.
Regosolic Turbic Cryosol Om, Cy, Cgy, Cz	These soils have developed on recently deposited soil materials and have very little or no soil development. They lack B horizons.
Histic Eutric Turbic Cryosol Om, Ah, Bmy or Bm (or both) or Cgy, Cz	These soils have a relatively high degree of base saturation, as indicated by a pH of 5.5 or greater in some or all of the B horizons, and thick (>15 cm) organic horizons in the upper 1 m of the solum.
Histic Dystric Turbic Cryosol Om, Ah, Bmy or Bm (or both) or Cgy, Cz	These soils have a low degree of base saturation, as indicated by a pH of less than 5.5 throughout the B horizons, and thick (>15 cm) organic horizons in the upper 1 m of the solum.
Histic Regosolic Turbic Cryosol Om, Cy, Cgy, Cz	These soils have developed on recently deposited soil materials and have very little or no soil development. They have thick (>15 cm) organic horizons in the upper 1 m of the solum. They lack B horizons.

*The diagnostic horizons are underlined in the horizon sequence.

Table 5.1.3. Brief description and classification of Static Cryosols (for more details, see Soil Classification Working Group, 1998).

Subgroup and Common Horizon Sequence*	**Description**
Orthic Eutric Static Cryosol Om, LFH, Bm, BCgj, Cz	These soils have a Bm horizon <10 cm thick and a relatively high degree of base saturation, as indicated by a pH of 5.5 or greater in some or all of the B horizons.
Orthic Dystric Static Cryosol Om, LFH, Bm, BCgj, Cz	These soils have a Bm horizon <10 cm thick and a low degree of base saturation, as indicated by a pH of less than 5.5 throughout the B horizons.
Brunisolic Eutric Static Cryosol LFH, Bm, BCgj, Cz	These soils have a Bm horizon ≥10 cm thick and a relatively high degree of base saturation, as indicated by a pH of 5.5 or greater in some or all of the B horizons.
Brunisolic Dystric Static Cryosol LFH, Bm, BCgj, Cz	These soils have a Bm horizon ≥10 cm thick and a low degree of base saturation, as indicated by a pH of less than 5.5 throughout the B horizons.
Luvisolic Static Cryosol LFH, Om, Ah or Ae, Bt, Cg, Cz	These soils have a clay accumulation in the B horizon (Bt). The Bt horizon is >10 cm thick.
Gleysolic Static Cryosol Om, Bg or Cg (or both), Cz	These soils have evidence of gleying in the form of low chromas or mottling to the mineral surface.
Regosolic Static Cryosol C, Cg, Cz	These soils have developed on recently deposited soil materials and have very little or no soil development. They lack B horizons.
Histic Eutric Static Cryosol Om, Ah, Bm, Cg, Cz	These soils have a relatively high degree of base saturation, as indicated by a pH of 5.5 or greater in some or all of the B horizons, and thick (>15 cm) organic horizons in the upper 1 m of the solum.
Histic Dystric Static Cryosol Om, Ah, Bm, Cg, Cz	These soils have a low degree of base saturation, as indicated by a pH of less than 5.5 throughout the B horizons, and thick (>15 cm) organic horizons in the upper 1 m of the solum.
Histic Regosolic Static Cryosol Om, C, Cg, Cz	These soils have developed on recently deposited soil materials and have very little or no soil development. They have thick (>15 cm) organic horizons in the upper 1 m of the solum.

*The diagnostic horizons are underlined in the horizon sequence.

Table 5.1.4. Brief description and classification of Organic Cryosols (for more details, see Soil Classification Working Group, 1998).

Subgroup and Common Horizon Sequence*	Description
Fibric Organic Cryosol Of or Om, Of, Ofz	These soils have organic layers >1 m thick and are composed dominantly of fibric materials in the control section below a depth of 40 cm.
Mesic Organic Cryosol Of or Om, Om, Omz	These soils have organic layers >1 m thick and are composed dominantly of mesic materials in the control section below a depth of 40 cm.
Humic Organic Cryosol Om or Oh, Ohz	These soils have organic layers >1 m thick and are composed dominantly of humic materials in the control section below a depth of 40 cm.
Terric Fibric Organic Cryosol Of, Ofz, Cz	These soils have organic layers <1 m thick and are composed dominantly of fibric materials above the mineral contact.
Terric Mesic Organic Cryosol Om, Omz, Cz	These soils have organic layers <1 m thick and are composed dominantly of mesic materials above the mineral contact.
Terric Humic Organic Cryosol Oh, Ohz, Cz	These soils have organic layers <1 m thick and are composed dominantly of humic materials above the mineral contact.
Glacic Organic Cryosol Of, Om or Oh, Wz	These soils have organic layers and a layer of ground ice >30 cm thick with an upper boundary within 1 m of the surface.

*The diagnostic horizons are underlined in the horizon sequence.

5. Conclusions

The main concepts for the development of the Cryosolic order in the Canadian System of Soil Classification are that soils with permafrost be placed at the highest order level of the classification and that cryogenic processes be recognized as soil-forming processes. Cold soil temperatures and the resulting cryogenic processes play the major role in the genesis of these soils.

The presence of permafrost had similar effects on the development of both mineral and organic soils and on their management and use. For these reasons, both perennially frozen mineral and organic soils are included in the Canadian Cryosolic order.

References

Canada Soil Survey Committee. 1978. The Canadian System of Soil Classification. Canada Department of Agriculture. Ottawa. Publication 1646. 164 pp.

Soil Classification Working Group. 1998. The Canadian System of Soil Classification, 3rd ed. Research Branch, Agriculture and Agri-Food Canada. Ottawa. Publication 1646. 187 pp.

Tarnocai, C. 1993. Chapter 71, Sampling frozen soils. In: M.R. Carter, ed. Soil Sampling and Methods of Analysis. Canadian Society of Soil Science. Lewis Publishers. Boca Raton, Florida. pp. 755-765

Tarnocai, C. 1998. The amount of organic carbon in various soil orders and ecological provinces in Canada. In: R. Lal, J.M. Kimble, R.L.F. Follett, and B.A. Stewart, eds. Soil Processes and the Carbon Cycle. Advances in Soil Science. CRC Press. New York. pp. 81-92

Tarnocai, C., W.W. Pettapiece, and S.C. Zoltai. 1973a. Cryoturbed Soils in Northern Canada. Canada Soil Survey. Department of Agriculture, Winnipeg. 64 pp.

Tarnocai, C., W.W. Pettapiece, and S.C. Zoltai. 1973b. Report on cryoturbed soils in northern Canada. In: J.H. Day and P.G. Lajoie, eds. Proceedings of the Ninth Meeting of the Canada Soil Survey Committee. University of Saskatchewan. Saskatoon, May 1973. pp. 96-116.

Chapter 2. Classification of Cryosols in Russia

Galena Mazhitova[1]

[1]Senior Researcher, Institute of Biology, Komi Science Center, Ural Division, Russian Academy of Sciences, 28 Kommunisticheskaya St., Syktyvkar, 167982 Russia; e-mail mazhitova@ib.komisc.ru.

1. Introduction

Russia contains the world's largest areas of permafrost and has a long-term tradition of soil classification. The traditional Russian genetic classification approach has been revised to change the relative role of ecology and soil properties. Still, most Russian classifications give priority to soil genesis and a subordinate place to arbitrary, rigid class definitions. Thus, after highly formalized world soil classification systems were adopted (firstly, *US Soil Taxonomy*), a gap between them and the Russian school appeared, and it persists.

In recent years, intense international collaboration has taken place in cryopedology. In particular, a large group of pedologists, mostly from the USA and Canada, proposed placing permafrost-affected soils at a high categorical level in world classification systems, *Soil Taxonomy* and the *World Reference Base for Soil Resources* (FAO/WRB). This was done and widened the gap between Russian and other world systems. Correlation of the various classification systems thus is needed.

The task is complicated by a large number of Russian classification systems differing more or less significantly from each other. Among them, only a few have been developed enough to fit practical needs. These systems have been adopted by State Ministries and All-Russia Prospecting Institutes, but in scientific publications any classification could be used without referring to the official one. This was especially evident with regard to permafrost-affected soils, as they were very poorly represented in the official classifications.

This chapter briefly reviews Russian classification approaches with regard to permafrost-affected soils and considers in more detail the newest classification system (Shishov et al., 1997). The latter evokes serious criticism, has not yet been approved by the All-Russia Congress of Soil Scientists, and, thus, still should be considered as a proposal. However, it is much better elaborated than previous classifications, particularly relative to permafrost-affected soils; it includes most recently obtained data; it takes into account foreign experience and it is now translated into English (*Shishov* et al., 2001). Thus, in our opinion, it deserves more close consideration here. We also provide a correlation of this system with *WRB* (FAO, 1998) and *Soil Taxonomy* (Soil Survey Staff, 1999) for permafrost-affected soils.

2. Permafrost-affected Soils in Russian Soil Classifications

For a long time, the ecological-genetic approach dominated most Russian soil classifications. Along with soil properties and genesis, geographical distribution of soils and soil-forming agents were used widely to classify soils. While permafrost-affected soils remained poorly studied, a landscape-based term *Tundra soil* was used. Later, a more specific term, *Tundra Gley*, appeared and, starting with Liverovsky (1937), most authors supported the concept of a tundra type of soil formation on the basis of soil properties and processes. For taiga soils, the name *Permafrost Taiga soil* was used, also landscape-based.

Intense studies of northern European and Siberian soils were conducted in the 1960s and 70s. A wide diversity of soils was found, mechanisms by which permafrost affects soil formation were studied, and summarizing concepts were developed. *Guidance on Classification and Diagnostics of Soils* (Ivanova and Nogina, 1967) was a USSR soil classification system adopted for the use of concerned ministries and institutions. It did not consider the soils of tundra and large permafrost areas in central and eastern Siberia because of a lack of data.

Among permafrost-affected soils, those used for agriculture in southern Siberia were included, called Permafrost Meadow-Forest soils. The general approach was to distinguish subzonal and facies subtypes within soil types. The specificity of soils of cold regions of Siberia was thus recognized.

The facies subtypes were supposed to reflect differences in soil genesis corresponding to environmental gradients along latitudes. However, names of facial subtypes were based on their thermal regime and not on soil features, indicating poor knowledge of the latter. Localization of gley and thickness, along with some chemical parameters, distinguish facial subtypes, which is not reflected in soil names. Thus, thermal features themselves were used to subdivide the soils. However, few data were available at that time, for permafrost-affected soils were touched on very briefly and these mostly were those of seasonally long-frozen subtypes of podzolic soils (they may have permafrost below the soil profile). Soils of large areas of permafrost-affected central and eastern Siberian taiga were not considered.

The Program of the USSR Soil Map at a scale 1:2.5 million, issued in 1971, included a much wider range of permafrost-affected soils. Ivanova (1976) developed a classification and diagnostic system for USSR soils. For the first time, tundra and permafrost taiga soils were considered over the whole area of their distribution within Russia. Three soil types were included in the list for tundra and 3 for permafrost-affected taiga.

Later, *Classification and Diagnostics of the USSR soils* (Egorov et al., 1977) became a widely used system. In contrast to Ivanova, its compilers stated again that permafrost-affected soils hardly could be considered because of poor data available. World soil classifications applicable outside of the USSR/Russia rarely were developed in Russia; one example is the classification based on a geochemical approach, by Glazovskaya (1971).

Meanwhile, various authors proposed a number of new soil types. A large group of tundra and taiga soils with morphologically undifferentiated profiles was divided into Podburs (Parapodzols) and Gleys (Targulian, 1971). Volkovintser (1978) supported genetic specificity of Steppe Cryoaridic Soils. Sokolov (1980) developed the concept of cryohydromorphic non-gley soil formation on loamy materials in continental areas. The lack of gley is due to drainage provided by frost cracks and due to enrichment of supra-permafrost water with oxygen.

Sokolov proposed the term *Cryozem* for these soils. He emphasized that effects of permafrost on soil genesis are manifested most strongly in them, and this, not the presence of permafrost, conditioned the choice of the name. However, few studies of water regime have been conducted to test his hypothesis and to confirm that the soils are saturated with water for long periods.

Sokolov and Bystryakov (1980) introduced a concept of *Pale soils* differing from the earlier adopted concept. Pale soils were considered metamorphic soils. The term *metamorphic* has a narrower meaning than *alteration* in the cambic horizon of *Soil Taxonomy*. Thus, subdivision of the soils with weakly developed profiles was continued based on genetic hypotheses, in spite of difficulties in their identification. Ignatenko (1980) included Homogenic Gleys (static soils underlain by permafrost) in his regional list for northeastern Asia, emphasizing subdivision of permafrost-affected soils into cryoturbated and static ones. The same term was used by other authors in the opposite sense, to designate cryogenic homogenization.

While Sokolov and other authors argued the idea that, for a genetic soil classification, the presence of permafrost itself is not a reason to set a special class at a high categorical level, Makeev (1981) considered that numerous types of permafrost-affected and seasonally long-frozen soils should be covered by two highest taxa, the permafrost and cold soils formations.

In the 1980s, important changes took place in dominant classification approaches. Under both the effect of western classifications and the national school development, a so-called *substantive-genetic approach* was adopted, based on soil properties rather than ecology (Fridland, 1982; Shishov, Sokolov, 1989; Sokolov, 1991).

A rather developed realization of the approach is the *Russian Soil Classification* (Shishov et al., 1997) discussed later in this chapter. The *Classification* also implements Sokolov's (1997) idea of the geoderma, which includes not only soils but also surface bodies of non-soil nature. These bodies can be included in a soil classification as no one other branch of science deals with them and because both soils and these bodies are products of exogenesis. Naturally, they should be separated from soils at the highest categorical level.

According to Sokolov, one of the earth's 5 soil sectors/types of exogenesis is a cryogenic one. It includes cryomorphic soils, non-cryomorphic soils, organic soils, cryopedolits and cryolits, and bedrock exposures. Non-cryomorphic soils mostly develop in coarsely textured materials and are characterized by weak effects of permafrost on soil formation.

3. The Minimum Classification Object and Patterned Ground

A pedon or another concept of the minimum object of soil classification has never been a part of the soil classification systems in Russia. Many authors, however, analyzed spacial soil variability at detail scales. Sokolov (1993, 1997) emphasized that soil classification is based on "central images" in a soil continuum. In this case, the minimum classification object need not be defined. He stresses, in particular, that soil classification systems are built "from top to bottom" and not bottom to top. However, the definition of the minimum object is necessary at the stage of identifying a classification unit, i.e., for soil-geographical purposes, in particular, soil mapping.

Traditionally, soil bodies which related to patterned ground components that regularly repeat themselves (earth hummocks, bare areas, etc.) were considered taxonomically different soils (Ivanova and Polyntseva, 1952). The term *soil microcomplex* was applied to such soil associations. Later, Fridland (1972) developed the concept of the structure of soil cover and introduced an *elementary soil areal*, an area of varying size, devoid of taxonomically significant boundaries within it.

Fridland also suggested that the minimum area which can be considered an elementary areal should be comparable in size with a plant. He gave numerous examples of soil microcomplexes connected to patterned grounds. Since then, a lateral dimension of 20 to 30 cm has been used by most researchers as a size of minimum object, though it never has been set officially. The microcomplexes were shown even on small-scale maps; for example, soils of tundra circles (bare frost-boiled areas) and frost cracks were included in the legend of the *USSR Soil Map* at a scale of 1:2.5 million.

Thus, soils of patterned ground always have been classified as a number of soils, not as a single anisotropic body, while a different approach is used in both *WRB* and *Soil Taxonomy* (Turbic or Ruptic-Histic units). Vasilevskaya (1979) suggested as an alternative that one could classify such soils as a single 3-dimensional body. She noticed that stable soil properties (*soil memory*) of the various components in a microcomplex are similar, whereas highly dynamic properties (*soil moment*) differ between components. Finally, however, she followed the traditional approach and separated soils of bare areas from those of vegetated surroundings at a level as high as a type.

4. Thermal Characteristics of Soils

As mentioned above, the 1967 soil classification included subdivision of soils into facies subtypes based mostly on their thermal regime. Dimo and Rosov (1974) elaborated the subdivision, and it has been used widely in soil classifications (Ivanova, 1976; Egorov et al., 1977).

Subdivision of soils into facies subtypes parallels their subdivision into genetic subtypes. Facies are large regions where climate affects soils in a similar way. Permafrost is not used directly because, in many cases, it does not affect soil genesis directly. Three indices were used to divide soils into the subtypes: Air Degree Days (>10°C), Soil Degree Days (>10°C) at a depth of 20 cm, and the duration of cold period at the same depth. The authors considered these indices the most important controls on soil formation and, in particular, on energy supply.

Four permafrost-affected subtypes, 6 seasonally long-frozen, 6 seasonally frozen, 3 seasonally short-frozen, and 1 periodically frozen subtype were distinguished. For the 4 permafrost-affected subtypes, the first and the second indices range from 0 to 1600°C, and the cold period is more than 8 months long in all 4 subtypes. Permafrost was included directly in the definition of types only when its effect was strong enough to single out a genetic subtype (for example, Podzolic Deep-Gleyed Permafrost soils with supra-permafrost gley).

The newest Russian classification also adopts the idea of introducing permafrost at different categorical levels of classification, depending of its actual effect on soil formation (Shishov et al., 1997). An approach has been proposed providing for the development of thermal and hydrological blocks separate from the main genetic block of classification (Fridland, 1982; Shishov, Sokolov, 1989). Permafrost-free, seasonally frozen, and permafrost soils were supposed to be included as separate classes in the hydrological block (Sokolov, 1991). So far, however, the 1997 classification does not contain such blocks.

5. 1997 Russian Soil Classification

5.1. General principles

As the compilers note, the work represents a new stage in the development of Russian soil classification. Soil-forming agents and ecological conditions no longer determine classification decisions; instead, soil properties and genesis do (Shishov et al., 1997). Traditional soil names are mostly preserved, yet landscape-based names no longer are used. Higher categories bear genetic sense, and diagnostic criteria for them are mostly qualitative.

Strict quantitative criteria, except for the thickness of the organic horizon and some other properties, are avoided at this level purposely, as they hardly can be combined with genetic logic. This is an obvious difference from *WRB* and *Soil*

Taxonomy. The lack of precise delineation of soil classes prevents construction of keys.

Quantitative criteria are used widely at lower levels. Type-diagnostic soil horizons are defined carefully, yet many of them have floating quantitative boundaries. For the soils of Russia, 32 such horizons are introduced, versus 34 in *WRB* and 27 in *Soil Taxonomy*, the world systems. This is partly due to a larger number of various organic horizons in the Russian system. However, genetic sequences of horizons rather than isolated horizons are diagnostic for soil types. Most horizon definitions are close to those in *WRB* and *Soil Taxonomy*. Also, a number of genetic soil properties have definitions similar to diagnostic properties in the other two systems. The 1997 Russian system gives correlations of the horizon definitions with the 1994 *WRB*.

The 1997 Russian system uses 8 categories, 2 of them higher and 5 lower than the soil type. The highest is a Trunk reflecting the relative importance of pedo- and lithogenesis. Post-lithogenic, sinlithogenic, and organogenic soils are distinguished. Organic soils, thus, are separated at the highest level from mineral soils. This explains why most Russian pedologists strongly opposed including organic soils together with mineral soils in the Cryosol *WRB* reference group.

The above-mentioned climatic facies subtypes are not included in this Russian system. Type is a central category, characterized by a specific genetic sequence of horizons and a combination of soil properties. Most subtypes are intergrades to other types.

An important feature of the classification is wide inclusion of agrogenic and humanly transformed soils, and an attached classification of technogenic surface bodies. Soil temperature and water regimes are not considered at all. Permafrost is not considered, despite its strong effect on soil properties (Cryozems). The concept of Cryosols/Gelisols adopted in *WRB* and *Soil Taxonomy* could not be considered consistently genetic and, hence, contradicts the basic principles of Russian classifications.

5.2. Handling permafrost-affected soils

A Cryoturbated horizon is included among type-diagnostic horizons in the set of horizons diagnostic for Cryozems. It is defined as follows.

> The cryoturbated horizon is dusky brown or grayish brown, lacks structure; and has distinct evidences of cryogenic distortions, such as translocation of large blocks of mineral and/or organic material, irregular wedge-like or broken boundaries. The horizon contains dispersed fragments of organic horizons and/or very dark carbonized organic residues. The reaction is nearly neutral. The material is excessively moist, usually thixotropic, and is frozen during part of the vegetative period.
>
> Shishov et al., 2001

Permafrost is not mentioned as a genetic soil property. Accordingly, no suffix symbol for a perennially frozen horizon or layer is introduced.

The Cryic horizon used in *WRB* in 1998 as diagnostic for Cryosols and Cryic Histosols has little in common with the Russian Cryoturbated (KR) horizon. Cryic is a perennially frozen mineral or organic horizon, whereas KR is a seasonally frozen, strongly cryoturbated mineral horizon. Besides, the KR covers far from all cryoturbated horizons occurring in mineral soils; it does not include cryoturbated horizons of other soil types (Gleyzems, Podburs, etc.) in the definition. It includes only cryoturbated horizons of Cryozems, which explains a relatively narrow property range, like the close-to-neutral reaction.

The concept of Gelic material in *Soil Taxonomy* is closer, yet wider than that of the KR. The 1997 Russian system does not introduce any pedon or another minimum unit of classification and does not discuss the issue. Thus, depending on the approach adopted, one can classify the soils of patterned grounds either as a number of soils or as a single soil, averaging, for example, organic horizon thickness to check it against diagnostic criteria.

5.3. Correlating classification systems

Various methods to correlate soil classification systems have been proposed. Rozhkov (1989), for example, proposed using statistical similarity criteria to compare taxa. The procedure requires compilation and analysis of a large database and is not possible within the framework of this chapter.

Stolbovoi (2000) published a monograph correlating the *Soil Map of Russian Federation* (Fridland, 1988) with *WRB*. Mazhitova et al. (1993) and Smith at al. (1995) have discussed correlating the Russian system with *Soil Taxonomy* and the Canadian system, with special regard to permafrost-affected soils, based on the materials of international expeditions.

This chapter presents correlations based on a direct check of Russian taxa against criteria of the taxa of the other two classifications. We tried to apply quantitative criteria of world classifications to Russian genetic soil types, using published analytical data and also unpublished data for northeastern Siberia, where a number of soil profiles had been analyzed in the USDA NRCS laboratory in Lincoln.

Soils of the Cryozem order of Russian classification always have permafrost at a shallow depth, whereas a number of other soil orders/types may or may not have permafrost, and its presence/absence is not specified at any categorical level. These orders/types classify partly as Cryosols/Gelisols and partly as Gelic or non-Gelic units of the groups other than Cryosols in *WRB*, or to the orders other than Gelisols in *Soil Taxonomy*.

Table 5.2.1 lists those soil orders/types from the Russian system which are represented widely in permafrost areas and may have permafrost within 2 m of the surface, and it correlates the permafrost-affected soils of these selected classes with *WRB* and *Soil Taxonomy*. The list is not complete, as it uses only main lower-level units, omitting units like Leptic or Thionic. We do not discuss Humus-Accumulative, Halomorphic, Alluvial, Lithozem, and some other soils which are less widespread or less frequently have permafrost within 2 m.

Table 5.2.1. Correlation of selected Russian taxa (permafrost-affected soils only) with *WRB* (1998) and *Soil Taxonomy* (1998).

Translated Russian name	Original Russian name	*WRB*, 1998	*Soil Taxonomy*, 1998
Trunk: POSTLITHOGENIC SOILS **Order:** Gleyzems/ Gleezemy **Types**:			
Gleyzems Raw-Humus Gleyzems	Gleezemy Gleezemy grubogu-musovyie	Gleyic Cryosols, Gleyi-Turbic Cryosols, Gelic Gleysols	Aquiturbels, Aquorthels
Peat Gleyzems	Torfiano-Gleezemy	Gleyi-Histic Cryosols, Turbi-Histic Cryosols (Gleyic), Fibri-Cryic Histosols, Geli-Histic Gleysols	Histoturbels, Historthels, Histels
Order: Cryozems/Kriozemy **Types**:			
Cryozems	Kriozemy	Oxiaqui-Turbic Cryosols, Turbi-Histic Cryosols (Oxiaquic)	Haploturbels, Histoturbels
Peat Cryozems	Torfiano-Kriozemy	Turbi-Histic Cryosols (Oxiaquic), Fibri-Cryic Histosols	Histoturbels, Histels
Order: Al-Fe-Humus/Alfegumusovyie **Types:**			
Podburs (Parapod-zols)	Podbury	Haplic Cryosols, Leptic Cryosols, Dystri-Gelic Cambisols	Haploturbels, Haplorthels
Gley Podburs	Podbury gleevyie	Haplic Cryosols, Gleyic Cryosols, Gleyi-Gelic Cambisols	Haploturbels, Aquiturbels, Haplorthels, Aquorthels
Dry Peaty Podburs	Sukhotorfiano-podbury	Haplic Cryosols, Foli-Cryic Histosols, Foli-Gelic Histosols, Gelic Cambisols	Histoturbels, Historthels, Folistels

(Continues)

Table 5.2.1. (cont.)

Translated Russian name	**Original Russian name**	***WRB*, 1998**	***Soil Taxonomy*, 1998**
Order: Al-Fe-Humus/Alfegumusovyie, cont. **Types:** cont.			
Peat-Gley Podburs	Torfiyno-podbury gleevyie	Gleyi-Histic Cryosols, Fibri-Cryic Histosols, Geli-Histic Gleysols	Fibristels, Hemistels Histoturbels, Historthels
Podzols	Podzoly	Haplic Cryosols, Hapli-Leptic Cryosols	Typic Haplorthels, Typic Haplocryods, Spodic Psammoturbels, Spodic Psammorthels
Gley Podzols	Podzoly gleevyie	Hapli-Gleyic Cryosols, Gleyi-Gelic Podzols	Aquiturbels, Haploturbels, Aquorthels, Haplorthels
Dry Peaty Podzols	Sukhotorfiano-podzoly	Haplic Cryosols, Foli-Cryic Histosols, Foli-Gelic Histosols, Gelic Podzols	Folistels, Historthels
Peat Podzols	Torfiano-podzoly	Histic Cryosols, Turbi-Histic Cryosols, Gleyi-Histic Cryosols, Gelic Podzols, Fibri-Cryic Histosols	Folistels, Histoturbels, Aquiturbels, Historthels, Aquorthels
Order:Metamorphic /Metamorficheskyie **Type:**			
Pale	Palevyie	Haplic Cryosols, Gelic Cambisols	Haplorthels, Eutrocryepts
Trunk: ORGANOGENIC **Order:** Peat/ Torfianyie **Type:**			
Oligotrophic Peat (high moor peat-bogs)	Torfianyie oligotrofnyie	Fibri-Cryic Histosols, Fibri-Gelic Histosols	Fibristels, Hemistels
Eutrophic Peat (low-moor peat-bogs)	Torfianyie eutrofnyie	Sapri-Cryic Histosols, Sapri-Gelic Histosols	Sapristels
Dry-Peaty	Sukhotorfianyie	Foli-Cryic Histosols, Foli-Gelic Histosols	Folistels

When strictly following quantitative criteria, rather often we should place a soil in a *WRB/Soil Taxonomy* unit which does not properly fit its genesis, though another unit does. For example, permafrost-affected Podburs are genetically close to Spodic Psammorthels. However, in most cases, their BHF horizon does not meet the quantitative criteria of a Spodic horizon or Psamm-Great Group; thus they should be classified in *Soil Taxonomy* as Typic Haploturbels (Smith, et al., 1995).

5.4. Gleyzems

"These soils are characterized by strongly pronounced gleying... evidenced by olive gray chromas over more than 80% of a vertical section of mineral horizons" (Shishov et al., 2001).

Gleyzems, according to the 1997 classification monograph, contain many permafrost-affected tundra soils. Yet Gleyzems occur in the permafrost-free or discontinuous permafrost environment of southern tundra, forest-tundra, or taiga. No genetic process except gleying is well manifest, in contrast to the gleyed soils of other orders (Peat-Gley Podzols, Peat-Solod Gley, etc.).

The Gley diagnostic horizon, as the Russian system defines it, cannot directly compare with Gleyic properties and Aquic conditions of the other two systems, although, with few reservations, one can draw the analogy. Table 5.2.2 shows quantitative aspects of the correlation between the three systems, mostly related to organic horizons. The Gleyzem type corresponds to Gleyic or Turbi-Gleyic Cryosols of *WRB*. In addition, it includes some Gelic and permafrost-free Gleysols. In *Soil Taxonomy*, the Gleyzem type corresponds to Aquorthels or Aquiturbels, yet some Gleyzems fall out of the Gelisol order.

The Russian type Peat-Gleyzem has a peat layer from 10 to 50 cm thick. A thickness of 40 cm separates Histic subunits and Histosols/Histels in both *WRB* and *Soil Taxonomy*, so representatives of this Russian type may belong to either one of these classes. The Russian Peat horizon, diagnostic for Peat Gleyzems, is the Histic horizon in terms of *WRB* and is organic soil material in terms of *Soil Taxonomy*, though the latter two have wider definitions.

The lowest organic matter content for all Russian peat horizons is 35%, whereas for the Histic horizon and organic soil material it is 20%. The Russian system considers horizons with organic matter content ranging from 20 to 35% as Mucky or Raw-humus horizons. Thus, Raw-Humus Gleyzems are Histic units of *WRB* and are Historthels/Histoturbels of *Soil Taxonomy*, but uncertainty still remains, as the Russian system does not set a lower limit of organic matter content for the Raw-humus horizon.

Table 5.2.2. Correlation of Russian Gleyzem order with *WRB* and *Soil Taxonomy*, with respect to the thickness of the organic horizon.

Organic horizon, thickness, cm	**No organic horizon**	**0-5**	**5-10**	**10-40**	**40-50**	**>50**
Russian, 1997		Raw-Humus Gleyzems Gleyzems		Peat Gleyzems		Peat
WRB, 1998	Gleyic/Gleyi-Turbic Cryosols			Gleyi-Histic/Turbi-Histic Cryosols	Fibri-Histic Histosols	
Soil Taxonomy, 1999	Aquirturbels, Aquorthels			Histoturbels, Historthels	Histels	

5.5. Cryozems

"The order is composed of soils that have beneath their organic horizons a homogeneous cryoturbated thixotropic mineral sequence of dusky grayish color which is underlain by ice-rich permafrost" (Shishov et al., 2001).

Due to similar spelling, the Russian term *Cryozems* is sometimes taken as an analogue of the term *Cryosols* either in the *WRB* sense or in a wider sense, as permafrost-affected soils in general. However, *Cryozems* covers a much smaller range of soils than *Cryosols*. Cryogenic processes affect soil genesis more strongly in Cryozems than in other permafrost-affected soils. This is the only soil type for which the diagnostic set of horizons includes the cryoturbated (KR) horizon.

These soils normally have permafrost within 100 cm. Their main area of occurrence is continental Siberia.

The best *WRB* correlative for Cryozems (as a type) is Oxiaqui-Turbic Cryosols. The formal definitions are similar, as are the genesis and volume of the two taxa. The Gelisol order of *Soil Taxonomy* does not contain an Oxiaquic unit; thus, Cryozems classify as Haploturbels (mostly Typic), though the latter covers a much wider variety of soils than Cryozems do (Figure 5.2.1). Peat Cryozems are divided between Histic Cryosols and Cryic Histosols or Histels and Histoturbels in the same way as are Peat Gleyzems.

5.6. Al-Fe-humus and metamorphic soils

These soils are distributed both within and beyond the limits of permafrost regions. Al-Fe-humus soils "are characterized by a morphologically and analytically distinct illuvial accumulation of Al-Fe-humus complex compounds producing a specific chemogenic Al-Fe-humus horizon." Metamorphic soils are "weakly differentiated morphologically. They consist of a mull or raw humus (moder) humus-accumulative horizon and a brown… metamorphic horizon which is formed by *in situ* ferrugination processes and pedogenic re-arrangement of the solid phase" (Shishov et al., 2001)

Genetic features of soil of these two orders are manifested regardless of the presence or absence of the permafrost. Thus, only some of the soils in these orders/types belong to the Cryosols/Gelic unit in *WRB* or to Gelisols in *Soil Taxonomy* (Figure 5.2.2). In contrast to Gleyzems and Cryosols, these orders include coarsely textured and stony soils.

Genesis of these soils, as presented in the Russian system, can not always be satisfactorily reflected by *WRB* and *Soil Taxonomy* units. In the *WRB*, Cryosols have no Spodic lower level units, so that Russian Podburs (Parapodzols) and Podzols, if they have permafrost within a 1-meter layer, are classified as Haplic Cryosols. This situation is common: *WRB/Soil Taxonomy* neglect genetic differences until they are quantitatively well manifested.

In the Gelisols Order of *Soil Taxonomy*, only Psammorthels and Psammoturbels have Spodic Subgroups, but Psamm-Great Groups are defined so narrowly that they do not include all Podburs and Podzols with permafrost within a 2-meter depth.

Correlating Dry Peaty Podburs and Dry Peaty Podzols with *WRB* is uncertain. In the definition of Russian Dry Peaty and Dry Peat horizons (the two horizons differ in thickness), saturation with water is not mentioned directly, and, based on the other horizons' features, saturation for long periods is unlikely. Taken as well aerated, the Dry Peaty horizon does not meet the requirement of Histic units, and Russian Dry Peat Podburs and Podzols should be classified as Haplic Cryosols, Dystri-Gelic Cambisols, or Haplic Podzols, in spite of the organic layer up to 50 cm thick. This peaty but unsaturated horizon brings to mind the Folic horizon, but for Cryosols and other mineral soil reference groups, only Histic, not Folic low level units, are included in this system.

Identification of Pale soils is based mainly on the presence of a metamorphic BM horizon, the closest analog of which is the Cambic horizon. Within both the Cryosol group and Gelisol order, these soils are classified as Haplic (or Leptic in *WRB*) units, together with permafrost-affected Podburs and, in *WRB*, Podzols.

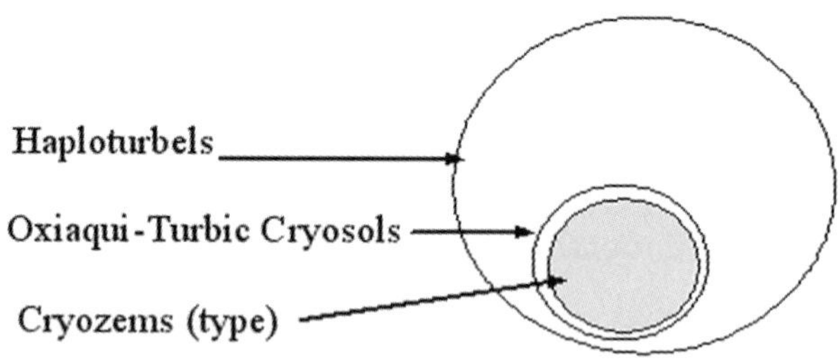

Figure 5.2.1. Correlation of Russian Cryozems with *WRB* and *Soil Taxonomy*.

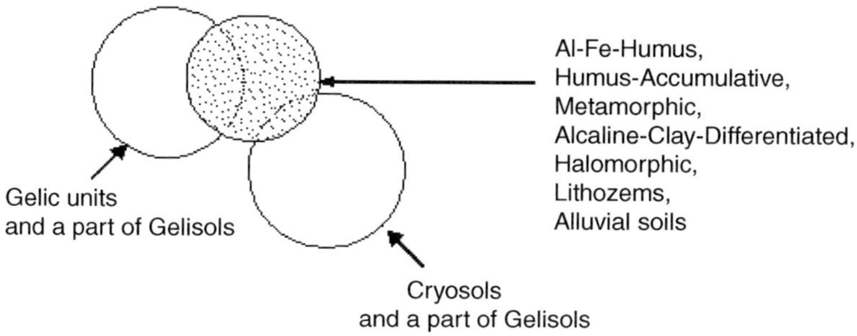

Figure 5.2.2. Correlation of Russian soil orders combining permafrost-affected and non-permafrost soils with *WRB* and *Soil Taxonomy*.

5.7. Organogenic soils

In the Russian system, organogenic soils start with a 50-cm peat layer, whereas a thickness of 40 cm (except for Lithic units) separates Histosols/Histels from mineral soils in *WRB* and *Soil Taxonomy*. Hence, some Peat Gleyzems, Peat Cryozems, Dry Peaty Podburs or Podzols, and Peat Podzols treated as mineral soils in the Russian system should be classified as Cryic Histosols and Histels. Soils with organic matter content from 20 to 35% also are treated as Raw humus or Mucky mineral soils in the Russian system, but they classify as Histosols/Histels in the two other systems. Figure 5.2.3 shows the correspondence between the Russian Peat order and *WRB/Soil Taxonomy* Histosols. This relationship is the same for both permafrost-free and permafrost-affected soils (Histosols/Histels in the latter case).

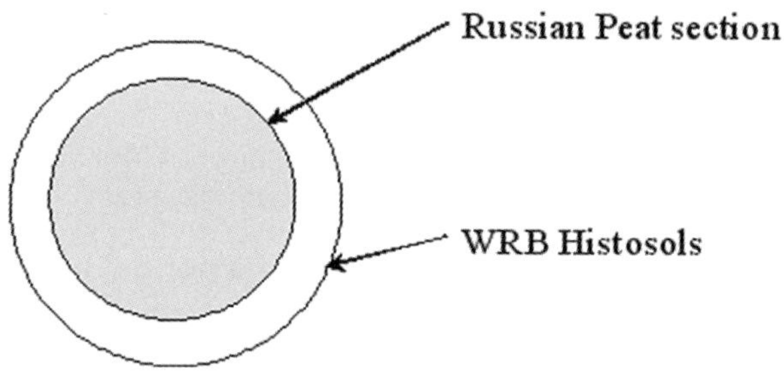

Figure 5.2.3. Correlation of Russian Peat order with *WRB*. Peat: organic matter content >35%, peat layer thickness > 50 cm. *WRB* Histosols: organic matter content ≥20% or ≥30%, depending of mineral clay content, peat layer thickness ≥40 cm.

6. Conclusion

The recently proposed version of the Russian classification system is closer than previous versions to the *WRB* and *Soil Taxonomy* with respect to general structure, definition of horizons, and soil features. However, differences due to different basic principles are still present.

A correlation problem not specific for only permafrost-affected soils is *WRB/Soil Taxonomy*'s neglect of soil genetic differences until they are quantitatively well manifested. As a result, a large number of Russian soil types can not be correlated properly with *WRB/Soil Taxonomy* soils, and they thus become Haplic/Typic units or mixed groups like Cambisols and orders like Inseptisols, where their genetic content is lost.

The gap between the Russian and other two systems became even wider recently with regard to permafrost-affected soils, because *WRB/Soil Taxonomy* gave permafrost a higher status, while the Russian system lowered it. A large number of Russian soil types contain both seasonally frozen and permafrost-affected soils, for which the depth of the permafrost table varies widely. No one category specifies the presence or absence of permafrost and a depth of the permafrost table. Thus, until hydrological and thermal blocks are included in the classification, many correlations of the Russian system to *WRB/Soil Taxonomy* will be uncertain.

7. Acknowledgment

I am grateful to participants in the IUSS-IPA Cryosol Working Group for their cooperation and useful discussions.

References

Dimo, V.N., and N.N. Rosov. 1974. Thermal criteria as a basis for facial-provincial subdivision of soils. Pochvovedenie. 5:12-22. (In Russian.)

Egorov, V.V., V.M. Fridland, E.N. Ivanova, et al., compilers. Classification and Diagnostics of the Soils of the USSR (Klassifikatsiya i diagnostika pochv SSSR). 1977. Kolos. Moscow. (In Russian.)

FAO. 1998. World Reference Base for Soil Resources. World Soil Resource Reports. 84. Rome.

Fridland, V.M. 1972. Structure of Soil Cover (Structura pochvennogo pokrova). Mysl. Moscow. (In Russian.)

Fridland, V.M. 1982. Major Principles and Elements of Basic Soil Classification and the Program of its Development (Osnovnyie printsipy i elementy bazovoi klassifikatsii i programma rabot po ee sozdaniyu). Dokuchaev Soil Institute RAAS, Moscow. 235 pp. (In Russian.)

Glazovskaya, M.A. 1972. Soils of the World (Pochvy mira). Vol. 1,2. Moscow State Univ. Publ. Moscow. (In Russian.)

Ignatenko, I.V. 1980. Classification, systematization, and nomenclature of the soils of Far North-East of the USSR. In: I.V. Ignatenko and D.I. Berman, eds. Geografiya i genezis poch Magadanskoi oblasti. Publ. Far East. Sci. Center, Russ. Acad. Sci. Vladivostok. pp. 55-92. (In Russian.)

Ivanova, E.N. 1976. USSR Soil Classification (Klassifikatsiya pochv SSSR). Nauka. Moscow. (In Russian.)

Ivanova, E.N., and N.A. Nogina, eds. 1967. Guidance on Classification and Diagnostics of Soils (Ukazaniya po klassifikatsii i diagnostike pochv). Kolos. Moscow. Issue 1.

Ivanova, E.N., and O.A. Polyntseva. 1952. Soils of European tundra. Trudy Komi filiala Akademii Nauk SSSR. Geogr. Series. 1:72-122. (In Russian.)

Liverovsky, Yu.A. 1937. Soils of the Tundra-Bog Strip (Pochvy tundrovo-bolotnoi polosy). Publ. Russ. Acad. Sci. Moscow-Leningrad. (In Russian.)

Mazhitova G.G., C.L. Ping, J.P. Moore, S.V. Gubin, and C.A.S. Smith. 1993. Correlation of soil classifications for north-eastern Russia with American and Canadian soil classifications. Pochvovedenie. 12: 26-33. (In Russian.)

Makeev, O.V. 1981. Facies of Soil Cryogenesis and Peculiarities of Soil Profiles (Fatsii pochvennogo kriogeneza i osobennosti organizatsii v nih pochvennyh profilei). Nauka. Moscow. (In Russian.)

Rozhkov, V.A. 1989. Soil Informatics (Pochvennaya informatika). VO Agropromizdat. Moscow. (In Russian.)

Shishov, L.L., and Sokolov, I.A. 1989. Genetic classification of the soils of the USSR. Pochvovedenie. 4:112-121. (In Russian.)

Shishov, L.L., V.D. Tonkonogov, and I.I. Lebedeva, compilers. 2001. Russian Soil Classification System. Dokuchaev Soil Institute, RAAS. Moscow. (Translation from Russian.)

Shishov, L.L., V.D. Tonkonogov, and I.I. Lebedeva, compilers. Russian Soil Classification (Klassifikatsia pochv Rossii). 1997. Dokuchaev Soil Institute, RAAS. Moscow. (In Russian.)

Smith, C.A.S., D.K. Swanson, J.P. Moore, R.J. Ahrens, J.G. Bockheim, J.M. Kimble, G.G. Mazhitova, C.L. Ping, and C. Tarnocai. 1995. A description and classification of soils and landscapes of the lower Kolyma River, northeastern Russia. Polar Geog. & Geol. 19, 2:107-126.

Soil Survey Staff. 1999. Soil Taxonomy, 2nd ed. Ag. Hndbk. 436. USDA, NRCS. Lincoln, NE.

Sokolov, I.A. 1980. Hydromorphic non-gley soil formation. Pochvovedenie. 1:21-33. (In Russian.)

Sokolov, I.A. 1991. Basic substantive-genetic soil classification. Pochvovedenie. 3:107-121. (In Russian.)

Sokolov, I.A. 1993. Theoretical Problems of Genetic Soil Science (Teoreticheskiye problemy geneticheskogo pochvovedeniya). Nauka. Novosibirsk. (In Russian.)

Sokolov, I.A. 1997. Soil Formation and Exogenesis (Pochvoobrazovaniye i ekzogenez). Dokuchaev Soil Sci. Institute. Moscow. (In Russian.)

Sokolov, I.A., and G.M. Bystryakov. 1980. Pale-yellow soils of the East Siberian northern taiga (Vestnik MGU, Seriya 17). Pochvovedeniye. 1:30-37. (In Russian.)

Stolbovoi, V. 2000. Soils of Russia: Correlated with the Revised Legend of the FAO Soil Map of the World and World Reference Base for Soil Resources. IIASA. Laxenburg, Austria.

Targulian, V.O. 1971. Soil Formation and Weathering in Cold Humid Regions (Pochvoobrazovaniye i vyvetrivaniye v holodnyh gumidnyh oblastyah). Nauka. Moscow. (In Russian.)

Vasilievskaya, V.D. 1979. Genetic features of soils in spotted tundra. Pochvovedenie. 7:20-32. (In Russian.)

Volkovintser, V.I. 1978. Steppe Cryoaridic Soils (Stepnyie crioaridnyie pochvy). Nauka. Novosibirsk. (In Russian.)

Chapter 3. The Gelisol Order in *Soil Taxonomy*

R.J. Ahrens[1], J.G. Bockheim[2], and C-L. Ping[3]

[1]Director, National Soil Survey Center, USDA-NRCS, 100 Centennial Mall North, Rm 152, Lincoln, Nebraska 68508, USA; e-mail bob.ahrens@nssc.nrcs.usda.gov.
[2]Professor, Univ. of Wisconsin, Soils Dept., 1525 Observatory Dr., Madison, Wisconsin 53706, USA; e-mail bockheim@facstaff.wisc.edu
[3]Professor, University of Alaska, Agricultural & Forest Experiment Station, 533 E. Fireweed, Palmer, Alaska 99645, USA; e-mail pfclp@uaa.alaska.edu.

1. Introduction

Although *Soil Taxonomy* attempts to account for all the soils in the world, taxa for permafrost regions of the world did not receive adequate attention until recently. This can be attributed to the low human population of permafrost regions and the limited suitability for traditional cultivated agriculture.

However, on a global scale permafrost is not unusual, and estimates of its extent on the earth's land surface range from 20 (Birkeland and Larson, 1989) to 26% (Black, 1954). Recently, the permafrost regions have attracted considerable interest because of the concerns for sources and sinks of greenhouse gases, the need to protect against terrestrial impacts of global warming, and the impact of transportation, development, and resource exploration in this fragile environment.

This chapter provides the rationale for the newest order in *Soil Taxonomy*, the Gelisols. It discusses some of the diagnostic features and taxa associated with the Gelisols, and it presents some of the suggestions to further improve the Gelisol order that have become apparent as more and more soil scientists familiarize themselves with it.

2. Background

Before 1998 and the new Gelisol order, *Soil Taxonomy* defined a pergelic soil temperature regime as having a mean annual soil temperature lower than 0°C at 50 cm below the soil surface. The assumption was that soils with mean annual soil temperatures below 0°C would contain permafrost. In reality, some soils experience cold winters and warm summers and have mean annual soil temperatures below 0°C, but the summer temperature is sufficiently high to prevent the formation of permafrost (Bockheim and Burns, 1991).

Before 1998, *Soil Taxonomy* had 18 pergelic subgroups distributed among the orders of Andisols, Entisols, Histosols, Inceptisols, Mollisols, and Spodosols. The subgroups were:

Pergelic Cryaquands
Pergelic Cryorthents
Pergelic Cryopsamments
Pergelic Cryofibrists
Pergelic Sphagnofibrists
Pergelic Cryohemists
Pergelic Cryosaprists
Histic Pergelic Cryaquepts
Pergelic Ruptic-Histic Cryaquepts
Humic Pergelic Cryaquepts
Pergelic Cryaquepts
Pergelic Cryochrepts
Pergelic Cryumbrepts
Pergelic Cryaquolls
Pergelic Cryoborolls
Pergelic Cryaquods
Pergelic Haplocryods
Pergelic Humicryods

An additional taxon, the Gelicryands, was recognized in the Andisol order at the great group level (Soil Survey Staff, 1996).

The priority key used in *Soil Taxonomy* was another constraint on properly identifying all the soils with a pergelic soil temperature regime. For example, Cryaquepts were defined with either a cryic or pergelic soil temperature regime, since pergelic was used at the subgroup level. Soils like Lithic Cryaquepts keyed ahead of Pergelic Cryaquepts, which implied that Lithic Cryaquepts could have a cryic or pergelic soil temperature regime.

In great groups without pergelic subgroups, like the Cryic Rendolls, the soil temperature regime could be either cryic or pergelic (Ahrens et al., 1994). Cryic and pergelic soil temperature regimes are not used at the family level and offered no way to separate these soils with different soil temperature regimes.

Before 1998, the United States had 50 soil series classified in pergelic subgroups. Ten of these series occur in the Rocky Mountains of the West, and forty occur in Alaska. Seventy-six percent of all soil series in pergelic subgroups were in the Inceptisol order, because many of them lack the features diagnostic of the other orders. Sixty-five percent of the permafrost-affected soils in Alaska occur in only two subgroups, the Pergelic Cryaquepts and the Ruptic-Histic Pergelic Cryaquepts (Moore et al., 1992).

3. International Committee

Because of the shortcomings mentioned above and the interest in global change and its impact on cold soils, an International Committee on Permafrost-Affected Soils (ICOMPAS) was established in 1993. Dr. James Bockheim of the University of Wisconsin chaired the committee.

ICOMPAS' mandate was simple, "improve the classification of permafrost-affected soils." The committee's first efforts concentrated on revising the definition of permafrost used in *Soil Taxonomy* and attempting to work with and improve existing taxa.

The committee first addressed the definition of permafrost. As indicated above, the previous definition proved unsatisfactory because of the extremes in soil temperatures between summer and winter that hindered the development of permafrost. The committee recommended that a soil with permafrost remain below 0°C for 2 or more continuous years (van Everdingen, 1998). The revised concept alleviates problems associated with the old definition.

At first, ICOMPAS focused its efforts on building taxa within the current system. The Gelicryands, introduced in 1990 by the International Committee on Andisols (ICOMAND), opened the door to potential "Geli" great groups in the other soil orders. The committee attempted to define Gelicryepts, Gelicryolls, etc., but soon realized that many of the taxa would have very few soils represented, and a few of the new taxa would contain most of the permafrost-affected soils.

Rather than Pergelic or Histic Pergelic Cryaquepts, most soils with permafrost would key into the proposed great group, Gelicryepts. To many pedologists, developing new taxa within the current system did little to improve the classification system because most of the soils with permafrost would still fall into one or two great groups.

4. A New Soil Order

The committee members next examined the feasibility of developing a new soil order. For the most part, soil orders in *Soil Taxonomy* reflect major soil-forming processes. The committee identified cryogenic processes associated with permafrost-affected soils.

Cryogenic processes in permafrost-affected soils are driven by the physical volume change of water to ice, by moisture migration along a thermal gradient in the frozen system, or by thermal contraction of the frozen material by continued rapid cooling (Bockheim et al., 1997). These processes leave markers in the soil and often are referred to as *cryoturbation* and/or *ice segregation*.

The indicators of cryoturbation include irregular or broken horizons, involutions, organic matter on the top or within the permafrost, oriented rock fragments, and silt-enriched layers. Cryoturbation is most evident in mineral soils and strongly depends on the number of freeze-thaw cycles and the texture and moisture content of the soil material (Van Vliet-Lano, 1985).

Two models attempt to explain cryoturbation. The first focuses on freezing fronts, from the surface and upward from the permafrost table, placing pressures on the unfrozen soil materials between the fronts (Nicholson, 1976). The second contends that heave-subsidence processes at the top of the active layer produce a net downward and outward movement of material, whereas heave processes at the bottom produce a net upward and outward movement (Mackay, 1980).

However, not all permafrost-affected soils exhibit cryoturbation. Ice segregation produces ice lenses, vein ice, segregated ice crystals, and ice wedges. Vein ice and segregated ice crystals promote strong granular structure, and ice lenses produce platy structure (Gubin, 1993). Both of these structures are important properties of these soils and greatly influence water movement.

In addition, the forces generated from freezing promote four fabrics that are common in permafrost-affected soils. These fabrics include a banded or lenticular fabric with a gradation in particle size from frost sorting, particularly from frequent freeze-thaw cycles; orbiculic fabric with circular or elliptical alignment of soil materials; suscitic fabric with a vertical alignment of grains and fines in spaces once occupied by ice lenses; and conglomeric fabric where pressures displace and move granular material within the soil (Tarnocai, 1983; Fox and Protz, 1981).

The committee used these soil markers to define a new term, *gelic materials*. These gelic materials are the dominant morphological features in permafrost-affected soils. Gelic materials, such as the features associated with cryoturbation, can occur in soils without permafrost or soils that had permafrost at one time. For this reason the committee also required the presence of permafrost to be included in the definition of the new soil order.

Permafrost itself hinders pedogenesis. It also is a barrier to roots and the movement of water. Permafrost-affected soils represent fragile environments with special concerns for use and management. All the above factors steered the committee toward defining a new soil order, the Gelisols.

The Canadian System of Soil Classification (1998) and the *World Resource Base for Soil Resources* (1998) use Cryosols to connote permafrost-affected soils. However, *Soil Taxonomy* has a different meaning for soils with a cryic soil temperature regime, and to use a similar word with a new meaning would cause considerable confusion for the users of *Soil Taxonomy*. The term, Gelisols, comes from the Greek word, *gelid*, which means very cold. The root has precedent in *Soil Taxonomy* in the terms *pergelic* and *Gelicryands*, and the formative element, *el*, can be used to form taxa.

5. Definition and Keying Sequence

The committee next struggled with defining Gelisols and keying the sequence of the soil orders. At first, Gelisols were required to have permafrost within 100 cm of the soil surface or gelic materials and permafrost within 100 cm of the soil surface.

After testing, the redundancy of the definition became apparent, because most soils with permafrost within 100 cm of the soil surface also had gelic materials. Also, soil scientists working in Alaska wanted permafrost recognized at deeper depths in the new soil order, even though it required extra effort to excavate beyond 100 cm.

The revised definition of the new order required permafrost within 100 cm of the soil surface, or permafrost within 200 cm of the soil surface and gelic materials within 100 cm of the soil surface. The revised definition generated little discussion and was incorporated into the eighth edition of the *Keys to Soil Taxonomy* (Soil Survey Staff, 1998).

Originally, the sequence for keying the orders included Histosols, Spodosols, and Andisols ahead of Gelisols. This sequence followed the idea that Histosols rely on anaerobic conditions to form, regardless of soil temperature. The soil-forming processes and subsequent markers in the soil are expressed strongly in Spodosols, and there are good reasons to maintain all like soils together, whether permafrost underlies them or not. The morphology of Andisols is more subdued when compared to Spodosols, but the common properties, including high phosphorous retention, make a strong case for grouping soils with like properties in the same soil order.

The literature describes Spodosols underlain by permafrost (Nicholson and Moore, 1977). However, the permafrost in these soils is well below the 200-cm limit that *Soil Taxonomy* uses to classify soils, or the permafrost is below 100 cm and the soils do not exhibit gelic features. The committee searched unsuccessfully for examples of Spodosols with permafrost within 200 cm of the soil surface and decided to key Gelisols ahead of Spodosols. However, recent research demonstrates that spodic horizons exist over permafrost in Antarctica (Beyer and Bolter, 2000; Beyer, et al., 2000)

Gelicryands, introduced by the International Committee on Andisols, suggested the presence of permafrost-affected Andisols. However, ICOMPAS was unable to find descriptions in the literature, and even scientists in Iceland were not aware of Andisols underlain by permafrost. Based on this information, ICOMPAS recommended moving the Gelisols ahead of the Andisols in the keying sequence of the soil orders.

Perhaps the most controversial decision facing the committee concerned the placement of organic soils underlain by permafrost. The committee considered two scenarios. The first kept all organic soils (Histosols) in the same soil order.

There are several logical reasons to include all organic soils in one soil order. Histosols require anaerobic conditions to facilitate the accumulation of organic materials. Anaerobic conditions occur in all soil temperature regimes, but soils with cold temperatures are more conducive to accumulating organic matter than warm soils, provided vegetation is adequate.

The second scenario, which was later approved for inclusion into *Soil Taxonomy*, advocated including all the permafrost-affected soils in the same soil order. Most committee members supported this concept because it made one soil order the focal point for management concerns and considerations associated with permafrost-affected soils. This second scenario also is in concert with some of the other classifications, notably the Canadian system.

6. Taxa

6.1. The Suborders

The committee next considered the suborders. Originally, Bockheim (1994) recommended five suborders, Histels, Turbels, Aquels, Aridels, and Statels. The Histels included the soils dominated by organic materials, and the Turbels contained the mineral soils with evidence of cryoturbation.

The Aquels tried to encompass the soils with aquic conditions and strong gleying or abundant redoximorphic features. This suborder generated debate with regard to keying sequence and the fact that many soils in cold regions commonly have aquic conditions because water perches above the permafrost. Several members of the committee felt it unimportant to have a property common to most Gelisols key high in the system. For this reason, the Aquel suborder was deleted and aquic conditions delegated to the great group level.

The Aridels, defined in terms of an aridic soil moisture regime, presented problems because the soils are never warm enough to meet the temperature requirements of an aridic soil moisture regime. The Aridels were intended for the extremely dry, cold soils of Antarctica. The committee decided to drop the Aridels and seek other means to accommodate these soils.

Three suborders were retained, the Histels, Turbels, and Statels, which were later named Orthels. The Histels represent the organic soils over permafrost and occur predominantly in the Subarctic and Low Arctic ecoclimatic regions in areas of continuous or widespread permafrost (Bockheim, 1994). The Turbels are the dominant suborder of Gelisols and are common in the High and Middle Arctic vegetative regions of North America and Eurasia (Bockheim, 1994).

Turbels require cryoturbation, which ranges in expression from involutions and organic material over the permafrost table to weakly oriented rock fragments. Moisture is a prerequisite to cryoturbation. Turbels that occur in drier areas occur on landscapes that receive run-on moisture or experienced moister conditions in the past. The Orthels occur in drier areas, are associated with coarsely textured soils, or occur in youthful parent materials where cryoturbation is inhibited.

The committee defined new diagnostic features for use in developing additional taxa. Anhydrous conditions accommodate soils common to cold desert regions. The Glacic defines layers largely composed of ice. Cryoturbation and gelic materials were mentioned above.

6.2. The Great Groups

The Histels are further divided into great groups based largely on the amount of decomposition of the organic materials. However, the Glacistels key high among the great groups to capture soils with layers of ice that present special engineering concerns. The amount of decomposition is least in the Fibristels, followed by the Hemistels and Sapristels, respectively.

The Turbels are divided into great groups based on the amount of organic materials, whether the soil is very dry or experiences aquic conditions, the nature of the epipedon, and the presence of sandy textures in the control section.

The great groups of the Orthels follow the Turbels closely, with one exception. Argillic horizons are recognized in the Orthels where lack of cryoturbation is more conducive to clay illuviation.

6.3. The Families

At the family level, temperature classes provide interpretative information about the Gelisols. The subgelic class (+1°C to -4°C) groups soils that are fragile and, if the organic mat is disturbed, tend to subside due to thawing of the upper permafrost table. The pergelic class (-4°C to -10°C) includes soils in the discontinuous and southern portions of the continuous permafrost zones. The hypergelic class (-10°C and lower) encompasses the continuous permafrost zone, where disturbances to the organic mat do not lower the permafrost table substantially.

The committee also attempted to add terms for features of patterned ground at the family level. Washburn, 1980, describes earth hummocks, sorted and non-sorted circles, sorted and non-sorted nets, sorted and non-sorted stripes, etc. However, these terms are difficult to define, and intergrades exist that have characteristics of more than one feature. For these reasons, terms for patterned ground were never incorporated into *Soil Taxonomy*, but these features are very useful as phases of map units.

7. Further Improvements

The committee's recommendations were accepted and incorporated into *Soil Taxonomy* in 1998 (Soil Survey Staff, 1998). Pedologists throughout the world have used the new soil order and in the process have noted items for improvement.

The cryic temperature regime is now defined to include soils with mean annual soil temperatures of lower than 8°C, but no permafrost. Some pedologists are finding this class too broad because it encompasses both the boreal, alpine, and tundra regions. Alternatives are being considered.

The discovery of spodic horizons underlain by permafrost means taxa should be developed to accommodate such soils. Spodorthels could be developed easily, as Beyer and Bolter (2000) and Beyer et al. (2000) suggested.

In the environment where Aquorthels occur, sometimes the organic surface horizons are not broken, nor frost cracked or mounded, and Hoefle and others (1998) have suggested that a Histic subgroup be added to Aquorthels to complement the already existing Ruptic-Histic subgroups. They also suggested adding a Turbic subgroup to Histels to accommodate soils in western Alaska.

Some pedologists have noted vegetative differences between the more acid and non-acid soils and want taxa developed to represent the contrasts in vegetation. The umbric and mollic intergrades make an attempt but are not adequate.

Cryoturbation is difficult to document adequately. The convoluted, broken horizons do not lend themselves well to our current pedon descriptions. The committee recommended that drawings be made to illustrate soils with cryoturbation. While drawings accurately portray these soils, they are not a medium readily transferred. Additional methods are needed to portray these soils accurately.

8. Summary

Originally, *Soil Taxonomy* inadequately defined and classified permafrost-affected soils. The international community under the leadership of Jim Bockheim devised taxa and diagnostic features to help us manage, communicate, and transfer knowledge about the fragile soils permafrost underlies.

The committee's success can be attributed to the chairperson's energy and leadership and the willingness of scientists worldwide to share their knowledge and data, especially our Canadian colleagues.

References

Ahrens, R.J., C.A.S. Smith, J.P. Moore, and L. Ratliff. 1994. Review of the classification of permafrost-affected soils in the United States and Canada. In: J.M. Kimble and R.J. Ahrens, eds. Proceedings of the meeting on the classification, correlation, and management of permafrost-affected soils. USDA-SCS, National Soil Survey Center. Lincoln, NE. pp. 3-15.

Beyer, L., and M. Bolter. 2000. Chemical and biological properties, formation, occurrence, and classification of Spodic Cryosols in a terrestrial ecosystem of East Antarctica (Wilkes Land). Catnea. 39:95-119.

Beyer, L., K. Pingpank, G. Wriedt, and M. Bolter, 2000. Soil formation in coastal Antarctica (Wilkes Land). Geoderma. 95:283-304.

Birkeland, P.W., and E.E. Larson. 1989. Putnam's Geology, 5th edition. Oxford Univ. Press. New York, NY.

Black, R.F. 1954. Permafrost, a review. Geol. Soc. Amer. Bull. 65:838-855.

Bockheim, J.G., and S.F. Burns. 1991. Pergelic soils of the western contiguous United States; distribution and taxonomy. Arct. & Alp. Res. 23:206-212.

Bockheim, J.G., C.L. Ping, J.P. Moore, and J.M. Kimble. 1994. Gelisols: A new proposed order for permafrost-affected soils. In: J.M. Kimble and R.J. Ahrens, eds. Proceedings of the meeting on the classification, correlation, and management of permafrost-affected soils. USDA-SCS. National Soil Survey Center. Lincoln, NE. pp. 25-44.

Bockheim, J.G., J.M. Kimble, C. Tarnocai, and C.L. Ping. 1997. Gelisols: A newly proposed order in Soil Taxonomy. Adv. of fundamental res earch of the earth cryosphere in arctic and subarctic. Proc. of international conf. Pushchino. Russia.

Fox, C.A., and R. Protz. 1981. Definition of fabric distributions to characterize the rearrangement of soil particles in the Turbic Cryosols. Can. J. of Soil Sci. 61:29-34.

Gubin, S.V. 1993. Late Pleistocene soil formation on coastal lowlands of northern Yakutia Pochvovedeniye 10:62-70. (In Russian.)

Hoefle, C.M., C.L. Ping, and J.M. Kimble. 1998. Properties of permafrost soils on northern Seward Peninsula, northwest Alaska. Soil Sci. Soc. Am. J. 6:1626-1639.

Mackay, J.R. 1980. The origin of hummocks, western Arctic coast, Canada. Can. J. Earth Sci. 13:889-897.

Moore, J.P., D. Swanson, and C.L. Ping. 1992. Warm and cold permafrost soils of interior Alaska. First Int. Conf. On Cryopedology, Nov. 10-14. Pushchino. Russia.

Nicholson, F.H. 1976. Patterned ground formation and description as suggested by low arctic and subarctic examples. Arct. & Alp. Res. 8:329-342.

Nicholson, H.M., and T.R. Moore. 1977. Pedogenesis in a subarctic iron-rich environment, Schefferville, Quebec. Can. J. Soil Sci. 57:35-45.

Soil Survey Staff. 1996. Keys to Soil Taxonomy. 7th edition. U.S. Gov't Printing Office.

Soil Survey Staff. 1998. Keys to Soil Taxonomy. 8th edition. U.S. Gov't Printing Office.

Tarnocai, C. 1983. Soils and vegetation. In: H.M. French and J.A. Heginbottom, eds. Northern Yukon Territory and Mackenzie Delta, Canada, Guidebook to permafrost and relative Features. IV International Conf. on permafrost, Div. of Geol. And Geophys. Surveys, Dept. of Natural Resources. Fairbanks, AK. pp. 21-34.

The Canadian system of soil classification. 1998. Third Edition. Research Branch and Agri-Food Canada. Pub. 1646.

Van Everdingen, R.O., ed. 1998. Multi-language glossary of permafrost and ground-ice terms. International Permafrost Association. The Arctic Institute of North America. The Univ. of Calgary. Calgary, Alberta. Canada.

Van Vliet-Lano, B. 1985. Frost effects in soils. In: J. Boardman, ed. Soils and Quaternary Landscape Evolution. J. Wiley. New York, NY. pp. 117-158.

Washburn, A.L. 1980. Geocryology: A survey of pereglacial processes and environments: Halsted Press, J. Wiley and Sons. New York, NY. 406 p.

World Reference Base for Soil Resources. 1998. World Soil Resources Report. 84.

Chapter 4. Classification of Permafrost-Affected Soils in the WRB

C. Tarnocai[1], G. Broll[2], and H.-P. Blume[3]

[1]Research Scientist, Agriculture and Agri-Food Canada, Research Branch (ECORC), K.W. Neatby Building, Rm. 1135, 960 Carling Avenue, Ottawa, Canada, K1A 0C6; e-mail tarnocaict@agr.gc.ca.
[2]Professor, Division of Geology and Agroecolgy, ISPA - University of Vechta, P.O. Box 1553, 49364 Vechta, Germany; e-mail gbroll@ispa.uni-vechta.de.
[3]Professor , Institute for Plant Nutrition and Soil Science, Christian-Albrechts-University Kiel, 24098 Kiel, Germany; e-mail hblume@soils.uni-kiel.de.

1. Introduction

The World Reference Base for Soil Resources (ISSS Working Group RB, 1998) classification includes permafrost-affected soils in three groups. The first is the Histosol major soil group, which includes both perennially frozen (Cryic, Glacic and Gelic lower-level units) (ISSS Working Group RB, 1998, pp. 13, 76) and unfrozen organic soils. The second is the Cryosol major soil group, which includes all of the perennially frozen mineral soils that have a cryic horizon (ISSS Working Group RB, 1998, pp. 14, 30). The third includes those soils in the Leptosol, Fluvisol, Solonchak, Gleysol, Podzol, Planosol, Albeluvisol, Umbrisol, Cambisol, Arenosol, and Regosol major soil groups (ISSS Working Group RB, 1998, pp. 76-78) that have permafrost at depths of 100 to 200 cm.

Regardless of how they are classified, all permafrost-affected soils have formed in a permafrost environment and have cryic, glacic, or gelic horizons. In addition, their soil water occurs primarily in the form of ice, and cryogenic processes are their dominant soil-forming processes.

This chapter describes the classification of permafrost-affected soils according to the 1998 edition of the WRB classification system and suggests improvements.

2. Classification Principles

The WRB classification system is based on four general principles (ISSS Working Group RB, 1998, pp. 4-9). First, the classification of soils is based on soil properties defined in term of diagnostic horizons and soil characteristics. Second, the selection of diagnostic horizons and characteristics takes into account their relationship with soil-forming processes. Third, climatic parameters are not applied in the classification of soils. (Note, however, that ISSS Working Group RB, 1998, does include temperature criteria in definitions of the cryic horizon [p. 30] and permafrost [p. 54] but emphasizes specific characteristics and properties that are a manifestation of a cold climate.) Fourth, the WRB classification system is composed of two tiers, the Reference Soil Groups (highest level; consists of 30 major soil groups) and the WRB Classification System (consists of a combination of prefixes as unique modifiers added to the reference soil groups).

Four specific principles apply to the WRB classification of permafrost-affected soils (ISSS Working Group RB, 1998). First, for Histosols, the material is placed at the highest level of the classification, thus maintaining the strict separation between organic and mineral major soil groups, regardless of soil properties. Second, for mineral soils, the presence of permafrost, together with cryogenic properties, enters into the highest level of the classification, with Cryosols being designated as a major soil group.

Third, for other mineral major soil groups (Leptosols, Fluvisols, Solonchaks, Podzols, Planosols, Albeluvisols, Umbrisols, Cambisols, Arenosols, and Regosols), the soil development of the particular group overrides the presence of permafrost and cryogenic processes if the cryic horizon is at a depth of more than one meter (otherwise they would have been classified as Cryosols, which key out earlier). In practice, this means that these mineral major soil groups do not have a cryic horizon within a depth of one meter. Permafrost-affected soils in these soil groups are identified by the Gelic lower-level unit of the group if the permafrost occurs at a depth of one to two meters.

Fourth, the diagnostic cryic horizon for permafrost-affected soils therefore is recognized only at the first level for the Cryosol major soil group. For Histosols and all mineral major soil groups other than Cryosols, the diagnostic cryic horizon is placed at the secondary level if this horizon occurs at a depth of one to two meters.

3. Subdivisions of the Major Soil Groups

Standardized subdivisions (lower-level units or subunits) were developed for each major soil group. The lower-level units closely relate to the diagnostic criteria but contain additional soil properties thought to be relevant to the lower level. The general rules for formulating the lower-level units are summarized in ISSS Working Group RB (1998, pp. 60-61). Table 5.4.1 summarizes the lower-level units used for permafrost-affected soils.

Table 5.4.1. List of lower-level units, their definitions, and the major soil groups to which they apply (ISSS Working Group RB, 1998).

No.	Lower-level Units	Definition	Major Soil Group
1	Histic	having a hystic horizon within 40 cm from the soil surface	Cryosols
2	Lithic	having continuous hard rock within 10 cm from the soil surface	Cryosols
3	Leptic	having continuous hard rock between 25 and 100 cm from the soil surface	Cryosols
4	Turbic	having cryoturbated features	Cryosols
5	Salic	having a salic horizon within 100 cm from the soil surface	Cryosols
6	Natric	having a natric horizon within 100 cm from the soil surface	Cryosols
7	Gleyic	having gleyic properties within 100 cm from the soil surface	Cryosols
8	Andic	having an andic horizon within 100 cm from the soil surface	Cryosols
9	Mollic	having a mollic horizon	Cryosols
10	Gypsic	having a gypsic horizon within 100 cm from the soil surface	Cryosols
11	Calcic	having a calcic horizon or secondary carbonates between 50 and 100 cm from the soil surface	Cryosols
12	Umbric	having an umbric horizon	Cryosols
13	Yermic	having a yermic horizon including a desert pavement	Cryosols
14	Aridic	having aridic properties without a takyric or yermic horizon	Cryosols
15	Glacic	having an ice layer that is >30 cm thick and contains >95% ice (by volume) within 100 cm of the soil surface	Cryosols
16	Thionic	having a sulfuric horizon within 100 cm of the soil surface	Cryosols
17	Oxyaquic	saturated with water during the thawing period and lacking redoximorphic features within 100 cm of the soil surface	Cryosols
18	Stagnic	having stagnic properties within 50 cm from the soil surface	Cryosols
19	Haplic	having no further or meaningful characterization	Cryosols
20	Cryic	having a cryic horizon within 100 cm from the soil surface	Histosols
21	Gelic	having permafrost within 200 cm from the soil surface	Others*

*Histosols, Leptosols, Fluvisols, Solonchaks, Podzols, Planosols, Albeluvisols, Umbrisols, Gleysols, Cambisols, Arenosols, and Regosols.

This table provides the names and brief descriptions of the lower-level units and indicates to which major soil groups they apply. The priority rules for the use of lower-level soil names should be followed to avoid confusion. Lower-level units numbered 1 to 18, which apply only to the Cryosol major soil group, are arranged in order of priority, with number 1 being the highest priority level.

For example, to classify a cryoturbated Cryosol with gleyic features, one would follow the priority list in Table 5.4.1 and note that lower-level unit 4 (cryoturbated features) is the first applicable unit, making the soil a Turbic Cryosol. Lower-level unit 7 (gleyic properties) also applies, however, so the soil is classified as a Gleyi-Turbic Cryosol.

4. Diagnostic Criteria

The diagnostic criterion for permafrost-affected soils in the WRB classification (ISSS Working Group RB, 1998) is the presence of a cryic horizon (pp. 14, 30). Specific properties of Cryosols, such as the presence of a high (>95%) amount of ice (Glacic) or cryoturbation features (Turbic), are used at the lower level so that cryoturbated Cryosols key out immediately after the organic Cryosols (Histic Cryosols), shallow soils over bedrock (Lithic Cryosols), and very gravelly Cryosols (Leptic Cryosols). The presence of permafrost within a depth of one to two meters in mineral soils other than Cryosols is indicated by the Gelic lower-level unit.

All *cryic horizons* are perennially frozen mineral or organic soil materials that must have:

1. Soil temperature at or below 0°C for two or more years in succession
2. a. in the presence of sufficient interstitial soil water, evidence of cryoturbation, frost heave, cryogenic sorting, thermal cracking, or ice segregation, <u>or</u>
   b. in the absence of sufficient interstitial soil moisture, evidence of thermal contraction of the frozen soil material
3. platy or blocky macrostructures resulting from vein ice development, and orbicular, conglomeratic, and banded microstructures resulting from sorting of coarse soil material.

If soil moisture is present, cryic horizons show evidence of perennial ice segregation and/or cryogenic processes (mixed soil material, disrupted soil horizons, involutions [swirl-like patterns in soil horizons], organic intrusions, frost heave, separation of coarse from fine soil materials, cracks, and patterned surface features such as earth hummocks, frost mounds, stone circles, nets, and polygons).

If insufficient interstitial soil water is present, the cryic horizons are dry but thermal contraction features occur, although more weakly developed than those in cryic horizons with higher moisture content.

The following properties are also diagnostic for permafrost-affected soils:

1. *Gelic*: having permafrost within 200 cm from the soil surface (ISSS Working Group RB, 1998, p. 66).

2. *Glacic*: having a horizon within 100 cm from the soil surface that is 30 cm or more thick and contains 95% or more (by volume) ice (ISSS Working Group RB, 1998, p. 66).
3. *Turbic*: having cryoturbative features (mixed soil material, disrupted soil horizons, involutions [swirl-like patterns in soil horizons], organic intrusions, frost heave, separation of coarse from fine soil materials, cracks, patterned surface features such as earth hummocks, frost mounds, stone circles, nets, and polygons), either at the surface or within 100 cm from the soil surface (*in Cryosols only*) (ISSS Working Group RB, 1998, p. 72).

5. Major Soil Groups of Permafrost-Affected Soils

5.1. Histosol major soil group

In the WRB classification, Histosols are defined as organic soils having a *histic* or *folic* horizon either 10 cm or more thick from the soil surface to a lithic or paralithic contact or 40 cm or more thick and starting within 30 cm from the soil surface, and lacking an *andic* or *vitric* horizon starting within 30 cm from the soil surface (ISSS Working Group RB, 1998, p. 13). The permafrost-affected Histosols are classified using the Cryic, Gelic, and Glacic lower-level units (ISSS Working Group RB, 1998, p. 76). Cryic, Gelic, and Glacic Histosols contain segregated ice and commonly are associated with patterned ground in the form of lowland polygons, peat plateaus, palsas, peat mounds, and ribbed and basin fens. Cryic Histosols, especially the shallow ones, are affected by cryoturbation.

5.2. Cryosol major soil group

In the WRB classification, Cryosols are defined as mineral soils having a *cryic* horizon within 100 cm from the soil surface (ISSS Working Group RB, 1998, p. 14). They contain segregated ice (except for those associated with dry permafrost), cryoturbated soil horizons, and macro- and microstructures resulting from cryogenic processes. In addition, Cryosols commonly are associated with patterned ground in the form of earth hummocks, circles, nets, steps, stripes, and ice-wedge polygons.

Cryosols were introduced into the draft version of the WRB as a major soil group in 1994 (Spaargaren, 1994). This draft version of the WRB was reviewed and revised during the following years, and the revised Cryosol major soil group appeared in the 1998 edition of the WRB (ISSS Working Group RB, 1998). Table 5.4.1 briefly defines the 20 lower-level units in the Cryosol major soil group.

5.3. Other major soil groups

The Gelic lower-level units of the Histosols, Leptosols, Fluvisols, Solonchaks, Gleysols, Podzols, Planosols, Albeluvisols, Umbrisols, Cambisols, Arenosols, and Regosols are characterized by cryoturbation features and/or the presence of permafrost within 1-2 m of the surface (ISSS Working Group RB, 1998). These permafrost-affected soils usually either are coarse-textured or have a deep active layer because of the removal of the vegetative cover and/or surface organic layer. This increase in depth of thaw could result from wildfires, agricultural practices, or activities such as pipeline and road construction.

6. Correlation with other Classifications

Permafrost-affected soils are classified by other countries as Cryosols (Canada: Soil Classification Working Group, 1998), Gelisols (USA: Soil Survey Staff, 1999), Cryozems and Cryomorphic soils (Russia: Mazhitova, this section, Chapter 2), and Polar desert soils (USA: Tedrow, 1958).

The Canadian Cryosol and American Gelisol soil orders both include all permafrost-affected soils (both mineral and organic) that have permafrost within 100 cm of the soil surface. These orders also contain all the cryoturbated soils that have permafrost within 200 cm of the soil surface.

Some of the Russian systems place permafrost-affected soils in a number of groups. For example, Makeev (1997) classifies all gleyed permafrost soils as various Tundra frozen soil types. He classified other mineral soils as Soddy frozen soils, Arctic stony frozen soils, Pale taiga frozen soils, Solonchaks, Podzols, Podburs, and Pseudogley frozen soils. All frozen organic soils are classified as Peaty frozen soils.

The Regic lower-level unit commonly is used in the classification systems mentioned above, but in the WRB system it is used only for the Anthrosol major soil group (ISSS Working Group RB, 1998, p. 76). Note also that the definition of Regic as lacking recognizable buried horizons (ISSS Working Group RB, 1998, p. 70) differs from that of the other classifications, which define Regic as soils lacking development (soils with A and C horizons).

Very dry, high salt content, permafrost soils (having temperatures below 0°C within 100 cm of the surface) that are not cryoturbated and contain no visible ice may seem to require classifying as Solonchaks in the WRB classification. As in other soil classifications, however, these soils should be classified as Cryosols, since they contain a cryic horizon.

7. Testing the WRB Classification of Permafrost-Affected Soils

The classification of permafrost-affected soils, and especially the Cryosol major soil group, was established only recently in the WRB. Any new classification system has to undergo a verification process, especially in the field, to test and validate the established classes, the criteria used to describe the classes described, and the definitions used for the various terms.

7.1. Changes to the Key to the Reference Soil Groups

If Histosols key out before Cryosols, Histic Cryosol subunits are characterized by a histic horizon that is 10 to 40 cm thick. Cryic Histosols have permafrost within 1 m of the surface and have a histic horizon >40 cm thick and cryoturbation features. Gelic Histosols lack cryoturbation features and/or permafrost within 1 to 2 m of the surface.

The second alternative would be to key out the Cryosols before the Histosols. Under these circumstances, all soils with a histic epipedon together with cryoturbation features and permafrost within 1 m of the surface are Histic (or Folic) Cryosols; those without cryoturbation features and/or permafrost within 1 to 2 m of the surface are Gelic Histosols.

The third alternative would be to key out the Cryosols before the Histosols and to classify all perennially frozen organic soils as either Organic Cryosols, as in the Canadian system (Soil Classification Working Group, 1998), or as Histic Cryosols, similar to the term used in the American system (Soil Survey Staff, 1998). In this alternative, there would be no Cryic Histosols, but Gelic Histosols would exist. Since the Cryic lower-level unit now exists only in the Histosols, it would no longer be needed in the WRB.

Since perennially frozen organic soils have morphological features that result from cryogenic processes such as cryoturbation and thermal cracking, they contain a cryic horizon (ISSS Working Group RB, 1998, p. 30). In addition, the use and management of perennially frozen organic soils are similar to those of perennially frozen mineral soils. Therefore, we recommend that Cryosols key out first in the Key to the Reference Soil Groups, followed by Histosols without permafrost. Such a change would make the WRB more consistent with other soil classifications that have been subjected to comprehensive testing.

7.2. Changes to horizon and lower-level unit definitions

7.2.1. Histic horizon

The histic horizon (ISSS Working Group RB, 1998, page 35), as now defined, has a thickness of 10 cm or more. The US *Soil Taxonomy* (Soil Survey Staff, 1999), however, requires a 20-cm minimum thickness for this horizon (Soil Survey Staff, 1998, p. 14). In *The Canadian System of Soil Classification* (Soil Classification Working Group, 1998, pp. 73-78), the minimum thickness criterion for Histic subgroups of Cryosols is 15 cm. The thickness is based on the average minimum thickness of the organic layer in the Canadian Arctic. We therefore suggest that the minimum thickness of the histic horizon be 15 cm.

7.2.2. Regic unit

A Regic lower-level unit should be included for Cryosols. The terms *Rego* or *Regosol* (*regolith* in geology) commonly are used in soil science and geology, and in some of the soil classifications they also are used for soils with very little or no soil development. The definition of *Regic* on page 70 (ISSS Working Group RB, 1998), "lacking recognizable buried horizons (in Anthrosols only)," is misleading; a term other than Regic should be used for this definition. The definition of Regic should be "Regic: lacking recognizable soil horizon development."

7.2.3. Organic unit

The Organic lower-level unit should be used only for Cryosols having an organic layer greater than 40 cm thick. This unit must have at least 18% (by weight) organic carbon (30% organic matter).

7.2.4. Cambic Horizon

This is the diagnostic horizon for the Cambic lower-level units. It should apply to the Cryosols as defined on page 28 (ISSS Working Group RB, 1998).

7.2.5. Luvic and Spodic units

Luvic and Spodic lower-level units should be included for Cryosols as defined on pages 68 and 72 (ISSS Working Group RB, 1998), respectively.

7.2.6. Cyric unit

The Cryic lower-level unit (having a cryic horizon within 100 cm from the soil surface) should be dropped from the WRB classification. The Gelic lower-level unit should be defined as having permafrost within 100 to 200 cm from the soil surface. The Gelic lower-level unit then should be used for all major soil groups except Cryosols.

7.2.7. Turbic unit

In the definition of *Turbic* (ISSS Working Group RB, 1998, p. 72), the statement that it applies *in Cryosols only* should be removed since it may apply to all major soil groups containing Gelic lower-level units.

7.3. Suggested third and fourth level qualifiers

Add third and fourth level qualifiers to the system. Third and fourth level qualifiers should be taken from lower portions of the classification hierarchy. The fourth level qualifier should be given in parentheses after the reference soil group name – *e.g.*, Umbri-Turbic Cryosol (Stagnic).

7.4. Suggested changes and additions to the priority listing of lower-level units

An asterisk (*) marks the proposed new units.

HISTOSOLS	CRYOSOLS
Cryic	Organic*
Glacic	Glacic
Gelic	Turbic
Salic	Thionic
	Stagnic
	Histic
	Leptic
	Salic
	Gleyic
	Andic
	Natric
	Gypsic
	Calcic
	Luvic*
	Spodic*
	Umbric
	Cambic*
	Regic*
	Lithic
	Mollic
	Yermic
	Aridic
	Oxyaquic
	Haplic

We also suggest using Calcic, Eutric, and Dystric as qualifiers for the Cryosol major soil group at only the third and fourth levels.

7.5. Suggestion relating to relict Cryosols

Paleosols that were formerly Cryosols but now are not associated with permafrost should not be classified as Cryosols even though they may contain cryogenic features such as cryoturbated soil horizons, sand wedges, and ice-wedge casts. The classification of these soils should be handled by using a turbic qualifier with the appropriate major soil group.

8. Conclusions

The WRB is intended to be the international reference base for classification of permafrost-affected soils. To achieve international acceptance, numerous compromises were made, but they led to contradictions in the definitions and concepts and, in some cases, negatively affected the scientific value of the classification system.

Cryopedologists should now rigorously test the WRB classification. We have begun this process (Tables 5.4.2, 5.4.3, 5.4.4, 5.4.5), but more extensive field testing and evaluation are needed. The classification also should be tested in a soil mapping project to determine how well the system works on a permafrost landscape. If these steps are carried out, the classification of permafrost-affected soils will truly fulfill the principles stated in ISSS Working Group RB (1998, pp. 4-5): that "WRB is meant to be a comprehensive classification system that enables people to accommodate their own national classification system" and that it will serve "as a link between existing classification systems" and "as a common denominator for communication at an international level."

Table 5.4.2. Classification of permafrost-affected soils according to WRB, American, and Canadian systems, with recommended changes.

No.	Site	Source	WRB Classification Current (1998)	Recommended	US Soil Taxonomy* / Canadian System **
1	22	1	Turbic Cryosol	Umbri-Turbic Cryosol	Orthic Dystric Turbic Cryosol**
2	Owl River 1	2	Turbic Cryosol	Cambi-Turbic Cryosol	Orthic Dystric Turbic Cryosol**
3	12-90-T2	3	Turbic Cryosol	Regi-Turbic Cryosol	Regosolic Turbic Cryosol**
4	2 NWT	1	Histi-Turbic Cryosol	Cambi-Turbic Cryosol	Orthic Eutric Turbic Cryosol**
5	17	1	Molli-Turbic Cryosol	Molli-Turbic Cryosol	Brunisolic Eutric Turbic Cryosol**
6	11	1	Umbri-Turbic Cryosol	Umbri-Turbic Cryosol (Stagnic)	Orthic Dystric Turbic Cryosol**
7	32	1	Umbri-Turbic Cryosol	Umbri-Turbic Cryosol (Gleyic)	Orthic Dystric Turbic Cryosol**
8	25a	1	Gleyi-Turbic Cryosol	Stagni- Stagni-Turbic Cryosol	Orthic Eutric Turbic Cryosol**
9	1YT	4	Gleyi-Turbic Cryosol	Stagni-Turbic Cryosol	Orthic Turbic Cryosol**
10	9	1	Turbi-Histic Cryosol	Histi-Turbic Cryosol (Umbric and Gleyic)	Gleysolic Turbic Cryosol**
11	12	1	Turbi-Histic Cryosol	Histi-Turbic Cryosol (Dystric)	Histic Dystric Turbic Cryosol**
12	28b	1	Hapli-Histic Cryosol	Histi-Regic Cryosol	Regosolic Turbic Cryosol**
13	Rundle	2	Haplic Cryosol	Hapli-Cambic Cryosol (Gleyic)	Gleysolic Static Cryosol**
14	12-90-T4	3	Haplic Cryosol	Hapli-Regic Cryosol (Salic)	Regosolic Static Cryosol**
15	12-89-83	3	Haplic Cryosol	Hapli-Regic Cryosol	Regosolic Static Cryosol**
16	12-89-08	3	Haplic Cryosol	Luvi-Haplic Cryosol	Luvisolic Static Cryosol**
17	94P0668	5	Haplic Cryosol	Cambi-Haplic Cryosol	Brunisolic Eutric Static Cryosol**
18	Landsc. 1-1	6	Gelic Gleysol	Gelic Gleysol	Gleysolic Turbic Cryosol**
19	Landsc. 1-2	6	Gelic Gleysol	Gelic Gleysol	Orthic Gleysol**
20	1	1	Cryic Histosol	Gelic Histosol	Mesic Organic Cryosol**
21	148	7	Gelic Histosol	Lepti-gelic Histosol	Lithic Fibristel*
22	137	7	Gelic Histosol	Lepti-gelic Histosol	Lithic Folistel*
23	202	7	Turbic Cryosol	Hapli-Turbic Cryosol	Typic Haploturbel*
24	2	7	Turbic Cryosol	Stagni-Turbic Cryosol	Lithic Aquiturbel*
25	25	7	Gelic Podzol	Lepti-Gelic Podzol	Lithic Haplorthel*
26	210	7	Gelic Podzol	Skeleti-Gelic Podzol (former penguin site)	Typic Haplorthel*
27	P7	8	Gelic Fluvisol	Distry-Gelic Fluvisol	Typic Psammorthel*
28	P1	8	Gelic Cambisol	Molli-Gelic Cambisol	Typic Mollorthel*
29	P13	8	Turbic Cryosol	Gleyi-Turbic Cryosol	Cumulic Umbriturbel*
30	P14	8	Umbri-Turbic Cryosol	Umbric Cryosol	Typic Haploturbel*

+ ISSS Working Group RB. (1998). * Soil Survey Staff (1998). ** Soil Classification Working Group (1998).

Sources: 1. Tarnocai et al. (1993); 2. Broll et al. (1998); 3. Tarnocai et al. (1991); 4. Ping et al. (1993); 5. Soil Survey Staff (1999); 6. Anonymous (1997); 7. Blume et al. (1997); 8. Kuhn (1997).

Table 5.4.3. Locations of sites.

No.	Site	Parent material	Latitude	Longitude	Location
1	22	Till	64° 32' 50" N	138° 14' 02" W	North Fork Pass, Yukon, Canada
2	Owl River 1	Till	66° 53' N	64° 40' W	Baffin Island, Nunavut, Canada
3	12-90-T2	Colluvium	81° 23' 41" N	76° 28' 00" W	Ellesmere Island, Nunavut, Canada
4	2 NWT	Till	68° 06' 39" N	133° 28' 29" W	Inuvik area, N.W.T., Canada
5	17	Colluvium over till	65° 04' 03" N	138° 16' 44" W	Ogilvie Mountains, Yukon, Canada
6	11	Mixed eolian and shale	66° 32' 27" N	136° 22' 28" W	Eagle Plain, Yukon, Canada
7	32	Colluvium	64° 04' 00" N	140° 57' 00" W	Little Gold Creek, Yukon, Canada
8	25a	Mixed loess and alluvium	64° 03' 13" N	138° 59' 51" W	Dawson City area, Yukon, Canada
9	1YT	Loess	66° 23' 20" N	136° 41' 44" W	Eagle Plain, Yukon, Canada
10	9	Colluvium	66° 42' 59" N	136° 21' 11" W	Richardson Mountains, Yukon, Canada
11	12	Mixed loess and colluvium	66° 23' 20" N	136° 41' 44" W	Eagle Plains, Yukon, Canada
12	28b	Colluvium	65° 56' 30" N	138° 43' 00" W	Too Much Gold Creek, Yukon, Canada
13	Rundle	Eolian	66° 31' N	65° 17' W	Baffin Island, Nunavut, Canada
14	12-90-T4	Eolian over fluvial	81° 24' 13" N	76° 51' 00" W	Ellesmere Island, Nunavut, Canada
15	12-89-83	Marine	81° 28' 50" N	76° 30' 30" W	Ellesmere Island, Nunavut, Canada
16	12-89-08	Fluvial	81° 24' 25" N	76° 38' 30" W	Ellesmere Island, Nunavut, Canada
17	94P0668	Loess	68° 43' 56" N	161° 31' 49" E	Lower Kolymar, Russia
18	Landscape 1-1	–	–	–	Vorkuta area, Russia
19	Landscape 1-2	–	–	–	Vorkuta area, Russia
20	1	Peat	68° 18' 57" N	133° 25' 51" W	Inuvik area, N.W.T., Canada
21	148	Peat over boulders	66° 17' 02" S	110° 32' 03" E	Casey, Wilkes Land, Antarctica
22	137	Peat over Leuco gneiss	66° 17' 04" S	110° 32' 03" E	Casey, Wilkes Land, Antarctica
23	202	Till	66° 17' 06" S	110° 32' 09" E	Casey, Wilkes Land, Antarctica
24	2	Gneiss to schist	66° 17' 06" S	110° 31' 58" E	Casey, Wilkes Land, Antarctica
25	25	Leuco gneiss	66° 17' 06" S	110° 31' 58" E	Casey, Wilkes Land, Antarctica
26	210	Leuco gneiss	66° 17' 04" S	110° 32' 05" E	Casey, Wilkes Land, Antarctica
27	P7	Glaciofluvial sand	62° 09' 44" S	58° 27' 45" W	Arctowski, King George Island, Antarctica
28	P1	Till (basaltic)	62° 09' 50" S	58° 28' 10" W	Arctowski, King George Island, Antarctica
29	P13	Glaciofluvial sand	62° 10' 00" S	58° 27' 43" W	Arctowski, King George Island, Antarctica
30	P14	Glaciofluvial sand	62° 10' 00" S	58° 27' 46" W	Arctowski, King George Island, Antarctica

Table 5.4.4. Morphological and analytical data for pedons classified in Table 5.4.2.

Site	Horizons[+]	Depth or thickness (cm)	Color (moist)	RMF[1]	Texture[2]	$CaCO_3$ Equiv. (%)	pH ($CaCl_2$)	Org. C (%)	El. cond. (ms/cm)	Fe d[3] (mg/g)
22	O	0-8	10YR 2/1	–	–	–	–	–	–	–
	C*1	0-25	10YR 3/1	–	gSL	–	4.8	0.5	–	–
	BC*	0-40	2.5Y 4/2	–	gSL	–	4.9	0.5	–	–
	B*	0-20	10YR 4/6	–	L	–	4.5	0.7	–	–
	C*2	10-40	10YR 3/1	–	vgSL	–	5.2	0.4	–	–
	C*3	–	10YR 3/1	–	vgSL	–	4.6	0.4	–	–
Owl River 1	Ah*	1-30	10YR 3/3	–	LS	–	4.5	3.4	–	0.43
	Bm*	2-38	10YR 3/4	–	S	–	5.0	0.6	–	0.15
	BC*	12-25	10YR 4/4	–	S	–	5.2	0.2	–	–
	C*	25-60	10YR 5/2	–	S	–	4.9	0.1	–	–
12-90-T2	C*1	0-20	10YR 4/2	–	SL	33.7	6.8	1.7	–	–
	C*2	20-60	10YR 4/2	–	SL	31.2	7.3	0.2	–	–
	C*i	60-80	10YR 4/2	–	SL	–	–	–	–	–
2 NWT	H	6-3	2.5YR 2.5/2	–	h1	–	4.1	43.0	–	–
	H	3-0	5YR 2.5/1	–	h10	–	5.7	35.9	–	–
	B	0-10	10YR 4/3	–	SiL–SiCL	–	4.6	3.0	–	–
	B*	10-35	10YR 3/2	–	SiL	–	6.1	1.8	–	–
	BC*	35-65	10YR 4/3	–	SiL–SiCL	–	6.5	2.7	–	–
	Ah*	up to 5	2.5Y 2.6/0	–	–	–	6.8	9.5	–	–
	C*i	65-110	2.5 Y 4/2	–	SiL	3.3	7.1	1.5	–	–
	Ah*	up to 5	2.5Y 2.5/0	–	–	–	–	–	–	–
	H*I	4-8	10YR 2/1	–	–	–	–	–	–	–

Table 5.4.4. (cont.)

Site	Horizons[+]	Depth or thickness (cm)	Color (moist)	RMF[1]	Texture[2]	$CaCO_3$ Equiv. (%)	pH ($CaCl_2$)	Org. C (%)	El. cond. (ms/cm)	Fe d[3] (mg/g)
17	O	6-0	5YR 2/1	–	–	–	6.8	32.4	–	–
	A*1	0-11	5YR 2/1	–	Si	–	6.8	12.2	–	–
	A*2	11-44	10YR 5/4	–	SiL	19.0	7.1	5.8	–	–
	B*1	44-61	10YR 3/3	–	SiL	19.0	7.2	1.9	–	–
	B1	61-89	2.5Y 4/4	–	SiL–SL	19.0	7.3	0.5	–	–
	B2	89-117	2.5Y 5/4	–	SL	44.0	7.4	0.4	–	–
	B3	117-127+	2.5 Y 4/2	–	SL	44.0	7.4	0.5	–	–
11	Ah1	0-10	10YR 4/2	–	SiC	–	3.4	9.2	–	–
	Ah2	0-30	10YR 5/3	–	C	–	3.5	4.2	–	–
	B*1	0-15	10YR 5/6	–	SiC	–	3.5	1.7	–	–
	B*2	0-30	2.5Y 4/4	–	SiC	–	3.5	2.0	–	–
	B*3	5-15	7.5YR 3/4	–	SiC	–	3.3	1.0	–	–
	Bg*	0-20	2.5Y 4/2	mottles	SiC	–	3.2	0.8	–	–
	C*	0-25	2.5Y 4/3	–	SiC	–	3.3	1.0	–	–
32	B	0-5	10YR 3/2	–	SiL	–	4.4	1.6	–	–
	BCg*1	5-35	5Y 4.5/1	mottles	SiL	–	4.2	1.0	–	–
	BC9*2	35-60+	10YR 3/2	mottles	SiL	–	4.3	1.5	–	–
25a	H	5-0	10YR 2/1	–	h1	–	4.5	44.6	–	–
	Ah	0-5	10YR 3/3	–	SiL	–	7.0	9.1	–	–
	Bg*1	5-23	2.5Y 3/2	mottles	SiL	–	6.6	1.8	–	–
	Bg*2	23-37	2.5Y 3/2	mottles	vfSL	–	6.9	1.4	–	–
	Bg*	37-54	2.5Y 3/2	mottles	vfSL	–	6.9	1.3	–	–
	Ah	54-58	10YR 2/1	–	vfSL	–	7.7	2.5	–	–
	Bg*4	58-63	5Y 3/2	mottles	SiL	–	7.1	1.6	–	–
	Bg*5	63-90	5Y 3/2	mottles	SiL	–	6.8	2.5	–	–
	Bg*I6	90-105	5Y 3/1	mottles	vfSL	–	6.8	2.1	–	–

Table 5.4.4. (cont.)

Site	Horizons[+]	Depth or thickness (cm)	Color (moist)	RMF[1]	Texture[2]	$CaCO_3$ Equiv. (%)	pH ($CaCl_2$)	Org. C (%)	El. cond. (ms/cm)	Fe d[3] (mg/g)
1YT	H	0-6	–	–	h5	–	–	–	–	–
	B*	6-26	10YR 3/2	–	SiL	–	–	–	–	–
	Bg*	26-62	2.5YR 5/3	Fe conc.	SiL	–	–	–	–	–
	Bg*	62-73	2.5YR 3/2	Fe conc.	SiL	–	–	–	–	–
	O*	–	–	–	–	–	–	–	–	–
	BCg*	73-92	2.5YR 3/2	Fe conc.	–	–	–	–	–	–
	C*I	92+	2.5YR 4/1	–	–	–	–	–	–	–
9	H	0-20	5YR 3/3	–	h1	–	3.8	43.2	–	–
	H	0-5	10YR 3/2	–	h5	–	4.4	25.0	–	–
	Bg*	3-9	5Y3/1	red. conc.	SiCL	–	4.2	5.4	–	–
	Bg*	3-27	5Y 5/1	red. conc.	SiCL	–	4.4	15.5	–	–
	Ah*1	0-7	7.5YR 4/4	–	SiCL	–	4.3	8.2	–	–
	Ah*2	15-20	7.5YR 4/4	–	SiCL	–	3.8	5.1	–	–
	O*I	40-45	10YR 4/2	–	–	–	4.1	20.7	–	–

Table 5.4.4. (cont.)

Site	Horizons[+]	Depth or thickness (cm)	Color (moist)	RMF[1]	Texture[2]	$CaCO_3$ Equiv. (%)	pH ($CaCl_2$)	Org. C (%)	El. cond. (ms/cm)	Fe d[3] (mg/g)
12	O	0-8	10YR 2/2	–	–	–	3.7	24.6	–	–
	H	0-27	2.5YR 3/6	–	h1	–	3.4	44.7	–	–
	H	0-15	7.5YR 3/2	–	h10	–	4.0	20.3	–	–
	B*	0-20	10YR 3/2	–	C	–	4.2	4.2	–	–
	Bg*	0-30	2.5Y 3/2	red. conc.	C	–	4.5	3.0	–	–
	Bg*	0-12	2.5Y 3/2	red. conc.	C	–	4.3	3.4	–	–
	O*I	0-25	5YR 3/3	–	–	–	4.2	23.5	–	–
	Bcgi	5-20	5Y 4/1	red. conc.	SiC	–	4.1	8.1	–	–
	Ci	–	5Y 4/1	–	SiC	–	4.7	1.8	–	–
28b	H	11-0	10YR 2/2	–	h5	–	3.9	44.9	–	–
	C*i1	0-28	10YR 4/1	–	SL	–	4.4	1.8	–	–
	Ahi	28-47	10YR 3/1	–	SL	–	4.8	2.1	–	–
	Ci	47-80+	10YR 4/1	–	gSL	–	5.1	1.2	–	–
Rundle	Ah	5-11	7.5YR 4/4	–	LS	–	4.7	5.0	–	–
	B1	12-17	7.5YR 3/3	–	S	–	4.5	3.4	–	–
	B2	7-15	7.5YR 3/4	–	S	–	4.6	2.7	–	–
	Cg	19-25	2.5Y 3/2	mottles	S	–	4.6	1.1	–	–
12-90-T4	C1	0-30	2.5Y 3/2	–	LS	11.5	7.0	0.8	–	–
	C2	30-54	2.5Y 3/2	–	LS	11.6	7.1	0.8	–	–
	Ci	54-74	2.5Y 3/2	–	LS	–	–	–	–	–
12-89-83	H	0-5	–	–	h1	5.3	7.4	28.8	–	–
	Ahg	5-10	10YR 5/1	–	L	2.0	6.7	9.6	–	–
	Cg	10-28	2.5Y 3/2	–	SL	0.7	6.2	1.3	–	–
	Cgi	28+	2.5Y 3/2	–	SL	–	–	–	–	–

Table 5.4.4. (cont.)

Site	Horizons[+]	Depth or thickness (cm)	Color (moist)	RMF[1]	Texture[2]	$CaCO_3$ Equiv. (%)	pH ($CaCl_2$)	Org. C (%)	El. cond. (ms/cm)	Fe d[3] (mg/g)
12-89-08	B	0-26	10YR 4/2	–	SL	10.9	7.1	0.6	–	–
	C	26-52	10YR 3.5/1	–	LS	9.9	6.8	0.1	–	–
	Ci	52+	10YR 3.5/1	–	LS	–	–	–	–	–
94P066 8	A	0-7	5YR 2/2	–	SiL	–	4.6	7.5	–	–
	AB	7-12	10YR 4/2	–	SiL	–	4.9	0.6	–	–
	B1	12-38	10YR 3/3	–	SiL	–	5.6	0.6	–	–
	B2	38-59	10YR 3/2	–	SiL	–	6.1	0.5	–	–
	B3	59-78	2.5Y 3/2	–	SiL	–	6.3	0.4	–	–
	B4	78-94	2.5Y 4/2	–	SiL	–	6.4	0.4	–	–
	Ci	94-110	2.5Y 4/2	–	SiL	–	6.5	0.4	–	–
Landsc. 1-1	H	0-12	–	–	h1	–	–	–	–	–
	H	12-20	–	–	h5	–	–	–	–	–
	A1	20-26	–	–	SiL	–	–	–	–	–
	Bg*1	26-46	–	–	SiL	–	–	–	–	–
	Bg*2	46-60	–	–	SiL	–	–	–	–	–
	BC*	60-88	–	–	–	–	–	–	–	–
Landsc. 1-2	O	0-1	–	–	–	–	–	–	–	–
	Bg	1-14	–	–	SiL	–	–	–	–	–
	Bg	14-127	–	–	SiL	–	–	–	–	–
1	H	0-35	2.5YR 2/5	–	h5	–	4.2	47.1	–	–
	Hi	35-39	7.5YR 4/4	–	h1	–	4.7	36.0	–	–
	Hi	39-110	5YR 3/3	–	h5	–	5.0	45.3	–	–
	Hi	110-175	5YR 2.5/2	–	h5	–	5.3	42.9	–	–
	Hi	175-240	10YR 2/1	–	h5	–	4.9	41.4	–	–
	Hi	240-265	10YR 2/1	–	h5	–	4.5	25.3	–	–
	IIAhi	265-300	–	–	SiL	–	4.1	18.8	–	–

Table 5.4.4. (cont.)

Site	Horizons+	Depth or thickness (cm)	Color (moist)	RMF[1]	Texture[2]	$CaCO_3$ Equiv. (%)	pH ($CaCl_2$)	Org. C (%)	EL cond. (ms/cm)	Fe d[3] (mg/g)
148	H1	0-1	5Y 6/1	oxidized	h1	0	6.0	27	1.8	–
	H2	1-8	5GY 5/1	reduced	h1–2	0	5.3	26	4.9	–
	H3	8-18	5GY 5/1	reduced	x4 gr3 h2	0	4.9	16	0.6	–
	R	>18	–	–	–	–	–	–	–	–
137	LO	0-3	–	–	h1	0	4.7	29	0.7	–
	O1	3-8	5YR 2.5/1.5	–	h3–5	0	3.9	20	0.6	–
	O2	8-16	5YR 2.5/5	–	h8–9	0	3.8	19	0.2	–
	OC	16-28	5YR 2.5/5	–	x4 h9	0	3.8	16	0.1	–
	R	>28	–	–	–	–	–	–	–	–
202	Ax	0-1	2.5Y 4/2	–	x3 gr3 SL	0	4.6	0.2	1.1	3.0
	AC	1-30	2.5Y 4/2	–	x3 gr3 SL	0	4.4	0.2	0.5	3.0
	Cfw	30-70	2.5Y 5/2	–	x4 gr3 LS	0	4.1	0.4	0.6	2.6
2	Azh	0-0.5	2.5Y 5/2	–	gr3 SL	0	4.4	1.5	1.9	3.5
	AC	0.5-5	2.5Y 6/4	–	gr2 CL	0	4.4	1.0	0.5	8.8
	Cg1	5-20	7.5YR 5/8; GY 6/1	mottles	gr2 CL	0	4.3	1.4	0.3	7.0
	Cg2	20-30	7.5YR 5/8; GY 6/1	mottles	gr2 C	0	4.4	0.6	0.2	15
25	AE	0-1	10YR 5/2	–	gr3 S	0	5.0	2.9	7.8	1.2
	Bh	1-5	5YR 3/1.5	–	gr3 S	0	4.2	2.1	0.4	2.9
	Bhs	5-26	10YR 3.5/3	–	gr3 SL	0	4.4	2.9	0.2	4.1

References

Anonymous. 1997. Guide to scientific soil excursion. Institute of Biology, Komi Scientific Centre, Ural Branch, Russian Academy of Sciences Pub. No. 33. 23 pp.

Blume, H.-P., L. Beyer, M. Boelter, H. Erlenheuser, E. Kalk, S. Kneesch, U. Pfisterer, and D. Schneider. 1997. Pedogenic zonation in soils of the southern circumpolar region. Adv. GeoEcology. 30:69-90.

Broll, G., G. Müller, and C. Tarnocai. 1998. Permafrost-affected soils in the Pangnirtung Pass area, Baffin Island, Canada. In: A.G. Lewkowicz and M. Allard, eds. Proceedings, Seventh International Conference on Permafrost. Centre d'études nordiques, Université Laval. Québec. pp. 89-94.

FAO-UNESCO-ISRIC. 1989. Soil map of the world. Revised legend. Technical Paper 20. International Soil Reference and Information Centre (ISRIC). Wageningen. Reprint of World Soil Resources Report 60. FAO. Rome.

ISSS Working Group RB. 1998. World Reference Base for Soil Resources. International Society of Soil Science (ISSS). International Soil Reference and Information Centre (ISRIC) and Food and Agriculture Organization of the United Nations (FAO). World Soil Resources Report 84. FAO. Rome. 91 pp. (http://www.fao.org/WHICENT/FAOINFO/AGRICULT/AGL/agll/wrb/wrbhome.htm)

Kuhn, D. 1997. Genese, Ökologie und Soziologie einer Bodengesellschaft in einem Periglazialgebiet der King-George-Insel (West-Antarktis). Schriftenr. Inst. Pflanzenern., Bodenkunde Univ. Kiel. Nr. 40.

Makeev, O.V. 1997. On classification of soils in permafrost region. In: D. Gilichinsky and J. Kimble, eds. Cryosols in Classification Hierarchy. Russian Academy of Sciences. Nauka. Novosibirsk. pp. 30-33.

Mazhitova, A. 2003. Classification of Cryosols in Russia. In: Kimble, ed. Cryosols: Permafrost-Affected Soils. pp. 609-624. (This book.)

Ping, C.L., W. Lynn, and C.A.S. Smith. 1993. Redoximorphic features in permafrost soils. In: D.A. Gilichinsky, ed. Post-Seminar Proceedings of the Joint Russian-American Seminar on Cryopedology and Global Change, November 15-16, 1992, Pushchino, Russia. Russian Academy of Sciences. pp. 233-244

Soil Classification Working Group. 1998. The Canadian System of Soil Classification, 3rd ed. Research Branch, Agriculture and Agri-Food Canada. Ottawa. Pub. 1646. 187 pp.

Soil Survey Staff. 1998. Keys to Soil Taxonomy, 8th ed. USDA-NRCS. Washington, D.C. 326 pp.

Soil Survey Staff. 1999. Soil Taxonomy, A basic system of soil classification of making and interpreting soil surveys, 2nd ed. USDA-NRCS. Ag. Hndbk. 436. Washington, D.C. 869 pp.

Spaargaren, O.C. 1994. World reference base of soil resources (Draft). International Society of Soil Science (ISSS), International Soil Reference and Information Centre (ISRIC) and Food and Agriculture Organization of the United Nations (FAO). Wageningen/Rome. 161 pp.

Tarnocai, C., D. Kroetsch, J. Gould, and H. Veldhuis. 1991. Soils, Vegetation and Trafficability, Tanquary Fiord and Lake Hazen areas, Ellesmere Island National Park Reserve. Canadian Parks Service, Prairie and Northern Region. Winnipeg, Man. Report RM91-1/ELS. 269 pp.

Tarnocai, C., C.A.S. Smith, and C.A. Fox. 1993. International Tour of Permafrost Affected Soils: The Yukon and Northwest Territories of Canada. Centre for Land and Biological Resources Research, Research Branch, Agriculture Canada. Ottawa. 197 pp.

Tedrow, J.C.F., and J.E. Cantlon. 1958. Concepts of soil formation and classification in Arctic regions. Arctic. 11:166-179.

Section 6. Management and Use of Cryosols

Management and Use of Cryosols: Introduction

I.B. Campbell[1]

[1]Research Scientist, Land and Soil Consultancy Services, 23 View Mount, Nelson 7001, New Zealand; e-mail iaincampbell@xtra.co.nz.

In recent decades, Cryosols of both the Arctic and Antarctic regions have been subjected to increasing anthropogenic activities originating from both local and lower-latitude, global actions. These regions are among the last relatively pristine areas remaining on our planet. Like all other areas, human actions threaten them.

Measurements have shown that atmospheric processes and ocean currents transport pollutants originating from mid latitudes to the Polar regions, more so to the Arctic than to the Antarctic region. Increasing levels of activities in the polar regions also have added materially to the degree of soil and environmental disturbance, while future effects on polar ecosystems from global warming and depletion of stratospheric ozone are expected to affect seasonal distribution of snow cover and sea and ground ice, as well as soils and associated ecosystems.

In recent times, awareness of the need for improved management of the Polar regions has increased. Eight countries, including Canada, Denmark, Finland, Iceland, Norway, Russia, Sweden, and the USA, agreed in 1991 to develop an Arctic monitoring and assessment programme (AMAP). The Antarctic Treaty established procedures that now largely govern activities in the Antarctic.

The contributions in this section's three chapters provide a useful insight into the extent and diversity of soil and land management issues for Cryosols of the polar regions. They demonstrate how, with increasing latitude and progressively colder conditions, the opportunities for use diminish, while the need for appropriate management grows. Greater knowledge and understanding of Cryosols will undoubtedly be the key to future management and use.

Chapter 1. Agricultural Use of Tundra Soils in the Vorkuta Area, Northeast European Russia

I. Archegova[1], N. Kotelina[2], and G. Mazhitova[3]

[1]Professor, Institute of Biology, Komi Science Center, Russian Academy of Sciences, 28 Kommunisticheskaya St., Syktyvkar, 167982 Russia; e-mail archegova@ib.komisc.ru.
[2]Senior Researcher, Institute of Biology, Komi Science Center, Russian Academy of Sciences, 28 Kommunisticheskaya St., Syktyvkar, 167982 Russia; e-mail archegova@ib.komisc.ru.
[3]Senior Researcher, Institute of Biology, Komi Science Center, Ural Division, Russian Academy of Sciences, 28 Kommunisticheskaya St., Syktyvkar, 167982 Russia; e-mail mazhitova@ib.komisc.ru.

1. Introduction

There are two centers of forage production in European Russian tundra, one near Murmansk on the Kola Peninsula and the other around Vorkuta, in the northeast of European Russia. Tundra on the Kola peninsula has no permafrost but around Vorkuta has massive-island permafrost. Vorkuta is a center of a large coal basin where coal has been developed since the 1930s. During the Soviet era, agricultural development of tundra lands was considered expedient around large towns (Vorkuta had pop. 101,000 in 1981; Murmansk, 394,000 in 1982) to supply them with food. Agriculture there included dairy husbandry, forage production, and greenhouse vegetable production.

Agronomists who were Gulag prisoners began research into the development of tundra soils in the 1940s, and the Institute of Biology, Komi Science Center, Russian Academy of Sciences, continued it since the late 1950s. In a relatively short period, the Institute found an expedient approach to tundra agriculture, establishing artificial meadows sown with perennials. Proper selection of native plants, along with proper management techniques during meadow establishment and maintenance, provides meadow stability, so that the meadow can function indefinitely (Khantimer, 1974; Archegova and Kotelina, 1979, 1983; Archegova et al., 1988). Cultivation of an annual pea-oat mixture widely adopted in the taiga zone also was tested in tundra but could not provide stable yields.

This chapter analyzes changes in tundra soil properties following these soils' development, by comparing virgin soils to those of annual and perennial agrophytocenoses.

2. Study Area

2.1. Landforms

The study area is a rolling plain with bogged depressions. It also has some hills, composed of bedrock and overlain with Quaternary deposits of varying thickness. All across the area, the uppermost layer of these deposits is a rather uniform covering of loess-like silty loam which serves as a soil-forming material and varies in thickness from 0.5 to 5 m, depending on the landform.

2.2. Climate

Mean annual air temperature in Vorkuta is -6.3°C. Snow cover persists from mid-October to early June. The mean duration of the season with an air temperature higher than 0°C is 125 days; higher than 5°C, 90 days; and higher than 10°C, 43 days. Mean annual precipitation is 550 mm, and the large share of it falls in late summer. Evaporation does not compensate for the precipitation.

The area belongs to the zone of massive-island permafrost (Ershov, 1988). Open and close taliks cover about 50% of the area. Permafrost temperatures vary around -1°C. Permafrost thickness varies greatly, from 10 to 150 m. Both an active layer and a seasonally frozen layer average 1.5 to 2.0 m thick, yet the shallowest active layer in mineral soils may be only 60 to 80 cm thick. Frost heave and thermokarst are characteristic.

2.3. Landscape

The area belongs to the southern tundra subzone characterized by the distribution of relatively high shrubs – birch and willow. Shrub communities cover about 60% of the watersheds. Two landscape types are commonly chosen for forage growth, though they do not cover the whole landscape diversity.

Landscape 1 is on nearly level hilltops and long gentle slopes. Shrubby birch *(Betula nana)*/willow (*Salix spp*.)/moss community is characteristic. Shrubs are 60 to 100 cm high. Also, dwarf shrubs like cowberry (*Vaccinium vitis-idaea*) and blueberry (*Vaccinium uliginosum*), as well as sparse grasses and sedges, occur in these communities. Green mosses and *Politrichum* dominate as ground cover. Organic hummocks occur. The sites are permafrost-free or with a deep position of permafrost table (more than 150 cm).

Landscape 2 is on convex landforms and some southeast facing slopes, where snow cover is shallow and winter conditions are most severe, resulting in a shallow (80 to 150 cm) active layer. Shrubby birch/moss/lichen communities are typical, sometimes with significant *Ledum* share. Birch is 30 to 40 cm high. Patterned grounds are well developed and include both earth hummocks and tundra circles (bare areas).

3. Methods

3.1. Field descriptions

Since the start of agricultural development of the tundra in the late 1950s, the Institute of Biology has monitored agrophytocenoces and their soils. The monitored lands are the property of two state farms belonging to the Vorkutaugol company. Two main monitoring locations are on the gentle slope of the Vorkuta River valley and at a Nerusovey-Musur hill (67°31' N 64°03 to 08'E).

During the first 10 years, the study was conducted annually. Later field observations were repeated at varying intervals but not rarer than once every three years. Studies monitored properties of the soils of artificial perennial meadows and of the pea-oat field (annual cenosis), their productivity, and the herbage composition of the meadows. At the same time, virgin soils were studied.

Monitored agrophytocenoses were under continuous economic use. Permanent transects were established, three in artificial meadows and three in the pea-oat field. Observations and sampling were conducted along the transects two or three times in a growing season, before the herbage harvesting in late July.

3.2. Geobotanical samples

Geobotanical descriptions were made, and herbage samples were taken to determine the yield separately for each species. Herbage was cut for this purpose in the plots 50x50 cm, at 3 cm above soil surface, in 16 replications per transect.

To determine root weight, soil monoliths were taken periodically (once in 3 to 5 years) from the upper 10- to 15-cm soil layer, in three to five stations per transect. The roots were washed out and passed through a round-hole 0.25-mm sieve; living and dead roots were treated separately.

3.3. Soil samples

Along the same transects, soil samples were taken in three to five stations per transect. Samples were taken from a cultivation-affected layer at a depth of 0 to 5 and 5 to 10 cm. Periodically, samples were taken from a deeper layer, down to a depth of 30 cm. Virgin soils were studied at Nerusovey-Musur in 2 permanent sites, representing the two landscape types described above. Four soil profiles per landscape were chosen, representing components of patterned ground (an earth hummock, a bare area, etc.). Sampling periodicity provided for characterization of interannual variability of the indexes.

All soil samples (from both virgin and cultivated soils) were analyzed for organic carbon, hydrolyzable nitrogen, available P and K, exchangeable Ca and Mg, and hydrolytic acidity, according to Russian standards (Frolov, 1975). A <1-mm fraction was used for most analyses, which is different from the western standards. The analyzed soils, however, practically do not contain a 1- to 2-mm fraction, which allows one to neglect the difference.

Organic carbon was determined by acid-dichromate digestion. Coefficient 1.72 was used to present data as humus content. Hydrolyzable *N* was measured by 0.5 *N* H_2SO_4 extraction, available P, by 0.2 *N* HCL extraction, available K, by atomic absorption, exchangeable Ca and Mg, by 1 *N* NH_4Cl, and hydrolytic acidity, by 1 *N* CH_3COONa extraction.

3.4. Water and temperature regimes

Soil water and temperature regimes of two virgin soils and two cultivated ones (under an artificial meadow and pea-oat field) were studied at Nerusovei-Musur in 1974-1980 (Kononenko, 1986). Mostly summer indexes were measured. Recently (from 1996 to the present) Mazhitova obtained new material on the soil moisture and thermal regime in the same area. Stress has been laid on year-round measurements. Miniature ONSET loggers were used to record temperature. Soil degree-days (SDD) were calculated by summarizing corresponding mean daily temperatures.

4. Characterization of Virgin Tundra Soils

Most of the important tundra soil properties are explainable from the soil's long preservation in a frozen state, slow thaw under heat-isolating moss cover, and short cool summers. Combined with low evaporation, these conditions cause saturation of the upper mineral soil layer with water for rather long periods, and thus the development of gley and thixotropy. The soil classification is:

Landscape Type 1 – Shallow-peatyfied Surface-gleyed Gleyzem (Klassifikatsiya pochv Rossii, 1997); Dystric Gleysol (FAO, 1998); Typic Cryaquept (USDA, 1998).

Landscape Type 2 – Shallow-peatyfied Surface-gleyed Gleyzem; Gleyi-Turbic Cryosol or Distry-Gelic Gleysols; Typic or Ruptic-Histic Aquiturbels (same sources, respectively).

A typical soil profile contains the horizons Oi-A-Bg-B-BC in Landscape 1 and Oi-A-Bgjj-Bjj-BCf in Landscape 2.

The Oi horizon thickness varies from 1 cm under bare areas (tundra circles) to 10 to 25 cm under the other components of patterned grounds. Its maximum thickness is under organic hummocks.

The A horizon is 2 to 5 cm thick. Humus accumulation is due to cryogenic coagulation (Archegova, 1972).

The Bg horizon when frozen has a network of ice microveins. This structureless thixotropic horizon, which is saturated for most of the time, is practically waterproof (Kononenko, 1986). Its thickness varies from 15 to 25 cm.

The B and BC horizons are morphologically non-gleyed or weakly gleyed and have well developed fine to medium angular structure or, in the lower part of the BC horizon, a medium prismatic one.

Morphological features of the soils correspond to nutrient distribution in their profiles (Table 6.1.1). Available N, P, and K and exchangeable Ca and Mg accumulate in the organic horizon and decrease in content in the Bg horizon. An increase in P and K farther down relates to the parent material's composition. Soil organic matter mainly occurs in a form of weakly decomposed plant residues in the organic horizon. Variations in nutrient content depend on the degree of the residue's decomposition.

Table 6.1.1. Selected chemical properties of virgin tundra soils.

Horizon	**Depth, cm**	**pH (H_2O)**	**Hydrolyt. acidity**	**Exchangeable bases**		**% humus**	**Available**		
				Ca	Mg		N	P	K
			cmol/kg soil				mg 100/g soil		
Landscape 1. Soil of an organic hummock.									
Oi	0-24	5.5	54.8	45.5	7.5	-	15.8	23.8	174.2
A	24-26	4.7	40.9	3.44	2.41	14.7	4.6	4.1	13.9
Bg	26-36	5.0	9.5	1.0	0.8	1.2	3.7	7.4	3.5
Bg	36-46	5.2	10.1	1.4	1.1	1.6	2.3	6.7	3.9
B	46-52	5.3	8.6	2.5	1.8	0.8	1.4	10.1	5.9
B	52-60	5.3	6.8	5.5	1.7	0.6	-	10.7	10.1
Landscape 2. Soil of an earth hummock.									
Oi	0-9	5.2	58.5	43.7	7.0	-	49.9	16.9	141.3
A	9-14	5.0	13.7	6.1	4.0	4.4	4.6	5.5	14.9
Bg	14-26	5.6	4.2	5.0	3.0	1.1	1.1	1.7	8.7
Bg	26-36	5.5	16.8	5.1	4.5	1.1	1.7	12.9	9.6
B	36-46	5.3	6.9	3.6	3.9	1.0	1.9	8.9	8.9
BC	46-56	5.4	6.5	4.9	2.8	0.8	1.7	10.8	9.6
BC	56-66	5.7	4.2	6.9	4.4	0.6	1.7	16.9	15.0
Landscape 2. Soil of a bare area.									
Oa	0-1.5	6.4	6.2	21.5	10.4	12.8	6.1	4.7	18.9
Bg	1.5-10	5.7	6.1	7.2	4.5	0.7	4.2	13.0	15.5
Bg	10-18	5.5	7.3	6.7	5.1	0.7	0.9	12.8	14.8
B1	18-30	5.7	6.9	8.5	5.1	0.6	1.7	12.9	17.2
B2	30-40	5.9	5.8	8.9	5.9	0.7	-	13.3	17.9
B2	40-50	6.1	4.4	8.9	6.7	0.6	2.9	15.1	17.8
BC	50-60	6.0	5.6	10.3	4.9	0.7	2.0	17.8	16.9

A boundary between organic and mineral soil layers is abrupt, and shrub roots do not go lower than a moss-peat layer, which is an obvious consequence of the unfavorable temperature and moisture regime of the mineral horizons. Shrubs have a widely spreading horizontal root system which pierces the organic layer, providing for the most efficient use of nutrients leached from plant residues. In the same layer, concentrated substrate microorganisms rework the residues. Thus the organic layer bears the soil's fertility. The biocenosis is highly autonomic and weakly connected with a mineral soil.

5. Cultivation of Tundra Soils

Cultivation starts with removing tundra vegetation and establishing grasses uniform in species composition. It also removes the fertility-bearing organic layer of virgin soils. The thixotropic Bg horizon, least favorable for plant growth, provides material for a cultivated soil formation.

Cultivation of tundra areas outside floodplains began in the mid-1950s and intensified in the early 1960s, when the Institute of Biology found expedient agrotechnical methods. Soon after that, the area of cultivated lands near Vorkuta reached several thousand hectares.

Climate and soil features limited the use of traditional agricultural techniques, and proper selection of forage species was especially important. Slow thaw, a short snowless period, long periods of soil saturation, and thixotropy in the upper mineral horizon made it practically impossible to till the soils with the upper layer inversion. Besides, the tillage required preliminary removal of tundra vegetation. Therefore, a tractor harrow was adopted as a main instrument of cultivation.

5.1. Tillage methods

A cultivated area was harrowed several times in different directions to provide for the disintegration of shrubs up to 80 cm high and for partial mixing of their residues with the Bg horizon. However, more common was preliminary bulldozing to remove shrubs and the moss-peat layer.

Later, cultivation techniques improved with rototilling, which does not require preliminary removal of shrubs up to 1.5 m. Repeated rototilling provided for the disintegration of mosses and other plants and mixed their remnants with a mineral layer, and it reduced the time to prepare a soil before sowing to one or two years, in contrast with the first method's two or three years (Archegova, et al., 1988). Expenses for primary cultivation also were lower.

5.2. Fertilizing

During the primary cultivation of a virgin soil, application of 80 to 100 t/ha of manure and 8 to 10 t/ha of lime was recommended. At sowing, a complete inorganic fertilizer also was required, at a rate of 60 to 120 kg/ha of primary nutrient. Later on, plant growth was supported by the annual application of nitrogen fertilizer and, every two or three years, additional application of P and K, not less than 60 kg/ha of primary nutrients.

Figure 6.1.1 shows the yield's fluctuations, depending on fertilization. It is important that plant residues mixed with the thixotropic Bg horizon remain weakly decomposed for more than 18 years, as our observations show. This improves the water regime by providing additional aeration, but it does not increase the nutrient pool.

5.3. Plant selection

We already have stressed the importance of plant species' selection to growing for forage in tundra. Establishing artificial perennial meadows based on the selection of native plant species has proved to be efficient in tundra areas. Tests showed that *Poa pratensis* and *Alopecurus pratensis* from local native populations were the best under cultivation. These species are well adapted to the local climate and efficiently reproduce themselves both by seeds and vegetatively. The latter is especially important for establishing and supporting an artificial meadow.

5.4. Meadow stages

An artificial meadow forms in three stages. First, a "weedy" stage persists for one or two years. Second is a stage which establishes the meadow as a cultural ecosystem; by the end of this stage, which lasts about 10 years, a soil which corresponds to the plant community has formed. The third stage is the stable functioning of the artificial meadow (agrocenosis); under proper meadow exploitation, this stage lasts indefinitely. One of the first artificial meadows has functioned successfully for over 40 years so far (Kotelina et al., 1998).

Studies also have been conducted on additional native plant species to enrich the herbage. From the species selected, grass, grass-leguminous, and herb-grass-leguminous perennial agrocenoses have been established which are still functioning in economic terms.

5.5. Yields

In the taiga zone, annual species (oats and peas) are sown traditionally for forage but were proved not expedient in the tundra. The rhythm of their biological development does not correspond with tundra climatic conditions, and economically significant yields can be obtained only in years with favorable weather conditions. For example, from 1959 to 1982, the yield of annual plants reached 10 t/ha only five times, whereas perennial grasses varied from 1.2 to 1.5 t/ha for more than half of this period. Even though difficulties with supplying farms with perennial seeds forced some of the farms to sow oats, establishing an artificial meadow proved ecologically the best method for using the tundra's limited soil and climate resources.

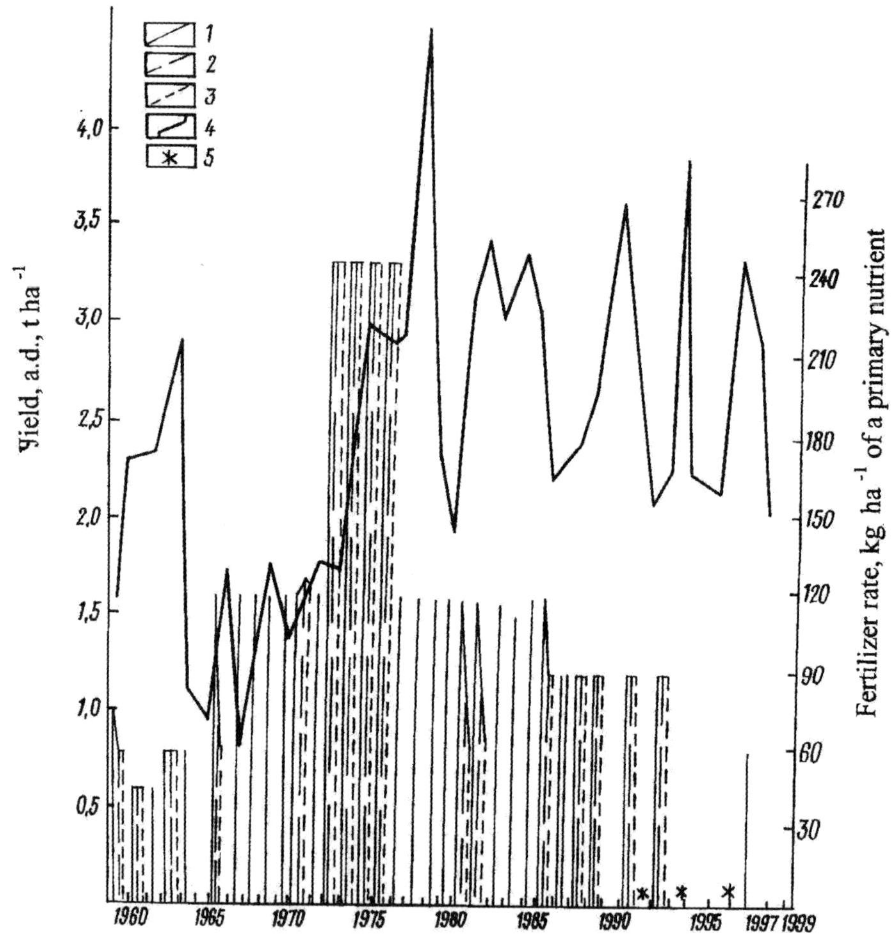

Figure 6.1.1. Yield dynamics in the artificial tundra meadow. Fertilizer rate: 1 - N, 2 - P, 3 - K, 4 - hay yield, 5 - years when manure has been applied.

6. Characterization of Cultivated Soils

Management, fertilizers, and grasses cause changes in soil properties. Soils under perennial and annual vegetation clearly differ.

6.1. Meadow Layer properties

Under an artificial meadow, in the third year after primary cultivation, the upper 10-cm soil layer was a silty loam, mottled due to alternation of reduced and oxidized areas, and with numerous coarse plant residues. In the sixth year, the cultivated layer was uniformly brown, and a roughly formed sod sublayer appeared, 3 cm thick. In the eighth year, this 3-cm sod was well formed and underlain by 7 cm of a humus-enriched layer with dense roots and residues of shrubs and mosses. A residual thixotropic Bg horizon and B horizon underlay this layer, and both of them preserved all features of virgin soils.

In the 17th and 37th years, the cultivated layer contained two horizons, a soddy one varying in thickness from 2 to 7 cm and an A horizon varying from 9 to 14 cm. Total depth of the newly formed layer ranged from 10 to 14 cm, which corresponded to the depth of cultivation. Deeper horizons repeated those of a virgin soil. Variations in the horizons' thickness reflected heterogeneity of the virgin soil.

A special study conducted in a chronosequence of meadows showed that the root density in a sod increased following differentiation of the cultivated layer into horizons. Root distribution corresponds with the horizonation. Thus, under a 38-year-old meadow, root weight in the 0- to 5-cm layer was 2.5 kg/m^2, whereas it was 1.1 kg/m^2 in the 5- to 10-cm layer.

6.2. Hydrology

Soil hydrological characteristics changed with changes in the cultivated layer properties. After 15 years from meadow establishment, total soil porosity reached 53 to 80%; at that level, pores of aeration totaled 39 to 66%. In virgin soils, pores of aeration in the thixotropic Bg horizon are practically absent or a total of 2 to 3% (Kononenko, 1986).

6.3. Nutrient content

Humus and nitrogen content in cultivated soils corresponds to biomass distribution within soil profiles. The thixotropic Bg horizon of a virgin soil has low humus content (see Table 6.1.1), the humus is of illuvial origin, and fulvic acids dominate its composition. Following cultivation, a humus-accumulative horizon develops with humic acids dominant. Nutrient content is affected by easy decomposition of grass residues, compared to the moss ones, but also by fertilizer application. Table

6.1.2 shows that nutrient content is maximum in a sod layer where the nutrients are taken from by roots, whereas, below there, their content decreases.

Thus, changes in a plant community and fertilizer application result in formation of a cultivated soil layer replacing partly the thixotropic Bg horizon. This layer contains a sod horizon typical for the soils of grasslands. It is important that formation of the horizon and transformation of a sown into a stable perennial community are completed simultaneously, i.e., that rhythms of the soil and biological components of the ecosystem coincide. The lower boundary of the agrophytocenosis corresponds to the depth of cultivation.

6.4. Classification

Soil classification after at least 10 years of cultivation is as follows: Haplic or Humic Gleysol, sometimes Gelic Gleysol (FAO, 1998); Typic or Humic Cryaquept, sometimes also Typic Aquiturbel (Soil Survey Staff, 1998).

The new Russian classification (Shishov and Dobrovolsky, 1997) does not provide a proper unit for classification of these soils because, at this stage, it does not consider the diversity of cultivated tundra soils. We propose including a Soddy subtype in the Surface-gleyed Agrozem type, the latter provided by the classification.

6.5. Annual plants

Soils under annual agrocenoses change in a different way. To support annual plants, a soil requires annual treatment, harrowing. The effect of biological factors is lower under an annual community than under the perennial one, but the effect of mechanical treatment and fertilizers is higher.

Soils of these communities also differ from those supporting perennial meadows. After 30 years of growing a pea-oat mixture, the soil had a cultivated Ap layer from 9 to 14 cm thick; this layer was loamy, dark gray, with very fine subangular structure and dense roots. It overlay a residual Bg horizon up to 17 cm thick, loamy, slightly thixotropic when saturated. The underlying B horizon, as in virgin soils, had fine to medium angular structure and was not gleyed and not thixotropic.

After 35 years of growing the pea-oat mixture, the soil had about the same profile. It had a 12- to 14-cm thick Ap horizon, dark gray, loose, with dense living and dead roots, remnants of an organic fertilizer, and an abrupt boundary with an underlying Bg horizon. The latter was hard, with rare roots, and represented a typical plough sole.

Table 6.1.2. Changes in chemical characteristics of the soil of an artificial meadow.

Horizon	Depth, cm	pH (H_2O)	% humus	Exchangeable		Available		
				Ca	Mg	N	P	K
				cmol/kg soil		mg/100g soil		
8th year after the meadow establishment (1965).								
A	0-10	6.1	6.9	14.8	2.4	13.6	6.2	18.3
Bg	10-20	5.6	1.1	4.4	1.5	9.3	3.9	3.9
19th year after the meadow establishment (1976).								
A (sod)	0-3	6.9	7.4	31.7	2.2	27.6	134.1	36.6
A	3-7	6.3	3.6	14.5	1.8	5.4	30.2	25.6
Bg	7-12	6.0	1.6	7.2	1.5	4.4	7.5	18.0
Bg	12-22	5.0	1.4	4.5	1.4	6.3	2.8	4.0
21st year after the meadow establishment (1978).								
A (sod)	0-5	6.3	5.5	8.2	1.2	11.5	50.3	18.9
A	5-10	6.2	3.2	5.9	1.4	8.1	20.5	9.8
Bg	10-15	6.3	2.8	5.6	1.5	8.3	18.5	11.4
30th year after the meadow establishment (1987).								
A (sod)	0-10	6.5	8.8	45.2	4.2	9.9	26.0	13.3
35th year after the meadow establishment (1992).								
A (sod)	0-5	6.4	18.9	-	-	18.2	509.5	44.6
A	5-10	5.9	5.8	-	-	6.1	216.2	17.7
39th year after the meadow establishment (1996).								
A (sod)	0-6	6.4	27.7	-	-	13.8	131.1	66.4
AB	6-17	5.9	16.3	-	-	4.4	50.8	23.6
Bg	17-27	5.6	2.9	-	-	3.7	4.1	8.6
BC	75-70	-	0.7	-	-	1.2	15.2	15.5

Chemical characteristics of a cultivated “arable” layer were conditioned by fertilizer application. Humus content in the Ap horizon was 5.5%, and available N, P, and K content was 16.0, 17.4, and 18.8 mg/100g soil, respectively. Irregular liming after the first years of cultivation caused a decrease in pH from 8.0 to 4.5, and that in exchangeable Ca, from 33.5 to 3.1 cmol/kg. Below the cultivated layer, nutrient and humus content dropped sharply.

Soil classification is as follows: Typical Surface-gleyed Agrogleyzem (Shishov and Dobrovolsky, 1997); Haplic or Humic Gleysol, sometimes Gelic (FAO, 1998); Typic or Humic Cryaquept, sometimes Typic Aquiturbel (Soil Survey Staff, 1998).

6.6. Changes in soil water and moisture regimes after cultivation

Shamanova (1970) is the only author who measured directly the depth of permafrost following tilling in the Vorkuta vicinity. A site tilled in July had a 195-cm active layer in October the same year, versus 160 cm in a virgin site located 150 m away. The experiment was conducted in Landscape Type 1. The same author reported that, in another site, also representing Landscape Type 1, six-year tilling resulted in a permafrost drop from 120 to 230 cm. However, the following should be taken into account.

Shamanova's research had been conducted in the period corresponding to the coldest part of a half-century natural climatic cycle. Data of comprehensive geocryological monitoring conducted in the same area by the Polar Ural Geology company show that the average depth of seasonal thaw increased sharply during the warming phase of the same cycle; in places, the permafrost table dropped down to several meters (Oberman, 1998).

Kononenko (1986) and Mazhitova did not find permafrost within at least 160 cm from the soil surface either under meadow and an arable site or in Landscape 1's virgin soils. Besides, they had no information on the depth of permafrost before cultivation on these sites. Thus, neither author could draw conclusions about the behavior of permafrost following cultivation. Kononenko's study was conducted in the coldest years of the climatic cycle and Mazhitova's, in its warmest years.

We also examined a number of meadows in various locations and found permafrost nowhere within 160 cm. Thus, either permafrost drops significantly (Landscape 2) or does not appear (Landscape 1) following cultivation. Still, we suppose that, several decades after establishing a meadow, thermal conditions may be in equilibrium and probably do not depend any more on the position of the permafrost table preceding cultivation.

Figure 6.1.2 shows a temperature regime of the soil of an artificial meadow in comparison with that of the soil of Landscape 1 and the coldest version of Landscape 2. Data for the period from September 1997 to October 1998 are presented as an example. The winter temperature regime of the meadow soil is very similar to that of Landscape 2, and the summer regime of this soil is closest to that of Landscape 1.

Table 6.1.3 shows some characteristics of the three soils. Snow depth is a principal index conditioning winter soil temperatures. According to Shamanova, values of about 50 cm separate permafrost-affected sites, on one hand, and permafrost-free sites or the sites with deep permafrost table position, on the other hand. Again, winter characteristics of the meadow soil (sums of negative temperatures and minimum temperatures but not the duration of cold period) are close to that of the soil of Landscape 2 (the "cold" one).

Summer characteristics of the meadow soil are even "warmer" than those of Landscape 1 (the "warm" one) (see, for example, sums of temperatures higher than 10°C). A wider range of variations in most indexes is characteristic for the meadow soil, compared to virgin soils. Mean annual soil temperature in the meadow is below zero, though permafrost (if it occurs) has a deep table position and seasonal freezing never reaches it. Negative MAST in the soils of close taliks can occur in the areas with island and discontinuous permafrost; for example, in Lapland (Raisanen et al., 1999).

Table 6.1.3. Selected characteristics of the temperature regime of virgin soils and a soil of artificial meadow (range of changes in 1996 through 1999).

Index	Artificial meadow	Landscape 1	Landscape 2
Snow depth, cm	40-50	78-112	35-45
MAST, °C, 50 cm	-0.4..-2.3	2.3*	-4.2*
Min temp., °C, 20 cm	-8.7..-16.0	-2.3...-3.3	-14.4..-15.3
SDD (<0°C), 20 cm	-577..-1529	-145..-187	-1406..-1650
SDD (>0°C), 20 cm	598-694	594-710	160-240
SDD(>10°C), 20 cm	0-82	0	0
Depth of seasonal freezing, cm	140-160	100-125	Not applicable
Depth of seasonal thaw, cm	Not applicable	Not applicable	63-67
Duration of period with temp. <0°C at 20 cm depth	234-248	143-231	220-250

* Data for only one year are available.

7. Conclusions

Forage can be grown in the tundra zone with massive-island permafrost. Grassing, i.e., establishing a meadow sown with perennials based on selection of native species, proved the most expedient method of forage production. Still, due to difficulties producing seeds of native species, annual plants (pea-oat mixture) traditional for the taiga zone continue to be grown in tundra.

Cultivating watershed tundra soils causes changes in the soils, depending on the type of agrophytocenosis. The most important changes are the formation of a soddy layer under perennial meadows and a ploughed horizon under annual communities. That is, given the same fertilizing, the difference between the soils depends on the type of agrophytocenosis (biological factor).

Soil hydrological and thermal regimes change into ones having no analogues among virgin soils. The soil classification reflects these changes.

Following cultivation, the spatial heterogeneity of a virgin soil disappears under the effect of management techniques and fertilizing, leading to a decrease in biological diversity.

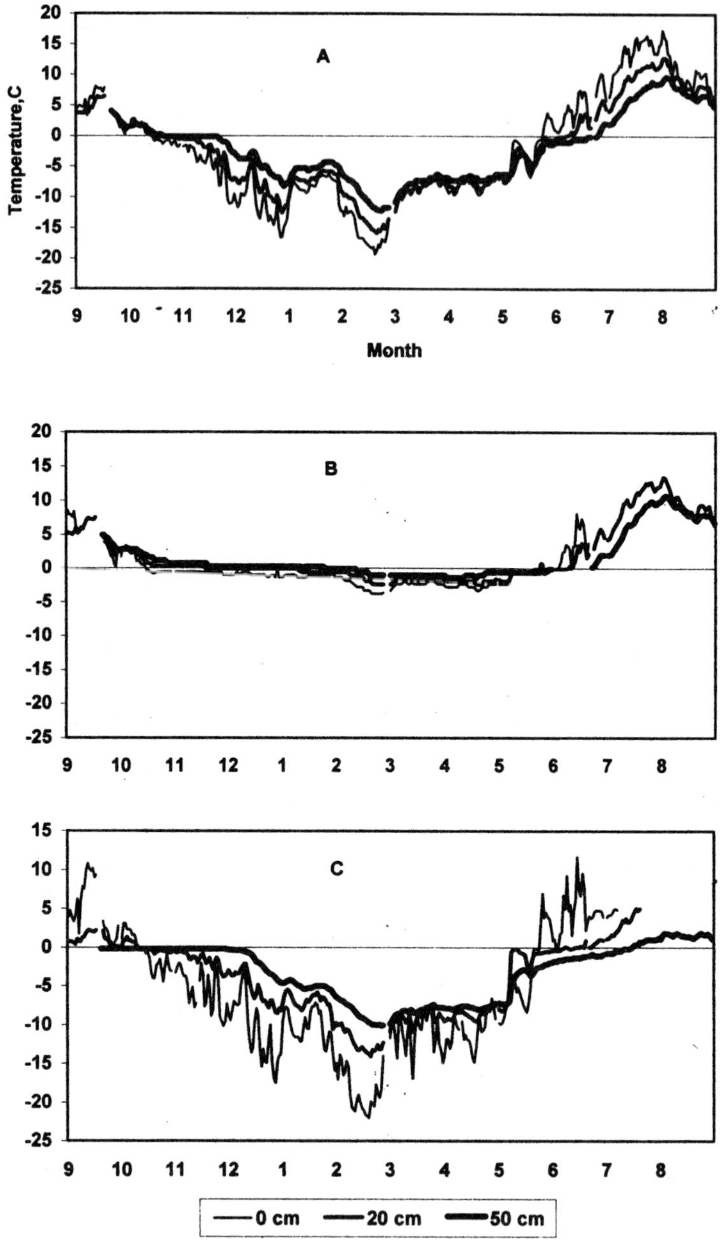

Figure 6.1.2. Temperature regime of a soil of artificial meadow in comparison with the regimes of virgin soils, 1997 through 1998. A – artificial meadow, B – landscape 1, C – landscape 2.

8. Acknowledgments

The Russian State Scientific Program, "Multidisciplinary research of oceans, seas, the Arctic and the Antarctic" (section 9.11.15), and a grant from the National Center for Scientific Programs (section 9.06.09, "Biological diversity") supported this study.

References

Archegova, I.B. 1972. Nature of humus-enriched horizon in loamy tundra soils in the northeast of European tundra (Harakter gumusirovannogo gorizonta v suglinistyh tundrovyh pochvah na severo-vostoke evropeiskoi tundry). Ecologiya,. 5:64-67. (In Russian.)

Archegova, I.B., and N.S. Kotelina. 1983. System of agriculture in the Komi Autonomy (Sistema vedeniya selskogo hoziaistva Komi ASSR). In: G.I. Gagiev, ed. Systema intensivnogo zemledeliya. Publ. of Komi Sci. Center. Syktyvkar. pp. 107-112. (In Russian.)

Archegova, I.B., N.A. Antonov, and A.F. Sharkov. 1988. Conditions and effectiveness of forage production in tundra (Usloviya i effectivnost kormoproizvodstva v tundre). Publ. of Komi Sci. Center. Syktyvkar. (In Russian.)

Archegova, I.B., and N.S. Kotelina, eds. 1979. Biogeocenological studies (Biogeotsenologicheskiye issledovaniya na seyanyh lugah v vostochnoevropeiskoi tundre). Nauka Publ. Leningrad. (In Russian.)

Ershov, E.D., ed. 1988. Geocryology of the USSR. European territory of the USSR (Geokriologiya SSSR. Evropeiskaya territoriya SSSR). Nedra Publ. Moscow. (In Russian.)

FAO. 1998. World Reference Base for Soil Resources. World Soil Resources Report 84. FAO. Rome.

Frolov, I.T., ed. 1975. Agrochemical research methods (Agrochimicheskiye metody issledovaniya). Nauka Publ. Moscow. (In Russian.)

Khantimer, I.S. 1974. Agricultural development of tundra (Selskohozyaistvennoie osvoenie tundry). Nauka Publ. Leningrad. (In Russian.)

Kononenko, A.B. 1986. Hydrothermal regime of taiga and tundra soils of the European northeast (Gidrotermicheskii rezhim taezhnyh i tundrivyh pochv evropeiskogo severo-vostoka). Nauka Publ. Leningrad. (In Russian.)

Kotelina, N.S. I.B. Archegova, G.G. Romanov, and L.P. Turubanova. 1998. Peculiarities of nature use and nature protection prospects in the Russian Far North (Osobennosti prirodopolzovaniya i perspectivy pripodovosstanovleniya na Krainem Severe Rossii). Publ. of Ural Division, Russ. Acad. Sci. Ekaterinburg. [In Russian]

Oberman, N.G. 1998. Permafrost and Cryogenic Processes in the Eats-European Surarctic. Eurasian Soil Sci. 5:540-550.

Shamanova, I.I. 1970. Effect of snow, vegetation and tillage on the ground temperatures in Vorkuta area (Vliyaniye snega, rastitelnogo pokrova i raspashki na teplovoie sostoianiye gruntov v raione Vorkuty). In: B.A. Tihomirov, ed. Biologicheskiye osnovy ispolzovaniya prirody Severa. Komi Publ. Syktyvkar. pp. 186-195. (In Russian.)

Shishov, L.L., and G.V. Dobrovolsky, eds. 1997. Russian soil classification (Klassifikatsiya pochv Rossii). Publ. of Dokuchaev Soil Sci. Inst. Moscow. (In Russian.)

Soil Survey Staff. Soil Taxonomy, 2nd ed. 1999. USDA, NRSC. Lincoln, NE. Ag. Hndbk # 436.

Räisänen, M.L., K. Löyttyjärvi, R. Pohjola, G. Broll, and G. Müller. 1999. Frost-affected soils in Finnish Lapland. Guide brochure, International excursion, August 24-31, 1999. Geological Survey of Finland, Munster University, Landscape Ecology.

Chapter 2. Disposal of Mine Tailings in Continuous Permafrost Areas: Environmental Aspects and Future Control Strategies

B. Elberling[1]

[1]Associate Professor, Institute of Geography, University of Copenhagen, Oster Voldgade 10, DK-1350 Copenhagen K., Denmark; e-mail be@geogr.ku.dk.

1. Introduction

The largest environmental concern facing the mining industry today is the occurrence of dissolved heavy metal contaminants and acidic effluents caused by sulfide oxidation (Paktunc, 1999), generally referred to as acid mine drainage (AMD). Numerous abandoned, operating, and proposed metal mines exist in permafrost-affected regions, and disposal of mine waste and AMD occur at several mine sites in the Arctic. Mine tailings result from the processing of ore and usually are transported to nearby disposal areas. Pollution in most cases spreads to the surroundings by acidic drainage to local watershed stream systems and marine environments.

Pyrite (FeS_2) is by far the most common sulfide mineral giving rise to AMD. The oxidation of pyrite is a complex biogeochemical process involving several redox reactions, hydrolysis and complex ion formation, solubility control, and microbial catalysis. Important aspects of this process are not understood clearly, but the following reaction commonly is given to describe the complete oxidation of the iron and sulfur moieties of pyrite:

$$FeS_{2(s)} + 3.75O_{2(g)} + 0.5H_2O \Rightarrow 2SO_4^{2-} + Fe^{3+} + H^+ \qquad (6.2.1)$$

In the reaction, oxidation forms ferric-iron and sulfuric acid, and heat is released. The above reaction generates 1 joule (J) per 7×10^{-7} moles of oxidized pyrite or about 12,000 kJ per kg pyrite (Dawson and Morin, 1996). Recent reviews on the process and controls are available (e.g., Evangelou, 1995).

It is well known that aerobic sulfide oxidizing bacteria can catalyze the oxidation of sulfide minerals, and numerous studies describe the catalyzing effects of chemolithotrophic bacteria such as *Thiobacillus ferrooxidans* (Evangelou, 1995; Nordstrom and Southam, 1997). These bacteria in particular can catalyze the oxidation of sulfide minerals (equation 6.2.1) under highly acidic conditions, thereby increasing the overall oxidation rate by up to six orders of magnitude, as compared to abiotic reactions (Nordstrom and Southam, 1997).

Besides the degree and type of microbial activity, these factors influence the generation of AMD: position of the sulfide body within the hydrological regime, types and abundances of sulfides as well as neutralizing gangue minerals, grain size distribution (specific surface area), climate, internal tailings' temperatures, and availability of oxygen.

The first part of this chapter discusses the influence of microbes, temperature, and oxygen on generation of AMD, as these factors play a unique role in permafrost regions. In addition, it considers depression of the freezing point, unfrozen water content, and redistribution of solutes, since these factors influence the environmental impact of AMD. The last part of the chapter focuses on literature reporting observed AMD in permafrost areas and discusses the current knowledge of strategies for controlling AMD.

2. Biotic Impact at Sub-Zero Temperatures

Within the permafrost region, soil temperatures are low for most of the year and temperature fluctuations can be extreme over short periods. Temperature is an important control on the physical-chemical reactions that are the basis for all biological processes. Thus, soil temperature is one of the most important parameters regulating the activity of microbes within the soil environment.

Bacteria adapted to live in cold conditions and to survive periods of extremely low temperatures have been known for more than 100 years (Forster, 1887). Since then, several surveys have documented the presence of viable bacteria in permafrost environments (e.g., Morita, 1975). Cold-adapted microbes are classified as either psychrophiles (cold-loving; optimal temperatures for growth below 15°C, active to below zero) or psychrotrophs (cold-tolerant; optimal temperatures for growth above 15°C, active to below zero).

One of the key assumptions about the presence of viable microbes at freezing temperatures is the presence of liquid water. In frozen soil, thin, intergranular water films allow chemical processes to be active even in winter and may represent survival niches for microbial life. The winter survival strategies for microorganisms in the Arctic are not understood well, but microbes are able to survive extremely low temperatures, and viable microorganisms occur in permafrost to depths of at least 18 meters and in material older than 70,000 years (Boyd and Boyd, 1964; Kjøller and Ødum, 1971).

2.1. Sulfide oxidizing bacteria in permafrost regions

A limited number of studies address the presence and potential activity of psychrophilic or psychrotrophic sulfide oxidizing bacteria in High Arctic environments. Kalin (1987) reported the presence of psychrophilic thiobacilli at the Nanisivik Mine, Baffin Island, Arctic Canada. Also at Nanisivik Mine, Elberling et al. (1999) detected psychrotrophic iron- and sulfur oxidizing bacteria including *Thiobacillus ferrooxidans* within alkaline tailings.

In the latter study, microcalorimetric measurements revealed that as much as 35% of the overall oxidation of tailings (measured as heat production) was due to the presence of the bacteria. Langdahl (1999) isolated a psychrotrophic acidophilic sulfide-oxidizing microbial population (dominated by *T. ferrooxidans*) from natural exposed sulfide ore material in North Greenland (83°N).

The presence of the well known sulfide oxidizing bacteria *T. ferrooxidans* in High Arctic environments indicates that these bacteria populations are likely at any location in the permafrost zone where reactive sulfide minerals are or have been present.

2.2. Temperature dependent biotic and abiotic sulphide oxidation

The Arrhenius Equation describes well how most biogeochemical reaction rates decrease with decreasing temperatures:

$$\ln (k_1/k_2) = E_a (T_1 - T_2) / RT_1T_2 \qquad (6.2.2)$$

where k_1 and k_2 are reaction rates at temperatures T_1 and T_2, E_a is the activation energy of the reaction, and R is the gas constant. Values of E_a for pyrite oxidation reported in the literature vary from 39 to 88 kJ/mol for a variety of experimental conditions (Nicholson et al., 1988).

Although the effects of temperature on sulfide oxidation rates can be complex, particularly under field conditions, several studies confirm that the simple relationship in the Arrhenius equation is a valid approximation for the chemical and the microbial reaction rates (Dawson and Morin, 1996). However, the Arrhenius equation considers only a single activation energy that is not temperature-dependent, whereas complex biological systems are presumed to have several temperature-dependent forms.

Figure 6.2.1 shows the temperature dependent biotic activity for sulfide oxidizing microbes collected from Arctic and temperate regions. The dashed line represents an equation fit (Jaynes et al., 1984) to relative activity measurements (X_t) of microbes collected at temperate latitudes and considered valid for the temperature range 4° to 55°C (T):

$$X_t = -1.2\times10^{-5}\,T^3 - 4.4\times10^{-4}\,T^2 + 0.066T - 0.25 \qquad (6.2.3)$$

The solid line represents the same equation fit to microbes collected in the Arctic. An overall difference in temperature optimum of 7 to 8°C occurs and indicates an important biological adaptation to cold climate.

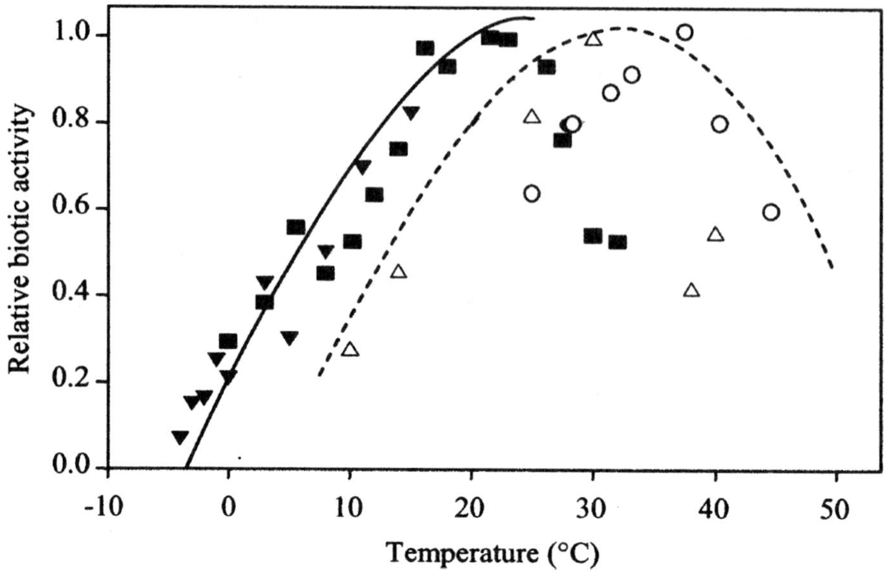

Figure 6.2.1. Relative changes in biological oxidation rates of sulfide minerals from temperate latitudes (open marks and dashed line) and Arctic latitudes (filled marks and solid line). Changes in oxidation are relative to the maximum oxidation observed for each species. Open marks represent data from Jaynes et al. (1984), filled squares are from Langdahl and Ingvorsen (1997), and filled triangles are from Elberling (2000).

Metabolic processes are active at temperatures as low as -10°C (Gilichinsky et al., 1992). However, no data are available on the growth of specific sulfide oxidizing bacteria at sub-zero temperatures. Langdahl and Ingvorsen (1997) studied the production rate of dissolved Fe^{2+} and Fe^{3+} of an enriched *T. ferrooxidans* culture from a naturally exposed sulfide ore in North Greenland. Based on the laboratory investigations, they concluded that the overall rate of microbial activity at 0°C constituted as much as 30% of the maximum activity at 21°C.

Oxygen uptake rates as a result of sulfide oxidation have been observed at temperatures as low as -4°C. These rates are based on the difference in oxygen uptake in ambient and sterilized tailing samples from Nanisivik Mine (Elberling, 2000). Results from that study indicate that both biotic and abiotic oxygen uptake in pyritic tailings drop with decreasing temperature (Figure 6.2.2).

The figure shows an Arrhenius relation, based on the activation energy and the maximum pyrite oxidation rate that Langdahl and Ingvorsen (1997) observed. The oxygen uptake rate was consistent with the Arrhenius relation in the temperature range above 0°C. Below 0°C, an increasing deviation occurs, and the linear correlation provides the best fit. Significant variation between the dependency of biological and chemical oxidation on temperature did not appear, indicating that the influence of temperature on the overall AMD is not sensitive to the percentage of biological oxidation.

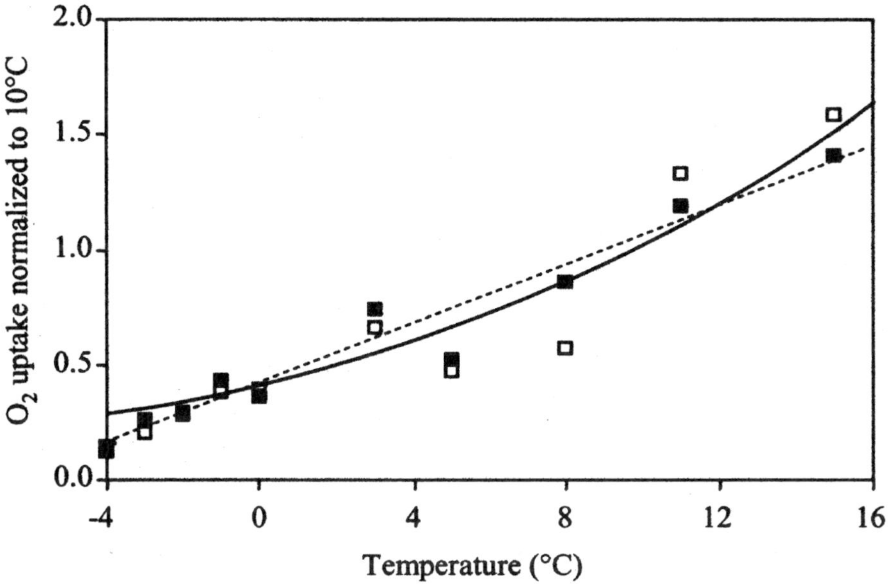

Figure 6.2.2. Relative oxygen uptake versus temperature in pyritic tailings from Nanisivik Mine. Biotic rates are shown as solid squares and chemical rates, as open squares. The solid line is generated from the Arrhenius equation, and the dashed line, from a linear regression equation (data from Elberling, 2000).

Using the same method to evaluate oxidation rates as mentioned above, Meldrum et al. (1999) studied the overall sulfide oxidation potential of tailings from Rankin Inlet, Arctic Canada, in freezer experiments at temperatures as low as -10°C. The authors conclude that oxygen consumption and heat generation due to oxidation are negligible at -10°C. The few studies mentioned above indicate that bacteria in natural systems to a large extent may be adapted to be active at sub-zero temperatures.

Although the observed bacterial activity indicates a significant reduction at 0°C compared to maximum activity, the results also suggest that biological sulfide oxidation may occur at temperatures as low as -4°C and that the activity becomes negligible as temperatures approach -10°C. Even though the kinetics of sulfide oxidation show significant potential AMD at sub-zero temperatures, sulfide oxidation will occur only if oxygen and liquid water are present. The next section therefore deals with the potential transport and redistribution of oxygen and water in frozen tailings.

3. Transport Phenomena in Frozen Ground

Frozen sediments consist of several phases, including solid particles, ice, unfrozen water, and gas. Within the frozen sediment, ions may be transported as a result of a concentration gradient (diffusion) or a pressure gradient (advection). Extreme differences in the diffusivity within the various phases play a major role in the control of the overall generation of AMD and spreading of contaminants.

Ion diffusion in pure monocrystalline ice is insignificant, as ion diffusion coefficients are about 5 orders of magnitude lower than in water (Barnall and Slotfeld-Ellington, 1983). For components available in the gas phase, e.g., oxygen, the diffusion coefficients are about 4 orders of magnitude higher in the gas phase than in water. Consequently, atmospheric oxygen will reach the depth where sulfide oxidation takes place primarily as a result of diffusion through the partially gas-filled pores (e.g., Pantelis and Ritchie, 1992; Elberling et al., 1993; 1994).

For waste material stored under nearly saturated or fully saturated conditions, the much lower oxygen diffusion in the water phase becomes important. The same is the case for contaminants from AMD, and even small quantities of unfrozen water may greatly enhance the overall contaminant migration in frozen grounds, by either ion diffusion in water or by water movement.

3.1. Transport of oxygen

As pyrite oxidizes in tailings, atmospheric oxygen is consumed, and the change in concentration with depth results in a concentration gradient, which is the driving force for the diffusive transport of oxygen. In tailing systems, one-dimensional molecular diffusion of oxygen gas can be described by Fick's First Law:

$$F_D = -D_e \, dC/dz \tag{6.2.4}$$

where F_D is the flux of O_2, D_e is the effective diffusion coefficient, and dC/dz is the vertical concentration gradient for oxygen. D_e is a function of the diffusion coefficient for oxygen in air and a factor that depends on the percentage of gas-filled pores (relative to the volumetric moisture content) and the tortuosity (Troeh et al., 1982).

Several models are available for estimating D_e but few are based on tailing material. Assuming that water-filled and air-filled pathways represent parallel resistance to diffusion in tailing material, Elberling and Nicholson (1996) showed that the effective diffusion coefficient (D_e) in tailings can be approximated well as the weighted sum of the diffusion coefficients in air and water (see dashed line in Figure 6.2.3).

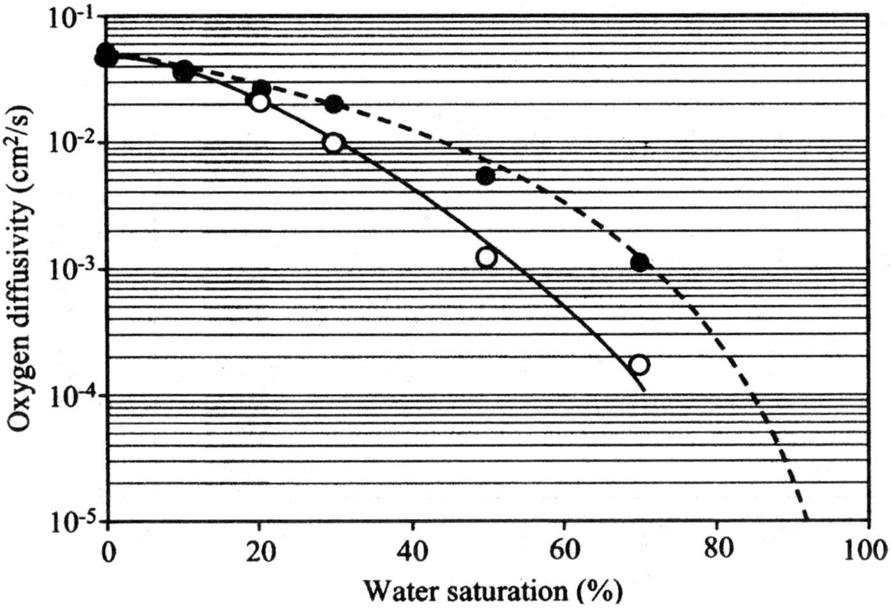

Figure 6.2.3. The effective diffusion coefficient (diffusivity) in non-frozen tailings (filled circles and dashed line) and frozen tailings (open circles and solid line) at water saturations up to 70%. The dashed line is an empirical equation for estimating D_e based on saturation values (Elberling and Nicholson, 1996), while the solid line is a free-drawn line representing the observed effect of freezing for comparison (modified from Elberling, 1998).

In frozen media, diffusion estimates require oxygen diffusion coefficients for super-cooled water as well as estimates on the overall change in gas pore volume fraction and connectivity during freezing. Such data presently are not available from the literature, and diffusion coefficients for oxygen gas in unfrozen and frozen tailings with variable water saturation are known only from experimental designs.

Based on measurements on unfrozen and frozen tailings from Nanisivik Mine (Figure 6.2.3; modified from Elberling, 1998), statistical tests suggest that freezing has no significant effect on D_e, in tailings with a water saturation less than 30%. Above 30% saturation, freezing lowers D_e, reaching a reduction factor of about 10 at 70% saturation.

Thus, freezing well drained, land-deposited tailings is not expected to decrease the overall availability of oxygen within tailings. The study points out that changes in the degree of saturation may be as important as freezing in permafrost areas and that a combination of saturation and freezing seems ideal for the control of oxygen availability within tailings deposited in the Arctic.

3.2. Unfrozen water

The amount of liquid water present in "partly" frozen soil and waste materials is critical in terms of controlling microbial activity and oxygen diffusion. It also will affect water and solute movement in frozen soils, which also influences soil strength, frost heaving, and contaminant transport.

Solutes lower the freezing point of water during freezing, and the relationship is given by the Van't Hoff equation. The freezing-point depression in dilute solutions can be calculated by using the following simplified equation (Marion, 1995):

$$\Delta T. \approx -1.86 \, v_{mb} \qquad (6.2.5)$$

where ΔT is the freezing-point depression, v is the number of aqueous species resulting from the dissolution of the solutes, and m_b is molarity of the solute. For pore water extracted from 6- to 7-year-old pyritic tailings in Nanisivik (unpublished data), concentrations of magnesium and sulfate alone result in a freezing point depression of about 1°, compared to the corresponding freezing point of pure water. However, deviations from equation 6.2.5 are apparent for high concentrations, and particularly for electrolytes that dissociate into ions of higher valance (e.g., $MgSO_4$).

A nearly pure ice phase will form when temperatures drop below the freezing point, concentrating solutes in the remaining unfrozen solution and thereby further decreasing the freezing point. If equilibrium is maintained between the solution and ice phases during freezing, then the concentrations of various ions in solution will follow the ice-solution equilibrium line, increasing in concentration as temperature decreases toward the eutectic composition (Banin and Anderson, 1974). For a pure $NaCl$-H_2O system with an initial NaCl molality greater than 5.17 mol/kg, the solution will not freeze until it reaches the eutectic temperature of -21.2°C; below this temperature, the solution will solidify as a mixture of ice and salt (Marion, 1995).

FREZCHEM can model the chemical equilibrium and the unfrozen water content during freezing (Mironenko et al., 1997), as this model includes the coefficients for the Pitzer-equations as well as solute redistribution during freezing. However, under natural conditions, and particularly for tailing water, site-specific characteristics will influence the actual ion exclusion and thereby the actual unfrozen water content. These characteristics, which the FREZCHEM model does not include, include texture, specific surface area, charged surfaces, freezing rate, and the initial distribution of salts (Marion, 1995).

Considering these limitations, FREZCHEM has been applied for simulating the observed unfrozen water content measured in columns filled with mine tailings from Nanisivik (Figure 6.2.4). Unfrozen water content was measured by time domain reflectometry (TDR), and tailings represent samples from the upper reactive zone and partly oxidized zone (20 cm), as well as the lower reduced zone near the permafrost (80 cm).

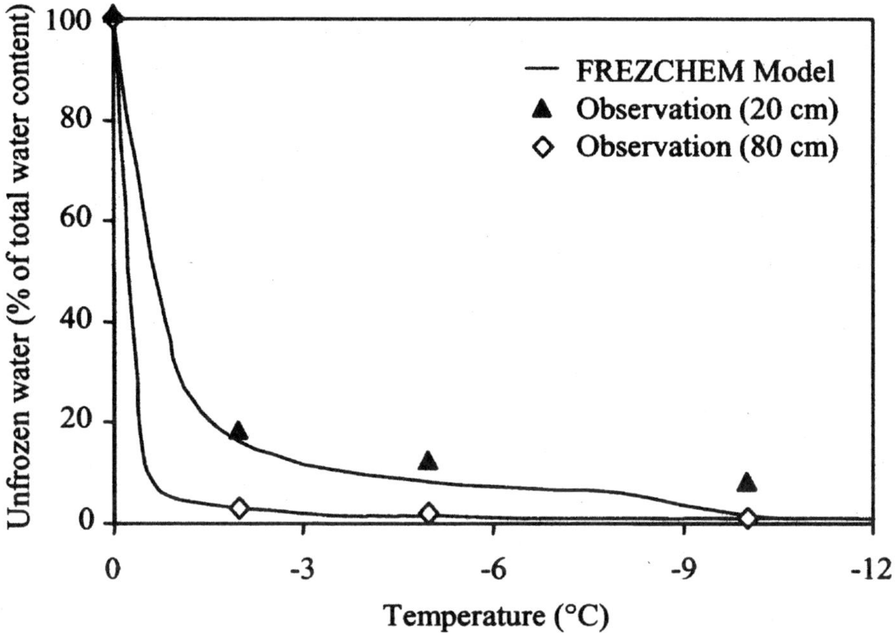

Figure 6.2.4. Unfrozen water content (% of total water content) in mine tailings collected in August from the upper oxidized zone (20 cm) and lower reduced zone (80 cm). Observations are compared to FREZCHEM simulations (Elberling, 2000).

The chemical composition of the pore water from the same site/depths has been used as input to FREZCHEM to predict the amount of unfrozen water content versus temperature (Elberling, 2000). The significant amount of unfrozen water, particularly within the reaction zone, suggests that liquid water is available for biological activity and contaminant transport well below 0°C.

3.3. Water movement in frozen soil

A potential environmental effect of unfrozen water is the diffusion of pollutants in the water phase and water movement within tailing profiles, the latter due to the fact that liquid films on particles provide a route for water and solutes in frozen tailings (Godwaldt et al., 1999). In general, water in frozen sediments moves from warm to cold regions, from regions with low concentration to regions with higher concentrations, and along water tension gradients (Marion, 1995). As tailings in permafrost areas freeze from the top down and from the permafrost table up, the thermal gradients induce a water flow toward the freezing fronts. Environmental factors such as water content, temperature, and snow cover determine the comparative importance of these two freezing fronts during the freezing process.

As solutes are largely excluded in the freezing process, maximum solute concentrations generally are close to the freezing front. The solute concentration in front of the freezing front may rise to 80 times the original concentration, if there is no solute inclusion in the frozen zone, and thereby cause the freezing front to leap over solute pockets (Hallet, 1978). These solute pockets ultimately may freeze, resulting in alternating bands of high and low ion concentrations in frozen tailings.

This potential ion exclusion in freezing sediment has received much attention and is shown to be controlled by temperature gradient (freezing rate), pore water composition, water content, mixing, and sediment type (e.g., Konrad and McCammon, 1990). The extent of ion exclusion increases with decreasing freezing rate (Marion, 1995). For freezing rates smaller than 0.1°C/day, corresponding to most field freezing conditions, more than 90% of the solutes have been observed to be rejected from the pore ice formed (Konrad and McCammon, 1990).

The ion exclusion may result in an increased environmental impact from AMD during spring where thawing starts from above and causes highly variable ion concentrations in the meltwater. Ion pulses (melt water with very high ion concentrations) are known from most Arctic areas (Melak and Sickman, 1995). Such pulses have a maximum impact on living organisms sensitive to toxic levels of heavy metals, for example, as well as on the overall flux of contaminants from tailing areas.

3.4. Soil strength

Frost heaving occurs as a result of the formation of segregated ice, for example, lenses and needles, as well as the volumetric expansion of water during freezing. Models are available to simulate and predict the mechanisms responsible for frost heaving (Chamberlain, 1981; Black and Hardenberg, 1991). Marion (1995) summarized the most important factors for the degree of frost heaving, including soil texture, pore size, freezing rate, temperature gradients, moisture, and overburden stress.

Frost heaving in tailings is considered a disadvantage, due to the formation of cracks and preferential pathways for infiltration of oxygen gas and oxygenated water. On the other hand, frost heaving seldom is observed in tailing areas, probably due to high ion concentrations that tend to reduce frost heaving. As Sheeran and Yong (1975) pointed out, this is expected because high ion concentrations increase the unfrozen water content and reduce the water available for forming segregated ice in tailings. Frost heaving may be important in cover materials provided as protection on top of tailings, and grain size and porosity should be considered in order to minimize frost heaving.

4. Acid Mine Drainage in Permafrost Regions

Annual temperature fluctuations in continuous permafrost regions cause the surface layer to thaw annually. The thaw depth, defined as the active layer, varies typically between <0.5 meter under thick organics to several meters under mineral soil. Pyrite oxidation occurs naturally in the active layer in permafrost areas where mining activity never has taken place.

Cameron (1977) studied oxidation of a massive sulfide body, named the *Agricola Lake prospect*, located northeast of Yellowknife, Canada. Apparently, intensive sulfide oxidation was occurring in the active layer both near the surface and at depth, and the lack of carbonate minerals in the area had resulted in natural acidic drainage (pH -3 to 4) as well as metal leaching.

Cameron (1979) observed the same phenomena for a massive sulfide deposit on Melville Peninsula in Northwest Territories, Canada. However, the presence of calcareous outcrops neutralized the acidic drainage and surface water. Newer observations in Yukon, Canada (Kwong, 1995), have confirmed active oxidation in permafrost areas in Canada.

Another example of a naturally exposed sulfide body is at Citronen Fjord in North Greenland, where the mean annual air temperature varies between -15 and -20°C and temperatures above zero are recorded only during a very short summer period. During this period, the actual sulfide oxidation rate has been evaluated based on *in-situ* oxygen uptake measurements made directly on well drained, ochre colored sediment below the exposed sulfide outcrops (Elberling and Langdahl, 1998). Oxygen uptake rates on the order of 0.2-0.4 moles/m^2 per day occurred during September, 1996. As organic carbon was absent, the amount of oxygen consumption was assumed to represent the overall rate of sulfide oxidation. The pore water and surface draining from the area were acidic and showed high concentrations of dissolved zinc and sulfate (Table 6.2.1).

The examples above illustrate that high sulfide oxidation rates occur naturally in continuous permafrost areas, at least during the short summer and, as a consequence, significant amounts of sulfate and heavy metal will appear in downstream aquatic systems. The environmental impact from natural exposed material is important, as the reactivity of such potential ore material can be used to provide minimum estimates of expected impact (potential generation of AMD) resulting from non-treated disposal of crushed ore material (tailings).

Table 6.2.1. Concentration (in ppm) of pore water and run-off from natural exposed sulfidic material in North Greenland (compiled from Elberling and Langdahl, 1998). The reference site is upstream from the area, with exposed sulfide outcrops.

	pH	SO_4	Fe	Zn	Pb	Cu	Ni	Cd
Pore water extracted from sulfidic material	1.39	66,780	53,700	6,150	35.8	0.6	3.1	2.4
Nearby surface runoff	2.54	2,300	195	139	1.8	0.04	0.4	0.5
Reference runoff	8.03	8	0.17	0.02	0.03	0.02	0.09	0.06

4.1. Oxidation of mine tailings deposited in continuous permafrost areas

The mining process produces different waste materials. Mine waste consists of large chunks of rocks or overburden which are transported by trucks or dragline to waste dumps. The ore is transported to the mill, where it is ground as part of ore processing, followed by extraction of the metals such as Zn, Pb, and Cu.

Mine tailings are the waste product resulting after metal extraction and consist of finely grained sulfidic ore with a reduced content of heavy metals. Mine tailings and waste rocks have very different hydrogeological, geochemical, and geotechnical characteristics, and therefore the two kind of materials must be considered separately in relation to AMD. This section evaluates only disposal of tailings; Dawson and Morin (1996) give a more thorough discussion of disposal of waste rocks in permafrost areas.

Tailings consist of ore material finer than 0.2 mm which results from the grinding procedure, which includes a large proportion of sulfide grains with a large fresh surface area. Results from ongoing investigations of oxygen uptake and release of heavy metals within pyritic tailings deposited near Nanisivik Mine on Baffin Island (NWT) in northern Canada (73°N) here illustrate the controlling parameters for generation of AMD in well drained tailings deposited in the continuous permafrost region (Elberling, 2000).

In Nanisivik, the annual mean temperature (1994 through 1998) is -15.6°C and the annual precipitation is approximately 125 mm. Since 1976, mine tailings have been deposited under water in a tailings impoundment. As the available subaqueous storage capacity was exhausted in 1991, disposal of tailings began on land. A test plot within this area was selected, and *in-situ* oxygen uptake rates have been estimated by changes in oxygen concentrations in closed gas chambers on top of the tailings (Elberling and Nicholson, 1996).

Measurements were made periodically throughout the year and were accompanied by climatic measurements. Figure 6.2.5 shows that, by the end of the summer, the active layer extended to a depth of approximately 2 meters and that this zone freezes during autumn. However, the freezing of the active layer varies significantly between years, depending on water content and snow accumulation that insulates the ground from low air temperatures.

During summer, oxygen uptake rates in 6-year-old well drained tailings were 0.2 to 0.5 moles O_2/m^2 per day (Elberling, 2000), which is of the same order of magnitude as rates observed in temperate climates (Elberling and Nicholson, 1996). During the following two-month period, air temperature dropped below 0°C, which resulted in significant effects on soil temperatures and the observed oxygen uptake (Figure 6.2.6).

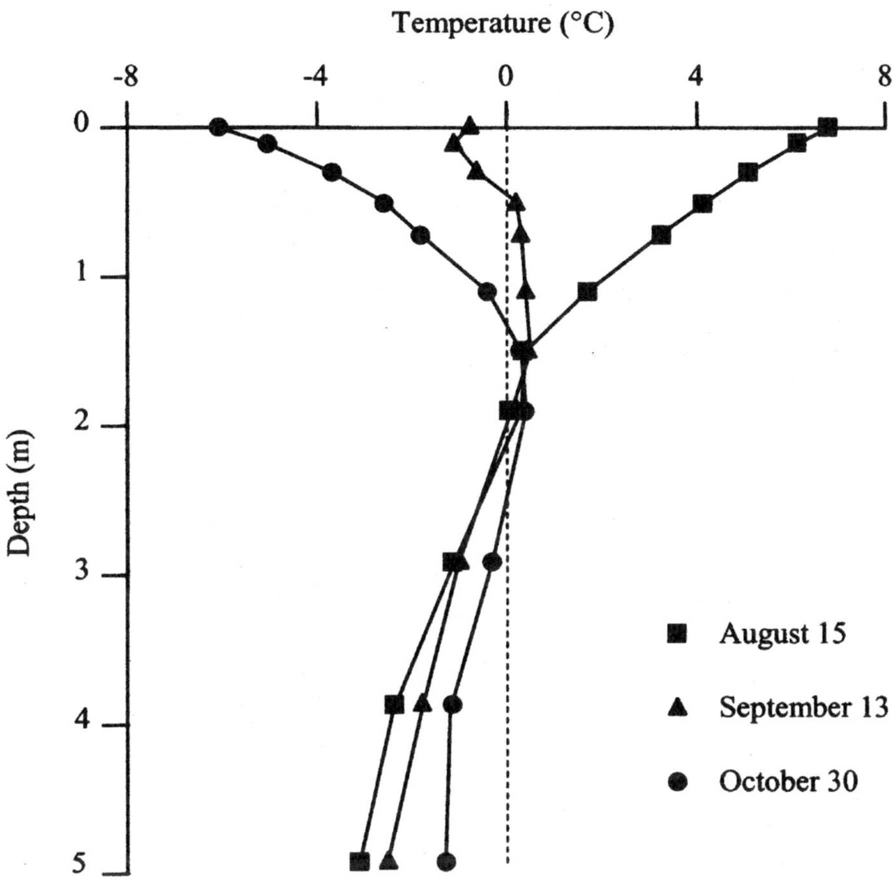

Figure 6.2.5. Soil temperature profile at an uncovered site within the tailings impoundment in Nanisivik. Freezing of tailings occurred during a period from August to October 1997. (Modified from Kyhn and Elberling, 2001).

Soil temperatures were used to model temporal variation in the reaction constant k (equation 6.2.2) and, subsequently, the expected variation in oxygen uptake over time, assuming that the diffusion coefficient within the reaction zone remained constant (Figure 6.2.6). The model failed to predict the oxygen uptake during the first days following major precipitation on Julian day 216 (11 mm over less than 10 hours). However, a site covered with plates in order to avoid infiltration by rain water but to allow free oxygen diffusion made it possible to differentiate between the effect of temperature and increasing water content on the overall oxygen uptake.

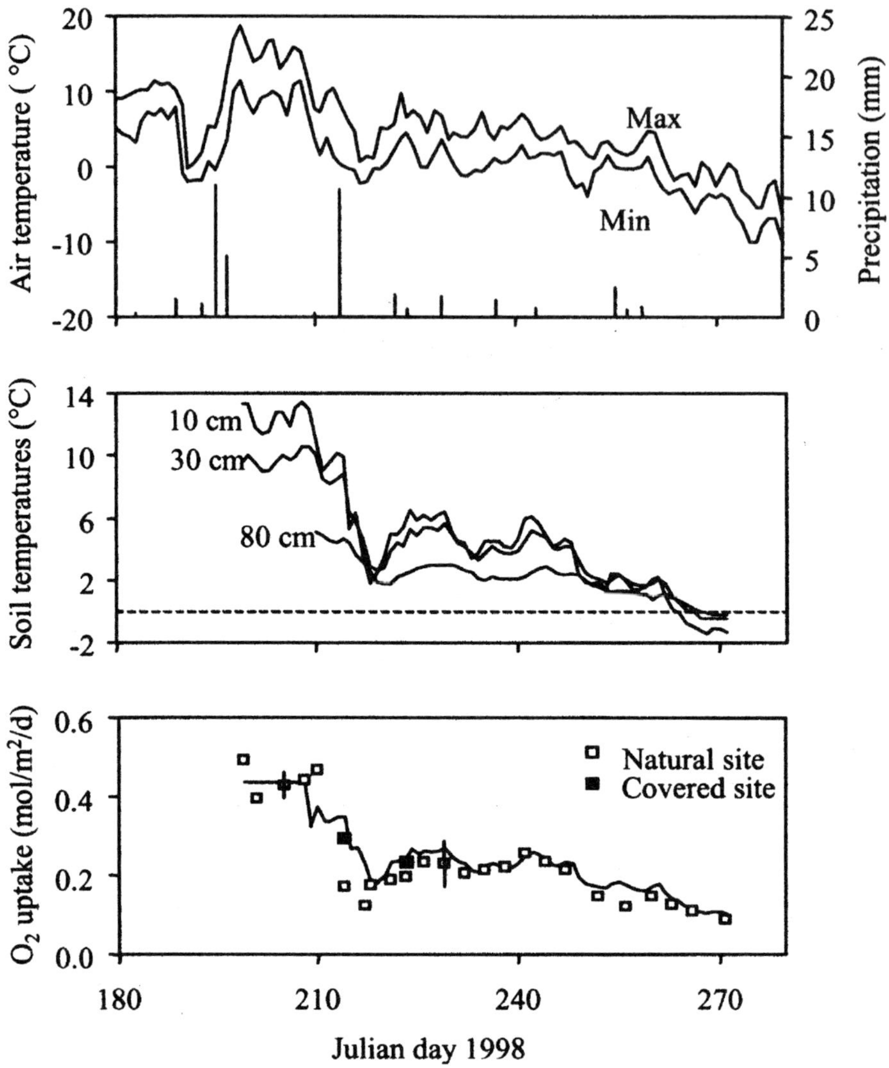

Figure 6.2.6. Air and soil temperatures (curves), precipitation (bars), and *in-situ* oxygen consumption rates at a natural site (open squares) and a covered site (solid squares) within well drained tailings measured in 1998. Bars through squares represent the SD of replicates (n=6). The simulated oxygen consumption is shown as a solid line. (Modified from Elberling, 2000).

Figure 6.2.6 shows that the model can predict the effect of temperature but not the effect of a decreasing diffusion of oxygen. The observations reveal the importance of both soil temperature and actual water content on AMD. Measurements during winter revealed that temperatures within the upper part of the snow-covered tailings were in the range of -1 to -3°C and the oxygen consumption was about 0.12 ± 0.02 moles of oxygen/m^2 per day or approximately 20 to 25% of the rates observed during summer. During early spring, temperatures within tailings were below -10°C and oxygen uptake was not detectable. The first oxygen uptake during spring was observed as the soil temperature rose to -4°C.

These detailed measurements of oxygen uptake throughout the year indicate that generation of AMD is highly sensitive to changes in temperature and that the effect of maintaining tailings below 0°C is predictable (Figure 6.2.2).

5. Strategies for Control of Acid Mine Drainage

Control strategies for AMD are based on either control of generation of AMD, control of contaminant migration, or collection and treatment of contaminated effluent. Although collection and treatment of contaminated effluent is a proven and demonstrated control strategy, maintaining such a facility at remote northern mine-sites, following closure, is costly and logistically difficult (Dawson and Morin, 1996). This section discusses generation of AMD and contaminant migration in relation to different control strategies, as these strategies aim to limit both generation of AMD and contaminant migration.

5.1. Frozen covers

Permafrost regions offer a unique possibility for the control of AMD that is not naturally available in warm climates: the freezing of mine tailings by covering the tailings with an inert material and thereby forcing the frost table to move up and maintain the tailings at sub-zero temperatures year-round. The application of frozen covers have been tested in few places in continuous permafrost areas.

At Lupin Mine, located 400 km NE of Yellowknife, Canada (65°N), sulfidic tailings (containing pyrrhotite, FeS) have been covered with 60 cm of gravel sand. Temperatures below the cover were above 0°C for more than 4 months each year, and tailing temperatures as high as 20°C have been observed during the summer (Dawson and Morin, 1996). At Nanisivik Mine, preliminary results (Figure 6.2.7, unpublished data) indicate that a 2-meter covering consisting of natural cover materials (shales, till and sand) without any sulfide minerals will adequately induce the permafrost table to move up, thereby maintaining reactive tailings below 0°C annually.

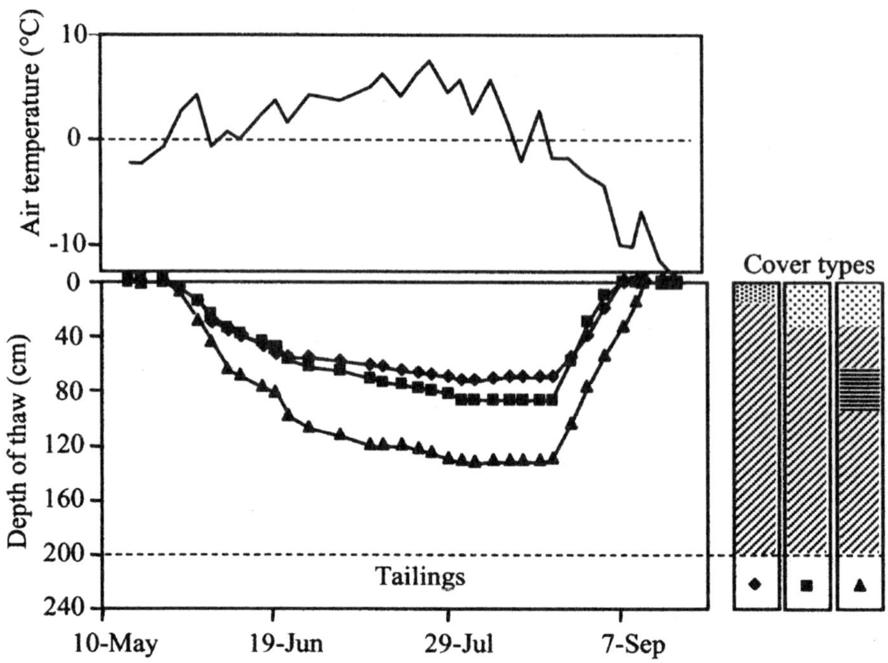

Figure 6.2.7. Active layer depths in tailings, 1997, covered by two meters of inert covers near Nanisivik Mine, Canada (unpublished data from Nanisivik Mine). Covers consist of sand and gravel (dots), till (horizontal lines), and shale (hatched lines).

However, as the test pads in Nanisivik during the test period were almost snow-free, additional snow in larger covered areas may result in increased insulation during autumn and higher early winter temperatures but a later start of thaw during spring. We discuss the impact of snow on ground temperatures and generation of AMD in general in greater detail in a later section of this chapter.

As figure 6.2.2 shows, temperatures slightly below zero will not guarantee negligible sulfide oxidation or completely frozen tailings. Therefore, the temperature at the tailing-cover interface becomes an important environmental parameter controlling generation of AMD. Temperature readings below the covers reveal that temperatures typically vary between -15 and a few degrees below zero. Thus, the thickness of covers recommended must be based on predictions of the sub-zero temperatures below the cover and the acceptance of the corresponding generation of AMD. Further laboratory studies of bacterial adaptation to freezing temperatures and of the sulfide oxidation rate in the temperature range 0 to -10°C are crucial in order to provide reasonable estimates for generation of AMD at sub-zero temperatures.

Although freezing is considered a method to control generation of AMD in permafrost areas, several aspects complicate the application of ground temperature as a fine-tuned control method for generation of AMD. As pyrite oxidation leads to a heat generation of approximately 12,000 kJ per kg pyrite (Dawson and Morin,

1996), the oxidation process may influence the rate of freezing of the ground and result in higher temperatures near sulfide minerals than the temperature measured in bulk material.

Assuming an annual average uptake of about 38 moles O_2 per m^2 as observed in Nanisivik 1998/99 (Elberling, 2000), more than 1 kg pyrite per m^2 are oxidizing annually, which generates about 32 kJ per m^2 per day. Compared to 334 kJ required to melt 1 kg of ice (Godwaldt et al., 1999), the rate of heat loss from oxidizing tailings to the surroundings may be enough to maintain elevated temperatures and generation of AMD for a period of time after bulk freezing.

5.2. Engineered dry and partly saturated covers

Engineered layered soil covers have been used to reduce oxygen and water movement within tailings effectively. Typically, cover systems are constructed from natural sediments and consist of layers with different hydraulic properties. The upper layers aim to prevent erosion and evaporation, the middle layers aim to retain water and act as a barrier for oxygen, and the lower layers aim to prevent drainage of overlying layers.

Using numerical simulation, Akindunni et al. (1991) showed that such barrier systems are possible when taking into account the air-entry value of the middle layer and the pressure head at which the lower layer approaches residual saturation. However, in permafrost areas, cracking due to seasonal freezing and reduced moisture caused by evaporation may alter the transport properties and even may enhance oxygen and water migration, due to preferential flow in crack systems.

5.3. Subaqueous disposal of tailings

Mine tailings have been placed under water in impoundments in many places in the world, primarily to minimize diffusion of oxygen into the tailings materials (Figure 6.2.3). This disposal strategy also has been applied in permafrost areas and may be attractive in areas where mine tailings can be placed in natural water bodies in a controlled fashion, as is the case at Polaris Mine, a lead-zinc mine on Little Cornwallis Island, Canada. Here, mine tailings have discharged 26 meters below the surface of the lake (Dawson and Morin, 1996).

In permafrost areas where a suitable natural water body is not present, construction of dams may be an alternative. The effect on AMD of covering tailings with water has not been tested in permafrost areas, but the potential effect can be illustrated by considering the maximum oxygen diffusion in tailings covered with 30 cm of saturated sand, again covered by a thin cover of oxygen-saturated water.

The maximum amount of oxygen which can diffuse through the 30-cm saturated cover at 5°C will be approximately 1.5×10^{-4} mol/m^2 per day (equation 6.2.4) or about 10,000 times less than the oxygen uptake observed in dry tailings at 5°C in Nanisivik (assuming instantaneous oxygen uptake at the tailing surface). These predictions are consistent with laboratory tests on the reactivity of tailings under water (Mugo et al., 1999; Elberling and Damgaard, 2001), indicating that storage of reactive tailings under water is effective in attenuating long-term fluxes of contaminants to the overlying water column.

The local hydrology and water balance are important factors to consider when designing a water-covered tailings impoundment. After mine closure, the water balance will be based on the precipitation and evaporation of the catchment area. Whether it is steady (i.e., no net change in pond level) will depend principally on the evaporation rate of water from the pond and on the input of water from meltwater. Strategies can be designed to maintain an equilibrium state, including installing snow fences to increase snow accumulation and meltwater and constructing channels to regulate the water input to the pond area. Difficulties in these strategies include sufficient storage capacity for extreme precipitation and the potential stability of dyke constructions in permafrost-affected areas.

Increasing oxygen solubility in cold water has been suggested as a concern about water covers (Dawson and Morin, 1996; Godwaldt et al., 1999). The amount of oxygen in the water increases by a factor of 1.4 as the temperature decreases from 15 to 0°C, while oxygen diffusivity decreases by a factor of 1.6 (Figure 6.2.8).

Comparing a saturated system at two different temperatures, using the above example with 30 cm saturated sand, the combined effect of increasing oxygen solubility and decreasing diffusivity will result in reducing the potential flux of oxygen at 2°C to approximately 90%, compared to the flux at 15°C. These simple calculations indicate that subaqueous disposal of tailings is a more effective remediation action in cold regions than elsewhere.

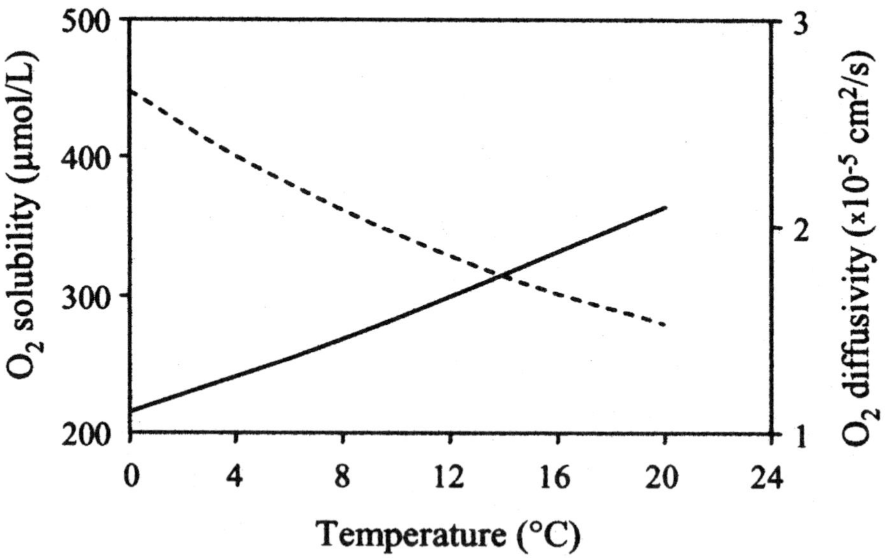

Figure 6.2.8. Solubility (dashed line) and diffusivity (solid line) of oxygen in water versus temperature. (Modified from Elberling and Damgaard, 2001.)

5.4. Snow fences and snow covers

In the Arctic, most precipitation falls as snow during winter. Snow covers are known to prevent wind erosion and redistribution of mine tailings during storms. Consequently, snow fences have been built, resulting in additional snow accumulation in tailing areas. The snow cover also may limit oxygen diffusion during winter and delay thawing in spring, thereby limiting the period of active layer development.

However, field observation over decades in Alaska and Canada have shown that snow has a very important influence on ground temperatures because of its insulating effects that reduce winter heat loss (e.g., Nicholson and Granberg, 1973). This insulating effect again depends on thickness and density of the snow layer, and variations in ground temperatures from year to year closely relate to annual variations in snow accumulation (Nicholson and Granberg, 1973).

Brooks et al. (1997; 1998) confirmed that snow covers insulate the frozen ground from the low winter air temperatures and result in surprisingly high subnivial levels of bacterial activity under seasonal snow cover.

Brooks et al. (1997) measured fluxes of CO_2 and N_2O as well as active microbial biomass when soil temperatures under a snow cover ranged from -5°C to 0°C. They suggested that subnivial bacterial activity may be a significant component of the annual C and N cycling in these environments. Similar conditions have been suggested for sulfidic tailings deposited in permafrost regions (Elberling, 2000). Further investigations are needed to quantify the effect of snow as a barrier for oxygen gas as compared to snow as a barrier of heat, controlling ground temperatures.

Although global changes in air temperatures have received much attention, global and local changes in snow cover (winter precipitation and wind regimes) may turn out to be the most critical parameter for long term generation of AMD. This is probably particularly important for facilities constructed for permanent frozen disposal of well drained tailings, whereas facilities constructed for permanent subaqueous disposal will be more sensitive to global changes affecting the water balance, including summer evaporation rates and annual precipitation.

6. Conclusions and Future Research

Currently there is strong interest in the general understanding of chemical and physical processes in frozen sediments in natural and engineered systems. This is due to a growing concern about contamination in the Arctic and to recent consideration of frozen barriers for preventing generation of AMD and, in more general terms, preventing the formation and dispersion of toxic chemicals at contaminated sites.

This review shows that AMD is not eliminated nor reduced to negligible levels at sub-zero temperatures, but it is reduced significantly. Very low temperatures during most of the year within the continuous permafrost zone inhibit generation of AMD. The efficiency of using ground temperature as the major control on tailings oxidation and release of pollutants in permafrost areas is understood only partly.

Reduction of AMD also may be achieved by limiting available oxygen to the tailings, by maintaining the reactive material at a nearly or fully water-saturated state. In the later case, natural freezing and the permafrost provide additional protection against generation of AMD and reduce movement of contaminants.

Future key questions relate to the overall limiting effects of freezing on AMD and to controls of oxidation processes and rates of migration of contaminants in aqueous solution within unsaturated and saturated frozen soils.

7. Acknowledgments

The Danish Environmental Protection Agency, Ministry of Environment and Energy, and the Danish Natural Sciences Research Council (no. 1235), Denmark, provided funding for this review. Special thanks to B.H. Jakobsen, G. Asmund, R. S. Sletten, K. LeDrew, and one anonymous reviewer for comments on the manuscript.

References

Akindunni, F., R.W. Gillham, and R.V. Nicholson. 1990. Numerical simulations to investigate moisture-retention characteristics in the design of oxygen-limiting covers for reactive mine tailings. Can. Geotech. J. 28: 446-451.

Banin, A., and D.M. Anderson. 1974. Effects of salt concentrations changes during freezing on the unfrozen water content of porous material. Water Resour. Res. 10:124-128.

Barnall, D., and D. Slotfeld-Ellington. 1983. Pulsed nuclear resonancestudies of doped ice Ih. J. Phys. Chem. 87:4321-4329.

Black, P.B., and M.J. Hardenberg. 1991. Historical perspectives in frost heave research: the early works of S. Taber and G. Beskow. USA Cold Regions Research and Engineering Laboratory. Special Report 91-23.

Boyd, W.L. and J.W. Boyd. 1964. The presence of bacteria in permafrost of the Alaskan Arctic. Can. J. Microbiol. 102:917-919

Brooks, P.D., S.K. Schmidt, and M.W. Williams. 1997. Winter production of CO_2 and N_2O from alpine tundra: environmental controls and relationship to inter-system C and N fluxes. Oecologia. 110:403-413.

Brooks, P.D., M.W. Williams, and S.K. Schmidt. 1998. Inorganic nitrogen and microbial biomass dynamics before and during spring snowmelt. Biogeochem. 43:1-15.

Cameron, E.M. 1977. Geochemical dispersion in mineralized soils of a permafrost environment. J. Geotechn. Explor. 7:301-326.

Cameron, E.M. 1979. Effect of graphite on the enhancement of surficial geochemical anomalies originating from the oxidation of sulphides. J. Geotechn. Explor. 12:35-43.

Dawson, R.F., and K.A. Morin. 1996. Acid mine frainage in permafrost regions: issues, control strategies and research requirements. Department of Indian and Northern Affairs, Canada. 65 pp.

Elberling, B. 2001. Environmental controls of the seasonal variations in oxygen uptake in sulfidic tailings deposited in a permafrost-affected area. Water Resour. Res. 37(1):99-107.

Elberling, B. 1998. Processes controlling oxygen uptake rates in frozen mine tailings in the Arctic. In: H.T. Shen, ed., Ice in Surface Waters. A.A. Balkema. Rotterdam/Brookfield. pp. 183-188.

Elberling, B., and L.R. Damgaard. 2001. Microscale measurements of oxygen diffusion and consumption in subaqueous sulfide tailings. Geochimica et Cosmochimic Acta. 65:1897-1905.

Elberling, B., R.V. Nicholson, and D.V. Darren. 1993. Field evaluation of sulphide oxidation rates. Nordic Hydrol. 245:323-338.

Elberling, B., R.V. Nicholson, and J.M. Scharer. 1994. A combined kinetic and diffusion model for pyrite oxidation in tailings: a change in control with time. J. Hydrol. 157:47-60.

Elberling, B., and R.V. Nicholson. 1996. Field measurements of oxygen diffusion and sulphide oxidation in mine tailings. Water Resour. Res. 326:1773-1784.

Elberling, B., and B.R. Langdahl. 1998. Natural heavy-metal release by sulphide oxidation in the High Arctic. Can. Geotech. J. 35:895-901.

Elberling, B., A. Schippers, and W. Sand. 1999. Bacterial and chemical oxidation of pyritic mine tailings at low temperatures. J. Cont. Hydrol. 41:225-238.

Evangelou, V.P. 1995. Pyrite oxidation and its control. CRC Press, Inc. 293 pp.

Forster, J. 1887. Ueber einige Eigenschaften leuchtender Bakterien. Zentr. Bakteriol. Parasitenk. Infekt. Hyg. 2:337-340.

Gilichinsky, D.A., E.A. Vorobyova, L.G. Erokhina, D.G. Fyodorov-Davydov, and N.R. Chaikovskaya. 1992. Long-term preservation of microbial ecosytems in permafrost. Adv. Space Res., 12:255-263.

Godwaldt, R.C., K.W. Biggar, and D.C. Sego. 1999. AMD generation at sub-zero temperatures. In: Assessment and Remediation of Contaminated sites in Arctic and Cold Climates, ARCSACC '99, Edmonton, May 3-4, 1999. pp. 75-82

Hallet, B. 1978. Solute redistribution in freezing ground. In: Proceedings of the Third International Conference on Permafrost, 1978, Edmonton, Canada. pp. 85-91.

Jaynes, D.B., A.S. Rogowski, and H.B. Pionke. 1984. Acid mine Drainage from reclaimed coal strip mines 1. Model description. Water Resour. Res. 20:233-242.

Kalin, M. 1987. Ecological engineering for gold and base metal mining operations in the North West Territories. In: Northern Affairs Program. Depart. of Indian Affairs and Northern Development, Environmental Studies no. 59. pp. 1-64

Kjøller, A., and S. Ødum. 1971. Evidence for longevity of seeds and microorganisms in permafrost. Arctic. 243:230-232.

Konrad, J.M., and A.W. McCammon. 1990. Solute partitioning in freezing soils. Can. Geotech. J. 67:726-736.

Kwong, Y.T.J. 1995. Geochemical, electrochemical and microbial control of sulphide oxidation and thier implication for the abatement of acid drainage. In: The International Workshop on Abatement of Geogenic Acidification in Mining Lakes, Sept. 4-6, 1995, Magdeburg, Germany.

Kyhn, C., and B. Elberling. 2001. Frozen cover actions limiting AMD from mine waste deposited on land in Arctic Canada. Cold Region Sci. and Tech. 32:133-142.

Langdahl, B.R. 1999. Acidophilic microorganisms in a High Arctic gossan environment. PhD. Thesis. Department of Microbial Ecology, Institute of Biological Sciences, Univ. of Aarhus, Denmark. 241 pp.

Langdahl, B.R., and K. Ingvorsen. 1997. Temperature characteristics of bacterial iron solubilisation and ^{14}C assimilation in naturally exposed sulfide ore material at Citronen Fjord, North Greenland (83°N). FEMS Microbiol. Ecol. 23:275-283.

Marion, G.M. 1995. Freeze-thaw processes and soil chemistry. CRREL special report 95-123. US Army Corps of Engineers, Cold Regions Research and Engineering Laboratory. pp. 1-28.

Melak, J.M., and J.O. Sickman. 1995. Snowmelt induced chemical changes in seven strems in the Sierra Nevada, California. In: Biogeochemistry of seasonally snow-covered Catchments (Proceedings of a Boulder Symposium, July 1995). IAHS Publ. 228. pp. 221-234.

Meldrum, J.L., H.E. Jamieson, and L.D. Dyke. 1999. Laboratory determination of sulphide oxidation potential in permafrost using tailings from Rankin Inlet, Nunavut. In: Mining and the Environment 2, Sudbury, Canada. September 12-16, 1999. pp. 119-126.

Mironenko, M.V., S.A. Grant, G.M. Marion, and R.E. Farren. 1997. FREZCHEM2. A chemical thermodynamic model for electrolyte solutions at subzero temperatures. In: Cold Regions Research & Engineering Laboratory, CRREL Report 97-5. pp. 1-40.

Morita, R.Y. 1975. Psychrophilic bacteria. Bacteriol. Rev. 39:144-167.

Mugo, R.K., D. McDonald, and G.W. Poling. 1999. Subaqueous tailings disposal in freshwater and marine environments - results of predictive geochemical testing using tailings with different compositions. In: Mining and the Environment 2, Sudbury, Canada, September 12-16, 1999. pp. 99-108.

Nicholson, F. H., and H.B. Granberg. 1973. Permafrost and snow cover relationships near Schefferville. In: Second International conference on Permafrost, Yakutsk, U.S.S.R., 13-28 July 1973. pp. 151-158.

Nicholson, R.V., R.W. Gillham, and E.J. Reardon. 1988. Pyrite oxidation in carbonate-buffered solution: 1 experimental kinetics. Geochim. Cosmochim. Acta. 52:1077-1085.

Nordstrom, D.K., and G. Southam. 1997. Geomicrobiology of sulfide mineral oxidation. Chapter 11 in Geomicrobiology. Mineralogical Society of America. Reviews in Mineralogy. 35:362-390.

Pantelis, G., and A.I.M. Ritchie. 1992. Macroscopic transport mechanisms as a rate limiting factor in dump leaching of pyritic ores. Appl. Mat. Model. 15:136-143.

Paktunc, A.D. 1999. Characterization of mine wastes for prediction of acid mine drainage. In: J. M. Azcue, ed. Environmental impacts of mining activity. Springer. pp. 19-40.

Price, W.S., H. Ide, and Y. Arata. 1999. Selfdiffusion of supercooled water to 238 Kusing PGSE NMR diffusion measurements. J. Phys. Chem. 103:448-450.

Ramsing, N., and J. Gundersen. 1994. Seawater and gases. Tabulated physical parameters, version 1.0.

Sheeran, D.E., and R.N. Yong. 1975. Water and slat redistribution in freezing soils. In: Conference on Soil Water Problems in Cold Regions, Division of Hydrology, AGU, Calgary, Alberta, Canada, May 6-7, 1975. pp. 58-69.

Troeh, R.F., J.D. Jabro, and D. Kirkham 1982. Gaseous diffusion equations for porous materials. Geoderma. 27:239-253.

Chapter 3. Environmental Impacts and Recovery from Human Activities on Cryosols of the Transantarctic Mountains

I.B. Campbell[1] and G.G.C. Claridge[2]

[1]Research Scientist, Land and Soil Consultancy Services, 23 View Mount, Nelson 7001, New Zealand; e-mail iaincampbell@xtra.co.nz.
[2]Research Scientist, 33 Kotari Road, Days Bay, Eastbourne, New Zealand; e-mail graeme.claridge@xtra.co.nz.

1. Introduction

Cryosols of the Transantarctic Mountains region of Antarctica are an important part of one of the earth's more extreme ecosystems. They have evolved in an environment dominated by extremely low temperatures, severe aridity, and the absence of a significant soil biological regime.

While these Cryosols have developed distinctive features related to the main weathering processes of oxidation and salinization on land surfaces that are in some places more than 10 million years old (Campbell and Claridge, 1978; Denton et al., 1993; Marchant et al., 1993), they are distinguished by a lack of structural development and a dominantly loose and incoherent form, due to the absence of colloidal clays and organic material (Bockheim, 1979, 1982; Campbell and Claridge, 1987; Claridge, 1965; Everett, 1971; Linkletter et al., 1973; Tedrow, 1977; Ugolini, 1963). In normal pedological terms, the soils would be described as *unweathered.* As a consequence of the long timescales for soil development, recovery from human disturbance is also extraordinarily slow.

Scientific research is the principal human activity in Antarctica and, since the current phase of intensive research activity began in 1956-57, the impacts on the soil environment are now evident and being recognized (Balks et al., 1995; Campbell et al., 1993, 1994, 1998; Claridge et al.; 1999, Sheppard et al., 1994). Recent estimates of the extent of Cryosol disturbances in the Transantarctic Mountains region (Campbell and Balks, 2001) suggest that less than 1% of the total soil area may have been disturbed through a variety of actions, ranging from base construction and logistical support to field research investigations. However, because these disturbances are chiefly in a few areas, particularly in the McMurdo

Sound region, they are very noticeable. They range from physical disturbance of the soil to contamination by foreign materials.

Increased understanding of the soil ecosystem's processes is now providing a basis for estimating the likely time scales for recovery from soil disturbances and provides a basis for better management decisions to minimize such impacts.

2. General Nature and Extent of Soil Disturbances

Environmental impacts on soils are greatest around established bases and facility centers where human activities are most intense. The most extensively disturbed areas are on southern Ross Island, the location of the New Zealand Scott Base and the McMurdo Station, and at Marble Point, a refuelling facility and site for a proposed all-weather aircraft landing field on the western side of McMurdo Sound (Figure 6.3.1). Lesser areas of disturbance are at the sites of a number of smaller stations.

Away from base areas where there are high concentrations of people, disturbances are primarily a result of research activities. Those within the McMurdo Dry Valley region are less intense than those around bases, but in total they have been estimated to cover a greater area than those around bases (Campbell and Balks, 2001).

3. Soil Disturbances around Bases

Logistical activities have catered for an estimated 80,000+ people at McMurdo Station and Scott Base since the late 1950s (Campbell et al., 1998). In this area have been repeated land scraping, quarrying, and earthworks for building sites, road construction, fuel storage, and wharf facilities. These activities have destroyed natural geomorphic features and caused permafrost retreat and associated landscape instability and thermokarst formation, widespread ground salinization, and extensive surface treading and track formation. Activity around this site also has spread dust over adjacent snow and ice-covered surfaces, causing changes in albedo and retreat of local snow and ice patches, along with accelerated stream flows and sedimentation.

Human activity around bases also has introduced a wide variety of contaminating foreign materials and wastes into the soils. These include organic and inorganic materials, fuels, garbage, and debris. Since natural soil biological or chemical activity is of very low intensity, the introduction of most of the contaminants is essentially permanent.

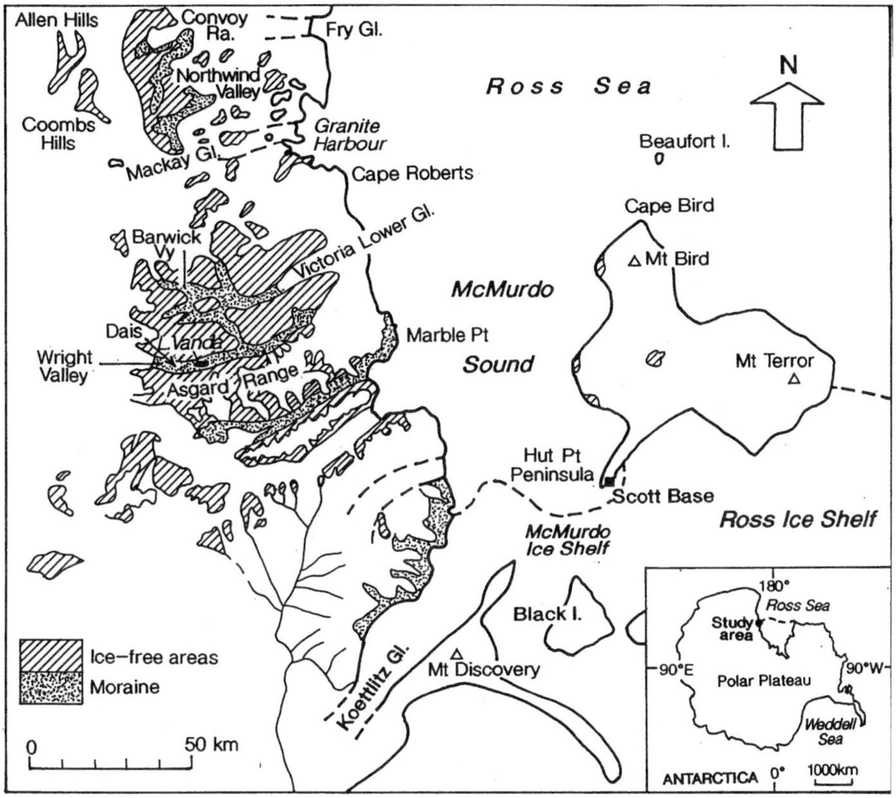

Figure 6.3.1. Map of the McMurdo Sound area showing ice-free areas and locations of sites the text mentions.

Another large area of disturbed soils near Marble Point (Figure 6.3.1) is a result of 1956-1958 construction for an aircraft runway. The site was abandoned and the buildings were trashed with little effort at site remediation. Marble Point has been used since then as a fuel storage and helicopter refuelling facility, and repeated fuel spills have contaminated soil. Hallett Station in North Victoria Land, like Marble Point, was abandoned, with little effort to remove foreign materials or rehabilitate the site.

Vanda Station was a small base by Lake Vanda in the McMurdo Dry Valleys (Figure 6.3.1). Its removal became necessary when the lake began to rise and threatened to flood the site (Waterhouse, 1997). Although most solids and other waste materials were removed as part of its operational management, wastewaters were disposed of locally and these, along with fuel spills and escaped liquids and solids, resulted in appreciable local soil contamination (Sheppard et al., 1994).

4. Soil Disturbances in Remote Areas

Soil disturbances are not restricted to bases but are widespread throughout the Transantarctic Mountains, as few areas have not been visited as part of some investigation. However, most activity has concentrated in the McMurdo Dry Valley region (Figure 6.3.1), with an estimated 10,000 people visiting here since 1957 (Sheppard et al., 1994; Vincent, 1996). The soil is disturbed by camps and activity around camp sites, such as treading and the formation of tracks between camp and work sites. Disturbances also result from activities such as pit digging, drilling, installing instruments, landing helicopters, disposing of human wastes, and minor littering and fuel spills. Compared with the soil disturbances around bases, the disturbances from field activities are less intense and more localized, often covering an area of a few hundred square meters, but in total they are estimated to cover an area greater than the combined area of disturbance at base sites (Campbell and Balks, 2001).

5. Soil Properties Affected by Large-scale Ground Scraping and Earthworks

Ground scraping and earthworks for building sites, storage, and roads are limited to the vicinity of the base areas, as there has been a general awareness of the destructive effects of heavy machinery and the intrusiveness of vehicle tracks within these landscapes. However, large areas of hillsides have been scraped to provide soil material for construction fill, with no thought to preserving natural landscape features. This destroys all small-scale and many larger-scale natural surface and topographic features. The resulting changes to the soil surface decrease the ground surface reflectivity and the albedo, causing changes in snow cover behavior, changes in the thermal regime of the soils, and disturbance of the soil's active layer and permafrost relationships (Balks et al., 1995).

When earth work and construction activity disturbs the ground surface, wind picks up dust and spreads it onto adjacent snow or ice-covered surfaces. At the southern end of Ross Island, the spread of fine particles and subsequent changes in albedo of snow or ice-covered surfaces have caused appreciable retreat of some snow and ice cover.

6. Permafrost Change and Thermokarst Formation following Disturbance

The soils in coastal sites are underlain by ice-cemented permafrost, often with a high frozen water content (Campbell et al., 1994, 1998a; Balks et al., 1995). Disturbances to the active layer typically result in some thawing of the underlying permafrost, a relatively rapid loss of water and accompanying ground instability, salinization, and increased water runoff and sedimentation. Permafrost disturbance is effectively permanent, with little recovery after 40 years (Campbell et al., 1994).

Adjacent to McMurdo Station, ground disturbance over a stagnant ice body has caused formation of an area of unstable thermokarst terrain (Figure 6.3.2). Stagnant or stranded glacial ice bodies are not uncommon in the Transantarctic Mountains and in some cases may be several million years old (Campbell and Claridge, 1987; Sugden et al., 1995).

7. Salinization

Arid Cryosols of the Transantarctic Mountains are saline, with a wide range of soluble salts, often in very high concentrations. The salts occur throughout the soil profiles, are transported by moisture, and are concentrated through evaporation to form a salty layer 5-15 cm below the soil surface. Salts also may form as surface efflorescences, more especially in the coastal regions, where evaporation may remove the higher amounts of moisture in these soils.

Figure 6.3.2. Thermokarst at Mcmurdo Station, Ross Island, caused by removal of surface cover by bulldozing and scraping, allowing the underlying ice-rich permafrost to thaw.

Disturbance causes a redistribution of existing soil salt patterns, particularly after large-scale earthmoving. However, it also occurs where digging profile pits disturbs soils. Salinity values in permafrost are often higher than in the active layer above (Campbell et al., 1994) and, in dry frozen permafrost, thick salty paleosol horizons are sometimes present.

Any soil surface or active layer disturbance which alters the soil's hydrological profile and increases surface evaporation may thaw permafrost and redistribute salts within the soil, sometimes causing the deposition of salts on the soil surface. Salt efflorescences are still forming on track surfaces formed 40 years ago at Marble Point and are widespread on the extensively scraped surfaces around McMurdo and Scott Base Stations.

8. Disturbances Caused by Pedestrian Traffic

Disturbance of the ground surface by pedestrian traffic is widespread on the arid Cryosols of the Transantarctic Mountains, and it is most clearly visible when the desert pavement is broken. Arid Cryosols are covered by a loose pavement of interlocking stones which protects the underlying finer soil from wind erosion. The nature of the desert pavement depends on the particle size of the soil's parent materials and the age of the ground surface. In soils with a sandy gravel range of particle size, the pavement consists of a surface coating of small pebbles overlying dominantly sandy textured soil, but in soils with coarser size grades, cobbles or boulders usually dominate. Because the underlying soils are typically loose and non-cohesive, human activities easily displace the surface clasts forming the desert pavement.

An experimental investigation of the rate at which tracks form in soils with markedly differing particle-size grades showed that distinct tracks formed in sandy gravel soils with fewer than 20 passes over the same surface, although, on some surfaces, a distinct and lasting effect was obvious after a single pass (Campbell et al., 1998) (Figure 6.3.3). On soils with increasingly coarser particle-size grades, the rate of track development was progressively slower. Where the surface cover included a high proportion of boulders, little surface displacement occurred.

At field camps, tent sites are sometimes levelled and larger stones moved aside. Extensive disturbance from treading may cover a few hundred square meters around the camp site. Research activities may involve sampling, drilling, or pit digging, which are often particularly disruptive to the soils, although usually limited to very small areas.

Figure 6.3.3. Track in alluvial sediments near Lake Vanda, formed after 20 passes. In inland regions where the climate is colder, the soils and the permafrost have very low moisture contents and the permafrost may be dry-frozen (Campbell et al., 1998a). In these situations, disturbances of the active layer of soil do not affect the permafrost's stability.

9. Introduction of Foreign Materials

Since the present phase of scientific activity began in the late 1950s, a very large amount of foreign material, especially building materials and domestic products, paper, cardboard, plastics, and wooden packages, etc., has been introduced into the environment and widely scattered about the landscape, especially in the vicinity of base sites.

Initially, there were no policies for removing wastes, including human waste, so much of it was either dumped, abandoned, or buried. Apart from the deliberate disposal of waste, wind has aided the wide dispersal of foreign materials, often many kilometers downwind from the source. Since biological decomposition of organic materials is extremely slow, many of these materials are expected to remain in the soil indefinitely.

Present codes of conduct aim to prevent or minimize the spread of foreign materials. Most currently produced wastes are removed from Antarctica and, while efforts are being made to remove previously deposited waste from some sites, it is impractical to remove smaller and most buried materials.

10. Soil Contamination

A wide variety of contaminants, including toxic metals (silver, cadmium, copper, lead, zinc, arsenic, etc.), petroleum products, deposits from incineration of waste, and landfill leachates have been detected at numerous sites, more especially in the vicinity of permanently occupied areas (Sheppard et al., 1994; Crumrine, 1992; Johnston and Stringer, 1988).

In the more remote areas away from base sites, the most common contaminant is urine, which is estimated to have added about 3000 kg of nitrogen to the soils of the Dry Valley region (Campbell and Balks, 2001). Some small fuel spills from generators and drilling equipment also have occurred. Most contaminants derive from the earlier absence of environmental protocols for management of wastes and from a lack of attention to environmentally sensitive practices.

In general, the nature, extent, and effects of contaminants on the arid Cryosols has been poorly investigated, and clean-up procedures have tended to be superficial. However, some disturbed and contaminated sites which were abandoned are now providing valuable scientific information on the behavior of contaminants in soils. Claridge et al. (1995) found that heavy metals moved less than 1 meter from a point-source of contamination over a 30-year period in a moist coastal environment. Experimental observations (Claridge et al., 1999) showed that the extent of leaching of an applied tracer salt was related to the soil moisture regime, with negligible leaching occurring in the most arid Dry Valley Cryosols.

11. Influence of Global Environmental Change

The main influences of global environmental change on the arid Cryosols are expected to be those driven by climate changes. Increasing global temperatures can be expected to cause increases in soil temperatures as they have in the Arctic (Koster et al., 1994), with consequent changes to the permafrost. Changes to the active layer (thaw depth) and soil biological activity, degradation to permafrost, increased thermokarst development, and thermo-erosion may occur in response to climatic changes.

The main changes to the soil chemistry are likely to be increased salinization and, at some sites, an increase in soil carbon content (Claridge et al., 1999). Given the Antarctic carbon budget, any increase in soil carbon may be offset by release of CO_2 from melting ice. Heavy metals, especially lead from motor fuel, and black carbon (soot) from worldwide industrial activity can be detected in polar snowfall and probably also falls on land, but in soils the effect is expected to be negligible.

12. Remediation and Recovery of Soil Disturbances

The time scales for formation of these Cryosols are very long, and recovery from disturbance can occur only at the same rate as the natural soil processes. The low intensity of soil ecosystems is significant for the very slow amelioration of disturbances of arid Cryosols.

For disturbances involving significant soil movement, no recovery of natural features can be expected for a very long time. Track and wheel marks from disturbances made 40 years ago by heavy machinery are still clearly visible, as are unfilled soil investigation pits (Campbell et al., 1998). Also, no accumulation of ice in the permafrost of disturbed soils has taken place (Campbell et al., 1994). On young (last glaciation) surfaces with unoxidized soils, disturbances may be less obvious than on old (Miocene) surfaces which have considerable color contrast between the surface and lower horizons.

For lower impact soil disturbances, the situation is rather different. In coastal areas where the soils are moist and there is some soil movement from freeze-thaw, light footprints or similar disturbances may disappear in less than one year. Similarly, where surface particles are distributed actively by water or wind, small disturbances soon are obliterated. On old, very dry, and stable surfaces, however, disturbances from footprints, etc., remain for a very long time (Campbell et al., 1998).

Little data is available about the rates of recovery from chemical contamination. Movement of soluble contaminants can be very slow in the absence of available liquid phases but, where there is a mobile liquid phase, subsurface transport of contaminants can be rapid (Sheppard et al., 2000). Some studies examining the biodegradation of hydrocarbons (Aislabie et al., 1998) indicate that rates of degradation are extremely slow. Spilt hydrocarbons have persisted in the soil for up to 40 years.

13. Protection of the Antarctic Environment

The management of Antarctica and the protection of its important intrinsic values have evolved through a unique series of international events and relationships, as Thomson (1990) outlined. In the early part of this century, the legal status of Antarctica was a result of discovery, exploration, and annexation by seven states, some of which other parties did not recognize. During the 1940s and 50s, the political situation deteriorated when some nations attempted to consolidate their sovereignty claims.

About this time, however, scientists planned the International Geophysical Year (IGY), a significant world-wide scientific program with priority for upper atmosphere, and polar research. This required the cooperation of a large number of Antarctic research stations and resulted in the establishment of the Scientific

Committee for Antarctic Research (SCAR). This committee was responsible for initiating, promoting, and coordinating scientific activity in Antarctica and for framing and reviewing scientific programs. SCAR provided, for the first time, a framework for managing essentially scientific Antarctic affairs.

When it became clear that IGY activity would continue in Antarctica for a long time and that cooperative science should be maintained, a formal structure was proposed, and, in 1961, the Antarctic Treaty came into existence. The Antarctic Treaty formalized IGY cooperative arrangements, but it also requires that Antarctica be used only for peaceful purposes, it bans nuclear waste, and it requires contracting parties to advise each other of planned activities. A consultative meeting system handles Antarctic matters.

Environmental concerns arose initially out of consideration for conserving and protecting living marine resources. However, with increasing pressure from scientific research and associated logistical activity, more especially in the McMurdo Dry Valley area of the Transantarctic Mountains, and with rising concern for the impacts of increased human activity on terrestrial ecosystems, various conservation measures soon appeared.

Among these was the designation of Specially Protected Areas (SPAs) and Sites of Special Scientific Interest (SSSI) to maintain significant areas of minimal human impact. At this time, few rules or procedures governed the management of sensitive values, and impacts from logistical activities were not considered seriously. Most waste from human activities, for example, was simply dumped and abandoned. The question of possible mineral resources also posed a problem, as interest arose from mineral exploration companies.

However, by 1988, a convention on the Regulation of Antarctic Mineral Resources was adopted. It places the highest priority on the fullest protection of the Antarctic environment. Since no mineral deposits of any significance are known (Behrendt, 1983), and in any case the costs of extraction would be exorbitant in this remote and harsh environment, the prospects of extractive mineral activity seem remote.

In 1991, the Protocol on the Environmental Protection to the Antarctic treaty was signed, providing for the comprehensive protection of the Antarctic environment. The protocol establishes Antarctica as a natural reserve devoted to peace and scientific research and it strengthens and confirms the Antarctic Treaty system.

The protocol covers all activities south of latitude 60°S and provides a framework for minimizing the impacts of all activities in Antarctica. For example, all activities must be planned and conducted to limit adverse impacts on the Antarctic environment and associated ecosystems, with sufficient information included in proposals to assess impacts and for monitoring to take place. Annexes to the protocol provide specific rules for implementing environmental impact assessment, waste management procedures, and monitoring.

14. Summary

The arid Cryosols of the Transantarctic Mountains have formed in an extreme environment with very slow weathering rates over an exceedingly long time. Because of their minimal development, they are highly vulnerable to human disturbances. Although human disturbance is estimated to have affected less than 1% of the Transantarctic Mountains Cryosols, the environmental disturbances are concentrated in the McMurdo Sound and Dry Valley region, an area of high use and therefore of appreciable visual impact.

The most severe disturbances are in the vicinity of bases where large-scale earth movement has taken place, while contamination has resulted from high levels of human activity. In more remote areas, the level of environmental disturbance is less and results primarily from surface soil displacement through walking, camping, and various scientific activities.

Physical disturbance to the soils results in changes to the soil's thermal, permafrost, and hydrological properties, destruction or disturbance of the desert pavement, and salinization. Contamination from solid waste and from chemical sources can be extremely long-lasting and is unlikely to be broken down or assimilated into the soil within meaningful time scales.

In the warmer coastal environments, there is some natural recovery from low-impact disturbances, but such disturbances persist in drier and older soils. Internationally agreed rules and procedures are now in place, under the Environmental Protocol to the Antarctic Treaty, to minimize further environmental damage and to safeguard the Antarctic environment.

References

Aislabie, J., M. McLeod, and R. Fraser. 1998. Potential for biodegradation of hydrocarbons in soil from the Ross Dependency. App. Microbiol. & Biotech. 49:210-214.

Balks, M.R., D.I. Campbell, I.B. Campbell, and G.G.C. Claridge. 1995. Interim results of 1993/94 soil climate, active layer and permafrost investigations at Scott Base, Vanda and Beacon Heights, Antarctica. Univ. of Waikato, Dept. of Earth Sciences Special report. pp. 1-64.

Bockheim, J.G. 1979. Relative age and origin of soils in eastern Wright Valley, Antarctica. Soil Science. 128:142-152.

Bockheim, J.G. 1982. Properties of a chronosequence of Ultraxerous soils in the Transantarctic Mountains. Geoderma. 28: 239-255.

Campbell, I.B., and G.G.C. Claridge. 1978. Soils and Late Cenozoic history of the upper Wright Valley, Antarctica. N.Z. J. Geol. & Geophys. 21:635-43.

Campbell, I.B., and G.G.C. Claridge. 1987. Antarctica: soils, weathering processes and environment. Elsevier Science Publishers. Amsterdam. pp. 1-368.

Campbell, I.B., G.G.C. Claridge, and M.R. Balks. 1993. A simple visual technique for estimating the effects of fieldwork on the terrestrial environment in an ice-free area of Antarctica. Polar Record. 29:321-328.

Campbell, I.B., G.G.C. Claridge, and M.R. Balks. 1994. The effect of human activities on the moisture content of soils and underlying permafrost from the McMurdo Sound region, Antarctica. Antarc. Sci. 6:307-314.

Campbell, I.B., G.G.C. Claridge, D.I. Campbell, and M.R. Balks. 1998. Short and long term impacts of human disturbances on snow-free surfaces in Antarctica. Polar Record. 34:17-26.

Behrendt, J.C. 1983. Petroleum and mineral resources of Antarctica. U.S Geological Survey Circular 909. 75 pp.

Campbell, I.B., G.G.C. Claridge, M.R. Balks, and D.I. Campbell. 1988a. Permafrost properties in the McMurdo Sound-Dry Valley region of Antarctica. Proceedings of the 7th International Conference on Permafrost, Yellowknife, Canada 1998. pp. 121-126.

Campbell. I.B., and M. Balks. 201. Soils and ice-free areas. In: E. Waterhouse, ed. A state of the environment report for the Ross Sea region of Antarctica. New Zealand Antarctic Institute. Christchurch. pp. 4.16-4.33.

Claridge, G.G.C. 1965. The clay mineralogy and chemistry of some soils from the Ross Dependency, Antarctica. N.Z. J. Geol. & Geophys. 8:186-220.

Claridge, G.G.C., I.B. Campbell, H.K.J. Powell, Z.H. Amin, and M.R. Balks. 1995. Heavy metal contamination in some soils of the McMurdo Sound region, Antarctica. Antarc. Sci. 7:9-14.

Claridge, G.G.C., I.B. Campbell, and M.R. Balks. 1999. Movement of salts in Antarctic soils: experiments using lithium chloride. Permafrost and Periglacial Processes. 10: 223-253.

Claridge, G.G.C., I.B.Campbell, and D.S. Sheppard. 1999. Carbon pools in Antarctica and their significance for global climate change. Advancements in Soil Science. (In press.)

Crumrine, K.Z. 1992. Surface remediation at McMurdo Station. Proceedings of the 5th Symposium on Antarctic Logistics and Operations. San Carlos de Bariloche, Argentina, June 1992. 41-60.

Denton, G.H., D.E Sugden, D.R. Marchant, B.L. Hall, and T.I. Wilch. 1993. East Antarctic Ice Sheet sensitivity to Pliocene climatic change from a Dry Valleys perspective. Geografiska Annaler. 75A:155-204.

Everett, K.R. 1971. Soils of the Meserve Glacier area, Wright Valley, Southern Victoria Land, Antarctica. Soil Sci. 112:425-438.

Johnston, P.A., and R.L. Stringer. 1988. Trace metals in soils and effluents from Antarctica. Greenpeace Inc and Queen Mary College, Univ. of London. Technical Note 11 Dec 1988. pp. 1-13.

Koster, E.A., M.E. Nienwenhuijzen, and A.S. Judge. 1994. Permafrost and climatic change: The annotated bibliography. Glaciological Data Report 27, WDC-A for Glaciology. Univ. of Colorado. Boulder, Colorado.

Linkletter, G., J. Bockheim, and F.C. Ugolini. 1973. Soils and glacial deposits in the Beacon Valley, Southern Victoria Land, Antarctica. N.Z. J. Geol. & Geophys. 16:90-108.

Marchant, D.R., G.H. Denton, G.E. Sugden, and C.C. Swisher 1993. Miocene Glacial stratigraphy and landscape evolution of the western Asgard Range, Antarctica. Geografisker Annaler. 75A. 303-330.

Sheppard, D.S., I.B. Campbell, G.G.C. Claridge, and J.M. Deely 1994. Contamination of soils about Vanda Station, Antarctica. Institute of Geological and Nuclear Sciences Report 94/20. Lower Hutt, New Zealand.

Sheppard, D.S., G.G.C. Claridge, and I.B. Campbell. 2000. Metal contamination of soils at Scott Base, Antarctica. Appl. Geochem. 5:513-530.

Sugden, D.E., D.R. Marchant, and N. Potter. 1995. Preservation of Miocene Glacier ice in East Antarctica. Nature 367:412-414.

Tedrow, J.C.F. 1977. Soils of the Polar Landscapes. Rutgers Univ. Press. New Brunswick. 638 pp.

Thomson, R.B. 1990. The international background. In: T. Hatherton, ed. Antarctica, the Ross Sea Region. DSIR Publishing. Wellington. pp. 259-275.

Ugolini, F.C. 1963. Soil investigations in the lower Wright Valley. Proceedings of the International Permafrost Conference. National Academy of Science. National Research Council Publication 1287. pp. 55-61.

Vincent, W.F. (ed) 1996. Environmental management of a Cold desert Ecosystem: the McMurdo Dry Valleys. Desert Research Institute, Univ. of Nevada, USA, special publication. 57 pp.

Waterhouse, E.J. 1997. Implementing the protocol on ice-free land: the New Zealand experience at Vanda Station. In: W.B. Lyons, C. Howard-Williams, and I. Hawes, eds. Ecosystem Processes in Antarctic ice-free Landscapes. Proceedings of an International workshop on Polar Desert Ecosystems, Christchurch New Zealand, 1-4 July 1996. Balkema. pp. 265-274.

Chapter 4. Soil Properties and Relationships in Cryosols of the Region of the Transantarctic Mountains in Antarctica

I.B. Campbell[1] and G.G.C. Claridge[2]

[1]Research Scientist, Land and Soil Consultancy Services, 23 View Mount, Nelson 7001, New Zealand; e-mail iaincampbell@xtra.co.nz.
[2]Research Scientist, 33 Kotari Road, Days Bay, Eastbourne, New Zealand; e-mail graeme.claridge@xtra.co.nz.

1. Introduction

The study of Antarctic soils is comparatively new, spurred in the late 1950s by the growing interest in earth and other sciences in Antarctica. Curiosity drove initial investigations, and the questions were whether soils actually existed there and, if so, what their properties were and what processes operated (McCraw, 1960; Ugolini, 1963; Claridge, 1965; Campbell and Claridge, 1966; Claridge and Campbell, 1968; Tedrow and Ugolini, 1966). As glaciological investigations intensified, weathering relationships became valuable for establishing glacial chronologies (Calkin, 1964; Ugolini and Bull, 1965; Everett, 1971; Campbell and Claridge, 1975; Bockheim, 1978, 1982) and attention focused on the spatial and temporal differences in and the broader significance of soil properties (Campbell and Claridge, 1987).

More recently, with the recognition of the fragility of the terrestrial ecosystems, the impacts from human activities, and local and global pressures, the focus of soil studies has shifted to consider soil properties of particular importance to ecosystem and environmental relationships and to management issues (Campbell et al., 1993, 1997, 1998; MacCulloch, 1996; Sheppard et al., 1994).

2. Occurrences of Soil

Small patches of ice-free ground are scattered throughout the Transantarctic Mountains on the western margin of the Ross Sea region (Figure 6.4.1). Their sporadic nature relates to the glaciological history of the area.

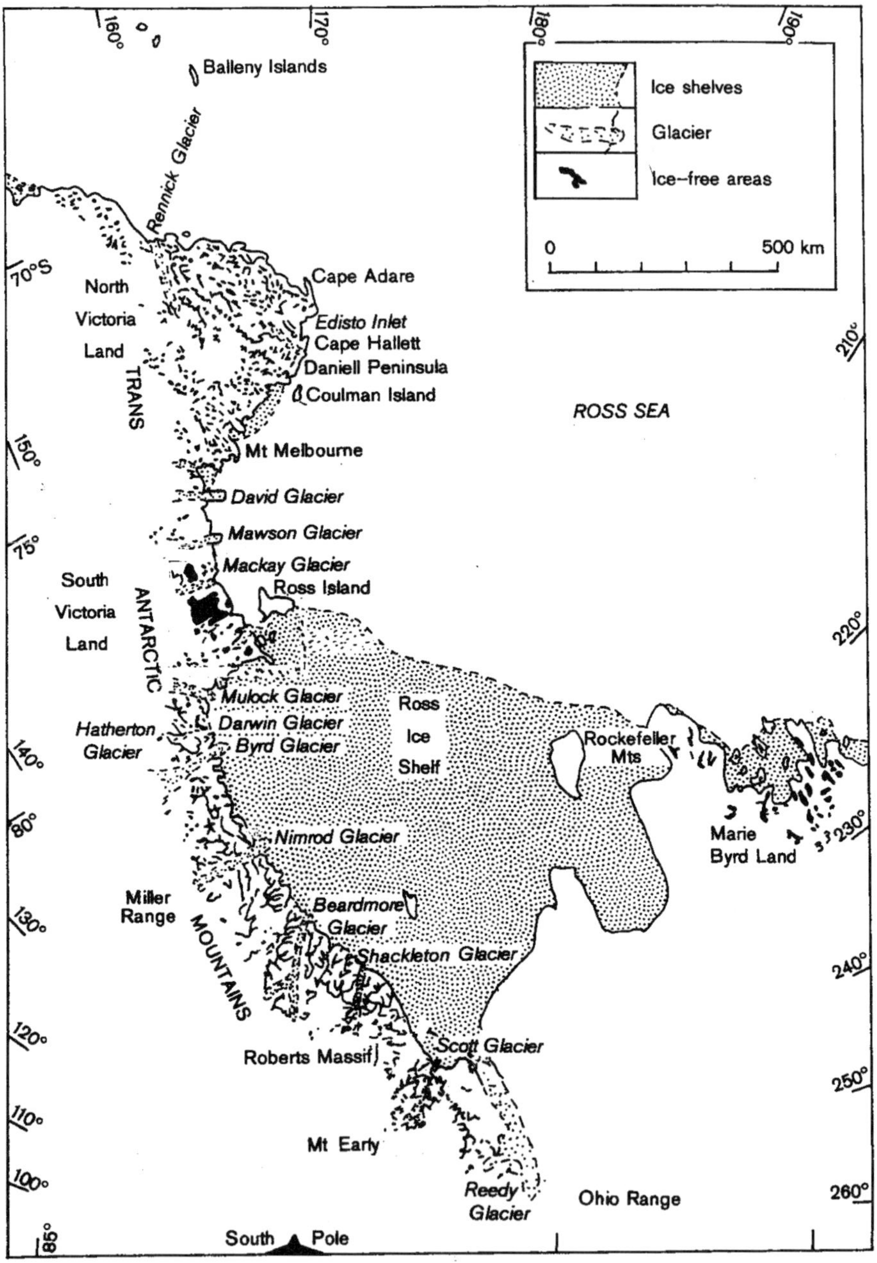

Figure 6.4.1. Ice-free areas in the Transantarctic Mountains.

In the northern locations, glaciation is predominantly alpine, and the bare ground and soil exposures largely occur only in small scattered areas of steep ground with numerous rock outcrops. Farther to the south, outlet glaciers from the Polar Plateau have cut through the mountains, and the resultant valleys have glacial deposits that record the history of glaciation in Antarctica. These land surfaces are the main sites of soils. The soil-landscape relationships are essentially a function of past complex glaciological interactions, including the effects of ancient (pre-Miocene) over-riding ice sheets (Denton et al., 1993), repeated invasions from ice-shelves grounded during periods of lower sea levels (Hollin, 1962), and alpine type glaciations responding to local variations in climate.

As glaciers occupy most valleys in the Transantarctic Mountains, the exposures of bare ground and soils in the valleys are largely discontinuous occurrences on valley sides and surfaces exposed by the retreat of ice. In the Dry Valley region of the McMurdo Sound area (Figure 6.4.2), however, extensive glacial retreat has given rise to a number of ice-free valleys which make up the largest contiguous area of bare ground within Antarctica; soils cover most of this 6,000 km^2 area.

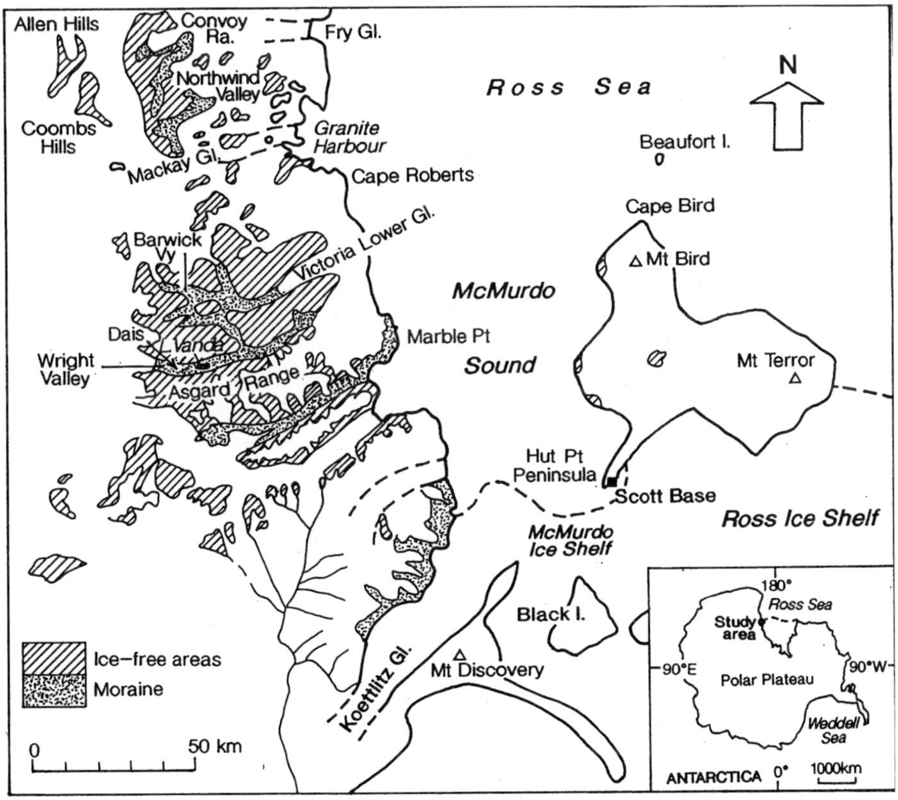

Figure 6.4.2. The ice-free areas of the McMurdo Sound region.

Glacial till is the predominant parent material for the soils on valley floors and sides, piedmonts, benches, and cirque basins. On some valley sides which have been ice-free for long periods, the soils formed from extensive deposits of scree, while, in some coastal areas, marine processes have reworked tills to form coastal beach deposits. Alluvium occurs only on a few valley floors and is of very limited extent, as sufficient water seldom is available to transport and sort sediment. Aeolian deposits are somewhat limited, usually restricted to comparatively small areas of coarse dune sands in scattered areas. Bare rock occurs on around 10% of the exposed land surfaces (Claridge et al., 2000), but even here is evidence of soil processes, with weathering and accumulation of fine material in fracture planes, as well as colonization and modification of rock surfaces by endolithic life forms.

Organic accumulations are rare in the region of the Transantarctic Mountains, since the biomass and metabolic activity of soils here are the lowest of any soil ecosystem. Algal peats occasionally occur adjacent to some small ponds, while moss, the most prolific vegetative form, usually occurs in patches of less than 5 square meters. The most significant organic accumulations are those associated with penguin rookeries, which a cover a total of probably less than 10 km^2 in coastal sites of the region.

3. Age Relationships of Soils

Although in normal pedological terms the Cryosols of the region of the Transantarctic Mountains can be described properly as unweathered or weakly weathered, a number of their features allow pedological distinctions. Chronosequence studies have been used widely to investigate Transantarctic Mountains Cryosols (Ugolini, 1964; Claridge and Campbell, 1968a, Everett, 1971; Linkletter et al., 1973; Bockheim, 1979, Pastor and Bockheim, 1980).

During the early investigations of these Cryosols, various workers recognized progressive weathering differences in soils of Dry Valley regions (Calkin, 1964; Tedrow and Ugolini, 1966) and correlated these within the framework of then-known glaciations (Wisconsin-Illinoian, etc.). Later work (Claridge and Campbell, 1968a) suggested soil age sequences spanned much greater time scales, while K/A dating (Everett, 1971) showed that soil time sequences indeed spanned several million years (Campbell and Claridge, 1975; Bockheim, 1982). Recent work, including dating widespread ash deposits in soils (Marchant et al., 1993; Denton et al., 1993), has confirmed that soil weathering on some surfaces extends to the Miocene (13.5 my) and probably beyond.

4. Soil Physical Properties

Soil weathering in the arid Cryosols involves relatively low levels of alteration of coarse materials. Soil descriptions therefore commonly use non-standard terminology to characterize the soils and highlight their significant and unusual features. Despite the number of soil scientists who have worked independently from each other in Antarctica, however, a considerable degree of uniformity exists among the methods and parameters used to describe the soils.

Weathering characteristics of the soil surface, including the oxidation state, shape, degree of polish and pitting, and the number and size of clasts, are important criteria used to distinguish weathering on different aged surfaces. Likewise, the angularity or roundness and weathering attributes of subsurface clasts, compared with those of surface clasts, provide an important measure of cryogenic movement of soil material and of ground surface stability.

Standard horizon designations are not widely used in descriptions of arid Cryosols, as the soil materials typically do not meet the usual definitions. In the absence of significant amounts of clay or organic matter, the arid Cryosols have no appreciable soil structural development, but small differences in cohesion are significant and are described. Moist soils occur to any extent only along water sources in summer, and soil color descriptions typically refer to dry soil.

4.1. Horizon nomenclature

As have most pedologists, we have been reluctant to use standard horizon nomenclature, since Antarctic soil materials do not fit the usual definitions. However, Bockheim (1997) has modified some definitions and defined new terms which can apply to arid Cryosols. For example, he uses *D* for the desert pavement, *Bw* for the underlying oxidized zone, and *Bz* for a salt-enriched horizon. These terms have not yet achieved universal usage.

4.2. Active layer

The active layer of the Transantarctic Mountains Cryosols varies in depth from around 70 cm to less than 10 cm. It clearly relates to climate and other factors that influence the soil thermal energy regime.

At Scott Base, for example, the active layer is around 35 cm, but at Marble Point, which has a similar mean annual temperature, it is about 65 cm (Balks et al., 1995). The difference is that the soils at Scott Base are more moist (owing to greater snow thaw from darker rocks) and much of the available energy is used for thawing ice in the soil rather than for heating the soil.

Accurate assessment of the active layer's thickness can be made only through measurements of soil temperature, as ice-cemented permafrost is not always present. When it is present, its position may not coincide with that of the annual thawing depth.

4.3. Permafrost

Permafrost is an important feature of the soils and exists either as hard ice-cemented material, where soil moisture content exceeds about 5% (gravimetric), or as loose uncemented soil, where the moisture content is less than about 5%. The moisture content is a function of the site's climate factors (Campbell et al., 1998a). In soils that have ice-cemented permafrost, soil weathering below the active layer/permafrost boundary is generally negligible and indicates that the active layer's thickness is a relatively stable feature. In dry frozen permafrost with low moisture content, alteration of soil continues at depth, irrespective of the depth to the permafrost.

4.4. Desert pavement

The desert or stone pavement is a layer of loose fine sandy gravel, stones, or boulders, and it occurs almost everywhere on the surface of Antarctic Cryosols (Figure 6.4.3). From an environmental perspective, it is one of the more important soil features since it is the part of the soil that human activity disturbs first and most easily.

The desert pavement forms largely through a combination of surface weathering and winnowing processes. Weathering breaks down surface clasts of all sizes over time, as wind sorts and shifts the smaller particles. Coarsely grained rocks typically disintegrate through granular disintegration assisted by salt weathering, often forming spectacular cavernous shapes. More finely grained rocks disintegrate by splitting along fracture planes and become rounded by exfoliation of oxidized surfaces. Abrasion is an important process which further reduces clasts to ventifacts. The end result of the surface weathering processes is the formation of an aerodynamically stable surface, with the wide range of particles present forming a well packed surface pavement.

Because of local rock types and environmental differences, the desert pavement is extremely variable in its appearance, in places consisting of a covering of well weathered, polished, oxidized or crumbled, wind faceted, or strongly pitted rocks, while in others it may consist of a loosely packed veneer of small pebbles.

4.5. Vesicular crusts

Beneath the desert pavement is the first soil horizon, which commonly consists of a pale, weakly cohesive layer, 1 to 2 cm thick, with a weakly developed vesicular structure. The vesicular structure forms through periodic moistening of the soil surface by snow melt, followed by freezing and evaporation of the moisture in the surface layer. The cycles of surface soil moistening followed by freezing and water loss can be rapid, with 70% of the frozen water which accumulated from a snow fall and thaw event being lost within 24 hours (Campbell and Claridge, 2000).

Figure 6.4.3. A well developed desert pavement forming a very stable surface. The surface rocks are well polished and stained.

Because of climatic and other site factors, the extent and degree to which surface soil moistening occurs vary considerably; hence the occurrence and development of the surface vesicular crust also vary.

4.6. Soil color

arid Cryosols are differentiated by their color. The youngest and least weathered soils have colors with high value and low chroma (for example, olive grey to light yellowish brown 5Y 6/2 - 2.5Y 6/4) without any distinction through the profile. The materials from which the soils are formed heavily influence the colors; the soils from sandstone, dolerite, granite, metasediment, and volcanic rocks all have distinctive colors. In an unweathered state, the soil color resembles the color of the fine fraction of the disaggregated rock from which the soil formed, but in a more weathered condition, the soil color relates to the relative abundance of iron-bearing minerals in the parent material.

With increasing age, the soil gradually reddens, through yellowish brown and strong brown (10YR 5/4, 7.5Y 5/6) to yellowish red and dark red (5YR 5/8,

2.5YR 3/6). The principle cause of the color change is the slow oxidation of iron-bearing minerals in the soil particles. In some older soils, aureoles of more intense color are sometimes present, surrounding individual weathering clasts, and appear to indicate that iron has migrated outward from the weathering source rock. Both chroma and value decrease with depth. Changes in color with increasing depth of soil generally are gradual and may continue for more than 1 meter in old soils.

4.7. Soil particle size

arid Cryosols are made largely of coarse particle-size material, although clast size varies widely. The frequency of boulders closely relates to the composition of the parent material, soils from dolerite generally having a higher proportion of boulders than soils with less resistant rocks. Between boulders, the soils are predominately gravelly sand, containing 50 to 90% of material >2mm in size. The fine earth fraction (<2 mm) seldom exceeds 5% of the soil and, of this, the silt fraction typically comprises <10% and the clay-size fraction (<0.002 mm) <1%. With increasing soil age, a small but consistent increase occurs in both the <2mm and the silt fraction, while the frequency of boulders, especially surface boulders, declines markedly (Figure 6.4.4).

Within the soil, marked changes in soil particle size may occur within a few centimeters, largely due to characteristics inherited from the parent materials. Cryic processes, such as the formation of patterned ground, also lead to some particle size sorting, giving rise to wide textural variations, especially adjacent to patterned ground cracks. Fluvial or aeolian sorted materials typically have high proportions of sand and low proportions of gravel.

4.8. Coarse particles

The description of the coarse particles (boulders, cobbles, pebbles, and granules) provides a useful means for evaluating both the extent of soil weathering and soil stability relationships. Systematic observation of surface clast size, shape, oxidation state, polish, pitting, and erosional characteristics provides useful data that distinguishes the soils on differing land surfaces. Similar observations of clasts within the profile give a clear indication of the extent of any mixing by cryic processes.

Clasts within the soil commonly contrast with those at the surface in having sharp edges and little surface oxidation or polish. Ghosts, or completely disaggregated clasts of some coarsely crystalline rocks, the shape of which remains intact, frequently occur in older soils. These features clearly indicate an absence of cryic movement. On the other hand, a layer of oxidized and faceted clasts in the soil commonly indicates the existence of a previous desert pavement with an underlying paleosol.

Figure 6.4.4. An arid Cryosol. A thin but distinct crust, formed by occasional moistening, is present at the surface. The lightly colored clast to the left of the tape is an *in situ* crumbled rock (ghost). Clasts within the profile are angular and contrast with the rounded surface rocks, indicating a lack of soil mixing by cryic movement.

4.9. Consistence

The absence of a significant clay and organic fraction gives little cohesion to these soils. They are typically incoherent and singly grained. Where the silt content is higher, they may be described as *pulverulent* or *powdery*.

Some small differences in consistence relate to increasing soil age, however. The least weathered soils are typically loose and totally uncohesive. With increasing soil age, cohesiveness increases slightly, and the side of a profile pit may remain intact if undisturbed. The increase in cohesion is partly due to the small increase in the fine fraction but may result from the accumulation of soluble salts in the soil. Where thick salt accumulations are present, the soil may be firmly cohesive or salt-cemented.

Figure 6.4.5. An inland soil on a young to intermediately aged surface, showing a distinct salt horizon between a 2- and 12-cm depth. Below this depth, the soil is less cohesive.

4.10. Soil salts

Since evaporation exceeds moisture supply in this environment, salt precipitation is a common feature in these soils (Campbell and Claridge, 1987). The most common occurrences are surface efflorescences, either as thin powdery accumulations or as thick encrustations, as concentrations beneath some surface clasts, a discrete horizon usually 5 to 15 cm below the soil surface, scattered flecks either within zones or through the subsurface soil, or accretions and aureoles associated with disintegrating subsurface clasts (Figure 6.4.5).

In coastal regions where the soils are young and generally moist, the salts are usually present as surface precipitates. Farther inland where soils are older and drier, the salts commonly occur within the profile in the various manifestations just mentioned. The salt horizons usually have developed best in the oldest and driest soils. They also occur in paleosols, as apparently intact or unmodified horizons. A distinct and well documented geographic pattern in the chemistry of the salts relates to the transport pathways, with nitrate being abundant especially in old soils far inland (Claridge and Campbell, 1968, 1977). The mechanisms of salt transport within the soils and the significance of different forms of precipitation, however, are understood poorly, although likely to relate to characteristics of soil moisture.

4.11. Organic accumulations

Since biological activity in the region's Cryosols is minimal, organic accumulations of any significance are rare. The main forms are pockets of algal material (algal peats) adjacent to small ponds, patches of moss seldom more than a few square meters in area, and the more extensive organic accumulations at penguin rookeries.

The algal peats commonly consist of a mixture of organic matter and sand-forming deposits, with well developed micro-polygonal structure about 10 to 15 cm in diameter. However, little evidence exists for organic decomposition or subsurface horizon alteration by organic compounds. Similarly, moss patches usually consist of a 1- to 2-cm layer of fibrous material without any visible sign of subsurface organic translocation.

In the much more extensive penguin rookeries, which cover around 20 km^2 for the whole of Antarctica, organic accumulations result from gradual accretion of guano, feathers, shells, bodies, and stones accumulated at the site for nesting. Investigation of the soils at abandoned rookery sites showed that gradual decomposition of the organic material and reductions in nitrate and phosphorous values occurred over several thousand years (Campbell and Claridge, 1966).

4.12. Other important properties

Recent ecological investigations in the McMurdo Dry Valleys have highlighted the importance of other soil properties in respect to functions of the ecosystem (Campbell et al., 1998). Soil moisture and soil temperature are especially significant, as they are essential components of soil processes and relationships within ecosystems.

4.13. Soil water content

Soil moisture is one of the most important properties of the arid Cryosols, since no chemical or biological interactions can occur without it. Recent studies (Balks et al., 1995; Campbell et al., 1997) have quantified soil moisture contents for a range of soils in the McMurdo Dry Valley region. In both the active layer and the permafrost, appreciable regional differences occur in the moisture content of soils. Highest soil moisture values are in the coastal areas, for example, while lowest values, averaging <1% over the whole profile, are in some of the Dry Valley soils. Rapid temporal changes are likely to limit the ability of the soil to sustain life.

4.14. Soil temperatures

Soil temperatures vary widely in the Transantarctic Mountains. The soils here are in an environment marked by large variations in incoming radiations, and differences in soil thermal properties are largely a function of the soil surface's capacity to absorb and radiate solar energy, along with the cooling effects of air (Campbell et al., 1997a). The active layer depth, frequency of freeze/thaw cycles, length of the thaw period, and maximum and minimum temperatures are important soil thermal characteristics influencing soil properties and relationships within ecosystems (Campbell and Claridge, 2000).

5. Conclusions

The arid Cryosols of the Transantarctic Mountains are soils that have formed under minimal weathering conditions in an extreme environment. The main pedological features, progressive oxidation, increasing salt accumulation, and decreasing particle size are a function of the very long periods of time through which the soil has weathered. Because of the extreme conditions, soil organic communities occur minimally and the soils are unmodified by the occasional accumulations of organic matter. Yet arid Cryosols have helped reveal the long and complex glacial history of Antarctica. Their significance, especially in terms of their soil thermal, moisture, and salinity regimes, is being recognized increasingly with respect to relationships within ecosystems and the study of life in extreme environments.

References

Balks, M.R., D.I. Campbell, I.B. Campbell, and G.G.C. Claridge. 1995. Interim results of, 1993/94 soil climate, active layer and permafrost investigations at Scott Base, Vanda and Beacon Heights, Antarctica. Univ. of Waikato, Depart. of Earth Sciences Special report. 64 pp.

Bockheim, J.G. 1977. Soil development in the Taylor Valley and McMurdo Sound area, Antarctica. Antarc. J. U.S. 12:105-108.

Bockheim, J.G. 1978. Soil weathering sequences in Wright Valley. Antarc. J. US. 13:36-39.

Bockheim, J.G. 1979. Relative age and origin of soils in eastern Wright Valley, Antarctica. Soil Sci. 128:142-152.

Bockheim, J.G. 1982. Properties of a chronosequence of ultraxerous soils in the Transantarctic Mountains. Geoderma. 28:239-255.

Bockheim, J.G. 1997. Properties and classification of cold desert soils from Antarctica. Soil Sci. Soc. Amer. J. 61: 224-231.

Calkin, P.E. 1964. Geomorphology and glacial geology of the Victoria Valley system, southern Victoria Land, Antarctica. Ohio State Univ. Inst. Polar Studies. Rept. 10. 66 pp.

Campbell, D.I., R.J.L. MacCulloch, and I.B. Campbell. 1997a. Thermal regimes of some soils in the McMurdo Sound region, Antarctica. In: W.B. Lyons, C. Howard-Williams, and I. Hawes, eds. Ecosystem Processes in Antarctic ice-free Landscapes. Proceedings of an International workshop on Polar Desert Ecosystems, Christchurch New Zealand, 1-4 July, 1996. Balkema. pp. 45-56

Campbell, I.B., and G.G.C. Claridge. 1966. A sequence of soils from a penguin rookery, Inexpressible Island, Antarctica. N.Z. J. Sci. 9:361-72.

Campbell, I.B., and G.G.C. Claridge. 1966. Site and soil differences in the Brown Hills region of the Darwin Glacier, Antarctica. N.Z. J. Sci. 10:563-577.

Campbell, I.B., and G.G.C. Claridge. 1975. Morphology and age relationships of Antarctic soils. In: R.P. Suggate and M.M. Cresswell, eds. Quaternary Studies. Royal Society of N.Z. Bull. 13:83-88.

Campbell, I.B., and G.G.C. Claridge. 1978. Soils and Late Cenozoic history of the upper Wright Valley, Antarctica. N.Z. J. Geol. Geophys. 21:635-43.

Campbell, I.B., and G.G.C. Claridge. 1987. Antarctica: soils, weathering processes and environment. Elesvier Science Publishers. Amsterdam. 368 pp.

Campbell, I.B., and G.G.C. Claridge. 2000. Temperature, moisture and salinity patterns in Transantarctic Mountain cold desert ecosystems. In: W. Davidson, C. Howard-Williams, and P. Broady, eds. Antarctic Ecosystems: Models for wider ecological understanding. Caxton Press. Christchurch. pp. 233-240.

Campbell, I.B., G.G.C. Claridge, M.R. Balks, and D.I. Campbell. 1997. Moisture content in soils of the McMurdo Sound Dry Valley region of Antarctica. In: W.B. Lyons, C. Howard-Williams, and I. Hawes, eds. Ecosystem Processes in Antarctic ice-free Landscapes. Proceedings of an International workshop on Polar Desert Ecosystems, Christchurch New Zealand, 1-4 July, 1996. Balkema. pp. 61-76.

Campbell, I.B., G.G.C. Claridge, D.I. Campbell, and M.R. Balks. 1998. The Soil Environment. In: J.C. Priscu, ed., Ecosystem Dynamics in a Polar Desert: the McMurdo Dry Valleys, Antarctica. Antarctic Research Series 72. American Geophysical Union. 297-322.

Campbell, I.B., G.G.C. Claridge, M.R. Balks, and D.I. Campbell. 1998a. Permafrost properties in the McMurdo Sound-Dry Valley region of Antarctica. In A Lewkowicz (ed), Proceedings of the 7th International Conference on Permafrost, Yellowknife, Canada, 1998. pp. 121-126.

Claridge, G.G.C. 1965. The clay mineralogy and chemistry of some soils from the Ross Dependency, Antarctica. N.Z. J. Geol. Geophy. 8:186-220.

Claridge, G.G.C., and I.B. Campbell. 1968. Origin of Nitrate deposits. Nature. 217:428-430.

Claridge, G.G.C., and I.B. Campbell. 1968a. Soils of the Shackleton Glacier region, Queen Maude Range, Antarctica. N.Z. J. Sci. 11:171-218.

Claridge, G.G.C., and I.B. Campbell. 1977. The salts in Antarctic soils, their distribution and relationship to soil processes. Soil Sci. 123: 377-384.

Claridge, G.G.C., I.B. Campbell., and D.S. Sheppard. 2000. Carbon Pools in Antarctica and their significance for global climate change. In: R Lal, J.M. Kimble and B.A Stewart, eds. Glogal Climate Change and Cold Regions Ecosystems. Lewis Publishers. Boca Raton, Fla. Pp. 59-78.

Denton, G.H., D.E Sugden, D.R. Marchant, B.L. Hall, and T.I. Wilch. 1993. East Antarctic Ice Sheet sensitivity to Pliocene climatic change from a Dry Valleys perspective. Geografiska Annaler. 75A:155-204.

Everett, K.R. 1971. Soils of the Meserve Glacier area, Wright Valley, Southern Victoria Land, Antarctica. Soil Sci. 112:425-438.

Hollin, J.T. 1962. On the Glacial history of Antarctica. J. Glaciology. 4:173-195.

Linkletter, G., J.G. Bockheim, and F.C. Ugolini. 1973. Soils and glacial deposits in the Beacon Valley, southern Victoria Land, Antarctica. N.Z. J. Geol. Geophys. 16:90-108.

MacCulloch, R.J.I. 1996. The microclimatology of Antarctic soils. Unpublished M.Sc. Thesis. Univer. of Waikato. Hamilton, New Zealand.

McCraw, J.D. 1960. Soils of the Ross Dependency, Antarctica. A preliminary note. N.Z. Soc. Soil Sci. Proc.4:30-35.

Marchant, D.R., G.H. Denton, D.E.Sugden, and C.C. Swisher. 1993. Miocene glacial stratigraphy and landscape evolution of the western Asgard Range, Antarctica. Geografiska Annaler. 75A:303-330.

Pastor, J., and J.G. Bockheim. 1980. Soil development on Moraines of Taylor Glacier, Lower Taylor Valley, Antarctica. Soil Sci. Soc. Am. J. 44:341-348.

Sheppard, D.S., I.B. Campbell, G.G.C. Claridge, and J.M. Deely. 1994. Contamination of soils about Vanda Station, Antarctica. Institute of Geological and Nuclear Sciences Report 94/20. Lower Hutt, New Zealand.

Tedrow, J.C.F., and F.C. Ugolini. 1966. Antarctic soils. In J.C.F. Tedrow (ed), Antarctic soils and soil forming processes. Antarc. Res. Ser. 8. American Geophysical Union.161-171.

Tedrow, J.C.F. 1977. Soils of the Polar Landscapes. Rutgers Univ. Press. New Brunswick. 638 pp.

Ugolini, F.C. 1963. Soil investigations in the lower Wright Valley. Proceedings of the International Permafrost Conference. Nat. Acad. Sci. Nat. Res. Council Pub. 1287: 55-61.

Ugolini, F.C., and C. Bull. 1965. Soil development and glacial events in Antarctica. Quaternaria. 7:251-269.

Printing: Mercedes-Druck, Berlin
Binding: Stein+Lehmann, Berlin